Stochastic Game Strategies and Their Applications

Stochastic Game Strategies and Their Applications

Bor-Sen Chen

CRC Press is an imprint of the
Taylor & Francis Group, an **informa** business

CRC Press
Taylor & Francis Group
6000 Broken Sound Parkway NW, Suite 300
Boca Raton, FL 33487-2742

First issued in paperback 2021

ISBN 13: 978-1-03-223795-4 (pbk)
ISBN 13: 978-1-138-36076-1 (hbk)

Publisher's Note
The publisher has gone to great lengths to ensure the quality of this reprint but points out that some imperfections in the original copies may be apparent.

Visit the Taylor & Francis Web site at
www.taylorandfrancis.com

and the CRC Press Web site at
www.crcpress.com

Contents

Preface

Game theory involves multi-person decision-making and has been extensively researched in the last few decades. Differential game has been applied to many fields such as financial markets, social economics, control systems, communication systems, and biological systems. In the cooperative differential game, the game strategy design for each player must be compromised with a common desired objective. Further, each player's game strategy depends on the information from other players. On the other hand, in the case of noncooperative game, the goal of each player is different from others and the information of each player is not exchanged. In general, the cooperative differential game strategy could quickly achieve their common objective which should be compromised beforehand while the noncooperative differential game strategy could achieve the results which are compromised based on the conflict multiobjective strategies. However, how to achieve their common objective is always a complicated and difficult compromised problem in the cooperative game strategy because some players need to sacrifice for the common objective in the long compromising process.

In recent decades, stochastic dynamic systems have been widely studied to model dynamic systems with intrinsic fluctuations and external disturbances, whether in engineering, economics, finance, or bio-science. In the field of control, the stochastic systems with random fluctuation due to Wiener process and Poisson process have been concerned and become a popular research topic in recent years. In general, the Wiener process can be regarded as a continuous random fluctuation and the Poisson process can be regarded as discontinuous random fluctuation. Due to the fact that the system dynamics are always disturbed by these random fluctuations, it is more realistic to consider these internal random fluctuations and external random disturbance into system model. In this situation, stochastic games become the new features of noncooperative and cooperative game problems in different research fields. At present, the conventional stochastic game problems focus on the iterative search for the solution of the corresponding Nash equilibrium point for some noncooperative quadratic game strategies, and several update algorithms one player by one player are developed for searching the Nash equilibrium solution of noncooperative game strategy of stochastic systems. However, it is still difficult to apply Nash equilibrium to solve multi-player game strategy problems of more complex systems efficiently. In this book, we focus on the recent developments of more practical stochastic noncooperative and cooperative H_2 and H_∞ games and their applications to control, signaling processing, communication, management, financial, and biological systems with random fluctuations and external disturbance. The exposition of material is quite novel, emphasizing concepts of stochastic game and their practical applications with efficient algorithms based on author's researches in the last two decades.

This book is divided into four parts. In the first part of this book, which comprises Chapters 1–3, the general concept and theory of multi-player stochastic games are introduced. Multi-player cooperative and noncooperative stochastic H_∞ games are discussed first in linear stochastic systems and then in nonlinear stochastic systems. In the second part of the book, which comprises Chapters 4–8, stochastic games are applied to control system designs, including robust robot tracking control design, adaptive vehicle tracking control design, mix H_2/H_∞ stochastic control design, and multi-player noncooperative and cooperative control designs in stochastic jump diffusion systems. In the third part of the book, which includes Chapters 9–11, the applications of stochastic games to signal processing and communication are introduced for robust deconvolution filter and equalizer design or robust mobile location estimation in wireless sensor network from the minmax Nash game perspective. In fourth part of the book, which includes Chapters 12–15, the stochastic cooperative and noncooperative games are introduced for management and financial systems. Stochastic H_∞ games are introduced for resource management strategy, noncooperative and cooperative strategies of power energy management in smart grid systems, and cooperative and noncooperative investment strategies in financial system. Finally, in the fifth part of the book, which includes Chapters 16–20, stochastic game strategies are applied to biological systems. The applications include the therapeutic drug strategy for the treatment of infectious diseases, robust design strategy of synthetic genetic circuits, stochastic evolutionary game strategy for a population of biological networks, and the noncooperative and cooperative game strategies of stochastic biological networks in the evolutionary process.

Since the effect of external disturbance is unpredictable, the conventional stochastic game based on the optimal H_2 payoff function (optimal quadratic performance) could not solve noncooperative and cooperative game problems of stochastic dynamic systems with external disturbance. Further, the present iterative schemes with updating algorithm have been widely employed to solve the Nash equilibrium for noncooperative game strategy indirectly. In general, it is not easy to solve the design problem of multiplayer noncooperative game strategies of stochastic system by these iterative schemes, especially in nonlinear stochastic systems. In this book, stochastic H_∞ noncooperative and cooperative game strategies are addressed to overcome the above two problems for nonlinear stochastic jump diffusion systems with external disturbance.

I grew up in a countryside at southern Taiwan. In my childhood, villagers would love to bet a gambling game with other people by cards, dice, Chinese chess, cricket, etc. in the slack season for farming, especially during the Lunar New Year. Under this environment, I bet with other children in the village when I was a kid. Sometimes, we were the curious spectators of assembling gambling games in the village. It was so exciting when you won a gambling and so depressed when you lost in gambling. The gambling game was a significant group activity in my childhood until I entered high school. From numerous gambling games, I had accumulated some feelings of stochastic game and human behaviors in game. It could be said that stochastic game was imprinted in my DNA when I was a kid. Further, in helping with my father's farming and fishing, we needed to face some unpredictable weather like typhoon, flood, and drought (external disturbance). Sometimes, faming and fishing seem to be a gambling game with heaven and to be on the safe side, you need to make a rational decision. When these uncertain environmental disturbances occurred, somethings needed to rescue for rehabilitation. These are why I have some good feeling in stochastic game theory with wide applications to engineering, social science, and biology in my research career even when I had not taken any course of game theory in school.

Last, but not least I would like to thank Dr. Gagandeep Singh, an editor of CRC, who has given me a lot of help during the course of this project. The author is also grateful to Ms. Chin Yin Wang for her careful typing of this book.

Bor-Sen Chen
Department of Electrical Engineering
National Tsing Hua University
Yuan Ze University

About the Author

Bor-Sen Chen received the B.S. degree from the Tatung Institute of Technology, Taipei, Taiwan, in 1970, the M.S. degree from National Central University, Chungli, Taiwan, in 1973, and the Ph.D. degree from the University of Southern California, Los Angeles, CA, USA, in 1982.

He was a Lecturer, an Associate Professor, and a Professor with the Tatung Institute of Technology from 1973 to 1987. He is currently a distinguished Tsing Hua chair Professor of electrical engineering and computer science with the National Tsing Hua University, Hsinchu, Taiwan. His current research interests include control engineering, signal processing, and systems biology. Dr. Chen was a recipient of the Distinguished Research Award from the National Science Council of Taiwan four times. He is a National Chair Professor of Ministry of Education of Taiwan. Prof. Chen is also a Life Fellow of IEEE.

Part I

General Theory of Stochastic Games

1

Introduction

1.1 Review of the Stochastic Games

Game theory involves multi-person decision-making and has been applied to many fields such as control engineering, communication engineering, economic and social science, and biology. At the beginning, the static game was proposed by Von Neumann and then extended to dynamic game by Nash. The concept of Nash equilibrium is central in analyzing the outcome of a game strategy in dynamic game systems. A Nash equilibrium for a dynamic game is a choice of strategies that tends to persist once the players are using it because neither player has an incentive to deviate from their strategy. In other words, no player could gain any benefit if he/she changes his/her strategy from Nash equilibrium. Even Nash equilibrium is the hallmark of noncooperative game strategy. However, at present, there still exists no effective way to solve stochastic multi-player noncooperative game problems for Nash equilibrium solution except some specific examples. In general, iterative schemes with updating algorithm are employed to search for the Nash equilibrium directly for multi-player noncooperative game strategy [1–3], which is not easily tackled by conventional methods, especially, for complex linear stochastic systems and nonlinear stochastic systems [3].

In the cooperative dynamic game, the game players must be compromised with each other a common desired target beforehand. At the same time, each player's game strategy needs the information from other players to achieve their common target. On the other hand, game players of noncooperative game have different and conflicting targets and the information of each player is not exchangeable. In recent decades, stochastic dynamic systems have been widely studied and applied to engineering, social, and bio-science to model complex and perturbative real systems. Therefore, the stochastic systems with random fluctuations due to Wiener process and Poisson process have been concerned in recent decades [2,4]. In general, the Wiener process is treated as a continuous random fluctuation and Poisson process can be regarded as a sequence of discontinuous random fluctuation in the stochastic dynamic systems to model intrinsic random fluctuations of a dynamic system in the real world.

Due to the fact that real system dynamics will be disturbed by these random intrinsic fluctuations and external disturbances, it is more realistic to consider these intrinsic random fluctuations and external disturbances into the stochastic system model. Since these intrinsic random fluctuations and external disturbances are unpredictable, we will introduce the stochastic game and their applications to multi-person decision-making problem of complex stochastic system with many agents in engineering, social, and bio-science.

In this book, we mainly introduce stochastic multi-player noncooperative H_∞ game strategy of linear and nonlinear stochastic systems with intrinsic random fluctuations and external disturbance. In general, external disturbance is unavailable for each player and is always considered as another player. If the stochastic system is free of external disturbance, the stochastic multi-player noncooperative H_2 game strategy will be also introduced to simplify the design procedure of noncooperative game strategy. Further, stochastic multi-player cooperative H_∞ game strategy of linear and nonlinear stochastic systems is employed for all players to achieve their common target which is compromised beforehand. If the external disturbance could be neglected, then the multi-player cooperative H_2 game strategy is also discussed for linear and nonlinear stochastic systems. In this book, not only the theoretical cooperative and noncooperative H_2 and H_∞ game strategy designs are introduced but also the practical design examples of cooperative and noncooperative H_2 and H_∞ games are introduced with applications to control engineering, signal processing and communication engineering, economic, financial, and management system as well as evolutionary biology.

1.2 The Scope of the Book

This book provides a perspective on a new framework that can help bring together the stochastic game theory and practical application to engineering, social, and bio-science, which have much offer each other. This book is divided into five parts. In part I, a general theory of stochastic m-player noncooperative and cooperative games is introduced from a more theoretical perspective. In part II, the applications of stochastic games to control system designs are given. In part III, the applications of stochastic game to signal processing and communication engineering are introduced. In part IV, the stochastic cooperative and noncooperative games of management and financial systems in social science are introduced. In part V, stochastic game strategies in biological systems are introduced.

Part I of this book investigates the general multi-player stochastic game theory in stochastic jump diffusion systems with continuous Wiener process and discontinuous Poisson process to model intrinsic random fluctuations and external disturbance to model the effect of environment in the real-world dynamic systems.

In Chapter 2, we discuss the m-player noncooperative and cooperative H_2 and H_∞ game strategies of linear stochastic jump diffusion systems. In the design of m-player noncooperative H_∞ game strategy, competitive strategies of other players and external disturbance are augmented as a competitive strategy vector to compete with each player based on a minmax H_∞ payoff function, i.e., each player tries to minimize the worst-case (maximum) effect of competitive strategy vector on the target tracking error to achieve his/her target with a parsimonious control effort (or the cost of strategy) simultaneously, i.e., the m-player noncooperative H_∞ game strategy design problem of linear stochastic system needs to solve m minmax stochastic H_∞ game problems simultaneously to achieve their respective targets.

In general, it is very difficult to solve m minmax stochastic H_∞ game problems simultaneously for m-player noncooperative H_∞ game strategy directly. An indirect method is proposed to transform the m-player noncooperative game problem to an equivalent multiobjective optimization problem (MOP) under the constraint of linear matrix inequalities (LMIs) from the suboptimal perspective. An LMIs-constrained multi-objective optimization evolution algorithm (MOEA) is also introduced to solve the m-player noncooperative H_∞ game strategy in a single round with the help of LMI toolbox in Matlab. If the linear stochastic system is free of external disturbance or the external disturbance could be neglected, then the m-player noncooperative H_2 game strategy design problem could be formulated for the simplicity of design procedure of the multi-person decision-making problem in linear stochastic system when each player has conflict objective without information exchange among each player. Following the similar procedure, an LMIs-constrained MOP is also derived for solving the m-player noncooperative H_2 game strategy of linear stochastic systems.

If all players in the linear stochastic system with external disturbance have compromised a common target with information exchange among each player, then m-player cooperative H_∞ game strategy design problem is formulated to minimize the worst-case effect of external disturbance on the common target tracking error to optimally achieve their common target with a parsimonious effort. In this case, all the strategies of players are augmented as an augmented strategy vector when external disturbance is considered as a competitive player. The m-player cooperative H_∞ game problem could be transformed to an LMIs-constrained single-objective optimization problem (SOP), which could be solved efficiently with the help of LMI toolbox in Matlab. Similarly, if the linear stochastic system is free of external disturbance, then the m-player cooperative game strategy design problem could be formulated as m-player cooperative stochastic H_2 game problem, which could be also transformed to LMIs-constrained SOP and solved efficiently by the help of LMI-toolbox in Matlab.

After introducing the general stochastic game theory of linear stochastic systems in Chapter 2, the stochastic game theory of nonlinear stochastic systems with random Wiener and Poisson process and external disturbance is introduced in Chapter 3. We first discuss the m-player noncooperative and cooperative stochastic H_∞ game strategies of nonlinear stochastic systems. The m-player noncooperative stochastic H_∞ game strategy of nonlinear stochastic systems could be transformed to an equivalent m Hamilton–Jaccobi–Issac inequalities (HJIIS)-constrained MOP by the proposed suboptimal method. Since there exists no analytic or numerical method to solve the MOP with m HJII constraints, the global linearization method is introduced to interpolate J local linearized stochastic systems at the vertices of polytope of global linearization to approximate the nonlinear stochastic system. Based on the interpolation of J local linearized stochastic systems for the nonlinear stochastic system, the HJIIs-constrained MOP for m-player noncooperative stochastic H_∞ game strategy could be transformed to an LMIs-constrained MOP, which could be solved efficiently by the LMIs-constrained MOEA in Chapter 2. The m-player cooperative H_∞ game could be formulated as a 2-tuple minmax stochastic H_∞ game, i.e., the total players are augmented as one player to minimize the worst-case effect of external disturbance, which is considered another player, to achieve their common target from the H_∞ tracking perspective. The m-player cooperative stochastic H_∞ game could be transformed to an equivalent HJII-constrained SOP, which could also be transformed to an LMIs-constrained SOP based on the global linearization method. Finally, the m-player noncooperative and cooperative stochastic H_2 game strategies are also discussed for nonlinear stochastic systems without external disturbance in Chapter 3. Similarly, the m-player noncooperative stochastic H_2 game strategy could be transformed to a corresponding HJIIs-constrained MOP, which in turn could be transformed to a corresponding LMIs-constrained MOP and solved by the proposed LMIs-constrained MOEA in Chapter 2. The m-player cooperative stochastic H_2 game problem of nonlinear stochastic systems without external disturbance could be solved indirectly by a HJII-constrained SOP from the suboptimal perspective, which could be also transformed to an LMIs-constrained SOP with the help of the global linearization method.

In Part II of the book, we introduce the applications of stochastic games to control system designs. In Chapter 4, a robust adaptive tracking control design of robot system based on minmax H_∞ dynamic game is introduced to enhance the robust tracking performance of robot system in spite of system uncertainties and external disturbance. At the beginning, a linear nominal robotic control design is specified via model reference tracking with desired eigenvalue assignment. Next, a fuzzy logic system is adaptively tuned to eliminate the nonlinear uncertainties as possibly as it can. Finally, a minmax H_∞ game strategy is specified to optimally attenuate the worst-case effect of both the residue due to adaptive fuzzy cancellation and external disturbance to achieve a minmax H_∞ tracking performance for robotic systems with large parameter perturbation and external disturbance [6].

In Chapter 5, a minmax H_∞ adaptive tracking control design is introduced for wheeled vehicles with trailer from the adaptive game perspective. The proposed minmax game control is equipped with adaptive fuzzy elimination scheme to achieve a robust H_∞ model reference tracking performance, despite the system uncertainties and external disturbance. In the design procedure, a nominal wheeled vehicle control design is obtained by model reference tracking with desired eigenvalue assignment. Then, a fuzzy logical system is subsequently tuned to enhance the tracking robustness and a minmax game control scheme is specified to optimally attenuate the worst-case effect of the residue of fuzzy elimination on the tracking error to be below a prescribed level to meet a robust H_∞ tracking performance, i.e., an adaptive fuzzy-based H_∞ optimization method is employed to solve the minmax adaptive H_∞ tracking control design of wheeled vehicle from the minmax game perspective [7].

H_2/H_∞ control design is a very important and popular control scheme to achieve both the optimal quadratic (H_2) control performance and a prescribed H_∞ attenuation level of external disturbance simultaneously [8]. In Chapter 6, the stochastic H_2/H_∞ control design for linear systems with state-dependent noise will be introduced through stochastic Nash game perspective. Based on the stabilization, exact observability, and stochastic detectability of linear stochastic systems, the infinite horizon stochastic H_2/H_∞ control design could be developed by the stochastic Nash game strategy. The finite horizon stochastic H_2/H_∞ control design could be also generalized from the deterministic one by the stochastic Nash game strategy. Finally, the observer-based stochastic H_2/H_∞ control design could be developed based on stochastic Nash game strategy for the linear stochastic systems when state variables could not be measured directly.

In Chapter 7, fuzzy multi-player stochastic H_2 games are introduced for nonlinear stochastic control systems with state variables unavailable for each player. Therefore, fuzzy-observer-based stochastic H_2 game strategies are developed for nonlinear stochastic control system. Based on T-S fuzzy scheme to interpolate several local linear stochastic systems to approximate the nonlinear stochastic system, a fuzzy-observer-based controller is introduced to deal with noncooperative stochastic H_2 game in the sense of Nash equilibrium strategies by an iterative update algorithm or with the cooperative game in the sense of Pareto-optimal strategies by an eigenvalue method [9].

In Chapter 8, multi-player stochastic minmax H_∞ cooperative and noncooperative game strategy design for a class of nonlinear stochastic jump diffusion control system with external disturbance is introduced. To attenuate the effect from competitive strategies of other players and unpredictable external disturbance on each player, a multi-player stochastic minmax H_∞ noncooperative game strategy design problem is transformed to a multi-tuple HJIIs-constrained MOP with a Nash equilibrium solution. Also, multi-player stochastic minmax H_∞ cooperative game strategy design problem is discussed and transformed to a HJII-constrained SOP. In order to overcome the difficult solution of HJII for the convenience of design, the Takagi–Sugeno (T-S) fuzzy model is employed to approximate the nonlinear stochastic control system so that the HJIIs-constrained MOP for the multi-player noncooperative stochastic minmax H_∞ game strategy could be transformed to an LMIs-constrained MOP. Therefore, the LMIs-constrained MOEA in Chapter 2 could be also employed to efficiently solve the multi-player stochastic minmax noncooperative H_∞ control game strategy design problem. In the case of stochastic minmax H_∞ cooperative game, by using T-S fuzzy interpolation method, the HJII-constrained SOP for stochastic minmax H_∞ cooperative game strategy could be transformed to an equivalent LMIs-constrained SOP.

In Part III, the application of stochastic game to signal processing and communication is introduced for robust signal reconstruction or equalization design in signal processing or signal transmission under channel uncertainty and interference.

In Chapter 9, a robust deconvolution filter design under stochastic parametric and channel noise uncertainties is introduced by the minmax game approach. In the signal transmission system, it is assumed that some large but bounded uncertainties exist in the driving and measurement noise covariance as well as the second-order statistics of stochastic parameters and initial condition. In this chapter, three kinds of minimax sensitivity criteria will be used to develop the minmax game strategy to the synthesis of minimax deconvolution filters under uncertain linear stochastic signal transmission systems, i.e., the robust filtering approach is based on the saddle-point theory and the sensitivity analysis to robust design of Kaman filter. The proposed minmax game design algorithms give the recursive realization of minimax deconvolution filter for the time-varying uncertain signal transmission systems [10].

In Chapter 10, the minmax stochastic H_2 game is applied to robust equalizer design of multi-input–multi-output (MIMO) communication systems with time-varying uncertain channel. Channel uncertainties within a neighborhood of the estimated channel matrix are considered within a bound on the spectral matrix norm of channel estimation errors. In this chapter, channel uncertainties are partitioned into finite Markov-transitioned channel uncertain states based on the least upper bound on matrix spectral norm of channel uncertainty. This robust equalizer design leads to a multiple model-based minmax stochastic H_2 game approach. On this basis, a feasible equalizer can be obtained from a weighted combination of multiple over-guaranteed cost-based equalizers, each of which is designed with respect to a channel uncertain state [11].

In Chapter 11, a minmax l_2 game approach is introduced for robust relative sensor location estimation problem in wireless sensor network (WSN), which is a prominent issue faced by several applications in WSNs. Therefore, how to determine the exact positions of location-unaware sensors is the primary focus. In this chapter, a robust minmax l_2 game method is proposed for the relative location estimation problem by minimizing the worst-case estimation error. The corresponding minmax l_2 optimization problem for robust relative sensor location estimation is originally noncovex but could be transformed to a convex semi-definite problem (SDP) and solved by the existing numerical techniques [12].

In Part IV, stochastic cooperative and noncooperative H_∞ game strategies for management and financial systems with intrinsic random fluctuations and external disturbance are introduced.

In Chapter 12, a stochastic H_∞ game approach is introduced to the robust resource management control for CO_2 emission and reduction of greenhouse effect. In the last decades, due to the increasingly severe global warning, investments in clean technology, reforestation, and political action have been studied to reduce CO_2 emission to help people protect their environment and resources. In this chapter, a nonlinear stochastic model is introduced to describe the stochastic dynamics of CO_2 emission with control inputs, including clean technology, reforestation, and carbon tax, under system uncertainties. For the efficient resources management, a robust tracking control based on stochastic H_∞ game is designed to force resources tracking a desired reference output. The worst-case effect of stochastic parametric fluctuations, i.e., external disturbances on the target tracking performance, is considered and minimized to achieve the robust tracking from the stochastic H_∞ game theory perspective. When one player (stochastic uncertainty) maximizes the tracking error and another player (management control) minimizes the tracking error, the stochastic H_∞ game could solve the robust minimax stochastic H_∞ tracking problem of resource management control for CO_2 emission and reduction of greenhouse effect. To avoid solving the HJII of nonlinear minmax H_∞ stochastic game for robust resource management control, a fuzzy model is proposed to approximate the nonlinear CO_2 emission model. Then, the nonlinear stochastic resource management control for nonlinear CO_2 emission and reduction of greenhouse effect could be easily solved by fuzzy stochastic H_∞ game approach via LMI technique [13].

In Chapter 13, noncooperative and cooperative H_∞ management strategies are introduced for dynamic energy transfer systems with renewable energy sources (RESs) in smart grid network. In this chapter, we will formulate a novel utility function for each manager (user) to design his/her noncooperative H_∞ management strategy according to their own consideration and the possible effect of other managers' strategies. Then, the noncoopearative H_∞ management strategy design problem in a smart grid network could be transformed to an LMIs-constrained MOP. Further an LMIs-constrained MOEA in Chapter 2 is developed to solve the multi-person noncooperative H_∞ management strategy in smart grid network. The cooperative H_∞ management strategy design problem of smart grid network is also introduced for smart grid network in this chapter.

In Chapter 14, we focus on the stochastic H_∞ model reference investment strategy of nonlinear stochastic financial systems with continuous and discontinuous (jumping) random intrinsic fluctuation and external disturbance. The proposed stochastic H_∞ model reference investment strategy could not only eliminate the effect of continuous and discontinuous intrinsic fluctuations but also attenuate the worst-case effect of external disturbance on the desired model reference tracking of a nonlinear stochastic financial system from the perspective of minimax stochastic H_∞ game strategy. Based on the Itô–Lévy lemma and stochastic Nash game, the minimax stochastic H_∞ model reference game strategy of stochastic nonlinear financial system is transformed to a nonlinear HJII-constrained optimization problem. Based on the global linearization, the HJII-constrained optimization problem is transformed to an equivalent LMIs-constrained optimization problem. Finally, a robust stochastic H_∞ model reference investment strategy of nonlinear stochastic financial system and a macroeconomic robust H_∞ reference control of the financial instability contagion due to international capital flow volatility are given as two simulation examples to illustrate the design procedure and to confirm the performance of the proposed stochastic H_∞ Nash game approach to the desired model reference control of the nonlinear stochastic financial system [14].

In Chapter 15, a multi-player noncooperative H_∞ stochastic game strategy is introduced to the stochastic financial system with random fluctuation through the multi-objective optimization method. In the financial market, most managers or investors hope their investment policies to be with high profit and low risk. Managers and investors involved pursue their own interests but are partly conflicting with others. Since financial systems are nonlinear with intrinsic random fluctuations and external disturbance, it still lacks an efficient method to solve the complex multi-player noncooperative game strategy design problem. In this chapter, a stochastic multi-player noncooperative H_∞ game strategy in nonlinear stochastic financial system is transformed to a multi-tuple HJIIs-constrained MOP. Then, the global linearization technique is employed to interpolate a set of local linearized stochastic systems to approximate the nonlinear stochastic financial system so that the HJIIs-constrained MOP could be transformed to an LMIs-constrained MOP, which could be easily solved by the LMIs-constrained MOEA in Chapter 2. Two financial systems are given with simulation results to validate the proposed noncooperative H_∞ game strategy.

Part V of this book introduces stochastic games in biological systems. Biological systems are very complex and evolve under intrinsic genetic variations and external disturbances, including competition with other species and environmental changes. Stochastic game theory has been widely applied to bioscience, especially evolutionary game of biological systems.

In Chapter 16, a robust model reference therapeutic control of immune response under exogeneous pathogenic attack is introduced based on the minimax dynamic H_∞ game strategy to match a prescribed immune response under uncertain initial states and environmental disturbances, including continuous intrusion of exogeneous pathogens. The worst-case effect of all possible environmental disturbances and uncertain initial states on the reference model matching for a desired reference immune response is minimized for the enhanced immune system, i.e., to design a robust therapeutic control to track a prescribed immune model reference response from the minimax H_∞ game perspective. This minimax H_∞ reference therapeutic control problem could herein be transformed to an equivalent dynamic H_∞ game problem. To be on the safe side, the exogeneous pathogens and environmental disturbances are considered as a player to maximize (worsen) the reference matching error when the therapeutic control agents are considered as another player to minimize the reference matching error. Since the innate immune system is highly nonlinear, it is not easy to solve the robust model reference therapeutic control problem by the nonlinear dynamic H_∞ game strategy directly. A fuzzy model is proposed to interpolate several local linearized immune systems to approximate the innate immune systems. Then, the H_∞ model reference therapeutic control problem of immune systems could be easily solved by the proposed fuzzy dynamic H_∞ game strategy via LMI technique [23].

In Chapter 17, a nonlinear stochastic system is introduced to model a synthetic gene network with intrinsic random fluctuation and environmental disturbance in the host cells. A stochastic H_∞ game strategy is introduced for robust design of synthetic gene network. Since the environmental disturbance is unavailable and considered as one player to distort the genetic network possible, the designer could select some adequate ribosomal binding site (RBS) components (parts) which are considered as another player. The minimax stochastic H_∞ game strategy is employed to treat the robust synthetic gene network design problem with a desired expression level at the steady state. Based on fuzzy approximation method, the minimax stochastic H_∞ game design problem could be transformed to an LMIs-constrained optimization problem, which could be efficiently solved by selecting adequate RBS components from RBS library [16].

In Chapter 18, the natural selection strategy in stochastic evolutionary game of a population of biological networks is introduced. In this chapter, a population of evolutionary biological network is modeled by a stochastic dynamic system with intrinsic random parameter fluctuations due to genetic variations and external disturbances due to environmental changes in the evolutionary process. Since the information on environmental changes is unavailable and their occurrences are unpredictable, they can be considered as a competitive game player to biological network with potential to destroy the phenotypic stability of the biological network. The biological network needs to develop an evolutionary strategy to improve its phenotypic stability as much as possible, so it can be considered as another game player in the evolutionary process, i.e., a stochastic minmax H_∞ evolutionary game. Based on the nonlinear stochastic minmax H_∞ evolutionary game strategy, some useful genetic variations could be used as raw materials by natural selection to construct negative feedback loop to efficiently improve robustness of biological network to resist environmental changes in order to maintain its phenotype. In this situation, the robust phenotype traits of stochastic biological network can be more frequently selected by natural selection in evolution. However, if the harbored neutral genetic variations are accumulated to a sufficiently large amount, and environmental disturbances are strong enough that the network robustness has no longer enough genetic robustness and environmental robustness, then the phenotype robustness might break down. In this case, a network phenotypic trait may be pushed from one equilibrium point to another, changing the phenotypic trait and starting a new phase of network evolution through the hidden neutral genetic variations harbored in network robustness by adaptive evolution [17].

Molecular biologists have long recognized carcinogenesis as an evolutionary process. In Chapter 19, the nonlinear stochastic H_∞ game strategy of evolutionary biological network in carcinogenic process is introduced. In this chapter, the evolution of somatic cancer cell lineages in carcinogenic process is modeled as an equilibrium point shifting during the process of a nonlinear stochastic evolutionary biological network. This evolutionary process is subject to intrinsic random fluctuations due to somatic genetic and epigenetic variations, as well as extrinsic disturbances because of carcinogens and stressors. In order to maintain the normal function (i.e., the phenotype at the nominal equilibrium point) of an evolutionary biological network under random intrinsic fluctuations and environmental disturbances, a network robustness scheme to incorporate with natural selection is developed based on stochastic H_∞ game strategy in this chapter. This can be accomplished by selecting certain genetic and epigenetic variations to modify the network structure to attenuate intrinsic fluctuations efficiently and to resist extrinsic disturbances in order to maintain the phenotype of the evolutionary biological network at an equilibrium point. However, during carcinogenesis, the remaining (or neutral) genetic and epigenetic variations accumulate on the basin of the equilibrium point, and the extrinsic disturbances become too large to maintain the normal phenotype at the desired equilibrium point of the nonlinear evolutionary network. Thus, the phenotype of biological network is shifted to a cancer phenotype at a new equilibrium point to begin a new evolutionary process. In this chapter, the natural selection scheme of an evolutionary biological network of somatic cells during carcinogenesis was derived from a robust negative feedback scheme based on nonlinear stochastic H_∞ evolutionary game strategy. The evolvability and phenotypic robustness criterion of evolutionary cancer network were estimated by solving a HJII-constrained optimization problem [18].

In Chapter 20, we discuss the noncooperative and cooperative stochastic H_∞ evolutionary game strategies of a population of evolutionary networks. In this chapter, the current static and dynamic evolutionary game strategies of biological networks are reviewed at first and then the roles of random genetic variations and stochastic environmental disturbances are discussed in these evolutionary biological system models. Therefore, a nonlinear stochastic biological model with Poisson-driven genetic variations and random fluctuations is proposed to mimic a population of evolving biological networks. To gain insight into the cooperative and noncooperative H_∞ evolutionary game theory of stochastic biological networks under natural selection, the phenotypic robustness and network evolvability of biological network with noncooperative and cooperative evolutionary H_∞ game strategies are discussed from the multi-player minmax stochastic H_∞ evolutionary game perspective. The noncooperative stochastic H_∞ evolutionary game strategy can be transformed into an equivalent MOP and is shown to display a significantly improved network robustness to tolerate genetic variations and buffer environmental disturbances, maintaining phenotypic traits for longer than the cooperative stochastic H_∞ evolutionary game strategy [19]. However, the noncooperative stochastic H_∞ game strategy requires more effort and more compromises between partly conflicting players. Finally, the global linearization is used to simplify the problem of solving these two nonlinear stochastic H_∞ evolutionary games.

1.3 Conclusion

In this chapter, some stochastic games, especially stochastic H_2 and H_∞ cooperative and noncooperative games, are reviewed and discussed at first. In order to help readers understand more easily, this book is divided into five parts according to their theoretical analysis and application. The scope of the book is also given chapter by chapter. Finally, the theoretical background of stochastic games and their applications to engineering, social, and bioscience are given in detail in the following chapters.

2

Introduction to Multi-player Stochastic Games in Linear Stochastic Systems

2.1 Introduction

Before we discuss the more general multiplayer stochastic game theory of nonlinear stochastic systems in Chapter 3, the multi-player game theory of linear stochastic systems is introduced in this chapter. In recent decades, stochastic dynamic systems have been widely studied, whether in control, signal processing, communication, economics, financial, or bioscience [1,2,9,20]. In these fields, the stochastic dynamic systems with random fluctuations due to Wiener process and Poisson process have been concerned in recent years [22,23]. In general, the Wiener process is treated as a continuous random fluctuation and Poisson process can be regarded as discontinuous jump fluctuation. For example, in the physical phenomena of control engineering, the system state jumps caused by failure of power supply or some unexpected events can be regarded as Poisson process, and the changes of loading and environmental situations could be considered as external disturbance. In the financial markets, stocks and social economics, all kinds of economic problems due to the interactions of factors like prices, investment, varying investment rate, and economic policy are dynamical and disturbed by these random fluctuations. Therefore, it is more realistic to consider these intrinsic random fluctuations and external disturbances into system model as linear stochastic system [22].

Game theory involves multi-person decision-making [1]. It is noncooperative if each player involved proposes his or her own interests which are partly conflicting with each other [1,9,19]. It is cooperative if all players have compromised a common target beforehand so that all of the interests of players could reach the same agreement [19,21]. In this chapter, the theory of multi-player noncooperative and cooperative game strategy will be introduced for linear stochastic systems with intrinsic random continuous and discontinuous fluctuations and external disturbance.

In this chapter, the stochastic m-player H_∞ noncooperative game design of linear stochastic systems with Wiener process and Poisson process is introduced at first. In this stochastic system, each player has his or her desired target and the targets and strategies of competitive players as well as external disturbance are unavailable for each player. For the convenience of strategy design $u_i(t)$ of each player, the competitive strategies and external disturbance (which is unpredictable and is considered as a strategy of an extra player) are augmented together as a competitive strategy vector $u_{-i}(t)$ so that the m-player H_∞ noncooperative game strategy design problem could be formulated as 2-tuple minmax H_∞ noncooperative game for each player from the perspective of minmax H_∞ target tracking, i.e., the worst-case (maximum) effect of competitive strategy of $u_{-i}(t)$ on the desired target tracking and control effort $u_i(t)$ of each player must be minimized as possible from the H_∞ control perspective [19]). Since it is not easy to simultaneously solve m minmax stochastic H_∞ game design problems, an indirect method is proposed to minimize the upper bounds of m minmax stochastic 2-tuple noncooperative H_∞ game problems simultaneously from the suboptimal perspective. In this situation, the *m*-player noncooperative stochastic H_∞ game design problem could be transformed to an equivalent m-objective multiobjective optimization problem (MOP). We also find the m-player noncooperative game problem of linear stochastic systems could be transformed to a corresponding linear matrix inequalities (LMIs)-constrained MOP, which could be shown as Nash equilibrium solution and solved efficiently by the proposed LMI-constrained multiobjective evolutionary algorithm (MOEA). Further, the special cases of m-player H_∞ noncooperative desired target tracking game problem, like the m-player noncooperative H_∞ stabilization game problem and the conventional optimal H_∞ tracking problem of linear stochastic systems with external disturbance (two-player H_∞ tracking problem with control and external disturbance as two competitive players), are also discussed.

If the linear stochastic system is free of external disturbance, the conventional multi-player noncooperative H_2 (quadratic) game strategy design problem is also discussed [1,2,9]. In general, *m*-player noncooperative H_2 strategy could not treat the multi-player noncooperative decision problem in the linear stochastic systems with external disturbance. If the effect of external disturbance is neglected, then multi-player noncooperative H_2 game strategy is very suitable for solving multi-player noncooperative decision problems. By the suboptimal methods, *m*-player noncooperative H_2 tracking game strategy could be solved by minimizing *m* players' upper bounds of minmax quadratic H_2 tracking performance to indirectly achieve the desired targets of each player simultaneously. This m-player noncooperative H_2 tracking game strategy design problem could be also transformed to an equivalent LMIs-constrained MOP, which could be also solved efficiently by the proposed LMIs-constrained MOEA algorithm. We find the solution of LMIs-constrained MOP for m-player noncooperative H_2 game strategy problem is also of Nash equilibrium solution.

If the *m* players in the linear stochastic system have compromised a common target with each other at the beginning, the m-player cooperative game strategy design could be employed for each player with information exchanges of strategies to achieve their common target efficiently. However, how to compromise with each other to achieve their common target is always a complicated and difficult process because some players need to sacrifice for the common target.

In this chapter, multi-player $H\infty$ cooperative game strategy in linear stochastic systems with intrinsic random fluctuation and external disturbance is also introduced for each player to achieve their common target. The worst-case effect of external disturbance and random fluctuation is minimized by the combined strategies of each player from the H_∞ common target tracking perspective to optimally and robustly achieve their common target with a parsimonious control effort. This multi-player stochastic H_∞ cooperative game strategy of linear stochastic system could be transformed to an equivalent LMIs-constrained single-objective optimization problem (SOP), which could be very easily solved with the help of LMI-toolbox in Matlab.

If the linear stochastic system is free of external disturbance, then multi-player stochastic H_2 cooperative game strategy design could be employed for each player to design its control strategies to achieve its common target. This stochastic H_2 cooperative game problem is reduced to a conventional optimal quadratic tracking design problem of linear stochastic system. This stochastic multi-player H_2 cooperative game strategy design problem could be also transformed to an LMIs-constrained SOP, which could be also very easily solved with the help of LMI-toolbox in Matlab.

2.2 Stochastic Multi-Player H_∞ Noncooperative Game Strategy in Linear Stochastic Systems

Consider the following linear stochastic jump-diffusion system:

$$dx(t) = (Ax(t) + v(t))dt + Cx(t)dw(t) + Dx(t)dp(t) \tag{2.1}$$

where $x(t) \in R^n$ denotes the state vector. $v(t) \in \mathbb{R}^n$ denotes the external disturbance, $w(t) \in \mathbb{R}$ denotes the standard Wiener process with zero mean and unit variance, and $p(t) \in \mathbb{R}$ denotes the Poisson counting process with $E\,p(t) = \lambda t$. $A \in \mathbb{R}^{n\times n}$, $C \in \mathbb{R}^{n\times n}$, and $D \in \mathbb{R}^{n\times n}$ are system matrices.

In the linear stochastic system, the diffusion term $Cx(t)dw(t) = Cx(t)n(t)dt$ denotes the intrinsic continuous random fluctuation, where $n(t)$ denotes the standard white noise. The jumping term $Dx(t)dp(t) = Dx(t)q(t)$ denotes the intrinsic discontinuous random fluctuation, where $q(t)$ denotes the Poisson point process with mean λ. $v(t)$ denotes the unpredictable external disturbance. The linear stochastic jump-diffusion system in (2.1) could be used to model many stochastic systems in engineering, economic, financial, and biological systems with both internal continuous and discontinuous fluctuations and external disturbance [1,2,19,22].

Some important properties of Wiener process $w(t)$ and Poisson process $p(t)$ in this chapter are given as follows [22]:

$$\begin{aligned}
&\text{(i) } E\{w(t)\} = 0\\
&\text{(ii) } E\{dw(t)\} = 0\\
&\text{(iii) } E\{dw(t)dw(t)\} = dt\\
&\text{(iv) } E\{dp(t)\} = \lambda dt, \text{ where } \lambda>0 \text{ denotes the Poisson jump intensity.}\\
&\text{(v) } E\{dp(t)dw(t)\} = 0, \text{ i.e., } dw(t) \text{ and } dp(t) \text{ are assumed independent.}\\
&\text{(vi) } E\{dp(t)dt\} = 0.
\end{aligned} \tag{2.2}$$

Remark 2.1 In the more engineering representation, (2.1) could be represented as follows:

$$\frac{d}{dt}x(t) = Ax(t) + v(t) + Cx(t)n(t) + Dx(t)q(t)$$

where $n(t)$ denotes the standard white noise with $dw(t) = n(t)dt$ and $q(t)$ denotes Poisson process with $dp(t) = q(t)dt$.

Suppose there are m players involved in multi-person decision-making of linear stochastic system in (2.1). It is noncooperative game if each player involved pursues his or her own interests which are partly conflicting with others. If all players have compromised a common target at the beginning and then all players pursue the common target together, it is a cooperative game. We will discuss the stochastic noncooperative game of linear stochastic system in (2.1) at first, and then the stochastic cooperative game will be discussed in the sequel.

Suppose m players are involved in the linear stochastic system in (2.1). In this situation, linear stochastic system should be modified as follows:

$$dx(t) = (Ax(t) + \sum_{i=1}^{m} B_i u_i(t) + v(t))dt + Cx(t)dw(t) + Dx(t)dp(t) \tag{2.3}$$

where $u_i(t) \in R^{l_i \times 1}$ denotes the control strategy of the ith player and $B^i \in R^{n \times l_i}$ denotes the input coupling (influencing) matrix of the ith player strategy $u_i(t)$ on the linear stochastic system.

In the noncooperative game problem of linear stochastic system in (2.3), the information of other players' control strategies $[u_1(t) \cdots u_{i-1}(t)u_{i+1}(t) \cdots u_m(t)]$ and external disturbance $v(t)$ are unavailable for the ith player. Since their effects are unpredictable, to be on the safe side, each player should consider them from the worst-case perspective. Further, the ith player has its own desired target x_d^i. Therefore, by augmenting other player's control strategies with the external disturbance as $u_{-i}(t) = [u_1^T(t) \cdots u_{i-1}^T(t)u_{i+1}^T(t) \cdots u_m^T(t)v^T(t)]^T$. Then, linear stochastic system in (2.3) could be represented by

$$dx(t) = (Ax(t) + B_i u_i(t) + B_{-i}u_{-i}(t))dt + Cx(t)dw(t) + Dx(t)dp(t) \tag{2.4}$$

where $B_{-i} = [B_1 \cdots B_{i-1}B_{i+1} \cdots B_m I]$

The utility function of m-player stochastic H_∞ noncooperative game for the ith player with desired target x_d^i is formulated as the following minmax stochastic H_∞ optimization problems:

$$\rho_i^* = \min_{u_i(t)} \max_{u_{-i}(t)} \frac{E\{\int_0^{t_p} ((x(t) - x_d^i)^T Q_i (x(t) - x_d^i) + u_i^T(t) R_i u_i(t) dt\}}{E\{(x(0) - x_d^i)^T (x(0) - x_d^i) + \int_0^{t_p} (u_{-i}^T(t) u_{-i}(t) dt\}} \tag{2.5}$$

$$i = 1, 2, \ldots, i, \ldots, m.$$

where $x(0)$ denotes the initial state, $Q_i \geq 0$ and $R_i > 0$ with appropriate dimension denote the corresponding weighting matrices for each player to tradeoff between the state tracking error $x(t) - x_d^i$ and the strategy effort of $u_i(t)$ from the perspective of player i. The value ρ_i^* in (2.5) denotes the tracking performance of the stochastic $H\infty$

noncooperative game with the corresponding optimal game strategy $u_i^*(t)$ of player i. The term $E\{(x(0)-x_d^i)^T(x(0)-x_d^i)\}$ in (2.5) denotes the effect of uncertain initial condition $x(0)-x_d^i$ on the stochastic noncooperative H∞ game strategy. t_p denotes the terminal time.

The main purpose of the noncooperative stochastic minmax H∞ game strategy $u_i(t)$ in (2.5) is to most attenuate the worst-case influence of the unpredictable external disturbance and other unavailable strategies on the individual goals to optimally achieve their individual targets with a parsimonious control effort $u_i(t)$, respectively, i.e., to be on the safe side, the noncooperative game strategy $u_i(t)$ of the ith player in (2.5) is designed to minimize the worst-case effect of unpredictable augmented competitive strategies and external disturbance $u_{-i}(t)$ with a parsimonious control strategy effort for all players simultaneously.

Since every player has its own desired target in the multiplayer noncooperative stochastic H∞ game in (2.5), for the simplicity of design, the shifted stochastic jump diffusion system in (2.3) is constructed as follows:

$$d\tilde{x}_i(t) = A\tilde{x}_i(t) + \sum_{i=1}^{m} B_i u_i(t) + v_i(t) + C\tilde{x}_i(t)dw(t) + D\tilde{x}_i(t)dp(t),\ i = 1,2,\ldots,m \tag{2.6}$$

where $\tilde{x}_i(t) = x(t) - x_d^i$ and $v_i(t) = v(t) + Ax_d^i + Cx_d^i dw + Dx_d^i dp(t)$. In the situation $u_{-i}(t)$ in (2.4) is modified as $u_{-i}(t) = [u_1^T(t)\cdots u_{i-1}^T(t)u_{i+1}^T(t)\cdots u_m^T(t)v_i^T(t)]^T$, then the linear stochastic systems in (2.6) could be represented by

$$d\tilde{x}_i(t) = (A\tilde{x}_i(t) + B_i u_i(t) + B_{-i}u_{-i}(t))dt + C\tilde{x}_i(t)dw(t) + D\tilde{x}_i(t)dp(t) \tag{2.7}$$

Then, the m-player of stochastic minmax H∞ noncooperative game in (2.5) could be reformulated for each player to solve the following m stochastic minmax H∞ games simultaneously:

$$\rho_i^* = \min_{u_i(t)} \max_{u_{-i}(t)} \frac{E\{\int_0^{t_p}(\tilde{x}_i^T(t)Q_i\tilde{x}_i(t) + u_i^T(t)R_i u_i^T(t))dt\}}{E\{\tilde{x}_i^T(0)\tilde{x}_i(0) + \int_0^{t_p} u_{-i}^T(t)u_{-i}(t)dt\}} \tag{2.8}$$

for $i = 1,2,\ldots,m$.

i.e., we need to solve the above 2-tuple m-player stochastic minmax H∞ noncooperative games $(\rho_1^*,\ldots\rho_i^*,\ldots\rho_m^*)$ simultaneously.

Remark 2.2 If only one control strategy is employed for linear stochastic system (2.6), i.e.,

$$d\tilde{x}(t) = (A\tilde{x}(t) + v'(t) + Bu(t))dt + C\tilde{x}(t)dw(t) + D\tilde{x}(t)dp(t) \tag{2.9}$$

where x_d is the desired target and $v'(t) = v(t) + Ax_d + Cx_d dw(t) + Dx_d dp(t)$, then stochastic H∞ noncooperative game in (2.8) becomes the following optimal H∞ stochastic control [2,23]

$$\rho^* = \min_{u(t)} \max_{v'(t)} \frac{E\{\int_0^{t_p}(\tilde{x}^T(t)Q\tilde{x}(t) + u^T(t)Ru(t))dt\}}{E\{\tilde{x}^T(0)\tilde{x}(0) + \int_0^{t_p} v'^T(t)v'(t)dt\}} \tag{2.10}$$

In the above optimal H∞ stochastic control problem, external disturbance $v'(t)$ is considered as one player because it is unavailable for control strategy $u(t)$ and only its worst-case effect is considered to be minimized in safety, i.e., the optimal H_∞ stochastic control problem in (2.10) is a two-player stochastic minmax H_∞ game.

Remark 2.3 In the conventional quadratic (i.e., H_2) noncooperative game control problems [1], i.e., noncooperative game with quadratic utility function [20], since it cannot treat the external disturbance $v(t)$, the external disturbances are always neglected. In general, external disturbances are unavoidable in real physical systems. In this chapter, only multi-player noncooperative stochastic H_∞ game problem in (2.6)–(2.8) is discussed for linear stochastic systems.

Remark 2.4 If the stabilization problem is considered in the m-player noncooperative stochastic H_∞ game control strategy, the following stochastic noncooperative H_∞ game problem in (2.6)–(2.8) is modified as the following for the linear stochastic system in (2.3):

$$\rho_i^* = \min_{u_i(t)} \max_{u_{-i}(t)} \frac{E\{\int_0^{t_p} (x^T(t)Q_i x(t) + u_i^T(t)R_i u_i(t))dt\}}{E\{x^T(0)x(0) + \int_0^{t_p} u_{-i}^T(t)u_{-i}(t)dt\}}$$

i.e., the desired target $x_d^i = 0$ for all $i = 1, \ldots, n$.

2.3 Suboptimal Method for Multi-Player Noncooperative H_∞ Game Strategy Design of Linear Stochastic Systems

In general, it is difficult to solve the multi-player noncooperative H_∞ game problem in (2.8) directly. Therefore, the following indirect suboptimal method is employed to solve the stochastic noncooperative H_∞ game problem in (2.8) by minimizing their corresponding upper bounds simultaneously

$$\rho_i^* = \min_{u_i(t)} \max_{u_{-i}(t)} \frac{E\{\int_0^{t_p} (\tilde{x}_i^T(t)Q_i\tilde{x}_i(t) + u_i^T(t)R_i u_i(t))dt\}}{E\{\tilde{x}_i^T(0)\tilde{x}_i(0) + \int_0^{t_p} u_{-i}^T(t)u_{-i}(t)dt\}} \leq \rho_i \quad (2.11)$$

$$i = 1, 2, \ldots, m.$$

where the positive value ρ_i denotes the upper bound of the ith player's H_∞ tracking performance ρ_i^*. The upper bound ρ_i of ρ_i^* in (2.11) could be given beforehand and then be decreased as small as possible to approach ρ_i^* for all players simultaneously to attain the real m-player noncooperative stochastic H_∞ game in (2.8). This indirect method is the so-called suboptimal approach to m-player noncooperative stochastic minmax H_∞ game design problem of linear stochastic game systems in (2.11), i.e., we could solve the following MOP by minimizing their upper bound ρ_i simultaneously:

$$(\rho_1^*, \ldots, \rho_i^*, \ldots, \rho_m^*) = \min_{(u_1(t)\ldots u_m(t))} (\rho_1, \ldots, \rho_i, \ldots, \rho_m) \quad (2.12)$$

subject to

$$\min_{u_i(t)}\max_{u_{-i}(t)} \frac{E\{\int_0^{t_p}(\tilde{x}_i^T(t)Q_i\tilde{x}_i(t)+u_i^T(t)R_iu_i(t))dt\}}{E\{\tilde{x}_i^T(0)\tilde{x}_i(0)+\int_0^{t_p}u_{-i}^T(t)u_{-i}(t)dt\}} \leq \rho_i\ ,\ i=1,2,\ldots,m. \tag{2.13}$$

where $\min(\rho_1,\ldots,\rho_i,\ldots,\rho_m)$ denotes the minimization of $\rho_1,\ldots,\rho_i,\ldots,$ and ρ_m simultaneously, i.e., an m-tuple MOP.

Remark 2.5 By the concept of Pareto dominance in the MOP in (2.12) and (2.13) [9,10], the vector $(\rho_1^*,\ldots,\rho_i^*,\ldots,\rho_m^*)$ dominates the vector $(\rho_1,\ldots,\rho_i,\ldots,\rho_m)$ if $\rho_i^* \leq \rho_i, \forall i \in \{1,2,\ldots,m\}$ and at least one of the inequalities is strict inequality. Hence, the MOP in (2.12) and (2.13) is well defined to find the Pareto optimal solution $(\rho_1^*,\ldots,\rho_i^*,\ldots,\rho_m^*)$ which cannot be dominated by other solutions. The formal definition of Pareto dominance can be referred to Definition 2.1 in the sequel.

Theorem 2.1
The solution of MOP in (2.12) and (2.13) is equivalent to the solution of the m-player noncooperative H_∞ game problem in (2.8).

Proof: Let $(\rho_1^*,\ldots,\rho_i^*,\ldots,\rho_m^*)$ be a solution of MOP in (2.12). If $(\rho_1^*,\ldots,\rho_i^*,\ldots,\rho_m^*)$ is not a solution H∞ noncooperative game problem in (2.8), then there at least exists a ρ_i' to be the solution of (2.8) such that

$$\min_{u_i(t)}\max_{u_{-i}(t)} \frac{E\{\int_0^{t_p}(\tilde{x}_i^T(t)Q_i\tilde{x}_i(t)+u_i^T(t)R_iu_i(t))dt\}}{E\{\tilde{x}_i^T(0)\tilde{x}_i(0)+\int_0^{t_p}u_{-i}^T(t)u_{-i}(t)dt\}} = \rho_i' \tag{2.14}$$

and $\rho_i'<\rho_i^*$ for some i. As a result, we immediately find the vector $(\rho_1^*,\ldots,\rho_i',\ldots,\rho_m^*)$ dominates the vector $(\rho_1^*,\ldots,\rho_i^*,\ldots,\rho_m^*)$ and this implies $(\rho_1^*,\ldots,\rho_i^*,\ldots,\rho_m^*)$ is not a solution of MOP in (2.12) and (2.13). This leads to a contradiction. Q.E.D.

The minmax constraints (2.13) on the MOP in (2.12) for the suboptimal m-player noncooperative H_∞ game problem in (2.11) are equivalent to m-constrained Nash minmax quadratic games because $u_{-i}(t)$ is independent of the minimization of $u_i(t)$ in (2.3) [1,8]

$$\begin{aligned}\min_{u_i(t)}\max_{u_{-i}(t)} E\int_0^{t_p}(\tilde{x}_i^T(t)Q_i\tilde{x}_i(t)+u_i^T(t)R_iu_i(t)-\rho_iu_{-i}^T(t)u_{-i}(t))dt \\ \leq \rho_iE\{\tilde{x}_i^T(0)\tilde{x}_i(0)\},\ \text{for } i=1,2,\ldots,m.\end{aligned} \tag{2.15}$$

Let us denote

$$J_i = E\int_0^{t_p}(\tilde{x}_i^T(t)Q_i\tilde{x}_i(t)+u_i^T(t)R_iu_i(t)-\rho_iu_{-i}^T(t)u_{-i}(t))dt,\ \text{for } i=1,2,\ldots,m. \tag{2.16}$$

We need two steps to solve the constrained minmax quadratic game problem in (2.15). The first step is to solve the following Nash minmax quadratic game problem:

$$J_i^* = \min_{u_i(t)}\max_{u_{-i}(t)} J_i \tag{2.17}$$

and the second step is to solve the following inequality constraints:

$$J_i^* \leq \rho_i E\{\tilde{x}_i^T(0)\tilde{x}_i(0)\} \tag{2.18}$$

By solving the constrained Nash minmax quadratic game problem in (2.17) and (2.18), the minmax H_∞ constraints in (2.13) of MOP of m-player noncooperative H_∞ game problem in (2.8) could be transformed to a set of LMIs so that the MOP with m minmax H_∞ constraints in (2.12) and (2.13) could be transformed to an MOP with m LMIs constraints to simplify the design procedure of m-player noncooperative H_∞ game problem of linear stochastic system.

If we could solve (2.17) and (2.18) for the m-constrained Nash minmax quadratic games in (2.15), then the MOP in (2.12) and (2.13) for m-player noncooperative H_∞ game in (2.8) could be transformed to an LMIs-constrained MOP in the following.

Before the description of main theorem, two following lemmas are given to help us solve the stochastic Nash minmax quadratic game in (2.15) or (2.17) and (2.18).

Lemma 2.1 [23] For any matrix X and Y with appropriate dimensions, we have

$$X^TY + Y^TX \leq X^TP^{-1}X + Y^TPY \tag{2.19}$$

where P is any positive definite symmetric matrix

Lemma 2.2 ([22,31]) Let $V : \mathbb{R}^{\kappa} \to \mathbb{R}$, $\mathbb{V}(\cdot) \in \mathbb{C}^{\not{k}}(\mathbb{R}^{\kappa})$ and $V(\cdot) \geq 0$. For the m-player linear stochastic jump diffusion system in (2.7), the Itô–Lévy formula of $V(\tilde{x}_i(t))$ is given as follows:

$$\begin{aligned} dV(\tilde{x}_i(t)) &= \left(\frac{\partial V(\tilde{x}_i(t))}{\partial \tilde{x}_i(t)}\right)^T (A\tilde{x}_i(t) + B_iu_i + B_{-i}u_{-i})dt + \left(\frac{\partial V(\tilde{x}_i(t))}{\partial \tilde{x}_i(t)}\right)^T C\tilde{x}_i(t)dw(t) \\ &+ \frac{1}{2}\tilde{x}^T(t)C^T\frac{\partial^2 V(\tilde{x}_i(t))}{\partial^2 \tilde{x}(t)}C\tilde{x}_i(t)dt + [V(\tilde{x}_i(t) + D\tilde{x}_i(t)) - V(\tilde{x}_i(t))]dp(t) \end{aligned} \tag{2.20}$$

Remark 2.6 The first two terms in the right-hand side of (2.20) are due to the chain rule in the conventional functional derivative. The Itô terms $\frac{1}{2}\tilde{x}_i^T(t)C^T\frac{\partial^2 V(\tilde{x}_i(t))}{\partial^2 \tilde{x}(t)}C\tilde{x}_i(t)$ and Lévy terms $[V(\tilde{x}_i(t) + D\tilde{x}_i(t)) - V(\tilde{x}_i(t))]$ are used to compensate the nondifferential of Wiener process and Poisson process in (2.6) in the conventional derivative through Itô and Lévy derivative, respectively [7,22].

Theorem 2.2
The constrained Nash minmax quadratic games in (2.15) for the suboptimal m-player noncooperative H∞ game in (2.11) or (2.13) or equivalently Nash minmax problems in (2.17) and (2.18) could be solved by

$$u_i^*(t) = -R^{-1}B_i^TP\tilde{x}_i(t) \tag{2.21}$$

$$u_{-i}^*(t) = \frac{1}{\rho_i}B_{-i}^TP\tilde{x}_i(t) \tag{2.22}$$

where the positive definite symmetric matrix P is the solution of the following Riccati-like inequalities

$$\begin{aligned}&PA + A^TP + Q_i + C^TPC + \lambda(PD + D^TP + D^TPD)\\&-PB_iR_i^{-1}B_i^TP + \frac{1}{\rho_i}PB_{-i}B_{-i}^TP \le 0, \quad i = 1, 2, \ldots, m\end{aligned} \tag{2.23}$$

with

$$P \le \rho_i I, \; i = 1, 2, \ldots, m. \tag{2.24}$$

Proof:

Let the Lyapunov function be $V(\tilde{x}) = \tilde{x}_i^T(t)P\tilde{x}_i(t)$. Then, (2.16) can be rewritten as

$$\begin{aligned}J_i = &E\tilde{x}^T(0)\tilde{x}(0) - E\tilde{x}^T(t_p)P\tilde{x}(t_p) + E\int_0^{t_p}(\tilde{x}_i^T(t)Q_i\tilde{x}_i(t) + u_i^T(t)R_iu_i(t)\\&-\rho_iu_{-i}^T(t)u_{-i}(t))dt + d\tilde{x}_i^T(t)P\tilde{x}_i(t)\end{aligned} \tag{2.25}$$

By applying Lemma 2.2 and using the fact that $Edw(t) = 0$ and $E[dp(t)] = \lambda dt$, (2.25) can be rewritten as

$$\begin{aligned}J_i = &E\tilde{x}_i^T(0)P\tilde{x}(0) - E\tilde{x}_i^T(t_p)P\tilde{x}(t_p) + E\int_0^{t_p}[(\tilde{x}_i^T(t)Q_i\tilde{x}_i(t) + u_i^T(t)R_iu_i(t)\\&-\rho_iu_{-i}^T(t)u_{-i}(t) + \tilde{x}_i^T(t)P(A\tilde{x}_i(t) + B_iu_i(t) + B_{-i}u_{-i}(t)) + (A\tilde{x}_i(t) + B_iu_i(t)\\&+B_{-i}u_{-i}(t))^TP\tilde{x}_i(t) + \tilde{x}_i^T(t)C^TP\tilde{x}_i(t) + \lambda\tilde{x}_i^T(t)(PD + D^TP + D^TPD)\tilde{x}_i(t)]dt\end{aligned} \tag{2.26}$$

Then, we get

$$\begin{aligned}J_i = &E\tilde{x}_i^T(0)P\tilde{x}(0) - E\tilde{x}_i^T(t_p)P\tilde{x}(t_p) + E\int_0^{t_p}\{(\tilde{x}_i^T(t)Q_i\tilde{x}_i(t) + \tilde{x}_i^T(t)PA\tilde{x}_i(t)\\&+\tilde{x}_i^T(t)A^TP\tilde{x}_i(t) + (u_i^T(t)R_i + \tilde{x}^T(t)PB_i)R_i^{-1}(R_iu_i(t) + B_i^TP\tilde{x}(t))\\&-(B_{-i}^TP\tilde{x}_i(t) - \rho_iu_{-i}(t))^T\rho_i^{-1}(B_{-i}^TP\tilde{x}_i(t) - \rho_iu_{-i}(t))\\&-\tilde{x}_i^T(t)PB_iR_i^{-1}B_i^TP\tilde{x}_i(t) + \frac{1}{\rho_i}\tilde{x}_i(t)PB_{-i}B_{-i}^TP\tilde{x}_i(t) + \tilde{x}_i^T(t)C^TPC\tilde{x}_i^T(t)\\&+\lambda\tilde{x}_i^T(t)(D^TP + PD + D^TPD)\tilde{x}_i(t)\}dt\end{aligned} \tag{2.27}$$

Then,

$$\begin{aligned}\min_{u_i(t)}\max_{u_{-i}(t)}J_i = &\min_{u_i(t)}\max_{u_{-i}(t)}\{(E\tilde{x}_i^T(0)P\tilde{x}_i(0) - E\tilde{x}_i^T(t_p)P\tilde{x}_i(t_p)\\&+E\int_0^{t_p}(\tilde{x}_i^T(t)Q_i\tilde{x}_i(t) + \tilde{x}_i^T(t)PA\tilde{x}_i(t) + \tilde{x}_i^T(t)A^TP\tilde{x}_i(t) - \tilde{x}_i^T(t)PB_iR_i^{-1}B_i^TP\tilde{x}_i(t)\\&+\frac{1}{\rho_i}\tilde{x}_i^T(t)PB_{-i}^TB_{-i}P\tilde{x}_i(t) + \tilde{x}_i^T(t)C^TPC\tilde{x}_i^T(t) + \lambda\tilde{x}_i^T(t)(D^TP\\&+PD + D^TPD)\tilde{x}_i(t) + (u_i^T(t)R_i + \tilde{x}^T(t)PB_i)R_i^{-1}(R_iu_i(t) + B_i^TP\tilde{x}(t))\\&-[(B_{-i}^TP\tilde{x}_i(t) - \rho_iu_{-i}(t))^T\rho_{-i}^{-1}(B_{-i}^TP\tilde{x}_i(t) - \rho_iu_{-i}(t))\}dt\end{aligned}$$

We get the solution $u_i^*(t)$ in (2.21) and $u_{-i}^*(t)$ in (2.22). Then,

$$\begin{aligned} J_i^* = \min_{u_i(t)} \max_{u_{-i}(t)} J_i = E\tilde{x}_i^T(0)P\tilde{x}_i(0) - E\tilde{x}_i^T(t_p)P\tilde{x}_i(t_p) + E\int_0^{t_p} \tilde{x}_i^T(t)[PA + A^TP + Q_i \\ + C^TPC + \lambda(PD + D^TP + D^TPD) - PB_iR_i^{-1}B_i^TP + \frac{1}{\rho_i}PB_{-i}B_{-i}^TP]\tilde{x}_i(t)dt \end{aligned} \tag{2.28}$$

If the Riccati-like inequalities in (2.23) hold then, we get

$$J_i^* = \min_{u_i(t)} \max_{u_{-i}(t)} J_i \leq E\tilde{x}^T(0)P\tilde{x}(0) - E\tilde{x}^T(t_p)P\tilde{x}(t_p) \leq E\tilde{x}^T(0)P\tilde{x}(0). \tag{2.29}$$

By (2.18) and (2.29), we get $E\tilde{x}^T(0)P\tilde{x}(0) \leq \rho_i E\tilde{x}_i^T(0)\tilde{x}_i(0)$. Then, we get (2.24).

Q.E.D.

The solutions $u_i^*(t)$ and $u_{-i}^*(t)$ in (2.21) and (2.22) are the solutions of constrained Nash minmax quadratic game problem in (2.17) and (2.18) or the m-constrained Nash minmax quadratic games in (2.15), which is just equivalent to the minmax H_∞ game constraints in (2.13) of the MOP in (2.12). Therefore, the MOP in (2.12) and (2.13) for the noncooperative minmax H_∞ game of linear stochastic system could be transformed into the following MOP to simultaneously minimize their upper bounds under the Riccati-like inequalities in (2.23) and (2.24) as follows:

$$(\rho_1^*, \dots, \rho_i^*, \dots, \rho_m^*) = \min_{P>0}(\rho_1, \dots, \rho_i, \dots, \rho_m) \tag{2.30}$$

subject to (2.23) and (2.24)

After solving MOP in (2.30) for the optimal solution P^*, the noncooperative minmax H_∞ game strategies of m players are constructed as $u_i^*(t) = -R^{-1}B_i^TP^*\tilde{x}_i(t)$ for i=1,2, ..., m. In general, the solution of MOP in (2.30) is not unique. There exist a set of solutions for MOP in (2.30). Some properties of MOP in (2.30) are given as follows:

Definition 2.1 Pareto Dominance: For the MOP in (2.30), suppose there have two multiobjective feasible solutions $(u_1^1(t), \dots, u_m^1(t))$ and $(u_1^2(t), \dots, u_m^2(t))$ with the corresponding objective values $(\rho_1^1, \dots, \rho_m^1)$ and $(\rho_1^2, \dots, \rho_i^2, \dots, \rho_m^2)$, respectively, the solution $(u_1^1(t), \dots, u_m^1(t))$ is said to dominate $(u_1^2(t), \dots, u_m^2(t))$ if $\rho_1^1 \leq \rho_1^2, \dots, \rho_i^1 \leq \rho_i^2, \dots, \rho_m^1 \leq \rho_m^2$ and at least one of the inequalities is a strict inequality.

Definition 2.2 Pareto optimality: The feasible solution $(u_1^*, \dots, u_m^*)$ with the corresponding objective values $(\rho_1^*, \dots, \rho_m^*)$ of MOP in (2.30) is said to be of the Pareto optimality with respect to the feasible solution set (i.e., the set of solutions satisfying with Riccati-like inequalities in (2.23) and (2.24)) if and only there does not exist another feasible solution that dominates it.

Remark 2.7 The optimal solution $(u_1^*(t), \dots, u_i^*(t), \dots, u_m^*(t))$ with the corresponding values $(\rho_1^*, \dots, \rho_i^*, \dots, \rho_m^*)$ of m-player noncooperative minmax H_∞ game in (2.8) or equivalently in (2.12) constitutes a Nash equilibrium solution if and only if [1,3]

$$
\begin{aligned}
&(\rho_1^*, ..., \rho_i^*, ..., \rho_m^*) \le (\rho_1, \rho_2^*, ..., \rho_m^*) \\
&\vdots \\
&(\rho_1^*, ..., \rho_i^*, ..., \rho_m^*) \le (\rho_1^*, ...\rho_{i-1}^*, \rho_i, \rho_{i+1}^*, ..., \rho_m^*) \\
&\vdots \\
&(\rho_1^*, ..., \rho_i^*, ..., \rho_m^*) \le (\rho_1^*, ..._i, \rho_{m-1}^*, \rho_m)
\end{aligned}
\tag{2.31}
$$

i.e., no player has anything to gain by changing their own strategy when Nash equilibrium solution is achieved.

Theorem 2.3
The multiobjective optimal solution $(u_1^*, \ldots, u_m^*)$ with the corresponding objective values $(\rho_1^*, \ldots, \rho_i^*, \ldots, \rho_m^*)$ of MOP in (2.30) is the Nash equilibrium solution of m-player noncooperative H_∞ game problem in (2.8) or equivalently the MOP in (2.12) for the linear stochastic system in (2.7) or (2.6).

Proof:
It could be proven by contradiction. If Nash equilibrium solution in (2.31) is violated, for example,

$$
(\rho_1^*, \ldots, \rho_i^*, \ldots, \rho_m^*) > (\rho_1^*, \ldots \rho_{i-1}^*, \rho_i, \rho_{i+1}^*, \ldots, \rho_m^*) \tag{2.32}
$$

for some i, it will violate the domination of $(\rho_1^*, \ldots, \rho_i^*, \ldots, \rho_m^*)$ of the solution of MOP in (2.30) for the noncooperative H_∞ game strategy of linear stochastic systems. Therefore, while MOP in (2.30) is solved, the inequalities (2.31) hold and MOP solution $(u_1^*(t), \ldots, u_i^*(t), \ldots, u_m^*(t))$ with the corresponding values $(\rho_1^*, \ldots, \rho_i^*, \ldots, \rho_m^*)$ is the Nash equilibrium solution. The proof is completed.

Q.E.D.

Since the Riccati-like inequalities in (2.23) are bilinear function of P, it is still difficult to solve it for P>0. It could be transformed to a set of equivalent LMIs by the following lemma.

Lemma 2.3 [23] Schur complements
The following LMI

$$
\begin{bmatrix} Q & S \\ S^T & R \end{bmatrix} < 0 \tag{2.33}
$$

where $Q = Q^T$, $R = R^T$ and S is arbitrary matrix, is equivalent to

$$
R > 0, Q - SR^{-1}S^T < 0 \tag{2.34}
$$

In other words, the bilinear inequality in (2.34) can be represented by an equivalent LMI in (2.33).

Let us denote $W = P^{-1}$ and perform the multiplication of W to both sides of (2.23), we get

$$
\begin{aligned}
&AW + WA^T + WQ_iW + WC^TW^{-1}CW - B_iR_i^{-1}B_i^T \\
&+ \lambda(DW + WD^T + WD^TW^{-1}DW) + \frac{1}{\rho_i}B_{-i}B_{-i}^T \le 0
\end{aligned}
\tag{2.35}
$$

By performing the Schur complements in Lemma 2.3 several times, we could obtain the following equivalent LMIs of Riccati-like inequalities in (2.35):

$$\begin{bmatrix} \Pi_i & W & CW & \lambda^{\frac{1}{2}}DW & B_{-i} \\ W & -Q_i^{-1} & 0 & 0 & 0 \\ (CW)^T & 0 & -W & 0 & 0 \\ \lambda^{\frac{1}{2}}(DW)^T & 0 & 0 & -W & 0 \\ B_{-i}^T & 0 & 0 & 0 & -\rho_i I \end{bmatrix} \leq 0, i = 1, \ldots, m. \tag{2.36}$$

where $\Pi_i \triangleq AW + WA^T + \lambda(DW + WD^T) - B_i R_i^{-1} B_i^T$.

Similarly, the inequalities in (2.24) are equivalent to

$$-\rho_i I + W^{-1} \leq 0, i = 1, 2, \ldots, m \tag{2.37}$$

or equivalent to the following LMIs by Schur complements

$$\begin{bmatrix} -\rho_i I & I \\ I & -W \end{bmatrix} \leq 0, i = 1, 2, \ldots, m \tag{2.38}$$

Remark 2.8 By the equivalence between (2.23), (2.24) and (2.36), (2.38), the Riccati-like inequalities-constrained MOP in (2.30) for the noncooperative stochastic H_∞ game strategy of m-player in linear stochastic system could be represented by the following LMIs-constrained MOP:

$$(\rho_1^*, \ldots, \rho_i^*, \ldots, \rho_m^*) = \min_{W>0, \rho_i} (\rho_1, \ldots, \rho_i, \ldots, \rho_m) \tag{2.39}$$

subject to (2.36) and (2.38)

Since Riccati-like inequalities (2.23) are transformed into equivalent LMIs in (2.36), the LMIs-constraints in (2.36) can be easily solved with the help of Matlab LMI toolbox. To solve MOP in (2.39), we always employ the MOEA algorithm because the MOEA can parallelly search the Pareto optimal solutions and escape from local optimal solution [24,25]. However, the conventional MOEA needs to search W directly to solve MOP in (2.39). Therefore, the dimension of search space $(w_{11}, w_{12}, \ldots, w_{nn})$ becomes very large, where w_{ij} is the ij component of W. In this situation, it becomes very complex when the dimension of state vector is large. With some modifications, the MOEAs could solve MOP in (2.39) indirectly. In our proposed LMIs-constrained MOEA method, we search feasible $(\rho_1, \ldots, \rho_2, \ldots, \rho_m)$ by MOEA algorithms, which must satisfy LMIs in (2.36) and (2.38). Then, we solve W>0 indirectly from (2.36) and (2.38) with the help of LMI toolbox in Matlab. Since MOEA algorithms could solve an MOP via a stochastic search method based on a "survival of the fittest" law to approach a set of compromised solutions called Pareto optimal solutions through the evolutionary algorithm (EA), including crossover, mutation, and elitist selection operator. Since the constraints of (2.36) and (2.38) on (2.39) are a set of LMIs, an LMIs-based MOEA is developed to efficiently solve the MOP in (2.39). Some important definitions about Pareto optimality of LMIs-constrained MOP in (2.39) are given as follows [9,10]:

Definition 2.3 Pareto optimal solution sets: For the given LMIs-constrained MOP in (2.39), the Pareto optimal solution set P_s is defined as $P_s \triangleq \{(u_i^*(t), \ldots, u_m^*(t)) |$. There does not exist another feasible solution $(u_1^0(t), \ldots, u_m^0(t))$ such that $(\rho_1^0, \ldots, \rho_n^0)$ dominates $(\rho_1^*, \ldots, \rho_m^*)\}$.

Definition 2.4 Pareto front: For the given LMIs-constrained MOP in (2.39), the Pareto front P_F is defined as $P_F \triangleq \{(\rho_1^*, \ldots, \rho_m^*)|(u_1^*(t), \ldots, u_m^*(t) \in P_s\}$.

For the m-player noncooperative H_∞ game of linear stochastic system in (2.3), based on the LMIs-constrained MOP in (2.39), a design procedure based on MOEA is introduced as follows:

Design procedure of LMIs-constrained MOEA for m-player noncooperative H∞ game strategy of linear stochastic systems

Step 1: Initialization

Step 2: Set the searching region $[\underline{\rho}_1, \bar{\rho}_1] \times \cdots \times [\underline{\rho}_m, \bar{\rho}_m] \triangleq S$, the maximum number N_g of individuals (population), the crowding distance ς_d, the crossover rate M_c, the mutation rate M_r, and the iteration number N_i in the LMIs-constrained MOEA, where $\underline{\rho}_i$ and $\bar{\rho}_i$ denote the lower and upper bounds of ρ_i, respectively.

Step 3: Set algorithm counter t=1. Randomly generate a feasible initial population $a_1, a_2, \ldots, a_{Ng} \in S$ and let $\bar{P}_t = \{a_1, a_2, \ldots, a_{Ng}\}$.

Step 4: Differential evolution

Step 4.1: Perform the crossover operation on $\bar{P}_t$ with crossover rate M_c.

Step 4.2: Perform the mutation operation on $\bar{P}_t$ with mutation rate M_r.

Step 4.3: Generate N_g feasible populations as the child population set P_t^c by examining whether their corresponding objective vectors $(\rho_1, \ldots, \rho_m)$ are feasible objective vectors satisfying the LMIs in (2.36) and (2.38), i.e., collects the feasible population generated by crossover operation and mutation operation.

Step 5: Updating.

Selecting N_g populations from the set $P_t^c \cup \bar{P}_t$ by using nondominated sorting method and crowded-comparing operator and the set $\bar{P}$ collects the selected population.

Set the algorithm counter $t = t + 1$ and $\bar{P}_t = \bar{P}$.

Step 6: Stopping criterion

Repeat Steps 4, 5 until the iteration number N_i is satisfied, then set the final population as the Pareto front P_F

Step 7: Knee selection

Select a preferable feasible objective individual $(\rho_1^*, \ldots, \rho_i^*, \ldots, \rho_m^*)$ and the corresponding solution $P^* = W^{*^{-1}}$ according to designer's own preference. In general, the knee point in the Pareto front is preferred.

Step 8: Obtain the Pareto optimal solution of knee point and construct the player strategy $u_i^*(t) = -R^{-1}B_i^T P^* \tilde{x}_i(t)$ from (2.21) for i=1,2, …, m.

Remark 2.9 If only one control strategy is employed to achieve a desired target x_d for linear stochastic system with external disturbance in (2.9), 2-player noncooperative H_∞ game problem becomes the optimal H_∞ robust control problem in (2.10). In this case, external disturbance is considered as another player because it is unavailable and unpredictable. This optimal H_∞ robust control design has been widely applied in linear stochastic systems to achieve a desired target despite intrinsic continuous and discontinuous random fluctuation and external disturbance.

By the similar indirect method, the optimal H_∞ robust control problem in (2.9) and (2.10) could be solved by the suboptimal H_∞ game method as follows:

$$\rho^* = \min_{u(t)} \max_{v'(t)} \frac{E\{\int_0^{t_p} (\tilde{x}^T(t)Q\tilde{x}(t) + u^T(t)Ru(t))dt\}}{E\{\tilde{x}^T(0)\tilde{x}(0) + \int_0^{t_p} v'^T(t)v'(t)dt\}} \leq \rho \tag{2.40}$$

subject to

$$d\tilde{x}(t) = (A\tilde{x}(t) + v'(t) + Bu(t))dt + C\tilde{x}(t)dw + D\tilde{x}(t)dp(t) \tag{2.41}$$

We need to solve the suboptimal 2-player H_∞ game problem in (2.40) and (2.41) with the upper bound ρ at first and then decrease ρ as small as possible to approach ρ^* for the optimal H∞ robust control design in (2.9) and (2.10).

Theorem 2.4
For the linear stochastic system (2.41), the suboptimal H_∞ game problem in (2.40) is solved by the following suboptimal control strategy $u^*(t)$ and the worst-case external disturbance $v'^*(t)$:

$$u^*(t) = -R^{-1}B^T P\tilde{x}(t) \tag{2.42}$$

$$v'^*(t) = \frac{1}{\rho} P\tilde{x}(t) \tag{2.43}$$

where the positive definite symmetry matrix P of the following Riccati-like inequality

$$\begin{aligned} &PA + A^T P + Q + C^T PC + \lambda(PD + D^T P + D^T PD) \\ &- PBR^{-1}B^T P + \frac{1}{\rho} PP \leq 0 \end{aligned} \tag{2.44}$$

with

$$P \leq \rho I \tag{2.45}$$

Proof: Similar to proof of Theorem 2.2.

By the similar derivation of LMIs in (2.36), the Riccati-like inequality in (2.44) could be transformed to the following equivalent LMI with $W = P^{-1}$

$$\begin{bmatrix} \Pi & W & CW & \lambda^{\frac{1}{2}}DW & I \\ W & -Q^{-1} & 0 & 0 & 0 \\ (CW)^T & 0 & -W & 0 & 0 \\ \lambda^{\frac{1}{2}}(DW)^T & 0 & 0 & -W & 0 \\ I & 0 & 0 & 0 & -\rho I \end{bmatrix} \leq 0 \tag{2.46}$$

where $\Pi \triangleq AW + WA^T + \lambda(DW + WD^T) - BR^{-1}B^T$. Similarly, (2.45) becomes the following LMI:

$$\begin{bmatrix} -\rho I & I \\ I & -W \end{bmatrix} \leq 0 \tag{2.47}$$

Since $u^*(t)$ in (2.42) and $v^{i*}(t)$ in (2.43) are the suboptimal solution of optimal H$_\infty$ robust control problem with the upper bound ρ in (2.40), the optimal stochastic H$_\infty$ control design problem becomes the following LMI-constrained optimization problem:

$$\begin{aligned}&\min_{W>0}\rho\\&\text{subject to (2.46) and (2.47)}\end{aligned} \tag{2.48}$$

The single-objective optimal problem (SOP) in (2.48) could be efficiently solved by decreasing ρ until no positive definite solution W exists with the help of LMI toolbox in Matlab. After solving SOP in (2.48), we could obtain ρ^* and $W^* = {P^*}^{-1}$. Then, the optimal H$_\infty$ robust control strategy for (2.10) of linear stochastic system (2.9) is given by $u^* = -R^{-1}B^TP^*\tilde{x}(t)$.

Remark 2.10 For the m-player noncooperative H$_\infty$ game strategy of the stabilization control problem in the linear stochastic system in (2.3) in Remark 2.4, i.e., the desired targets $x_d^i = 0$ for i=1, 2, ...,m, we don't need to shift the origin of the linear stochastic system (2.3) to x_d^i for each player as (2.6). In this situation, the suboptimal m-player noncooperative H$_\infty$ game (2.11) for the stabilization control problem of linear stochastic system (2.3) should be modified as follows:

$$\begin{aligned}&\rho_i^* = \min_{u_i(t)}\max_{u_{-i}(t)}\frac{E\{\int_0^{t_p}(x^T(t)Q_ix(t)+u_i^T(t)R_iu_i(t))dt\}}{E\{x^T(0)x(0)+\int_0^{t_p}u_{-i}^T(t)u_{-i}(t)dt\}} \le \rho_i\\&i=1,2,\ldots,m.\end{aligned} \tag{2.49}$$

Then, the suboptimal m-player noncooperative H$_\infty$ tracking control game strategy in Theorem 2.2 should be modified for m-player noncooperative H$_\infty$ stabilization control game strategy as follows:

Theorem 2.5

The suboptimal problem of m-player noncooperative H$_\infty$ stabilization game strategy of (2.49) for linear stochastic system in (2.3) could be solved by

$$\begin{aligned}u_i^*(t) &= -R_i^{-1}B_i^TP^*x(t), \quad i=1,\ldots,m\\u_{-i}^*(t) &= \frac{1}{\rho_i^*}B_{-i}^TP^*x(t), \quad i=1,\ldots,m\end{aligned} \tag{2.50}$$

where ρ_i^* and the positive definite symmetric matrix P^* are the solution of the following Riccati-like inequalities-constrained MOP

$$(\rho_1^*,\ldots,\rho_i^*,\ldots,\rho_m^*) = \min_P(\rho_1,\ldots,\rho_i,\ldots,\rho_m) \tag{2.51}$$

subject to

$$\begin{aligned}&PA + A^TP + Q_i + C^TPC + \lambda(PD + D^TP + D^TPD)\\&-PB_iR_i^{-1}B_i^TP + \frac{1}{\rho}PB_i^TB_{-i}P \le 0 \quad i=1,2,\ldots,m.\end{aligned} \tag{2.52}$$

with

$$P \leq \rho_i \ \ Ii = 1, 2, \ldots, m. \tag{2.53}$$

Proof: Similar to the proof of Theorem 2.2.

By the Schur complements and with $W = P^{-1}$, the Riccati-like inequalities in (2.52) and (2.53) could be equivalently transformed to the following LMIs, respectively

$$\begin{bmatrix} \Pi_i & W & CW & \lambda^{\frac{1}{2}}DW & B_{-i} \\ W & -Q_i^{-1} & 0 & 0 & 0 \\ (CW)^T & 0 & -W & 0 & 0 \\ \lambda^{\frac{1}{2}}(DW)^T & 0 & 0 & -W & 0 \\ B_{-i}^T & 0 & 0 & 0 & -\rho_i I \end{bmatrix} \leq 0 \ \ i = 1, 2, \ldots, m \tag{2.54}$$

where $\Pi_i \triangleq AW + WA^T + \lambda(DW + WD^T) - B_i R_i^{-1} B_i^T$.

$$\begin{bmatrix} -\rho_i I & I \\ I & -W \end{bmatrix} \leq 0 \tag{2.55}$$

Therefore, the Riccati-like inequalities-based MOP in (2.51) for the m-player H_∞ game strategy of (2.49) for the stabilization control problem of linear stochastic system in (2.3) becomes how to solve the following LMIs-constrained MOP:

$$(\rho_1^*, \ldots, \rho_i^*, \ldots, \rho_m^*) = \min_{W>0}(\rho_1, \ldots, \rho_i, \ldots, \rho_m) \tag{2.56}$$

subject to (2.54) and (2.55)

The proposed LMIs-constrained MOEA could be also employed to solve ρ_i^* and $P^* = (W^*)^{-1}$ from the above MOP for $u_i^*(t) = -R_i^{-1} B_i^T P^* x(t)$ as m-player H_∞ strategy in (2.49) for the stabilization of linear stochastic system in (2.3).

2.4 Stochastic Multi-Player H_2 Noncooperative Game Strategy in Linear Stochastic Systems

If the linear stochastic system in (2.6) is free of external disturbance $v_i(t)$ or the external disturbance $v_i(t)$ of linear stochastic system in (2.6) could be neglected, i.e.,

$$d\tilde{x}_i(t) = (A\tilde{x}_i(t) + \sum_{i=1}^{m} B_i u_i(t))dt + C\tilde{x}_i(t)dw(t) + D\tilde{x}_i(t)dp(t), \ \ i = 1, \ldots, m \tag{2.57}$$

which could be represented by

$$d\tilde{x}_i(t) = (A\tilde{x}_i(t) + B_i u_i(t) + B_{-i} u_{-i}(t))dt + C\tilde{x}_i(t)dw(t) + D\tilde{x}_i(t)dp(t) \tag{2.58}$$

where $B_{-i} = [B_1 \cdots B_{i-1} \ B_{i+1} \cdots B_m]$ and $u_{-i}(t) = [u_1^T(t) \cdots u_{i-1}^T(t) \ u_{i+1}^T(t) \cdots u_m^T(t)]^T$.

The following m-player noncooperative stochastic H_2 game problem is formulated by the following minimax quadratic Nash games simultaneously:

$$J_i^* = \min_{u_i(t)} \max_{\substack{u_j(t) \\ j=1,2,\dots,m \\ j\neq i}} E\int_0^{t_p} [(x(t)-x_d^i)^T Q_i (x(t)-x_d^i) + u_i^T(t) R_i u_i(t) - \sum_{\substack{j=1 \\ j\neq i}}^{m} u_j^T(t) R_j u_j(t)]dt \tag{2.59}$$

$i=1,2,\dots,m.$

which could be represented by

$$J_i^* = \min_{u_i(t)} \max_{u_{-i}(t)} E\int_0^{t_p} [\tilde{x}_i^T(t) Q_i \tilde{x}_i(t) + u_i^T(t) R_i u_i(t) - u_{-i}^T(t) R_{-i} u_{-i}(t)]dt \tag{2.60}$$

$i = 1, 2, \dots, m.$

where $R_{-i} = \begin{bmatrix} R_1 & & & & & 0 \\ & \ddots & & & & \\ & & R_{i-1} & & & \\ & & & R_{i+1} & & \\ & 0 & & & \ddots & \\ & & & & & R_m \end{bmatrix}$

In general, it is not easy to solve the m minmax H_2 noncooperative games in (2.60) for linear stochastic system in (2.58) to obtain $(J_1^*, \dots, J_i^*, \dots, J_m^*)$ directly. An indirect method is introduced to simultaneously solve m minmax H_2 games in (2.60) by minimizing their upper bounds in the following MOP from the suboptimal perspective:

$$(J_1^*, \dots, J_i^*, \dots, J_m^*) = \min_{u_1(t),\dots,u_m(t)} (J_1, \dots, J_i, \dots, J_m) \tag{2.61}$$

subject to

$$\min_{u_i(t)} \max_{u_{-i}(t)} E\int_0^{t_p} [\tilde{x}_i^T(t) Q_i \tilde{x}_i(t) + u_i^T(t) R_i u_i(t) - u_{-i}^T(t) R_{-i} u_{-i}(t)]dt \leq J_i \tag{2.62}$$

$i = 1, 2, \dots, m.$

where $\min(J_1, \dots, J_i, \dots, J_m)$ denotes the minimization of $J_1, \dots, J_i, \dots,$ and J_m simultaneously.

Theorem 2.6
The solution of MOP in (2.61) and (2.62) is equivalent to the solution of m-player noncooperative stochastic H_2 game problem (2.60)

Proof:
Let $(J_1^*, \dots, J_i^*, \dots, J_m^*)$ be the solution of MOP in (2.61). If $(J_1^*, \dots, J_i^*, \dots, J_m^*)$ is not the solution of m-player H_2 noncooperative game problem in (2.60), then there at least exists a J_i' to be the solution of (2.60) such that

$$\min_{u_i(t)}\max_{u_{-i}(t)} E\int_0^{t_p}[\tilde{x}_i^T(t)Q_i\tilde{x}_i(t)+u_i^T(t)R_iu_i(t)-u_{-i}^T(t)R_{-i}u_{-i}(t)]dt=J_i^{'} \tag{2.63}$$

and $J_i^{'}<J_i^*$ for some i. As a result, we immediately find the vector $(J_1^*,\ldots,J_i^{'},\ldots,J_m^*)$ dominates the vector $(J_1^*,\ldots,J_i^*,\ldots,J_m^*)$ and implies $(J_1^*,\ldots,J_i^*,\ldots,J_m^*)$ is not the solution of MOP in (2.61) and (2.62). This leads to a contradiction.

Q.E.D.

Theorem 2.7
The multiobjective optimal solution $(u_1^*(t),\ldots,u_i^*(t),\ldots,u_m^*(t))$ with the corresponding objective values $(J_1^*,\ldots,J_i^*,\ldots,J_m^*)$ of MOP in (2.61) and (2.62) is the Nash equilibrium solution of m-player stochastic H_2 noncooperative game problem of linear stochastic systems in (2.57).

Proof: Similar to Theorem 2.3.

Lemma 2.4 Let $V:\mathbb{R}^n\rightarrow\mathbb{R}$, $V(\cdot)\in C^2(\mathbb{R}^n)$ and $V(\cdot)\geq 0$. For the m-player linear stochastic jump diffusion system in (2.58), the Itô–Lévy formula of $V(\tilde{x}(t))$ is given as follows [32,33]:

$$\begin{aligned}dV(\tilde{x}_i(t)) &= \left(\frac{\partial V(\tilde{x}(t))}{\partial\tilde{x}_i(t)}\right)^T(A\tilde{x}_i(t)+B_iu_i(t)+B_{-i}u_{-i}(t))dt\\&+\left(\frac{\partial V(\tilde{x}(t))}{\partial\tilde{x}_i(t)}\right)^T C\tilde{x}_i(t)dw(t)+\frac{1}{2}\tilde{x}^T(t)C^T\frac{\partial^2V(\tilde{x}(t))}{\partial\tilde{x}^2(t)}C\tilde{x}(t)dt\\&+[V(\tilde{x}(t)+D\tilde{x}(t))-V(\tilde{x}(t))]dp(t)\end{aligned} \tag{2.64}$$

Theorem 2.8
The MOP in (2.61) and (2.62) for m-player noncooperative H_2 game strategy problem in (2.60) could be solved by

$$u_i^*(t)=-R_i^{-1}B_i^TP^*\tilde{x}_i(t) \tag{2.65}$$

$$u_{-i}^*(t)=R_{-i}^{-1}B_{-i}P^*\tilde{x}_i(t) \tag{2.66}$$

where the positive definite matrix P^* is the solution of the following MOP:

$$(J_1^*,\ldots,J_i^*,\ldots,J_m^*)=\min_{P>0}(J_1,\ldots,J_i,\ldots,J_m) \tag{2.67}$$

subject to

$$\begin{aligned}&PA+A^TP+Q_i+C^TPC+\lambda(D^TP+PD+D^TPD)\\&-PB_iR_i^{-1}B_i^TP+PB_{-i}R_{-i}^{-1}B_i^TP\leq 0\end{aligned} \tag{2.68}$$

$$PR_0^i\leq\frac{J_i}{n}I_n\ i=1,2,\ldots,m \tag{2.69}$$

where $R_0^i=E[\tilde{x}_i(0)\tilde{x}_i^T(0)]$, and I_n denotes the identity matrix of n dimension.

Proof:
From the constrained minmax quadratic game in (2.62) and Itô–Lévy formula in (2.64), by the fact $Edw=0$ and Lyapunov function $dV(\tilde{x}_i(t))=\tilde{x}_i^T(t)P\tilde{x}_i(t)$, we get

$$\begin{aligned}
&\min_{u_i(t)}\max_{u_{-i}(t)} E\int_0^{t_p}[\tilde{x}_i^T(t)Q\tilde{x}_i(t)+u_i^T(t)R_iu_i(t)-u_{-i}^T(t)R_{-i}u_{-i}(t)]dt\\
&=\min_{u_i(t)}\max_{u_{-i}(t)} E\{V(\tilde{x}_i(0))-V(\tilde{x}_i(t_p))+\int_0^{t_p}[\tilde{x}_i^T(t)Q_i\tilde{x}_i(t)\\
&\quad+u_i^T(t)R_iu_i(t)-u_{-i}^T(t)R_{-i}u_{-i}(t)]dt+dV(\tilde{x}_i(t))\\
&=\min_{u_i(t)}\max_{u_{-i}(t)} E\{V(\tilde{x}_i(0))-V(\tilde{x}_i(t_p))+\int_0^{t_p}[\tilde{x}_i^T(t)Q_i\tilde{x}_i(t)+u_i^T(t)R_iu_i(t)\\
&\quad-u_{-i}^T(t)R_{-i}u_{-i}(t)+\left(\frac{\partial V(\tilde{x}_i(t))}{\partial\tilde{x}_i(t)}\right)^T(A\tilde{x}_i(t)+B_iu_i(t)+B_{-i}u_{-i}(t)\\
&\quad+\frac{1}{2}\tilde{x}_i(t)C^T\frac{\partial^2V(\tilde{x}_i(t))}{\partial\tilde{x}_i^2(t)}C\tilde{x}_i(t))+[V(\tilde{x}_i(t)+D\tilde{x}_i(t))-V(\tilde{x}_i(t))]\\
&=\min_{u_i(t)}\max_{u_{-i}(t)} E\{\tilde{x}_i^T(0)P\tilde{x}_i(0)-\tilde{x}_i^T(t_p)P\tilde{x}_i(t_p)+\int_0^{t_p}[\tilde{x}_i^T(t)Q_i\tilde{x}_i(t)+u_i^T(t)R_iu_i(t)\\
&\quad-u_{-i}^T(t)R_{-i}u_{-i}(t)+\tilde{x}_i^T(t)P(A\tilde{x}_i(t)+B_iu_i(t)+B_{-i}u_{-i}(t))\\
&\quad+(A\tilde{x}_i(t)+B_iu_i(t)+B_{-i}u_{-i}(t))^TP\tilde{x}_i(t)+\tilde{x}_i^T(t)C^TPC\tilde{x}_i(t)\\
&\quad+\lambda\tilde{x}_i^T(t)(D^TP+PD+D^TPD)\tilde{x}_i(t)]dt\}\\
&=\min_{u_i(t)}\max_{u_{-i}(t)} E\{\tilde{x}_i^T(0)P\tilde{x}_i(0)-\tilde{x}_i^T(t_p)P\tilde{x}_i(t_p)+\int_0^{t_p}[\tilde{x}_i^T(t)Q_i\tilde{x}_i(t)\\
&\quad+\tilde{x}_i(t)PA\tilde{x}_i(t)+\tilde{x}_i^T(t)AP\tilde{x}_i(t)+\tilde{x}_i^T(t)C^TPC\tilde{x}_i(t)\\
&\quad+\lambda\tilde{x}_i^T(t)(D^TP+PD+D^TPD)\tilde{x}_i(t)-\tilde{x}_i^T(t)PB_iR_i^{-1}B_i^TP\tilde{x}_i(t)\\
&\quad+\tilde{x}_i^T(t)PB_{-i}R_{-i}^{-1}B_{-i}^TP\tilde{x}_i(t)+(R_iu_i(t)+B_i^TP\tilde{x}_i(t))^TR_i^{-1}(R_iu_i(t)\\
&\quad+B_i^TP\tilde{x}_i(t))-(R_{-i}u_{-i}(t)+B_{-i}^TP\tilde{x}_i(t))^TR_{-i}^{-1}(R_{-i}u_{-i}(t)+B_{-i}^TP\tilde{x}_i(t))]dt\}\\
&=E\{\tilde{x}_i^T(0)P\tilde{x}_i(0)-\tilde{x}_i^T(t_p)P\tilde{x}_i(t_p)+\int_0^{t_p}\tilde{x}_i^T(t)[PA+A^TP+Q_i+C^TPC\\
&\quad+\lambda(D^TP+PD+D^TPD)-PB_iR_i^{-1}B_i^TP+PB_{-i}R_{-i}^{-1}B_i^TP]\tilde{x}_i(t)dt
\end{aligned} \tag{2.70}$$

and the noncooperative minmax H_2 game strategy in (2.65) and (2.66). By (2.68) and (2.62), (2.70) leads to

$$E\tilde{x}_i^T(0)P\tilde{x}_i(0)\le J_i \tag{2.71}$$

By the fact, $R_0^i=E[\tilde{x}_i(0)\tilde{x}_i^T(0)]$, we get $T_rPR_0^i<J_i$, i.e., $PR_0^i\le\frac{J_i}{n}I_n$ in (2.69).

Q.E.D.

Since (2.68) is a bilinear matrix inequality (BMI) of P, for the convenience of solving MOP in (2.67)–(2.69), it should be transformed to an equivalent LMI. Let us denote $W=P^{-1}$ and then perform the multiplication of W to both sides of (2.68), then we get

$$\begin{aligned}
&AW+A^TW+WQ_iW+WC^TW^{-1}CW+\lambda(WD^T\\
&+DW+WD^TW^{-1}DW)-B_iR_i^{-1}B_i^T+B_{-i}R_{-i}^{-1}B_{-i}^T\le 0
\end{aligned} \tag{2.72}$$

By performing the Schur complements in Lemma 2.3 on (2.72) several times, we could obtain the following LMIs:

$$\begin{bmatrix} \Pi_i' & W & CW & \lambda^{\frac{1}{2}}DW \\ W & -Q_i^{-1} & 0 & 0 \\ (CW)^T & 0 & -W & 0 \\ \lambda^{\frac{1}{2}}(DW)^T & 0 & 0 & -W \end{bmatrix} \leq 0 \; i = 1, 2, \ldots, m \tag{2.73}$$

where $\Pi_i' = AW + A^T W + \lambda(WD^T + DW) - B_i R_i^{-1} B_i^T + B_{-i} R_{-i}^{-1} B_{-i}^T$.

The LMI in (2.69) is equivalent to

$$-\frac{J_i}{n}(R_0^i)^{-1} + W^{-1} \leq 0, \; i = 1, 2, \ldots, m \tag{2.74}$$

By Schur complements, it is equivalent to

$$\begin{bmatrix} -\frac{J_i}{n}(R_0^i)^{-1} & I \\ I & -W \end{bmatrix} \leq 0, \; i = 1, 2, \ldots, m \tag{2.75}$$

Therefore, the MOP in (2.67)–(2.69) for m-player noncooperative stochastic H_2 game strategy could be transformed to the following LMIs-constrained MOP:

$$(J_1^*, \ldots, J_i^*, \ldots, J_m^*) = \min_{W>0}(J_1, \ldots, J_i, \ldots, J_m) \tag{2.76}$$

$$\text{subject to LMIs (2.73) and (2.75)} \tag{2.77}$$

The LMIs-constrained MOEA proposed to solve MOP in (2.39) for m-player noncooperative stochastic H_∞ game strategy in the previous section could be also employed to solve the MOP in (2.76) for m-player noncooperative stochastic H_2 game strategy.

Remark 2.11 For the m-player noncooperative stocahstic H_2 game strategy of the stabilization control problem in the linear stochastic system (2.57), i.e., the desired targets $x_d^i = 0$ for all players, we don't need to shift the origin of linear stochastic to x_d^i for each player as (2.58).

In this case, the suboptimal m-player noncooperative H_2 game problem in (2.61) for linear stochastic system (2.57) should be modified as the following MOP:

$$(J_1^*, \ldots, J_i^*, \ldots, J_m^*) = \min_{(u_1(t), \ldots, u_m(t))}(J_1, \ldots, J_i, \ldots, J_m) \tag{2.78}$$

subject to

$$\min_{u_i(t)} \max_{u_{-i}(t)} E \int_0^{t_p} (x^T(t)Qx(t) + u_i^T(t)R_i u_i(t) - u_{-i}^T(t)R_{-i}u_{-i}(t))dt \leq J_i, \; i = 1, 2, \ldots, m \tag{2.79}$$

Then, we get the following suboptimal result of m-player noncooperative H_2 game strategy in the stabilization control problem.

Theorem 2.9
The linear stochastic m-player noncooperative H_2 stabilization game strategy problem in (2.78) and (2.79) could be solved by

$$u_i^*(t) = -R_i^{-1}B_i^T P^* x(t) \tag{2.80}$$

$$u_{-i}^*(t) = R_{-i}^{-1}B_{-i}P^* x(t) \tag{2.81}$$

where $P^* = W^{*^{-1}}$ is the solution of the following MOP:

$$(J_1^*, \ldots, J_i^*, \ldots, J_m^*) = \min_W (J_1, \ldots, J_i, \ldots, J_m) \tag{2.82}$$

subject to

$$\begin{bmatrix} \Pi_i' & W & CW & \lambda^{\frac{1}{2}}DW \\ W & -Q_i^{-1} & 0 & 0 \\ (CW)^T & 0 & -W & 0 \\ \lambda^{\frac{1}{2}}(DW)^T & 0 & 0 & -W \end{bmatrix} \leq 0 \;\; i = 1, \ldots, m \tag{2.83}$$

where Π_i' is defined in (2.73) and

$$\begin{bmatrix} -\frac{l_i}{n} R_0^{i^{-1}} & I \\ I & -W \end{bmatrix} \leq 0, \;\; i = 1, \ldots, m \tag{2.84}$$

Proof: Similar to the proof procedure in Theorem 2.8 and the Schur complement procedure (2.72)–(2.76).

Therefore, the proposed LMIs-constrained MOEA could be also employed to solve the MOP in (2.82)–(2.84) for m-player noncooperative stochastic H_2 stabilization game strategy.

2.5 Stochastic Multi-Player Stochastic H_∞ Cooperative Game Strategy in Linear Stochastic Systems

If the m players in linear stochastic system (2.3) have compromised a common desired state x_d (target) with each other at the beginning, then the cooperative H_∞ game strategy of linear stochastic system in (2.3) becomes how to specify $u_1, u_2, \ldots, u_m$ to achieve the following m-player stochastic H_∞ game tracking strategy

$$\rho_0 = \min_{u(t)} \max_{V(t)} \frac{E\{\int_0^{t_p}((x(t)-x_d)^T Q(x(t)-x_d) + \sum_{i=1}^m u_i^T(t)R_i u_i(t))dt\}}{E\{(x(0)-x_d)^T(x(0)-x_d) + \int_0^{t_p} v^T(t)v(t)dt\}} \tag{2.85}$$

Let us denote $\tilde{x}(t) = x(t) - x_d$ as the tracking error between the state $x(t)$ and the common desired state x_d, $u(t) = [u_1^T(t), u_2^T(t), \ldots, u_m^T(t)]^T$ denotes the combination of all strategies, $Q \geq 0$ denotes the weighting (punishment) matrix on tracking errors, and $R_i > 0$ denotes the weighting matrix on the strategy effort of each player.

In the m-player cooperative stochastic H∞ game problem, for the convenience of design, the origin of linear stochastic system in (2.3) is shifted to x_d as follows:

$$d\tilde{x}(t) = (A\tilde{x}(t) + Bu(t) + V'(t))dt + C\tilde{x}(t)dw(t) + D\tilde{x}(t)dp(t) \tag{2.86}$$

where $v'(t) = v(t) + Ax_d + Cx_d dw(t) + Dx_d dp(t)$, $B = [B_1, \ldots, B_m]$, $u(t) = [\, u_1(t) \quad \cdots \quad u_m(t)\,]$.

Then, the utility function of m-player cooperative stochastic H∞ game tracking system in (2.85) is modified as

$$\rho_0 = \min_{u(t)} \max_{V(t)} \frac{E\{\int_0^{t_p} (\tilde{x}^T(t)Q\tilde{x}(t) + u^T(t)Ru(t))dt\}}{E\{(\tilde{x}^T(0)\tilde{x}(t) + \int_0^{t_p} v^{T'}(t)v'(t)dt\}} \tag{2.87}$$

where $R = \begin{bmatrix} R_1 & & & 0 \\ & R_2 & & \\ & & \ddots & \\ 0 & & & R_m \end{bmatrix}$.

In general, it is not easy to solve m-player cooperative stochastic H∞ game tracking strategy in (2.87). An indirect method to solve m-player cooperative stochastic H∞ game tracking strategy is introduced from the suboptimal perspective as follows:

$$\rho_0 = \min_{u(t)} \max_{V(t)} \frac{E\{\int_0^{t_p} (\tilde{x}^T(t)Q\tilde{x}(t) + u^T(t)Ru(t))dt\}}{E\{(\tilde{x}^T(0)\tilde{x}(0) + \int_0^{t_p} v'^T(t)v'(t)dt\}} \leq \rho \tag{2.88}$$

subject to linear stochastic system (2.86).

Therefore, the suboptimal m-player cooperative stochastic H∞ game tracking strategy could be formulated as SOP

$$\rho_0 = \min_{u(t),\rho} \rho \tag{2.89}$$

subject to (2.88), i.e.,

$$\min_{u(t)} \max_{v'(t)} \frac{E\{\int_0^{t_p} (\tilde{x}^T(t)Q\tilde{x}(t) + u^T(t)Ru(t))dt\}}{E\{(\tilde{x}^T(0)\tilde{x}(0) + \int_0^{t_p} v'^T(t)v'(t)dt\}} \leq \rho \tag{2.90}$$

Theorem 2.10
When the optimal solution ρ_0 of SOP in (2.89) and (2.90) is achieved, it is equivalent to solving the m-player cooperative stochastic H∞ game tracking strategy in (2.87) for the linear stochastic system in (2.3).

Proof: Similar to the proof procedure by contradiction in Theorem 2.1.

Theorem 2.11
For the linear stochastic system (2.3), the m-player cooperative stochastic H∞ game tracking strategy in (2.89) could be solved by the following:

$$u^0(t) = -R^{-1}B^TP^0\tilde{x}(t) \tag{2.90}$$

$$v^{\rho^0}(t) = -\frac{1}{\rho^0}P^0\tilde{x}(t) \tag{2.91}$$

where ρ^0 and P^0 are the solution of the following SOP:

$$\rho_0 = \min_{P,\rho} \rho \tag{2.92}$$

subject to

$$\begin{aligned} PA + A^TP + Q + C^TPC + \lambda(PD + D^TP + D^TPD) \\ - PBR^{-1}B^TP + \tfrac{1}{\rho}PP \leq 0 \end{aligned} \tag{2.93}$$

with

$$P \leq \rho I \tag{2.94}$$

Proof: Similar to Theorem 2.4.

Since (2.93) is still a Riccati-like inequality, it is not easy to solve the SOP in (2.92)–(2.94). Let us denote $W = P^{-1}$ and perform the multiplications of W to both sides of (2.93) to obtain

$$\begin{aligned} AW + WA^T + WQW + WC^TW^{-1}CW \\ + \lambda(DW + WD^T + WD^TW^{-1}DW) - BR^{-1}B^T + \tfrac{1}{\rho}I \leq 0 \end{aligned} \tag{2.95}$$

then, we get the following LMI by the use of several Schur complements of (2.95)

$$\begin{bmatrix} \Pi & W & CW & \lambda^{\frac{1}{2}}DW & I \\ W & -Q^{-1} & 0 & 0 & 0 \\ (CW)^T & 0 & -W & 0 & 0 \\ \lambda^{\frac{1}{2}}(DW)^T & 0 & 0 & -W & 0 \\ I & 0 & 0 & 0 & -\rho I \end{bmatrix} \leq 0 \tag{2.96}$$

where $\Pi \triangleq AW + WA^T + \lambda(DW + WD^T) - BR^{-1}B^T$.

Similarly, (2.94) is equivalent to

$$\begin{bmatrix} -\rho I & I \\ I & -W \end{bmatrix} \leq 0 \tag{2.97}$$

Therefore, we get the following result for m-player cooperative stochastic H_∞ game strategy for the desired target tracking of linear stochastic system.

Theorem 2.12

For linear stochastic system (2.3), the m-player cooperative stochastic H_∞ game tracking strategy in (2.87) is solved by

$$u^0(t) = -R^{-1}B^T P^0 \tilde{x}(t)$$

$$v'^0(t) = -\frac{1}{\rho^0} P^0 \tilde{x}(t)$$

where ρ_0 and $P^0 = W^{0^{-1}}$ are the solution of the following SOP

$$\rho_0 = \min_{W,\rho} \rho \tag{2.98}$$

subject to LMIs in (2.96) and (2.97)

The solution of W^0 and ρ_0 of SOP in (2.98) could be solved by decreasing ρ until no positive symmetric matrix $W > 0$ is solved for LMIs in (2.96) and (2.97) with the help of LMI toolbox in Matlab.

Remark 2.11 For the m-player stochastic H$_\infty$ cooperative game strategy in the following robust H$_\infty$ stabilization problem of linear stochastic system

$$\rho^0 = \min_{u(t)} \max_{V(t)} \frac{E\{\int_0^{t_p} (x^T(t)Qx(t) + \sum_{i=1}^{m} u_i^T(t)R_i u_i(t))dt\}}{E\{(x^T(0)x(0) + \int_0^{t_p} v^T(t)v(t)dt\}} \tag{2.99}$$

subject to

$$\begin{aligned} dx(t) &= (Ax(t) + \sum_{i=1}^{m} B_i u_i(t) + v(t)) + Cx(t)dw(t) + Dx(t)dp(t) \\ &= (Ax(t) + Bu(t) + v(t))dt + Cx(t)dw(t) + Dx(t)dp(t) \end{aligned} \tag{2.100}$$

then we get the following result.

Theorem 2.13
For the linear stochastic system in (2.100), the m-player cooperative stochastic H$_\infty$ game strategy for the robust H$_\infty$ stabilization problem in (2.99) is solved by

$$u^0(t) = -R^{-1}B^T P_0 x(t) \tag{2.101}$$

$$v^0(t) = -\frac{1}{\rho^0} P_0 x(t) \tag{2.102}$$

where ρ^0 and P_0 are the solution of the following SOP:

$$\rho_0 = \min_{P>0} \rho \tag{2.103}$$

$$\begin{aligned} &PA + A^T P + Q + C^T PC + \lambda(PD + D^T P + D^T PD) \\ &- PBR^{-1}B^T P + \tfrac{1}{\rho} PP \le 0 \end{aligned} \tag{2.104}$$

with

$$P \le \rho I \tag{2.105}$$

Proof: Similar to Theorem 2.11.

Let us denote $W = P^{-1}$ and perform the multiplications of W to both sides of (2.104) to obtain $AW + WA^T + WQW + WC^TW^{-1}CW + \lambda(DW + WD^T + WD^TW^{-1}DW) - BRB^T + \frac{1}{\rho}I \le 0$

Then, we get the following LMI by the use of several Schur complements:

$$\begin{bmatrix} \Pi & W & CW & \lambda^{\frac{1}{2}}DW & I \\ W & -Q^{-1} & 0 & 0 & 0 \\ (CW)^T & 0 & -W & 0 & 0 \\ \lambda^{\frac{1}{2}}(DW)^T & 0 & 0 & -W & 0 \\ I & 0 & 0 & 0 & -\rho I \end{bmatrix} \le 0 \tag{2.106}$$

where $\Pi \triangleq AW + WA + \lambda(DW + WD^T) - BR^{-1}B^T$.

Similarly, (2.105) is equivalent to

$$\begin{bmatrix} -\rho I & I \\ I & -W \end{bmatrix} \le 0 \tag{2.107}$$

Therefore, we get the following result for m-player cooperative H_∞ game strategy for the robust H_∞ stabilization problem of linear stochastic system.

Theorem 2.14

The m-player stochastic H_∞ cooperative game strategy for the robust H_∞ stabilization problem is solved by the following:

$$u^0(t) = -R^{-1}B^TP_0x(t)$$
$$v^0(t) = -\frac{1}{\rho}P_0x(t)$$

where ρ^0 and $P_0 = W_0^{-1}$ is solved by the following SOP:

$$\rho_0 = \min_W \rho \tag{2.108}$$

subject to LMIs in (2.106) and (2.107)

2.6 Stochastic Multi-Player H_2 Cooperative Game Strategy in Linear Stochastic Systems

Suppose m players in the linear stochastic system (2.57) has a compromise of common desired target x_d. Since the control strategies of all players are available and the cooperative H_2 game strategy could be formulated as the following linear quadratic tracking problem

$$J^0 = \min_{u(t)} E \int_0^{t_p} ((x(t) - x_d)^T Q(x(t) - x_d) + u^T(t)Ru(t))dt \tag{2.109}$$

subject to

$$dx(t) = (Ax(t) + Bu(t))dt + Cx(t)dw(t) + Dx(t)dp(t) \tag{2.110}$$

where $u(t) = \begin{bmatrix} u_1(t) \\ \vdots \\ u_i(t) \\ \vdots \\ u_m(t) \end{bmatrix}$, $B = [B_1, \ldots, B_i, \ldots, B_m]$, $R = \begin{bmatrix} R_1 & & & & 0 \\ & \ddots & & & \\ & & R_i & & \\ & & & \ddots & \\ 0 & & & & R_m \end{bmatrix}$.

For the convenience of design in the multi-player H_2 cooperative tracking control in (2.109) and (2.110), the origin of linear stochastic system in (2.110) should be shifted to x_d as the following tracking error dynamic system:

$$\begin{aligned} d\tilde{x}(t) = {} & (A\tilde{x}(t) + Bu(t) + Ax_d)dt + C\tilde{x}(t)dw(t) + D\tilde{x}(t)dp(t) \\ & + Cx_d dw(t) + Dx_d dp(t) \end{aligned} \tag{2.111}$$

Then, the suboptimal m-player cooperative stochastic H_2 tracking problem in (2.109) and (2.110) is formulated with the upper bound J

$$J^0 = \min_{u(t)} E \int_0^{t_p} ((\tilde{x}^T(t)Q\tilde{x}(t) + u^T(t)Ru(t))dt \leq J \tag{2.112}$$

subject to (2.111).

Then, we will solve the suboptimal m-player cooperative H_2 tracking problem with the upper bound J and then decrease J to approach J^0 to achieve the m-player cooperative H_2 tracking control solution.

Theorem 2.15
For the linear stochastic system (2.110), the m-player cooperative H_2 game tracking strategy could be solved by the following:

$$u^0(t) = -R^{-1}B^T P^0 \tilde{x}(t), \tag{2.113}$$

where P^0 is the solution of the following SOP:

$$J^0 = \min_P J \tag{2.114}$$

$$\begin{bmatrix} PA + A^T P + Q + \lambda(D^T P + PD) \\ + C^T PC + \lambda D^T PD - PBR^{-1}B^T P & 0 \\ 0 & C^T PC + \lambda D^T PD \end{bmatrix} \leq 0 \tag{2.115}$$

$$P \leq \frac{J}{n} R_0^{-1} \tag{2.116}$$

Proof:

$$\begin{aligned} J^0 &= \min_{u(t)} E\int_0^{t_p} ((\tilde{x}^T(t)Q\tilde{x}(t) + u^T(t)Ru(t))dt \le J \\ &= \min_{u(t)} E\{\tilde{x}^T(0)P\tilde{x}(0) - \tilde{x}^T(t_p)P\tilde{x}(t_p) \\ &+ \int_0^{t_p} (\tilde{x}^T(t)Q\tilde{x}(t) + u^T(t)Ru(t))dt + d\tilde{x}^T(t)P\tilde{x}(t) \end{aligned} \tag{2.117}$$

By the fact $Edw(t) = 0$, we get

$$\begin{aligned} J^0 &= \min_{u(t)} E\{\tilde{x}^T(0)P\tilde{x}(0) - \tilde{x}^T(t_p)P\tilde{x}(t_p) \\ &+ \int_0^{t_p} [\tilde{x}^T(t)Q\tilde{x}(t) + u^T(t)Ru(t) + \tilde{x}^T(t)P(A\tilde{x}(t) + Bu(t) + Ax_d) \\ &+ (Ax_d^T C^T \frac{\partial^2(\tilde{x}^T(t)P\tilde{x}(t))}{\partial^2\tilde{x}(t)} Cx_d + Bu(t) + Ax_d)^T P\tilde{x}(t) \\ &+ \tilde{x}^T(t)C^T \frac{\partial^2(\tilde{x}^T(t)P\tilde{x}(t))}{\partial^2\tilde{x}(t)} C\tilde{x}(t) + \lambda[(\tilde{x}(t) + D\tilde{x}(t))^T P(\tilde{x}(t) \\ &+ D\tilde{x}(t)) - \tilde{x}^T(t)P\tilde{x}(t)] + \lambda[(\tilde{x}(t) + Dx_d)^T P(\tilde{x}(t) + Dx_d) - \tilde{x}^T(t)P\tilde{x}(t)]\}dt \end{aligned} \tag{2.118}$$

By the assumption $E\tilde{x}(t) = 0$, we get

$$\begin{aligned} J^0 &= \min_{u(t)} E\{\tilde{x}^T(0)P\tilde{x}(0) - \tilde{x}^T(t_p)P\tilde{x}(t_p) \\ &+ \int_0^{t_p} [\tilde{x}^T(t)Q\tilde{x}(t) + u^T(t)Ru(t) + \tilde{x}^T(t)PA\tilde{x}(t) \\ &+ \tilde{x}^T(t)PBu(t) + \tilde{x}^T(t)PAx_d + \tilde{x}^T(t)A^T P\tilde{x}(t) \\ &+ u^T(t)B^T P\tilde{x}(t) + x_d^T A^T P\tilde{x}(t) + \tilde{x}^T(t)C^T PC\tilde{x}(t) \\ &+ x_d^T C^T PCx_d + \lambda\tilde{x}^T(t)(D^T P + PD + D^T PD)\tilde{x}(t) \\ &+ \lambda x_d^T D^T PDx_d]dt\} \\ &= \min_{u(t)} E\{\tilde{x}^T(0)P\tilde{x}(0) - \tilde{x}^T(t_p)P\tilde{x}(t_p) + \int_0^{t_p} [(\tilde{x}^T(t)Q\tilde{x}(t) \\ &+ \tilde{x}^T(t)PA\tilde{x}(t) + \tilde{x}^T(t)PAx_d) + \tilde{x}^T(t)C^T PC\tilde{x}(t) \\ &+ x_d^T C^T PCx_d + \lambda\tilde{x}^T(t)(D^T P + PD + D^T PD)\tilde{x}(t) \\ &+ \lambda x_d^T D^T PDx_d - \tilde{x}^T(t)PBR^{-1}B^T P\tilde{x}(t) \\ &+ (Ru(t) + B^T P\tilde{x}(t))R^{-1}(Ru(t) + B^T P\tilde{x}(t))]dt\} \\ &= E\{\tilde{x}^T(0)P\tilde{x}(0) - \tilde{x}^T(t_p)P\tilde{x}(t_p) \\ &+ \int_0^{t_p} [\tilde{x}^T(t) \;\; x_d^T] \begin{bmatrix} \begin{matrix} PA + A^T P + Q \\ +\lambda(D^T P + PD) \\ +C^T PC + \lambda D^T PD \\ -PBR^{-1}B^T P \end{matrix} & 0 \\ 0 & C^T PC + \lambda D^T PD \end{bmatrix} \begin{bmatrix} \tilde{x}(t) \\ x_d \end{bmatrix} dt \end{aligned} \tag{2.119}$$

From m-player H_2 game tracking strategy (2.113) and by (2.115), we get

$$J^0 \leq E\{\tilde{x}^T(0)P\tilde{x}(0)\} = T_r PR_0 \leq J \tag{2.120}$$

where $R_0 = E\{\tilde{x}(0)\tilde{x}^T(0)\}$.

From (2.120), we get (2.116). Q.E.D.

Because (2.115) is a Riccati-like inequality, it is not easy to solve MOP, for J^0 and P^0 from MOP in (2.114)–(2.116) for m-player cooperative H_2 game tracking strategy.

Let us denote $W = P^{-1}$ and perform the multiplication of the following matrix to both sides of (2.115):

$$\begin{bmatrix} W & 0 \\ 0 & W \end{bmatrix} \tag{2.121}$$

then, we get the following LMI by several Schur complements:

$$\begin{bmatrix} \Pi & W & CW & \lambda^{\frac{1}{2}}DW & CW & \lambda^{\frac{1}{2}}DW \\ W & -Q^{-1} & 0 & 0 & 0 & 0 \\ (CW)^T & 0 & -W & 0 & 0 & 0 \\ \lambda^{\frac{1}{2}}(DW)^T & 0 & 0 & -W & 0 & 0 \\ (CW)^T & 0 & 0 & 0 & -W & 0 \\ \lambda^{\frac{1}{2}}(DW)^T & 0 & 0 & 0 & 0 & -W \end{bmatrix} \tag{2.122}$$

where $\Pi \triangleq AW + WA + \lambda(DW + WD^T) - BR^{-1}B^T$

Similarly, (2.116) is equivalent to

$$\begin{bmatrix} -\frac{J}{n}R_0^{-1} & I \\ I & W \end{bmatrix} \leq 0 \tag{2.123}$$

Then, we get the following result.

Theorem 2.16

For the linear stochastic system (2.110) or (2.111), the m-player cooperative H_2 game tracking strategy could be solved by

$$u^0(t) = -R^{-1}B^TP^0\tilde{x}(t)$$

where $P^0 = W^{0^{-1}}$ is the solution of the following LMIs-constrained SOP:

$$J^0 = \min_{W>0} J \tag{2.124}$$

subject to (2.122) and (2.123)

Remark 2.12 For the m-player H_2 cooperative game strategy in the following H_2 quadratic stabilization problem of linear stochastic system

$$J^0 = \min_{u(t)} E \int_0^{t_p} ((x^T(t)Qx(t) + \sum u_i^T(t)R_i u_i(t))dt \tag{2.125}$$

subject to

$$\begin{aligned} dx(t) &= (Ax(t) + \sum B_i u_i(t))dt + Cx(t)dw(t) + Dx(t)dp(t) \\ &= (Ax(t) + Bu(t))dt + Cx(t)dw(t) + Dx(t)dp(t) \end{aligned} \tag{2.126}$$

where $u(t) = \begin{bmatrix} u_1(t) \\ \vdots \\ u_m(t) \end{bmatrix}$, $B = [B_1, \ldots, B_m]$, $R = \begin{bmatrix} R_1 & & 0 \\ & R_2 & \\ 0 & & \ddots \\ & & & R_m \end{bmatrix}$.

Then, we get the following result:

Theorem 2.17
For the linear stochastic system in (2.126), the m-player H_2 cooperative game strategy for the stochastic quadratic stabilization in (2.125) could be solved by

$$u^0(t) = -R^{-1}B^T P^0 x(t) \tag{2.127}$$

where P^0 is the solution of the following SOP:

$$J^0 = \min_{P>0} J \tag{2.128}$$

subject to

$$PA + A^T P + Q + \lambda(D^T P + PD) + C^T PC + \lambda D^T PD - PBR^{-1}B^T P \leq 0 \tag{2.129}$$

$$P \leq \frac{J}{n} R_0^{-1} \tag{2.130}$$

where $R_0 = E[x(0)x^T(0)]$.

Let $W = P^{-1}$ and perform W to both sides of (2.129), with the help of Schur complement in Lemma 2.3, we get

$$\begin{bmatrix} \Pi & W & CW & \lambda^{\frac{1}{2}}DW \\ W & -Q^{-1} & 0 & 0 \\ (CW)^T & 0 & -W & 0 \\ \lambda^{\frac{1}{2}}(DW)^T & 0 & 0 & -W \end{bmatrix} \tag{2.131}$$

where Π is defined in (2.122). By Schur complement, (2.130) could be transformed to LMI in (2.123).

Theorem 2.18
m-player H_2 cooperative game strategy design problem for the stochastic H_2 quadratic stabilization problem in (2.125) and (2.126) could be solved by

$$u^0(t) = -R^{-1}B^T P^0 x(t)$$

where $P^0 = (W^0)^{-1}$ is the solution of the following SOP:

$$J^0 = \min_{W>0} J$$

subject to LMIs in (2.131) and (2.123)

Proof: Similar to the procedure is Theorem 2.16

2.7 Conclusion

In this chapter, the multiplayer stochastic H_∞ noncooperative game strategy design problem of linear stochastic systems with intrinsic fluctuation due to Wiener process and Poisson process and external disturbance is introduced at first when the interests of each player are partly conflicting with others and the competitive strategies of other players are unavailable. The multi-player stochastic H_∞ noncooperative game strategy design problem is transformed to an equivalent LMIs-constrained MOP with Nash equilibrium solution to the multi-player noncooperative stochastic H_∞ game problem. An LMIs-constrained MOEA is also proposed to efficiently solve the LMIs-constrained MOP to obtain the Nash equilibrium solution for the m-player stochastic H_∞ noncooperative game strategy in a single run with the help of Matlab LMI toolbox. If the stochastic linear system is free of external disturbance, then multiplayer stochastic H_2 noncooperative game strategy design problem is also discussed. The multiplayer H_2 noncooperative game strategy design problem is transformed to a simpler LMIs-constrained MOP with Nash equilibrium solution too. The proposed LMIs-constrained MOEA could be also employed to solve the MOP for H_2 noncooperative game strategy of linear stochastic systems without external disturbance.

If the m players have their common target and the strategies of each player are available for all players, then m-player cooperative stochastic H_∞ and H_2 game strategy design problems are employed for multi-person decision-making for common interests and could be transformed to the corresponding LMIs-constrained SOPs. Then, LMIs-constrained SOPs for H_2 and H_∞ cooperative game strategies could be easily solved with the help of LMI toolbox in Matlab. More detailed applications of cooperative and noncooperative stochastic H_2 and H_∞ cooperative game strategies of linear stochastic systems will be given in the following chapters.

3

Introduction to Multi-player Stochastic Games in Nonlinear Stochastic Systems

3.1 Introduction

After the introduction of multi-player noncooperative and cooperative game strategies of linear stochastic systems in Chapter 2, this chapter will discuss the multi-player noncooperative and cooperative game strategies of nonlinear stochastic systems. In real physical systems, stochastic dynamics are always nonlinear [1,24]. In this chapter, we consider multi-player noncooperative and cooperative game strategy design problem of nonlinear stochastic jump diffusion systems with external disturbance. The nonlinear stochastic systems are always with continuous intrinsic random fluctuation due to Wiener process and discontinuous random fluctuation due to Poisson process as well as external disturbance to model complex and uncertain dynamic systems with multiple players [5,29,31]. We found the present game strategies for the nonlinear dynamic systems are lack of random fluctuation and environmental disturbance [1,9,27,28]. Further, the present noncooperative game strategies of nonlinear dynamic systems have focused on the discussions on the stabilization problem and the existence of Nash equilibrium solutions instead of the design methods of noncooperative game strategy of nonlinear dynamic systems [26,28]. Since the external disturbances were always neglected in the nonlinear dynamic game problems, only quadratic (H_2) payoff functions were considered in the conventional nonlinear game strategy problems [2,26,28]. In this chapter, multi-player noncooperative and cooperative H_∞ and H_2 game strategy design problems will be introduced for nonlinear stochastic systems with continuous Wiener process and discontinuous Poisson process as well as external disturbance from the more practical application perspective.

First, multi-player noncooperative H_∞ game strategy design problem is introduced for nonlinear stochastic systems with Wiener and Poisson process as well as external disturbance. Each player has his/her desired target and payoff function. The multi-player noncooperative H_∞ game strategy design problem cannot be solved directly and is first transformed to an equivalent Hamilton-Jacobi-Issac inequalities (HJIIs)-constrained MOP by an indirect suboptimal method. We will prove that the equivalent HJIIs-constrained MOP has Nash equilibrium solution for the multi-player noncooperative game problem of nonlinear stochastic systems. Since it is very difficult to solve the HJIIs-constrained MOP for the multi-player noncooperative H_∞ game strategy design problem of nonlinear stochastic systems, the global linearization method [2,12] (or T-S fuzzy method [9]) will be employed to interpolate several local linearized stochastic systems to approach the nonlinear stochastic system so that the HJIIs-constrained MOP could be transformed to an equivalent LMIs-constrained MOP. Since the conventional multiobjective evolutionary algorithms (MOEAs) search control gain matrices directly to solve MOP and cannot be applied to solving the complex LMIs-constrained MOP directly, an LMIs-constrained MOEA [30] is also proposed to solve the LMIs-constrained MOP for the multi-player noncooperative H_∞ game strategy of nonlinear stochastic system. In the special cases, the multi-player noncooperative H_∞ game strategy of the stabilization of nonlinear stochastic system is also discussed [34].

If the nonlinear stochastic systems are free of external disturbances or external disturbances are neglected in the design procedure, then the m-player noncooperative H_2 game strategy design problem is discussed based on the quadratic payoff (utility) function [2,32]. By the proposed indirect method via minimizing the upper bound of H_2 (quadratic) payoff function of each player, the m-player noncooperative H_2 game strategy design problem of nonlinear stochastic systems is transformed to an equivalent HJIIs-constrained MOP whose solution is guaranteed to be the Nash equilibrium solution of m-player noncooperative H_2 game strategy. In order to overcome the difficulty of solving HJIIs-constrained MOP for m-player noncooperative stochastic H_2 game strategy, the global linearization technique [2,12] is also employed to interpolate several local linearized stochastic linear systems to approximate the nonlinear stochastic system so that the HJIIs-constrained MOP for m-player noncooperative stochastic H_2 game problem could be also transformed to an equivalent LMIs-constrained MOP, which could be easily solved by the proposed LMIs-constrained MOEA to search the Pareto optimal solutions [30] for m-player noncooperative stochastic H_2 game strategies in a single round. Further, the m-player noncooperative H_2 game strategy for the stabilization problem of nonlinear stochastic system is also discussed in the same vein.

In the nonlinear stochastic system with multiple players, if these players have compromised one another a common target with the same payoff function, then the multi-player cooperative game strategy design problem needs to be solved to achieve their purpose while the control strategy information are exchangeable among these players. The multi-player cooperative H_∞ game strategy is introduced for nonlinear stochastic system with external disturbance. Since the external disturbance is unavailable for all players, it is considered as a competitive player to these players and all players are combined as an augmented player. Therefore, the multi-player cooperative stochastic H_∞ game problem could be formulated as a two-tuple (one is augmented player and another is external disturbance) minmax stochastic H_∞ game problem, which could be transformed to an equivalent HJII-constrained SOP and then be transformed to an LMIs-constrained SOP with the help of global linearization technique, which could be efficiently solved through the LMI toolbox in Matlab.

As a special case, the multi-player cooperative H_∞ game strategy for the stabilization of nonlinear stochastic system could be solved by a similar procedure. In the nonlinear stochastic systems without external disturbance, the multi-player cooperative H_2 game strategy could be employed for these players to achieve their common target compromised beforehand through a same payoff function. The multi-player cooperative stochastic H_∞ game strategy could be transformed to an equivalent HJII-constrained SOP and then transformed to an LMIs-constrained SOP, which could be efficiently solved with the help of LMI toolbox in Matlab. Further, the multi-player cooperative H_2 game strategy for the stabilization problem of nonlinear stochastic system is also discussed as a special case in the same vein.

After the introduction of theoretical backgrounds of m-player noncooperative and cooperative H_∞/H_2 game strategy design problems of nonlinear stochastic system in this chapter, their applications to engineering, social, and bio-science will be discussed in the following chapters.

3.2 Stochastic Multi-Player Noncooperative H_∞ Game Strategy in Nonlinear Stochastic Systems

In general, most physical systems are always nonlinear stochastic systems because they may interact between nonlinear factors and may suffer from random intrinsic fluctuation and external disturbance [2,3]. For the discussion of m-player noncooperative H_∞ game strategy, let us consider the following nonlinear stochastic system with m players (m-person decision-makers):

$$dx(t) = [f(x(t)) + \sum_{i=1}^{m} g_i(x(t))u_i(t) + h(x(t))v(t)]dt + l(x(t))dw(t) + n(x(t))dp(t) \tag{3.1}$$

where $x(t) = \begin{bmatrix} x_1(t) \\ \vdots \\ x_i(t) \\ \vdots \\ x_n(t) \end{bmatrix}$ and $u_i(t) = \begin{bmatrix} u_{i1}(t) \\ \vdots \\ u_{ik}(t) \end{bmatrix}, i = 1, 2, \ldots, m$

denote the state vector and control strategies of m players, respectively. $v(t)$ denotes the external disturbance. $w(t)$ is the standard 1-dimensional Wiener process. $l(x(t))dw(t)$ denotes the continuous random fluctuation. $p(t)$ denotes Poisson counting process with mean λ in a unit time and $n(x(t))dp(t)$ is regarded as intrinsic discontinuous random fluctuation (jumping process). The nonlinear interaction vector $f(x(t))$ must satisfy with Lipschitz continuity and $g_i(x(t))$ denotes the input matrix of the ith control strategy $u_i(t)$ of the ith player [34].

For the nonlinear stochastic system in (3.1), we want to design noncooperative game strategies for each player to achieve their desired targets as possible as they can, despite the effect of intrinsic continuous and jumping random fluctuations as well as external disturbance and competitive strategies of other players.

For the ith game strategy $u_i(t)$, it assumes that all other competitive game strategies $u_1(t), \ldots, u_{i-1}(t), u_{i+1}(t), \ldots, u_m(t)$ and environmental disturbance are unavailable for the ith player and can be bunched together as a competitive game strategy $u_{-i}(t)$ to $u_i(t)$ to simplify the representation and the corresponding analysis. Since external disturbance $v(t)$ plays an important role and is unpredictable to each player in the stochastic control system, to be on the safe side, its influence on the design of noncooperative game strategy must be considered from the worst-case perspective and each player tends to minimize the worst-case effect of external disturbance from the minmax viewpoint. Therefore, the external disturbance $v(t)$ could be considered as a player in the noncooperative game of the nonlinear stochastic system in (3.1). Thus, the nonlinear stochastic system in (3.1) could be represented by

$$\begin{aligned} dx(t) =& [f(x(t)) + g_i(x(t))u_i(t) + g_{-i}(x(t))u_{-i}(t)]dt \\ & + l(x(t))dw(t) + n(x(t))dp(t) \end{aligned} \tag{3.2}$$

where $g_{-i}(t) = [g_1(x(t)...g_{i-1}(x(t)g_{i+1}(x(t)...h(x(t))]$ denotes input matrices of competitive strategies and $u_{-i}(t) = [u_1^T(t)...u_{i-1}^T(t)u_{i+1}^T(t)...v^T(t)]^T$ denotes competitive strategies including external disturbance.

In the noncooperative nonlinear stochastic system in (3.2), since other competitive strategies and external disturbance are unavailable for each player, to be on the safe side, the game strategy $u_i(t)$ of each player i attempts to simultaneously minimize the worst-case effect of the combined competitive strategy and external disturbance $u_{-i}(t)$ to achieve his or her desired target x_d^i as possible with a tradeoff of control effort, in spite of continuous and discontinuous intrinsic random fluctuation, i.e., each strategy needs to simultaneously minimize the deviation to their target $\tilde{x}_i(t) = x(t) - x_d^i$ with a parsimonious control effort $u_i(t)$ under the worst-case (maximum) effect of other competitive strategies and external disturbance as the following stochastic m-player H_∞ noncooperative game:

$$\begin{aligned} \rho_i^* =& \min_{u_i(t)} \max_{u_{-i}(t)} \frac{E\int_0^{t_p} [(x(t) - x_d^i)^T Q_i (x(t) - x_d^i) + u_i^T(t) R_i u_i(t)]dt}{E(x(0) - x_d^i)^T (x(0) - x_d^i) + E\int_0^{t_p} u_{-i}^T(t) u_{-i}(t)dt} \\ & \text{for} i = 1, 2, \ldots, m \end{aligned} \tag{3.3}$$

where x_d^i denotes the desired target of player i; Q_i and R_i indicate the corresponding weighting matrices selected by each player to compromise between the deviation $\tilde{x}_i(t)$ to the desired target and the strategy effort of $u_i(t)$ toward the goal of player i; t_p denotes the terminal time of the game; ρ^* denotes the performance of minmax H_∞ game strategy of the ith player to achieve their target.

For the convenience of further analysis and the simplicity of design procedure of stochastic multi-player H_∞ game strategy, we get the following shifted nonlinear stochastic system by the fact $\tilde{x}_i(t) = x(t) - x_d^i$:

$$\begin{aligned} d\tilde{x}_i(t) =& [f(\tilde{x}_i(t) + g_i(\tilde{x}_i(t))u_i(t) + g_{-i}(\tilde{x}_i(t))u_{-i}(t) + l(\tilde{x}_i(t))]dw(t) \\ & + n(\tilde{x}_i(t))dP(t) \text{ for } i = 1, 2, \ldots, m \end{aligned} \tag{3.4}$$

where $\begin{array}{l} f(\tilde{x}_i(t)) = f(\tilde{x}_i(t) + x_d^i), g_i(\tilde{x}_i(t)) = g_i(\tilde{x}_i(t) + x_d^i), h_i(\tilde{x}_i(t)) = h_i(\tilde{x}_i(t) + x_d^i), \\ g_{-i}(\tilde{x}_i(t)) = g_{-i}(\tilde{x}_i(t) + x_d^i), l(\tilde{x}_i(t)) = l(\tilde{x}_i(t) + x_d^i), n(\tilde{x}_i(t)) = n(\tilde{x}_i(t) + x_d^i) \end{array}$

for the simplicity of notation.

Then, the m-player noncooperative H_∞ game in (3.3) for nonlinear stochastic system (3.2) is modified as

$$\rho_i^* = \min_{u_i(t)} \max_{u_{-i}(t)} \frac{E\int_0^{t_p} [\tilde{x}_i^T(t) Q_i \tilde{x}_i(t) + u_i^T(t) R_i u_i(t)]dt}{E\tilde{x}_i^T(0)\tilde{x}_i(0) + E\int_0^{t_p} u_{-i}^T(t) u_{-i}(t)dt}, \text{for} i = 1, 2, \ldots, m \tag{3.5}$$

The physical meaning of stochastic multi-player H_∞ noncooperative game strategy in (3.5) is that the worst-case effect of uncertainties due to the initial condition, competitive game strategies, and external disturbance on each target tracking performance must be minimized in safety by player i with a parsimonious control effort $u_i(t)$ according to the tradeoff of two weighting matrices Q_i and R_i.

Remark 3.1

(i) The m-player noncooperative H_∞ game strategy $(u_1^*, u_2^*, \ldots, u_m^*)$ in (3.5) constitutes a Nash equilibrium solution if and only if [1,4]

$$\begin{gathered}(\rho_1^*, \rho_2^*, \ldots, \rho_m^*) \leq (\rho_1, \rho_2^*, \ldots, \rho_m^*) \\ \vdots \\ (\rho_1^*, \ldots, \rho_i^*, \ldots, \rho_m^*) \leq (\rho_1^*, \ldots, \rho_{i-1}^*, \rho_i, \rho_{i+1}^*, \ldots, \rho_m^*) \\ \vdots \\ (\rho_1^*, \rho_2^*, \ldots, \rho_m^*) \leq (\rho_1^*, \rho_2^*, \ldots, \rho_m)\end{gathered} \tag{3.6}$$

where $(\alpha_1, \beta_1) \leq (\alpha_2, \beta_2)$ if and only if $\alpha_1 \leq \alpha_2$ and $\beta_1 \leq \beta_2$, i.e., (α_1, β_1) dominates (α_2, β_2), i.e., no player could gain any benefit if he/she changes his/her strategy from Nash equilibrium solution.

(ii) The solution vector $(\rho_1^*, \rho_2^*, \ldots, \rho_m^*)$ corresponds to the H_∞ tracking performance of each player with m-player noncooperative H_∞ game strategy of the H_∞ noncooperative game problem in (3.5) of the nonlinear stochastic tracking system in (3.4). Different choices of weighting matrices Q_i, R_i, and target x_d^i in (3.5) could give different strategies with different degrees of noncooperative characteristic among players. Since the involved players' purposes of these strategies in (3.5) are not fully in conflict with one another, a certain degree of cooperation still exists in the multi-player noncooperative game strategies. All players have to assign their weighting matrices Q_i, R_i, and target x_d^i according to their purposes.

(iii) In the linear stochastic system case, (3.1) is reduced to (2.3), i.e., the linear stochastic system in (2.3) is the local linearized system (3.1) at an operation point.

(iv) The primary purpose of each player is to minimize the state tracking error $\tilde{x}(t)$ from his/her target x_d^i by parsimonious effort $u_i(t)$. $x_d^i = 0$ for all i are considered in the traditional noncooperative game strategy [1,13]. Namely, only the stabilization issue is considered in most traditional noncooperative game strategies.

(v) If the nonlinear system in (3.1) is free of random fluctuation, i.e., the nonlinear system is deterministic, then the expectation E in (3.3) and (3.4) should be neglected.

Since the Wiener process $w(t)$ and Poisson counting process $p(t)$ are non-differentiable, the following lemma is crucial for the solution of stochastic noncooperative game strategy of nonlinear stochastic system in (3.4).

Lemma 3.1 (Itô–Lévy Lemma [22,33]). For the nonlinear stochastic system in (3.4), the Itô–Lévy formula for a Lyapunov function $V(\tilde{x}(t))>0$ is given as the following:

$$
\begin{aligned}
dV(\tilde{x}(t)) = & \left(\frac{\partial V(\tilde{x}_i(t))}{\partial \tilde{x}_i(t)}\right)^T [(f(\tilde{x}_i(t) + g_i(\tilde{x}_i(t))u_i(t) + g_{-i}(\tilde{x}_i(t))u_{-i}(t))]dt \\
& + \frac{\partial V(\tilde{x}_i(t))}{\partial \tilde{x}_i(t)} l(\tilde{x}_i(t))dw(t) + \frac{1}{2} l^T(\tilde{x}_i(t)) \frac{\partial^2 V(\tilde{x}_i(t))}{\partial^2 \tilde{x}_i(t)} l(\tilde{x}_i(t))dt \\
& + [V(\tilde{x}_i(t) + n(\tilde{x}_i(t))) - V(\tilde{x}_i(t))]dp(t)
\end{aligned} \tag{3.7}
$$

Basically, it is hard to solve the m minmax H_∞ game problems in (3.5) simultaneously for the nonlinear stochastic systems in (3.4) directly. Consequently, the m-player stochastic H_∞ noncooperative game problem in (3.5) could be solved by the indirect method as Chapter 2 in the following:

$$
\rho_i^* = \min_{u_i(t)} \max_{u_{-i}(t)} \frac{E \int_0^{t_p} \tilde{x}_i^T(t) Q_i \tilde{x}_i(t) + u_i^T(t) R_i u_i(t)}{E\tilde{x}_i^T(0)\tilde{x}_i(0) + E \int_0^{t_p} u_{-i}^T(t) u_{-i}(t) dt} \leq \rho_i, \text{ for } i = 1, 2, \ldots, m \tag{3.8}
$$

where $\rho_i > 0$ indicates the upper bound of ρ_i^* by the ith strategy.

In the indirect method, an initial upper bound ρ_i of ρ_i^* is provided at first. To simultaneously approach ρ_i^* for all i, all the m initial upper bounds in (3.8) need to be minimized simultaneously as small as possible. For the stochastic nonlinear system in (3.4), the so-called indirect approach to the m-player noncooperative stochastic H_∞ game is constituted in (3.8) from the suboptimal game perspective, i.e., we finally realize the real m minmax H_∞ optimization problems of m-player noncooperative H_∞ game strategy in (3.8) simultaneously. Namely, we must solve the following multiobjective optimization problem (MOP) by simultaneously minimizing the relevant upper bounds in (3.8) as follows:

$$
(\rho_1^*, \ldots, \rho_i^*, \ldots, \rho_m^*) = \min_{(u_1(t), \ldots, u_i(t), \ldots, u_m(t))} (\rho_1, \ldots, \rho_i, \ldots, \rho_m) \tag{3.9}
$$

subject to

$$
\begin{aligned}
& \min_{u_i(t)} \max_{u_{-i}(t)} \frac{E \int_0^{t_p} [\tilde{x}_i^T(t) Q_i \tilde{x}_i(t) + u_i^T(t) R_i u_i(t)]dt}{E[\tilde{x}_i^T(0)\tilde{x}_i(0) + \int_0^{t_p} u_{-i}^T(t) u_{-i}(t) dt]} \leq \rho_i, \\
& i = 1, 2, \ldots, m
\end{aligned} \tag{3.10}
$$

Therefore, the m-player noncooperative stochastic H_∞ game problem in (3.5) or (3.8) is transformed to a multiobjective problem in (3.9) and (3.10) from the suboptimal perspective and we need to design game strategies $u_1(t), \ldots, u_i(t), \ldots, u_m(t)$ for each player to simultaneously minimize the objective vector $(\rho_1, \ldots, \rho_i, \ldots, \rho_m)$ in the Pareto optimal sense [30]. In general, the solution of MOP in (3.9) and (3.10) is not unique and called Pareto optimal solution. Before we further investigate the solution of MOP in (3.9) and (3.10) for m-player noncooperative H_∞ game strategy of nonlinear stochastic systems, some fundamental concepts of MOP in (3.9), (3.10) are introduced in the following.

Definition 3.1 (Pareto dominance) [21,24,25]. For two feasible solutions $(u_1^1(t), \ldots, u_m^1(t))$ and $(u_1^2(t), \ldots, u_m^2(t))$ corresponding to objective vectors $(\rho_1^1, \ldots, \rho_m^1)$ and $(\rho_1^2, \ldots, \rho_m^2)$ of MOP in (3.9) and (3.10), respectively, the solution $(u_1^1(t), \ldots, u_m^1(t))$ is said to dominate $(u_1^2(t), \ldots, u_m^2(t))$ if $\rho_1^1 \leq \rho_1^2, \ldots, \rho_m^1 \leq \rho_m^2$ and at least one of the inequalities is of strict inequality.

Definition 3.2 (Pareto optimality) [21,24,25]. The feasible solution $(u_1(t), \ldots, u_m(t))$ is said to be Pareto optimality with respect to the feasible solution set if and only if there does not exist another feasible solution that dominates it.

Definition 3.3 (Pareto optimal solution set) [21,24,25] For the MOP in (3.9) and (3.10), which has feasible solution $(u_1(t), \ldots, u_m(t))$ with the corresponding objective vector $(\rho_1, \ldots, \rho_m)$, Pareto optimal solution set σ^* is defined as $\sigma^* = \{(u_1^*(t), \ldots, u_m^*(t))| (u_1^*(t), \ldots, u_m^*(t))$ is the Pareto optimality$\}$, where $(u_1^*(t), \ldots, u_m^*(t))$ denotes the Pareto optimal solution of the objective vector $(\rho_1^*, \ldots, \rho_m^*)$.

Definition 3.4 (Pareto front) [21,24,25]. For the MOP in (3.9) and (3.10) with feasible solution $(u_1(t), \ldots, u_m(t))$, the corresponding objective vector $(\rho_1, \ldots, \rho_m)$, and the Pareto optimal solution set σ^*, the Pareto front is defined as $P_F = \{(\rho_1^*, \ldots, \rho_m^*)|(u_1^*, \ldots, u_m^*) \in \sigma^*\}$.

Theorem 3.1
The solution $(\rho_1^*, \ldots, \rho_m^*)$ of MOP in (3.9) and (3.10) is equivalent to the solution of m-player noncooperative H_∞ game problem in (3.5) of nonlinear stochastic system in (3.4).

Proof: The proof of Theorem 3.1 only needs to prove the inequalities in (3.8) or (3.10) become equalities for the Pareto optimal solutions by contradiction. First, the MOP problem defines the Pareto optimal solution $(\rho_1^*, \ldots, \rho_i^*, \ldots, \rho_m^*)$ in (3.9) and (3.10). Suppose a strict inequality at the optimality holds for any one of inequalities in (3.8) or (3.10). Without loss of generality, suppose $\min_{u_i(t)} \max_{u_{-i}(t)} \frac{E\int_0^{t_p} [\tilde{x}_i^T(t)Q_i\tilde{x}_i(t)+u_i^T(t)R_iu_i(t)]dt}{E\tilde{x}_i^T(0)\tilde{x}_i(0)+E\int_0^{t_p} u_{-i}^T(t)u_{-i}(t)dt} = \rho_i'$ such that $\rho_i' < \rho_i^*$, which means $(\rho_1^*, \ldots, \rho_i', \ldots, \rho_m^*)$ dominates Pareto optimal solution $(\rho_1^*, \ldots, \rho_i^*, \ldots, \rho_m^*)$ of MOP in (3.9) and (3.10). The above consequence results in a contradiction. Thus, when the Pareto optimal solution of MOP in (3.9) and (3.10) is reached, we can conclude that the m strict inequalities in (3.10) became equalities. Therefore, the MOP in (3.9) and (3.10) is equivalent to m-player noncooperative H_∞ game problem in (3.5).

Q.E.D.

Since $u_{-i}(t)$ is independent of $u_i(t)$, the m-constrained H_∞ game problem in (3.10) is equivalent to the following m-constrained Nash quadratic games [1]

$$\min_{u_i(t)} \max_{u_{-i}(t)} E\int_0^{t_p} [\tilde{x}_i^T(t)Q_i\tilde{x}_i(t) + u_i^T(t)R_iu_i(t) - \rho_i u_{-i}^T(t)u_{-i}(t)]dt \le \rho_i E\tilde{x}_i^T(0)\tilde{x}_i(0), \quad i = 1, 2, \ldots, m \tag{3.11}$$

Before further discussion, we denote

$$J_i = E\int_0^{t_p} [\tilde{x}_i^T(t)Q_i\tilde{x}_i(t) + u_i^T(t)R_iu_i(t) - \rho_i u_{-i}^T(t)u_{-i}(t)]dt \quad i = 1, 2, \ldots, m \tag{3.12}$$

Two steps are needed to solve the m Nash quadratic game constraints in (3.11) for the m H_∞ game constraints in (3.10). In the first step, we need to solve the following m Nash quadratic game problem:

$$\begin{aligned} J_i^* &= \min_{u_i(t)} \max_{u_{-i}(t)} J_i \\ &= \min_{u_i(t)} \max_{u_{-i}(t)} E \int_0^{t_p} [\tilde{x}_i^T(t) Q_i \tilde{x}_i(t) + u_i^T(t) R_i u_i(t) - \rho_i u_{-i}^T(t) u_{-i}(t)] dt, i = 1, 2, \ldots, m \end{aligned} \tag{3.13}$$

In the second step, we need to solve the following upper constrained problems:

$$J_i^* \leq \rho_i E \tilde{x}_i^T(0) \tilde{x}_i(0), i = 1, 2, \ldots, m \tag{3.14}$$

By the above two-step procedure for m-constrained Nash quadratic game problem is (3.11) or (3.10), solving (3.13) and (3.14) brings the following result in preparation for solving the MOP in (3.9) and (3.10) for m-player stochastic H_∞ game strategy design problem in (3.5).

Theorem 3.2

The constrained Nash games in (3.13) and (3.14) or m-constrained Nash games in (3.11) could be solved by

$$u_i^*(t) = -\frac{1}{2} R_i^{-1} g_i^T(\tilde{x}_i(t)) \left(\frac{\partial V(\tilde{x}_i(t))}{\partial \tilde{x}_i(t)} \right), \; i = 1, 2, \ldots, m \tag{3.15}$$

$$u_{-i}^*(t) = \frac{1}{2\rho_i} g_{-i}^T(\tilde{x}_i(t)) \left(\frac{\partial V(\tilde{x}_i(t))}{\partial \tilde{x}_i(t)} \right), \; i = 1, 2, \ldots, m \tag{3.16}$$

where the positive Lyapunov function $V(\tilde{x}_i(t)) > 0$ is the solution of the following Hamilton-Jacobi Issac inequalities (HJIIs):

$$\begin{aligned} &\tilde{x}_i^T(t) Q_i \tilde{x}_i(t) + \left(\frac{\partial V(\tilde{x}_i(t))}{\partial \tilde{x}_i(t)} \right)^T f(\tilde{x}_i(t)) \\ &- \frac{1}{4} \left(\frac{\partial V(\tilde{x}_i(t))}{\partial \tilde{x}_i(t)} \right)^T g_i(\tilde{x}_i(t)) R_i^{-1} g_i^T(\tilde{x}_i(t)) \left(\frac{\partial V(\tilde{x}_i(t))}{\partial \tilde{x}_i(t)} \right) \\ &+ \frac{1}{4\rho_i} \left(\frac{\partial V(\tilde{x}_i(t))}{\partial \tilde{x}_i(t)} \right)^T g_{-i}(\tilde{x}_i(t)) g_{-i}^T(\tilde{x}_i(t)) \left(\frac{\partial V(\tilde{x}_i(t))}{\partial \tilde{x}_i(t)} \right) \\ &+ \frac{1}{2} l^T(\tilde{x}_i(t)) \frac{\partial^2 V(\tilde{x}_i(t))}{\partial \tilde{x}_i^2(t)} l(\tilde{x}_i(t)) + \lambda [V(\tilde{x}_i(t) + n(\tilde{x}_i(t))) - V(\tilde{x}_i(t))] \leq 0 \\ &i = 1, 2, \ldots, m. \end{aligned} \tag{3.17}$$

with

$$E\{V(\tilde{x}_i(0))\} \leq \rho_i E\{\tilde{x}_i^T(0) \tilde{x}_i(0)\}, i = 1, \ldots, m. \tag{3.18}$$

Remark 3.2 (i) The last two terms $\frac{1}{2} l^T(\tilde{x}_i(t)) \frac{\partial^2 V(\tilde{x}_i(t))}{\partial \tilde{x}_i^2(t)} l(\tilde{x}_i(t))$ and $\lambda[V(\tilde{x}_i(t) + n(\tilde{x}_i(t))) - V(\tilde{x}_i(t))]$ in (3.17) are the effect of intrinsic fluctuation of Wiener process and Poisson process, respectively. (ii) The minmax H_∞ game constraints in (3.10) are transformed to HJIIs in (3.17) and (3.18).

Proof: From (3.12), we get

$$J_i = E\int_0^{t_p} (\tilde{x}_i^T(t)Q_i\tilde{x}_i(t) + u_i^T(t)R_iu_i(t) - \rho_i u_{-i}^T(t)u_{-i}(t))dt$$
$$= E\{V(\tilde{x}_i(0)) - V(\tilde{x}_i(t_p))\} + E\{V(\tilde{x}_i(t_p)) - V(\tilde{x}_i(0))\} + E\int_0^{t_p} [\tilde{x}_i^T(t)Q_i\tilde{x}_i(t) + u_i^T(t)R_iu_i(t)$$
$$- \rho_i u_{-i}^T(t)u_{-i}(t)]dt$$
$$= E\{V(\tilde{x}_i(0)) - V(\tilde{x}_i(t_p))\} + E\{\int_0^{t_p} [(\tilde{x}_i^T(t)Q_i\tilde{x}_i(t) + u_i^T(t)R_iu_i(t) - \rho_i u_{-i}^T(t)u_{-i}(t))dt$$
$$+ dV(\tilde{x}_i(t))]\}$$

By applying the Itô–Lévy formula in (3.7), we get

$$\begin{aligned}
J_i = {} & E\{V(\tilde{x}_i(0)) - V(\tilde{x}_i(t_p))\} + E\{\int_0^{t_p} [(\tilde{x}_i^T(t)Q_i\tilde{x}_i(t) + u_i^T(t)R_iu_i(t) \\
& - \rho_i u_{-i}^T(t)u_{-i}(t) + \left(\frac{\partial V(\tilde{x}_i(t))}{\partial \tilde{x}_i(t)}\right)^T f(\tilde{x}_i(t)) + \left(\frac{\partial V(\tilde{x}_i(t))}{\partial \tilde{x}_i(t)}\right)^T g_i(\tilde{x}_i(t))u_i(t) \\
& + \left(\frac{\partial V(\tilde{x}_i(t))}{\partial \tilde{x}_i(t)}\right)^T g_{-i}(\tilde{x}_i(t))u_{-i}(t) + \frac{1}{2}l^T(\tilde{x}_i(t))\frac{\partial^2 V(\tilde{x}_i(t))}{\partial \tilde{x}_i^2(t)}l(\tilde{x}_i(t))]dt \\
& + \left(\frac{\partial V(\tilde{x}_i(t))}{\partial \tilde{x}_i(t)}\right)^T l(\tilde{x}_i(t))dw_i(t) + [V(\tilde{x}_i(t) + n(\tilde{x}_i(t))) - V(\tilde{x}_i(t))]dp(t)\}
\end{aligned} \tag{3.19}$$

Completing the square for $u_i(t)$ and $u_{-i}(t)$, i.e.

$$\begin{aligned}
& u_i^T(t)R_iu_i(t) + \left(\frac{\partial V(\tilde{x}_i(t))}{\partial \tilde{x}_i(t)}\right)^T g_i(\tilde{x}_i(t))u_i(t) \\
& = (u_i(t) + \frac{1}{2}R_i^{-1}g_i^T(\tilde{x}_i(t))\left(\frac{\partial V(\tilde{x}_i(t))}{\partial \tilde{x}_i(t)}\right))^T R_i(u_i(t) + \frac{1}{2}R_i^{-1}g_i^T(\tilde{x}_i(t))\left(\frac{\partial V(\tilde{x}_i(t))}{\partial \tilde{x}_i(t)}\right)) \\
& - \frac{1}{4}\left(\frac{\partial V(\tilde{x}_i(t))}{\partial \tilde{x}_i(t)}\right)^T g_i(\tilde{x}_i(t))R_i^{-1}g_i^T(\tilde{x}_i(t))\left(\frac{\partial V(\tilde{x}_i(t))}{\partial \tilde{x}_i(t)}\right)
\end{aligned} \tag{3.20}$$

and

$$\begin{aligned}
& - \rho_i u_{-i}^T(t)u_{-i}(t) + \left(\frac{\partial V(\tilde{x}_i(t))}{\partial \tilde{x}_i(t)}\right)^T g_{-i}(\tilde{x}_i(t))u_{-i}(t) \\
& = - \rho_i\left(u_{-i}(t) - \frac{1}{2\rho_i}g_{-i}^T(\tilde{x}_i(t))\left(\frac{\partial V(\tilde{x}_i(t))}{\partial \tilde{x}_i(t)}\right)\right)^T\left(u_{-i}(t) - \frac{1}{2\rho_i}g_{-i}^T(\tilde{x}_i(t))\left(\frac{\partial V(\tilde{x}_i(t))}{\partial \tilde{x}_i(t)}\right)\right) \\
& + \frac{1}{4\rho_i}\left(\frac{\partial V(\tilde{x}_i(t))}{\partial \tilde{x}_i(t)}\right)^T g_{-i}(\tilde{x}_i(t))g_{-i}^T(\tilde{x}_i(t))\left(\frac{\partial V(\tilde{x}_i(t))}{\partial \tilde{x}_i(t)}\right)
\end{aligned} \tag{3.21}$$

By the fact that $Edw(t) = 0$, $Edp(t) = \lambda dt$ and substituting (3.20) and (3.21) into (3.19), we get

$$
\begin{aligned}
J_i^* &= \min_{u_i(t)} \max_{u_{-i}(t)} J_i \\
&= \min_{u_i(t)} \max_{u_{-i}(t)} E\{V(\tilde{x}_i(0)) - V(\tilde{x}_i(t_p)) + \int_0^{t_p} (\tilde{x}_i^T(t)Q_i\tilde{x}_i(t) + \left(\frac{\partial V(\tilde{x}_i(t))}{\partial \tilde{x}_i(t)}\right)^T f(\tilde{x}_i(t)) \\
&+ \frac{1}{2} l^T(\tilde{x}_i(t)) \frac{\partial^2 V(\tilde{x}_i(t))}{\partial \tilde{x}_i^2(t)} l(\tilde{x}_i(t)) + \lambda[V(\tilde{x}_i(t) + n(\tilde{x}_i(t))) - V(\tilde{x}_i(t))] \\
&+ \frac{1}{4\rho_i} \left(\frac{\partial V(\tilde{x}_i(t))}{\partial \tilde{x}_i(t)}\right)^T g_{-i}(\tilde{x}_i(t)) g_{-i}^T(\tilde{x}_i(t)) \left(\frac{\partial V(\tilde{x}_i(t))}{\partial \tilde{x}_i(t)}\right) \\
&- \frac{1}{4} \left(\frac{\partial V(\tilde{x}_i(t))}{\partial \tilde{x}_i(t)}\right)^T g_i(\tilde{x}_i(t)) R_i^{-1} g_i^T(\tilde{x}_i(t)) \left(\frac{\partial V(\tilde{x}_i(t))}{\partial \tilde{x}_i(t)}\right) \\
&+ \left(u_i(t) + \frac{1}{2} R_i^{-1} g_i^T(\tilde{x}_i(t) \left(\frac{\partial V(\tilde{x}_i(t))}{\partial \tilde{x}_i(t)}\right)\right)^T R_i \left(u_i(t) + \frac{1}{2} R_i^{-1} g_i^T(\tilde{x}_i(t)) \left(\frac{\partial V(\tilde{x}_i(t))}{\partial \tilde{x}_i(t)}\right)\right) \\
&- \rho_i \left(u_{-i}(t) - \frac{1}{2\rho_i} g_i^T(\tilde{x}_i(t)) \left(\frac{\partial V(\tilde{x}_i(t))}{\partial \tilde{x}_i(t)}\right)\right)^T \left(u_{-i}(t) - \frac{1}{2\rho_i} g_i^T(\tilde{x}_i(t)) \left(\frac{\partial V(\tilde{x}_i(t))}{\partial \tilde{x}_i(t)}\right)\right)
\end{aligned}
\tag{3.22}
$$

It is obvious that the suboptimal strategies $u_i^*(t)$ and $u_{-i}^*(t)$ for the minmax of J_i in (3.22) are given as follows:

$$
u_i^*(t) = -\frac{1}{2} R_i^{-1} g_i^T(\tilde{x}_i(t)) \left(\frac{\partial V(\tilde{x}_i(t))}{\partial \tilde{x}_i(t)}\right) \tag{3.23}
$$

$$
u_{-i}^*(t) = \frac{1}{2\rho_i} g_{-i}^T(\tilde{x}_i(t)) \left(\frac{\partial V(\tilde{x}_i(t))}{\partial \tilde{x}_i(t)}\right) \tag{3.24}
$$

which are (3.15) and (3.16), respectively.

Substituting (3.23) and (3.24) into (3.22), we have

$$
\begin{aligned}
J_i^* = E\Bigg\{ & V(\tilde{x}_i(0)) - V(\tilde{x}_i(t_p)) + \int_0^{t_p} \Bigg[(\tilde{x}_i^T(t)Q_i\tilde{x}_i(t) + \left(\frac{\partial V(\tilde{x}_i(t))}{\partial \tilde{x}_i(t)}\right)^T f(\tilde{x}_i(t)) \\
& + \frac{1}{2} l^T(\tilde{x}_i(t)) \frac{\partial^2 V(\tilde{x}_i(t))}{\partial \tilde{x}_i^2(t)} l(\tilde{x}_i(t)) + \lambda[V(\tilde{x}_i(t)) + n(\tilde{x}_i(t)) - V(\tilde{x}_i(t))] \\
& - \frac{1}{4} \left(\frac{\partial V(\tilde{x}_i(t))}{\partial \tilde{x}_i(t)}\right)^T g_i(\tilde{x}_i(t)) R_i^{-1} g_i^T(\tilde{x}_i(t)) \left(\frac{\partial V(\tilde{x}_i(t))}{\partial \tilde{x}_i(t)}\right) \\
& + \frac{1}{4\rho_i} \left(\frac{\partial V(\tilde{x}_i(t))}{\partial \tilde{x}_i(t)}\right)^T g_{-i}(\tilde{x}_i(t)) g_{-i}^T(\tilde{x}_i(t)) \left(\frac{\partial V(\tilde{x}_i(t))}{\partial \tilde{x}_i(t)}\right) \Bigg] dt \Bigg\}
\end{aligned}
\tag{3.25}
$$

By the HJIIs in (3.17) and (3.14), we get

$$
J_i^* \le E[V(\tilde{x}_i(0)) - V(\tilde{x}_i(t_p))] \le EV(\tilde{x}_i(0)) \le \rho_i E\{\tilde{x}_i^T(0)\tilde{x}_i(0)\} \tag{3.26}
$$

which is (3.18).

Q. E. D.

In Theorem 3.2, the suboptimal $u_i^*(t)$ in (3.15) and $u_{-i}^*(t)$ in (3.16) are the constrained game solution in (3.11) or (3.10) with respect to upper bound ρ_i. Therefore, m-player noncooperative H_∞ game strategy design problem could be solved based on Theorem 3.2 and MOP in (3.9) and (3.10) by minimizing these ρ_i simultaneously in the following.

Theorem 3.3

'The nonlinear stochastic m-player noncooperative H_∞ game strategy design problem could be solved by

$$u_i^*(t) = -\frac{1}{2}R_i^{-1}g_i^T(\tilde{x}_i(t))\left(\frac{\partial V^*(\tilde{x}_i(t))}{\partial \tilde{x}_i(t)}\right) \tag{3.27}$$

$$u_{-i}^*(t) = \frac{1}{2\rho_i^*}g_{-i}^T(\tilde{x}_i(t))\left(\frac{\partial V^*(\tilde{x}_i(t))}{\partial \tilde{x}_i(t)}\right) \tag{3.28}$$

for $i = 1, 2, \ldots, m$.

where ρ_i^* and $V^*(\tilde{x}_i(t))$ are the solution of the following MOP:

$$(\rho_1^*, \ldots, \rho_i^*, \ldots, \rho_m^*) = \min_{V(\tilde{x}_i(t))} (\rho_1, \ldots, \rho_i, \ldots, \rho_m) \tag{3.29}$$

subject to HJIIs in (3.17) and (3.18).

Proof: Since the solution of (3.15)–(3.18) in Theorem 3.2 is only the suboptimal solution with upper bounds ρ_i in (3.10), we still need to minimize $(\rho_1, \ldots, \rho_i, \ldots, \rho_m)$ simultaneously in (3.9) to solve MOP for $(\rho_1^*, \ldots, \rho_i^*, \ldots, \rho_m^*)$ as shown in (3.29).

Theorem 3.4

The solution and the corresponding noncooperative game strategy $(u_1^*(t), \ldots, u_i^*(t), \ldots, u_m^*(t))$ in Theorem 3.3 is the Nash equilibrium solution of m-player noncooperative H_∞ game problem in nonlinear stochastic system in (3.4).

Proof: It will be proven by contradiction. According to the definition of Nash equilibrium solution in (3.6) of noncooperative H_∞ game problem for nonlinear stochastic system in (3.4), the inequalities in (3.6) must be all satisfied. If any inequality in (3.6) is violated, for example,

$$\begin{aligned}&(\rho_1^*, \ldots, \rho_i^*, \ldots, \rho_m^*) \geq (\rho_1^*, \ldots, \rho_{i-1}^*, \rho_i, \rho_{i+1}^*, \ldots, \rho_m^*)\\ &i = 1, 2, \ldots, m.\end{aligned} \tag{3.30}$$

It will violate the Pareto domination of $(\rho_1^*, \ldots, \rho_i^*, \ldots, \rho_m^*)$ of MOP in (3.29). Therefore, when MOP in (3.29) is solved, the inequalities in (3.6) all hold and $(\rho_1^*, \ldots, \rho_i^*, \ldots, \rho_m^*)$ and the corresponding $(u_1^*(t), \ldots, u_i^*(t), \ldots, u_m^*(t))$ in Theorem 3.3 are the Nash equilibrium solution.

Q.E.D.

Remark 3.3 Unlike the conventional iterative methods [1,26] with recursive algorithm like gradient searching method one player by one player to solve Nash equilibrium solution of noncooperative game strategy, the H_∞ noncooperative game strategy is to systematically solve MOP in (3.29) by all players $(u_1^*(t), \ldots, u_m^*(t))$ in (3.27) simultaneously. However, it is extremely difficult to solve HJIIs-constrained MOP in (3.29) for m-player H_∞ game strategy design in nonlinear stochastic systems because it still lacks an efficient method for solving HJIIs.

In this chapter, the global linearization with a set of local linearized systems at the vertices of polytope is employed to interpolate a nonlinear stochastic system in (3.4) with a set of local linear stochastic systems to overcome the difficulty in solving HJIIs of MOP in (3.29). This leads to bounding all local linearized stochastic systems by a polytope C_0 with J vertices [23,32]

$$\begin{bmatrix} \frac{\partial f(\tilde{x}_i(t))}{\partial \tilde{x}_i(t)} \\ \frac{\partial g_i(\tilde{x}_i(t))}{\partial \tilde{x}_i(t)} \\ \frac{\partial g_{-i}(\tilde{x}_i(t))}{\partial \tilde{x}_i(t)} \\ \frac{\partial l(\tilde{x}_i(t))}{\partial \tilde{x}_i(t)} \\ \frac{\partial n(\tilde{x}_i(t))}{\partial \tilde{x}_i(t)} \end{bmatrix} \in C_0 \left[\begin{bmatrix} A_{i1} \\ B_{i1} \\ B_{-i1} \\ L_{i1} \\ N_{i1} \end{bmatrix} \cdots \begin{bmatrix} A_{ij} \\ B_{ij} \\ B_{-ij} \\ L_{ij} \\ N_{ij} \end{bmatrix} \cdots \begin{bmatrix} A_{iJ} \\ B_{iJ} \\ B_{-iJ} \\ L_{iJ} \\ N_{iJ} \end{bmatrix} \right], \forall \tilde{x}_i(t) i = 1,2,\ldots m. \quad (3.31)$$

Namely, we can say that the trajectory $\tilde{x}_i(t)$ of the nonlinear stochastic system in (3.4) will be represented by the convex combination of the trajectories of the following J local linearized stochastic systems at J vertices of the polytope if the convex hull C_0 consists of all the local linearized systems at all [23,32]

$$\begin{aligned} d\tilde{x}_i(t) &= [A_{ij}\tilde{x}_i(t) + B_{ij}u_i(t) + B_{-ij}u_{-j}(t)]dt + L_{ij}\tilde{x}_i(t)dw(t) + N_{ij}\tilde{x}_i(t)dp(t) \\ i &= 1,2,\ldots,m, \ j = 1,2,\ldots,J \end{aligned} \quad (3.32)$$

where $B_{-ij} = [B_{1j}\ldots B_{i-1j}B_{i+1j}\ldots B_{mj}]$.

According to the global linearization theory [23], the trajectory $\tilde{x}_i(t)$ of the nonlinear stochastic system in (3.4) can be represented by a convex combination of the trajectories of J local linearized stochastic systems in (3.32) as follows:

$$\begin{aligned} d\tilde{x}_i(t) = \sum_{j=1}^{J} \alpha_j(\tilde{x}_i(t))[(A_{ij}\tilde{x}_i(t) + B_{ij}u_i(t) + B_{-ij}u_{-j}(t)dt + L_{ij}\tilde{x}_i(t)dw(t) \\ + N_{ij}\tilde{x}_i(t)dp(t)] \end{aligned} \quad (3.33)$$

where $\alpha_j(\tilde{x}_i(t))$ denotes the interpolation functions, $0 \leq \alpha_j(\tilde{x}_i(t)) \leq 1$ and $\sum_{j=1}^{J} \alpha_j(\tilde{x}_i(t)) = 1$, i.e., we can replace the trajectory of nonlinear stochastic system in (3.4) by the trajectory of the interpolated stochastic system in (3.33).

Remark 3.4 Except the global linearization method, there are other interpolation methods to interpolate some local linearized stochastic systems through different interpolation functions to efficiently approximate a nonlinear stochastic system, for example, T-S fuzzy method through the interpolatory fuzzy bases [5], which will be also employed in the following chapters to treat the m-player noncooperative or cooperative game strategy design problem in nonlinear stochastic systems.

The following lemma is necessary for the proof of the following theorem.

Lemma 3.2 [5] For any matrix S_i with appropriate dimension and interpolation function $\alpha_i(\tilde{x}_j(t))$ with $0 \leq \alpha_i(\tilde{x}_j(t)) \leq 1$ and $\sum_{i=1}^{J} \alpha_i(\tilde{x}_j(t)) = 1$, then for the matrix $P>0$, we have

$$\left(\sum_{i=1}^{J}\alpha_i(\tilde{x}_j(t))S_i^T\right)P\left(\sum_{i=1}^{J}\alpha_i(\tilde{x}_j(t))S_i\right)\leq\sum_{i=1}^{J}\alpha_i(\tilde{x}_j(t))S_i^TPS_i \tag{3.34}$$

If the nonlinear stochastic system in (3.4) is represented by the global linearization system in (3.33), then Theorem 3.2 for solving the constrained Nash games in (3.13) and (3.14) could be replaced by the following theorem.

Theorem 3.5
Based on the global linearization system (3.33), the m-constrained Nash quadratic games in (3.13) and (3.14) or m-constrained Nash quadratic games in (3.11) could be solved by

$$u_i^*(t)=-\sum_{j=1}^{J}\alpha_j(\tilde{x}_i(t))R_i^{-1}B_{ij}^TP\tilde{x}_i(t) \tag{3.35}$$

$$\begin{aligned}u_{-i}^*(t)&=\frac{1}{\rho_i}\sum_{j=1}^{J}\alpha_j(\tilde{x}_i(t))B_{-ij}^TP\tilde{x}_i(t)\\ i&=1,2,\ldots,m\end{aligned} \tag{3.36}$$

where $P>0$ is the common solution of the following Riccati-like inequalities

$$\begin{aligned}&Q_i+A_{ij}^TP+PA_{ij}-PB_{ij}R_i^{-1}B_{ij}^TP+\frac{1}{\rho_i}PB_{-ij}B_{-ij}^TP+L_{ij}^TPL_{ij}\\&+\lambda(N_{ij}^TPN_{ij}+N_{ij}^TP+PN_{ij})\leq0, i=1,2,\ldots,mj=1,2,\ldots,J\end{aligned} \tag{3.37}$$

with

$$0\leq P\leq\rho_iI, i=1,2,\ldots,m \tag{3.38}$$

Proof: Based on Theorem 3.2, we select Lyapunov function $V(\tilde{x}_i(t))=\tilde{x}_i^T(t)P\tilde{x}_i(t)$. By the global linearization technique in (3.31)–(3.33), we get

$$\begin{aligned}f(\tilde{x}_i(t))&=\sum_{j=1}^{J}\alpha_j(\tilde{x}_i(t))A_{ij}\tilde{x}_i(t), l(\tilde{x}_i(t))=\sum_{j=1}^{J}\alpha_j(\tilde{x}_i(t))L_{ij}\tilde{x}_i(t)\\ g_i(\tilde{x}_i(t))&=\sum_{j=1}^{J}\alpha_j(\tilde{x}_i(t))B_{ij}, n(\tilde{x}_i(t))=\sum_{j=1}^{J}\alpha_j(\tilde{x}_i(t))N_{ij}\tilde{x}_i(t)\\ g_{-i}(\tilde{x}_i(t))&=\sum_{j=1}^{J}\alpha_j(\tilde{x}_i(t))B_{-ij}\end{aligned} \tag{3.39}$$

Then,

$$\begin{aligned}u_i^*(t)&=-\frac{1}{2}R_i^{-1}g_i^T(\tilde{x}_i(t))\left(\frac{\partial V^*(\tilde{x}_i(t))}{\partial\tilde{x}_i(t)}\right)\\&=-\sum_{j=1}^{J}\alpha_j(\tilde{x}_i(t))R_i^{-1}B_{ij}^TP\tilde{x}_i(t)\end{aligned} \tag{3.40}$$

and

$$\begin{aligned}u_{-i}^*(t)&=\frac{1}{2\rho_i^*}g_{-i}^*(\tilde{x}_i(t))\left(\frac{\partial V^*(\tilde{x}_i(t))}{\partial\tilde{x}_i(t)}\right)\\&=\frac{1}{\rho_i}\sum_{j=1}^{J}\alpha_j(\tilde{x}_i(t))B_{-ij}^TP\tilde{x}_i(t)\end{aligned} \tag{3.41}$$

which are (3.35) and (3.36), respectively.

The HJIIs in (3.17) and (3.18) have the following result after using Lemma 3.2:

$$\begin{aligned}
&\left(\frac{\partial V(\tilde{x}_i(t))}{\partial \tilde{x}_i(t)}\right)^T f(\tilde{x}_i(t)) + \tilde{x}_i^T(t)Q_i\tilde{x}_i(t) - \frac{1}{4}\left(\frac{\partial V(\tilde{x}_i(t))}{\partial \tilde{x}_i(t)}\right)^T g_i(\tilde{x}_i(t))R_i^{-1}g_i^T(\tilde{x}_i(t))\left(\frac{\partial V(\tilde{x}_i(t))}{\partial \tilde{x}_i(t)}\right)\\
&+\frac{1}{4\rho_i}\left(\frac{\partial V(\tilde{x}_i(t))}{\partial \tilde{x}_i(t)}\right)^T g_{-i}(\tilde{x}_i(t))g_{-i}^T(\tilde{x}_i(t))\left(\frac{\partial V(\tilde{x}_i(t))}{\partial \tilde{x}_i(t)}\right) + \frac{1}{2}l^T(\tilde{x}_i(t))\frac{\partial^2 V(\tilde{x}_i(t))}{\partial \tilde{x}_i^2(t)}l(\tilde{x}_i(t))\\
&+\lambda[V(\tilde{x}_i(t)+n(\tilde{x}_i(t))) - V(\tilde{x}_i(t))]\\
&=\sum\nolimits_{j=1}^{J}\alpha_j(\tilde{x}_i(t))\tilde{x}_i^T(t)[Q_i + A_{ij}^TP + PA_{ij} - PB_{ij}R_i^{-1}B_{ij}^TP + \frac{1}{\rho_i}PB_{-ij}B_{-ij}^TP + L_{ij}^TPL_{ij}\\
&+\lambda(N_{ij}^TPN_{ij} + N_{ij}^TP + PN_{ij})]\tilde{x}_i(t) \leq 0
\end{aligned} \tag{3.42}$$

According to (3.42), once the Riccati-like inequalities in (3.37) hold, the HJIIs in (3.17) can also be satisfied. For the inequalities in (3.18), we have

$$0 \leq E\{\tilde{x}_i^T(t)P\tilde{x}_i(t)\} \leq \rho_i\{\tilde{x}_i^T(0)\tilde{x}_i(0)\} \tag{3.43}$$

The sufficient and necessary condition to satisfy (3.43) is the following LMIs:

$$0 \leq P \leq \rho_i I, \text{for } i = 1, 2, \ldots, m \tag{3.38}$$

which is LMIs in (3.18)

Q.E.D.

Remark 3.5 (i) The Riccati-like inequalities in (3.37) are considered as local linearized HJIIs in (3.17) of the polytope at J vertices in (3.31). It still needs much effort to solve Riccati-like inequalities in (3.37) and (3.38). For the simplicity of design, the Riccati-like inequalities in (3.37) can be readily transformed into the following equivalent LMIs through applying Schur complement transformation in Lemma 2.3 several times after multiplying $W = P^{-1}$ to both sides of (3.37)

$$\begin{bmatrix}
\Pi_{ij} & WL_{ij}^T & \lambda^{1/2}WN_{ij}^T & W & B_{-ij}\\
L_{ij}W & -W & 0 & 0 & 0\\
\lambda^{1/2}N_{ij}W & 0 & -W & 0 & 0\\
W & 0 & 0 & -Q_i^{-1} & 0\\
B_{-ij}^T & 0 & 0 & 0 & -\rho_i I
\end{bmatrix} \leq 0,\ i = 1, 2, \ldots, m, j = 1, \ldots, J \tag{3.44}$$

where $\Pi_{ij} = WA_{ij}^T + A_{ij}W - B_{ij}R_i^{-1}B_{ij}^T + \lambda(WN_{ij}^T + N_{ij}W)$

(ii) Similarly, the inequalities in (3.38) are equivalent to

$$\begin{bmatrix} -\rho_i I & I\\ I & W \end{bmatrix} \leq 0, i = 1, 2, \ldots, m \tag{3.45}$$

(iii) The LMIs in (3.44) and (3.45) could be easily solved for $W>0$ by the LMI toolbox in Matlab.

(iv) Obviously, the LMIs in (2.36) and (2.37) in Chapter 2 are the special case of (3.44) and (3.45), respectively, at a local linearization.

Based on Theorem 3.5 through the global linearization, the solution of m-player noncooperative H_∞ game strategy design problem in Theorem 3.3 could be simplified in the following theorem.

Theorem 3.6

The nonlinear stochastic m-player noncooperative H_∞ game strategy design problem of nonlinear stochastic system in (3.4) and (3.5) could be solved based on the global linearization method in the following:

$$u_i^*(t) = -\sum_{j=1}^{J} \alpha_j(\tilde{x}_i(t))R_i^{-1}B_{ij}^T P^* \tilde{x}_i(t) \tag{3.46}$$

$$u_{-i}^*(t) = \frac{1}{\rho_i^*}\sum_{j=1}^{J} \alpha_j(\tilde{x}_i(t))B_{ij}^T P^* \tilde{x}_i(t), \quad i = 1, 2, \ldots, m \tag{3.47}$$

where ρ_i^* and $P^* = W^{*^{-1}}$ are the solution of following LMIs-constrained MOP:

$$(\rho_1^*, \ldots, \rho_i^*, \ldots, \rho_m^*) = \min_{W>0}(\rho_1, \ldots, \rho_i, \ldots, \rho_m) \tag{3.48}$$

subject to LMIs in (3.44) and (3.45)

From Theorem 3.6, the main work for m-player noncooperative H_∞ game strategy design problem of nonlinear stochastic system is to solve the LMIs-constrained MOP in (3.48). There are two steps to achieve LMIs-constrained MOP for m-player noncooperative stochastic H_∞ game strategy. The first step is to perform the global linearization (3.31) of nonlinear stochastic system (3.4) to obtain (3.33). The second step is to employ the proposed LMIs-constrained MOEA to search for ρ_i^* and $W^* = P^{*^{-1}}$ in (3.48) for the noncooperative H_∞ game strategies $u_i^*(t) = -\sum_{j=1}^{J} \alpha_j(\tilde{x}_i(t))R_i^{-1}B_{ij}^T P^* \tilde{x}_i(t)$ for each player in (3.46) for $i = 1, 2, \ldots, m$. Different from the traditional MOEAs in [25,35,36] to search for W^* or P^* for MOP in (3.48), the proposed LMI-constrained MOEA is to search $(\rho_1^*, \ldots, \rho_m^*)$ and then find the corresponding W^* or P^* from LMIs in (3.44) and (3.45) by LMI toolbox in Matlab. The searching region of LMIs-constrained MOEA is the set of objective values $(\rho_1, \ldots, \rho_m)$ under LMIs constraints in (3.44) and (3.45). First, the MOEA encodes the individuals $(\rho_1^n, \ldots, \rho_m^n)$ for all $n = 1, 2, \ldots, N_p$ as the so-called initial parent population P_r is a feasible set with the initial number $r = 1$, where P_r is the population at the rth iteration. In other words, the nth individual $(\rho_1^n, \ldots, \rho_m^n)$ must satisfy the LMIs in (3.44) and (3.45) with $\rho_i^L \leq \rho_i^n \leq \rho_i^v$, $i = 1, 2, \ldots, m$ for all $n = 1, 2, \ldots, N_p$, where N_p is the population size. ρ_i^v and ρ_i^L denote the upper and lower bounds of each ρ_i^n, respectively. Second, the candidate Pareto dominance solution can be obtained by the crowded-comparison operator [13,30]. Then, the child population can also be generated by crossover and mutation. Note that if some individuals are not feasible, i.e., if they cannot satisfy with the requirement of LMIs in (3.44) and (3.45), these individuals should be deleted from the candidate set. The proposed algorithm is executed iteratively while the final population is achieved.

At last, once the set of Pareto front P_F is reached, the Pareto optimal solution $(u_1^*(t), \ldots, u_i^*(t), \ldots, u_m^*(t))$ of MOP in (3.46)–(3.48) for the m-player noncooperative game strategy of nonlinear stochastic system can also be obtained. According to the above analyses, the algorithm of LMIs-constrained MOEA algorithm to solve the MOP in (3.48) for m-player H_∞ noncooperative game strategy of nonlinear stochastic system is proposed as follows:

LMIs-constrained MOEA algorithm for noncooperative H_∞ game strategy of nonlinear stochastic systems

Step 1: Choose the searching region $(\rho_1^L, \ldots, \rho_m^L) \times (\rho_1^v, \ldots, \rho_m^v)$ for the feasible objective vector $(\rho_1, \ldots, \rho_m)$ and provide the population number N_p, the iteration number N_i, the crossover rate C_r, and mutation ratio m_r in the proposed MOEA. Set iteration number $i = 1$.

Step 2: Choose N_p feasible individuals (chromosomes) as the initial population P_1.

Step 3: Employ EA and produce $2N_p$ feasible individuals by checking whether their corresponding objective vectors $(\rho_1, \ldots, \rho_m)$ are feasible or not (i.e., satisfying the LMIs in (3.44) and (3.45)).

Step 4: Set the iterative index $i = i + 1$ and choose N_p elite individuals from $2N_p$ feasible individuals in step 3 via the non-dominated sorting scheme and the crowed comparison method to the population P_{i+1}.

Step 5: Select a preferable feasible objective $(\rho_1^*, \ldots, \rho_m^*) \in P_F$ according to designer's own preference with the optimal $W^* = (P^*)^{-1}$, where W^* is the solution of MOP in (3.48). Once the preferable solution is selected as the control strategy $(u_1^*(t), \ldots, u_i^*(t), \ldots, u_m^*(t))$ with $u_i^*(t) = -\sum_{j=1}^{J} \alpha_j(\tilde{x}_i(t)) R_i^{-1} B_{ij}^T P^* \tilde{x}_i(t)$, $i = 1, 2, \ldots, m$, the m-player noncooperative H_∞ game strategy of nonlinear stochastic system in (3.1) could be solved.

Remark 3.6 In the case of m-player noncooperative H_∞ game strategy for the stabilization problem of nonlinear stochastic system in (3.1) as the following

$$\rho_i^* = \min_{u_i(t)} \max_{u_{-i}(t)} \frac{E\{\int_0^{t_p} [(x^T(t)Q_i x(t) + u_i^T(t) R_i u_i(t)] dt\}}{E\{x(0)^T x(0) + E\int_0^{t_p} u_{-i}^T(t) u_{-i}(t) dt\}} \quad (3.49)$$
$$i = 1, 2, \ldots, m$$

subject to

$$\begin{aligned} dx(t) &= [f(x(t)) + \sum_{i=1}^{m} g_i(x(t)) u_i(t) + h(x(t)) v(t)] dt + l(x(t)) dw(t) + n(x(t)) dp(t) \\ &= [f(x(t)) + g_i(x(t)) u_i(t) + g_{-i}(x(t)) u_{-i}(t)] dt + l(x(t)) dw(t) + n(x(t)) dp(t) \end{aligned} \quad (3.50)$$

where $g_{-i}(x(t)) = [g_1(x(t)) \ldots g_{i-1}(x(t))\ g_{i+1}(x(t)) \ldots g_m(x(t))\ h(x(t))]$ and the total competitive strategies are $u_{-i}(t) = [u_1^T(t) \ldots u_{i-1}^T(t)\ u_{i+1}^T(t) \ldots u_m^T(t)\ v^T(t)]$

Then, the suboptimal m-player noncooperative H_∞ game strategy for the stabilization problem in (3.49) and (3.50) could be transformed to the following MOP:

$$(\rho_1^*, \ldots, \rho_i^*, \ldots, \rho_m^*) = \min_{(u_1(t), \ldots, u_i(t), \ldots, u_m(t))} (\rho_1, \ldots, \rho_i, \ldots, \rho_m) \quad (3.51)$$

subject to

$$\min_{u_i(t)}\max_{u_{-i}(t)} \frac{E\{\int_0^{t_p}[(x^T(t)Q_ix(t)+u_i^T(t)R_iu_i(t)]dt\}}{E\{x(0)^Tx(0)+E\int_0^{t_p}u_{-i}^T(t)u_{-i}(t)dt\}} \le \rho_i$$
$$i=1,2,\ldots,m \tag{3.52}$$

Following the previous results, we get

Theorem 3.7

The m-player noncooperative H_∞ game strategy for the stabilization problem in (3.49) and (3.50) could be solved by

$$u_i^*(t) = -\frac{1}{2}R_i^{-1}g_i^T(\tilde{x}_i(t))\left(\frac{\partial V^*(\tilde{x}_i(t))}{\partial \tilde{x}_i(t)}\right),\ i=1,2,\ldots,m \tag{3.53}$$

$$u_{-i}^*(t) = \frac{1}{2\rho_i^*}g_{-i}^*(\tilde{x}_i(t))\left(\frac{\partial V^*(\tilde{x}_i(t))}{\partial \tilde{x}_i(t)}\right),\ i=1,2,\ldots,m \tag{3.54}$$

where ρ^* and $V^*(x(t))$ are the solution of the following HJIIs-constrained MOP

$$(\rho_1^*,\ldots,\rho_i^*,\ldots,\rho_m^*) = \min_{V(x(t))>0}(\rho_1,\ldots,\rho_i,\ldots,\rho_m) \tag{3.55}$$

subject to

$$\begin{aligned}&x^T(t)Q_ix(t)+\left(\frac{\partial V(x(t))}{\partial x(t)}\right)^Tf(x(t))-\frac{1}{4}\left(\frac{\partial V(x(t))}{\partial x(t)}\right)^Tg_i(x(t))R_i^{-1}g_i^T(x(t))\left(\frac{\partial V(x(t))}{\partial x(t)}\right)\\&+\frac{1}{4\rho_i}\left(\frac{\partial V(x(t))}{\partial x(t)}\right)^Tg_{-i}(x(t))g_{-i}^T(x(t))\left(\frac{\partial V(x(t))}{\partial x(t)}\right)+\frac{1}{2}l^T(x(t))\frac{\partial^2V(x(t))}{\partial x^2(t)}l(x(t))\\&+\lambda[V(\tilde{x}_i(t)+n(\tilde{x}_i(t)))-V(\tilde{x}_i(t))]\le 0,\ i=1,2,\ldots,m\end{aligned} \tag{3.56}$$

with

$$EV(x(0))\le\rho_iE\{x^T(0)x(0)\} \tag{3.57}$$

Since it is not easy to solve the HJIIs-constrained MOP in (3.55)–(3.57), the global linearization method at J vertices of polytope C_0 is employed as follows [23]:

$$\begin{bmatrix}\frac{\partial f(x(t))}{\partial x(t)}\\ \frac{\partial g_i(x(t))}{\partial x(t)}\\ \frac{\partial g_{-i}(x(t))}{\partial x(t)}\\ \frac{\partial l(x(t))}{\partial x(t)}\\ \frac{\partial n(\tilde{x}_i(t))}{\partial \tilde{x}_i(t)}\end{bmatrix}\in C_0\left[\begin{bmatrix}A_1\\B_{i1}\\B_{-i1}\\L_1\\N_1\end{bmatrix}\cdots\begin{bmatrix}A_j\\B_{ij}\\B_{-ij}\\L_j\\N_j\end{bmatrix}\cdots\begin{bmatrix}A_J\\B_{iJ}\\B_{-iJ}\\L_J\\N_J\end{bmatrix}\right],\forall x(t)\ i=1,\ldots,m \tag{3.58}$$

Then, the nonlinear stochastic system in (3.50) could be represented by the following global linearization system [23,32]:

$$dx(t) = \sum_{j=1}^{J} \alpha_j(x(t))[(A_j x(t) + B_{ij}u_i(t) + B_{-ij}u_{-i}(t))dt \\ + L_j x(t)dw(t) + N_j x(t)dp(t)] \tag{3.59}$$

Based on the global linearization system in (3.59), we get

Theorem 3.8

The m-player noncooperative H_∞ game strategy for the stabilization problem in (3.49) and (3.50) could be solved by

$$u_i^*(t) = R_i^{-1}g_i P^* x(t) = \sum_{j=1}^{J} \alpha_j(x(t))B_{ij}^T P^* x(t) \tag{3.60}$$

$$u_{-i}^*(t) = \frac{1}{\rho_i^*}g_i(x(t))P^* x(t) = \frac{1}{\rho_i^*}\sum_{j=1}^{J} \alpha_j(x(t))B_{-ij}^T P^* x(t) \tag{3.61}$$

where ρ^* and P^* are the solution of the following Riccati-like inequalities-constrained MOP

$$(\rho_1^*, \ldots, \rho_i^*, \ldots, \rho_m^*) = \min_{P>0}(\rho_1, \ldots, \rho_i, \ldots, \rho_m) \tag{3.62}$$

subject to

$$Q_i + A_j^T P + PA_j - PB_{ij}R_i^{-1}B_{ij}^T P + \frac{1}{\rho_i}PB_{-ij}B_{-ij}^T P + L_j^T PL_j \\ + \lambda(N_j^T PN_j + N_j^T P + PN_j) \le 0, i = 1, 2, \ldots, m\ j = 1, 2, \ldots, J \tag{3.63}$$

$$0 \le P \le \rho_i I, i = 1, 2, \ldots, m \tag{3.64}$$

Since it is still not easy to solve the Riccati-like inequalities-based MOP in (3.62)–(3.64), we let $W = P^{-1}$ and perform the multiplication of W to both sides of (3.63) in the following:

$$WQ_i W + WA_j^T + A_j W - B_{ij}R_i^{-1}B_{ij}^T + \frac{1}{\rho_i}B_{ij}B_{ij}^T + WL_j^T W^{-1}L_j W \\ + \lambda(WN_j^T W^{-1}N_j W + WN_j^T + N_j W) \le 0, i = 1, 2, \ldots, m\ j = 1, 2, \ldots, J \tag{3.65}$$

By performing Schur complements several times in the above inequalities, we get the following equivalent LMIs:

$$\begin{bmatrix} \Pi_{ij} & W & WL_j^T & WN_j^T & B_{-ij} \\ W & -Q_i^{-1} & 0 & 0 & 0 \\ L_j W & 0 & -W & 0 & 0 \\ N_j W & 0 & 0 & -W & 0 \\ B_{-ij}^T & 0 & 0 & 0 & -\rho_i I \end{bmatrix} \le 0\ i = 1, 2, \ldots, m\ j = 1, \ldots, J \tag{3.66}$$

where $\Pi_{ij} = WA_{ij}^T + A_{ij}W - B_{ij}R_i^{-1}B_{ij}^T + \lambda(WN_{ij}^T + N_{ij}W)$

and the inequalities in (3.64) is equivalent to

$$\begin{bmatrix} -\rho_i I & I \\ I & -W \end{bmatrix} \leq 0 \tag{3.67}$$

Theorem 3.9
The m-player noncooperative H_∞ game strategy for the stabilization of nonlinear stochastic system could be solved by

$$u_i^*(t) = -R_i^{-1} g_i(x(t))P^* x(t)$$

$$u_{-i}^*(t) = \frac{1}{\rho_i} g_i(x(t))P^* x(t)$$

where ρ_i^* and $P^* = W^{*^{-1}}$ are the solution of the following LMIs-constrained MOP:

$$(\rho_1^*, \ldots, \rho_i^*, \ldots, \rho_m^*) = \min_{W>0}(\rho_1, \ldots, \rho_i, \ldots, \rho_m) \tag{3.68}$$

subject to LMIs in (3.66) and (3.67)

3.3 Multi-Player Noncooperative H_2 Game Strategy in Nonlinear Stochastic Systems without External Disturbance

If the nonlinear stochastic system in (3.1) is free of external disturbance or the external disturbance $v(t)$ is neglected as follows

$$dx(t) = [f(x(t)) + \sum\nolimits_{i=1}^{m} g_i(x(t)u_i(t)]dt + l(x(t))dw(t) + n(x(t))dp(t) \tag{3.69}$$

then, the following m-player noncooperative H_2 (quadratic) game strategy design problem is formulated as follows:

$$J_i^* = \min_{u_i(t)} \max_{\substack{u_j(t) \\ j=1,\ldots,m \\ j\neq i}} E\{\int_0^{t_p} ((x(t) - x_d^i)^T Q_i (x(t) - x_d^i) + u_i^T(t) R_i u_i(t) - \sum_{\substack{j=1 \\ j\neq i}}^{m} u_j^T(t) R_j u_j(t))dt\} \tag{3.70}$$

Let us denote $\tilde{x}_i(t) = x(t) - x_d^i, g_{-i}(t) = [g_1(t) \quad \cdots \quad g_{i-1}(t) \quad g_{i+1}(t) \quad \cdots \quad g_m(t)]$,

$$u_{-i}(t) = \begin{bmatrix} u_1(t) \\ \vdots \\ u_{i-1}(t) \\ u_{i+1}(t) \\ \vdots \\ u_m(t) \end{bmatrix}, R_{-i} = \begin{bmatrix} R_1 & & & & & 0 \\ & \ddots & & & & \\ & & R_{i-1} & & & \\ & & & R_{i+1} & & \\ & 0 & & & \ddots & \\ & & & & & R_m \end{bmatrix}$$

Then, nonlinear stochastic system in (3.70) could be represented by

$$\begin{aligned} d\tilde{x}_i(t) =& [f(\tilde{x}_i(t)) + g_i(\tilde{x}_i(t))u_i(t) + g_{-i}(\tilde{x}_i(t))u_{-i}(t)]dt \\ &+ l(\tilde{x}_i(t))dw(t) + n(\tilde{x}_i(t))dp(t) \ i = 1,2,\dots,m. \end{aligned} \tag{3.71}$$

where $f(\tilde{x}_i(t)) = f(\tilde{x}_i(t) + x_d^i), g_i(\tilde{x}_i(t)) = g_i(\tilde{x}_i(t) + x_d^i), g_{-i}(\tilde{x}_i(t)) = g_{-i}(\tilde{x}_i(t) + x_d^i), l(_i\tilde{x}(t)) = l(\tilde{x}_i(t) + x_d^i), n(\tilde{x}_i(t)) = n(\tilde{x}_i(t) + x_d^i)$ for the simplicity of notation.

Then, the m-player noncooperative H_2 game in (3.70) for nonlinear stochastic system (3.71) could be modified as [1,4]

$$\begin{aligned} J_i^* = \min_{u_i(t)} \max_{u_{-i}(t)} E\{\int_0^{t_p} ((\tilde{x}_i^T(t)Q_i\tilde{x}_i(t) + u_i^T(t)R_iu_i(t) - u_{-i}^T(t)R_{-i}u_{-i}(t))dt\} \\ i = 1,2,\dots,m \end{aligned} \tag{3.72}$$

In general, it is not easy to solve the above m-player noncooperative H_2 game strategy in (3.72) simultaneously for nonlinear stochastic system in (3.71) to obtain $(J_1^*,\dots,J_i^*,\dots,J_m^*)$ directly. An indirect method is employed to solve m minmax H_2 Nash quadratic games in (3.72) by minimizing their upper bounds as the following MOP from the suboptimal perspective:

$$(J_1^*,\dots,J_i^*,\dots,J_m^*) = \min_{(u_1(t),\dots,u_i(t),\dots,u_m(t))} (J_1,\dots,J_i,\dots,J_m) \tag{3.73}$$

subject to

$$\begin{aligned} \min_{u_i(t)} \max_{u_{-i}(t)} E\{\int_0^{t_p} ((\tilde{x}_i^T(t)Q_i\tilde{x}_i(t) + u_i^T(t)R_iu_i(t) - u_{-i}^T(t)R_{-i}u_{-i}(t))dt\} \le J_i \\ i = 1,2,\dots,m \end{aligned} \tag{3.74}$$

where $\min(J_1,\dots,J_i,\dots,J_m)$ denotes the minimization of $J_1,\dots,J_i,\dots,J_{m-1}$ and J_m simultaneously.

Theorem 3.10
The solution of MOP in (3.73) and (3.74) is equivalent to the solution of m-player noncooperative H_2 game strategy design problem in (3.72).

Proof: Similar to the proof procedure in Theorem 3.1.

Lemma 3.3 Let $V : \mathbb{R}^{\kappa} \to R$, $V(\cdot) \in C^2(\mathbb{R}^{\kappa})$ and $V(\cdot) \ge 0$. For the m-player nonlinear stochastic system in (3.71), the Itô–Lévy formula of $V(\tilde{x}(t))$ is given as follows [33,34]:

$$\begin{aligned} dV(\tilde{x}_i(t)) =& \left(\frac{\partial V(\tilde{x}_i(t))}{\partial \tilde{x}_i(t)}\right)^T (f(\tilde{x}_i(t) + g_i(\tilde{x}_i(t))u_i(t) + g_{-i}(\tilde{x}_i(t))u_{-i}(t))dt \\ &+ \left(\frac{\partial V(\tilde{x}_i(t))}{\partial \tilde{x}_i(t)}\right)^T l(\tilde{x}_i(t)dw(t) + \frac{1}{2} l^T(\tilde{x}_i(t)) \frac{\partial^2 V(\tilde{x}_i(t))}{\partial \tilde{x}_i^2(t)} l(\tilde{x}_i(t))dt \\ &+ [V(\tilde{x}_i(t) + n(\tilde{x}_i(t))) - V(\tilde{x}_i(t))]dp(t) \end{aligned} \tag{3.75}$$

Based on Theorem 3.10 and Itô–Lévy formula in Lemma 3.3, we get the following result for m-player noncooperative H_2 game strategy.

Theorem 3.11
The m-player noncooperative H_2 game strategy in (3.72) for the nonlinear stochastic system in (3.71) could be solved by the following:

$$u_i^*(t) = -\frac{1}{2}R_i^{-1}g_i^T(\tilde{x}_i(t))\left(\frac{\partial V^*(\tilde{x}_i(t))}{\partial \tilde{x}_i(t)}\right) \tag{3.76}$$

$$u_{-i}^*(t) = \frac{1}{2}g_{-i}^T(\tilde{x}_i(t))\left(\frac{\partial V^*(\tilde{x}_i(t))}{\partial \tilde{x}_i(t)}\right) \quad \text{for } i = 1, 2, \ldots, m \tag{3.77}$$

where $V^*(\tilde{x}_i(t))$ is the solution of the following HJIIs-constrained MOP:

$$(J_1^*, \ldots, J_i^*, \ldots, J_m^*) = \min_{V(\tilde{x}_i(t))>0}(J_1, \ldots, J_i, \ldots, J_m) \tag{3.78}$$

subject to

$$\begin{aligned}
&\tilde{x}_i^T(t)Q_i\tilde{x}_i(t) + \left(\frac{\partial V(\tilde{x}_i(t))}{\partial \tilde{x}_i(t)}\right)^T f(\tilde{x}_i(t)) - \frac{1}{4}\left(\frac{\partial V(\tilde{x}_i(t))}{\partial \tilde{x}_i(t)}\right)^T g_i(\tilde{x}_i(t))R_i^{-1}g_i^T(\tilde{x}_i(t))\left(\frac{\partial V(\tilde{x}_i(t))}{\partial \tilde{x}_i(t)}\right) \\
&-\frac{1}{4}\left(\frac{\partial V(\tilde{x}_i(t))}{\partial \tilde{x}_i(t)}\right)^T g_{-i}(\tilde{x}_i(t))g_{-i}^T(\tilde{x}_i(t))\left(\frac{\partial V(\tilde{x}_i(t))}{\partial \tilde{x}_i(t)}\right) + \frac{1}{2}l^T(\tilde{x}_i(t))\frac{\partial^2 V(\tilde{x}_i(t))}{\partial \tilde{x}_i^2(t)}l(\tilde{x}_i(t)) \\
&+ \lambda[V(\tilde{x}_i(t) + n(\tilde{x}_i(t))) - V(\tilde{x}_i(t))] \leq 0
\end{aligned} \tag{3.79}$$

$$V(\tilde{x}_i(0)) \leq J_i \tag{3.80}$$

Proof: Similar to Theorem 3.5 and Theorem 3.6.

Theorem 3.12
The solution $(u_1^*(t), \ldots, u_i^*(t), \ldots u_m^*(t))$ with the corresponding objective vector $(J_1^*, \ldots, J_i^*, \ldots, J_m^*)$ of m-player noncooperative H_2 game strategy in Theorem 3.11 for nonlinear stochastic system in (3.71) is Nash equilibrium solution.

Proof: Similar to Theorem 3.4.

In general, it is not easy to solve HJIIs-constrained MOP in (3.78)–(3.80) for the m-player noncooperative H_2 game strategy of nonlinear stochastic systems. Therefore, the global linearization method in (3.31) is also introduced to interpolate several local linearized systems at the vertices of the polytope to approximate the nonlinear stochastic system in (3.71) as the following convex combination of J local linearized stochastic systems:

$$\begin{aligned}
d\tilde{x}_i(t) = \sum\nolimits_{j=1}^{J} \alpha_j(\tilde{x}_i(t))[A_i\tilde{x}(t) + B_{ij}u_i(t) + B_{-ij}u_{-i}(t) \\
+ L_{ij}\tilde{x}_i(t)dw(t) + N_{ij}\tilde{x}(t)dp(t)]
\end{aligned} \tag{3.81}$$

where $\alpha_j(\tilde{x}_i(t)), i = 1, \ldots, J$ denote the interpolation functions with $0 \leq \alpha_j(\tilde{x}_i(t)) \leq 1$ and $\sum_{j=1}^{J} \alpha_j(\tilde{x}_i(t)) = 1$.

Based on the global linearization system in (3.81), we get the following result.

Theorem 3.13

The m-player noncooperative H_2 game strategy in (3.72) for the nonlinear stochastic system in (3.81) could be solved by the following:

$$u_i^*(t) = -\sum_{j=1}^{J} \alpha_j(\tilde{x}_i(t)) R_i^{-1} B_{ij}^T P^* \tilde{x}_i(t) \quad (3.82)$$

$$\begin{aligned} &u_{-i}^*(t) = \sum_{j=1}^{J} \alpha_j(\tilde{x}_i(t)) B_{-ij}^T P^* \tilde{x}_i(t) \\ &i = 1, 2, \ldots, m \end{aligned} \quad (3.83)$$

where $P^* > 0$ is the solution of the following Riccati-like inequalities-constrained MOP:

$$\min_{P>0}(J_1, \ldots, J_i, \ldots, J_m) \quad (3.84)$$

subject to

$$\begin{aligned} &PA_{ij} + A_{ij}^T P + Q_i - PB_{ij} R_i^{-1} B_{ij}^T P + PB_{-ij} B_{-ij}^T P + L_{ij}^T P L_{ij} \\ &+ \lambda(N_{ij}^T P N_{ij} + N_{ij}^T P + P N_{ij}) \leq 0, i = 1, 2, \ldots, mj = 1, 2, \ldots, J \end{aligned} \quad (3.85)$$

$$0 \leq Tr P R_0 \leq J_i I, \ i = 1, 2, \ldots, m \quad (3.86)$$

Proof: Similar to m-player noncooperative H_2 game strategy in Theorem 3.5 and Theorem 3.6.

It is still not easy to solve the Riccati-like inequalities-constrained MOP in (3.84)–(3.86) for m-player noncooperative H_2 game strategy in (3.82). Let us denote $W = P^{-1}$ and perform the multiplication of W to both sides of (3.85), we get

$$\begin{aligned} &A_{ij} W + W A_{ij}^T + W Q_i W - B_{ij} R_i^{-1} B_{ij}^T + B_{-ij} B_{-ij}^T + W L_{ij}^T W^{-1} L_{ij} W \\ &+ \lambda(W N_{ij}^T W^{-1} N_{ij} W + W N_{ij}^T + N_{ij} W) \leq 0 \end{aligned} \quad (3.87)$$

By perform Schur complement in Lemma 2.1 on (3.87) several times, we get the following equivalent LMIs:

$$\begin{aligned} &\begin{bmatrix} \Pi_{ij} & W & W L_{ij}^T & \lambda^{1/2} W N_{ij}^T \\ W & -Q_i^{-1} & 0 & 0 \\ L_{ij} W & 0 & -W & 0 \\ \lambda^{1/2} N_{ij} W & 0 & 0 & -W \end{bmatrix} \leq 0 \\ &\Pi_{ij} = A_{ij} W + W A_{ij}^T - B_{ij} R_i^{-1} B_{ij}^T + B_{-ij} B_{-ij}^T + \lambda(W N_{ij}^T + N_{ij} W) \end{aligned} \quad (3.88)$$

and (3.86) becomes

$$\begin{bmatrix} -\frac{J_i}{n} R_0^{-1} & I \\ I & -W \end{bmatrix} \leq 0 \quad (3.89)$$

where $R_0 = E[\tilde{x}(0)\tilde{x}^T(0)]$.

Therefore, we get the following result for m-player noncooperative H_2 game strategy in (3.72) for the nonlinear stochastic system in (3.81)

Theorem 3.14
Based on the global linearization, m-player noncooperative H_2 game strategy problem in (3.72) of nonlinear stochastic system in (3.81) could be solved by

$$u_i^*(t) = -\sum_{j=1}^{J} \alpha_j(\tilde{x}_i(t)) R_i^{-1} B_{ij}^T P^* \tilde{x}_i(t) \tag{3.90}$$

$$u_{-i}^*(t) = \sum_{j=1}^{J} \alpha_j(\tilde{x}_i(t)) B_{-ij}^T P^* \tilde{x}_i(t) \tag{3.91}$$

where $P^* = (W^*)^{-1}$ is the solution of the following LMIs-based MOP:

$$\min_{W>0}(J_1, \ldots, J_i, \ldots, J_m) \tag{3.92}$$

subject to LMIs in (3.88) and (3.89).

The proposed LMIs-constrained MOEA algorithm for m-player noncooperative stochastic H_∞ game strategy in Theorem 3.6 in the previous section could be also employed to efficiently solve the LMIs-constrained MOP in (3.92) for m-player noncooperative H_2 game strategy design problem.

3.4 Stochastic M-Player Cooperative H_∞ Game Strategy in Nonlinear Stochastic Systems

For the nonlinear stochastic system with m-person decision-makers in (3.1), if these decision-makers have compromised with a common target x_d with each other at the beginning by some sacrifice of their own objectives, then the cooperative game strategy of nonlinear stochastic system in (3.1) becomes how to design $u_1(t), u_2(t), \ldots, u_m(t)$ altogether to solve the following cooperative H_∞ game problem:

$$\rho^0 = \min_{u(t)} \max_{v(t)} \frac{E\{\int_0^{t_p} [(x(t)-x_d)^T Q (x(t)-x_d) + \sum_{i=1}^{m} u_i^T(t) R_i u_i(t)] dt\}}{E\{(x(0)-x_d)^T (x(0)-x_d) + \int_0^{t_p} v^T(t) v(t) dt\}} \tag{3.93}$$

subject to

$$dx(t) = (f(x(t)) + G(x(t))u(t) + h(x(t))v(t))dt + l(x(t))dw(t) + n(x(t)dp(t) \tag{3.94}$$

where $G(x(t)) = [g_1(x(t)) \ldots g_m(x(t))]$, $u(t) = [u_1^T(t) \ldots u_m^T(t)]^T$ and R denotes

$$R = \begin{bmatrix} R_1 & & & & 0 \\ & \ddots & & & \\ & & R_i & & \\ & & & \ddots & \\ 0 & & & & R_m \end{bmatrix}$$

Let us denote $\tilde{x}(t) = x(t) - x_d$, then we get the following H_∞ tracking performance of cooperative game in (3.93) and (3.94):

$$\rho^0 = \min_{u(t)} \max_{v(t)} \frac{E\{\int_0^{t_p} [(\tilde{x}^T(t)Q\tilde{x}(t) + u^T(t)Ru(t)]dt\}}{E\{(\tilde{x}^T(0)\tilde{x}(0) + \int_0^{t_p} v^T(t)v(t)dt\}} \tag{3.95}$$

subject to

$$d\tilde{x}(t) = (f(\tilde{x}(t)) + G(\tilde{x}(t))u(t) + h(\tilde{x}(t))v(t))dt + l(\tilde{x}(t))dw(t) + n(\tilde{x}(t))dp(t) \tag{3.96}$$

where $f(\tilde{x}(t)) \triangleq f(\tilde{x}(t) + x_d), G(\tilde{x}(t)) \triangleq G(\tilde{x}(t) + x_d), h(\tilde{x}(t)) \triangleq h(\tilde{x}(t) + x_d),$
$l(\tilde{x}(t)) \triangleq l(\tilde{x}(t) + x_d), \text{and} n(\tilde{x}(t)) \triangleq n(\tilde{x}(t) + x_d).$

Remark 3.7 In the m-player cooperative H_∞ game strategy design problem (3.93) and (3.94) or (3.95) and (3.96), even m players $u_1(t), \ldots, u_m(t)$ cooperate together as an augmented player $u(t) = [u_1^T(t), \ldots, u_m^T(t)]^T$; with the exchange of information among players, the environmental disturbance $v(t)$ is still unavailable and unpredictable for these players. Therefore, external disturbance is considered as a competitive player against $u(t)$.

Since it is still not easy to solve cooperative stochastic H_∞ game strategy in (3.95) and (3.96) directly, the following suboptimal cooperative H_∞ game strategy by minimizing the upper bound ρ is proposed to solve (3.95) and (3.96) indirectly

$$\rho^0 = \min \rho \tag{3.97}$$

subject to

$$\min_{u(t)} \max_{v(t)} \frac{E\{\int_0^{t_p} [(\tilde{x}^T(t)Q\tilde{x}(t) + u^T(t)Ru(t)]dt\}}{E\{(\tilde{x}^T(0)\tilde{x}(0) + \int_0^{t_p} v^T(t)v(t)dt\}} \leq \rho \tag{3.98}$$

Then, we get the following result for multi-player cooperative H_∞ game strategy for nonlinear stochastic system.

Theorem 3.15
The multi-player cooperative H_∞ game strategy in (3.95) for nonlinear stochastic system in (3.96) could be solved by the following:

$$u^0(t) = -\frac{1}{2}R^{-1}G^T(\tilde{x}(t))\left(\frac{\partial V^0(\tilde{x}(t))}{\partial \tilde{x}}\right) \tag{3.99}$$

$$v^0(t) = \frac{1}{2\rho_0}h^T(\tilde{x}(t))\left(\frac{\partial V^0(\tilde{x}(t))}{\partial \tilde{x}}\right) \tag{3.100}$$

where ρ^0 and the positive Lyapunov function $V^0(\tilde{x}(t))>0$ are the solution of the following HJII-constrained SOP:

$$\rho^0 = \min_{V(\tilde{x}(t))} \rho \tag{3.101}$$

subject to

$$\begin{aligned}
&\tilde{x}^T(t)Q\tilde{x}(t) + \left(\frac{\partial V(\tilde{x}(t))}{\partial \tilde{x}(t)}\right)^T f(\tilde{x}(t)) - \frac{1}{4}\left(\frac{\partial V(\tilde{x}(t))}{\partial \tilde{x}(t)}\right)^T G(\tilde{x}(t))R^{-1}G^T(\tilde{x}(t))\left(\frac{\partial V(\tilde{x}(t))}{\partial \tilde{x}(t)}\right) \\
&+\frac{1}{4\rho}\left(\frac{\partial V(\tilde{x}(t))}{\partial \tilde{x}(t)}\right)^T h(\tilde{x}_i(t))h^T(\tilde{x}(t))\left(\frac{\partial V(\tilde{x}(t))}{\partial \tilde{x}(t)}\right) + \frac{1}{2}l^T(\tilde{x}(t))\frac{\partial^2 V(\tilde{x}(t))}{\partial \tilde{x}^2(t)}l(\tilde{x}(t)) \\
&+ \lambda[V(\tilde{x}(t) + n(\tilde{x}(t))) - V(\tilde{x}(t))] \leq 0
\end{aligned} \tag{3.102}$$

with

$$EV(\tilde{x}(0)) \leq E\tilde{x}^T(0)\tilde{x}(0) \tag{3.103}$$

Proof: According to the independence of $v(t)$ with $u(t)$ [1,9], (3.98) is equivalent to

$$\min_{u(t)}\max_{v(t)} E\int_0^{t_p} (\tilde{x}^T(t)Q\tilde{x}(t) + u^T(t)Ru(t) - \rho v^T(t)v(t)]dt \leq \rho E\tilde{x}^T(0)\tilde{x}(0) \tag{3.104}$$

Following the procedure of proof in Theorem 3.2, the constrained Nash minmax quadratic game in (3.98) is equivalent to the following:

$$\begin{aligned}
&\min_{u(t)}\max_{v(t)} E\{V(\tilde{x}(0)) - V(\tilde{x}(t_p)) + \int_0^{t_p} (\tilde{x}^T(t)Q\tilde{x}(t) + \left(\frac{\partial V(\tilde{x}(t))}{\partial \tilde{x}(t)}\right)^T f(\tilde{x}(t)) \\
&+\frac{1}{2}l^T(\tilde{x}(t))\frac{\partial^2 V(\tilde{x}(t))}{\partial \tilde{x}^2(t)}l(\tilde{x}(t)) - \frac{1}{4}\left(\frac{\partial V(\tilde{x}(t))}{\partial \tilde{x}(t)}\right)^T G(\tilde{x}(t))R^{-1}G^T(\tilde{x}(t))\left(\frac{\partial V(\tilde{x}(t))}{\partial \tilde{x}(t)}\right) \\
&+\frac{1}{4\rho}\left(\frac{\partial V(\tilde{x}(t))}{\partial \tilde{x}(t)}\right)^T h(\tilde{x}))h^T(\tilde{x}(t))\left(\frac{\partial V(\tilde{x}(t))}{\partial \tilde{x}(t)}\right) \\
&+\left(u(t) + \frac{1}{2}R^{-1}G^T(\tilde{x}(t))\left(\frac{\partial V(\tilde{x}(t))}{\partial \tilde{x}(t)}\right)\right)^T R\left(u(t) + \frac{1}{2}R^{-1}G^T(\tilde{x}(t))\left(\frac{\partial V(\tilde{x}(t))}{\partial \tilde{x}(t)}\right)\right) \\
&-\rho\left(v(t) - \frac{1}{2\rho}h^T(\tilde{x}(t))\left(\frac{\partial V(\tilde{x}(t))}{\partial \tilde{x}(t)}\right)\right)^T\left(v(t) - \frac{1}{2\rho}h^T(\tilde{x}(t))\left(\frac{\partial V(\tilde{x}(t))}{\partial \tilde{x}(t)}\right)\right) \\
&+ \lambda[V(\tilde{x}(t)) + n(\tilde{x}(t)) - V(\tilde{x}(t))]\} \leq \rho E\tilde{x}^T(0)\tilde{x}(0)
\end{aligned} \tag{3.105}$$

The suboptimal H_∞ game strategy in (3.105) is obtained as

$$u^0(t) = -\frac{1}{2}R^{-1}G(\tilde{x}(t))\left(\frac{\partial V^0(\tilde{x}(t))}{\partial \tilde{x}}\right), v^* = \frac{1}{2\rho}h^T(\tilde{x}(t))\left(\frac{\partial V(\tilde{x}(t))}{\partial \tilde{x}(t)}\right) \tag{3.106}$$

which are (3.99) and (3.100), respectively, and the constrained Nash minmax quadratic game in (3.105) becomes

$$\begin{aligned}
&E\{V(\tilde{x}(0)) - V(\tilde{x}(t_p)) + \int_0^{t_p} (\tilde{x}^T(t)Q\tilde{x}(t) + \left(\frac{\partial V(\tilde{x}(t))}{\partial \tilde{x}(t)}\right)^T f(\tilde{x}(t)) \\
&-\frac{1}{4}\left(\frac{\partial V(\tilde{x}(t))}{\partial \tilde{x}(t)}\right)^T G(\tilde{x}(t))R^{-1}G^T(\tilde{x}(t))\left(\frac{\partial V(\tilde{x}(t))}{\partial \tilde{x}(t)}\right) \\
&+\frac{1}{2}l^T(\tilde{x}(t))\frac{\partial^2 V(\tilde{x}(t))}{\partial \tilde{x}^2(t)}l(\tilde{x}(t)) + \frac{1}{4\rho}\left(\frac{\partial V(\tilde{x}(t))}{\partial \tilde{x}(t)}\right)^T h(\tilde{x}(t))h^T(\tilde{x}(t))\left(\frac{\partial V(\tilde{x}(t))}{\partial \tilde{x}(t)}\right) \\
&+ \lambda[V(\tilde{x}(t) + n(\tilde{x}(t))) - V(\tilde{x}(t))] \leq \rho E\tilde{x}^T(0)\tilde{x}(0)
\end{aligned} \tag{3.107}$$

By the HJII in (3.102), the above inequality is reduced to

$$E\{V(\tilde{x}(0))\} \leq \rho E\tilde{x}^T(0)\tilde{x}(0) \tag{3.108}$$

which is (3.103). Therefore, the suboptimal m-player cooperative H_∞ game strategy of (3.97) and (3.98) could be solved by (3.99) and (3.100) through SOP in (3.101)–(3.103). Q.E.D.

However, it is still difficult to solve the HJII-constrained SOP in (3.101)–(3.103) for the m-player cooperative H_∞ game strategy because of no analytic or numerical method at present for HJII in (3.102). Therefore, the global linearization method in (3.31)–(3.33) is employed to treat the HJII in (3.102). By the global linearization method [23], the nonlinear stochastic system in (3.96) could be represented by the interpolation of J local linearized linear stochastic systems at J vertices of the polytope as follows [23,32]:

$$d\tilde{x}(t) = \sum_{j=1}^{J} \alpha_j(\tilde{x}(t))[A_j\tilde{x}(t) + B_j u(t) + H_j v(t)]dt + L_j\tilde{x}(t)dw + N_j\tilde{x}(t)dp(t) \tag{3.109}$$

where $\alpha_j(\tilde{x}(t))$ denotes the interpolation functions with $0 \leq \alpha_j(\tilde{x}(t)) \leq 1$ and $\sum_{j=1}^{J} \alpha_j(\tilde{x}(t)) = 1$, i.e., the nonlinear stochastic system in (3.96) could be represented by the interpolated stochastic system (3.109).

Based on the above analysis, we could get the following result.

Theorem 3.16 Based on the global linearization method in (3.109), the above multi-player cooperative H_∞ game strategy design problem could be solved as follows:

$$u^0(t) = -\sum_{j=1}^{J} \alpha_j(\tilde{x}(t))R^{-1}B_j^T P^0(\tilde{x}(t)) \tag{3.110}$$

$$v^0(t) = -\frac{1}{\rho_0}\sum_{j=1}^{J} \alpha_j(\tilde{x}(t))H_j^T P^0(\tilde{x}(t)) \tag{3.111}$$

where ρ^0 and P^0 are the solution of the following Riccati-like inequalities-constrained SOP

$$\rho^0 = \min_{P>0} \rho \tag{3.112}$$

subject to

$$\begin{aligned} &PA_j + A_j^T P + Q - PB_jR^{-1}B_j^T P + \frac{1}{\rho}PH_jH_j^T P + L_j^T PL_j \\ &+ \lambda(N_j^T PN_j + N_j^T P + PN_j) \leq 0, j = 1,2,\ldots,J \end{aligned} \tag{3.113}$$

with

$$0 \leq P \leq \rho I \tag{3.114}$$

It is still difficult to solve the Riccati-like inequalities-constrained SOP in (3.112)–(3.114) for the multi-player cooperative H_∞ game strategy in (3.110) and (3.111). Let us denote $W = P^{-1}$ in (3.112) and perform the multiplication of W to both sides of (3.113), we get

$$\begin{aligned} &A_jW + WA_j^T + WQW - B_jR^{-1}B_j^T + \frac{1}{\rho_i}B_{ij}B_{ij}^T P + \frac{1}{\rho}H_jH_j^T \\ &+ WL_j^T W^{-1}L_jW + \lambda(WN_j^T W^{-1}N_jW + WN_j^T + N_jW) \leq 0,\ j = 1,2,\ldots,J \end{aligned} \tag{3.115}$$

By performing the Schur complements several times to the above inequalities, we get the following equivalent LMIs:

$$\begin{bmatrix} \Pi_j & W & WL_j^T & \lambda^{1/2}WN_j^T & H_j \\ W & -Q^{-1} & 0 & 0 & 0 \\ L_jW & 0 & -W & 0 & 0 \\ \lambda^{1/2}N_jW & 0 & 0 & -W & 0 \\ H_j^T & 0 & 0 & 0 & -\rho I \end{bmatrix} \leq 0, \mathrm{j} = 1, 2, \ldots, \mathrm{m} \tag{3.116}$$

where $\Pi_j = A_jW + WA_j^T - B_jR^{-1}B_j^T + \lambda(WN_j^T + N_jW)$ and (3.114) is equivalent to

$$\begin{bmatrix} -\rho I & I \\ I & -W \end{bmatrix} \leq 0 \tag{3.117}$$

Therefore, we get the following result.

Theorem 3.17

The multiplayer cooperative H_∞ game strategy in (3.93) and (3.94) could be solved by

$$u^0(t) = -\sum_{j=1}^{J} \alpha_j(\tilde{x}(t))R^{-1}B_j^T P^0 \tilde{x}(t) \tag{3.118}$$

$$v^0(t) = -\frac{1}{\rho^0}\sum_{j=1}^{J} \alpha_j(\tilde{x}(t))H_j^T P^0 \tilde{x}(t) \tag{3.119}$$

where ρ^0 and $P^0 = (W^0)^{-1}$ are the solution of the following LMIs-constrained SOP:

$$\rho^0 = \min_{W>0} \rho \tag{3.120}$$

subject to LMIs in (3.116) and (3.117).

The LMIs-constrained SOP in (3.120) could be efficiently solved by decreasing ρ until no existence of $W>0$ with the help of LMI toolbox in Matlab.

Remark 3.8 In the case of m-player cooperative H_∞ game strategy for the stabilization problem of nonlinear stochastic system in the following

$$\rho_0 = \min_{u(t)} \max_{v(t)} \frac{E\{\int_0^{t_p} [(x^T(t)Qx(t) + u^T(t)Ru(t)]dt\}}{E\{(x^T(0)x(0) + \int_0^{t_p} v^T(t)v(t)dt\}} \tag{3.121}$$

subject to

$$dx(t) = (f(x(t)) + G(x(t))u(t) + h(x(t))v(t))dt + l(x(t))dw(t) + n(x(t))dp(t) \tag{3.122}$$

where $G(x(t))$, $u(t)$ and R are defined in (3.94), respectively.

The indirect suboptimal method by minimizing the upper bound of (3.121) is given by

$$\rho_0 = \min \rho \tag{3.123}$$

subject to

$$\min_{u(t)}\max_{v(t)}\frac{E\{\int_0^{t_p}[(x^T(t)Rx(t)+u^T(t)Ru(t)]dt\}}{E\{(x^T(0)x(0)+\int_0^{t_p}v^T(t)v(t)dt\}}\leq\rho \tag{3.124}$$

Then, we get the following result.

Theorem 3.18

The multi-player cooperative H_∞ game strategy for the stabilization of nonlinear stochastic system in (3.121) and (3.122) could be solved

$$u^0(t)=-\frac{1}{2}R^{-1}G^T(x)\left(\frac{\partial V^0(x(t))}{\partial x(t)}\right),v^0(t)=\frac{1}{2\rho_0}h^T(x(t))\left(\frac{\partial V^0(x(t))}{\partial x(t)}\right) \tag{3.125}$$

where ρ_0 and $V^0(x(t))$ are the solution of the following HJII-constrained SOP:

$$\rho_0=\min_{V(x(t))>0}\rho \tag{3.126}$$

subject to

$$\begin{aligned}&(x^T(t)Qx(t)+\left(\frac{\partial V(x(t))}{\partial x(t)}\right)^T f(x(t))-\frac{1}{4}\left(\frac{\partial V(x(t))}{\partial x(t)}\right)^T G(x(t))R^{-1}G^T(x(t))\left(\frac{\partial V(x(t))}{\partial x(t)}\right)\\&+\frac{1}{4\rho}\left(\frac{\partial V(x(t))}{\partial x(t)}\right)^T h(x(t))h^T(x(t))\left(\frac{\partial V(x(t))}{\partial x(t)}\right)+\frac{1}{2}l^T(x(t))\frac{\partial^2 V(x(t))}{\partial x^2(t)}l(x(t))\\&+\lambda[V(x(t)+n(x(t)))-V(x(t))]\leq 0\end{aligned} \tag{3.127}$$

with

$$EV(x(0))\leq\rho Ex^T(0)x(0) \tag{3.128}$$

Proof: Similar to Theorem 3.15.

In order to overcome the difficult of solving HJII in (3.127) for the m-player cooperative H_∞ game strategy for the stabilization of nonlinear stochastic system in (3.121) and (3.122), the global linearization method is employed to interpolate several local linearized stochastic systems at the J vertices of polytope to approach the nonlinear stochastic system as follows:

$$dx(t)=\sum_{j=1}^J\alpha_j(x(t))[A_jx(t)+B_ju(t)+H_jv(t)]dt+L_jx(t)dw+N_jx(t)dp(t) \tag{3.129}$$

where $\alpha_j(x(t))$ denotes the interpolation functions with $0\leq\alpha_j(x(t))\leq 1$ and $\sum_{j=1}^J\alpha_j(x(t))=1$, i.e., the nonlinear stochastic system in (3.122) could be represented by the global linearization system in (3.129).

Based on the global linearization in (3.129), we get the following result.

Theorem 3.19
The multi-player cooperative H_∞ game strategy for the stabilization of nonlinear stochastic system in (3.121) and (3.122) could be solved by the following:

$$u^0(t) = -\sum_{j=1}^{J} \alpha_j(x(t))R^{-1}B_j^T P^0 x(t) \tag{3.130}$$

$$v^0(t) = \frac{1}{\rho^0}\sum_{j=1}^{J} \alpha_j(x(t))H_j^T P^0 x(t) \tag{3.131}$$

where ρ_0 and P^0 are the solution of the following Riccati-like inequalities-constrained SOP

$$\rho_0 = \min_{P>0} \rho \tag{3.132}$$

subject to

$$\begin{aligned} &PA_j + A_j^T P + Q - PB_jR^{-1}B_j^T P + \frac{1}{\rho}PH_jH_j^T P + L_j^T PL_j \\ &+ \lambda(N_j^T PN_j + N_j^T P + PN_j) \le 0, j = 1, 2, \ldots, J \end{aligned} \tag{3.133}$$

with

$$0 \le P \le \rho I \tag{3.134}$$

For the convenience of solving SOP in (3.132)–(3.134), we let $W = P^{-1}$ in (3.133) and perform the multiplication of W to both sides of (3.133) as follows:

$$\begin{aligned} &A_jW + WA_j^T + WQW - B_jR^{-1}B_j^T + \frac{1}{\rho}H_jH_j^T + WL_j^T W^{-1}L_jW \\ &+ \lambda(WN_j^T W^{-1}N_jW + WN_j^T + N_jW) \le 0,\ j = 1, 2, \ldots, J. \end{aligned} \tag{3.135}$$

By performing Schur complements several times to the above inequalities, we get the following LMIs:

$$\begin{bmatrix} \Pi_j & W & WL_j^T & \lambda^{1/2}WN_j^T & H_j \\ W & -Q_i^{-1} & 0 & 0 & 0 \\ L_jW & 0 & -W & 0 & 0 \\ \lambda^{1/2}N_jW & 0 & 0 & -W & 0 \\ H_j^T & 0 & 0 & 0 & -\rho I \end{bmatrix} \le 0, \mathrm{j} = 1, 2, \ldots, \mathrm{m} \tag{3.136}$$

where $\Pi_j = A_jW + WA_j^T - B_jR^{-1}B_j^T + \lambda(WN_j^T + N_jW)$
And the inequality in (3.134) is equivalent to

$$\begin{bmatrix} -\rho I & I \\ I & -W \end{bmatrix} \le 0 \tag{3.137}$$

Then, we get the following result:

Theorem 3.20

The multi-player cooperative H_∞ game strategy for the stabilization of nonlinear stochastic system in (3.121) and (3.122) could be solved by the following:

$$u^0(t) = -\sum_{j=1}^{J} \alpha_j(x(t))R^{-1}B_j^T P^0 x(t) \quad (3.138)$$

$$v^0(t) = \frac{1}{\rho^0}\sum_{j=1}^{J} \alpha_j(x(t))H_j^T P^0 x(t) \quad (3.139)$$

where ρ_0 and P^0 are the solution of the following LMIs-constrained SOP:

$$\rho_0 = \min_{W>0} \rho \quad (3.140)$$

subject to LMIs in (3.136) and (3.137).

3.5 Stochastic Multiplayer Cooperative H_2 Game Strategy in the Nonlinear Stochastic Systems without External Disturbance

Consider the following nonlinear stochastic system with m players and without external disturbance:

$$dx(t) = [f(x(t)) + \sum_{i=1}^{m} g_i(x(t))u_i(t)]dt + l(x(t))dw(t) + n(x(t))dp(t) \quad (3.141)$$

Then, the m-player cooperative game strategy design problem is formulated as

$$J^0 = \min_{u_1(t),\ldots,u_m(t)} E\int_0^{t_p} [(x(t)-x_d)^T Q(x(t)-x_d) + \sum_{i=1}^{m} u_i^T(t)R_i u_i(t)]dt \quad (3.142)$$

where x_d denotes the compromised common target of these players beforehand.
Let us denote $\tilde{x}(t) = x(t) - x_d$ and $G(x(t)) = [g_1(x(t)), \ldots, g_m(x(t))]$,

$$u(t) = \begin{bmatrix} u_1(t) \\ \vdots \\ u_i(t) \\ \vdots \\ u_m(t) \end{bmatrix}, R = \begin{bmatrix} R_1 & & & & 0 \\ & \ddots & & & \\ & & R_i & & \\ & & & \ddots & \\ 0 & & & & R_m \end{bmatrix}$$

Then, the multiplayer cooperative H_2 game strategy in (3.142) of nonlinear stochastic system in (3.141) could be formulated as

$$J^0 = \min_{u(t)} E \int_0^{t_p} [\tilde{x}^T(t)Q\tilde{x}(t) + u^T(t)Ru(t)]dt \tag{3.143}$$

subject to

$$d\tilde{x}(t) = [f(\tilde{x}(t)) + G(\tilde{x}(t))u(t)]dt + l(\tilde{x}(t))dw(t) + n(\tilde{x}(t))dp(t) \tag{3.144}$$

where
$f(\tilde{x}(t)) = f(\tilde{x}(t) + x_d), G(\tilde{x}(t)) = G(\tilde{x}(t) + x_d), l(\tilde{x}(t)) = l(\tilde{x}(t) + x_d), n(\tilde{x}(t)) = n(\tilde{x}(t) + x_d)$.

By the suboptimal method through minimizing the upper bound J of J^0 indirectly, we get

$$J^0 = \min J \tag{3.145}$$

subject to

$$\min_{u(t)} E \int_0^{t_p} [\tilde{x}^T(t)Q\tilde{x}(t) + u^T(t)Ru(t)]dt \leq J \tag{3.146}$$

Theorem 3.21
The multi-player cooperative H_2 game strategy in (3.145) and (3.146) could be solved by

$$u^0(t) = -\frac{1}{2}R^{-1}G^T(\tilde{x}(t))\left(\frac{\partial V^0(\tilde{x}(t))}{\partial \tilde{x}}\right) \tag{3.147}$$

where $V^0(\tilde{x}(t))$ is the solution of the following HJII-constrained SOP:

$$\min_{V(\tilde{x}(t))} J \tag{3.148}$$

subject to

$$\begin{aligned} &\tilde{x}^T(t)Q\tilde{x}(t) + \left(\frac{\partial V(\tilde{x}(t))}{\partial \tilde{x}(t)}\right)^T f(\tilde{x}(t)) - \frac{1}{4}\left(\frac{\partial V(\tilde{x}(t))}{\partial \tilde{x}(t)}\right)^T G(\tilde{x}(t))R^{-1}G^T(\tilde{x}(t))\left(\frac{\partial V(\tilde{x}(t))}{\partial \tilde{x}(t)}\right) \\ &+ \frac{1}{2}l^T(\tilde{x}(t))\frac{\partial^2 V(\tilde{x}(t))}{\partial^2 \tilde{x}(t)}l(\tilde{x}(t)) + \lambda[V(\tilde{x}(t) + n(\tilde{x}(t))) - V(\tilde{x}(t))] \leq 0 \end{aligned} \tag{3.149}$$

with

$$0 \leq V(\tilde{x}(0)) \leq J \tag{3.150}$$

Proof: The inequality in (3.146) becomes

$$\min_{u(t)} E[V(\tilde{x}(0)) - V(\tilde{x}(t_p)) + \int_0^{t_p} [\tilde{x}^T(t)Q\tilde{x}(t) + u^T(t)Ru(t)]dt + dV(\tilde{x}(t)) \leq J \tag{3.151}$$

By Itô–Lévy formula [11,13] of nonlinear stochastic system in (3.144), we get

$$dV(\tilde{x}(t)) = \left(\left(\frac{\partial V(\tilde{x}(t))}{\partial \tilde{x}(t)} \right)^T f(\tilde{x}(t) + G(\tilde{x}(t))u(t) \right) dt + \left(\frac{\partial V(\tilde{x}(t))}{\partial \tilde{x}(t)} \right)^T l(\tilde{x}(t))dw(t)$$
$$+ \frac{1}{2} l^T(\tilde{x}(t)) \frac{\partial^2 V(\tilde{x}(t))}{\partial^2 \tilde{x}(t)} l(\tilde{x}(t))dt + [V(\tilde{x}(t) + n(\tilde{x}(t))) - V(\tilde{x}(t))]dp(t) \quad (3.152)$$

Then, the inequality in (3.151) becomes

$$\min_{u(t)} E[V(\tilde{x}(0)) - V(\tilde{x}(t_p)) + \int_0^{t_p} [\tilde{x}^T(t)Q\tilde{x}(t) + u^T(t)Ru(t)$$
$$+ \left(\frac{\partial V(\tilde{x}(t))}{\partial \tilde{x}(t)} \right)^T l(\tilde{x}(t))]dw(t) + \left(\frac{\partial V(\tilde{x}(t))}{\partial \tilde{x}(t)} \right)^T (f(\tilde{x}(t) + G(\tilde{x}(t))u(t)) \quad (3.153)$$
$$+ \frac{1}{2} l^T(\tilde{x}(t)) \frac{\partial^2 V(\tilde{x}(t))}{\partial^2 \tilde{x}(t)} l(\tilde{x}(t))dt + \lambda[V(\tilde{x}(t) + n(\tilde{x}(t))) - V(\tilde{x}(t))]dt] \leq J$$

By the fact $Edw(t) = 0$,

$$\min_{u(t)} E[V(\tilde{x}(0)) - V(\tilde{x}(t_p)) + \int_0^{t_p} [\tilde{x}^T(t)Q\tilde{x}(t) + \left(\frac{\partial V(\tilde{x}(t))}{\partial \tilde{x}(t)} \right)^T f(\tilde{x}(t))$$
$$+ \frac{1}{2} l^T(\tilde{x}(t)) \frac{\partial^2 V(\tilde{x}(t))}{\partial^2 \tilde{x}(t)} l(\tilde{x}(t))dt - \frac{1}{4} \left(\frac{\partial V(\tilde{x}(t))}{\partial \tilde{x}(t)} \right)^T G(\tilde{x}(t))R^{-1}G^T(\tilde{x}(t)) \left(\frac{\partial V(\tilde{x}(t))}{\partial \tilde{x}(t)} \right) \quad (3.154)$$
$$+ \left((Ru(t) + \frac{1}{2} G^T(x(t)) \left(\frac{\partial V(\tilde{x}(t))}{\partial \tilde{x}(t)} \right) \right)^T R^{-1} \left((Ru(t) + \frac{1}{2} G^T(x(t)) \left(\frac{\partial V(\tilde{x}(t))}{\partial \tilde{x}(t)} \right) \right)$$
$$+ \lambda[V(\tilde{x}(t)) + n(\tilde{x}(t)) - V(\tilde{x}(t))]dt \leq J$$

From the minimization of the l.h.s of (3.154), we get the optimal strategy $u^0(t)$ as (3.147) and then (3.154) becomes

$$V(\tilde{x}(0)) + \int_0^{t_p} [\tilde{x}^T(t)Q\tilde{x}(t) + \left(\frac{\partial V(\tilde{x}(t))}{\partial \tilde{x}(t)} \right)^T f(\tilde{x}(t))$$
$$- \frac{1}{4} \left(\frac{\partial V(\tilde{x}(t))}{\partial \tilde{x}(t)} \right)^T G(\tilde{x}(t))R^{-1}G^T(\tilde{x}(t)) \left(\frac{\partial V(\tilde{x}(t))}{\partial \tilde{x}(t)} \right) \quad (3.155)$$
$$+ \frac{1}{2} l^T(\tilde{x}(t)) \frac{\partial^2 V(\tilde{x}(t))}{\partial^2 \tilde{x}(t)} l(\tilde{x}(t)) + \lambda[V(\tilde{x}(t) + n(\tilde{x}(t))) - V(\tilde{x}(t))]dt \leq J$$

By the HJII in (3.149), we get

$$V(\tilde{x}(0)) \leq J \quad (3.156)$$

which is (3.150).

Q.E.D.

In order to simplify the HJII-constrained SOP in (3.148)–(3.150) for multi-player cooperative H_2 game strategy in Theorem 3.21, the global linearization method is employed to interpolate J' local linearized stochastic systems at the vertices of polytope to approach nonlinear stochastic system in (3.144) as follows:

$$d\tilde{x}(t) = \sum_{j=1}^{J'} \alpha_j(\tilde{x}(t))[A_j\tilde{x}(t) + B_ju(t) + L_j\tilde{x}(t)dw(t) + N_j\tilde{x}(t)dp(t)] \tag{3.157}$$

With the global linearization system in (3.157) to replace the nonlinear stochastic system in (3.144), we get the following result.

Theorem 3.22
The multiplayer cooperative H_2 game strategy in (3.143) and (3.144) could be solved by

$$u^0(t) = -R^{-1}G^T(\tilde{x}(t))P^0\tilde{x}(t) \tag{3.158}$$

where P^0 is the solution of the following Riccati-like inequalities-constrained SOP:

$$\min_{P>0} J \tag{3.159}$$

subject to

$$PA_j + A_j^TP + Q - PB_jR^{-1}B_j^TP + L_j^TPL_j + \lambda(N_j^TPN_j + N_j^TP + PN_j) \leq 0,\ j = 1, 2, ..., J' \tag{3.160}$$

with

$$E\tilde{x}^T(0)P\tilde{x}(0) \leq J \tag{3.161}$$

Proof: Similar to previous Theorems.

For the convenience of solving Riccati-like inequalities constrained-SOP in (3.159)-(3.161) for multiplayer cooperative H_2 game strategy, we let $W = P^{-1}$ and perform the multiplication of W to both sides of (3.160):

$$\begin{aligned} &A_jW + WA_j^T + WQW - B_jR^{-1}B_j^T + WL_j^TW^{-1}L_jW \\ &+ \lambda(WN_j^TW^{-1}N_jW + WN_j^T + N_jW) \leq 0,\ j = 1, 2, ..., J'. \end{aligned} \tag{3.162}$$

By performing Schur complements several times on (3.162), we could get the following equivalent LMIs:

$$\begin{bmatrix} \Pi_j & W & WL_j^T & WN_j^T \\ W & -Q^{-1} & 0 & 0 \\ L_jW & 0 & -W & 0 \\ N_jW & 0 & 0 & -W \end{bmatrix} \leq 0, j = 1, 2, \ldots, J' \tag{3.163}$$

where Π_j is defined in (3.136)and the inequality in (3.161) becomes

$$\begin{bmatrix} -\frac{J}{n}R_0 & I \\ I & -W \end{bmatrix} \leq 0 \tag{3.164}$$

where $R_0 = E[\tilde{x}(0)\tilde{x}^T(0)]$

Then we get the following result.

Theorem 3.23
The multi-player cooperative H_2 game strategy in (3.143) and (3.144) could be solved by

$$u^0(t) = -R^{-1}G(\tilde{x}(t))P^0\tilde{x}(t) \tag{3.165}$$

where $P^0 = (W^0)^{-1}$ is the solution of the following LMIs-constrained SOP:

$$\min_{W>0} J \tag{3.166}$$

subject to LMIs in (3.163) and (3.164)

Based on Theorem 3.23, we could solve LMIs-constrained SOP in (3.166) with the help of LMI toolbox in Matlab to obtain W^0. Then, we obtain $P^0 = (W^0)^{-1}$ for the m-player cooperative H_2 game strategy $u^0(t)$ in (3.165).

Remark 3.9 In the case of m-player cooperative H_2 game strategy for the stabilization problem of nonlinear stochastic system without external disturbance in the following

$$J^0 = \min_{u(t)} E\{\int_0^{t_p} [x^T(t)Qx(t) + \sum_{i=1}^{m} u_i^T(t)R_iu_i(t)]dt\} \tag{3.167}$$

subject to

$$dx(t) = [f(x(t)) + \sum_{i=1}^{m} g_i(x(t))u_i(t)]dt + l(x(t))dw(t) + n(x(t))dP(t)$$

Let us denote $G(x(t)) = [g_1(x(t))\ldots g_i(x(t))\ldots g_m(x(t))]$ and

$$u(t) = \begin{bmatrix} u_1(t) \\ \vdots \\ u_i(t) \\ \vdots \\ u_m(t) \end{bmatrix}, R = \begin{bmatrix} R_1 & & & & 0 \\ & \ddots & & & \\ & & R_i & & \\ & & & \ddots & \\ 0 & & & & R_m \end{bmatrix}$$

Then the m-player cooperative H_2 game strategy for the stabilization of nonlinear stochastic system without external disturbance is reformulated as follows

$$J^0 = \min_{u(t)} E\{\int_0^{t_p} [x^T(t)Qx(t) + u^T(t)Ru(t)dt\} \tag{3.168}$$

subject to

$$dx(t) = (f(x(t)) + G(x(t))u(t))dt + l(x(t))dw(t) + n(x(t))dp(t) \tag{3.169}$$

By the suboptimal method, we solve the H_2 optimal problem indirectly in (3.168) and (3.169) as follows.

$$\min J \tag{3.170}$$

subject to

$$\min_{u(t)} E\{\int_0^{t_p} [x^T(t)Qx(t) + u^T(t)Ru(t)\} \leq J \tag{3.171}$$

Following the result in Theorem 3.21, we get

Theorem 3.24
The multi-player cooperative H_2 game strategy for the stabilization problem of nonlinear stochastic system in (3.167) and (3.168) could be solved by

$$u^0(t) = -\frac{1}{2}R^{-1}G^T(x(t))\left(\frac{\partial V^0(x(t))}{\partial x}\right) \tag{3.172}$$

where $V^0(x)$ is the solution of the following HJII-constrained SOP

$$\min_{V(x(t))>0} J \tag{3.173}$$

subject to

$$\begin{aligned}[x^T(t)Qx(t) &+ \left(\frac{\partial V(x(t))}{\partial x(t)}\right)^T f(x(t)) - \frac{1}{4}\left(\frac{\partial V(x(t))}{\partial x(t)}\right)^T G(x(t))R^{-1}G^T(x(t))\left(\frac{\partial V(x(t))}{\partial x(t)}\right) \\ &+ \frac{1}{2}l^T(x(t))\frac{\partial^2 V(x(t))}{\partial^2 x(t)}l(x(t)) + \lambda[V(x(t) + n(x(t))) - V(x(t))] \leq 0\end{aligned} \tag{3.174}$$

with

$$0 \leq V(x(0)) \leq J \tag{3.175}$$

Proof: Similar to Theorem 3.21

Since it is very difficult to solve the HJII-constrained HJII in (3.173)-(3.175) for the multi-player cooperative H_2 game strategy in (3.172). The global linearization method is employed to interpolate several local linearized stochastic systems at J vertices of polytope to approach the nonlinear stochastic system in (3.169) in the following:

$$d(x(t)) = \sum_{j=1}^{J'} \alpha_j(x(t))(A_j x(t) + B_j u(t))dt + L_j x(t)dw(t) + N_j x(t)dp(t) \tag{3.176}$$

Based on the global linearization system in (3.176) for the nonlinear stochastic system in (3.169), we could then obtain the following result for the multi-player cooperative H_2 game strategy for the nonlinear stochastic system.

Theorem 3.25
The multi-player cooperative H_2 game strategy for the stabilization problem of nonlinear stochastic system in (3.168) and (3.169) could be solved by

$$u^0(t) = -R^{-1}G^T(x(t))P^0x(t) \tag{3.177}$$

where P^0 is the solution of the following Riccati-like inequalities-constrained SOP:

$$\min_{P>0} J \tag{3.178}$$

subject to

$$PA_j + A_j^T P + Q - PB_j R^{-1} B_j^T P + L_j^T PL_j + \lambda(N_j^T PN_j + N_j^T P + PN_j) \leq 0, j = 1, 2, \ldots, J' \tag{3.179}$$

with

$$x^T(0)Px(0) \leq J \tag{3.180}$$

Proof: Similar to Theorem 3.22.

Since it is still complex to solve Riccati-like inqualities-constrained SOP in (3.178)–(3.180) for P^0 in (3.177), we let $W = P^{-1}$ and perform the multiplication of W to both sides of (3.179), we get

$$\begin{aligned} &A_j W + WA_j^T + WQW - B_j R^{-1} B_j^T + WL_j^T W^{-1} L_j W \\ &+ \lambda(WN_j^T W^{-1} N_j W + WN_j^T + N_j W) \leq 0,\ j = 1, 2, \ldots, J'. \end{aligned} \tag{3.181}$$

Performing the Schur complements to (3.181) several times, we could get the following equivalent LMIs:

$$\begin{bmatrix} \Pi_{ij} & W & WL_j^T & WN_j^T \\ W & -Q^{-1} & 0 & 0 \\ L_j W & 0 & -W & 0 \\ N_j W & 0 & 0 & -W \end{bmatrix} \leq 0,\ j = 1, 2, \ldots, J' \tag{3.182}$$

And the inequality (3.180) is equivalent to

$$\begin{bmatrix} -\frac{J}{n} R_0 & I \\ I & -W \end{bmatrix} \leq 0 \tag{3.183}$$

where $R_0 = E[x(0)x^T(0)]$.

Therefore, we get the following result.

Theorem 3.26

The multi-player cooperative H_2 game strategy for the stabilization of nonlinear stochastic system in (3.167) or (3.168) could be solved as

$$u^0(t) = -R^{-1} G^T(x(t)) P^0 x(t) \tag{3.184}$$

where $P^0 = (W^0)^{-1}$ is the solution of the following LMIs-constrained SOP:

$$\min_{W>0} J \tag{3.185}$$

subject to LMIs (3.182) and (3.183).

3.6 Conclusion

In this chapter, the stochastic games are extended from linear stochastic systems in Chapter 2 to nonlinear stochastic systems with multi-players. Unlike the previous studies, each player in this chapter has his/her target and utility (payoff) function in noncooperative game strategy or common target and utility function in cooperative game strategy. We first discussed multi-player noncooperative H_∞ game strategy for nonlinear stochastic systems with external disturbance and noncooperative H_2 game strategy for nonlinear stochastic systems free of external disturbance. Basically, multi-player H_∞ and H_2 noncooperative game strategy of nonlinear stochastic systems could be all transformed to through the global linearization an equivalent HJIIs-constrained MOP and then transformed to a corresponding LMIs-constrained MOP, which could be solved by the proposed LMIs-constrained MOEA to search for the Pareto optimal solutions in a single run to obtain the Nash equilibrium solution of the corresponding game strategy. We then discussed multi-player cooperative H_∞ and H_2 game strategy of nonlinear stochastic systems with a common goal and the same payoff function. These multi-player cooperative game strategies could be transformed based on the suboptimal method to a corresponding HJIIs-constrained SOP and then transformed to LMIs-constrained SOP through the global linearization, which could be easily solved with the help of LMI toolbox in Matlab. The applications of multi-player H_∞ and H_2 game strategies in this chapter will be introduced in the following chapters.

Part II

Stochastic Game Strategies in Control System Designs

4

Robust Adaptive Tracking Control Design of Robot Systems under Parameter Perturbation and External Disturbance: A Minimax H_∞ Dynamic Game Approach

4.1 Introduction

In the literature of robotics, robotic dynamic systems are always described by a second-order differential equation at each joint of arm with the acting torque (or force) as the control input. Extensive control approaches had been proposed for the feedback control of robot arms [37–40,43,44]. A total robot dynamic model (arm dynamics plus motor (actuator) dynamics and the interaction between motors and joints) have recently been considered in robotic control design [37,39,41,42]. Under this circumstance, a third-order robot dynamic model must be developed to include the actuator dynamics. Owing to factors such as high-velocity moment, highly varying loads, friction, and saturation of actuator, the robotic system becomes uncertain. In [41], a robust control is developed on the basis of the robust stabilization technique in [52,53]. In these approaches, the nonlinear uncertain dynamics are assumed to be norm bounded. A robust controller based on a small gain theorem is then derived to guarantee the global boundedness of robotic system. The result may be conservative if the plant perturbations are large. Furthermore, if only the stability is guaranteed, no further tracking performance could be achieved, leading to the limitation of practical applications of robotic system.

Robust H_∞ control design has been widely studied in the recent decade for desired disturbance attenuation [48–50]. An H_∞ tracking control design has been developed for robotic systems to attenuate the worst-case effect of parameter perturbations and external noises [51]. However, if the magnitudes of uncertain dynamics and external noises are large, the effect on tracking error may still be obvious or even lead to system instability. In this situation, before an H_∞ attenuation is applied, an effective elimination of uncertain dynamics is an appealing strategy for enhancing the tracking performance.

Conventionally, adaptive control schemes can be employed to match and then cancel this uncertain nonlinear dynamics. However, linear parametrization must be assumed in conventional adaptive control schemes, i.e., the unknown parameters must be of linear structure. In nonlinear uncertain robotic systems, this assumption may not be true. Since fuzzy logic system can be tuned to approximate any nonlinear dynamic with model free, in this chapter, a fuzzy adaptive scheme is used to efficiently eliminate the plant uncertainties so that tracking performance can be enhanced via an adaptive learning method.

Fuzzy approximation techniques have been recently used to effectively approximate unknown nonlinear dynamics [45–47]. However, in the conventional adaptive fuzzy control, the desired tracking performance cannot be guaranteed from the more theoretical control perspective. More recently [61], an H_∞ adaptive tracking control has been proposed for single-input–single-output (SISO) unknown nonlinear systems via an adaptive fuzzy control based on the feedback linearization technique and H_∞ control scheme. However, the limitation of this adaptive fuzzy control method is that it employs feedback linearization-based adaptive fuzzy control, which contains an inverse term of adaptive fuzzy logic system. Therefore, it is sensitive in the adaptive control process, particularly, in the case of approaching a small value. Furthermore, only SISO systems have been discussed, but robotic tracking systems are multi-input–multi-output (MIMO) nonlinear systems. Unlike the conventional fuzzy control schemes for the nonlinear robotic system, the proposed robust H_∞ adaptive scheme is used to cancel the nonlinear part of robotic system so that the robotic system becomes a linear system with an adaptive cancelation residue. Then a linear H_∞ tracking control scheme is employed based on minimax H_∞ game to achieve the robust tracking. Recently, a neural-based adaptive H_∞ control has been applied to robotic systems via adaptive feedback linearization method in [63].

Unlike the conventional adaptive control design methods, in this chapter, a linear nominal robotic system is given and only the robust H_∞ tracking problem for uncertain dynamics and external disturbance is addressed. A fuzzy-based minimax H_∞ game scheme is employed to cancel the effects of nonlinear uncertainties to enhance the robust adaptive tracking performance of uncertain robotic systems. The conventional adaptive feedback linearization technique for uncertain nonlinear system control design is unnecessary in this design.

The proposed adaptive robust H_∞ game controller contains two terms, one contains a self-tuning fuzzy logic system which is developed to optimally cancel the uncertain dynamics and external disturbance to enhance the tracking robustness, and the other contains a minimax H_∞ game strategy control algorithm that is employed to optimally attenuate the worst-case effect of both the residue due to adaptive fuzzy cancelation and the exogenous disturbance below a desired level from a minimax H_∞ game perspective.

The proposed H_∞ adaptive tracking control design method attempts to combine the dynamic game technique and an adaptive fuzzy cancelation algorithm to guarantee a robust H_∞ adaptive tracking performance for uncertain robotic control systems including motor dynamics and external disturbances. In the proposed robust H_∞ adaptive tracking control, the adaptive fuzzy cancelation technique plays a role of rough tuning and the minimax H_∞ game attenuation technique plays a role of fine tuning. Unlike the conventional adaptive control in robotic systems, the uncertain dynamics do not need to have a linear parameterized structure in this design, i.e., uncertain dynamics can be free of structure in the proposed fuzzy-based minimax H_∞ adaptive control systems. Since the adaptive feedback linearization technique is avoided, an inverse adaptive matrix is unnecessary. Hence, the computational complexity and sensitivity of the proposed H_∞ adaptive tracking control algorithm are significantly reduced. Therefore, the proposed minimax H_∞ adaptive game strategy for robust control design method is appropriate for the robust tracking control design of robotic systems with large uncertainties and external disturbances.

In this chapter, Section 4.2 presents a dynamic model of the robotic manipulator and joint motors first. The robotic system has a state vector composed of the errors in joint position, velocity, and acceleration, and a control vector composed of the armatures' input voltages. In Section 4.3, the robust minimax H_∞ adaptive tracking control problem for uncertain robotic system via fuzzy-based adaptive control scheme is described. The uncertainty is matched and canceled by an adaptive fuzzy scheme. In this approach, we propose a fuzzy-based minimax H_∞ dynamic game scheme to treat the nonlinear robust adaptive tracking control design of robotic manipulators, including actuator dynamics. The proposed scheme is simpler and more robust than the approaches proposed in [39–42]. In Section 4.4, the proof of the adaptive minimax H_∞ game tracking control problem via adaptive fuzzy cancelation scheme is given and the design procedures is proposed. In Section 4.5, a simulation example is provided to demonstrate the design procedure's effectiveness and to confirm the performance of the proposed robust minimax H_∞ adaptive tracking control design for robotic systems with consideration of motor drives. Simulation results indicate that a desired robust H_∞ adaptive tracking performance can be guaranteed for an uncertain robotic system via the proposed fuzzy cancelation-based minimax H_∞ game strategy under large time-varying parameter perturbations and external disturbance. Concluding remarks are finally made in Section 4.6.

4.2 Nonlinear Dynamics of Robotic Systems with External Disturbance

A nonlinear dynamic model of robotic manipulator and joint motors is first described. Following Tarn *et al.* [39], Beekman and Lee [42], and Mahmoud [41], the joints of robotic manipulators driven by dc motor are considered and a combined dynamic model of the robot manipulator plus joint motors is developed herein. An n-joints robot dynamic (including actuators) is described as follows. First, the motion equations of a robotic manipulator with revolute joints can be expressed as

$$M'(q)\ddot{q} + C(q,\dot{q})\dot{q} + G(q) = \tau + d' \tag{4.1}$$

where $q, \dot{q}, \ddot{q} \in R^n$ are the vectors of joint positions, velocities, and accelerations, respectively; $M'(q) \in R^{n\times n}$ is the matrix of the moment inertia; $C(q,\dot{q})\dot{q} \in R^n$ is the vector of the centripetal and Coriolis forces; $G(q) \in R^n$ is the vector of gravitational force; $\tau \in R^n$ is the vector of torques developed at the joint side of gear box; $d' \in R^n$ is the external disturbance such as loading and environment changes.

Remark 4.1 The matrix M'in the robot model (4.1) is symmetric positive-definite.

The relation between the joint position q and the motor-shaft position q_m is given by

$$q_m = Nq \tag{4.2}$$

where $N \in R^{n\times n}$ is a diagonal matrix of the gear ratios for the n joints and $N>0$ (which means that the matrix N is positive-definite). By armature-controlled dc motors, the electrical model of the jth motor is characterized by

$$R_j i_j + L_j \frac{di_j}{dt} + K_{b_j} \frac{dq_{m_j}}{dt} = u_j, \text{for } j = 1, 2, \ldots, n \tag{4.3}$$

where R_j is the resistance of the armature circuit, L_j is the inductance of the armature circuit, K_{b_j} is the back electromotive force (EMF) constant of the motor, i_j is the armature current, q_{m_j} is the motor shaft position, and u_j is the armature input voltage. Let us define

$$R = diag[R_j], L = diag[L_j], K_b = diag[K_{b_j}] \tag{4.4}$$

$$u = \begin{bmatrix} u_1 \\ u_2 \\ \vdots \\ u_n \end{bmatrix}, \; i = \begin{bmatrix} i_1 \\ i_2 \\ \vdots \\ i_n \end{bmatrix}, u = \begin{bmatrix} q_{m_1} \\ q_{m_2} \\ \vdots \\ q_{m_n} \end{bmatrix} \tag{4.5}$$

where $diag[\cdot]$ denotes a diagonal matrix of dimension n. Then, the electrical models of armature-controlled dc motors in (4.3) can be grouped in a compact form

$$Ri + L\frac{di}{dt} + K_b\dot{q}_m = u. \tag{4.6}$$

Furthermore, the developed torque at the joint side is related to the armature current by

$$\tau = NK_t i \tag{4.7}$$

where $K_t \in R^{n\times n}$ is the diagonal matrix of motor torque constants and $K_t > 0$. Substituting (4.7) into (4.6) yields

$$R_n\tau + L_n\dot{\tau} + K_{bn}\dot{q} = u \tag{4.8}$$

where

$$R_n = R(NK_t)^{-1}, L_n = L(NK_t)^{-1}, K_{bn} = K_b N. \tag{4.9}$$

Remark 4.2 The indicated inverse in (4.9) exists due to the physical nature of N and K_t. Now, to obtain the combined robot dynamic model, we substitute (4.1) into (4.8). Thus, we get

$$\begin{aligned} &L_n M'(q)q^{(3)} + (R_n M'(q) + L_n[\dot{M}'(q) + C(q,\dot{q})])\ddot{q} + (R_n C(q,\dot{q}) \\ &+ L_n\dot{C}(q,\dot{q}) + K_{bn})\dot{q} + R_n G(q) + L_n\dot{G}(q) = u + R_n d' + L_n\dot{d}' \end{aligned} \tag{4.10}$$

where $q^{(3)}$ denotes the third partial time derivative of q. The dynamic model in (4.10) can be written in the following compact form:

$$M(q)q^{(3)} + D(q,\dot{q},\ddot{q}) = u + R_n d' + L_n\dot{d}' \tag{4.11}$$

where

$$M(q) = L_n M'(q) \tag{4.12}$$

$$\begin{aligned} D(q,\dot{q},\ddot{q}) = &(R_n M'(q) + L_n[\dot{M}'(q) + C(q,\dot{q})])\ddot{q} + (R_n C(q,\dot{q}) \\ &+ L_n\dot{C}(q,\dot{q}) + K_{bn})\dot{q} + R_n G(q) + L_n\dot{G}(q) \end{aligned} \tag{4.13}$$

By introducing the state vector

$$x = \begin{bmatrix} x_1 \\ x_2 \\ x_3 \end{bmatrix} \triangleq \begin{bmatrix} q \\ \dot{q} \\ \ddot{q} \end{bmatrix}. \tag{4.14}$$

Equation (4.11) is transferred into the following standard form:

$$\dot{x} = \begin{bmatrix} x_2 \\ x_3 \\ -M^{-1}(x_1)D(x) \end{bmatrix} + \begin{bmatrix} 0 \\ 0 \\ M^{-1}(x_1) \end{bmatrix} u + \begin{bmatrix} 0 \\ 0 \\ M^{-1}(x_1)(R_n d' + L_n \dot{d}') \end{bmatrix}. \tag{4.15}$$

Here we denote $d = M^{-1}(x_1)(R_n d' + L_n \dot{d}')$ and assume $d \in L_2[0, t_f]$. Therefore, (4.15) can be rewritten as

$$\dot{x} = \begin{bmatrix} x_2 \\ x_3 \\ -M^{-1}(x_1)D(x) \end{bmatrix} + \begin{bmatrix} 0 \\ 0 \\ M^{-1}(x_1) \end{bmatrix} u + \begin{bmatrix} 0 \\ 0 \\ I \end{bmatrix} d \tag{4.16}$$

which represents a typical nonlinear dynamical robotic system with an affine nonlinear input function. In this chapter, we assume the state vector x can be measured.

The state vector in (4.16) is composed of joint positions, velocities, and accelerations of robotic dynamic system, and the input vector u consists of the armatures' input voltages. Because of the model's complexity and nonlinearity of robotic dynamic system in (4.16), a direct design of control laws is not easy. This situation is further compounded by the drift incurred in online measurements of acceleration, the frequent changes in load and model parameter, and the corruption of external disturbances.

Given a task of a continuously differentiable and uniformly bounded trajectory in the joint space q_d for which we wish the robot manipulator to follow. Therefore, we define

$$\tilde{q} = q - q_d = x_1 - q_d \tag{4.17}$$

as the joint position error. Tarn *et al.* have developed a feedback linearization plus decoupling technique based on differential geometric control theory to provide a nonlinear feedback control law for the regulation of robotic arms [39]. However, this design is possible only while the dynamics of the robotic dynamic are well known. Assume that $M_0(\cdot)$ and $D_0(\cdot,\cdot,\cdot)$ are the nominal estimates of $M(\cdot)$ and $D(\cdot,\cdot,\cdot)$, respectively. We follow the design procedure in [41] and use the following control law:

$$u = M_0(q_d^{(3)} - K_1\tilde{q} - K_2\dot{\tilde{q}} - K_3\ddot{\tilde{q}} + u_0) + D_0(\dot{q}, \ddot{q}, \dddot{q}) \tag{4.18}$$

where K_1, K_2, K_3 are diagonal matrices to be designed and u_0 is an auxiliary control signal yet to be specified.

Substituting (4.17) and (4.18) into (4.16) leads to

$$\dot{x} = \begin{bmatrix} x_2 \\ x_3 \\ M^{-1}(D_0 - D) + M^{-1}M_0 v + d \end{bmatrix} \tag{4.19}$$

where

$$v = q_d^{(3)} - K_1\tilde{q} - K_2\dot{\tilde{q}} - K_3\ddot{\tilde{q}} + u_0. \tag{4.20}$$

For further development, let us define the system uncertainties as

$$f(x) = M^{-1}[(D_0 - D) + (M_0 - M)v]. \tag{4.21}$$

For design purpose, let us denote

$$e = \begin{bmatrix} \tilde{q} \\ \dot{\tilde{q}} \\ \ddot{\tilde{q}} \end{bmatrix} = \begin{bmatrix} x_1 - q_d \\ x_2 - \dot{q}_d \\ x_3 - \ddot{q}_d \end{bmatrix} \tag{4.22}$$

as the state error vector. From (4.21), if our nominal estimates of M and D are exact (i.e., $M_0 = M$ and $D_0 = D$), then the system uncertainties will be zero except the external noise. This is impossible in practical robot systems. In order to enhance the robust tracking performance, an adaptive fuzzy logic system (Fig. 4.1) is introduced herein to eliminate the uncertainties.

In the form of (4.20) and (4.21), we obtain a simple form of (4.19)

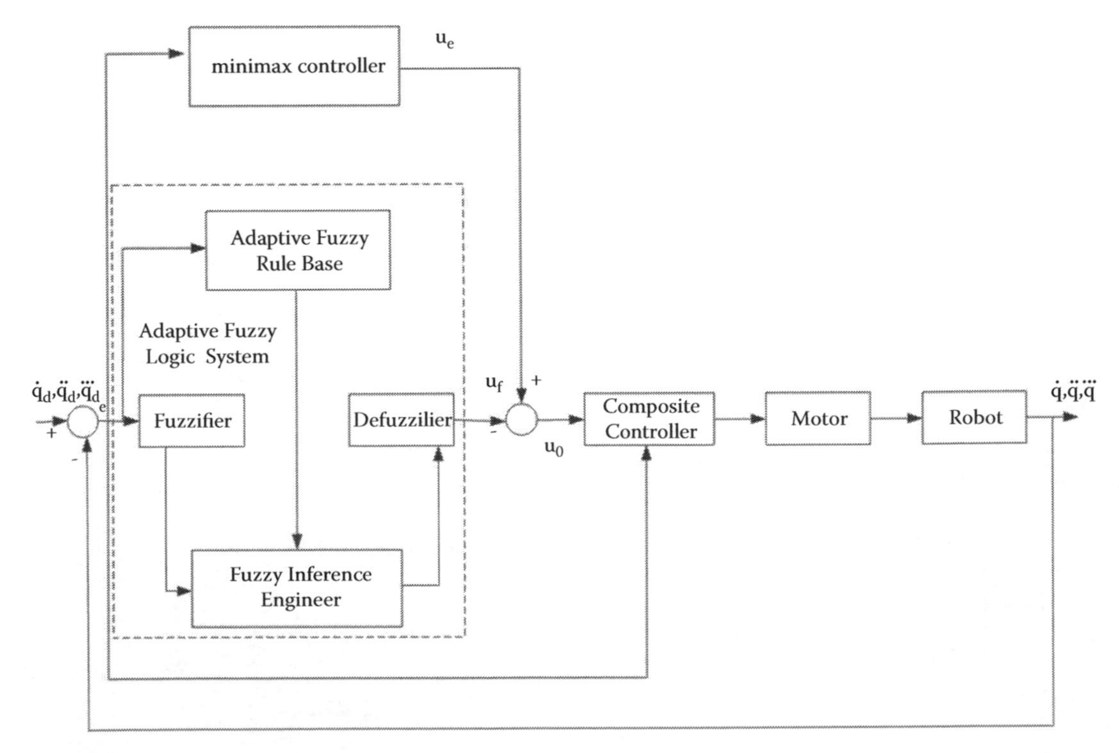

FIGURE 4.1
Fuzzy logic system for robot tracking control.

$$\dot{x} = \begin{bmatrix} x_2 \\ x_3 \\ v + f(x) + d(t) \end{bmatrix}. \tag{4.23}$$

From (4.17), (4.19), and (4.22), we could obtain

$$\begin{aligned} \begin{bmatrix} \dot{\tilde{q}} \\ \ddot{\tilde{q}} \\ \tilde{q}^{(3)} \end{bmatrix} &= \begin{bmatrix} \tilde{q} \\ \dot{\tilde{q}} \\ -K_1\tilde{q} - K_2\dot{\tilde{q}} - K_3\ddot{\tilde{q}} + u_0 + f(x) + d \end{bmatrix} \\ &= \begin{bmatrix} 0_{n\times n} & I_{n\times n} & 0_{n\times n} \\ 0_{n\times n} & 0_{n\times n} & I_{n\times n} \\ -K_1 & -K_2 & -K_3 \end{bmatrix} \begin{bmatrix} \tilde{q} \\ \dot{\tilde{q}} \\ \ddot{\tilde{q}} \end{bmatrix} + \begin{bmatrix} 0_{n\times n} \\ 0_{n\times n} \\ I_{n\times n} \end{bmatrix} u_0 + \begin{bmatrix} 0_{n\times n} \\ 0_{n\times n} \\ I_{n\times n} \end{bmatrix} (f(x) + d). \end{aligned} \tag{4.24}$$

The system in (4.24) represents the robotic tracking error dynamics. In terms of the terminology of Garofalo and Leitmann [54], it is a nominally linear uncertain system. The nonlinear uncertainties are modeled by the vector $f(x)$. A more convenient form of (4.24) would be

$$\dot{e} = Ae + Bu_0 + Bf(x) + Bd(t) \tag{4.25}$$

where

$$A = \begin{bmatrix} 0 & I & 0 \\ 0 & 0 & I \\ -K_1 & -K_2 & -K_3 \end{bmatrix}, B = \begin{bmatrix} 0 \\ 0 \\ I \end{bmatrix}. \tag{4.26}$$

Remark 4.3 The control parameters K_1, K_2, and K_3 are specified so that A has desired eigenvalues and the tracking dynamic in (4.25) has a desired response while the robotic system is free of uncertainty $f(x)$ and external disturbance $d(t)$.

4.3 Adaptive Minimax H∞ Game Tracking Control Design Problem in Robotic Systems

The tracking error dynamics of uncertain robotic manipulator with actuator dynamics are described by (4.25) in the above section. If the robotic system is free of uncertainties and external disturbance (i.e., $f(x) = 0$ and $d(t) = 0$), by the nonlinear control law in (4.18) with adequate choice of linear control $u_0(t)$, the tracking error $e(t)$ in (4.25) will asymptotically converge to zero as $t \to \infty$ [37]. In practical robotic systems, however, uncertainties due to parameter perturbations, unmodeled dynamics, and external noises are inevitable. These uncertainties deteriorate the tracking performance or even lead to system instability in the worst case. Hence, the effect of uncertainties $f(x)$ and $d(t)$ on tracking error in (4.25) must be eliminated in practical design. Since they are uncertain, several robust design algorithms [52–54] have been employed as a robust controller to override the upper norm bound of $\|f(x)\|$. Due to high nonlinearity and uncertainty of $f(x)$, the estimation of its upper norm bound is a difficult task. Furthermore, an effective suppression on the effect of the external noise $d(t)$ is not easy, for example, by variable structure system (VSS) control or dead-zone control based on an upper bound of external disturbance to treat this problem. All of these robust control methods may lead to a conservative and imprecise result.

In this chapter, the control signal u_0 is divided into two parts as follows:

$$u_0 = u_e - u_f(x, \Theta). \tag{4.27}$$

Under this circumstance, the tracking error dynamics in (4.25) is of the following form:

$$\dot{e} = Ae + Bu_e + B(f(x) - u_f(x, \Theta)) + Bd \tag{4.28}$$

where $u_f(x, \Theta)$ is a fuzzy logic system and $\Theta(t)$ is a parameter matrix to be tuned.

In (4.28), the fuzzy logic system $u_f(x, \Theta)$ will be tuned to approximate $f(x)$ as closely as possible. Furthermore, the control signal $u_e(t)$ will be used based on the minimax H_∞ dynamic game to attenuate the total effect of the residue of $f(x) - u_f(x, \Theta)$ and external disturbance $d(t)$ on the tracking error $e(t)$ from a minimax H_∞ tracking perspective. In this chapter, since the fuzzy logic system $u_f(x, \Theta)$ is employed to adaptively cancel the uncertain $f(x)$ in (4.28) to enhance the robust tracking, the fuzzy logic system $u_f(x, \Theta)$ in Fig. 4.1 is described in the following paragraphs.

The fuzzy logic systems can perform universal approximation from the perspective of human experts and can uniformly approximate nonlinear continuous functions to an arbitrary accuracy [55,56]. The fuzzy logic systems in Fig. 4.1 are qualified as building blocks of $u_f(x, \Theta)$ for adaptive cancelation of nonlinear function $f(x)$ in (4.28). The fuzzy logic systems are constructed from the fuzzy If–Then rules using some specific inference, fuzzification, and defuzzification strategies. Therefore, the linguistic information from human experts and the information from state measurement can be incorporated into fuzzy logic systems to adjust their parameters to achieve the optimal approximation [60].

The basic configuration of a fuzzy logic system is depicted in Fig. 4.1 as a fuzzy logic system. The fuzzy logic system $u_f(x, \Theta)$ in this work performs a mapping from $U \in R^{3n}$ to $V \in R^n$. Let $U = U_1 \times U_2 \times U_3$, where $U_i \in R^n$, for $i = 1, 2, 3$. A fuzzy rule base consists of a collection of fuzzy If–Then rules [55,56,60]

$$\begin{aligned} &R^{(l)} : \text{If } x_1 \text{is } F_1^l, \text{and} \cdots \text{and, } x_{3n} \text{ is } F_{3n}^l \\ &\text{Then } u_f \text{ is } G^l \end{aligned} \tag{4.29}$$

where $x = (x_1, x_2, \cdots, x_{3n})^T \in U$ and $u_f \in V \subset R^n$ are the input and output of the fuzzy logic system, respectively, and $l = 1, 2, \ldots, M$. The fuzzy inference engine performs a mapping from fuzzy sets in $U \in R^{3n}$ to fuzzy sets in R^{3n} based upon the fuzzy If–Then rules in the fuzzy rule base and the compositional rule of inference. The fuzzifier maps a crisp point $x = (x_1, x_2, \cdots, x_{3n})^T \in U$ into a fuzzy set A_x in U. The defuzzifier maps a fuzzy set in U to a crisp point in V. More information can be found in [55,56,60].

The fuzzy logic systems of Fig. 4.1 comprise an extremely rich class of static systems mapping from $U \subset R^{3n}$ to $V \subset R^n$ because many different choices are available within each block. In addition, many combinations of these choices can result in a useful subclass of fuzzy logic systems. One subclass of fuzzy logic systems is used here as building blocks of our adaptive fuzzy approximation (cancelation) controller and is described by the following important result.

Lemma 4.1 [60]: The fuzzy logic systems with center-average defuzzifier, product inference, and singleton fuzzifier are in the following form:

$$u_{f_i}(x,\theta_i) = \frac{\sum_{l=1}^{M} \theta_{il}\left(\prod_{j=1}^{3n} \mu_{F_j^l}(x_i)\right)}{\sum_{l=1}^{M}\left(\prod_{j=1}^{3n} \mu_{F_j^l}(x_j)\right)} \tag{4.30}$$

where θ_{il} is the point at which the given membership function $\mu_{F_j^l}(x_j)$ achieves its maximum value, and we assume that $\mu_{F_j^l}(\theta_{il}) = 1$.

Let us denote the fuzzy basis functions as

$$\varsigma_l(x) = \frac{\prod_{j=1}^{3n} \mu_{F_j^l}(x_i)}{\sum_{l=1}^{M}\left(\prod_{j=1}^{3n} \mu_{F_j^l}(x_j)\right)} \tag{4.31}$$

for $l = 1, 2, \ldots, M$ and denote

$$\varsigma(x) \triangleq [\varsigma_1(x) \quad \varsigma_2(x) \quad \cdots \quad \varsigma_M(x)]. \tag{4.32}$$

Consequently, $u_{f_i}(x,\theta_i)$ in (4.30) is of the following form:

$$u_{f_i}(x,\theta_i) = [\varsigma_1(x) \quad \varsigma_2(x) \quad \cdots \quad \varsigma_M(x)] \times \begin{bmatrix} \theta_{i_1}(t) \\ \theta_{i_2}(t) \\ \vdots \\ \theta_{i_M}(t) \end{bmatrix} \tag{4.33}$$
$$\triangleq \varsigma(x)\theta_i(t)$$

where $\theta_i(t) = [\theta_{i_1}(t) \quad \theta_{i_2}(t) \quad \cdots \quad \theta_{i_M}(t)]^T$ for $i = 1, 2, \ldots, n$. Therefore, the fuzzy logic system for the n input and n output of n-link robotic system with the same fuzzy basis functions is of the following form:

$$u_f(x,\Theta) = \begin{bmatrix} u_{f_1}(x,\theta_1) \\ u_{f_2}(x,\theta_2) \\ \vdots \\ u_{f_n}(x,\theta_n) \end{bmatrix} = \begin{bmatrix} \varsigma(x)\theta_1(t) \\ \varsigma(x)\theta_2(t) \\ \vdots \\ \varsigma(x)\theta_n(t) \end{bmatrix}$$
$$= \begin{bmatrix} \varsigma(x) & 0 & \cdots & 0 \\ 0 & \varsigma(x) & \ddots & \vdots \\ \vdots & \ddots & \ddots & \vdots \\ 0 & \cdots & 0 & \varsigma(x) \end{bmatrix} \times \begin{bmatrix} \theta_1(t) \\ \theta_2(t) \\ \vdots \\ \theta_n(t) \end{bmatrix} \tag{4.34}$$

i.e.,

$$u_f(x,\Theta) = \xi(x)\Theta(t) \tag{4.35}$$

where $\xi(x) = \begin{bmatrix} \varsigma(x) & 0 & \cdots & 0 \\ 0 & \varsigma(x) & \ddots & \vdots \\ \vdots & \ddots & \ddots & \vdots \\ 0 & \cdots & 0 & \varsigma(x) \end{bmatrix}, \Theta(t) = \begin{bmatrix} \theta_1(t) \\ \theta_2(t) \\ \vdots \\ \theta_n(t) \end{bmatrix}$.

Remarks 4.4

1) The membership function $\mu_{F_j^l}$ can be triangular or any type of membership functions.
2) In general, as the number M of fuzzy basis functions approaches infinite [60], fuzzy logic system $u_f(x,\Theta)$ can approximate any uncertain function by adequately selecting parameter matrix Θ in (4.35).
3) In this robust adaptive design, the membership functions are specified by the designer's experience or knowledge regarding the uncertainties of a robotic system and the parameter Θ is to be tuned according to the tracking error $e(t)$.

In practical robust adaptive control design, for the convenience of computation and implementation, the number of fuzzy basis functions of fuzzy logic system is chosen as small as possible. Under this circumstance, an adaptive law must be developed to tune the parameter $\Theta(t)$ to construct a fuzzy logic system $u_f(x,\Theta)$ with an adequate dimension to approximate to $f(x)$ as closely as possible.

Now define the following optimal fuzzy approximation [60]:

$$\Theta^* = \arg\min_{\Theta\in\Omega}\max_{x\in\Omega_x}\|f(x) - \xi\Theta\| \tag{4.36}$$

where $\|\cdot\|$ denotes the Euclidean norm, i.e., $\|x\| = \sqrt{x^T x}$. Ω and Ω_x denote the sets of suitable bounds on $\Theta(t)$ and x, respectively. We assume that $\Theta(t)$ and x never reach the boundary of Ω and Ω_x and $f(x) = \xi\Theta^* + \varepsilon(t)$, $\varepsilon(t) \le \bar{\varepsilon}$ in Ω and Ω_x for some $\bar{\varepsilon}>0$. Otherwise, the projection algorithm such as described in the remark following the proof of Theorem 4.1 must be introduced to prevent the divergence of $\Theta(t)$.

Consequently, the tracking error dynamic (4.28) can be rewritten as follows:

$$\begin{aligned} \dot{e} &= Ae + Bu_e + B(\xi\Theta^* - \xi\Theta) + B(f(x) - \xi\Theta^*) + Bd(t) \\ &= Ae + Bu_e + B\xi\tilde{\Theta} + Bw(t) \end{aligned} \tag{4.37}$$

where $w(t) \triangleq f(x) - \xi\Theta^* + d(t)$ denotes the sum of the optimal fuzzy cancelation error via fuzzy logic system and external disturbance and $\tilde{\Theta}(t)$ is defined as

$$\tilde{\Theta}(t) \triangleq \Theta^* - \Theta(t). \tag{4.38}$$

Our design procedure is divided into two steps. In the first step, the adaptive fuzzy algorithm $u_f(x,\Theta)$ is tuned via $\Theta(t)$ to optimally cancel the uncertain term $f(x)$. After cancelation, the term $\xi\Theta^* - \xi\Theta = \xi\tilde{\Theta}$ will finally vanish. If the effect of $w(t)$ cannot be eliminated efficiently, the tracking performance will be deteriorated, particularly, in the case of a small number of membership functions in (4.32). This is a weak point to overcome in fuzzy-based control design methods. However, the cancelation error and external disturbance (i.e., $w(t)$) are uncertain and cannot be estimated perfectly and canceled completely. If such a situation arises, to be on the safe side, their worst-case effect on tracking error must be attenuated as fully as possible. Therefore, in the second step, the control signal $u_e(t)$ should be specified such that the worst-case effect of $w(t)$ on the tracking error $e(t)$ must be attenuated as much as possible and below a prescribed level ρ, i.e., the following minimax H∞ game strategy tracking performance must be satisfied for the tracking error dynamics in (4.37) (see [48,49,62]):

$$\begin{aligned}&\min_{u_e(t)\in L_2[0,t_f]}\max_{w(t)\in L_2[0,t_f]}\int_0^{t_f}\left(e^T(t)Qe(t)+u_e^T(t)Ru_e(t)-\rho^2 w^T(t)w(t)\right)dt\\&\leq e^T(0)Pe(0)+\frac{1}{\gamma}\tilde{\Theta}^T(0)\tilde{\Theta}(0)\end{aligned} \tag{4.39}$$

where $\gamma>0$ is a weighting factor, $Q=Q^T>0$, $P=P^T>0$ and $R>0$ are some positive-definite weighting matrices and the final time $t_f>0$ [48–51].

Our design objective involves how to tune the parameters of the fuzzy logic system $\Theta\xi$ to eliminate the term $\Theta^*\xi$ and then to specify an adequate minimax control law $u_e(t)$ in (37) so that the worst-case effect of $w(t)$ on $e(t)$ is guaranteed to be less than or equal to a prescribed attenuation level ρ. Under the case $e(0)=0$ and $\tilde{\Theta}(0)=0$, (4.39) can be rewritten as the following minimax H∞ tracking game problem because $w(t)$ is independent on $u_e(t)$:

$$\min_{u_e(t)\in L_2[0,t_f]}\max_{w(t)\in L_2[0,t_f]}\frac{\left\|\left[Q^{1/2}e(t)\; R^{1/2}u_e(t)\right]\right\|_{L_2}}{\|w(t)\|_{L_2}}\leq\rho \tag{4.40}$$

where $\|w(t)\|_{L_2}\triangleq\sqrt{\int_0^{t_f}w^T(t)w(t)dt}$ and

$$\left\|\left[Q^{1/2}e(t)\; R^{1/2}u_e(t)\right]\right\|_{L_2}\triangleq\sqrt{\int_0^{t_f}e^T(t)Qe(t)+u_e^T(t)Ru_e(t)dt}.$$

However, in the case $e(0)\neq 0$ and $\tilde{\Theta}(0)\neq 0$, the H∞ tracking control problem in (4.40) must be modified to (4.39) to consider $e(0)$ and $\tilde{\Theta}(0)$ as some kind of disturbance [48,49].

At present [48–51], the minimax H∞ tracking control is the most efficient method of eliminating the worst-case effect of the uncertain $w(t)$ on $e(t)$ in (4.37). Therefore, it will be employed to attenuate the effect of $w(t)$ to achieve the minimax H∞ game strategy in (4.40) for the robust tracking in our fuzzy-based control design of robotic system.

The expression in (4.40) implies that the minimax H∞ game strategy tracking performance from $w(t)$ to $Q^{1/2}e(t)$ and $R^{1/2}u_e(t)$ must be less than or equal to ρ. The physical meaning is that the worst-case influence of $w(t)$ on $e(t)$ and $u(t)$ must be less than or equal to ρ for any $w(t)\in L_2[0,t_f]$ from an energy perspective. Owing to this reasoning, we are actually dealing with the minimax H∞ game strategy tracking problem for the model reference robotic control systems via an adaptive fuzzy cancelation scheme. The primary difference between the proposed H_∞ game tracking algorithm and the conventional H_∞ tracking control is that, to be on the safe side, an adaptive fuzzy control $u_f(x,\Theta)$ is employed in our design to eliminate the uncertain term $f(x)$ as much as possible to enhance the tracking robustness before the minimax H_∞ tracking control $u_e(t)$ is used. In the conventional H_∞ control design, only H_∞ control $u_e(t)$ is employed to directly attenuate the uncertain term $f(x)$. Since $f(x)$ is generally very large, it may lead to system instability. Furthermore, its effect on tracking error will be still very large even when H_∞ attenuation is used. The study in [51] is a case of H_∞ tracking control design of robotic manipulator without employing the adaptive fuzzy cancelation scheme to eliminate the uncertainties. Therefore, the result is conservative and more control effort is deemed necessary. In this minimax H_∞ game adaptive tracking design, the use of an adaptive fuzzy logic system $u_f(x,\Theta)$ allows eliminating the uncertain term $f(x)$ as much as possible to enhance the tracking robustness under large system uncertainties and external disturbance. Employing $u_e(t)$ to attenuate the effect of $w(t)$ could allow achieving a prescribed tracking performance. The role of adaptive fuzzy cancelation in the proposed minimax H_∞ game tracking control design of uncertain robotic systems is discussed in the simulation example of Section 4.5.

Remark 4.5

1. In general, the cancelation error and external disturbance $w(t)$ always exist; $\rho<1$ is necessary for the attenuation of $w(t)$ to achieve robust tracking. If $\rho \to \infty$, then the robust minimax H_∞ adaptive tracking performance design is reduced to a conventional H_2 optimal adaptive tracking control design without considering the attenuation of $w(t)$(see [48–51]). In this case, the H_2 optimal adaptive tracking performance is deteriorated by $w(t)$ and is indicated in our simulation example in Section 4.5.
2. The performance in (4.39) or (4.40) is of finite-time minimax H_∞ adaptive tracking control so $w(t)$ can be finite in $[0,t_f]$. If the terminal time $t_f \to \infty$, the integrations $\int_0^\infty e^T(t)Qe(t)+u_e^T(t)Ru_e(t)dt$ and $\int_0^\infty w^T(t)w(t)dt$ may increase to ∞. However, the integral inequality (4.40) always holds true.

Our design objective involves in specifying an updated law of $\Theta(t)$ for fuzzy logic system $\xi(x)\Theta(t)$ and a control law $u_e(t)$ for the tracking error dynamic in (4.37) such that the robust minimax H_∞ adaptive tracking performance (4.39) or (4.40) is guaranteed. The updated law for $u_e(t)$ plays a prominent role in rough tuning and the specification of $u_e(t)$ plays a role in fine tuning for the robust tracking control of the uncertain robotic system in (4.11). A more detailed description is given in the next section.

4.4 Minimax H_∞ Game Tracking Control of Robotic Systems via Adaptive Fuzzy Cancelation Scheme

From the analysis in the above section, the robust tracking control design of uncertain robotic systems is formulated as a minimax H_∞ game tracking control problem in (4.40) via an adaptive fuzzy control $u_f(x,\Theta)=\xi(x)\Theta(t)$ to eliminate uncertainties $f(x)$ as fully as possible and then via a minimax H_∞ control $u_e(t)$ to attenuate the worst-case influence of $w(t)$ on tracking error below a prescribed level ρ. The first step in our minimax H_∞

game strategy design involves specifying an updated law for $\Theta(t)$ and a control law $u_e(t)$ such that for any $w(t) \in L_2[0, t_f]$, the following equivalent minimax quadratic game problem is achieved, i.e.,

$$\min_{u_e(t)\in L_2[0,t_f]} \max_{w(t)\in L_2[0,t_f]} \int_0^{t_f} (e^T(t)Qe(t) + u_e^T(t)Ru_e(t) - \rho^2 w^T(t)w(t))dt$$
$$\leq e^T(0)Pe(0) + \frac{1}{\gamma}\tilde{\Theta}^T(0)\tilde{\Theta}(0) \tag{4.41}$$

subject to the tracking error dynamic equation in (4.37)

Let us define the cost function

$$J(e, u_e, w) = \int_0^{t_f} e^T(t)Qe(t) + u_e^T(t)Ru_e(t) - \rho^2 w^T(t)w(t))dt \tag{4.42}$$

Following some arrangement, we obtain

$$\begin{aligned} J(e, u_e, w) &= e^T(0)Pe(0) - e^T(t_f)Pe(t_f) + \frac{1}{\gamma}\tilde{\Theta}^T(0)\tilde{\Theta}(0) \\ &\quad - \frac{1}{\gamma}\tilde{\Theta}^T(t_f)\tilde{\Theta}(t_f) + \int_0^{t_f} [e^T(t)Qe(t) + u_e^T(t)Ru_e(t) \\ &\quad - \rho^2 w^T(t)w(t) + \frac{d}{dt}(e^T(t)Pe(t)) + \frac{1}{\gamma}\frac{d}{dt}(\tilde{\Theta}^T(t)\tilde{\Theta}(t))]dt \\ &= e^T(0)Pe(0) - e^T(t_f)Pe(t_f) + \frac{1}{\gamma}\tilde{\Theta}^T(0)\tilde{\Theta}(0) \\ &\quad - \frac{1}{\gamma}\tilde{\Theta}^T(t_f)\tilde{\Theta}(t_f) + \int_0^{t_f} [e^T(t)Qe(t) + u_e^T(t)Ru_e(t) \\ &\quad - \rho^2 w^T(t)w(t) + \dot{e}^T(t)Pe(t) + e^T(t)P\dot{e}(t) \\ &\quad + \frac{1}{\gamma}\dot{\tilde{\Theta}}^T(t)\tilde{\Theta}(t) + \frac{1}{\gamma}\tilde{\Theta}^T(t)\dot{\tilde{\Theta}}(t)]dt. \end{aligned} \tag{4.43}$$

Substituting (4.37) into the above equation yields

$$\begin{aligned} J(e, u_e, w) &= e^T(0)Pe(0) - e^T(t_f)Pe(t_f) + \frac{1}{\gamma}\tilde{\Theta}^T(0)\tilde{\Theta}(0) \\ &\quad - \frac{1}{\gamma}\tilde{\Theta}^T(t_f)\tilde{\Theta}(t_f) + \int_0^{t_f} [e^T(t)(A^TP + PA + Q)e(t) \\ &\quad + u_e^T(t)Ru_e(t) - \rho^2 w^T(t)w(t) + u_e^T(t)B^TPe(t) \\ &\quad + e^T(t)PBu_e(t) + e^T(t)PB\xi(x)\tilde{\Theta}(t) \\ &\quad + \tilde{\Theta}^T(t)\xi^T(x)B^TPe(t) + e^T(t)PBw(t) \\ &\quad + w^T(t)B^TPe(t) + \frac{1}{\gamma}\dot{\tilde{\Theta}}^T(t)\tilde{\Theta}(t) + \frac{1}{\gamma}\tilde{\Theta}^T(t)\dot{\tilde{\Theta}}(t)]dt \end{aligned} \tag{4.44}$$

then, we get the following main result of minimax H_∞ game tracking control in (4.41) or (4.39) and (4.40).

Theorem 4.1
For the uncertain robotic system (4.15) or (4.16), if the adaptive control $u(t)$ is chosen as

$$u(t) = M_0(q_d^{(3)} - K_1\tilde{q} - K_2\dot{\tilde{q}} - K_3\ddot{\tilde{q}} + u_e^*(t) - \xi\Theta(t)) + D_0(q, \dot{q}, \ddot{q}) \tag{4.45}$$

with

$$\dot{\Theta}(t) = \gamma\xi^T(x)B^T Pe(t) \tag{4.46}$$

$$u_e^*(t) = -R^{-1}B^T Pe(t) \tag{4.47}$$

where $R = R^T > 0$ is a weighting matrix and $P = P^T > 0$ is the solution of the following algebraic Riccati-like equation:

$$PA + A^T P + Q - PB(R^{-1} - \frac{1}{\rho^2}I)B^T P = 0 \tag{4.48}$$

then, the minimax H_∞ tracking game performance in (4.39) or (4.40) is guaranteed for a prescribed ρ and the corresponding worst case $w^*(t)$ to deteriorate the tracking performance is of the form

$$w^*(t) = \frac{1}{\rho^2}B^T Pe(t). \tag{4.49}$$

Remark 4.6 $w^*(t)$ and $u_e^*(t)$ are the strategies of two players of the minimax H_∞ game in (4.40) or (4.41): $w^*(t)$ denotes a possible disturbance in $L_2[0, t_f]$, which leads to the worst-case influence on tracking error from the H_∞ perspective and $u_e^*(t)$ denotes the minimax H_∞ robust tracking control under the worst-case disturbance $w^*(t)$.

Proof: From (4.44) and after some rearrangement, we get

$$\begin{aligned} J(e, u_e, w) = {} & e^T(0)Pe(0) - e^T(t_f)Pe(t_f) + \frac{1}{\gamma}\tilde{\Theta}^T(0)\tilde{\Theta}(0) \\ & - \frac{1}{\gamma}\tilde{\Theta}^T(t_f)\tilde{\Theta}(t_f) + \int_0^{t_f} [e^T(t)(A^T P + PA + Q)e(t) \\ & + u_e^T(t)Ru_e(t) + u_e^T(t)B^T Pe(t) + e^T(t)PBu_e(t) \\ & + \xi(x)e^T(t)PB\tilde{\Theta}(t) + \frac{1}{\gamma}\dot{\tilde{\Theta}}^T(t)\tilde{\Theta}(t) + \frac{1}{\gamma}\tilde{\Theta}^T(t)\dot{\tilde{\Theta}}(t) \\ & + \tilde{\Theta}^T(t)B^T Pe(t)\xi^T(x) + w^T(t)B^T Pe(t) \\ & + e^T(t)PBw(t) - \rho^2 w^T(t)w(t)]dt. \end{aligned} \tag{4.50}$$

From (4.46)–(4.49), we get

$$\begin{aligned}
J(e,u_e,w) &= e^T(0)Pe(0)+\frac{1}{\gamma}\tilde{\Theta}^T(0)\tilde{\Theta}(0)-e^T(t_f)Pe(t_f) \\
&-\frac{1}{\gamma}\tilde{\Theta}^T(t_f)\tilde{\Theta}(t_f)+\int_0^{t_f}[e^T(t)(PA+A^TP+Q)e(t) \\
&+u_e^T(t)Ru_e(t)-\rho^2w^T(t)w(t)]dt \\
&= e^T(0)Pe(0)+\frac{1}{\gamma}\tilde{\Theta}^T(0)\tilde{\Theta}(0)-\frac{1}{\gamma}\tilde{\Theta}^T(t_f)\tilde{\Theta}(t_f) \\
&+\int_0^{t_f}[e^T(t)(PA+A^TP+Q-PB(R^{-1}-\frac{1}{\rho^2}I)B^TP)e(t) \\
&+u_e^T(t)Ru_e(t)+u_e^T(t)B^TPe(t)+e^T(t)PBu_e(t) \\
&+e^T(t)PBR^{-1}Pe(t)-\rho^2w^T(t)w(t)+w^T(t)B^TPe(t) \\
&+e^T(t)PBw(t)-\frac{1}{\rho^2}e^T(t)PBB^TPe(t)]dt.
\end{aligned} \tag{4.51}$$

From (4.48) and using the technique of completion of the squares, we get

$$\begin{aligned}
J(e,u_e,w) &= e^T(0)Pe(0)+\frac{1}{\gamma}\tilde{\Theta}^T(0)\tilde{\Theta}(0)-e^T(t_f)Pe(t_f) \\
&-\frac{1}{\gamma}\tilde{\Theta}^T(t_f)\tilde{\Theta}(t_f)+\int_0^{t_f}[(Ru_e(t)+B^TPe(t)^TR^{-1} \\
&\times(Ru_e(t)+B^TPe(t))-(\rho w(t)-\frac{1}{\rho}B^TPe(t))^T \\
&\times(\rho w(t)-\frac{1}{\rho}B^TPe(t))]dt.
\end{aligned} \tag{4.52}$$

From (4.52) and the dynamic minimax game theory [48–50,62], we obtain the optimal control $u_e^*(t)$ as (4.47) and the worst case $w^*(t)$ as (4.49). Then,

$$\begin{aligned}
\min_{u_e(t)}\max_{w(t)} J(e,u_e,w) &= e^T(0)Pe(0)+\frac{1}{\gamma}\tilde{\Theta}^T(0)\tilde{\Theta}(0) \\
&-e^T(t_f)Pe(t_f)-\frac{1}{\gamma}\tilde{\Theta}^T(t_f)\tilde{\Theta}(t_f) \\
&\le e^T(0)Pe(0)+\frac{1}{\gamma}\tilde{\Theta}^T(0)\tilde{\Theta}(0).
\end{aligned} \tag{4.53}$$

The above inequality holds by the fact that $P=P^T>0$, $R=R^T>0$ and $\gamma>0$. From (4.42), it is seen

$$\begin{aligned}
&\min_{u_e(t)}\max_{w(t)} J(e,u_e,w) = \int_0^{t_f}\min_{u_e(t)}\max_{w(t)}(e^T(t)Qe(t)+u_e^T(t)Ru_e(t)-\rho^2w^T(t)w(t))dt \\
&\le e^T(0)Pe(0)+\frac{1}{\gamma}\tilde{\Theta}^T(0)\tilde{\Theta}(0).
\end{aligned} \tag{4.54}$$

This is (4.41): if $e(0)=0$ and $\tilde{\Theta}(0)=0$, then (54) is reduced to (4.40).

Remark 4.7

1. To guarantee the positive-definite solution of P in the algebraic Riccati-like equation (4.48), the following constraint must hold [49]:

$$R^{-1} - \frac{1}{\rho^2} I \geq 0 \text{ or } \rho^2 I \geq R. \tag{4.55}$$

For a prescribed attenuation level ρ, the weighting matrix R on the control $u_e(t)$ in (47) must satisfy the above constraint to guarantee the solvability of minimax H_∞ tracking control of the uncertain robotic system in (4.11). Hence, a robust H_∞ tracking design with an arbitrary attenuation of $w(t)$ is possible by the proposed method with an adequate choice of R. However, if the attenuation level ρ is specified as an extremely small value, from (4.55), R must be of an extremely small value. Under this circumstance, $u_e(t)$ may require a large control energy. There is a tradeoff between ρ and $u_e(t)$ in the minimax H_∞ robust tracking control.

1. Notably, Ω and Ω_e do not need to be known or specified beforehand. Since $w(t)$ is bounded by the universal approximation theorem [60] and the assumption of bounded external disturbances, is also given $e(t)$ is bounded too. Moreover, additional tools concerning projection algorithm [57] can be used to analyze the bounded problems of $\Theta(t)$. Assume that the constraint set Ω is specified as $\Omega \triangleq \{\Theta | \|\Theta\| \leq M\}$, where M is a positive constant. Then, the parameter update law in (4.46) must be modified as [57]

$$\dot{\Theta} = \gamma \xi^T(x) B^T P e(t) - U(\Theta) \tag{4.56}$$

where

$$U(\Theta) = \begin{cases} 0 \\ \text{if} \|\Theta\| < M \text{ or } (\|\Theta\| \text{ and } \Theta^T \xi^T B^T P e(t) \leq 0) \\ \gamma \Theta \frac{\Theta^T \xi^T B^T P e(t)}{\|\Theta\|^2} \\ \text{if} \|\Theta\| \geq M \text{ and } \Theta^T \xi^T B^T P e(t) > 0. \end{cases}$$

Since $(e^T(t) P B \xi(x) \tilde{\Theta}(t)) + (1/\gamma)(\dot{\tilde{\Theta}}^T(t) \tilde{\Theta}(t)) = (e^T(t) P B \xi^T(x) \Theta(t) / \|\Theta(t)\|^2) \Theta^T(t) \Theta(t)$, it can be shown by geometric argument that the angle between $\tilde{\Theta}(t)$ and $\Theta(t)$ is larger than $\frac{\pi}{2}$ at projection case. So, $\tilde{\Theta}^T(t)\Theta(t) < 0$. We obtain $(e^T(t) P B \xi(x) \tilde{\Theta}(t)) + (1/\gamma)(\dot{\tilde{\Theta}}^T(t) \tilde{\Theta}(t)) \leq 0$ and $\tilde{\Theta}^T(t) \xi^T B^T(x) P e(t) + (1/\gamma) \tilde{\Theta}^T(t) \dot{\tilde{\Theta}}(t) \leq 0$. By the same procedure as the proof in Theorem 4.1, the minimax H_∞ tracking performance (4.41) can also be guaranteed via the updated law (4.56).

From the above analysis, a design procedure for the adaptive minimax H_∞ game tracking control of uncertain robotic systems with motor driving is outlined as follows.

Design procedure of adaptive minimax H_∞ game tracking control of robotic systems:

Step 1: Specify K_1, K_2 and K_3 to determine matrix A with desired eigenvalues.

Step 2: Determine the fuzzy architecture $\xi(x)$ and specify the desired attenuation level ρ.

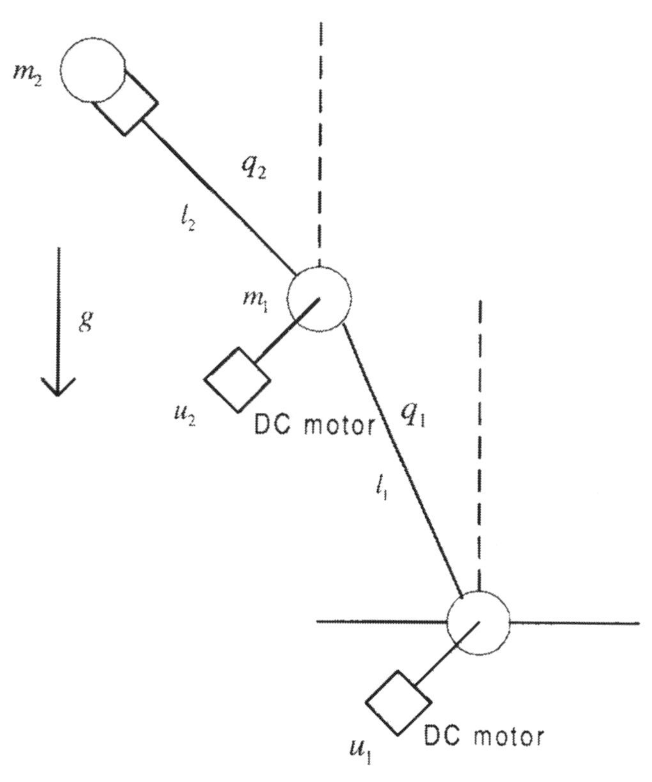

FIGURE 4.2
The two-link robotic manipulator driven by dc motors.

Step 3 : Select positive-definite weighting matrices Q and R with $R \leq \rho^2 I$ to guarantee the solvability of the minimax tracking.

Step 4: Solve the positive-definite matrix P from the Riccati-like equation (4.48).

Step 5: Compute the minimax tracking control law (4.47) and the parameter update law in (4.46) and then $u(t)$ in (4.45).

In the next section, a design example of adaptive minimax H_∞ game tracking control of robotic system is given according to the above design procedure to demonstrate the effectiveness of the proposed design method.

4.5 Design Example

In this section, we test our proposed adaptive fuzzy-based minimax H_∞ game tracking control on the robust tracking design of a two-link robot by using a computer. Consider a two-link manipulator driven by dc motor as Fig. 4.2 with system parameters as: link mass m_1, m_2 (kg), lengths l_1, l_2 (m), and angular positions q_1, q_2 (rad).

The systematic parameters for the dynamic equation of motion (4.1) are

$$M'(q) = \begin{bmatrix} (m_1+m_2)l_1^2 & m_2l_1l_2(s_1s_2+c_1c_2) \\ m_2l_1l_2(s_1s_2+c_1c_2) & m_2l_2^2 \end{bmatrix}$$

$$C(q,\dot{q}) = m_2l_1l_2(c_1s_2-s_1c_2)\begin{bmatrix} 0 & -\dot{q}_2 \\ -\dot{q}_1 & 0 \end{bmatrix}$$

$$G(q) = \begin{bmatrix} -(m_1+m_2)l_1gs_1 \\ -m_2l_2gs_2 \end{bmatrix}$$

where $q \in R^2$ and the shorthand notations $c_1 = \cos(q_1), c_2 = \cos(q_2), s_1 = \sin(q_1)$ and $s_2 = \sin(q_2)$ are used.

Assume that the desired trajectory planning problem for a weight-lifting operation is considered and the two-link manipulator suffers from time-varying parametric uncertainties and exogenous disturbances.

- The resistance matrix of the armature circuit is

$$R_t = \begin{bmatrix} 1.6 & 0 \\ 0 & 1.6 \end{bmatrix}.$$

- The inductance matrix of the armature circuit is

$$L = \begin{bmatrix} 0.048 & 0 \\ 0 & 0.048 \end{bmatrix}.$$

- The back EMF constant of the motor is

$$K_b = \begin{bmatrix} 0.19 & 0 \\ 0 & 0.19 \end{bmatrix}.$$

- The motor torque constant is

$$K_t = \begin{bmatrix} 0.2613 & 0 \\ 0 & 0.2613 \end{bmatrix}.$$

- The gear ratio is

$$N = \begin{bmatrix} 62.55 & 0 \\ 0 & 107.81 \end{bmatrix}.$$

The proposed robust H_∞ adaptive fuzzy-based model reference tracking control with the minimax H_∞ game strategy is employed to treat this robust robotic trajectory planning problem. For the convenience of simulation, the nominal parameters of the robotic system are given as $m_1 = 4.6(\text{kg}), m_2 = 2.3(\text{kg})$, $l_1 = 0.5(\text{m})$, $l_2 = 0.2(\text{m})$, $g = 9.8\text{m/s}^2$, and the initial conditions $q_1(0) = 2, q_2(0) = 1, \dot{q}_1(0) = \dot{q}_2(0) = \ddot{q}_1(0) = \ddot{q}_2(0) = 0$. The desired reference trajectories are $q_{1d}(t) = \sin(t), q_{2d}(t) = \cos(t)$, respectively.

Assume that the parameters m_1 and m_2 be perturbed in the following form:

$$\Delta m_1(t) = 0.1\sin(2t) \text{ and } \Delta m_2(t) = 0.1\sin(2t)$$

respectively. Moreover, the exogenous disturbances d_1 and d_2 are assumed as

$$d_1(t) = 0.1\sin(2t),\ d_2(t) = 0.05\sin(2t).$$

Obviously, the parameter uncertainties and exogenous disturbances are extremely large. Therefore, the proposed adaptive fuzzy-based minimax H_∞ game tracking control algorithm is employed to treat this robust tracking control design of robotic systems with parameter uncertainties and exogeneous disturbance.

Now, following the design procedure in the above section, the robust adaptive minimax H_∞ game tracking control design is given by the following steps:

Step 1: Specify

$$K_1 = \begin{bmatrix} 9 & 0 \\ 0 & 7 \end{bmatrix}, K_2 = \begin{bmatrix} 10 & 0 \\ 0 & 14 \end{bmatrix}, K_3 = \begin{bmatrix} 5 & 0 \\ 0 & 8 \end{bmatrix}$$

such that the following desired eigenvalues of the nominal tracking system are given with desired systematic characteristics $-1.3037+1.4359i$, $-1.3037-1.4359i$, -2.3926, -5.7913, -1.2087, and -1.0000.

Step 2: The following membership functions are selected:

$$\mu_{F_i^1} = \exp\left[-\left(\frac{x_i - 3\times a_i}{b_i}\right)^2\right], \mu_{F_i^2} = \exp\left[-\left(\frac{x_i - 2\times a_i}{b_i}\right)^2\right],$$

$$\mu_{F_i^3} = \exp\left[-\left(\frac{x_i - 1\times a_i}{b_i}\right)^2\right], \mu_{F_i^4} = \exp\left[-\left(\frac{x_i}{b_i}\right)^2\right]$$

$$\mu_{F_i^5} = \exp\left[-\left(\frac{x_i + 1\times a_i}{b_i}\right)^2\right], \mu_{F_i^6} = \exp\left[-\left(\frac{x_i + 2\times a_i}{b_i}\right)^2\right],$$

$$\mu_{F_i^7} = \exp\left[-\left(\frac{x_i + 3\times a_i}{b_i}\right)^2\right]$$

where $i = 1,2,3$, and $a_1 = 0.5, a_2 = 1, a_3 = 2, b_1 = 10, b_2 = 20, b_3 = 30$. Because the system has six state variables, forty-two fuzzy rules of the following form are included in the fuzzy rule bases. Denote

R^l : If x_1 is F_1^j, then u_f is G_1^j for $j = 1,2,\ldots,7$ and $l = 1,2,\ldots,7$.

R^l : If x_2 is F_2^j, then u_f is G_2^j for $j = 1,2,\ldots,7$ and $l = 8,9,\ldots,14$.

R^l : If x_3 is F_3^j, then u_f is G_3^j for $j = 1,2,\ldots,7$ and $l = 15,16,\ldots,21$.

R^l : If x_4 is F_4^j, then u_f is G_4^j for $j = 1,2,\ldots,7$ and $l = 22,23,\ldots,28$.

R^l : If x_5 is F_5^j, then u_f is G_5^j for $j = 1,2,\ldots,7$ and $l = 29,30,\ldots,35$.

R^l : If x_6 is F_6^j, then u_f is G_6^j for $j = 1,2,\ldots,7$ and $l = 36,37,\ldots,42$.

Denote

$$D_m = \sum_{j=1}^{7} \prod_{i=1}^{3} \mu_{F_i^j}(x_i),\ \varsigma(x) = \begin{bmatrix} \frac{\prod_{i=1}^{3} \mu_{F_i^1}(x_i)}{D_m} \cdots \frac{\prod_{i=1}^{3} \mu_{F_i^7}(x_i)}{D_m} \end{bmatrix},$$

$$\xi(x) = \begin{bmatrix} \varsigma(x) & 0 \\ 0 & \varsigma(x) \end{bmatrix}, \Theta(t) = [\theta_{11} \quad \theta_{12} \quad \cdots \quad \theta_{17} \quad \theta_{21} \quad \theta_{22} \quad \cdots \quad \theta_{27}].$$

The attenuation level ρ is selected as $\rho = \infty, \rho = 0.1$, and $\rho = 0.05$, respectively. In the case of $\rho = \infty$ (i.e., the H_2 optimal tracking control is used but without attenuation of $w(t)$), we only want to reveal the deterioration of tracking performance by $w(t)$.

Step 3: Select weighting matrix $Q = diag[100I_2, 10I_2, I_2]$, and $R = \rho^2 I$ to guarantee the solvability of the minimax H_∞ adaptive tracking problem [see (4.55)].

Step 4: Solve the Riccati-like equation (4.48). In the $\rho = \infty$ case, we can solve

$$P = \begin{bmatrix} 17.35 & 0 & 10.05 & 0 & 0.031 & 0 \\ 0 & 17.35 & 0 & 10.05 & 0 & 0.031 \\ 10.05 & 0 & 17.41 & 0 & 0.054 & 0 \\ 0 & 10.05 & 0 & 17.41 & 0 & 0.054 \\ 0.031 & 0 & 0.054 & 0 & 0.031 & 0 \\ 0 & 0.031 & 0 & 0.054 & 0 & 0.031 \end{bmatrix}.$$

In the cases of $\rho = 0.1$ and $\rho = 0.05$, we can solve

$$P = \begin{bmatrix} 23.97 & 0 & 15.46 & 0 & 0.556 & 0 \\ 0 & 18.05 & 0 & 11.10 & 0 & 0.714 \\ 15.46 & 0 & 11.10 & 0 & 2.046 & 0 \\ 0 & 11.10 & 0 & 19.24 & 0 & 1.150 \\ 0.556 & 0 & 2.046 & 0 & 1.409 & 0 \\ 0 & 0.714 & 0 & 1.409 & 0 & 0.769 \end{bmatrix}.$$

In this simulation, we use the function "are" in the WINDOW MATLAB to solve the algebraic Riccati- like equation.

By using the control algorithms (4.45)–(4.47) to obtain the minimax H_∞ tracking controller $u_e^*(t)$, the voltage control input $u(t)$ and the update law for $\Theta(t)$ are obtained as follows:

Step 5: In the $\rho = \infty$ case, the minimax H_∞ game tracking controller is of the following form:

$$\dot{\Theta} = \xi^T(t) \begin{bmatrix} \Theta_1 \\ \Theta_2 \end{bmatrix} - U(\Theta)$$

$$\Theta_1 = -0.0307(q_1 - q_{1d}) - 0.0539(\dot{q}_1 - \dot{q}_{1d}) - 0.0313(\ddot{q}_1 - \ddot{q}_{1d})$$

$$\Theta_2 = -0.0309(q_2 - q_{2d}) - 0.0539(\dot{q}_2 - \dot{q}_{2d}) - 0.0310(\ddot{q}_2 - \ddot{q}_{2d})$$

$$u_f(t) = \begin{bmatrix} u_{f_1}(t) \\ u_{f_2}(t) \end{bmatrix} = \xi(x)\Theta(t)$$

and

$$\begin{aligned} u(t) &= \begin{bmatrix} 0.0051 & 0.0004 \\ 0.0002 & 0.0002 \end{bmatrix} \begin{bmatrix} u_1(t) \\ u_2(t) \end{bmatrix} + D_0 \\ u_1(t) &= q_{1d}^{(3)} - 9(q_1 - q_{1d}) - 10(\dot{q}_1 - \dot{q}_{1d}) - 5(\ddot{q}_1 - \ddot{q}_{1d}) - u_{f_1} \\ u_2(t) &= q_{2d}^{(3)} - 7(q_2 - q_{2d}) - 14(\dot{q}_2 - \dot{q}_{2d}) - 8(\ddot{q}_2 - \ddot{q}_{2d}) - u_{f_2} \end{aligned}$$

where D_0 is the nominal value of $D(\dot{q}, \ddot{q}, \dddot{q})$ in (4.13).
In the $\rho = 0.1$ case, the minimax H∞ game tracking controller is obtained as

$$\begin{aligned} \dot{\Theta} &= \xi^T(t) \begin{bmatrix} \Theta_1 \\ \Theta_2 \end{bmatrix} - U(\Theta) \\ \Theta_1 &= -0.5556(q_1 - q_{1d}) - 2.0461(\dot{q}_1 - \dot{q}_{1d}) - 1.4092(\ddot{q}_1 - \ddot{q}_{1d}) \\ \Theta_2 &= -0.7143(q_2 - q_{2d}) - 1.1497(\dot{q}_2 - \dot{q}_{2d}) - 0.7687(\ddot{q}_2 - \ddot{q}_{2d}) \\ u_{e_1}(t) &= -55.5556(q_1 - q_{1d}) - 204.6070(\dot{q}_1 - \dot{q}_{1d}) - 140.9214(\ddot{q}_1 - \ddot{q}_{1d}) \\ u_{e_2}(t) &= -71.4286(q_2 - q_{2d}) - 114.9660(\dot{q}_2 - \dot{q}_{2d}) - 76.8707(\ddot{q}_2 - \ddot{q}_{2d}) \\ u_f(t) &= \begin{bmatrix} u_{f_1}(t) \\ u_{f_2}(t) \end{bmatrix} = \xi(x)\Theta(t) \end{aligned}$$

and

$$\begin{aligned} u(t) &= \begin{bmatrix} 0.0051 & 0.0004 \\ 0.0002 & 0.0002 \end{bmatrix} \begin{bmatrix} u_1(t) \\ u_2(t) \end{bmatrix} + D_0 \\ u_1(t) &= q_{1d}^{(3)} - 9(q_1 - q_{1d}) - 10(\dot{q}_1 - \dot{q}_{1d}) - 5(\ddot{q}_1 - \ddot{q}_{1d}) - u_{f_1} \\ u_2(t) &= q_{2d}^{(3)} - 7(q_2 - q_{2d}) - 14(\dot{q}_2 - \dot{q}_{2d}) - 8(\ddot{q}_2 - \ddot{q}_{2d}) - u_{f_2}. \end{aligned}$$

In the $\rho = 0.05$ case, the minimax H∞ game tracking controller is of the following form:

$$\begin{aligned} \dot{\Theta} &= \xi^T(t) \begin{bmatrix} \Theta_1 \\ \Theta_2 \end{bmatrix} - U(\Theta) \\ \Theta_1 &= -0.5556(q_1 - q_{1d}) - 2.0461(\dot{q}_1 - \dot{q}_{1d}) - 1.4092(\ddot{q}_1 - \ddot{q}_{1d}) \\ \Theta_2 &= -0.7143(q_2 - q_{2d}) - 1.1497(\dot{q}_2 - \dot{q}_{2d}) - 0.7687(\ddot{q}_2 - \ddot{q}_{2d}) \\ u_{e_1}(t) &= -222.2222(q_1 - q_{1d}) - 818.4282(\dot{q}_1 - \dot{q}_{1d}) - 563.6856(\ddot{q}_1 - \ddot{q}_{1d}) \\ u_{e_2}(t) &= -285.7143(q_2 - q_{2d}) - 459.8639(\dot{q}_2 - \dot{q}_{2d}) - 307.4830(\ddot{q}_2 - \ddot{q}_{2d}), \\ u_f(t) &= \begin{bmatrix} u_{f_1}(t) \\ u_{f_2}(t) \end{bmatrix} = \xi(x)\Theta(t) \end{aligned}$$

and

$$u(t) = \begin{bmatrix} 0.0051 & 0.0004 \\ 0.0002 & 0.0002 \end{bmatrix} \begin{bmatrix} u_1(t) \\ u_2(t) \end{bmatrix} + D_0$$

$$u_1(t) = q_{1d}^{(3)} - 9(q_1 - q_{1d}) - 10(\dot{q}_1 - \dot{q}_{1d}) - 5(\ddot{q}_1 - \ddot{q}_{1d}) - u_{f_1}$$

$$u_2(t) = q_{2d}^{(3)} - 7(q_2 - q_{2d}) - 14(\dot{q}_2 - \dot{q}_{2d}) - 8(\ddot{q}_2 - \ddot{q}_{2d}) - u_{f_2}.$$

Figs. 4.3–4.8 present the simulation results. Results obtained from the H_2 optimal tracking case (i.e., $\rho = \infty$) are shown in Fig. 4.3. In Figs. 4.5 and 4.7, simulation results of the proposed adaptive fuzzy-based minimax H_∞ game tracking control have demonstrated the tracking performance for attenuation levels $\rho = 0.1$ and 0.05, respectively. Figs. 4.4, 4.6, and 4.8 show the control inputs under different attenuation levels. Figs. 4.9 and 4.10 show the fluctuations of the adaptive parameters for the cases of $\rho = 0.1$ and 0.05, respectively.

According to the simulation results of the above three attenuation level cases, a specification of smaller attenuation level ρ may yield a better tracking performance. In the case of H_2 optimal tracking, the effect of $w(t)$ on tracking error has not been attenuated and subsequently leads to a poor tracking performance. According to these results, the tracking performance is unsatisfactory and even unstable. The effect of the

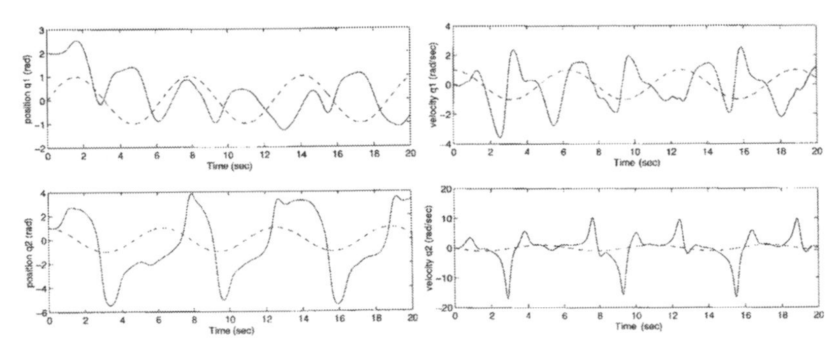

FIGURE 4.3
The H_2 optimal tracking results without attenuation of $w(t)$ (desired trajectory: ","; actual trajectory: "______").

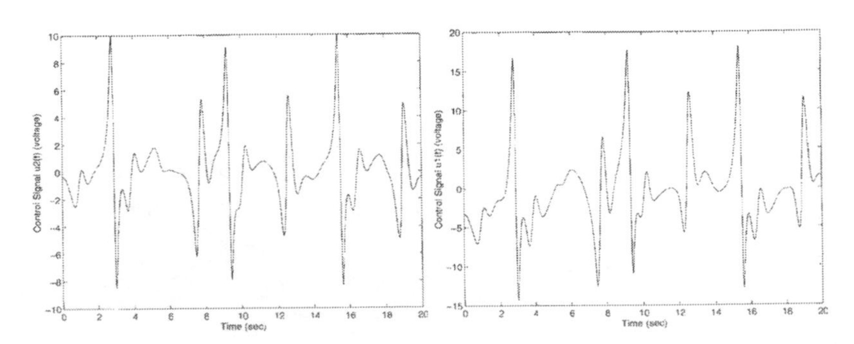

FIGURE 4.4
Control signals of the H_2 optimal tracking case.

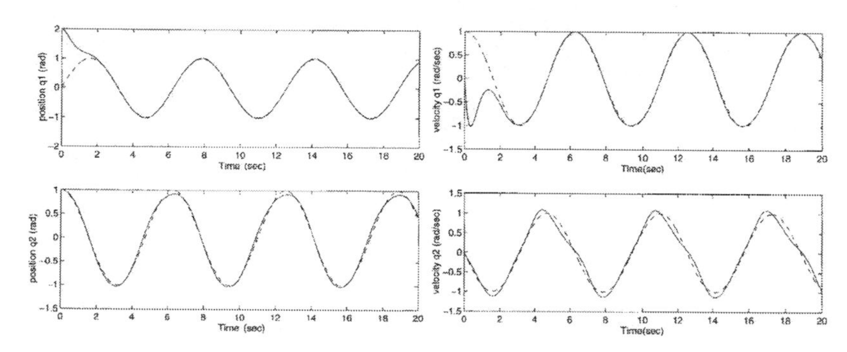

FIGURE 4.5
Robust H_∞ game tracking results with attenuation level $\rho = 0.1$ (desired trajectory: ","; actual trajectory: "____").

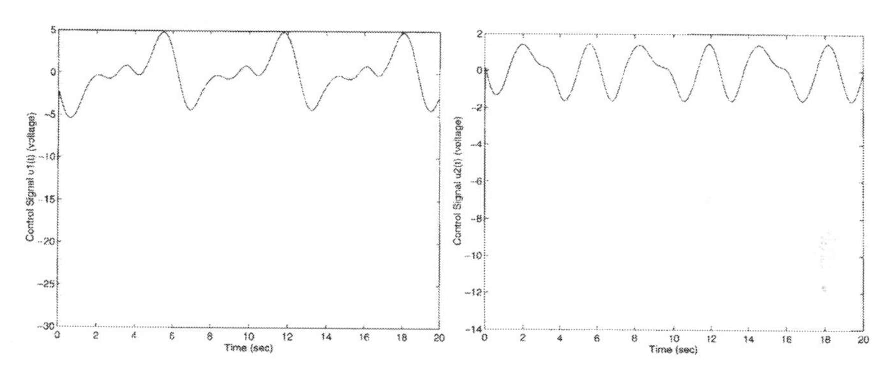

FIGURE 4.6
Control signals with case $\rho = 0.1$.

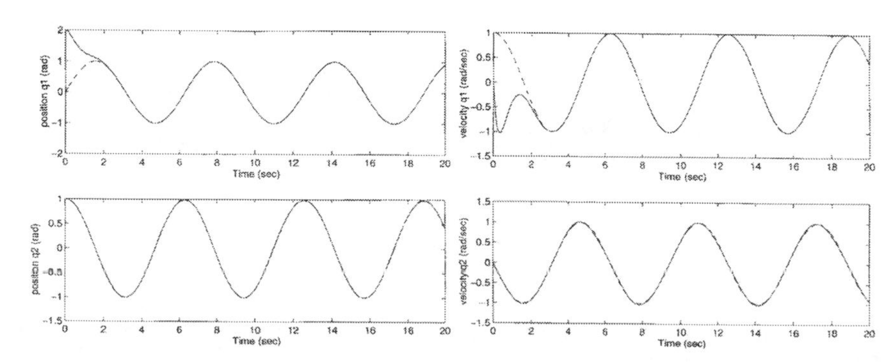

FIGURE 4.7
Robust H_∞ game tracking results with attenuation level $\rho = 0.05$ (desired trajectory: ","; actual trajectory: "____").

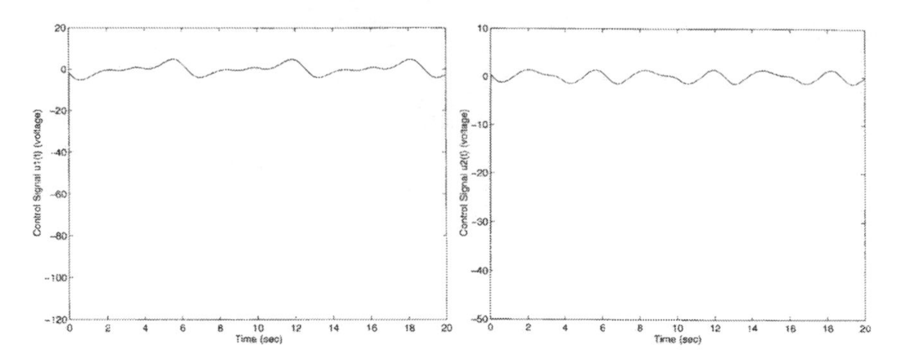

FIGURE 4.8
Control signals with case $\rho = 0.05$.

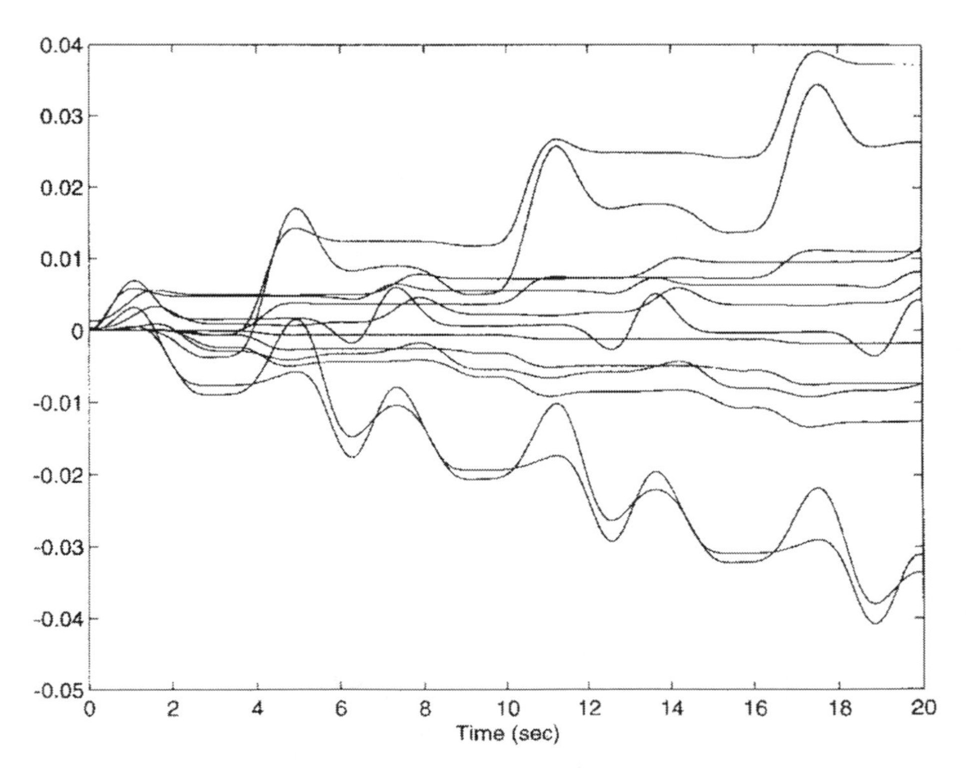

FIGURE 4.9
The fluctuations of the adaptive parameters θ_i for the case of $\rho = 0.1$.

combined disturbance $w(t)$ on the tracking error is attenuated evidently as ρ decreases. However, the voltage control inputs also indicate a high-gain effect of the designed controller as ρ decreases, particularly, at the beginning of control. In practical control engineering design, this effect due to the residue of fuzzy cancelation and exogenous disturbance must be diminished by the minimax H_∞ game tracking design in fuzzy-based adaptive control robotic systems. Otherwise, it will damage the entire system,

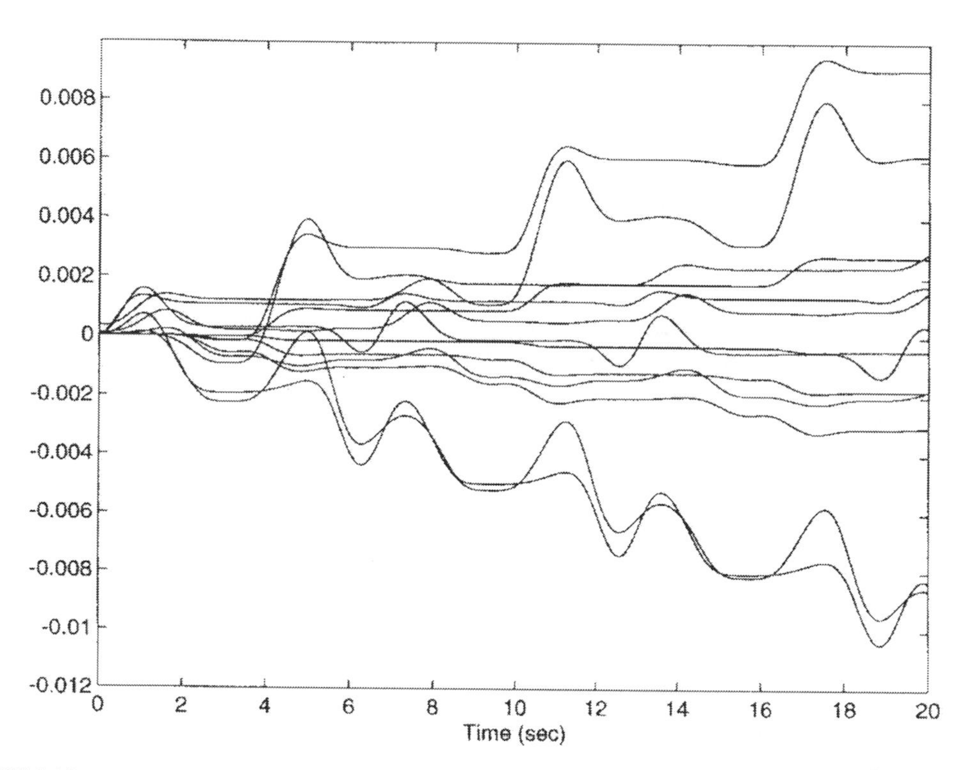

FIGURE 4.10
The fluctuations of the adaptive parameters θ_i for the case of $\rho = 0.05$.

particularly, in using a small number of membership functions. However, the attenuation level cannot be decreased without considering the increase of the control input. This is a tradeoff between the amplitude of control signal and the tracking performance.

In [41], the following control is employed for robust tracking of robot system with dc motor:

$$u_0[e(t)] = -\rho B^T P_n e(t) - \mu(e(t)) \frac{B^T P_n e(t)}{\|B^T P_n e(t)\| + \varepsilon}, \varepsilon > 0 \tag{4.57}$$

where $\mu(e(t)) = \alpha_1 \|e(t)\| + \alpha_2 \|e(t)\|^2 + \alpha_3 \|e(t)\|^3 + \alpha_4$ and P_n is solved by the following Riccati-like equation:

$$P_n(A + \beta_n I) + (A + \beta_n I)^T P_n + Q - \rho P_n B B^T P_n = 0. \tag{4.58}$$

Some comparisons are given as follows:

1. In control algorithm of [41], the second term in (4.57) is employed to override the uncertainties or disturbance directly through a high-gain control that is proportional to $\|e(t)\|$, $\|e(t)\|^2$, and $\|e(t)\|^3$. If the initial $e(0)$ is not very small, it will lead to a very high-control signal and the control system will diverge. By applying the control algorithm in [41] to our simulation example, we have found the tracking system diverges even with small initial $e(0)$. Obviously, it is not a good design to

override the uncertainties directly. In our method, an adaptive fuzzy scheme is employed to optimally cancel the uncertainties beforehand. Therefore, we only need $u_e^*(t)$ to eliminate the effect of residue with a little effort.

2. The choice of control parameters $\alpha_1, \alpha_2, \alpha_3$, and α_4 in [41] are complex and case by case. However, in our case, a simple design procedure is proposed by a systematic method. Furthermore, a comparison with conventional optimal H_2 tracking control without uncertain cancelation is shown in Figs. 4.3 and 4.4. From the results in Figs. 4.5–4.8, it is seen that the proposed method has much better performance than that of the conventional H_2 optimal tracking control. The reason is that the effort of uncertain parameters and external disturbance is efficiently eliminated by the proposed fuzzy-based minimax H_∞ game cancelation strategy.

4.6 Conclusion

In this chapter, an adaptive fuzzy cancelation technique and a minimax H_∞ game attenuation technique have functioned in the roles of a rough tuning and a fine tuning, respectively, and are combined together to enhance the robust tracking performance of uncertain robotic systems including motor dynamics. The solvability of this robust minimax H_∞ game tracking problem for uncertain robot system is also investigated. According to our results, a desired minimax H_∞ game tracking performance can be achieved if the weighting matrix on control signal is adequately specified. Actually, the proposed adaptive fuzzy-based minimax H_∞ game tracking method can be applied to any robust adaptive control design of an uncertain nonlinear mechanical system of the form in (4.11) or (4.15). With the aid of adaptive fuzzy cancelation algorithm, the minimax H_∞ game tracking control design can be extended from exactly known linear systems toward nonlinear uncertain systems with disturbances. Furthermore, by employing the minimax H_∞ game attenuation technique, the robust tracking performance of fuzzy-based adaptive control design for uncertain nonlinear mechanical systems can be significantly improved. Therefore, the proposed adaptive minimax H_∞ game tracking control design algorithm is appropriate for practical control design of mechanical systems with large parameter perturbations and external disturbances. The proposed minimax H_∞ game design method is simple, and the number of membership functions for the proposed control law can be extremely small. However, because of the use of both fuzzy cancelation technique and minimax H_∞ tracking game strategy, the results are less conservative than the other robust control methods. Simulation results have indicated that the desired robust tracking performance of uncertain dc motor-driven robotic systems with external disturbance can be achieved via the proposed adaptive minimax H_∞ game tracking control method.

5

A Robust Adaptive Tracking Control Design Based on Minimax H_∞ Game Strategy for Wheeled Vehicles with Trailer

5.1 Introduction

Recently, the growing use of wheeled vehicles with trailer in numerous applications, such as transportation vehicles, surveillance and ferry mobile robots, is necessary for the development of strategies for auto-parking, auto-docking, as well as auto-path tracking and correction in higher speeds, in both structured and unstructured environments. The nonlinear dynamics of the wheeled vehicle with trailer is one kind of nonholonomic systems.

During the last five years, the published literature pertaining to the nonholonomic control problem has grown enormously. In [64], an invariant manifold technique and a control strategy has been proposed, which steers the nonholonomic systems onto manifolds of the global and relative local-local controllability, to bring the nonholonomic vehicles under kinematic constraints. In [65,66], the invariant manifold technique has been introduced to solve the tracking problem under dynamic and kinematic constraints. In [67], the controllability of the general multibody mobile robotic system has been discussed. In [68], a summary of recent developments in control of nonholonomic systems, including tractors with trailers, robot manipulation, etc. has been presented. In [69], a control design for speed and azimuth of the vehicles without trailer is solved by using the fuzzy-Gaussian neural-network controller. However, the parameter uncertainty and external disturbance have not been considered in this work, while the control performance can also not be guaranteed from a more theoretical perspective.

This chapter concerns with the tracking control problem of a wheeled vehicle driven by two independent wheels. In the autonomous mobile robot system, an optimal control, or a model-based approach, can be adopted for achieving such control, if the mathematical model is completely known. However, it is not easy to measure some physical parameters, such as viscous friction factor and the moment of inertia around the center of gravity (c.g.) for the wheeled vehicle. Therefore, a robust controller, not based on precise mathematical model, is recommendable for such control design since it allows the simplification of modeling and also considers parameter variation, load change, the elasticity of the wheels, and road disturbance.

In the last decades, H_∞ attenuation control design has been well developed for attenuating the effects of uncertain dynamics and external noises [6,51]. However, if the magnitudes of uncertain dynamics and external noises of the dynamic system to be controlled are large, their effect on tracking error may still be obvious and even lead to system instability. In such situations, an effective elimination of uncertain dynamics and external noise, prior to the application of H_∞ attenuation scheme, is necessary for the enhancement of the tracking robustness. Conventionally, adaptive control schemes are employed to cancel the uncertain nonlinear dynamics. However, parameters are linear in conventional adaptive control schemes, i.e., the unknown parameters are considered to be of linear model form. In nonlinear uncertain wheeled vehicle systems, this may not be true. Since the fuzzy logic system can be tuned to approximate any nonlinear dynamic system, a fuzzy control technique has been used in this chapter to efficiently eliminate the plant uncertainties, via an adaptive learning method.

Recently, fuzzy logic techniques have been effectively used to approximate unknown nonlinear dynamics [47,61,70]. However, in the conventional adaptive fuzzy control, the desired tracking performance cannot be guaranteed from a more theoretical control point of view. In this chapter, a fuzzy-based dynamic H_∞ game scheme has been employed to treat the robust H_∞ tracking design problem of uncertain wheeled vehicle systems. The proposed robust controller comprises two terms, one being a self-tuning fuzzy logic system, which is employed to optimally eliminate the uncertain dynamics in order to enhance the tracking robustness, and the other being a minimax H_∞ game control strategy, which is employed to attenuate the worst-case effects of both the residue of fuzzy elimination and the exogenous disturbance on the tracking error within a desired level. The proposed adaptive H_∞ game design method attempts to combine the dynamic game technique and adaptive fuzzy algorithm to guarantee a robust H_∞ tracking performance for uncertain wheeled vehicle control systems.

In the proposed adaptive robust H_∞ tracking control, the adaptive fuzzy elimination technique is used as a rough tuning while the minimax H_∞ game attenuation technique serves as a fine-tuning. Unlike the conventional adaptive control in nonlinear robust control systems, the uncertain dynamics need not be of linearly parameterized structure for this approach, i.e., uncertain dynamics can be free of structure in the proposed adaptive fuzzy-based robust H_∞ game control systems. Since the adaptive feedback linearization technique is avoided, an inverse of adaptive matrix becomes unnecessary. Hence, the computational complexity and the sensitivity of the proposed adaptive H_∞ game control strategy are significantly reduced. Therefore, the proposed minimax H_∞ game tracking control design method is more suitable for robust tracking performance control design for wheeled vehicle systems with uncertainties and external disturbance.

5.2 Dynamic Model of the Wheeled Mobile Robot

The following notations are used in this chapter.

1) I_v: moment of inertia around the c.g. of mobile robot.
2) I_w: moment of inertia of the wheel.
3) M_1: mass of the front-wheeled vehicle.
4) M_2: mass of the rear-wheeled vehicle.
5) l: distance between left or right wheel and c.g. of the wheeled vehicle.
6) L: length of the link bar.

7) d_L: distance between force T and c.g. of the wheeled vehicle.
8) ϕ_1: azimuth of the front-wheeled vehicle.
9) ϕ_2: azimuth of the rear-wheeled vehicle.
10) (x_1, y_1): position of the front-wheeled vehicle in absolute coordinate $O - XY$.
11) (x_2, y_2): position of the rear-wheeled vehicle in absolute coordinate $O - XY$.
12) v_1: velocity of the front-wheeled vehicle.
13) v_2: velocity of the rear-wheeled vehicle.
14) r: radius of the wheel.
15) θ_l, θ_r: rotational angles of the left and right wheels, respectively.
16) c: viscous friction factor.
17) k: driving gain factor.
18) u_l, u_r: driving input.
19) T: tension force acting on the rear mobile vehicle.
20) D_l, D_r: left and right driving forces.

Let the wheeled vehicle with two independent driving wheels be rigid, moving on the plane as shown in Fig. 5.1, where $O - XY$ indicates the absolute (fixed) coordinates on the plane. In this chapter, we deal with the wheeled vehicle under the following assumptions.

Assumptions 5.1:

1. The running wheeled vehicle satisfies both the pure rolling and nonslipping conditions.
2. The link bar of the wheeled vehicle crosses the centers of gravity of the front and rear wheeled vehicles, i.e., $d_L = 0$.

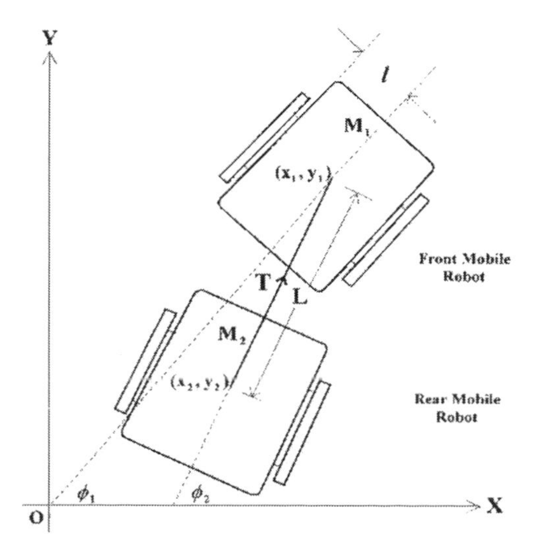

FIGURE 5.1
A wheeled mobile robot with trailer and two independent wheels.

Under the above assumptions, the model of such wheeled vehicle can be composed of the following equations [65,69].

1) The dynamic equation of the front mobile robot

$$\begin{aligned} I_v\ddot{\phi}_1 - Td_L &= D_r l - D_l l \\ M_1\dot{v}_1 - T\cos(\phi_2 - \phi_1) &= D_r + D_l. \end{aligned} \tag{5.1}$$

2) The dynamic equation of the driving systems for right and left wheels of the front mobile robot

$$I_w\ddot{\theta}_r + c\dot{\theta}_r = ku_r - rD_r \tag{5.2}$$

$$I_w\ddot{\theta}_l + c\dot{\theta}_l = ku_l - rD_l. \tag{5.3}$$

3) Pure rolling constraints

$$r\dot{\theta}_r = v_1 + l\dot{\phi}_1 \tag{5.4}$$

$$r\dot{\theta}_l = v_1 - l\dot{\phi}_1. \tag{5.5}$$

4) Nonslipping constraints

$$\dot{x}_1 \sin\phi_1 - \dot{y}_1 \cos\phi_1 = 0 \tag{5.6}$$

$$\dot{x}_2 \sin\phi_2 - \dot{y}_2 \cos\phi_2 = 0 \tag{5.7}$$

or

$$\dot{x}_1 = v_1 \cos\phi_1 \tag{5.8}$$

$$\dot{y}_1 = v_1 \sin\phi_1 \tag{5.9}$$

$$\dot{x}_2 = v_2 \cos\phi_2 \tag{5.10}$$

$$\dot{y}_2 = v_2 \sin\phi_2 \tag{5.11}$$

5) Relationships between front- and rear-wheeled vehicles

$$T = M_2\dot{v}_2 \tag{5.12}$$

$$v_2 = v_1 \cos(\phi_2 - \phi_1) \tag{5.13}$$

$$\dot{\phi}_2 = -\frac{1}{L} v_1 \sin(\phi_2 - \phi_1) \tag{5.14}$$

For the convenience of analysis, we combine the above equations in the following. First, substituting (5.4) and (5.5) into (5.2) and (5.3), we get

$$\frac{I_w}{r}(\dot{v}_1 + l\ddot{\phi}_1) + \frac{c}{r}(v_1 + l\dot{\phi}_1) = ku_r - rD_r \tag{5.15}$$

$$\frac{I_w}{r}(\dot{v}_1 + l\ddot{\phi}_1) + \frac{c}{r}(v_1 + l\dot{\phi}_1) = ku_l - rD_l. \tag{5.16}$$

From (5.1) and (5.12)–(5.14), we obtain

$$D_r = \frac{[lM_1\dot{v}_1 - lT\cos(\phi_2 - \phi_1) + I_v\ddot{\phi}_1]}{2l} \tag{5.17}$$

$$D_r = \frac{[lM_1\dot{v}_1 - lT\cos(\phi_2 - \phi_1) - I_v\ddot{\phi}_1]}{2l} \tag{5.18}$$

and

$$T = [M_2\cos(\phi_2 - \phi_1) \quad 0]\begin{bmatrix} \dot{v}_1 \\ \ddot{\phi}_1 \end{bmatrix} \tag{5.19}$$

$$+[\tfrac{1}{L}M_2 v_1 \sin(\phi_2 - \phi_1) \quad M_2 v_1 \sin(\phi_2 - \phi_1)] \times \begin{bmatrix} v_1 \\ \dot{\phi}_1 \end{bmatrix} \tag{5.20}$$

respectively. Substituting (5.17), (5.18), and (5.19) into (5.15) and (5.16), we get the following differential equation:

$$M(\phi_1, \phi_2)\begin{bmatrix} \dot{v}_1 \\ \ddot{\phi}_1 \end{bmatrix} + C(v_1, \phi_1, \phi_2)\begin{bmatrix} v_1 \\ \dot{\phi}_1 \end{bmatrix} = \begin{bmatrix} u_r \\ u_l \end{bmatrix} \tag{5.21}$$

where $M(\phi_1, \phi_2) = \begin{bmatrix} \frac{I_w}{kr} + \frac{rM_1}{2k} - \frac{rM_2}{2k}\cos^2(\phi_2 - \phi_1) & \frac{I_w l}{kr} + \frac{rI_V}{2lk} \\ \frac{I_w}{kr} + \frac{rM_1}{2k} - \frac{rM_2}{2k}\cos^2(\phi_2 - \phi_1) & \frac{I_w l}{kr} - \frac{rI_V}{2lk} \end{bmatrix}$

and $C(v_1, \phi_1, \phi_2)$ is shown as

$$C(v_1, \phi_1, \phi_2) = \begin{bmatrix} \frac{c}{kr} - \frac{rM_2}{2kL} v_1 \cos(\phi_2 - \phi_1)\sin^2(\phi_2 - \phi_1) & \frac{cl}{kr} - \frac{rM_2}{2k} v_1 \cos(\phi_2 - \phi_1)\sin(\phi_2 - \phi_1) \\ \frac{c}{kr} - \frac{rM_2}{2kL} v_1 \cos(\phi_2 - \phi_1)\sin^2(\phi_2 - \phi_1) & -\frac{cl}{kr} - \frac{rM_2}{2k} v_1 \cos(\phi_2 - \phi_1)\sin(\phi_2 - \phi_1) \end{bmatrix}.$$

Remark 5.1 For the wheeled vehicle system as shown in Fig. 5.1, the position of the interaction through the link bar between the front-wheeled vehicle and the rear-wheeled vehicle has the following holonomic constraints in the X–Y directions:

$$x_1 - x_2 = L\cos\phi_2 \tag{5.22}$$

$$y_1 - y_2 = L\sin\phi_2. \tag{5.23}$$

Therefore, the trajectory planning for the wheeled vehicle systems must satisfy the requirements of the above constraints.

However, in practical control cases, some parameter variations (or uncertainties) will occur on I_v, I_w, M_1, M_2 and c, i.e., these parameters can be composed of nominal and perturbed terms. In this situation, the coefficient matrices in (5.21) can be expressed as

$$M(\phi_1,\phi_2) = M_0(\phi_1,\phi_2) + \Delta M(\phi_1,\phi_2) \tag{5.24}$$

$$C(v_1,\phi_1,\phi_2) = C_0(v_1,\phi_1,\phi_2) + \Delta C(v_1,\phi_1,\phi_2) \tag{5.25}$$

where the nominal coefficient matrices M_0 and C_0 are given by

$$M_0(\phi_1,\phi_2) = \begin{bmatrix} \frac{I_{w0}}{kr} + \frac{rM_{10}}{2k} - \frac{rM_{20}}{2k}\cos^2(\phi_2-\phi_1) & \frac{I_{w0}l}{kr} + \frac{rI_{V0}}{2lk} \\ \frac{I_{w0}}{kr} + \frac{rM_{10}}{2k} - \frac{rM_{20}}{2k}\cos^2(\phi_2-\phi_1) & \frac{I_{w0}l}{kr} - \frac{rI_{V0}}{2lk} \end{bmatrix},$$

$$C_0(v_1,\phi_1,\phi_2) = \begin{bmatrix} \frac{c_0}{kr} - \frac{rM_{20}}{2kL}v_1\cos(\phi_2-\phi_1)\sin^2(\phi_2-\phi_1) & \frac{c_0 l}{kr} - \frac{rM_{20}}{2k}v_1\cos(\phi_2-\phi_1)\sin(\phi_2-\phi_1) \\ \frac{c_0}{kr} - \frac{rM_{20}}{2kL}v_1\cos(\phi_2-\phi_1)\sin^2(\phi_2-\phi_1) & -\frac{c_0 l}{kr} - \frac{rM_{20}}{2k}v_1\cos(\phi_2-\phi_1)\sin(\phi_2-\phi_1) \end{bmatrix}.$$

and the perturbed coefficient matrices ΔM and ΔC are given by

$$\Delta M(\phi_1,\phi_2) = \begin{bmatrix} \frac{\Delta I_w}{kr} + \frac{r\Delta M_1}{2k} - \frac{r\Delta M_2}{2k}\cos^2(\phi_2-\phi_1) & \frac{\Delta I_w l}{kr} + \frac{r\Delta I_V}{2lk} \\ \frac{\Delta I_w}{kr} + \frac{r\Delta M_1}{2k} - \frac{r\Delta M_2}{2k}\cos^2(\phi_2-\phi_1) & \frac{\Delta I_w l}{kr} - \frac{r\Delta I_V}{2lk} \end{bmatrix},$$

$$\Delta C(v_1,\phi_1,\phi_2) = \begin{bmatrix} \frac{\Delta c}{kr} - \frac{r\Delta M_2}{2kL}v_1\cos(\phi_2-\phi_1)\sin^2(\phi_2-\phi_1) & \frac{\Delta c l}{kr} - \frac{r\Delta M_2}{2k}v_1\cos(\phi_2-\phi_1)\sin(\phi_2-\phi_1) \\ \frac{\Delta c}{kr} - \frac{r\Delta M_2}{2kL}v_1\cos(\phi_2-\phi_1)\sin^2(\phi_2-\phi_1) & -\frac{\Delta c l}{kr} - \frac{r\Delta M_2}{2k}v_1\cos(\phi_2-\phi_1)\sin(\phi_2-\phi_1) \end{bmatrix}.$$

Furthermore, external disturbances are also unavoidable. If r_1 denotes the position of the front-wheeled vehicle, i.e., $\dot{r}_1 = v_1$ and $d' = [d_1', d_2']^T$ denotes the vector of external disturbance, then the dynamic equation of motion of the uncertain wheeled vehicle system can be expressed as

$$(M_0 + \Delta M)\ddot{q} + (C_0 + \Delta C)\dot{q} = u + d' \tag{5.26}$$

where $q = [r_1, \phi_1]^T$ and $u = [u_r, u_l]^T$.

Let us consider the following nominal control law:

$$u = M_0[\ddot{q}_d - K_1\tilde{q} - K_2\dot{\tilde{q}} + u_0] + C_0\dot{q} \tag{5.27}$$

where K_1 and K_2 are diagonal matrices to be designed, u_0 is an auxiliary control signal to be specified, $q_d = [r_{1d}, \phi_{1d}]^T$ denotes the desired trajectory vector of the wheeled vehicle, which could be specified beforehand through GPS, and $\tilde{q}$ is defined as

$$\tilde{q} \triangleq q - q_d = \begin{bmatrix} r_1 - r_{1d} \\ \phi_1 - \phi_{1d} \end{bmatrix}. \tag{5.28}$$

By substituting (5.27) into (5.26), we get

$$M_0(\ddot{\tilde{q}} + K_1\tilde{q} + K_2\dot{\tilde{q}}) + \Delta M\ddot{q} + \Delta C\dot{q} = M_0u_0 + d' \tag{5.29}$$

or

$$\ddot{\tilde{q}} + K_1\tilde{q} + K_2\dot{\tilde{q}} + M_0^{-1}(\Delta M\ddot{q} + \Delta C\dot{q}) = u_0 + M_0^{-1}d'. \tag{5.30}$$

If we define

$$e = \begin{bmatrix} \tilde{q} \\ \dot{\tilde{q}} \end{bmatrix} = \begin{bmatrix} r_1 - r_{1d} \\ \phi_1 - \phi_{1d} \\ \dot{r}_1 - \dot{r}_{1d} \\ \dot{\phi}_1 - \dot{\phi}_{1d} \end{bmatrix} \tag{5.31}$$

and

$$f(\eta) = -M_0^{-1}(\Delta M\ddot{q} + \Delta C\dot{q}) \tag{5.32}$$

$$d(t) = M_0^{-1}d'(t) \tag{5.33}$$

with $\eta = (\dot{q}, \ddot{q})^T$, then we get the tracking error dynamic equation of the wheeled vehicle system as follows:

$$\dot{e}(t) = Ae(t) + B(u_0 + f(\eta) + d(t)). \tag{5.34}$$

where

$$A = \begin{bmatrix} O_{2\times2} & I_{2\times2} \\ -K_1 & -K_2 \end{bmatrix} = \begin{bmatrix} 0 & 0 & 1 & 0 \\ 0 & 0 & 0 & 1 \\ -k_{11} & 0 & -k_{21} & 0 \\ 0 & -k_{12} & 0 & -k_{22} \end{bmatrix}$$

$$B = \begin{bmatrix} O_{2\times2} \\ I_{2\times2} \end{bmatrix} = \begin{bmatrix} 0 & 0 \\ 0 & 0 \\ 1 & 0 \\ 0 & 1 \end{bmatrix}.$$

Assumption 5.2
The parameter variations $\Delta I_v, \Delta I_w, \Delta M_1, \Delta M_2$ and Δc are bounded, then the induced norm of the perturbed coefficient matrices ΔM and ΔC in (5.24) and (5.25), and $f(\eta)$ in (5.32) are bounded also.

Remarks 5.2

1) Note that the feedback control law in (5.27) for the nonlinear system (5.26) is obtained via the nominal parameter matrices M_0 and C_0. It leads to the tracking error dynamic equation in (5.34).
2) The control parameters k_{11}, k_{12}, k_{21} and k_{22} in (5.34) are specified so that all eigenvalues of A have their desired values. Thus, by setting $u_0 = 0$, the tracking error $e(t)$ of dynamic equation (5.34) will approach zero as $t \to \infty$ when the wheeled vehicle system is free of uncertainty $f(\eta)$ and external disturbance $d(t)$.

5.3 A Description of Fuzzy Adaptive System

The tracking error dynamics of the uncertain wheeled vehicle is described by (5.34). If the wheeled vehicle is free of uncertainties and external disturbances (i.e., $f(\eta) = 0$ and $d(t) = 0$), the tracking error $e(t)$ in (5.34) will asymptotically converge to zero as $t \to \infty$ owing to the nonlinear control law in (5.27), with adequate choice of linear control $u_0(t)$. In practical wheeled vehicle systems, however, uncertainties due to parameter perturbations, unmodeled dynamics, elasticity of the wheels, change of load, and external disturbance of air resistance are unavoidable. Hence, it is important to eliminate the effect of uncertainties $f(\eta)$ and $d(t)$ on the tracking error in (5.34). Since $f(\eta)$ is uncertain, several robust design algorithms [6,51–54] can be employed as robust controllers to override the upper norm bound of $\|f(\eta)\|$. Due to high nonlinearity and uncertainty of $f(\eta)$, it is, however, not easy to estimate the upper norm bound. Furthermore, it is also not easy to suppress the effect of external disturbance $d(t)$ by using, for example, a VSS control or a dead-zone control based on the upper bound of external disturbance to treat this robust tracking problem. All of these robust control methods may lead to conservative and imprecise results.

In this chapter, the control signal u_0 in (5.34) is divided into two parts as follows:

$$u_0 = u_e - u_f(\eta, \Theta). \tag{5.35}$$

In this situation, the tracking error dynamics in (5.34) is of the following form:

$$\dot{e} = Ae + Bu_e + B[f(\eta) - u_f(\eta, \Theta)] + Bd \tag{5.36}$$

where $u_f(\eta, \Theta)$ is a fuzzy logic system and Θ is the update parameter matrix which will be discussed in the following paragraph. In (5.36), the fuzzy logic system $u_f(\eta, \Theta)$ is tuned to approximate $f(\eta)$ as closely as possible. Furthermore, the control signal $u_e(t)$ is used to attenuate the effects of residue of $f(\eta) - u_f(\eta, \Theta)$ as well as the external disturbance $d(t)$ on the tracking error, from the minimax H_∞ game perspective.

The basic configuration of the fuzzy logic system for wheeled vehicle tracking control is shown in Fig. 5.2. The fuzzy logic system $u_f(\eta, \Theta)$ in this work performs a mapping from $U \in R^4$ to $V \in R^2$. Let $U = U_1 \times U_2 \times U_3 \times U_4$ where $U_i \subset R$, for $i = 1,2,3,4$. The fuzzy rule base consists of a collection of fuzzy If–Then rules as follows:

$$
\begin{aligned}
&R^{(l)}: \text{If } \eta_1 \text{ is } F_1^l, \eta_2 \text{ is } F_2^l, \eta_3 \text{ is } F_3^l \text{ and } \eta_4 \text{ is } F_4^l, \\
&\text{Then } u_f \text{ is } G^l, \text{for } l = 1, 2, \ldots, M
\end{aligned} \tag{5.37}
$$

where $\eta = (\eta_1, \eta_2, \eta_3, \eta_4)^T \in U$ and $u_f \in V \subset R^2$ are the input and output of the fuzzy logic system, respectively. In our design, the fuzzy inference engine performs a mapping from fuzzy sets in $U \in R^4$ to fuzzy sets in R^4, based upon the fuzzy If–Then rules in the fuzzy rule base and the compositional rule of inference. The fuzzifier maps a crisp point $\eta = (\eta_1, \eta_2, \eta_3, \eta_4)^T \in U$ into a fuzzy set A_x in U. The defuzzifier maps a fuzzy set in V to a crisp point in V. More information can be found in [55,56,60].

The fuzzy logic system contained in Fig. 5.2 comprises a very rich class of static system mapping from $U \subset R^4$ to $V \subset R^2$, since many different choices are available within each block, and in addition, many combinations of these choices can result in a useful subclass of fuzzy logic systems. One subclass of fuzzy logic systems is used here as building blocks of the fuzzy approximation controller $u_f(\eta, \Theta)$ for adaptive cancelation of nonlinear function $f(\eta)$ and is described by the following important result.

Lemma 5.1 [60]: If the fuzzy basis functions are defined as

$$
\xi_{il}(\eta) = \frac{\prod_{j=1}^{4} \mu_{F_j^{il}}(\eta_j)}{\sum_{k=1}^{M} \left[\prod_{j=1}^{4} \mu_{F_j^{ik}}(\eta_j)\right]} \tag{5.38}
$$

for $i = 1, 2$ and $l = 1, 2, \ldots, M$, then the fuzzy logic systems with center-average defuzzifier, product inference, and singleton fuzzifier for wheeled mobile robotic tracking systems are of the following form [61]:

$$
u_f(\eta, \Theta) = \begin{bmatrix} u_{f_1}(\eta, \Theta) \\ u_{f_2}(\eta, \Theta) \end{bmatrix} = \begin{bmatrix} \xi_1^T \Theta_1 \\ \xi_2^T \Theta_2 \end{bmatrix} = \Xi(\eta)\Theta \tag{5.39}
$$

where

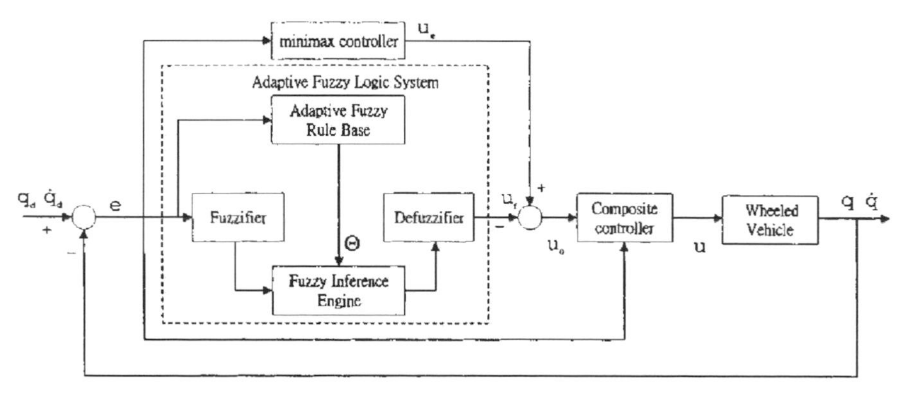

FIGURE 5.2
Fuzzy adaptive system for wheeled vehicle H_∞ game tracking control.

$$\Theta = \begin{bmatrix} \Theta_1 \\ \Theta_2 \end{bmatrix}, \Xi(\eta) = \begin{bmatrix} \xi_1^T(\eta) & 0 \\ 0 & \xi_2^T(\eta) \end{bmatrix}$$

and $\Theta_i = [\theta_{i1} \cdots \theta_{iM}]^T$, $\xi_i(\eta) = [\xi_{i1}(\eta) \cdots \xi_{iM}(\eta)]^T$, for $i = 1, 2$ with that θ_{il} is the point at which the given membership function $\mu_{F_j^{il}}(\theta_{il})$ achieves its maximum value, and we assume that $\mu_{F_j^{il}}(\theta_{il}) = 1$.

Remarks 5.3

1) There are two main reasons for using the fuzzy logic system in (5.39) as a basic building block of adaptive fuzzy controller. First, the fuzzy logic systems (5.39) are constructed from the fuzzy If–Then rules of (5.37), using some specific fuzzy inference, fuzzification, and defuzzification strategies. Therefore, the linguistic information from a human expert can be directly incorporated into the controllers. Second, the fuzzy logic systems of the form in (5.39) have been proven in [70] and [60] to be universal approximations, i.e., for any given real continuous function $f(\eta)$ defined on the compact set U, there exists a fuzzy adaptive system of the form in (5.39) such that it can uniformly approximate $f(\eta)$ over U to an arbitrary accuracy by adequate choice of parameter matrix Θ as $M \to \infty$. Therefore, the fuzzy logic systems (5.39) are qualified as building blocks for the adaptive controller for an elimination of nonlinear uncertainties of wheeled vehicle systems.
2) In this design, the membership functions $\mu_{F_j^{il}}$ are specified according to the designer's experience or knowledge about the uncertainties of the wheeled mobile systems and the parameter Θ is to be tuned according to the tracking error e.

However, in practical control design, the number of fuzzy basis functions of fuzzy logic system is chosen as small as possible for the convenience of computation and implementation. In this situation, an adaptive law must be developed to tune the parameter Θ, in order to make a fuzzy logic system $u_f(\eta, \Theta)$ with an adequate dimension to eliminate $f(\eta)$ as fully as possible.

Let us define the following minimax update parameter of elimination [60]:

$$\Theta^* = \arg \min_{\Theta \in \Omega_\Theta} \max_{\eta \in \Omega_\eta} \| f(\eta) - \Xi(\eta)\Theta \| \tag{5.40}$$

where $\| \cdot \|$ denotes the Euclidean norm, i.e., $\|\eta\| = \sqrt{\eta^T \eta}$ and Ω_Θ and Ω_η denote the sets of Θ and η that have suitable bounds, respectively. Obviously, Θ^* exists and is a constant. Furthermore, in Ω_Θ and Ω_η, $f(\eta) = \Xi(\eta)\Theta^* + \varepsilon(t)$ with $\varepsilon(t) \leq \bar{\varepsilon}$ for some $\bar{\varepsilon} > 0$. To guarantee that Θ lies inside the bounded set Ω_Θ, a projection algorithm is introduced to prevent the divergence of Θ in the next section.

Then, the tracking error dynamic equation (5.36) can be rewritten as follows:

$$\begin{aligned} \dot{e} &= Ae + Bu_e + B(\Xi(\eta)\Theta^* - \Xi(\eta)\Theta) + B\varepsilon(t) + Bd \\ &= Ae + Bu_e + B\Xi(\eta)\tilde{\Theta} + Bw \end{aligned} \tag{5.41}$$

where $w = \varepsilon(t) + d$ denotes the sum of optimal elimination error due to the fuzzy adaptive system and the external disturbance, and $\tilde{\Theta}(t)$ is defined as

$$\tilde{\Theta}(t) = \Theta^* - \Theta(t). \tag{5.42}$$

The proposed design procedure is divided into two steps. In the first step, the adaptive fuzzy logic system $u_f(\eta, \Theta)$ is tuned by way of $\Theta(t)$ to optimally approximate and eliminate the uncertain term $f(\eta)$. In this situation, the term $\Xi(\eta)\Theta^* - \Xi(\eta)\Theta = \Xi(\eta)\tilde{\Theta}$ vanishes as $t \to \infty$. However, the sum of approximation error and external disturbance w cannot be estimated and eliminated. Therefore, in the second step, the control signal $u_e(t)$ is specified such that the worst-case effect of the uncertainty $w(t)$ on the tracking error $e(t)$ must be attenuated as much as possible, and below a prescribed level ρ, i.e., the following minimax H_∞ game (optimal H_∞) tracking performance must be satisfied for the error dynamics in (5.41):

$$\begin{aligned} &\min_{u_e(t)\in L_2[0,t_f]} \max_{w(t)\in L_2[0,t_f]} \int_0^{t_f} [e^T(t)Qe(t) + u_e^T(t)Ru_e(t) - \rho^2 w^T(t)w(t)]dt \\ &\le e^T(0)Pe(0) + \frac{1}{\gamma}[\tilde{\Theta}^T(0)\tilde{\Theta}(0)] \end{aligned} \tag{5.43}$$

where $\gamma > 0$ is a weighting factor, $Q = Q^T > 0$, $P = P^T > 0$ and $R > 0$ are some positive-definite weighting matrices and the final time $t_f > 0$ [48–51].

For the case $e(0) = 0$ and $\tilde{\Theta}(0) = 0$, since $w(t)$ is independent of $u_e(t)$, (5.43) is reduced to the following minimax H_∞ game tracking strategy [48–51]:

$$\min_{u_e(t)\in L_2[0,t_f]} \max_{w(t)\in L_2[0,t_f]} \frac{\left\| [Q^{\frac{1}{2}}e(t) \quad R^{\frac{1}{2}}u_e(t)] \right\|_{L_2}}{\|w(t)\|_{L_2}} \le \rho \tag{5.44}$$

where

$$\left\| [Q^{1/2}e(t) \quad R^{1/2}u_e(t)] \right\|_{L_2} = \sqrt{\int_0^{t_f} e^T(t)Qe(t) + u_e^T(t)Ru_e(t)dt}$$

and

$$\|w(t)\|_{L_2} = \sqrt{\int_0^{t_f} w^T(t)w(t)dt}.$$

However, in general, if $e(0) \neq 0$ and $\tilde{\Theta}(0) \neq 0$, the minimax H_∞ game tracking control strategy in (5.44) must be modified to (5.43) by considering $e(0)$ and $\tilde{\Theta}(0)$ as some kind of disturbances [48,49], i.e., (5.43) is the more general formulation of minimax H_∞ game tracking control design problem than (5.44). Our design objective is to tune the parameters of the fuzzy adaptive system to eliminate the uncertain term $\Xi(\eta)\Theta^* - \Xi(\eta)\Theta$, and then to derive an adequate minimax H_∞ game control law $U_e(t)$ so that the worst-case effect of $w(t)$ on $e(t)$ is guaranteed to be less than or equal to ρ.

At present [48–51], the minimax H_∞ game control (or the optimal H_∞ control) is the most efficient method to eliminate the effect of uncertainty $w(t)$ on $e(t)$. Therefore, this method has been employed to treat the residue $w(t)$, and achieve robust H_∞ tracking control of the wheeled vehicle system by fuzzy-based H_∞ minimax game strategy design.

The expression in (5.44) means that the worst-case influence (induced L_2 norm) from $w(t)$ to $Q^{1/2}e(t)$ and $R^{1/2}u_e(t)$ must be minimized, and brought below a prescribed ρ. For achieving this, we are actually dealing with the minimax H_∞ game tracking control problem for the model reference control of wheeled vehicle with the robustness enhancement by using adaptive fuzzy logic system. The main difference between the proposed adaptive H_∞ game tracking control and the conventional H_∞ control is that an adaptive fuzzy logic system $u_f(\eta, \Theta)$ is employed in the present design to eliminate the uncertain term $f(\eta)$ as fully as possible before the minimax H_∞ game control $u_e(t)$ is applied. In the conventional H_∞ control design, only H_∞ control $u_e(t)$ is employed to directly attenuate the uncertain terms $f(\eta)$ and . If $f(\eta)$ and $d(t)$ are very large, the stability may not be guaranteed. Even if the stability is guaranteed, the effect of $f(\eta)$ and $d(t)$ on tracking error will be still very large even if H_∞ attenuation is used. Therefore, the result may be conservative and more control effort will be needed. The role of robustness enhancement via adaptive fuzzy logic system in the design of adaptive minimax H_∞ game tracking controller on the uncertain wheeled vehicle system will be discussed in the simulation example in Section 5.4.

Remarks 5.4

1) In general, $\rho < 1$ is necessary for the attenuation of the effect of $w(t)$ to achieve robust tracking. If $\rho \to \infty$ [48–51], then the robust H_∞ tracking performance design is reduced to an H_2 optimal tracking control design, without considering the attenuation of the effect of $w(t)$.
2) The performance in (5.43) or (5.44) is of finite time minimax tracking control. In this situation, $w(t)$ can be finite in $[0, t_f]$. If $t_f \to \infty$, the integrations $\int_0^\infty [e^T(t)Qe(t) + u_e^T(t)Ru_e(t)]dt$ and $\int_0^\infty [w^T(t)w(t)]dt$ may be increased to ∞. However, the integral inequality always holds.

5.4 Minimax H_∞ Game Tracking Control of Mobile Robot Systems via Adaptive Fuzzy Cancellation Scheme

Based on the analysis in the above section, the robust tracking control design of uncertain wheeled vehicle systems is formulated as a fuzzy-based minimax H_∞ game tracking control design problem. An adaptive fuzzy control $u_f(\eta, \Theta) = \Xi(\eta)\Theta$ is used to eliminate the uncertainty $f(\eta)$ as fully as possible and then a minimax H_∞ game control u_e is specified to attenuate the worst-case influence of $w(t)$ on tracking error below aprescribed level ρ. The main work in this design is to specify an update law for Θ and a control law $u_e(t)$ such that for any $w(t)$ the minimax H_∞ game tracking performance in (5.43) or (5.44) is achieved.

Subject to the tracking error dynamic equation in (5.41), let us define the cost functional

$$J(e,u_e,w)=\int_0^{t_f}[e^T(t)Qe(t)+u_e^T(t)Ru_e(t)-\rho^2w^T(t)w(t)]dt. \tag{5.45}$$

Then we get the following main result.

Theorem 5.1

For the uncertain wheeled vehicle system with external disturbance in (5.26), if the adaptive tracking control is chosen as

$$u(t)=M_0(\ddot{q}_d-K_1\tilde{q}-K_2\dot{\tilde{q}}+u_e-\Xi\Theta)+C_0\dot{q} \tag{5.46}$$

with

$$\dot{\Theta}=\gamma\Xi^T(\eta)B^TPe(t) \tag{5.47}$$

$$u_e=-R^{-1}B^TPe(t) \tag{5.48}$$

where $R=R^T$ is a weighting matrix and $P=P^T>0$ is the solution of the following algebraic Riccati-like equation:

$$PA+A^TP+Q-PB(R^{-1}-\frac{1}{\rho^2}I)B^TP=0. \tag{5.49}$$

Then, the minimax H_∞ game tracking performance in (5.43) is guaranteed for a prescribed ρ and the corresponding worst-case $w^*(t)$ deteriorating the tracking performance is of the form

$$w^*(t)=\frac{1}{\rho^2}B^TPe(t) \tag{5.50}$$

Proof: By the fact that

$$\begin{aligned}&e^T(0)Pe(0)-e^T(t_f)Pe(t_f)+\frac{1}{\gamma}\tilde{\Theta}^T(0)\tilde{\Theta}(0)-\frac{1}{\gamma}\tilde{\Theta}^T(t_f)\tilde{\Theta}(t_f)\\&+\int_0^{t_f}[\dot{e}^T(t)Pe(t)+e^T(t)P\dot{e}(t)+\frac{1}{\gamma}\dot{\tilde{\Theta}}^T(t)\tilde{\Theta}(t)+\frac{1}{\gamma}\tilde{\Theta}^T(t)\dot{\tilde{\Theta}}(t)]dt\\&=0\end{aligned}$$

Eq. (5.45) is equivalent to

$$\begin{aligned}&J(e,u_e,w)\\&=e^T(0)Pe(0)-e^T(t_f)Pe(t_f)+\frac{1}{\gamma}\tilde{\Theta}^T(0)\tilde{\Theta}(0)-\frac{1}{\gamma}\tilde{\Theta}^T(t_f)\tilde{\Theta}(t_f)\\&\quad+\int_0^{t_f}[e^T(t)Qe(t)+u_e^T(t)Ru_e(t)-\rho^2w^T(t)w(t)+\dot{e}^T(t)Pe(t)\\&\quad+e^T(t)P\dot{e}(t)+\frac{1}{\gamma}\dot{\tilde{\Theta}}^T(t)\tilde{\Theta}(t)+\frac{1}{\gamma}\tilde{\Theta}^T(t)\dot{\tilde{\Theta}}(t)]dt.\end{aligned} \tag{5.51}$$

Substituting (5.41) into the above equation, we get

$$
\begin{aligned}
&J(e, u_e, w) \\
&= e^T(0)Pe(0) - e^T(t_f)Pe(t_f) + \frac{1}{\gamma}\tilde{\Theta}^T(0)\tilde{\Theta}(0) - \frac{1}{\gamma}\tilde{\Theta}^T(t_f)\tilde{\Theta}(t_f) \\
&+ \int_0^{t_f} [e^T(t)(A^TP + PA + Q)e(t) + u_e^T(t)Ru_e(t) - \rho^2 w^T(t)w(t) \\
&+ u_e^T(t)B^TPe(t) + e^T(t)PBu_e(t) + \tilde{\Theta}^T(t)\Xi^T(\eta)B^TPe(t) + e^T(t)PBw(t) \\
&+ \frac{1}{\gamma}\dot{\tilde{\Theta}}(t)\tilde{\Theta}(t) + \frac{1}{\gamma}\tilde{\Theta}^T(t)\dot{\tilde{\Theta}}(t)]dt.
\end{aligned} \tag{5.52}
$$

By the adaptive law in (5.47) and the fact that $\dot{\tilde{\Theta}} = -\dot{\Theta}$ from (5.42), (5.52) can be rewritten as

$$
\begin{aligned}
&J(e, u_e, w) \\
&= e^T(0)Pe(0) + \frac{1}{\gamma}\tilde{\Theta}^T(0)\tilde{\Theta}(0) - e^T(t_f)Pe(t_f) - \frac{1}{\gamma}\tilde{\Theta}^T(t_f)\tilde{\Theta}(t_f) \\
&+ \int_0^{t_f} [e^T(t)(A^TP + PA + Q)e(t) + u_e^T(t)Ru_e(t) - \rho^2 w^T(t)w(t) \\
&+ u_e^T(t)B^TPe(t) + w^T(t)B^TPe(t) + e^T(t)PBu_e(t) + e^T(t)PBw(t)]dt.
\end{aligned} \tag{5.53}
$$

Using (5.49) and applying the technique of completing the squares, we get

$$
\begin{aligned}
&J(e, u_e, w) \\
&= e^T(0)Pe(0) + \frac{1}{\gamma}\tilde{\Theta}^T(0)\tilde{\Theta}(0) - e^T(t_f)Pe(t_f) - \frac{1}{\gamma}\tilde{\Theta}^T(t_f)\tilde{\Theta}(t_f) \\
&+ \int_0^{t_f} [(Ru_e(t) + B^TPe(t))^T \times R^{-1}(Ru_e(t) + B^TPe(t)) \\
&- (\rho w(t) - \frac{1}{\rho}B^TPe(t))^T \times (\rho w(t) - \frac{1}{\rho}B^TPe(t))]dt.
\end{aligned} \tag{5.54}
$$

From the game theory [49,50,62], we obtain the minimax H_∞ game control $u_e(t)$ as in (5.48) and the worst case $w^*(t)$ as in (5.50). Hence,

$$
\begin{aligned}
&\min_{u_e(t)} \max_{w(t)} J(e, u_e, w) \\
&= e^T(0)Pe(0) + \frac{1}{\gamma}\tilde{\Theta}^T(0)\tilde{\Theta}(0) - e^T(t_f)Pe(t_f) - \frac{1}{\gamma}\tilde{\Theta}^T(t_f)\tilde{\Theta}(t_f) \\
&\le e^T(0)Pe(0) + \frac{1}{\gamma}\tilde{\Theta}^T(0)\tilde{\Theta}(0).
\end{aligned} \tag{5.55}
$$

The above inequality holds due to the fact, $P = P^T > 0$, and $\gamma > 0$. By the definition (5.45), (5.55) is equivalent to the minimax H_∞ game tracking performance in (5.43). Moreover, if $e(0) = 0$ and $\tilde{\Theta}(0) = 0$, then (5.55) is reduced to (5.44).

Remark 5.5 denotes apossible disturbance in $L_2[0, t_f]$, which makes the worst-case influence on the tracking error $e(t)$ from the L_2-norm point of view.

Remarks 5.6

1) In order to guarantee the positive-definite solution P of the algebraic Riccati-like equation (5.49), the following constraint must hold [49]:

$$R^{-1} - \frac{1}{\rho^2} I > 0 \tag{5.56}$$

or

$$\rho^2 I > R \tag{5.57}$$

i.e., for a prescribed attenuation level ρ, the weighting matrix R on the control $u_e(t)$ in (5.48) must satisfy the above constraint to guarantee the solvability of minimax H_∞ game tracking control of the uncertain wheeled vehicle system in (5.26). Hence, a robust H_∞ game tracking control design with arbitrary attenuation of the influence of $w(t)$ is possible by the proposed minimax H_∞ game tracking control design method. However, if the attenuation level ρ is specified to be a very small value, by the limit of (5.57), the singular value of R^{-1} must be very large. In this situation, it will require large control energy. Therefore, there is a tradeoff between the attenuation level and the control energy.

2) Notice that the bounds on Θ and e need not be known or be specified. Moreover, additional tools from the projection algorithm [57] can be used to analyze the problem of the bound of Θ. Assume that the constraint set Ω_Θ is specified as $\Omega_\Theta \equiv \{\Theta | \|\Theta\| \leq M_f$, where M_f is a positive constant. Then, the parameter update law in (5.47) must be modified as [57]

$$\dot{\Theta} = \begin{cases} \gamma \Xi^T(\eta) B^T P e(t), \\ \text{if } \|\Theta\| < M_f \text{ or } (\|\Theta\| = M_f \text{ and } \Theta^T \Xi^T(\eta) B^T P e(t) < 0) \\ \gamma \Xi^T(\eta) B^T P e(t) - \gamma \frac{\Theta^T \Xi^T(\eta) B^T P e(t)}{\|\Theta\|^2} \Theta, \\ \text{if } \|\Theta\| \geq M_f \text{ and } \Theta^T \Xi^T(\eta) B^T P e(t) \geq 0. \end{cases} \tag{5.58}$$

Since $\tilde{\Theta}^T(t)[\Xi^T(\eta) B^T P e(t) + (1/\gamma)\dot{\tilde{\Theta}}(t)] \leq 0$ in (5.52) due to the update law in (5.58), the minimax H_∞ game tracking performance in (5.43) can also be guaranteed.

3) The boundedness of $e(t)$ and $\eta(t)$ is discussed in the Appendix.

5.4.1 Discussion

1) The proposed minimax H_∞ game control scheme in this chapter can deal with the following general nonlinear uncertain systems:

$$(M_0 + \Delta M)\ddot{q} + (C_0 + \Delta C)\dot{q} + (G_0 + \Delta G) = u + d$$

with bounded uncertainties $\Delta M, \Delta C$, and ΔG, and disturbance d if the number of membership function is large enough.

2) There are two advantages of this approach over other fuzzy controllers: a) The fuzzy controller in this chapter is updated to approximate only the unmodeled term $f(\eta)$ in (5.32) rather than the true model terms; it produces less approximation errors than the traditional fuzzy controller which is updated to approximate the whole model. b) The proposed nonlinear adaptive controller in (5.46) is a hierarchical control algorithm and is divided into three parts to achieve their design purposes, respectively. The first part contains K_1 and K_2, and are designed to achieve the asymptotical tracking of the nominal wheeled vehicle system. The second part, i.e., fuzzy adaptive controller $\Xi\Theta$, is tuned to cancel the uncertain part $f(\eta)$ as much as possible. Finally, the minimax H_∞ game tracking control u_e is designed to attenuate the effects of the residue of cancelation $f(\eta) - u_e$ as well as the external disturbance $d(t)$ on the tracking error.
3) In our design method, the fuzzy adaptive controller $\Xi\Theta$ is tuned to cancel the uncertain part $f(\eta)$ as much as possible. This resolves most problems on other control methods in which the bounds of the system uncertainties are needed to be known and which are difficult to compute. On this viewpoint, the simulation results using the sliding-mode control method are discussed in Section 5.5 to compare with our control method.

5.5 Design Procedure

Based on the above analysis, a design procedure for the fuzzy-based minimax H_∞ game tracking control of wheeled vehicle systems can be outlined as follows:

Step 1) Specify k_{11}, k_{12}, k_{21}, and k_{22} to determine matrix A with desired eigenvalues.

Step 2) Decide the fuzzy architecture $\Xi(\eta)$ and specify the desired attenuation level ρ.

Step 3) Select positive-definite weighting matrices Q and R with $R<\rho^2 I$ to guarantee the solvability of the minimax H_∞ tracking control.

Step 4) Solve the positive-definite matrix P from the Riccati-like equation in (5.49).

Step 5) Apply the control $u(t)$ in (5.46) to the uncertain wheeled vehicle system in (5.26), by using the minimax H_∞ game tracking control law in (5.48) and the parameter update law in (5.47).

In the next section, a design example is presented, given according to the above design procedure, to illustrate the performance of the proposed minimax H_∞ game tracking design method.

5.6 Simulation Example

In this section, the proposed fuzzy-based adaptive minimax H_∞ game control on the tracking design of a wheeled vehicle is tested by using a computer. For the convenience of simulation, we consider a wheeled vehicle system as shown in Fig. 5.1, with the nominal parameters of this wheeled vehicle system as follows [69]:

$$\begin{aligned}
&I_v = 10(\text{kgm}^2),\ I_w = 0.005(\text{kgm}^2),\\
&M_1 = 200(\text{kg}),\ M_2 = 100(\text{kg}),\\
&l = 0.3(\text{m}),\ c = 0.05(\text{kgm}^2/\text{s}),\\
&r = 0.1(\text{m}),\ k = 5, L = 1.2(\text{m}).
\end{aligned}$$

Suppose that the desired reference trajectories in the Cartesian coordinate are $x_{1d} = t$ and $y_{1d} = 8\sin(0.4t)$. Assume the parameters I_v, I_w, c, M_1 and M_2 to be perturbed in the following form:

$$\Delta I_v = 0.1\sin(t), \Delta I_w = 0.0001\cos(\text{t}),$$
$$\Delta M_1 = 2\cos(t), \Delta M_2 = 1.5\cos(t),$$

and

$$\Delta c = 0.001\sin(t)$$

respectively. Moreover, the exogenous disturbances d_1' and d_2' are sinusoidal waves with period $T_p = 2\pi$, i.e.,

$$d_1' = -0.3 + 0.3\sin t$$
$$d_2' = 0.2 + 0.2\sin t.$$

Obviously, the parameter uncertainties and exogenous disturbances are significant. Therefore, the proposed fuzzy-based minimax H_∞ game tracking control algorithm is employed to treat this robust adaptive tracking control design. Now, following the design procedure as given in the above section, the minimax H_∞ game tracking control design is achieved by the following steps:

Step 1) Specify $K_1 = \begin{bmatrix} 5 & 0 \\ 0 & 5 \end{bmatrix}, K_2 = \begin{bmatrix} 5 & 0 \\ 0 & 5 \end{bmatrix}$ such that the eigenvalues of the nominal tracking system are 1.3820, 3.6180, 1.3820, and 3.6180.

Step 2) Because there are two outputs of fuzzy logic system with each corresponding to seven fuzzy rules and four state variables. The following 56 membership functions (i.e., $M = 7$) are selected:

$$\mu_{F_j^{i1}} = \exp[-((\eta_j - 3 \times a_j)/b_j)^2]$$
$$\mu_{F_j^{i2}} = \exp[-((\eta_j - 2 \times a_j)/b_j)^2]$$
$$\mu_{F_j^{i3}} = \exp[-((\eta_j - 1 \times a_j)/b_j)^2]$$
$$\mu_{F_j^{i4}} = \exp[-(\eta_j/b_j)^2], \text{ for } i = 1,2,\ j = 1,2,3,4$$
$$\mu_{F_j^{i5}} = \exp[-((\eta_j + 1 \times a_j)/b_j)^2]$$
$$\mu_{F_j^{i6}} = \exp[-((\eta_j + 2 \times a_j)/b_j)^2]$$
$$\mu_{F_j^{i7}} = \exp[-((\eta_j + 3 \times a_j)/b_j)^2]$$

where $a_1 = 0.5, a_2 = 0.5, a_3 = 1,\ a_4 = 1, b_1 = 100, b_2 = 100, b_3 = 1000, b_4 = 1000$. The fuzzy rules in the following form are included in the fuzzy rule bases:

$$R^{(i1)}: \text{If } \eta_1 \text{ is } F_1^{i1}, \eta_2 \text{ is } F_2^{i1}, \eta_3 \text{ is } F_3^{i1}, \eta_4 \text{ is } F_4^{i1}, \text{ Then } \mu_{fi} \text{ is } G^{i1}$$
$$R^{(i2)}: \text{If } \eta_1 \text{ is } F_1^{i2}, \eta_2 \text{ is } F_2^{i2}, \eta_3 \text{ is } F_3^{i2}, \eta_4 \text{ is } F_4^{i2}, \text{ Then } \mu_{fi} \text{ is } G^{i2}$$
$$R^{(i3)}: \text{If } \eta_1 \text{ is } F_1^{i3}, \eta_2 \text{ is } F_2^{i3}, \eta_3 \text{ is } F_3^{i3}, \eta_4 \text{ is } F_4^{i3}, \text{ Then } \mu_{fi} \text{ is } G^{i3}$$
$$R^{(i4)}: \text{If } \eta_1 \text{ is } F_1^{i4}, \eta_2 \text{ is } F_2^{i4}, \eta_3 \text{ is } F_3^{i4}, \eta_4 \text{ is } F_4^{i4}, \text{ Then } \mu_{fi} \text{ is } G^{i4}, \text{ for } i=1,2.$$
$$R^{(i5)}: \text{If } \eta_1 \text{ is } F_1^{i5}, \eta_2 \text{ is } F_2^{i5}, \eta_3 \text{ is } F_3^{i5}, \eta_4 \text{ is } F_4^{i5}, \text{ Then } \mu_{fi} \text{ is } G^{i5}$$
$$R^{(i6)}: \text{If } \eta_1 \text{ is } F_1^{i6}, \eta_2 \text{ is } F_2^{i6}, \eta_3 \text{ is } F_3^{i6}, \eta_4 \text{ is } F_4^{i6}, \text{ Then } \mu_{fi} \text{ is } G^{i6}$$
$$R^{(i7)}: \text{If } \eta_1 \text{ is } F_1^{i7}, \eta_2 \text{ is } F_2^{i7}, \eta_3 \text{ is } F_3^{i7}, \eta_4 \text{ is } F_4^{i7}, \text{ Then } \mu_{fi} \text{ is } G^{i7}$$

Denote

$$D = \sum_{k=1}^{7} \prod_{j=1}^{4} \mu_{F_j^{ik}}(\eta_j)$$

and we have

$$\Theta_1 = [\theta_{11}\ \theta_{12}\ \cdots\ \theta_{17}\ ]^T, \Theta_2 = [\theta_{21}\ \theta_{22}\ \cdots\ \theta_{27}\ ]^T$$

$$\xi_i(\eta) = \left[\frac{\left(\prod_{j=1}^{4} \mu_{F_j^{i1}}\right)}{D} \cdots \frac{\left(\prod_{j=1}^{4} \mu_{F_j^{i7}}\right)}{D}\right], \text{for } i = 1, 2.$$

The attenuation level ρ is chosen to be $\rho = 0.5, \rho = 0.2$, and $\rho = 0.1$, respectively.

Step 3) Select weighting matrix $Q = 10I$, and $R = 0.25\rho^2 I$ to guarantee the solvability of the minimax H_∞ game tracking control problem.

Step 4) Solve the Riccati-like equation in (5.49). In case of $\rho = 0.5$, we have the solution

$$P = \begin{bmatrix} 10.5732 & 0 & 0.5868 & 0 \\ 0 & 10.5732 & 0 & 0.5868 \\ 0.5868 & 0 & 0.6344 & 0 \\ 0 & 0.5868 & 0 & 0.6344 \end{bmatrix}.$$

In case of $\rho = 0.2$, we have the solution

$$[P = \begin{bmatrix} 10.3002 & 0 & 0.3045 & 0 \\ 0 & 10.3002 & 0 & 0.3045 \\ 0.3045 & 0 & 0.3153 & 0 \\ 0 & 0.3045 & 0 & 0.3153 \end{bmatrix}]$$

In case of $\rho = 0.1$, we have the solution

$$P = \begin{bmatrix} 10.1653 & 0 & 0.1667 & 0 \\ 0 & 10.1653 & 0 & 0.1667 \\ 0.1667 & 0 & 0.1697 & 0 \\ 0 & 0.1667 & 0 & 0.1697 \end{bmatrix}.$$

In this simulation, the software WINDOWS MATLAB 5.0 is used to solve the algebraic Riccati equation.

Step 5) Use (5.46)–(5.48) to obtain the minimax H_∞ game tracking control $u_e(t)$, control input $u(t)$, and the adaptive law for Θ, respectively. The simulation results are presented in Figs. 5.3–5.8. The position and applied torques of the wheeled vehicle are presented in Figs. 5.3 and 5.4 when $\rho = 0.5$, Figs. 5.5 and 5.6 when $\rho = 0.2$, and Figs. 5.7 and 5.8 when $\rho = 0.1$.

Fig. 5.9 shows the state errors of the wheeled vehicle. Fig. 5.10 shows the minimax H_∞ game tracking performances of the wheeled vehicle. Fig. 5.11 shows the fuzzy control approximation errors. From the simulation results of the above three cases, it can be seen that the smaller attenuation level ρ may yield a better minimax H_∞ game tracking performance. The effect of $w(t)$, including exogenous disturbances and approximation error due to adaptive fuzzy elimination, is evidently attenuated as ρ decreases. But the control inputs also indicate a high-gain effect of the designed controller as ρ decreases. In practical control engineering design, this effect due to plant uncertainties and exogenous disturbances must be diminished, which will otherwise damage the tracking performance of the whole system. However, the attenuation level ρ cannot be decreased without considering the increase in the control input. This is a tradeoff between the amplitude of control signal $u_e(t)$ and the tracking performance.

To compare with our design method, simulation results using sliding-mode control on the same vehicle system are shown in Figs. 5.12 and 5.13. From these results, more chattering exists in the control values and it causes larger tracking errors. Besides, in the sliding-mode control, the bounds of the system uncertainties are needed to know and are difficult to get. By the adaptive minimax H_∞ game control method designed in this

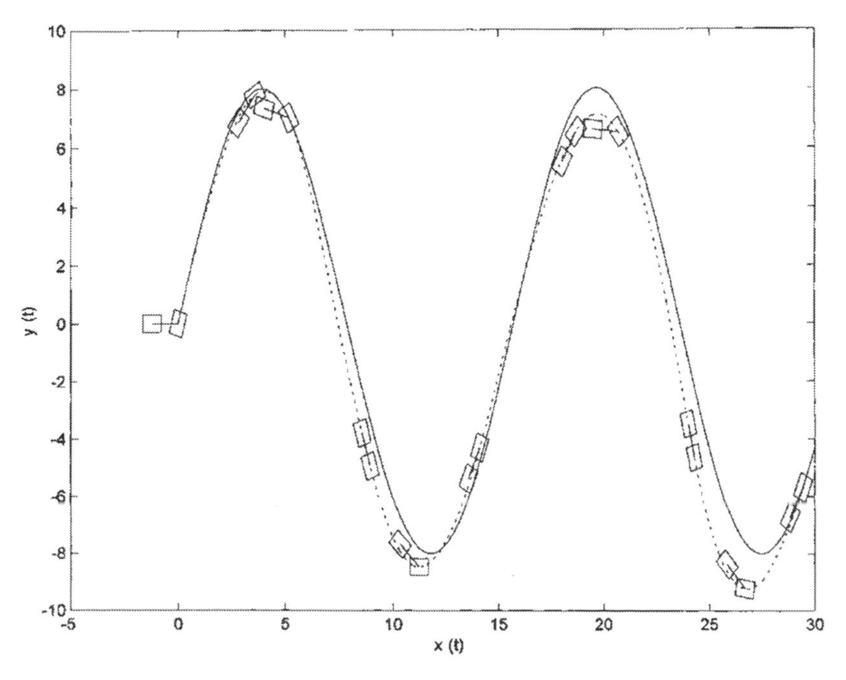

FIGURE 5.3
Trajectory of the wheeled vehicle in X–Y plane when $\rho = 0.5$.

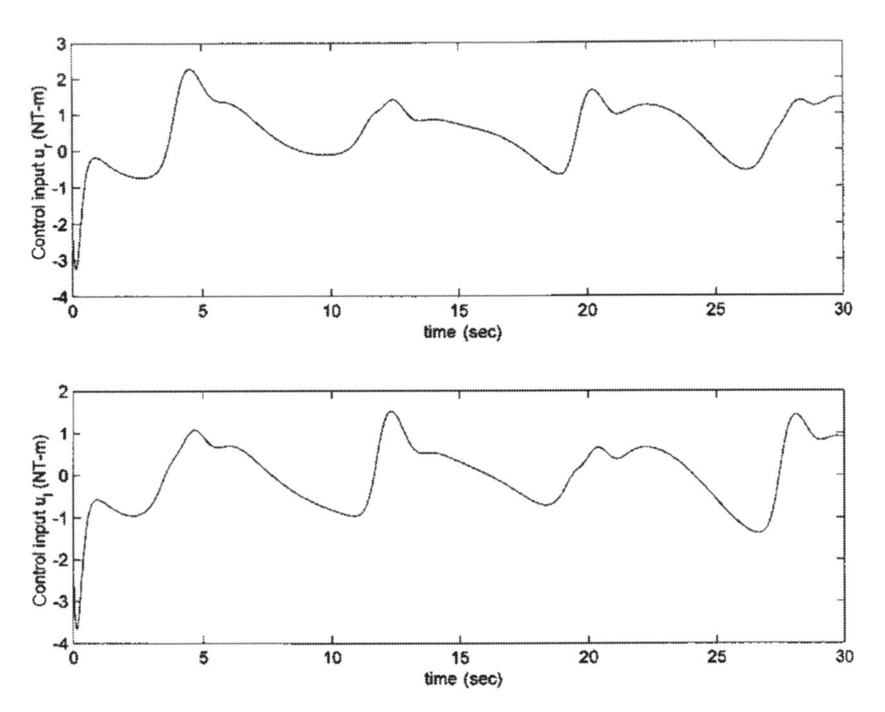

FIGURE 5.4
Control inputs applied to the wheeled vehicle when $\rho = 0.5$.

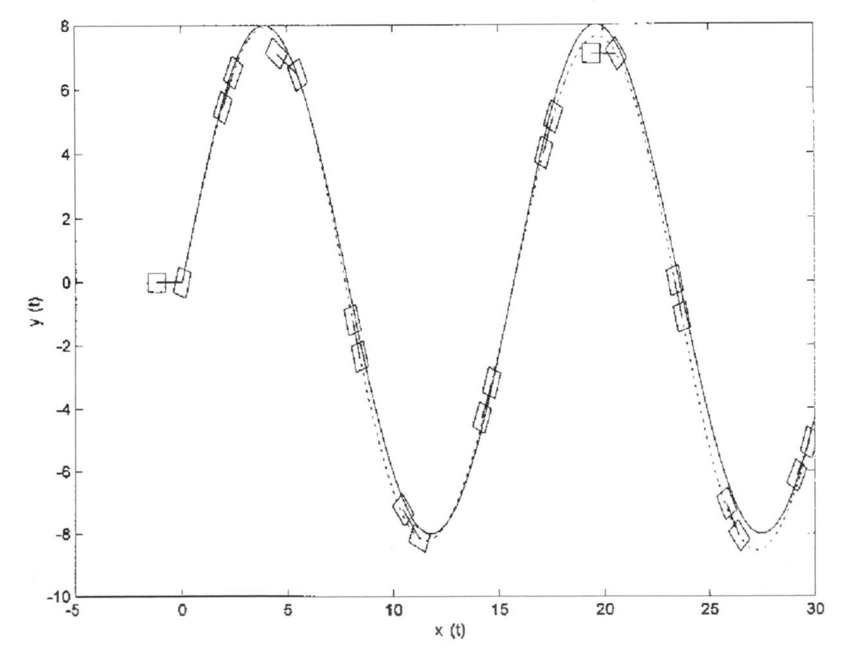

FIGURE 5.5
Trajectory of the wheeled vehicle in X–Y plane when $\rho = 0.2$.

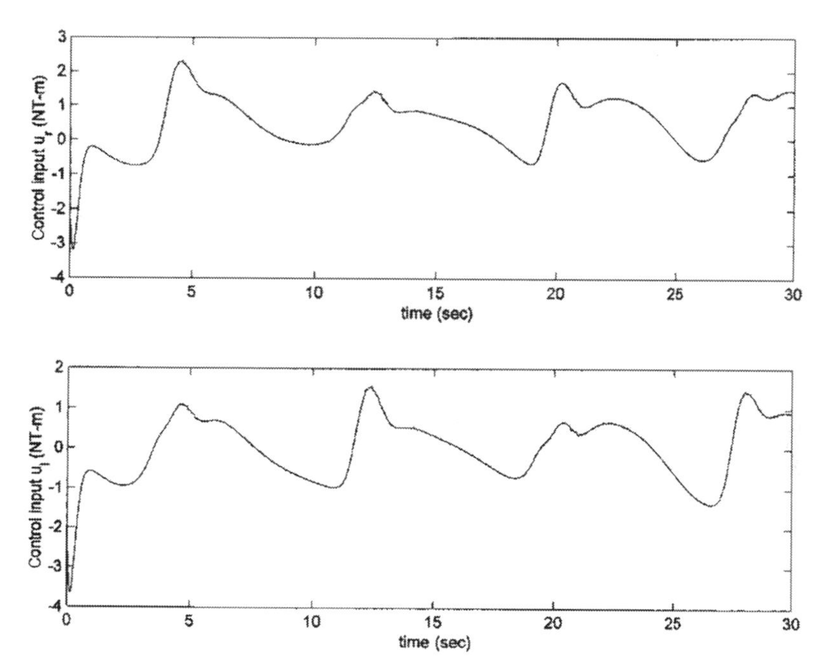

FIGURE 5.6
Control inputs applied to the wheeled vehicle when $\rho = 0.2$.

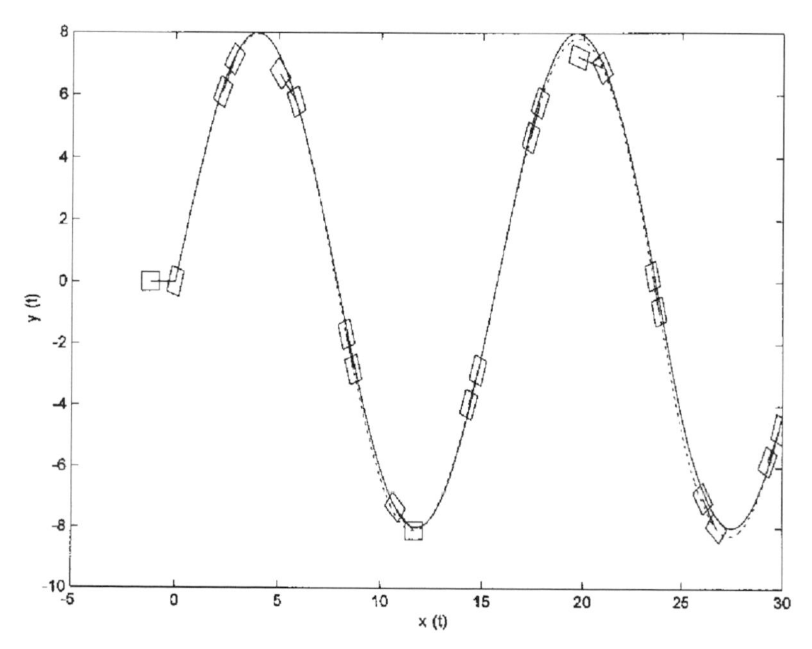

FIGURE 5.7
Trajectory of the wheeled vehicle in X–Y plane when $\rho = 0.1$.

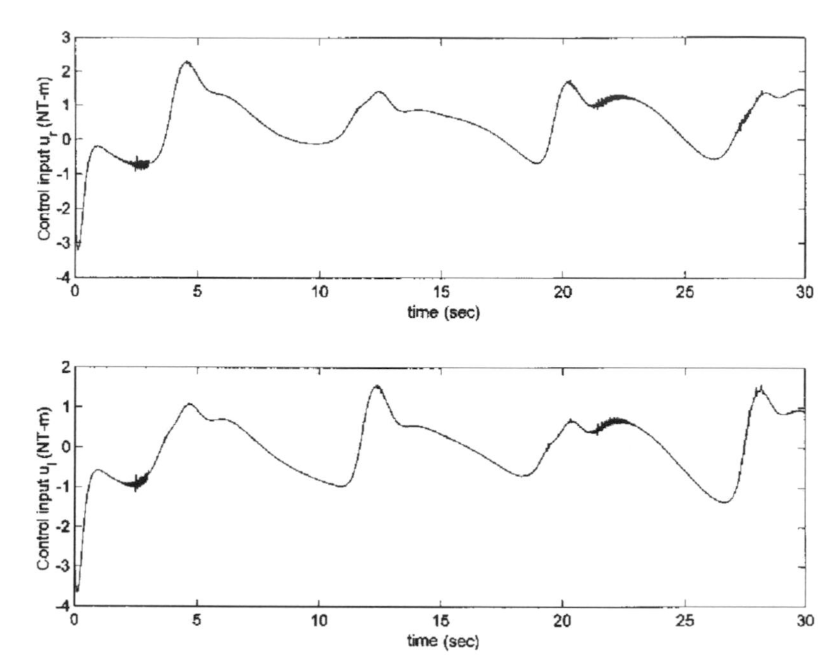

FIGURE 5.8
Control inputs applied to the wheeled vehicle when $\rho = 0.1$.

chapter, the bounds of the system uncertainties are not needed and the control values are smoother and acceptable.

5.7 Conclusion

In this chapter, an adaptive fuzzy cancelation technique and aminimax H_∞ game (or minimum H_∞) attenuation technique are used for rough-tuning and fine-tuning, respectively, to treat the robust H_∞ adaptive tracking control of uncertain wheeled vehicle systems. The enhancement of robust tracking performance is guaranteed by the proposed adaptive fuzzy-based minimax H_∞ game tracking control scheme from a more theoretical design perspective. Unlike the conventional nonlinear adaptive control, the uncertain dynamics need not be of linear parametrization in this chapter. Furthermore, the adaptive feedback linearization scheme is avoided and an inverse of adaptive matrix becomes unnecessary in our proposed control algorithm. The solvability of this robust adaptive minimax H_∞ game tracking control problem is also illustrated. It has been observed that a desired minimax H_∞ game tracking performance can be achieved if weighting matrix R on control signal is adequately specified.

Actually, the proposed adaptive fuzzy-based minimax H_∞ game tracking method can be applied to robust adaptive tracking control design of any uncertain nonlinear nonholonomic constraint system. With the aid of adaptive fuzzy elimination algorithm, the minimax H_∞ game tracking control design can be extended from exactly known

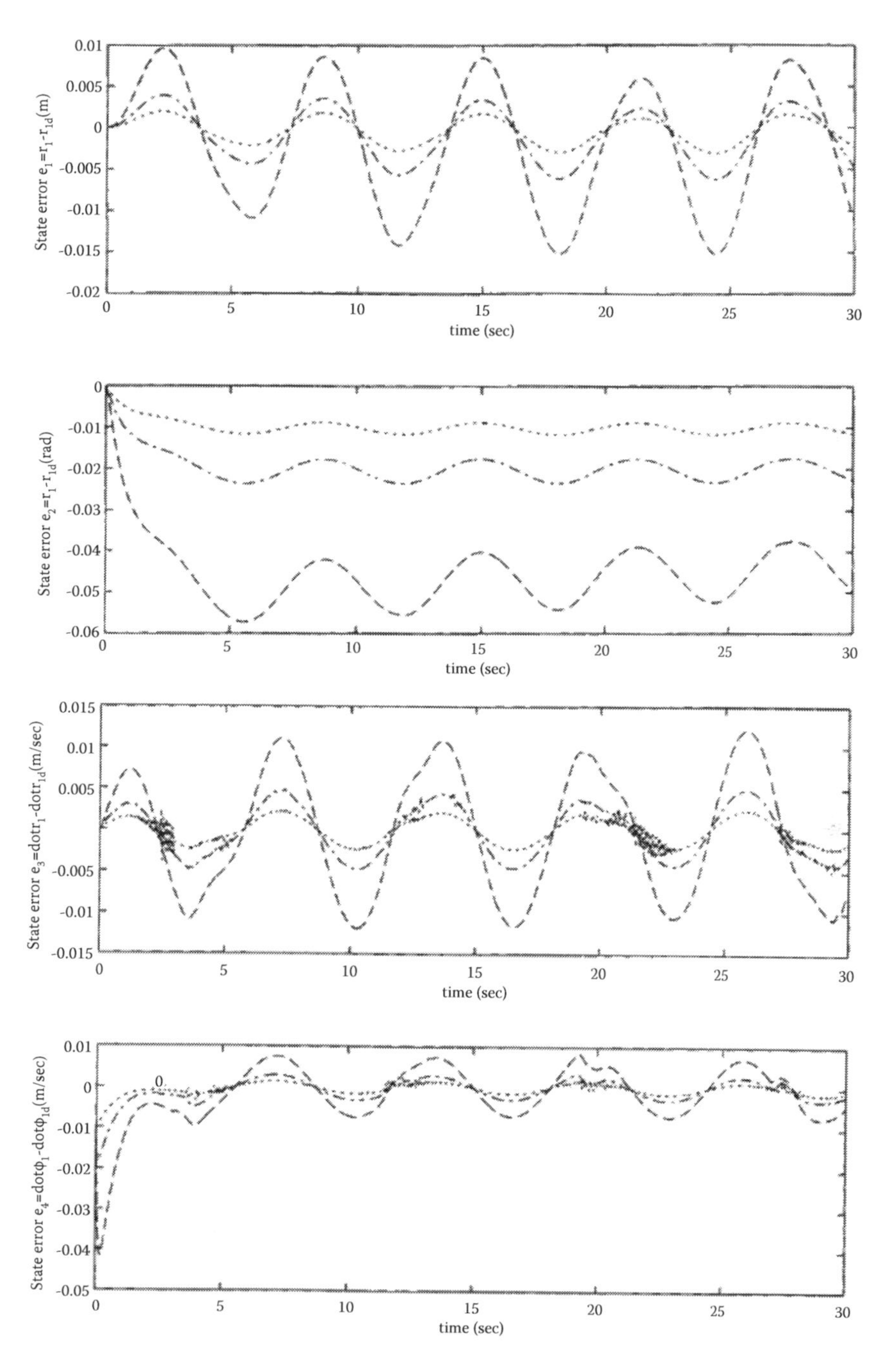

FIGURE5.9
State errors of the wheeled vehicle: "–" for $\rho = 0.5$, "-.-." for $\rho = 0.2$, and "…" for $\rho = 0.1$.

systems to uncertain systems. Furthermore, with the use of minimax game attenuation technique, the H_∞ game tracking performance of the fuzzy adaptive control design for uncertain nonlinear systems can be significantly improved. Therefore, the proposed adaptive minimax H_∞ game design algorithm is suitable for practical control design of

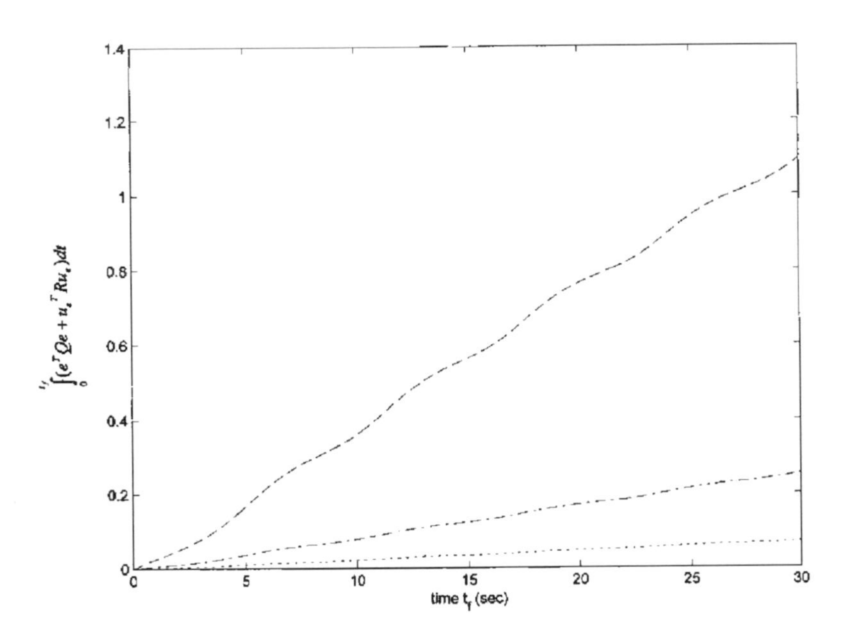

FIGURE 5.10
H_∞ game tracking performances of the wheeled vehicle: "–" for $\rho = 0.5$, "-.-." for $\rho = 0.2$, and "..." for $\rho = 0.1$.

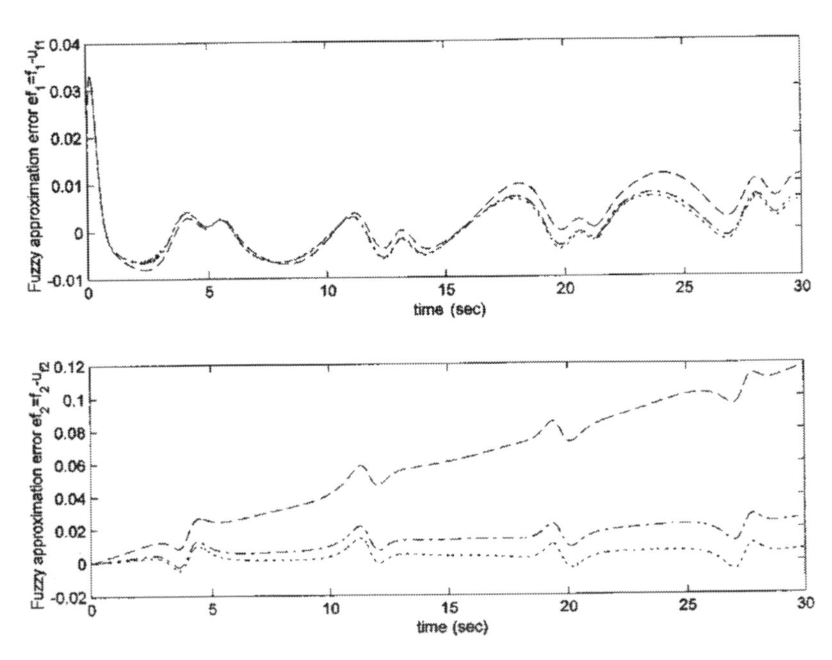

FIGURE 5.11
Fuzzy adaptive approximation errors: "–" for $\rho = 0.5$, "-.-." for $\rho = 0.2$, and "..." for .

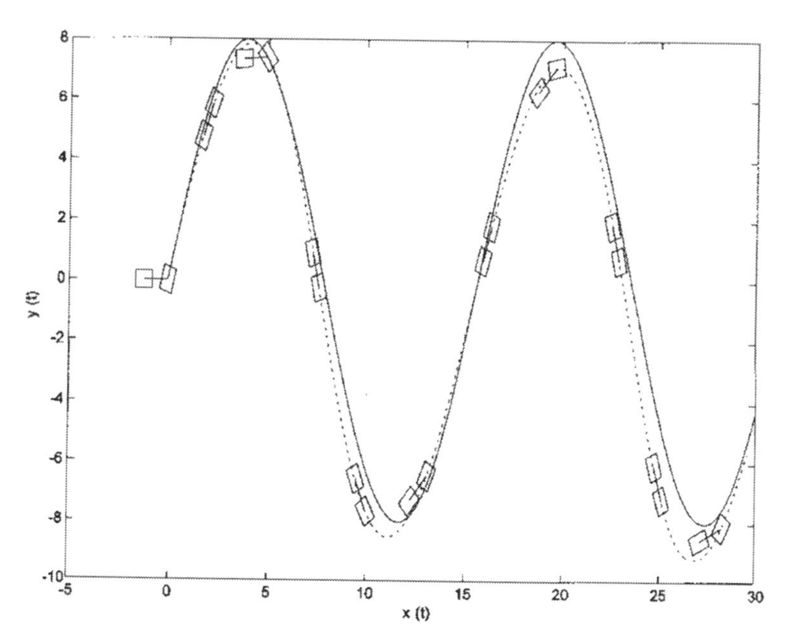

FIGURE 5.12
Trajectory of the wheeled vehicle in X–Y plane using the sliding-mode control method.

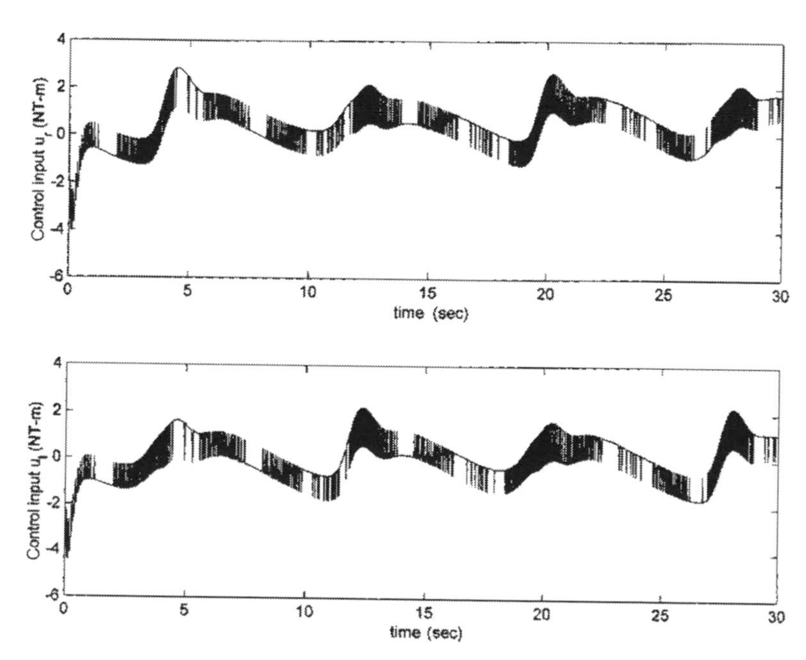

FIGURE 5.13
Control inputs applied to the wheeled vehicle using the sliding-mode control method.

uncertain wheeled vehicle systems. The proposed method is simple and the results are less conservative than the other methods. The simulation results clearly reveal that the robust tracking performance of uncertain wheeled vehicle systems can be enhanced by using the proposed adaptive minimax H_∞ game control method.

5.8 Appendix

Boundedness of $e(t)$ and $\eta(t)$

From (5.41) and (5.48), we get

$$\begin{aligned}\dot{e}(t) &= Ae(t) + Bu_e(t) + B\Xi^T\tilde{\Theta}(t) + Bw(t)\\ &= A_r e(t) + B\Xi^T\tilde{\Theta}(t) + Bw(t)\end{aligned} \tag{5.59}$$

where $A_r = A - BR^{-1}B^TP$. The solution of (5.59) can be obtained as

$$e(t) = e^{A_r t}e(0) + \int_0^t [e^{A_r(t-\tau)}(B\Xi^T\tilde{\Theta}(\tau) + Bw(\tau))]d\tau. \tag{5.60}$$

By the asymptotical stability of A_r, we have

$$\|e^{A_r t}\| \le \beta\exp(-\alpha t) \tag{5.61}$$

for some positive scales α and β, where $\|\cdot\|$ denotes the induced norm operator. Taking the norm operation on the both sides of (5.60), we obtain

$$\|e(t)\| \le \|e^{A_r t}\| \cdot \|e(0)\| \int_0^t [\|e^{A_r(t-\tau)}\| \|B\|(\|\Xi\| \cdot \|\tilde{\Theta}(\tau)\| + \|w(\tau)\|)]d\tau. \tag{5.61}$$

By the inequality in (5.61), we get

$$\begin{aligned}\|e(t)\| &\le \beta e^{-\alpha t}\|e(0)\| + \int_0^t [\beta e^{-\alpha(t-\tau)}\|B\| \times (\|\Xi\| \cdot \|\tilde{\Theta}(\tau)\| + \|w(\tau)\|)]d\tau\\ &\le \beta e^{-\alpha t}\|e(0)\| + \sup_{\tau\in[0,\infty]} [\|B\|(\|\Xi\| \cdot \|\tilde{\Theta}(\tau)\| + \|w(\tau)\|)] \times \int_0^t [\beta e^{-\alpha(t-\tau)}]d\tau\\ &\le \beta e^{-\alpha t}\|e(0)\| + \sup_{\tau\in[0,\infty]} [\|B\|(\|\Xi\| \cdot \|\tilde{\Theta}(\tau)\| + \|w(\tau)\|)] \times \left[\frac{\beta}{\alpha}(1 - e^{-\alpha t})\right].\end{aligned} \tag{5.62}$$

Then, the tracking error $e(t)$ is bounded by the following inequality:

$$\|e(t)\| \le \beta\|e(0)\| + \frac{\beta}{\alpha} \times \sup_{\tau\in[0,\infty]} [\|B\|(\|\Xi\| \cdot \|\tilde{\Theta}(\tau)\| + \|w(\tau)\|)]$$

for all t since $\|\Xi\|$ and $\|\tilde{\Theta}(t)\|$ are bounded (see (5.39), (5.40), (5.42) and the projection algorithm in (5.58)) and the norm $\|w(t)\|$ is assumed to be bounded also (see Assumption 5.2). Furthermore, by the fact

$$\eta = e + \eta_d \tag{5.63}$$

and the bounded assumption of the reference trajectory η_d, the state vector η is bounded.

6

Stochastic Nash Game Approach to Stochastic H_2/H_∞ Robust Control Design for Linear Systems with State-Dependent Wiener Noise and External Disturbance

6.1 Introduction

In the past decades, the mixed H_2/H_∞ control problem for deterministic systems has become a popular research topic for robust stabilization and optimal control. It has attracted much attention and has been widely applied to various fields; see [50,71–75] and the references therein. In recent years, some researchers have turned their attentions to the stochastic H_∞ robust control problem. For example, on the systems modeled by Itô's stochastic differential equation, a class of very general linear stochastic H_∞ problem with state- and control-dependent Wiener noise was studied by [76], and a stochastic bounded real lemma was derived, which has important applications in the robust stochastic H_∞ filtering design with state-dependent Wiener noise (see, e.g., [77]). Authors in [78] treated with the robust H_∞ control in the presence of stochastic uncertainty. H_2 and H_∞-control (H_2 and H_∞ filtering) are discussed in [79] for Markov jump linear systems, and H_∞ control (filtering) for discrete-time stochastic systems is discussed in [80]. In this chapter, we will introduce a kind of stochastic H_2/H_∞ control design with state-dependent Wiener noise for Itô's differential systems.

In the last decades, H_∞ control is an important robust control design for eliminating efficiently the effect of disturbance $v(t)$, and has been widely employed to deal with robust control design problem for the systems with uncertain disturbance. Obviously, there may be more than one solutions to H_∞ control problem with a desired robustness. In engineering practice, we want the control $u(t)$ not only to eliminate the effect of disturbance, but also to minimize a desired control performance when the worst-case disturbance $v^*(t,x)$ is imposed. Since the optimal H_2 performance is more appealing for control engineering design, it naturally leads to the mixed H_2/H_∞ robust control problem [50,71,74]. If the solution $(u^*(t,x), v^*(t,x))$ of the above H_2/H_∞ design exists, then we say the H_2/H_∞ robust control problem has a pair of solutions $(u^*(t,x), v^*(t,x))$.

In deterministic H_∞ theory, H_∞ norm is defined by a norm of the rational transfer matrix, which cannot be directly generalized to nonlinear or stochastic systems [76,81]. However, from the view of time-domain, a norm of the transfer function is the same as L_2-induced norm of the input–output operator with zero initial state zero, and this important feature makes it possible to develop the nonlinear or stochastic H_∞ control theory. In [76], the stochastic H_∞ norm is given by a norm of the perturbation operator L, which measures the worst-case effect that the stochastic disturbance may have on the controlled output z.

One of the important approaches solving H_2/H_∞ robust control problems belongs to the Nash game theory [71,82,83]. By constructing two performances $J_1(u,v)$ and $J_2(u,v)$ associated with the H_∞ robustness and H_2 regulation optimization, respectively, the H_2/H_∞ robust control design can be converted into finding the Nash equilibria point (u^*,v^*), such that [71]

$$J_1(u^*,v^*) \leq J_1(u^*,v),\ J_2(u^*,v^*) \leq J_2(u,v^*).$$

In [71], by using a Nash game approach, the mixed H_2/H_∞ robust control problem of deterministic linear systems was solved, and the necessary and sufficient conditions were presented in terms of the existence of solutions of a cross-coupled Riccati equations. Because the results of [71], on the one hand, are very elegant in theory, and on the other hand, the cross-coupled Riccati equations may be solved by a standard numerical integration, it has become a popular method in the mixed H_2/H_∞ control design. The method used in [71] has been generalized to the nonlinear [84] and output feedback H_2/H_∞ robust control [85].

Up to now, few results have been obtained on the stochastic problem with state- or control-dependent Wiener noise. This chapter will make an elementary discussion on the stochastic H_2/H_∞ robust control, especially, extending the results of [71] on the deterministic H_2/H_∞ control problem to the stochastic H_2/H_∞ control problem of the stochastic systems governed by Itô differential equations with state-dependent Wiener noise. The stabilization, estimation, and control problems for the systems with state-dependent noise have been studied by [86] and [87]. To investigate our stochastic H_2/H_∞ control problem, some concepts such as stabilization, exact observability, and stochastic detectability should be introduced for stochastic system with state-dependent Wiener noise.

In order to develop a parallel stochastic H_2/H_∞ robust control theory to that of [71], two essential difficulties arise, i.e., how to extend Lemma 4 in [50] and Lemma 2.2 in [71] in deterministic system to stochastic systems. By utilizing a comparison theorem on the algebraic Riccati equation (ARE) [88], stochastic bounded real lemma [76], and the standard theory of differential equations, we could overcome these two difficulties to obtain two more general results (Lemmas 6.5 and 6.7). Mainly based on Lemmas 6.5 and 6.7, the infinite and finite horizon stochastic H_2/H_∞ state feedback control problems are solved, respectively. When the state variables cannot be measured directly, how to design a stochastic H_2/H_∞ controller based on the available information is very valuable in practice, and has been studied in the stochastic H_∞ [76] and multiobjective H_2/H_∞ control [85]. By solving a two-step convex optimization problem in this chapter, an observer-based suboptimal stochastic H_2/H_∞ output feedback control design is developed for the stochastic systems with state-dependent Wiener noise and external disturbance.

This chapter consists of six sections. In Section 6.2, some definitions and lemmas are introduced. Section 6.3 describes the infinite horizon stochastic H_2/H_∞ robust control problem, which could be solved by a pair of coupled AREs. Section 6.4 discusses the finite-horizon stochastic H_2/H_∞ robust control problem. Section 6.5 treats with the observer-based design for the stochastic H_2/H_∞ robust control problem. Finally, Section 6.6 ends this chapter with some discussion remarks.

For convenience, we adopt the following notations:

A^T	transpose of a matrix or vector A;
$A \geq 0 (A > 0)$	positive semidefinite (positive definite) symmetric matrix A;
$L^2_F(R_+, R^l)(L^2_F[0,T], R^l)$	space of nonanticipative stochastic processes $y(t) \in R^l$ with respect to an increasing σ-algebras $F_t (t \geq 0)$ satisfying $E\int_0^\infty \|y(t)\|^2 dt < \infty (E\int_0^T \|y(t)\|^2 dt < \infty)$;
I	identity matrix.

In addition, we make the following assumption:

Assumption 6.1 All matrices mentioned in this chapter are real constant except being especially pointed out.

6.2 System Descriptions and Preliminaries

We first introduce the stabilizability, which is an essential assumption in this chapter.

Definition 6.1 The following stochastic controlled system governed by Itô's equation

$$dx = (Fx + G_1 u)dt + Gxdw_1 \quad x(0) = x_0 \tag{6.1}$$

is called stabilizable (in the mean square sense), if there exists a feedback control $u(t) = Kx(t)$, such that for any $x_0 \in \Re^n$, the closed-loop system

$$dx = (F + G_1 K)x(t)dt + Gx(t)dw_1(t) \quad x(0) = x_0 \tag{6.2}$$

is asymptotically mean square stable, i.e.,

$$\lim_{t\to\infty} Ex(t)x^T(t) = 0$$

where K is a constant matrix.

In (6.1), $x(t)$ and $u(t)$ are called the state and input, respectively. Without loss of generality, throughout this chapter, we assume w_1 to be one-dimensional standard Wiener process defined on the filtered probability space (Ω, F, F_t, P). When system (6.1) is stabilizable, we also call (F, G_1, G) stabilizable. We deem (F, G) stable, if $dx = Fxdt + Gxdw_1$ is asymptotically mean square stable. From the result of [89], (F, G) is stable $\Rightarrow F$ is stable. Conversely, it is not true.

Exact observability [90] and stochastic detectability are also essential concepts, which play important roles in this chapter.

Definition 6.2 Consider the following stochastic system with measurement equation:

$$\begin{cases} dx(t) = Ax(t)dt + A_1 x(t)dw_1(t) \quad x(0) = x_0 \in \Re^n \\ y(t) = Cx(t). \end{cases} \tag{6.3}$$

We call (6.3) or $[A, A_1|C]$ exactly observable, if

$$y(t) \equiv 0 \quad \forall t \geq 0 \Rightarrow x_0 = 0. \tag{6.4}$$

From Th. 4.2 in [90], the complete observability of (A, C) implies the exact observability of $[A, A_1|C]$. However, the inverse is not true.

Definition 6.3 System (6.3) or $[A, A_1|C]$ is called stochastically detectable, if there exists a real matrix H of suitable dimension, such that

$$dx = (A + HC)xdt + A_1 x dw_1 \tag{6.5}$$

is asymptotically mean square stable.

Definition 6.3 is the generalization of the detectability of deterministic systems to stochastic context. We should note that Definitions 6.2 and 6.3 are not equivalent; therefore, we cannot say that one of them implies the other one.

Remark 6.1 In (6.5), let $X(t) = Ex(t)x^T(t)$, then by Itô's formula, we have

$$\dot{X}(t) = (A + HC)X(t) + X(t)(A + HC)^T + A_1 X(t) A_1^T.$$

Obviously, $\lim_{t\to\infty} Ex(t)x^T(t) = 0 \Leftrightarrow \lim_{t\to\infty} X(t) = 0$. According to the results of [91], stochastic detectability of $[A, A_1|C]$ is equivalent to (C, A) MS-detectability with $\Gamma(X(t)) = A_1 X(t) A_1^T$ in [91, eq. (3.1)].

Lemma 6.1 [88]: If $[A, A_1|C]$ is exactly observable, then (A, A_1) is stable if the following Lyapunov-type equation

$$PA + A^T P + A_1^T P A_1 = -C^T C \tag{6.6}$$

has a unique positive-definite solution $P > 0$.

Proof: See Appendix A.

Lemma 6.1 extends the related property on Lyapunov equation $PA + A^T P = -C^T C$. Corresponding to Lemma 6.1, under the condition of stochastic detectability, we have the following result.

Lemma 6.2 If $[A, A_1|C]$ is stochastically detectable, then (A, A_1) is stable if (6.6) has a unique solution $P \geq 0$.

Proof: Due to Remark 6.1, this lemma is a direct corollary of [91, Lemma 3.2].

Lemma 6.3 Assume $\gamma \neq 0$

$$\tilde{A}_2 = \begin{bmatrix} C \\ \gamma^{-1} B_1^T P_1 \\ B_2^T P_2 \end{bmatrix} \qquad \tilde{A}_3 = \begin{bmatrix} C \\ B_2^T P_2 \end{bmatrix}.$$

i) If $[A, A_1|C]$ is exactly observable, then so is $[A - B_2 B_2^T P_2, A_1|\tilde{A}_2]$.
ii) If $[A - \gamma^{-2} B_1 B_1^T P_1, A_1|C]$ is exactly observable, then so is $[A - \gamma^{-2} B_1 B_1^T P_1 - B_2 B_2^T P_2, A_1|\tilde{A}_3]$.

Proof: See Appendix B.

Lemma 6.4 Under the conditions of **Lemma 6.3**, we have the following result:

1) If $[A, A_1|C]$ is stochastically detectable, then so is $[A - B_2B_2^TP_2, A_1|\tilde{A}_2]$
2) If $[A - \gamma^{-2}B_1B_1^TP_1, A_1|C]$ is stochastically detectable, then so is $[A - \gamma^{-2}B_1B_1^TP_1 - B_2B_2^TP_2, A_1|\tilde{A}_3]$

Proof: By Definition 6.3, there exists a real matrix H, such that $(A + HC, A_1)$ is stable by the stochastic detectability of $[A, A_1|C]$. If we set $\tilde{H} = [H \quad 0 \quad B_2]$, then $A - B_2B_2^TP_2 + \tilde{H}\tilde{A}_2 = A + HC$. So $(A - B_2B_2^TP_2 + \tilde{H}\tilde{A}_2, A_1)$ is stable, and then (6.1) is proved. Repeating the same procedure, (6.2) can be shown.

Consider the following stochastic perturbed system:

$$\begin{cases} dx(t) = (A_{11}x(t) + B_{11}v(t))dt + A_{12}x(t)dw_1 \\ z_1(t) = C_{11}x(t, v, x_0) \end{cases} \tag{6.7}$$

where $x(0) = x_0 \in \Re^n$, $z_1 \in L_F^2(\Re_+, \Re^{n_{z_1}})$. Define the perturbation operator of (6.7) as

$$\tilde{L}(v) = C_{11}x(t, v, 0), \quad t \geq 0, \quad v \in L_F^2(\Re_+, \Re^{n_v})$$

$$\begin{aligned} \|\tilde{L}\|_\infty &= \sup_{\substack{v \in L_F^2(\Re_+, \Re^{n_v}) \\ v \neq 0, x_0 = 0}} \frac{\|z_1\|_2}{\|v\|_2} \\ &= \sup_{\substack{v \in L_F^2(\Re_+, \Re^{n_v}) \\ v \neq 0, x_0 = 0}} \frac{\{E\int_0^\infty x^TC_{11}^TC_{11}x dt\}^{1/2}}{\{E\int_0^\infty v^Tv dt\}^{1/2}}. \end{aligned}$$

The following lemma generalizes Lemma 4 in [50], which is a very important result in the deterministic H_∞ control theory, to a stochastic case.

Lemma 6.5 For stochastic system (6.7), if it is internally stable [76] (in the mean square sense with $v = 0$) and $\|\tilde{L}\|_\infty < \gamma$ for some $\gamma > 0$, then there exists a solution $P_1 \leq 0$ to the following ARE:

$$PA_{11} + A_{11}^TP + A_{12}^TPA_{12} - \gamma^{-2}PB_{11}B_{11}^TP - C_{11}^TC_{11} = 0. \tag{6.8}$$

Moreover, $(A_{11} - \gamma^{-2}B_{11}B_{11}^TP_1, A_{12})$ is stable.

Proof: This is a direct corollary of Cor. 2.14 in [76] and a comparison theorem on the generalized AREs [87]. For convenience, see Appendix C for the proof.

Lemma 6.6 In the stochastic system of (6.7), if (A_{11}, B_{11}, A_{12}) is stabilizable, then we have the following result.

1) ARE

$$PA_{11} + A_{11}^TP + A_{12}^TPA_{12} + C_{11}^TC_{11} - PB_{11}B_{11}^TP = 0 \tag{6.9}$$

has a maximal solution $P_2 \geq 0$.
2)

$$\min_{v \in L_F^2(\Re_+, \Re^{n_v})} \left\{ E \int_0^\infty (x^T C_{11}^T C_{11} x + v^T v) dt \right\} = x_0^T P_2 x_0 \tag{6.10}$$

and the optimal value v^* is given by

$$v^*(t, x) = -B_{11}^T P_2 x(t)$$

3) If $[A_{11}, A_{12} | C_{11}]$ is exactly observable (stochastically detectable), then $P_2 > 0 (P_2 \geq 0)$ is the unique real solution of (6.9).

Proof: 1) and 2) are immediately derived from Corollary 4 and Th. 12 in [92]. Note that (6.9) can be written as

$$P(A_{11} - B_{11}B_{11}^T P) + (A_{11} - B_{11}B_{11}^T P)^T P + A_{12}^T P A_{12} = -[C_{11}^T \quad PB_{11}] \begin{bmatrix} C_{11} \\ B_{11}^T P \end{bmatrix}.$$

$[A_{11} - B_{11}B_{11}^T P, A_{12} | [C_{11}^T PB_{11}]^T]$ is also exactly observable (stochastically detectable) due to the exact observability (stochastic detectability) of $[A_{11}, A_{12} | C_{11}]$. So, 3) is a direct corollary of Lemmas 6.1 and 6.2.

6.3 Stochastic Nash Game Approach to Infinite Horizon Stochastic H_2/H_∞ Control

Consider the following stochastic linear system:

$$\begin{cases} dx(t) = (Ax(t) + B_2 u(t) + B_1 v(t))dt + A_1 x(t) dw_1 \\ z(t) = \begin{bmatrix} Cx(t) \\ Du(t) \end{bmatrix} \end{cases} \tag{6.11}$$

where $x(0) = x_0$ and $D^T D = I$. For $0 < T < \infty$, by [93], when $(u, v, x_0) \in L_F^2([0, T], \Re^{n_v}) \times \Re^n$, there exists a unique solution $x(t) = x(t, u, v, x_0) \in L_F^2([0, T], \Re^n)$.

We view $v(t)$ and $z(t)$ as the disturbance and controlled output of the system (6.11), respectively. The infinite horizon stochastic H_2/H_∞ control problem can be stated as follows.

Definition 6.4: Given disturbance attenuation $\gamma > 0$, to find $u^*(t, x) \in L_F^2(\Re_+, \Re^{n_v})$, such that

1)

$$\|L\|_\infty = \sup_{\substack{v \in L_F^2(\Re_+, \Re^{n_v}) \\ v \neq 0, x_0 = 0}} \frac{\|z\|_2}{\|v\|_2}$$

$$:= \sup_{\substack{v \in L_F^2(\Re_+, \Re^{n_v}) \\ v \neq 0, x_0 = 0}} \frac{\{E \int_0^\infty (x^T C^T C x + u^{*T} u^*) dt\}^{1/2}}{\{E \int_0^\infty v^T v dt\}^{1/2}} < \gamma$$

where

$$L(v) = \begin{bmatrix} Cx(t,u^*,v,0) \\ Du^* \end{bmatrix}$$

is called the perturbation operator of (6.11).

2) $u^*(t,x)$ stabilizes system (6.11) internally, i.e, $\lim_{t\to\infty} E\|x(t,u^*,0,x_0)\|^2 = 0$.

3) When the worst-case disturbance $v^*(t,x) \in L_F^2(\Re_+, \Re^{n_v})$, if it exists, is applied to system (6.11), $u^*(t,x)$ minimizes the output energy [84]

$$J_2^\infty(u,v^*) = \|z\|_2^2 = E\int_0^\infty (x^T C^T x + u^T u)dt.$$

Here, $v^*(t,x)$ is called a worst-case disturbance in the sense that

$$v^*(t,x) = \arg\min_v J_1^\infty(u^*,v) \qquad \forall x_0 \in \Re^n$$
$$\forall v \in L_F^2(\Re_+, \Re^{n_v})$$

with

$$J_1^\infty(u^*,v) = E\int_0^\infty (\gamma^2 v^T v - z^T z)dt.$$

If an admissible control $u(t,x) \in L_F^2(\Re_+, \Re^{n_v})$ only satisfies 1) and 2), then this $u(t,x)$ is called a solution to H_∞ control of (6.11). Obviously, there may be more than one solution to the H_∞ control problem. If the previous (u^*,v^*) exists, then we say that the infinite horizon H_2/H_∞ control admits a pair of solutions.

Remark 6.2 The infinite horizon stochastic H_2/H_∞ control is associated with the two-player, nonzero-sum Nash game strategies (u^*,v^*) solved by the following Nash equilibrium solution (see [71,83,84]):

$$J_1^\infty(u^*,v^*) \le J_1^\infty(u^*,v) \tag{6.12}$$

and

$$J_2^\infty(u^*,v^*) \le J_2^\infty(u,v^*) \tag{6.13}$$

To guarantee the unique global Nash game solution in (6.12) and (6.13), the both players are only allowed to use linear, memoryless, constant state feedback control. From our following theorems, one can see that if the above u^* also stabilizes (6.11) internally, and $J_1^\infty(u^*,v^*) \ge 0$, then u^* is a solution to the stochastic H_2/H_∞ control, and v^* is the corresponding worst-case disturbance.

In this section, we discuss the infinite horizon stochastic H_2/H_∞ robust control problem. Under the condition of exact observability, two results of stochastic H_2/H_∞ robust control for stochastic linear system in (6.11) are obtained, which extend [73, Th. 3.1] to the stochastic case.

Theorem 6.1
For the linear stochastic system in (6.11), suppose the following coupled AREs

$$A^T P_1 + P_1 A + A_1^T A_1 - C^T C - [P_1 \quad P_2] \begin{bmatrix} \gamma^{-2} B_1 B_1^T & B_2 B_2^T \\ B_2 B_2^T & B_2 B_2^T \end{bmatrix} \times \begin{bmatrix} P_1 \\ P_2 \end{bmatrix} = 0 \tag{6.14}$$

$$A^T P_2 + P_2 A + A_1^T P_2 A_1 + C^T C - [P_1 \quad P_2] \begin{bmatrix} 0 & \gamma^{-2} B_1 B_1^T \\ \gamma^{-2} B_1 B_1^T & B_2 B_2^T \end{bmatrix} \times \begin{bmatrix} P_1 \\ P_2 \end{bmatrix} = 0 \tag{6.15}$$

have a pair of solutions $(P_1, P_2), P_1 < 0, P_2 > 0$. If $[A, A_1|C]$ and $[A - \gamma^{-2} B_1 B_1^T P_1, A_1|C]$ are exactly observable, then the stochastic H_2/H_∞ robust control problem admits a pair of solutions

$$u^*(t,x) = -B_2^T P_2 x(t), \qquad v^*(t,x) = -\gamma^{-2} B_1^T P_1 x(t).$$

That is, i) $(u^*, v^*) \in L_F^2(\Re_+, \Re^{n_u}) \times L_F^2(\Re_+, \Re_v^n)$, and $(A - B_2 B_2' P_2, A_1)$ is stable. ii) $\|L\|_\infty < \gamma$. iii) u^* minimizes the output energy $\|z\|_2^2$ when v^* is applied in the stochastic linear system in (6.11), i.e.,

$$u^* = \arg\min_u J_2^\infty(u, v^*) \ \forall u \in L_F^2(\Re_+, \Re^{n_u})$$

Proof: Note that (6.14) and (6.15) can be rearranged as

$$P_1(A - B_2 B_2^T P_2) + (A - B_2 B_2^T P_2)^T P_1 + A_1^T P_1 A_1 = \tilde{A}_2^T \tilde{A}_2 \tag{6.16}$$

and

$$P_2(A - B_2 B_2^T P_2 - \gamma^{-2} B_1 B_1^T P_1) + (A - B_2 B_2^T P_2 - \gamma^{-2} B_1 B_1^T P_1)^T P_2 + A_1^T P_2 A_1 = -\tilde{A}_3^T \tilde{A}_3 \tag{6.17}$$

respectively, where $\tilde{A}_2$ and $\tilde{A}_3$ are defined in Lemma 6.3. By Lemma 6.3, $[A - B_2 B_2^T P_2 - \gamma^{-2} B_1 B_1^T P_1, A_1 | \tilde{A}_3]$ is exactly observable. So from Lemma 6.1, (6.17) yields $(A - B_2 B_2^T P_2 - \gamma^{-2} B_1 B_1^T P_1, A_1)$ being stable. Hence, the stochastic Nash game solution $(u^*(t,x), v^*(t,x)) = (-B_2^T P_2 x(t), -\gamma^{-2} B_1^T P x(t)) \in L_F^2(\Re_+, \Re^{n_u}) \times L_F^2(\Re_+, \Re^{n_v})$. Second, from Lemmas 6.1 and 6.3, (6.16) yields $(A - B_2 B_2^T P_2, A_1)$ being stable, i.e., (6.11) is internally stabilizable by $u(t) = u^*(t,x) = -B_2^T P_2 x(t)$; i) is proved.

Substituting $u(t) = u^*(t,x) = -B_2^T P_2 x(t)$ into (6.11) gives

$$\begin{cases} dx(t) = ((A - B_2 B_2^T P_2) x(t) + B_1 v(t)) dt + A_1 x(t) dw_1 \\ z(t) = \begin{bmatrix} Cx(t) \\ D B_2^T P_2 x(t) \end{bmatrix} \end{cases} \tag{6.18}$$

where $x(0) = x_0$. Since $(A - B_2 B_2^T P_2, A_1)$ is stable, and $v \in L_F^2(\Re_+, \Re^{n_v})$, we have $x(t) \in L_F^2(\Re_+, \Re^n)$ from Rem. 2.6 in [76], where $x(t)$ is the trajectory of (6.18). Applying Itô's formula to (6.18) and considering (6.14), by completing the square, we have

$$
\begin{aligned}
J_1^\infty(u^*, v) &= E\int_0^\infty (\gamma^2 v^T v - z^T z)dt \\
&= x_0^T P_1 x_0 + \gamma^2 E\int_0^\infty (v - v^*)^T (v - v^*)dt \\
&\geq J_1^\infty(u^*, v^*) = x_0^T P_1 x_0.
\end{aligned} \tag{6.19}
$$

From which we could see the worst-case disturbance corresponding to $u^*(t,x) = -B_2^T P_2 x(t)$ is given by $v^*(t,x) = -\gamma^{-2} B_1^T P_1 x(t)$. Along the lines of [71] and [82], we define the operator $L_1 : L_F^2(\Re_+, \Re^{n_v}) \mapsto L_F^2(\Re_+, \Re^{n_v})$, $L_1 v(t) = v(t) - v^*(t,x)$ with the realization in the linear stochastic system of (6.11)

$$
\begin{aligned}
&dx(t) = ((A - B_2 B_2^T P_2)x(t) + B_1 v(t))dt + A_1 x(t) dw_1 \quad x(0) = 0 \\
&v(t,x) - v^*(t,x) = v(t,x) - \gamma^{-2} B_1^T P_1 x(t)
\end{aligned}
$$

then, L_1^{-1} exists, which is determined by

$$
\begin{aligned}
&dx(t) = ((A - B_2 B_2^T P_2 - \gamma^{-2} B_1^T P)x(t) + B_1(v(t) - v^*(t,x)))dt + A_1 x(t) dw_1 \\
&x(0) = 0 \\
&v = \gamma^{-2} B_1^T P_1 x(t) + (v - v^*).
\end{aligned}
$$

From (6.19), we have $(x(0) = 0)$

$$
J_1^\infty(u^*, v) = \gamma^2 \|v\|_2^2 - \|z\|_2^2 = \gamma^2 \|L_1 v\|^2 \geq \varepsilon \|v\|^2 > 0
$$

for some $\varepsilon > 0$, which yields $\|L\|_\infty < \gamma$.

Finally, when the worst-case disturbance $v = v^*(t,x) = -\gamma^{-2} B_1^T P_1 x(t)$ is applied to (6.11), we have

$$
\begin{cases}
dx(t) = ((A - \gamma^{-2} B_1 B_1^T P_1)x(t) + B_2 u)dt + A_1 x(t) dw_1 \\
z(t) = \begin{bmatrix} Cx(t) \\ Du(t) \end{bmatrix}
\end{cases} \tag{6.20}
$$

where $x(0) = 0$. Now, the H_2 optimization problem becomes a standard stochastic linear quadratic optimal control problem, that is how to minimize the H_2 performance $\min_{u \in L_F^2(\Re_+, \Re^{n_u})} J_2^\infty(u, v^*)$ under the constraint of (6.20). From the stability of $(A - B_2 B_2^T P_2 - \gamma^{-2} B_1 B_1^T P_1, A_1)$, we conclude that $(A - \gamma^{-2} B_1 B_1^T P_1, B_2, A_1)$ is stabilizable. In addition, (6.15) can be written as

$$
P_2(A - \gamma^{-2} B_1 B_1^T P_1) + (A - \gamma^{-2} B_1 B_1^T P_1)^T P_2 + A_1^T P_2 A_1 - P B_2 B_2^T P_2 + C^T C = 0.
$$

Lemma 6.6 immediately yields

$$
\min_{u \in L_F^2(\Re_+, \Re^{n_u})} J_2^\infty(u, v^*) = J_2^\infty(u^*, v^*) = x_0^T P_2 x_0.
$$

This ends the proof.

In fact, by Lemma 6.6, under the condition of exact observability, the cross-coupled AREs (6.14) and (6.15) have at most one pair of solutions (P_1, P_2) with $P_1 < 0$ and $P_2 > 0$.

Theorem 6.2
If the stochastic H_2/H_∞ robust control problem has a pair of Nash equilibrium solutions (u^*, v^*) as follows:

$$u^*(t,x) = K_2 x(t), \quad v^*(t,x) = K_1 x(t)$$

where $u^*(t,x)$ and $v^*(t,x)$ are time-invariant feedback laws. If $[A + B_1 K_1, A_1 | C]$ is exactly observable, then the coupled AREs (6.14) and (6.15) admit solutions $P_1 \leq 0$ and $P_2 > 0$, respectively. Moreover, we can take Nash equilibrium solution as

$$u^*(t,x) = -B_2^T P_2 x(t), \quad v^*(t,x) = -\gamma^{-2} B_1^T P_1 x(t).$$

Proof: By means of Lemma 6.5, this theorem can be proved via following the line of [71]; the details are given as follows. Implementing $u(t,x) = u^*(t,x) = K_2 x(t)$ in (6.11) gives

$$dx(t) = ((A + B_2 K_2) x(t) + B_1 v(t)) dt + A_1 x(t) dw_1 \tag{6.21}$$

$$z(t) = \begin{bmatrix} Cx(t) \\ DK_2 x(t) \end{bmatrix} \tag{6.22}$$

where $x(0) = 0$. By definition, $(A + B_2 K_2, A_1)$ is stable, $\|L\|_\infty < \gamma$. By Lemma 6.5, the following equation

$$P(A + B_2 K_2) + (A + B_2 K_2)^T P + A_1^T P A_1 - C^T C - K_2^T K_2 - \gamma^{-2} P B_1 B_1^T P = 0 \tag{6.23}$$

has a solution $P_1 \leq 0$, and $(A + B_2 K_2 - \gamma^{-2} B_1 B_1^T P_1, A_1)$ is stable.

In addition, since $(A + B_2 K_2, A_1)$ is stable, for any $v \in L_F^2(\Re_+, \Re^{n_v})$, by [78, Rem. 2.6], $x(t) \in L_F^2(\Re_+, \Re^n)$. Consider (6.21) and (6.23), completing the square yields

$$\begin{aligned} J_1^\infty(u^*, v) &= E\int_0^\infty (\gamma^2 v^T v - z^T z) dt \\ &= x_0^T P_1 x_0 + \gamma^2 E\int_0^\infty (v + \gamma^{-2} B_1^T P_1 x)^T (v + \gamma^{-2} B_1^T P_1 x) dt \\ &\geq J_1^\infty(u^*, v^*) \end{aligned}$$

where $v^*(t,x) = -\gamma^{-2} B_1^T P_1 x(t)$, which belongs to $L_F^2(\Re_+, \Re^{n_v})$ due to Lemma 6.5. So, we get

$$K_1 = -\gamma^{-2} B_1^T P_1.$$

Implementing $v(t,x) = v^*(t,x) = -\gamma^{-2} B_1^T P_1 x(t)$ in system (6.11) yields

$$
\begin{aligned}
dx(t) &= ((A - \gamma^{-2}B_1B_1^TP_1)x(t) + B_2u(t))dt + A_1x(t)dw_1 \\
x(0) &= x_0.
\end{aligned}
\tag{6.24}
$$

Since $(A + B_2K_2 - \gamma^{-2}B_1B_1^TP_1, A_1)$ is stable, $(A - \gamma^{-2}B_1B_1^TP_1, B_2, A_1)$ is stabilizable. From the given condition and Lemma 6.3, we know that $[A + B_1K_1, A_1|C]$ =$[A - \gamma^{-2}B_1B_1^TP_1, A_1|C]$ is exactly observable. Applying Lemma 6.6, the following ARE

$$
P_2(A - \gamma^{-2}B_1B_1^TP_1) + (A - \gamma^{-2}B_1B_1^TP_1)^TP_2 + A_1^TP_2A_1 - PB_2B_2^TP_2 + C^TC = 0 \tag{6.25}
$$

has a unique solution $P_2 > 0$, which satisfies

$$
\begin{aligned}
&\min_{u \in L_F^2(\Re_+, \Re^{n_u})} J_2^\infty(u, v^*) = J_2^\infty(u^*, v^*) = x_0^TP_2x_0 \\
&u^*(t, x) = -B_2^TP_2x(t)
\end{aligned}
\tag{6.25}
$$

So, $K_2 = -B_2^TP_2$. It is easy to test that (6.25) is the same as (6.15).

Substituting $K_2 = -B_2^TP_2$ into (6.23) yields (6.14). This ends the proof of Theorem 6.2.

By the same procedure as before, we can present the counterparts of Theorems 6.1 and 6.2 under the condition of stochastic detectability, respectively, as follows.

Theorem 6.3

For (6.11), suppose the coupled AREs in (6.14) and (6.15) have a pair of solutions $(P_1, P_2), P_1 \leq 0, P_2 \geq 0$. If $[A, A_1|C]$ and $[A - \gamma^{-2}B_1B_1^TP_1, A_1|C]$ are stochastically detectable, then the stochastic H_2/H_∞ control problem admits a pair of Nash game solutions

$$
u^*(t, x) = -B_2^TP_2x(t), \quad v^*(t, x) = -\gamma^{-2}B_1^TP_1x(t).
$$

Theorem 6.4

If the stochastic H_2/H_∞ control problem has a pair of Nash equilibrium solutions (u^*, v^*) as follows:

$$
u^*(t, x) = K_2x(t), \quad v^*(t, x) = K_1x(t)
$$

where $u^*(t, x)$ and $v^*(t, x)$ are time-invariant state feedback laws. If $[A + B_1K_1, A_1|C]$ is stochastically detectable, then the coupled Riccati-like equations in Nash game solutions are given by (6.14) and (6.15) and admit solutions $P_1 \leq 0$ and $P_2 \geq 0$, respectively. Moreover,

$$
u^*(t, x) = -B_2^TP_2x(t), \quad v^*(t, x) = -\gamma^{-2}B_1^TP_1x(t).
$$

To obtain the optimal solution u^* for stochastic H_2/H_∞ control problem, we need to solve the coupled Riccati like equations in (6.14) and (6.15), which can be solved by standard numerical integration as pointed out in [71].

6.4 Stochastic Nash Game Approach to Finite Horizon Stochastic H_2/H_∞ Control

In this section, we discuss the finite horizon stochastic H_2/H_∞ control problem of linear stochastic system. In particular, the finite horizon stochastic H_∞ filtering on discrete-time stochastic systems with multiplicative noise was considered in [80].

Consider the following stochastic linear system:

$$\begin{aligned} dx(t) =&(A(t)x(t) + B_2(t)u(t) + B_1(t)v(t))dt \\ &+ A_1(t)x(t)dw_1 + B_0(t)dw_2x(0) = x_0 \\ z(t) =& \begin{bmatrix} C(t)x(t) \\ D(t)u(t) \end{bmatrix}, \ D^T(t)D(t) = I \end{aligned} \tag{6.27}$$

where various matrices are continuous functions of time with suitable dimensions. Without loss of generality, we assume w_1 and w_2 to be mutually independent, one-dimensional standard Wiener processes. The finite horizon stochastic H_2/H_∞ control problem of system (6.27) can be stated as follows:

Definition 6.5 Given the disturbance attenuation $\gamma > 0, 0 \leq T < \infty$, to find a state feedback control $u_T^*(t,x) \in L_F^2([0,T], \Re^{n_u})$, such that with the constraint

$$\begin{aligned} &dx(t) = (A(t)x(t) + B_2(t)u(t) + B_1(t)v(t))dt + A_1(t)x(t)dw_1 \\ &x(0) = x_0 \\ &z(t) = \begin{bmatrix} C(t)x(t) \\ D(t)u(t) \end{bmatrix}, \quad D^T(t)D(t) = I \end{aligned} \tag{6.28}$$

we have

$$1) \ \|L\|_{[0,T]} = \sup_{\substack{v \in L_F^2([0,T],\Re^{n_v}) \\ v \neq 0, x_0 = 0}} \frac{\|z\|_{[0,T]}}{\|v\|_{[0,T]}}$$

$$:= \sup_{\substack{v \in L_F^2([0,T],\Re^{n_v}) \\ v \neq 0, x_0 = 0}} \frac{\{E\int_0^T (x^T C^T Cx + u_T^{*T} u_T^*)dt\}^{1/2}}{\{E\int_0^T v^T v dt\}^{1/2}}$$

$< \gamma$.

2) When the worst-case disturbance $v_T^*(t,x) \in L_F^2([0,T], \Re^{n_v})$, if it exists, is applied to (6.27), u_T^* minimizes the output energy

$$J_2^T(u, v_T^*) = E\int_0^T (x^T C^T Cx + u^T u)dt.$$

Here, the so-called worst-case disturbance v_T^* means that

$$\begin{aligned} v_T^*(t,x) &= \arg\min_v J_1^T(u_T^*, v) \\ &= \arg\min_v E\int_0^T (\gamma^2 v^T v - z^T z)dt, \quad \forall x_0 \in \Re^n \end{aligned}$$

$\forall v \in L_F^2([0,T], \Re^{n_v})$. If the previous (u_T^*, v_T^*) exists, then we say that the finite horizon H_2/H_∞ control has a pair of stochastic Nash equilibrium solutions (u_T^*, v_T^*).

As before, the finite horizon stochastic H_2/H_∞ control problem can be formulated as a stochastic Nash game [71]. In this section, we shall give a necessary and sufficient condition for the existence of the linear, memoryless state feedback pair (u_T^*, v_T^*), which is equivalent to finding the following Nash equilibrium solution:

$$J_1^T(u_T^*, v_T^*) \leq J_1^T(u_T^*, v) \qquad J_2^T(u_T^*, v_T^*) \leq J_2^T(u, v_T^*) \tag{6.29}$$

The following lemma is necessary in the derivation which follows.

Lemma 6.7 For stochastic system (6.7), $\|\tilde{L}\|_{[0,T]} < \gamma$ for some $\gamma > 0$ if the following differential Riccati equation (DRE) (with the time argument suppressed)

$$\begin{cases} \dot{P} + PA_{11} + A_{11}^T P + A_{12}^T P A_{12} - \gamma^{-2} P B_{11} B_{11}^T P - C_{11}^T C_{11} = 0 \\ P(T) = 0 \end{cases} \tag{6.30}$$

has a unique solution $P_1(t) \leq 0$ on $[0,T]$.

Proof: The sufficient part can be proved in the same way as [71]. Next, we prove the necessary condition, i.e., $\|\tilde{L}\|_{[0,T]} < \gamma$ implies (6.30) having a solution $P_1(t)$ on $[0,T]$. Otherwise, by the standard theory of differential equations, there exists a unique solution $P_T(t)$ backward in time on a maximal interval $[T_0, T], T_0 \geq 0$, and as $t \to T_0$, $P_T(t)$ becomes unbounded, i.e., (6.30) exhibits the phenomenon of a finite escape time. We shall show that the existence of a finite escape time will lead to a contradiction. Let $0 < \varepsilon < T - T_0, x(T_0 + \varepsilon) := x_{T_0,\varepsilon} \in \Re^n$

$$\begin{aligned} J_1^T(x, v, x_{T_0,\varepsilon}, T_0 + \varepsilon) : &= E\int_{T_0+\varepsilon}^T (\gamma^2 \|v\|^2 - \|z_1\|^2)dt \\ &= E\int_{T_0+\varepsilon}^T (\gamma^2 \|v\|^2 - x^T C_{11}^T C_{11})dt \end{aligned}$$

then by completing the square

$$\begin{aligned} &J_1^T(x, v, x_{T_0,\varepsilon}, T_0 + \varepsilon) \\ &= x_{T_0,\varepsilon}^T P_1(T_0+\varepsilon) x_{T_0,\varepsilon} - Ex^T(T) P_1(T) x(T) + E\int_{T_0+\varepsilon}^T (\gamma^2\|v\|^2 - \|z_1\|^2)dt + E\int_{T_0+\varepsilon}^T d(x^T P_1 x) \\ &= x_{T_0,\varepsilon}^T P_1(T_0+\varepsilon) x_{T_0,\varepsilon} + E\int_{T_0+\varepsilon}^T (\gamma v + \gamma^{-1} B_{11}^T P_1 x)^T (\gamma v + \gamma^{-1} B_{11}^T P_1 x)dt \\ &+ E\int_{T_0'}^T x^T(\dot{P}_1 + P_1 A_{11} + A_{11}^T P_1 + A_{12}^T P_1 A_{12} - \gamma^{-2} P_1 B_{11} B_{11}^T P_1 - C_{11}^T C_{11}) x dt \\ &= E\int_{T_0+\varepsilon}^T (\gamma v + \gamma^{-1} B_{11}^T P_1 x)^T (\gamma v + \gamma^{-1} B_{11}^T P_1 x)dt + x_{T_0,\varepsilon}^T P_1(T_0+\varepsilon) x_{T_0,\varepsilon}. \end{aligned}$$

So,

$$\begin{aligned}\min_{v\in L_F^2([T_0+\varepsilon,T],R^{n_v})} J_1^T(x,v,x_{T_{0,\varepsilon}},T_0+\varepsilon) &= J_1^T(x,v_T^*,x_{T_{0,\varepsilon}},T_0+\varepsilon) = x_{T_{0,\varepsilon}}^T P_1(T_0+\varepsilon)x_{T_{0,\varepsilon}} \\ &\le J_1^T(x,0,x_{T_{0,\varepsilon}},T_0+\varepsilon) = E\int_{T_0+\varepsilon}^T(-\|z_1\|^2)dt \le 0\end{aligned} \quad (6.31)$$

and the corresponding minimum $v_T^*(t,x)$ is given by

$$v_T^*(t,x) = K_1(t)x(t) = -\gamma^{-2}B_{11}^TP(t)x(t).$$

Inequality (6.31) follows

$$P_1(T_0+\varepsilon) \le 0 \quad (6.32)$$

for any $0<\varepsilon<T-T_0$. Here, we shall show that there exists $C>0$, such that

$$J_1^T(x,v,x_{T_{0,\varepsilon}},T_0+\varepsilon) \ge -C\|x_{T_{0,\varepsilon}}\|^2 \quad \forall v\in L_F^2([T_0+\varepsilon,T],R^{n_v}). \quad (6.33)$$

This can be done by following the line of Lemma 2.10 in [76], but needs a little modification. For convenience, we give a simple illustration in the sequel. By the linearity, the solution $x(t,v,x_{T_{0,\varepsilon}},T_0+\varepsilon)$ of (7) with initial state $x_{T_{0,\varepsilon}}$ satisfying

$$x(t,v,x_{T_{0,\varepsilon}},T_0+\varepsilon) = x(t,0,x_{T_{0,\varepsilon}},T_0+\varepsilon) + x(t,v,0,T_0+\varepsilon).$$

Denote by $X_T(t)$ the solution of

$$\begin{aligned}&\dot X(t) + X(t)A_{11} + A_{11}^TX(t) + A_{12}^TX(t)A_{12} - C_{11}^TC_{11} = 0 \\ &X(T) = 0.\end{aligned} \quad (6.34)$$

It is easy to check that for $x_{T_{0,\varepsilon}} \in \Re^n$,

$$\begin{aligned}&J_1^T(x,v,x_{T_{0,\varepsilon}},T_0+\varepsilon) - J_1^T(x,v,0,T_0+\varepsilon) \\ &= x'_{T_{0,\varepsilon}}X_T(T_0+\varepsilon)x_{T_{0,\varepsilon}} + E\int_{T_0+\varepsilon}^T (v^TB_{11}^TX_T(t)x(t,0,x_{T_{0,\varepsilon}},T_0+\varepsilon) \\ &+ x'(t,0,x_{T_{0,\varepsilon}},T_0+\varepsilon)X_T(t)B_{11}v)dt.\end{aligned}$$

Take $0<\varepsilon^2<\gamma^2-\|\tilde L\|_{[0,T]}^2$, then

$$\begin{aligned}J_1^T(x,v,0,T_0+\varepsilon) &\ge \gamma^2\|\bar v\|_{[0,T]}^2 - \|z_1\|_{[0,T]}^2 \\ &\ge \varepsilon^2\|\bar v\|_{[0,T]}^2 = \varepsilon^2\|v\|_{[T_0+\varepsilon,T]}^2\end{aligned} \quad (6.35)$$

where $\bar v$ is the extension of v from $[T_0+\varepsilon,T]$ to $[0,T]$ by 0. Therefore, by completing the square [76]

$$
\begin{aligned}
&J_1^T(x, v, x_{T_{0,\varepsilon}}, T_0+\varepsilon) \\
&\geq x_{T_{0,\varepsilon}}^T X_T(T_0+\varepsilon) x_{T_{0,\varepsilon}} - E\int_{T_0+\varepsilon}^T \left\|\varepsilon^{-1} B_{11}^T X_T(t) x(t, 0, x_{T_{0,\varepsilon}}, T_0+\varepsilon)\right\|^2 dt.
\end{aligned}
\tag{6.36}
$$

It is well known that there exists $\alpha > 0$ such that

$$
E\int_{T_0+\varepsilon}^T \left\|x(t, 0, x_{T_{0,\varepsilon}}, T_0+\varepsilon)\right\|^2 dt \leq \alpha \left\|x_{T_{0,\varepsilon}}\right\|^2.
$$

As seen from the aforementioned information, there exist $\beta > 0$ and $\beta_1 > 0$ satisfying

$$
\begin{aligned}
&x_{T_{0,\varepsilon}}^T X_T(T_0+\varepsilon) x_{T_{0,\varepsilon}} \\
&= -E\int_{T_0+\varepsilon}^T x^T(t, 0, x_{T_{0,\varepsilon}}, T_0+\varepsilon) C_{11}^T C_{11} \cdot x(t, 0, x_{T_{0,\varepsilon}}, T_0+\varepsilon) dt \\
&\geq -\beta \left\|x_{T_{0,\varepsilon}}\right\|^2
\end{aligned}
\tag{6.37}
$$

and

$$
E\int_{T_0+\varepsilon}^T \left\|\varepsilon^{-1} B_{11}^T X_T(t) x(t, 0, x_{T_{0,\varepsilon}}, T_0+\varepsilon)\right\|^2 dt \leq \beta_1 \left\|x_{T_{0,\varepsilon}}\right\|^2.
$$

So, from (6.36),

$$
J_1^T(x, v, x_{T_{0,\varepsilon}}, T_0+\varepsilon) \geq -(\beta+\beta_1)\left\|x_{T_{0,\varepsilon}}\right\|^2 := -C\left\|x_{T_{0,\varepsilon}}\right\|^2
$$

i.e., (6.33) holds. Inequality (6.33), together with (6.31) and (6.32), yields, for any $0 < \varepsilon < T - T_0$

$$
-CI_{n\times n} \leq P_1(T_0+\varepsilon) \leq 0.
$$

So, $P_1(T_0+\varepsilon)$ cannot tend to ∞ as $\varepsilon \to 0$, showing that (6.30) has a unique solution $P_1(t) \leq 0$ on $[0, T]$. Lemma 6.7 is proved.

The following theorem generalizes Th. 2.3 in [71]:

Theorem 6.5

Finite horizon H_2/H_∞ control has Nash game solution $(u_T^*(t,x), v_T^*(t,x))$, where $u_T^*(t,x)$ and $v_T^*(t,x)$ are the following time-variant feedback strategies:

$$
\begin{aligned}
u_T^*(t,x) &= K_2(t)x(t) \\
v_T^*(t,x) &= K_1(t)x(t)
\end{aligned}
$$

respectively, iff the coupled Riccati equations

$$
\begin{aligned}
&-\dot{P}_1 = A^T P_1 + P_1 A + A_1^T P_1 A_1 - C^T C - [P_1, P_2]\begin{bmatrix} \gamma^{-2} B_1 B_1^T & B_2 B_2^T \\ B_2 B_2^T & B_2 B_2^T \end{bmatrix}\begin{bmatrix} P_1 \\ P_2 \end{bmatrix} \\
&P_1(T) = 0
\end{aligned}
\tag{6.38}
$$

$$-\dot{P}_2 = A^T P_2 + P_1 A + A_1^T P_2 A_1 + C^T C - [P_1, P_2] \begin{bmatrix} 0 & \gamma^{-2} B_1 B_1^T \\ \gamma^{-2} B_1 B_1^T & B_2 B_2^T \end{bmatrix} \begin{bmatrix} P_1 \\ P_2 \end{bmatrix} \tag{6.39}$$
$$P_2(T) = 0$$

have solutions $P_1(t) \leq 0, P_2(t) \geq 0$ on $[0, T]$. In this case,

1. $u_T^*(t,x) = -B_2^T P_2(t)x(t), v_T^*(t,x) = -\gamma^{-2} B_1^T P_1(t)x(t);$
2. $J_2^T(u_T^*, v_T^*) = x_0^T P_2 x_0 + tr(\int_0^T B_0^T(t)P_2(t)B_0(t)dt).$

Proof: Sufficiency: Applying the standard completion of square argument and Itô's formula, under the constraint of (6.28), we have

$$\begin{aligned} &J_1^T(u,v) \\ &= x_0^T P_1 x_0 - Ex^T(T)P_1(T)x(T) + E\int_0^T [(\gamma^2 u^T u - z^T z)dt + d(x^T P_1(t)x)] \\ &= x_0^T P_1 x_0 + E\int_0^T [(\gamma^2 u^T u - z^T z)dt + x^T P_1(t)x + (dx)^T P_1(t)x + x^T P_1(t)dx + (dx)^T P_1(t)dx]. \end{aligned}$$

By a series of simple computation together with (6.38), this can be written as

$$\begin{aligned} &J_1^T(u,v) - x_0^T P_1 x_0 \\ &= E\int_0^T [\gamma^2 (v - v_T^*)^T (v - v_T^*) - u^T u + u_T^{*T} u_T^* + 2x^T P_1(t) B_2(u(t) - u_T^*(t,x))]dt. \end{aligned} \tag{6.40}$$

Implementing $u(t) = u_T^*(t,x) = -B_2' P_2(t)x(t)$ in (6.40) follows:

$$J_1^T(u_T^*, v) \geq J_1^T(u_T^*, v_T^*) = x_0^T P_1(0) x_0$$

where $v_T^*(t,x) = -\gamma^{-2} B_1^T P_1(t)x(t)$. So v_T^* is the worst-case disturbance with respect to u_T^*. When $x_0 = 0$, then

$$J_1^T(u_T^*, v) = E\int_0^T \gamma^2 (v - v_T^*)^T (v - v_T^*)dt.$$

By the same discussion as in the proof of Theorem 6.1, one can obtain $\|L\|_{[0,T]} < \gamma$.

In addition, under the constraint of (6.27), by completing the square and considering (6.39), we have

$$\begin{aligned} \min_{u \in L_F^2(\Re_+, \Re^{n_u})} J_2^T(u, v_T^*) &= J_2^T(u_T^*, v_T^*) \\ &= x_0^T P_2 x_0 + \int_0^T B_0^T(t) P_2(t) B_0(t) dt. \end{aligned} \tag{6.41}$$

By Definition 6.5, this shows that the finite horizon H_2/H_∞ control has a pair of solutions (u_T^*, v_T^*) with

$$u_T^*(t,x) = K_2(t)x(t) = -B_2^T P_2(t)x(t)$$
$$v_T^*(t,x) = K_1(t)x(t) = -\gamma^{-2} B_1^T P_1(t)x(t).$$

Necessity: Implement $u_T^*(t,x) = K_2(t)x(t)$ in (6.28), then

$$\begin{cases} dx(t) = (A(t) + B_2(t)K_2(t))x(t) + B_1(t)v(t))dt + A_1(t)xdw_1, \quad x(0) = x_0 \\ z(t) = \begin{bmatrix} C(t)x(t) \\ D(t)K_2(t)x(t) \end{bmatrix} \quad D'(t)D(t) = I. \end{cases} \tag{6.42}$$

By Definition 6.5, $\|L\|_{[0,T]} < \gamma$. By Lemma 6.7, the following DRE:

$$\begin{cases} \dot{P} + P(A + B_2K_2) + (A + B_2K_2)^T P + A_1^T P A_1 - \gamma^{-2} P B_1 B_1^T P - C^T C - K_2^T K_2 = 0 \\ P(T) = 0 \end{cases} \tag{6.43}$$

has a unique solution $P_1(t) \leq 0$ on $[0,T]$. From the proof of Lemma 6.7, we can see $v_T^*(t,x) = K_1(t)x(t) = -\gamma^{-2} B_1^T P_1(t)x(t)$, where $P_1(t)$ solves (6.43). Substituting $v = v_T^* = -\gamma^{-2} B_1^T P_1(t)x(t)$ into (6.27) yields

$$\begin{cases} dx(t) = (A(t) + \gamma^{-2} B_1(t) B_1^T(t) P_1(t))x(t) + B_2(t)u(t))dt \\ \qquad + A_1(t)xdw_1 + B_0(t)dw_2, x(0) = x_0 \\ z(t) = \begin{bmatrix} C(t)x(t) \\ D(t)u(t) \end{bmatrix} \quad D^T(t)D(t) = I. \end{cases} \tag{6.44}$$

How to minimize $J_2^T(u, v_T^*)$ under the constraint of (6.44) is a standard optimal regulator problem. By [94], the following DRE

$$\begin{cases} \dot{P} + P(A - \gamma^{-2} B_1 B_1^T P_1) + (A - \gamma^{-2} B_1 B_1^T P_1)^T P + A_1^T P A_1 - P B_2 B_2^T P + C^T C = 0 \\ P(T) = 0 \end{cases} \tag{6.45}$$

has a positive-semidefinite solution $P_2(t) \geq 0$ on $[0,T]$. Substituting $P_2(t)$ into (6.45) follows (6.39).

Additionally, by completing the square with a series of computations, we have

$$J_2^T(u, v_T^*) = x_0^T P_2 x_0 + E\int_0^T [(u + B_2^T P_2 x)^T (u + B_2^T P_2 x) + B_0^T P_2 B_0] dt.$$

Therefore, $u_T^*(t,x) = K_2(t)x(t) = -B_2^T P_2(t)x(t)$ solves the following stochastic H_2 optimization problem:

$$J_2^T(u_T^*, v_T^*) = \min_{u \in L_F^2(\Re_+, \Re^{n_u})} J_2^T(u, v_T^*) = x_0^T P_2(0) x_0 + \int_0^T B_0^T(t) P_2(t) B_0(t) dt.$$

Substituting $P_1(t) \leq 0$ and $K_2 = -B_2^T P_2(t)$ into (6.43), (6.38) is obtained. The proof of Theorem 6.5 is complete.

Remark 6.3 Th. 2.3 in [71] only obtained a sufficient part of Theorem 6.5 with $A_1 = 0$ in (6.27). Here, we not only generalize the result of [71] to the more general stochastic models, but also give a necessary and sufficient condition for the finite horizon stochastic H_2/H_∞ control problems.

6.5 Stochastic Nash Game to Observer-Based Design for Stochastic H_2/H_∞ Control

The previous sections assume that the state variables in stochastic H_2/H_∞ control are available, which does not always hold in practice. In this case, one should estimate x from the measurement output y. Some analogous problems have also been considered by [95,96] in the static output feedback for deterministic linear systems, [85] in the dynamic output feedback for multiobjective H_2/H_∞ control, and [74] in the output feedback for deterministic nonlinear systems. In this section, we study observer-based mixed H_2/H_∞ control design for the stochastic systems of the form

$$\begin{cases} dx(t) = (Ax(t) + B_2 u(t) + B_1 v(t))dt + A_1 x dw_1. \\ y(t) = C_1 x(t) + D_1 v(t) \end{cases} \tag{6.46}$$

where $y(t)$ represents the vector-valued measurement variable, which is corrupted by noise $v(t)$. Different from the above discussion, this section deals with the nominal disturbance $v(t)$, i.e., assuming that the exogenous input $v(t)$ has bounded total energy, i.e., $E\int_0^\infty v(t)v^T(t)dt = R_0 \geq 0$. By minimizing an upper bound of the H_2 performance, an observer-based suboptimal H_2/H_∞ control design is given. We refer the reader to [97] and [98] for various guaranteed cost bounds, robust stability, and performance analysis of the state-space systems. We choose, as usual, a finite-dimensional compensator as follows:

$$\begin{cases} d\hat{x} = (A\hat{x} + B_2 u)dt + L(y - C_1\hat{x})dt. \\ u = K\hat{x}, \ \hat{x}(0) = \hat{x}_0 \in \Re^n \end{cases} \tag{6.47}$$

If we set $\tilde{x} = x - \hat{x}$, $\bar{x} = [\hat{x}, \tilde{x}^T]$, then the resulting closed-loop system becomes

$$d\bar{x} = \begin{bmatrix} A + B_2 K & LC_1 \\ 0 & A - LC_1 \end{bmatrix} \bar{x}dt + \begin{bmatrix} LD_1 \\ B_1 - LD_1 \end{bmatrix} vdt + \begin{bmatrix} 0 & 0 \\ A_1 & A_1 \end{bmatrix} \bar{x}dw_1. \tag{6.48}$$

Given $\gamma > 0$ and weighting matrices $\tilde{Q}_1, \tilde{Q}_2, \tilde{R}_2$, our aim is to determine whether there exists acompensator (6.47) stabilizing system (6.48) internally, achieving the H_∞ control performance (with $\bar{x}(0) = 0$)

$$E\int_0^\infty \bar{x}^T\tilde{Q}_1\bar{x}dt < \gamma^2 E\int_0^\infty v^T v dt \tag{6.49}$$

and minimizing the H_2 performance

$$J(\bar{x}, u) = E\int_0^\infty (\bar{x}^T\tilde{Q}_2\bar{x}dt + u^T(t)\tilde{R}_2 u(t))dt$$

simultaneously. $\tilde{Q}_1, \tilde{Q}_2$ and $\tilde{R}_2$ are specified beforehand according to the design purpose. For simplicity, we assume $\tilde{Q}_1 > 0, \tilde{Q}_2 > 0$ and $\tilde{R}_2 > 0$. Otherwise, exact observability or stochastic detectability should be imposed. Suppose $\bar{x}(t, u, v, \bar{x}(0))$ is the solution of (6.48), the following result is a sufficient condition for the H_∞ control performance. Let

$$\bar{A} = \begin{bmatrix} A + B_2K & LC_1 \\ 0 & A - LC_1 \end{bmatrix} \quad \bar{B} = \begin{bmatrix} LD_1 \\ B_1 - LD_1 \end{bmatrix} \quad \bar{C} = \begin{bmatrix} 0 & 0 \\ A_1 & A_1 \end{bmatrix}.$$

Proposition 6.1: If the following matrix inequalities

$$P\bar{A} + \bar{A}^T P + \bar{C}^T P\bar{C} + \frac{1}{\gamma^2} P\bar{B}\bar{B}^T P + \tilde{Q}_1 < 0 \tag{6.50}$$

and

$$P^2 + P\bar{A} + \bar{A}^T P + \bar{C}^T P\bar{C} + \tilde{Q}_2 + \tilde{K}^T\tilde{R}_2\tilde{K} < 0 \tag{6.51}$$

have a common solution $P > 0$, where $\tilde{K} = [K, 0]$. Then, i) the system in (6.48) is internally stable; ii) the H_∞ performance in (6.49) holds; and iii) the H_2 performance has an upper bound, i.e.,

$$J(\bar{x}, u = K\hat{x}) \le \|\bar{x}(0)\|^2 tr(P) + tr(\bar{B}\bar{R}_0\bar{B}^T). \tag{6.52}$$

Proof: By Lemma 6.1, the inequality in (6.50) yields the system in (6.48) being internally stable. To prove ii), by Itô's formula, we have ($\bar{x}(0) = 0$)

$$\begin{aligned}
&E\int_0^\infty \bar{x}^T\tilde{Q}_1\bar{x}dt - \gamma^2 E\int_0^\infty v^T v dt \\
&= E\int_0^\infty [(\bar{x}^T\tilde{Q}_1\bar{x} - \gamma^2 v^T v)dt + d(\bar{x}^T P_1\bar{x})] - Ex^T(\infty)Px(\infty) \\
&< E\int_0^\infty [(d\bar{x})^T P_1\bar{x} + \bar{x}^T P_1 d\bar{x} + (d\bar{x})^T P_1 d\bar{x}] + E\int_0^\infty (\bar{x}^T\tilde{Q}_1\bar{x} - \gamma^2 v^T v)dt \\
&= E\int_0^\infty \begin{bmatrix} \bar{x} \\ v \end{bmatrix}^T \Xi \begin{bmatrix} \bar{x} \\ v \end{bmatrix} dt
\end{aligned} \tag{6.53}$$

where

$$\Xi = \begin{bmatrix} P_1\bar{A} + \bar{A}^T P_1 + \bar{C}^T P_1 \bar{C} + \tilde{Q}_1 & P\bar{B} \\ \bar{B}' P & -\gamma^2 I \end{bmatrix}.$$

By Schur's complement [23], the inequality in (6.50) having a solution is equivalent to $P>0$, so (6.49) holds, ii) is proved.

To prove iii), note that with the constraint of (6.48), for a solution $P>0$ to (6.51), by completing the square

$$\begin{aligned}
&J(\bar{x}, u = K\hat{x}) \\
&= E\int_0^\infty (\bar{x}^T \tilde{Q}_2 \bar{x} dt + u^T \tilde{R}_2 u)dt \\
&= -\lim_{T\to\infty} E\bar{x}(T)^T P\bar{x}(T) + \bar{x}(0)^T P\bar{x}(0) \\
&\quad + E\int_0^\infty (\bar{x}^T \tilde{Q}_2 \bar{x} dt + u^T \tilde{R}_2 u)dt + E\int_0^\infty d(\bar{x}^T P\bar{x}) \\
&= \bar{x}(0)^T P\bar{x}(0) + E\int_0^\infty \bar{x}(t)^T (P\bar{A} + \bar{A}^T P + \bar{C}^T P\bar{C} + \tilde{Q}_2 + \tilde{K}^T \tilde{R}_2 \tilde{K})\bar{x} dt \\
&\quad + E\int_0^\infty (v^T \bar{B}^T P\bar{x} + \bar{x}^T P\bar{B} v)dt - \lim_{T\to\infty} E\bar{x}(T)^T P\bar{x}(T) \\
&\le \bar{x}(0)^T P\bar{x}(0) - \lim_{T\to\infty} E\bar{x}(T)^T P\bar{x}(T) + E\int_0^\infty v^T \bar{B}^T \bar{B} v dt \\
&\quad + E\int_0^\infty \bar{x}^T (P^2 + P\bar{A} + \bar{A}^T P + \bar{C}^T P\bar{C} + \tilde{Q}_2 + \tilde{K}^T \tilde{R}_2 \tilde{K})\bar{x} dt
\end{aligned} \tag{6.54}$$

where the following inequality is used [99] with $\varepsilon = 1$: For any real matrices X and Y of appropriate dimensions, we have

$$X^T Y + Y^T X \le \varepsilon X^T X + \varepsilon^{-1} Y^T Y \quad \forall \varepsilon > 0.$$

By Lemma 6.1, (6.51) having a positive solution $P>0$ yields $(\bar{A}, \bar{C})$ being stable, so $\bar{x} \in L_F^2(R_+, R^{2n})$ if $v \in L_F^2(R_+, R^{n_v})$. Hence, $\lim_{T\to\infty} E\bar{x}(T)^T P\bar{x}(T) = 0$. iii) is immediately followed from (6.54).

By minimizing the upper bound of $J(\bar{x}, u)$ subject to (6.50) and (6.51), we can obtain a suboptimal H_2/H_∞ control design, which is equivalent to minimizing $tr(P)$ under the same constraints. In what follows, we focus our attention on the suboptimal stochastic H_2/H_∞ control problem, which can be formulated as follows:

$$\min_{P>0} tr(P) \tag{6.55}$$

subject to (6.50) and (6.51).

It is difficult to solve analytically the suboptimal control problem, because it is not a convex optimization problem. But if we let P, $\tilde{Q}_1$, and $\tilde{Q}_2$ take the following special forms:

$$P = \begin{bmatrix} P_{11} & 0 \\ 0 & P_{22} \end{bmatrix} \quad \tilde{Q}_1 = \begin{bmatrix} Q_{11}^{(1)} & 0 \\ 0 & Q_{22}^{(1)} \end{bmatrix} \quad \tilde{Q}_2 = \begin{bmatrix} Q_{11}^{(2)} & 0 \\ 0 & Q_{22}^{(2)} \end{bmatrix} \tag{6.56}$$

then (6.50) and (6.51) become (6.57) and (6.58), respectively; see (6.57)–(6.64).

$$\begin{bmatrix} \{P_{11}(A+B_2K)+(A+B_2K)^T P_{11} & \{{}^1/_{\gamma^2} P_{11}LD_1(B_1-LD_1\}^T P_{22} \\ +{}^1/_{\gamma^2} P_{11}LD_1D_1^TL^TP_{11}+Q_{11}^{(1)}\} & +P_{11}LC_1+A_1^TP_{22}A_1\} \\ \{{}^1/_{\gamma^2} P_{22}(B_1-LD_1)\}D_1^TL^TP_{11} & \{P_{22}(A-LC_1)+(A-LC_1)^TP_{22}+A_1^TP_{22}A_1 \\ +A_1^TP_{22}A_1+C_1^TL^TP_{11}\} & +{}^1/_{\gamma^2} P_{22}(B_1-LD_1)(B_1-LD_1)^TP_{22}+Q_{22}^{(1)}\} \end{bmatrix} < 0 \tag{6.57}$$

$$\begin{bmatrix} \{P_{11}^2+P_{11}(A+B_2K)+(A+B_2K)^TP_{11} & P_{11}LC_1+A_1^TP_{22}A_1 \\ +A_1^TP_{22}A_1+Q_{11}^{(2)}+K^T\tilde{R}_2K\} & \\ C_1^TL^TP_{11}+A_1^TP_{22}A_1 & \{P_{22}^2P_{22}(A-LC_1)+(A-LC_1)^TP_{22} \\ & +Q_{22}^{(2)}+A_1^TP_{22}A_1\} \end{bmatrix} < 0 \tag{6.58}$$

$$\begin{bmatrix} \{(A+B_2K)W_{11}+W_{11}(A+B_2K)^T & \{{}^1/_{\gamma^2} LD_1(B_1-LD_1)^TP_{22} \\ +W_{11}A_1^TP_{22}A_1W_{11}+{}^1/_{\gamma^2} LD_1D_1^TL^T & +LC_1+W_{11}A_1^TP_{22}A_1\} \\ +W_{11}Q_{11}^{(1)}W_{11}\} & \\ \{{}^1/_{\gamma^2} P_{22}(B_1-LD_1)\}D_1^TL^T & \{P_{22}(A-LC_1)+(A-LC_1)^TP_{22} \\ +A_1^TP_{22}A_1W_{11}+C_1^TL^T\} & +{}^1/_{\gamma^2} P_{22}(B_1-LD_1)(B_1-LD_1)^TP_{22} \\ & +Q_{22}^{(1)}+A_1^TP_{22}A_1\} \end{bmatrix} < 0 \tag{6.59}$$

$$\begin{bmatrix} \{(A+B_2K)W_{11}+W_{11}(A+B_2K)^T & \\ +W_{11}K^T\tilde{R}_2KW_{11}+W_{11}A_1^TP_{22}A_1W_{11} & LC_1+W_{11}A_1^TP_{22}A_1 \\ +W_{11}Q_{11}^{(1)}W_{11}+I\} & \\ C_1^TL^T+A_1^TP_{22}A_1W_{11} & \{P_{22}(A-LC_1)+(A-LC_1)^TP_{22} \\ & +Q_{22}^{(2)}+A_1^TP_{22}A_1+P_{22}^2\} \end{bmatrix} < 0 \tag{6.60}$$

$$\begin{bmatrix} \begin{matrix} AW_{11} + W_{11}A^T + B_2Y + Y^TB_2^T \\ +W_{11}A_1^TP_{22}A_1W_{11} + {}^1\!/_{\gamma^2}LD_1D_1^TL^T \\ +W_{11}Q_{11}^{(1)}W_{11}\} \end{matrix} & \begin{matrix} LC_1 + {}^1\!/_{\gamma^2}LD_1(B_1 - LD_1)^TP_{22} \\ +W_{11}A_1^TP_{22}A_1 \end{matrix} \\ \begin{matrix} C_1^TL^T + {}^1\!/_{\gamma^2}P_{22}(B_1 - LD_1)\}D_1^TL^T \\ +A_1^TP_{22}A_1W_{11} \end{matrix} & \begin{matrix} \{P_{22}(A - LC_1) + (A - LC_1)^TP_{22} + Q_{22}^{(1)} \\ +{}^1\!/_{\gamma^2}P_{22}(B_1 - LD_1)(B_1 - LD_1)^TP_{22} \\ +A_1^TP_{22}A_1\} \end{matrix} \end{bmatrix} < 0 \quad (6.61)$$

$$\begin{bmatrix} \begin{matrix} \{(AW_{11} + W_{11}A^T + B_2Y + Y^TB_2^T + Y^T\tilde{R}_2Y \\ +W_{11}A_1^TP_{22}A_1W_{11} + W_{11}Q_{11}^{(2)}W_{11} + I\} \end{matrix} & LC_1 + W_{11}A_1^TP_{22}A_1 \\ C_1^TL^T + A_1^TP_{22}A_1W_{11} & \begin{matrix} \{P_{22}A + A^TP_{22} - ZC_1 - C_1^TZ^T \\ +Q_{22}^{(2)} + A_1^TP_{22}A_1 + P_{22}^2\} \end{matrix} \end{bmatrix} < 0 \quad (6.62)$$

$$\begin{bmatrix} \begin{matrix} \{AW_{11} + W_{11}A^T \\ +B_2Y + Y^TB_2^T\} \end{matrix} & \begin{matrix} {}^1\!/_{\gamma^2}LD_1(B_1 - LD_1)^TP_{22} \\ +LC_1 + W_{11}A_1^TP_{22}A_1 \end{matrix} & W_{11}^TA_1^T & W_{11}^T & LD_1 \\ \begin{matrix} \{{}^1\!/_{\gamma^2}P_{22}(B_1 - LD_1)\}D_1^TL^T \\ +A_1^TP_{22}A_1W_{11} + C_1^TL^T\} \end{matrix} & \begin{matrix} \{P_{22}A + A^TP_{22} - ZC_1 \\ -C_1^TZ^T + Q_{22}^{(1)} + A_1^TP_{22}A_1 \\ +{}^1\!/_{\gamma^2}P_{22}(B_1 - LD_1) \\ \cdot(B_1 - LD_1)^TP_{22}\} \end{matrix} & 0 & 0 & 0 \\ A_1W_{11} & 0 & -S & 0 & 0 \\ W_{11} & 0 & 0 & -(Q_{11}^{(1)})^{-1} & 0 \\ D_1^TL^T & 0 & 0 & 0 & -\gamma^2I \end{bmatrix} < 0 \quad (6.63)$$

$$\begin{bmatrix} \begin{matrix} \{AW_{11} + W_{11}A^T \\ +B_2Y + Y^TB_2^T + I\} \end{matrix} & LC_1 + W_{11}A_1^TP_{22}A_1 & Y^T & W_{11}^T & W_{11}^TA_1^T \\ C_1^TL^T + A_1^TP_{22}A_1W_{11} & \begin{matrix} \{P_{22}^2 + P_{22}A + A^TP_{22} - ZC_1 \\ -C_1^TZ^T + Q_{22}^{(2)} + A_1^TP_{22}A_1\} \end{matrix} & 0 & 0 & 0 \\ Y & 0 & -(\tilde{R}_2)^{-1} & 0 & 0 \\ W_{11} & 0 & 0 & -(Q_{11}^{(2)})^{-1} & 0 \\ A_1W_{11} & 0 & 0 & 0 & -S \end{bmatrix} < 0 \quad (6.64)$$

Let $W = diag(W_{11}, I) = diag(P_{11}^{-1}, I)$, and multiplying W on both sides of (6.57) and (6.58), we have (6.59) and (6.60), respectively. Let $Z = P_{22}L$, $Y = KW_{11}$, then (6.59) and (6.60) become (6.61) and (6.62) respectively. Let $S = P_{22}^{-1}$, by Schur's complement [23], (6.61) and (6.62) can be written as (6.63) and (6.64), respectively. Now, the original optimization problem becomes solving (6.55) subject to (6.63) and (6.64). Inequalities in (6.63) and (6.64) are not the forms of LMIs, because the variables L, P_{22} and W_{11} are coupled together in some components. So in the sequel, we look for another kind of suboptimal solution for the observer-based mixed stochastic H_2/H_∞ robust control design. Note that from (6.63) and (6.64), one must have (6.65) and (6.66).

$$P_{22}A + A^T P_{22} - ZC_1 - C_1^T Z^T + \frac{1}{\gamma^2} P_{22}(B_1 - LD_1)(B_1 - LD_1)^T P_{22} + Q_{22}^{(1)} + A_1^T P_{22} A_1 < 0 \tag{6.65}$$

$$P_{22}^2 + P_{22}A + A^T P_{22} - ZC_1 - C_1^T Z^T + Q_{22}^{(2)} + A_1^T P_{22} A_1 < 0. \tag{6.66}$$

Inequalities in (6.65) and (6.66) can be written in the LMIs forms as (6.67) and (6.68), respectively.

$$\begin{bmatrix} P_{22}A + A^T P_{22} - ZC_1 - C_1^T Z^T + A_1^T P_{22} A_1 + Q_{22}^{(1)} & P_{22}B_1 - ZD_1 \\ (P_{22}B_1 - ZD_1)^T & -\gamma^2 I \end{bmatrix} < 0 \tag{6.67}$$

$$\begin{bmatrix} P_{22}A + A^T P_{22} - ZC_1 - C_1^T Z^T & P_{22} \\ P_{22} & -I \end{bmatrix} < 0. \tag{6.68}$$

Based on the above analyses, the observer-based suboptimal stochastic H_2/H_∞ control design is divided into two steps. In the first step, we solve the convex optimization problem

$$\min_{\text{subject to } P_{22} > 0, (6.67), (6.68)} tr(P_{22}) \tag{6.69}$$

to determine P_{22} and Z, accordingly, $L = P_{22}^{-1} Z$, which is for the observer design in (6.47). The convex optimization problem (6.69) can be easily treated with efficient tools [23,100]. After P_{22}, Z and L are obtained, substituting them into (6.63), (6.64), they both then become LMIs. In the second step, by solving the following convex optimization problem

$$\max_{\text{subject to } W_{11} > 0, (6.63), (6.64)} tr(W_{11}) \tag{6.70}$$

feedback gain K and $P_{11} = W_{11}^{-1}$ can also be solved. Hence, a suboptimal solution $P = diag(P_{11}, P_{22})$ and compensator (6.47) are obtained. Summarizing the aforementioned discussion, we have the following result.

Theorem 6.6
If the two-step convex optimization problems (6.69), (6.70) are feasible, then there is an observer-based compensator (6.47), which solves the suboptimal stochastic H_2/H_∞ control problem with

$$K = YW_{11}^{-1}L = P_{22}^{-1}Z.$$

The suboptimal control and H_2 performance index are given, respectively, as follows:

$$u^*(t) = YW_{11}^{-1}\hat{x}(t)$$
$$J^*(\bar{x}, u^*) = \mathrm{tr}(W_{11}^{-1} + P_{22})\|\bar{x}(0)\|^2 + \mathrm{tr}(\bar{B}\bar{R}_0\bar{B}').$$

6.6 Conclusion

This chapter has discussed stochastic Nash game approach to the stochastic H_2/H_∞ control with state-dependent noise, including both finite and infinite horizon cases. It has been shown that finite (infinite) horizon stochastic H_2/H_∞ control has close relation to a pair of coupled DREs (a pair of coupled AREs, correspondingly). We believe that Lemmas 6.5 and 6.7 are of independent interest, and may play important roles in some other applications, especially in the H_2/H_∞ robust control of nonlinear stochastic systems. An observer-based compensator design for suboptimal stochastic H_2/H_∞ control is also presented with a feasible algorithm which needs to solve two convex optimization problems (6.69) and (6.70) by the LMI optimization toolbox in Matlab [100]. There are still some open problems, which are not only interesting but also valuable. For example, what happens when control u enters into the diffusion term? It remains to study in future work.

6.7 Appendix

6.7.1 Appendix A

A Proof of Lemma 6.1

If (A, A_1) is stable, $PA + A^TP + A_1^TPA_1 = -C^TC - \varepsilon I$ has a solution $P_\varepsilon > 0$ for any $\varepsilon > 0$; see [76]. Let $\varepsilon \to 0$; it is easy to show $P_\varepsilon \to P \geq 0$, which is the solution of (6.6). Now, we show $P > 0$; otherwise, there exists $x_0 \neq 0$, such that $Px_0 = 0$. Applying Itô's formula to x^TPx and considering (6.6) yield

$$E\int_0^T x^TC^TCxds = x_0^TPx_0 - Ex^T(T)Px(T), \quad \forall T > 0. \tag{6.71}$$

From (6.71), for any $T > 0$, we have

$$0 \leq E\int_0^T x^TC^TCxds = -Ex^T(T)Px(T) \leq 0$$

which follows $y(t) = Cx(t) \equiv 0, \forall t \in [0, T]$, but this is impossible because of exact observability, so $P > 0$. The uniqueness of the solutions of (6.6) is obvious. In fact, under the stability of (A, A_1), for any real solutions P_1 and P_2 of (6.6), by completing the square, we have

$$E \int_0^T x^T C^T C x ds = x_0^T P_1 x_0 = x_0^T P_2 x_0, \quad \forall x_0 \in \Re^n$$

which yields $P_1 = P_2$, the necessity part is proved.

If (6.6) has a solution $P > 0$, from (6.71), we know $V(x(t)) := Ex^T(t) P x(t)$ is monotonically decreasing and bounded from below with respect to t, so $\lim_{t \to \infty} V(x(t))$ exists. For any fixed T, if we let $t_n = nT, n = 0, 1, 2, \ldots$, then by (6.71)

$$V(x(t_{n+1})) - V(x(t_n)) = -E \int_{t_n}^{t_{(n+1)}} x^T(t) C^T C x(t) dt.$$

From [94], we know that the following equation

$$-\dot{H}(t) = H(t) A + A^T H(t) + A_1^T H(t) A_1 + C^T C, \quad H(T) = 0$$

has a unique solution $H(\cdot) \geq 0$ on $[0, T]$. By completing the square, we have

$$V(x(t_1)) - V(x(t_0)) = -E \int_0^T x^T(t) C^T C x(t) dt = -x_0^T H(0) x_0.$$

As in the necessity proof, one has $H(0) > 0$ because of the exact observability. Due to the time-invariance of (6.3), it is easy to verify

$$V(x(t_{n+1})) - V(x(t_n)) = -Ex^T(t_n) H(0) x(t_n).$$

Taking limit in this, we have $\lim_{t \to \infty} V(x(t_n)) = \lim_{n \to \infty} E\|x(t_n)\|^2 = 0$. Because $\lim_{t \to \infty} V(x(t))$ exists and $P > 0$, we have $\lim_{t \to \infty} Ex(t) x^T(t) = \lim_{t \to \infty} V(x(t)) = 0$, i.e., (A, A_1) is stable. The proof of this lemma is ended.

6.7.2 Appendix B

A Proof of Lemma 6.3

Th. 4.2 in [90] asserted that $[\tilde{A}, A_1 | \tilde{A}_2]$ is exactly observable if $Ker(P_0) = \{\theta\}$, where

$$P_0 = [\tilde{A}_2^T, \tilde{A}^T \tilde{A}_2^T, A_1^T \tilde{A}_2^T, \tilde{A}^T A_1^T \tilde{A}_2^T, A_1^T \tilde{A}^T \tilde{A}_2^T, (\tilde{A}^T)^2 \tilde{A}_2^T, (A_1^T)^2 \tilde{A}_2^T, \ldots]^T.$$

The above fact can be easily proved; for the reader's convenience, we give a simple illustration for Th. 4.2 in [90]. In fact, any initial state $x(0)$ corresponding to $y(t) \equiv 0$ must belong to $Ker(P_0)$. First, for any $t \geq 0$, from

$$\begin{aligned} y(t) &= \tilde{A}_2 x(t) \\ &= \tilde{A}_2 x(0) + \int_0^t \tilde{A}_2 \tilde{A} x(s) ds + \int_0^t \tilde{A}_2 A_1 x(s) dw_1(t) \\ &\equiv 0 \end{aligned}$$

we have $\tilde{A}_2 x(0) = 0, \tilde{A}_2 \tilde{A} x(t) \equiv 0, \tilde{A}_2 A_1 x(t) \equiv 0$.

Second, repeating the above procedure, one has $x(0) \in Ker(P_0)$. Th. 4.2 in [90] is accordingly derived by Definition 6.2.

Let $\tilde{A} = A - B_2 B_2^T P_2$. If $[\tilde{A}, A_1 | \tilde{A}_2]$ is not exactly observable, then there exists a nonzero vector $\xi \in Ker(P_0)$. From $P_0 \xi = 0$, it yields

$$\tilde{A}_2 \tilde{A}^i A_1^j \xi = 0, \tilde{A}_2 A_1^j \tilde{A}^i \xi = 0, \quad i, j = 0, 1, 2, \ldots$$

which leads to

$$C A^i A_1^j \xi = 0, C A_1^j A^i \xi = 0, \quad i, j = 0, 1, 2, \ldots.$$

This is equivalent to

$$Ker(P_0^*) := Ker([C^T, A^T C^T, A_1^T C^T, A^T A_1^T C^T, A_1^T A^T C^T, (A^T)^2 C^T, (A_1^T)^2 C^T, \ldots]^T) \neq \{\theta\}. \quad (6.72)$$

Again, by [90, Th. 4.2], (6.72) shows that $[A, A_1 | C]$ is not exactly observable. This is a contradiction with the given condition, so $[\tilde{A}, A_1 | \tilde{A}_2]$ is exactly observable. In the same way, we can prove $[A - \gamma^{-2} B_1 B_1^T P_1 - B_2 B_2^T P_2, A_1 | \tilde{A}_3]$ to be exactly observable. Lemma 3 is proved.

6.7.3 Appendix C

A Proof of Lemma 6.5

By a comparison theorem on generalized AREs [88], this lemma is immediately derived. For the reader's convenience, we present a sketch for the proof. By Corr. 2.14 in [76], there exists a sufficiently small $\delta > 0$, such that ARE

$$P A_{11} + A_{11}^T P + A_{12}^T P A_{12} + C_{11}^T C_{11} - \delta^2 I - \gamma^{-2} P B_{11} B_{11}^T P = 0 \quad (6.73)$$

has a solution $P_\delta < 0$. Applying Lemma 6.6 and Ths. 10 and 12 in [92], there exists a maximal solution P_1 to (6.8), such that $P_1 > P_\delta$ satisfies

$$\begin{aligned} &\min_{v \in L_F^2(\Re_+, \Re^{n_v})} \{\tilde{J}_1^\infty(x_0, 0) := E \int_0^\infty (\gamma^2 v^T v - z_1^T z_1) dt \lim_{T \to \infty} E \|x(t)\|^2 = 0\} \\ &= x_0^T P_1 x_0 \leq \tilde{J}_1^\infty(x_0, 0) \leq 0. \end{aligned} \quad (6.74)$$

Because x_0 is arbitrary, from before, $P_1 \le 0$. The rest is to prove $(A_{11} - \gamma^{-2}B_{11}B_{11}^T P_1, A_{12})$ stable. Denote $\hat{A} = A_{11} + B_{11}K_\delta$, where

$$K = -\gamma^{-2}B_{11}^T P_1 K_\delta = -\gamma^{-2}B_{11}^T P_\delta.$$

Note that the (6.8) and (6.73) can be written, respectively, as

$$P_1\hat{A}_{11} + \hat{A}_{11}^T P_1 + A_{12}^T P_1 A_{12} + \gamma^2 K^T K - C_{11}^T C_{11} = 0 \tag{6.75}$$

$$P_\delta A_\delta + A_\delta^T P_\delta + A_{12}^T P_\delta A_{12} + \gamma^2 K_\delta^T K_\delta - C_{11}^T C_{11} - \delta^2 I = 0. \tag{6.76}$$

Subtracting (6.75) from (6.76), by a series of computations, we have $(P_1 - P_\delta)\hat{A} + \hat{A}^T(P_1 - P_\delta) + A_{12}^T(P_1 - P_\delta)A_{12} = -\gamma^{-2}(P_\delta B_{11} + \gamma^2 K^T)(P_\delta B_{11} + \gamma^2 K^T)^T - \delta^2 I < 0$. This equation shows that Lyapunov-type inequality

$$P\hat{A} + \hat{A}^T P + A_{12}^T P A_{12} < 0$$

has a positive solution $P > 0$, so $(\hat{A}, A_{12})$ is stable from Lemma 6.1. Lemma 6.5 is proved.

7

Fuzzy Multi-player Stochastic Noncooperative and Cooperative H_2 Game Strategies for Nonlinear Stochastic Observer-based Control Systems

7.1 Introduction

In general large-scale systems are often controlled by more than one controller or decision-maker with each using an individual control strategy. These decentralized controllers may operate in a group as a team with a common objective function or in a conflicting manner with multiple-objective goals as a game problem [101]. Stochastic game theory has been widely applied to multiperson decision-making problems, stimulated by a vast number of applications, including those in economics, management, communication networks, power networks, and in the design of complex stochastic systems. In this stochastic game problem, many decision-makers are present or many possible conflicting objectives should be taken into account in order to reach some form of stochastic performance optimality [1,102]. Typically, n-person (or n-player) stochastic games are divided into two classes: a noncooperative type of stochastic game in the sense of Nash equilibrium solution and a cooperative one in the sense of Pareto optimization solution. In the noncooperative stochastic game with n players, each participant pursues an individual goal which may partly conflict with others. The players in the cooperative stochastic game work together and act as one player seeking their maximum common profit. In this chapter, both noncooperative and cooperative stochastic H_2 game problems of nonlinear stochastic control systems are considered.

In general, in the nonlinear n-person stochastic H_2 game problems, one needs to solve n-simultaneous Hamilton–Jacobi–Issac (HJI) equations, which are all nonlinear partial differential equations [1]. At present, it is very difficult to solve the nonlinear n-person stochastic game problems, except for very special cases. For this reason, it is not easy to apply nonlinear n-person stochastic H_2 game theory to address the practical design problems. The purpose of this chapter is to find a simple and feasible method to deal with the general design problem of nonlinear n-person stochastic H_2 games so the results can be applied in a practical design problem.

Recently, fuzzy models have been used to efficiently approximate nonlinear systems [4,103–105]. In this chapter, in order to avoid solving n-simultaneous HJI equations, the Takagi–Sugeno fuzzy model [4] is employed to approximate the nonlinear stochastic dynamic systems in the nonlinear stochastic H_2 game problem. Therefore, the n-person nonlinear stochastic H_2 game problem is transformed to a n-person fuzzy stochastic game problem. Based on the fuzzy model, the n-person fuzzy stochastic H_2 game problems are characterized in terms of a minimization problem subject to some Riccati-like inequalities.

Since the state variables are not all available in practice, a state estimation algorithm is needed to estimate the state variables for the control design. In this chapter, a suboptimal fuzzy observer is proposed to estimate the states for controller design in these H_2 (quadratic) fuzzy stochastic game problems when state variables are unavailable. Using a separation method, the solution of the observer-based fuzzy stochastic H_2 game problem is also characterized in terms of a minimization problem subject to some Riccati-like inequalities.

Solving the minimization problem subject to some Riccati-like inequalities in n-person fuzzy stochastic H_2 game is still a challenging task. Fortunately, using the techniques of Schur complements, certain form of Riccati-like inequalities can be transformed into equivalent linear matrix inequalities (LMIs) [23,107]. Therefore, the fuzzy stochastic H_2 game problems are reduced to solving the minimization problem subject to LMIs, which is known as an eigenvalue problem (EVP) [23]. The EVP can be solved very efficiently by convex optimization techniques using interior-point methods with the aid of a toolbox in Matlab [100].

This chapter consists of five sections: the problem formulation is presented in Section 7.2, while fuzzy observer combined with the fuzzy control for both noncooperative and cooperative stochastic H_2 games are described in Section 7.3. In Section 7.4, simulation examples are provided to demonstrate the design procedures and indicate the performance of the proposed methods. Finally, concluding remarks are made in Section 7.5.

7.2 Problem Description

Consider the following nonlinear stochastic system:

$$\begin{aligned} \dot{x}(t) &= f_1(x(t), u_1(t), \ldots, u_n(t)) + g_1(x(t))w(t) \\ y(t) &= f_2(x(t)) + g_2(x(t))v(t) \end{aligned} \tag{7.1}$$

where $x(t) = (x_1(t), \ldots, x_m(t))^T$ denotes state variables, $u(t) = (u_1(t), \ldots, u_n(t))^T$ denotes control inputs of n players, $y(t) = (y_1(t), \ldots, y_r(t))^T$ denotes output of the system, and external disturbance $w(t)$ and measurement noise $v(t)$ are assumed to be uncorrelated, zero-mean, white noises with identity power density without loss of generality.

We assume that the action of the kth player ($k \in \{1, 2, \ldots, n\}$) is determined by a control policy $u_k(t)$ and denotes the class of all such policies for the kth player by U_k, i.e., $u_k(t) \in U_k$.

For the noncooperative stochastic H_2 game of the nonlinear stochastic system (7.1), the individual H_2 payoff (utility) function to be minimized by the kth player $u_k(t)$ is [1]

$$J_k^{NC}(u_k) = \lim_{t_f \to \infty} \frac{1}{t_f} E\left\{ x^T(t_f)Sx(t_f) + \int_0^{t_f} (x^T(t)Qx(t) + u_k^T(t)R_k u_k(t))dt \right\} \tag{7.2}$$

where E denotes expectation, $S > 0, Q = Q^T > 0$ and $R_k > 0$ for $k = 1, 2, \ldots, n$. The solution for noncooperative stochastic H_2 game problem in (7.2) is the Nash equilibrium solution. In other words, we seek a multipolicy $u^*(t) = (u_1^*(t), \ldots, u_n^*(t))^T$ that no controller has incentive to deviate from [102], i.e.,

$$J_k^{NC}(u^*(t)) = \inf_{u_k(t) \in U_k} J_k^{NC}([u_k(t) | u_{-k}^*(t)]) \tag{7.3}$$

where $[u_k(t)|u^*_{-k}(t)]$ is the policy obtained when for each $j \neq k$, player j uses policy $u_j^*(t)$, and player k uses $u_k(t)$, i.e.,

$$[u_k(t)|u^*_{-k}(t)] = [u_k(t)|u_1^*(t), \ldots, u_{k-1}^*(t), u_{k+1}^*(t), \ldots, u_n^*(t)]. \tag{7.4}$$

For an n-person differential game, an n-tuple of strategies provides a feedback Nash equilibrium solution for the noncooperative stochastic H_2 game.

On the other hand, for the cooperative stochastic H_2 game (i.e., team), the common H_2 payoff function to be minimized is [1,103]

$$\begin{aligned} J^C(u) &= \lim_{t_f \to \infty} \frac{1}{t_f} E\left\{ x^T(t_f)Sx(t_f) + \int_0^{t_f} (x^T(t)Qx(t) + \sum_{k=1}^{n} u_k^T(t)R_k u_k(t))dt \right\} \\ &= \lim_{t_f \to \infty} \frac{1}{t_f} E\left\{ x^T(t_f)Sx(t_f) + \int_0^{t_f} [x^T(t)Qx(t) + u^T Ru(t)]dt \right\} \end{aligned} \tag{7.5}$$

where $S > 0, Q = Q^T > 0, R = diag(R_1, R_2, \ldots, R_n)$ and $R_k > 0$ for $k = 1, 2, \ldots, n$.

For an n-person cooperative stochastic H_2 game, we seek a cooperative strategy $u^*(t)$ to provide a feedback Pareto-optimal solution for the cooperative stochastic game in (6.5), i.e.,

$$J^C(u^*) - \inf_{u(t)} \lim_{t_f \to \infty} \frac{1}{t_f} E\left\{ x^T(t_f)Sx(t_f) + \int_0^{t_f} [x^T(t)Qx(t) + u^T Ru(t)dt] \right\} \tag{7.6}$$

The fuzzy linear model is described by fuzzy If–Then rules and will be employed here to deal with the stochastic H_2 game control design problem for nonlinear stochastic systems. The ith rule of the fuzzy linear model for the nonlinear stochastic system in (7.1) is of the following form [104,105,108]:

Plant Rule i:

If $z_1(t)$ is F_{i1} and $\cdots$ and $z_g(t)$ is F_{ig}

Then,

$$\begin{aligned} \dot{x}(t) &= A_i x(t) + \sum_{k=1}^{n} (B_{ik} u_k(t)) + G_i w(t) \\ y(t) &= C_i x(t) + D_i v(t) \end{aligned} \tag{7.7}$$

for $i = 1, 2, \ldots, L$ where F_{ij} the fuzzy set, $A_i \in R^m, B_{ik} \in R^{m \times 1}, G_i \in R^{m \times p}, C_i \in R^r, D_i \in R^{r \times q}$; L is the number of If–Then rules; $z_1(t), z_2(t), \ldots, z_g(t)$ are the premise variables.

Assumption: (A_i, B_i) are controllable and (A_i, C_i) are observable for $i = 1, 2, \ldots, L$. The fuzzy system is inferred as follows [104,105,108]:

$$\dot{x}(t) = \sum_{i=1}^{L} h_i(z(t))[A_i x(t) + \sum_{k=1}^{n} (B_{ik} u_k(t)) + G_i w(t)] \tag{7.8}$$

$$y(t) = \sum_{i=1}^{L} h_i(z(t))[C_i x(t) + D_i v(t)] \tag{7.9}$$

where

$$\begin{aligned} h_i(z(t)) &= \frac{\mu_i(z(t))}{\sum_{i=1}^{L} \mu_i(z(t))} \\ \mu_i(z(t)) &= \prod_{i=1}^{g} F_{ij}(z_j(t)) \\ z(t) &= [z_1(t), z_2(t), \ldots, z_g(t)] \end{aligned} \tag{7.10}$$

and where $F_{ij}(z_j(t))$ is the grade of membership of $z_j(t)$ in F_{ij}. The normalized membership functions in (7.10) satisfy

$$\sum_{i=1}^{L} h_i(z(t)) = 1 \tag{7.11}$$

where $h_i(z(t)) \in [0, 1]$ [109].

Suppose the following fuzzy controller of the kth player is employed to deal with the above fuzzy control system design

Control Rule i:

If $z_1(t)$ is F_{i1} and $\cdots$ and $z_g(t)$ is F_{ig}

Then,

$$u_k(t) = -K_{ik}x(t) \tag{7.12}$$

for $i = 1, 2, \ldots, L$, and $k = 1, 2, \ldots, n$.

Hence, the fuzzy controller is given by

$$u_k(t) = -\sum_{i=1}^{L} h_i(z(t))K_{ik}x(t) \tag{7.13}$$

where the control parameters K_{ik} (for $i = 1, 2, \ldots, L$, and $k = 1, 2, \ldots, n$) are to be specified later to achieve the desired control purpose of noncooperative or cooperative stochastic H_2 game strategy.

In this chapter, we define $\sum_{k=1}^{n} B_{ik}u_k = B_i u$ where $B_i = [\, B_{i1} \quad B_{i2} \quad \cdots \quad B_{in} \,]$; $B_{ik} \in R^{m \times 1}$ for $i = 1, 2, \ldots, L$, and $k = 1, 2, \ldots, n$, and $u(t)$ is represented as follows:

$$\begin{aligned} u(t) &= [u_1(t), u_2(t), \ldots, u_n(t)]^T \\ &= -\sum_{i=1}^{L} h_i(z(t)) \left(\begin{bmatrix} K_{i1} \\ \vdots \\ K_{in} \end{bmatrix} x(t) \right) \\ &= -\sum_{i=1}^{L} h_i(z(t)) \left(\begin{bmatrix} k_{i11} & \cdots & k_{i1m} \\ \vdots & \ddots & \vdots \\ k_{in1} & \cdots & k_{inm} \end{bmatrix} x(t) \right) \\ &= -\sum_{i=1}^{L} h_i(z(t))(K_i x(t)) \end{aligned} \tag{7.14}$$

where

$$\begin{aligned} u_k(t) &= -\sum_{i=1}^{L} h_i(z(t))[K_{ik}x(t)] \\ &= -\sum_{i=1}^{L} h_i(z(t))([k_{ik1} \quad k_{ik2} \quad \cdots \quad k_{ikm}]x(t)) \end{aligned} \tag{7.15}$$

and

$$K_i = \begin{bmatrix} K_{i1} \\ \vdots \\ K_{in} \end{bmatrix} = \begin{bmatrix} k_{i11} & \cdots & k_{i1m} \\ \vdots & \ddots & \vdots \\ k_{in1} & \cdots & k_{inm} \end{bmatrix} \tag{7.16}$$

for $i = 1, 2, ..., L$, and $k = 1, 2, ..., n$.

Substituting (7.13) into (7.8), the fuzzy control system is obtained as follows:

$$\dot{x}(t) = \sum_{i=1}^{L}\sum_{j=1}^{L} h_i(z(t))h_j(z(t)) \times [(A_i - B_iK_j)x(t) + G_iw(t)]. \tag{7.17}$$

7.3 Stochastic H_2 Games via Combined Fuzzy Observer and Control

In practice, state variables are not all available. For this situation, we need to estimate the state vector $x(t)$ from the output $y(t)$ for state feedback control. Suppose the following fuzzy observer is proposed to deal with the state estimation for the nonlinear stochastic system (7.7).

Observer Rule i:

If $z_1(t)$ is F_{i1} and $\cdots$ and $z_g(t)$ is F_{ig}

Then

$$\dot{\hat{x}}(t) = [A_i\hat{x}(t) + B_iu(t) + L_i(y(t) - \hat{y}(t))] \tag{7.18}$$

where L_i is the observer gain for the ith observer rule and is to be specified later to achieve the desired control purpose of noncooperative or cooperative H_2 stochastic game strategy and $\hat{y}(t) = \sum_{i=1}^{L} h_i(z(t))(C_i\hat{x}(t))$.

The overall fuzzy observer is represented as follows:

$$\dot{\hat{x}}(t) = \sum_{i=1}^{L} h_i(z(t))\{A_i\hat{x}(t) + B_iu(t) + L_i[y(t) \quad \hat{y}(t)]\} \tag{7.19}$$

and the fuzzy observer-based controller is modified by

$$u(t) = -\sum_{i=1}^{L} h_i(z(t))[K_i\hat{x}(t)] \tag{7.20}$$

Then, the augmented system is of the following form:

$$\begin{bmatrix} \dot{x}(t) \\ \dot{\hat{x}}(t) \end{bmatrix} = \begin{bmatrix} \left\{ \sum_{j=1}^{L} h_i(z(t))[A_i x(t) + B_i u(t) + G_i w(t)] \right\} \\ \left\{ \sum_{i=1}^{L} h_i(z(t))[A_i \hat{x}(t) + B_i u(t) + L_i(y(t) - \hat{y}(t))] \right\} \end{bmatrix}. \tag{7.21}$$

Let us denote the estimation error as

$$\tilde{x}(t) = x(t) - \hat{x}(t). \tag{7.22}$$

By differentiating (7.22) and after some manipulation, we get

$$\dot{\tilde{x}}(t) = \sum_{i=1}^{L} \sum_{j=1}^{L} h_i(z(t)) h_j(z(t)) \times [(A_i - L_i C_j)\tilde{x}(t) + G_i w(t) - L_i D_j v(t)] \tag{7.23}$$

The design purpose in this chapter is to specify the fuzzy control in (7.13) and the fuzzy observer in (7.19) to achieve noncooperative control performance in (7.3) and cooperative control performance in (7.6), respectively.

A. Fuzzy noncooperative H_2 stochastic game design:

Let us consider the noncooperative stochastic H_2 game performance index in (7.2) at first. The design purpose of the noncooperative stochastic H_2 game strategy is to specify the control gain K_{ik} and the estimator gain L_i (for $i = 1, 2, \ldots, L$) such that the individual H_2 quadratic cost function in (7.2) is minimized for the noncooperative stochastic fuzzy H_2 game strategy. We now use the well-known relation [110,111]

$$E\{E\{x(t)|Y(t)\}\} = E\{x(t)\} \tag{7.24}$$

to describe (7.2) in a form more suitable for the analysis to follow where

$$Y(t) = \{y(\tau)|0 \le \tau \le t\}.$$

Equations (7.2) and (7.24) imply that

$$\begin{aligned} J_k^{NC}(u_k) = \lim_{t_f \to \infty} \frac{1}{t_f} E\left\{ E\{x^T(t_f) S x(t_f)|Y(t)\} + \int_0^{t_f} E\{x^T(t) Q x(t)|Y(t)\} dt \right\} \\ + E\left\{ \int_0^{t_f} u_k^T(t) R_k u_k(t) dt \right\}. \end{aligned} \tag{7.25}$$

By the fact that

$$\begin{aligned} E\{x^T(t) Q x(t)\} &= \bar{x}^T(t) Q \bar{x}(t) + \text{tr}\{QE\{[x - \bar{x}][x - \bar{x}]^T\}\} \\ &= \bar{x}^T(t) Q \bar{x}(t) + \text{tr}\{Q \text{cov}[x(t)]\} \end{aligned} \tag{7.26}$$

where $\bar{x}(t) = E\{x(t)\}$, (7.25) can be rewritten as follows [110,111]:

$$\begin{aligned}J_k^{NC}(u_k) &= \lim_{t_f\to\infty}\frac{1}{t_f}E\Big\{\hat{x}^T(t_f)S\hat{x}(t_f)+\mathrm{tr}(S\sum\nolimits_{\tilde{x}}(t_f))+\int_0^{t_f}[\hat{x}^T(t)Q\hat{x}(t)\\&\quad+u_k^T(t)R_ku_k(t)]dt\Big\}+\lim_{t_f\to\infty}\frac{1}{t_f}E\Big\{\int_0^{t_f}\mathrm{tr}(QE\{[x-\bar{x}][x-\bar{x}]^T|Y(t)\})dt\Big\}\\&=\lim_{t_f\to\infty}\frac{1}{t_f}E\Big\{\hat{x}^T(t_f)S\hat{x}(t_f)+\int_0^{t_f}[\hat{x}^T(t)Q\hat{x}(t)+u_k^T(t)R_ku_k(t)]dt\Big\}\\&\quad+\lim_{t_f\to\infty}\frac{1}{t_f}\Big\{\mathrm{tr}(S\sum\nolimits_{\tilde{x}}(t_f))+\int_0^{t_f}\mathrm{tr}(Q\sum\nolimits_{\tilde{x}}(t))dt\Big\}\\&=J_k^{NC1}(u_k)+J_k^{NC2}(\tilde{x})\end{aligned}\tag{7.27}$$

where $\hat{x}(t)=E\{x(t)|Y(t)\}$ [106,111] and $\sum_{\tilde{x}}(t)=E\{[x-\bar{x}][x-\bar{x}]^T\}=\mathrm{cov}(\tilde{x}(t),\tilde{x}(t))$ [106]

$$J_k^{NC1}(u_k)=\lim_{t_f\to\infty}\frac{1}{t_f}E\Big\{\hat{x}^T(t_f)S\hat{x}(t_f)+\int_0^{t_f}(\hat{x}^T(t)Q\hat{x}(t)+u_k^T(t)R_ku_k(t)]dt\Big\}\tag{7.28}$$

and

$$J_k^{NC2}(\tilde{x})=\lim_{t_f\to\infty}\frac{1}{t_f}\Big\{\mathrm{tr}(S\sum\nolimits_{\tilde{x}}(t_f))+\int_0^{t_f}\mathrm{tr}(Q\sum\nolimits_{\tilde{x}}(t))dt\Big\}.\tag{7.29}$$

It is found that $J_k^{NC2}(\tilde{x})$ depends on the observer gain L_i only. Therefore, the minimization for $J_k^{NC}(u_k)$ can be done by minimizing $J_k^{NC2}(\tilde{x})$ first and then minimizing $J_k^{NC1}(u_k)$.

First, we work on the estimator gain L_i (for $i=1,2,\dots,L$) such that $J_k^{NC2}(\tilde{x})$ (for $k=1,2,\dots,n$) is minimized. The matrix differential equation for $\sum_{\tilde{x}}(t)$ can be determined as follows:

$$\begin{aligned}\dot{\sum}_{\tilde{x}}(t)&=\sum_{i=1}^{L}\sum_{j=1}^{L}h_i(z(t))h_j(z(t))[(A_i-L_iC_j)\Sigma_{\tilde{x}}(t)\\&\quad+\Sigma_{\tilde{x}}(t)(A_i-L_iC_j)^T+G_iG_i^T+L_iD_jD_j^TL_i^T].\end{aligned}\tag{7.30}$$

For a steady-state solution

$$\sum_{\tilde{x}}(t)=\mathrm{cov}(\tilde{x}(t),\tilde{x}(t))=\tilde{P}\tag{7.31}$$

for all $t>0$ where $\tilde{P}$ is a symmetry positive–semidefinite constant matrix.

Hence,

$$\dot{\sum}_{\tilde{x}}(t)=0\tag{7.32}$$

and

$$\sum_{i=1}^{L}\sum_{j=1}^{L} h_i(z(t))h_j(z(t))[(A_i - L_iC_j)\tilde{P} + \tilde{P}(A_i - L_iC_j)^T + G_iG_i^T + L_iD_jD_j^TL_i^T] = 0. \tag{7.33}$$

Therefore, the optimal quadratic performance for $J_k^{NC2}(\tilde{x})$ is obtained as

$$J_k^{NC2}(\tilde{x}) = \mathrm{tr}(Q\tilde{P}). \tag{7.34}$$

Note that a sufficient condition for (7.33) implies that

$$(A_i - L_iC_j)\tilde{P} + \tilde{P}(A_i - L_iC_j)^T + G_iG_i^T + L_iD_jD_j^TL_i^T] = 0. \tag{7.35}$$

If the observer parameters are chosen as follows:

$$L_i = \tilde{P}C_i^T$$

we obtain

$$A_i\tilde{P} + \tilde{P}A_i^T - \tilde{P}C_i^TC_j\tilde{P} - \tilde{P}C_j^TC_i\tilde{P} + G_iG_i^T + \tilde{P}C_i^TD_jD_j^TC_i\tilde{P} = 0 \tag{7.36}$$

for $i = 1, 2, \ldots, L$.

Next, we work on the control gain K_{ik} such that $J_k^{NC1}(u_k)$ (for $k = 1, 2, \ldots, n$) is minimized. From the stochastic Hamilton–Jacobi–Bellman equation, we define

$$V_k(\hat{x}(t), t) = E\left\{\hat{x}^T(t_f)S\hat{x}(t_f) + \int_0^{t_f} [(\hat{x}(\tau)^TQ\hat{x}(\tau) + u_k^T(\tau)R_ku_k(\tau)]d\tau|\hat{x}(t)\right\}. \tag{7.37}$$

The stochastic Hamilton–Jacobi–Bellman equation then implies that

$$\frac{\partial V_k}{\partial t} + \hat{x}^TQ\hat{x} + u_k^TR_ku_k + \left(\frac{\partial V_k}{\partial \hat{x}}\right)^T\left\{\sum_{i=1}^{L} h_i(z(t))[A_i\hat{x}(t) + B_iu(t)]\right\} + \frac{1}{2}\mathrm{tr}\left(\sum_{i=1}^{L}\sum_{j=1}^{L} h_i(z(t))h_j(z(t)) \times \left(L_iD_jD_j^TL_i^T\frac{\partial^2 V_k}{\partial \hat{x}^2}\right)\right) = 0 \tag{7.38}$$

with endpoint condition $V_k(\hat{x}(t_f), t_f) = \hat{x}^T(t_f)S\hat{x}(t_f)$. Assuming that a solution of the above equation is of the following form:

$$V_k(\hat{x}(t), t) = \hat{x}^T(t)P_k^*(t)\hat{x}(t) + \eta_k(t) \tag{7.39}$$

For a steady-state solution, let $P_k^*(t) = P_k^*$ for all $t \geq 0$. Substituting (7.39) into (7.38), we obtain

$$\begin{aligned}
&\sum_{i=1}^{L} h_i(z(t))\{\hat{x}^T Q\hat{x} + u_k^T R_k u_k + (2P_k^*\hat{x})^T A_i x + (2P_k^*\hat{x})^T B_i u\} \\
&= \sum_{i=1}^{L} h_i(z(t))\{\hat{x}^T Q\hat{x} + u_k^T R_k u_k + (P_k^*\hat{x})^T A_i\hat{x} + (A_i\hat{x})^T P_k^*\hat{x} + (P_k^*\hat{x})^T B_i u + (B_i u)^T P_k^*\hat{x}\} \\
&= 0.
\end{aligned} \tag{7.40}$$

Then, (7.40) can be rewritten as

$$\begin{aligned}
&\sum_{i=1}^{L} h_i(z(t))\{\hat{x}^T Q\hat{x} + u_k^T R_k u_k + (P_k^*\hat{x})^T A_i\hat{x} + (A_i\hat{x})^T P_k^*\hat{x} \\
&+ \hat{x}^T P_k^*\left(\sum_{l=1,l\neq k}^{n} B_{il}u_l^* + B_{ik}u_k\right) + \left(\sum_{l=1,l\neq k}^{n} B_{il}u_l^* + B_{ik}u_k\right)^T P_k^*\hat{x}\} \\
&= \sum_{i=1}^{L} h_i(z(t))\sum_{j=1}^{L} h_j(z(t))\{\hat{x}^T[Q + P_k^*A_i + A_i^T P_k^* - \frac{1}{2}(P_k^*B_{ik}R_k^{-1}B_{jk}^T P_k^* \\
&+ P_k^*B_{jk}R_k^{-1}B_{ik}^T P_k^*)]\hat{x}\} + \hat{x}^T P_k^*\left(\sum_{l=1,l\neq k}^{n}(B_{il}u_l^*)\right) + \left(\sum_{l=1,l\neq k}^{n}(B_{il}u_l^*)\right)^T P_k^*\hat{x} \\
&+ u_k^T R_k u_k + \hat{x}^T P_k^*\left(\sum_{i=1}^{L} h_i(z(t))B_{ik}u_k\right) + \left(\sum_{i=1}^{L} h_i(z(t))B_{ik}u_k\right)^T P_k^*\hat{x} \\
&+\frac{1}{2}\sum_{i=1}^{L} h_i(z(t))\sum_{j=1}^{L} h_j(z(t))[\hat{x}^T(P_k^*B_{ik}R_k^{-1}B_{jk}^T P_k^* + P_k^*B_{jk}R_k^{-1}B_{ik}^T P_k^*)\hat{x}] = 0.
\end{aligned} \tag{7.41}$$

By the fact that $\sum_{i=1}^{L} h_i(z(t)) = \sum_{j=1}^{L} h_j(z(t))$, (7.41) can be rewritten as

$$\begin{aligned}
&\sum_{i=1}^{L} h_i(z(t))\sum_{j=1}^{L} h_j(z(t))\{\hat{x}^T[Q + P_k^*A_i + A_i^T P_k^* - \frac{1}{2}(P_k^*B_{ik}R_k^{-1}B_{jk}^T P_k^* \\
&+ P_k^*B_{jk}R_k^{-1}B_{ik}^T P_k^*)]\hat{x}\} + \hat{x}^T P_k^*\left(\sum_{l=1,l\neq k}^{n}(B_{il}u_l^*)\right) + \left(\sum_{l=1,l\neq k}^{n}(B_{il}u_l^*)\right)^T P_k^*\hat{x} \\
&+\left(R_k u_k + \sum_{i=1}^{L} h_i(z(t))(B_{ik}P_k^*\hat{x})\right)^T R_k^{-1}\left(R_k u_k + \sum_{i=1}^{L} h_i(z(t))(B_{ik}P_k^*\hat{x})\right)
\end{aligned} \tag{7.42}$$

It is found that if we let

$$u_k = -\sum_{i=1}^{L} h_i(z(t))(K_{ik}^*\hat{x}) \tag{7.43}$$

be denoted as u_k^*, where $K_{ik}^* = R_k^{-1}B_{ik}^T P_k^*$ for $i = 1,2,\ldots,L$, and $k = 1,2,\ldots,n$, then by substituting (7.43) into (7.42), we obtain

$$\sum_{i=1}^{L}\sum_{j=1}^{L}h_i(z(t))h_j(z(t))\{\hat{x}^T[Q+P_k^*A_i+A_i^TP_k^*-\frac{1}{2}(P_k^*B_{ik}R_k^{-1}B_{jk}^TP_k^*$$
$$+P_k^*B_{jk}R_k^{-1}B_{ik}^TP_k^*)-P_k^*\left(\sum_{l=1,l\neq k}^{n}(B_{il}u_l^*R_l^{-1}B_{il}^TP_l^*)\right) \tag{7.44}$$
$$-\left(\sum_{l=1,l\neq k}^{n}(B_{il}u_l^*R_l^{-1}B_{il}^TP_l^*)\right)^TP_k^*]\hat{x}\}=0.$$

A sufficient condition for (7.44) implies that

$$P_k^*A_i+A_i^TP_k^*-\frac{1}{2}(P_k^*B_{ik}R_k^{-1}B_{jk}^TP_k^*+P_k^*B_{jk}R_k^{-1}B_{ik}^TP_k^*)+Q$$
$$-P_k^*\left(\sum_{l=1,l\neq k}^{n}(B_{il}u_l^*R_l^{-1}B_{il}^TP_l^*)\right)-\left(\sum_{l=1,l\neq k}^{n}(B_{il}u_l^*R_l^{-1}B_{il}^TP_l^*)\right)^TP_k^*=0. \tag{7.45}$$

Therefore, the optimal H_2 quadratic performance for $J_k^{NC1}(u^*)$ is obtained as

$$J_k^{NC1}(u^*)=\lim_{t_f\to\infty}\frac{1}{t_f}\{V_k(\hat{x}(0),0)\}$$
$$=\lim_{t_f\to\infty}\frac{1}{t_f}\left\{\int_0^{t_f}\mathrm{tr}\left(\sum_{i=1}^{L}h_i(z(\tau))\sum_{j=1}^{L}h_j(z(\tau))[L_iD_jD_j^TL_i^TP_k^*]d\tau\right)\right\}. \tag{7.46}$$

Furthermore, the noncooperative optimal H_2 performance is obtained as

$$J_k^{NC}(u^*(t))=\mathrm{tr}(Q\tilde{P})+\lim_{t_f\to\infty}\frac{1}{t_f}\left\{\int_0^{t_f}\mathrm{tr}\left(\sum_{i=1}^{L}h_i(z(\tau))\sum_{j=1}^{L}h_j(z(\tau))(L_iD_jD_j^TL_i^TP_k^*)\right)d\tau\right\}. \tag{7.47}$$

In general, it is very difficult to get common solutions $\tilde{P}$ from a set of Riccati-like equations defined in (7.36). The following suboptimal solution is dealt with this problem. From (7.34), we get the upper bound of $J_k^{NC2}(\tilde{x})$ as

$$J_k^{NC2}(\tilde{x})=\mathrm{tr}(Q\tilde{P})<\mathrm{tr}(Q\hat{P}) \tag{7.48}$$

for any $\hat{P}=\hat{P}^T>0$ such that

$$A_i\hat{P}+\hat{P}A_i^T-\hat{P}C_i^TC_j\hat{P}-\hat{P}C_j^TC_i\hat{P}+G_iG_i^T+\hat{P}C_i^TD_jD_j^TC_i\hat{P}<0. \tag{7.49}$$

There are many feasible solutions for $\hat{P}$ in (7.49), in which a solution to minimize the upper bound $\mathrm{tr}(Q\hat{P})$ is the suboptimal solution for $\tilde{P}$ in (7.36).

With $\hat{W}=\hat{P}^{-1}$, we get

$$\hat{W}A_i + A_i^T\hat{W} - C_i^TC_j - C_j^TC_i + \hat{W}G_iG_i^T\hat{W} + C_i^TD_jD_j^TC_i < 0. \tag{7.50}$$

By the Schur complements [23], (7.50) is equivalent to the following LMIs:

$$\begin{bmatrix} \hat{W}A_i + A_i^T\hat{W} - C_i^TC_j - C_j^TC_i + C_i^TD_jD_j^TC_i & \hat{W}G_i \\ G_i^T\hat{W} & -I \end{bmatrix} < 0 \tag{7.51}$$

for $i = 1,2,\ldots,L$.

In other words, we seek the estimator gain $L_i = \hat{P}C_i^T = \hat{W}^{-1}C_i^T$ (for $i = 1,2,\ldots,L$) such that $\mathrm{tr}(Q\hat{W}^{-1})$ is minimized subject to (7.51). Since Q is symmetric positive, there exists a symmetric $\tilde{Q}$ such that $Q = \tilde{Q}\tilde{Q}^T$, i.e., $\tilde{Q} = Q^{1/2}$. We obtain $\mathrm{tr}(Q\hat{W}^{-1}) = \mathrm{tr}(\tilde{Q}\tilde{Q}^T\hat{W}^{-1}) = \mathrm{tr}(\tilde{Q}^T\hat{W}^{-1}\tilde{Q})$. Consider a new matrix variable $\Phi > \tilde{Q}^T\hat{W}^{-1}\tilde{Q}$, then $\mathrm{tr}(\Phi) > \mathrm{tr}(\tilde{Q}^T\hat{W}^{-1}\tilde{Q})$. Also, by Schur complement [23], $\Phi > \tilde{Q}^T\hat{W}^{-1}\tilde{Q}$ is equivalent to

$$\begin{bmatrix} \Phi & \tilde{Q}^T \\ \tilde{Q} & \hat{W} \end{bmatrix} > 0. \tag{7.52}$$

Therefore, the suboptimal fuzzy observer can be obtained by solving the following EVP:

$$\begin{aligned} &\min_{\hat{W}} \mathrm{tr}(\Phi) \\ &\text{subject to } \hat{W} = \hat{W}^T > 0, \begin{bmatrix} \Phi & \tilde{Q}^T \\ \tilde{Q} & \hat{W} \end{bmatrix} > 0 \text{ and (7.51).} \end{aligned} \tag{7.53}$$

Similarly, it is very difficult to solve P_k^* from the Riccati-like equations in (7.45). By the same argument as before, we can take a suboptimal approach for P_k^*. From (7.46) and (7.45), we get

$$J_k^{NC1}(u^*) < \lim_{t_f \to \infty} \frac{1}{t_f} \left\{ \int_0^{t_f} \mathrm{tr}\left(\sum_{i=1}^{L} h_i(z(\tau)) \sum_{j=1}^{L} h_j(z(\tau))[L_iD_jD_j^TL_i^TP_k]d\tau \right) \right\} \tag{7.54}$$

for any $P_k = P_k^T > 0$ such that

$$\begin{aligned} &P_kA_i + A_i^TP_k - \frac{1}{2}(P_kB_{ik}R_k^{-1}B_{jk}^TP_k + P_kB_{jk}R_k^{-1}B_{ik}^TP_k) \\ &- P_k\left(\sum_{l=1, l\neq k}^{n} (B_{il}R_l^{-1}B_{il}^TP_l) \right) - \left(\sum_{l=1, l\neq k}^{n} (B_{il}R_l^{-1}B_{il}^TP_l) \right)^T P_k + Q < 0 \end{aligned} \tag{7.55}$$

i.e., the suboptimal solution is to find a P_k from all feasible solutions of the inequality constraints in (7.55) such that the upper bound of $J_k^{NC1}(u^*)$ in the right-hand side of (7.54) is the smallest one.

With $W_k = P_k^{-1}$, (7.55) is equivalent to

$$A_iW_k + W_kA_i^T - \frac{1}{2}(B_{ik}R_k^{-1}B_{jk}^T + B_{jk}R_k^{-1}B_{ik}^T) - \left(\sum_{l=1,l\neq k}^{n}(B_{il}R_l^{-1}B_{il}^TP_l)\right)W_k$$
$$- W_k\left(\sum_{l=1,l\neq k}^{n}(B_{il}R_l^{-1}B_{il}^TP_l)\right)^T + W_kQW_k < 0. \tag{7.56}$$

By the Schur complements [23], (7.56) is equivalent to the following LMIs:

$$\begin{bmatrix} G_{ki} & W_k \\ W_k & -Q^{-1} \end{bmatrix} < 0 \tag{7.57}$$

where G_{ki} is designed as

$$G_{ki} = A_iW_k + W_kA_i^T - \frac{1}{2}(B_{ik}R_k^{-1}B_{jk}^T + B_{jk}R_k^{-1}B_{ik}^T) - \left(\sum_{l=1,l\neq k}^{n}(B_{il}R_l^{-1}B_{il}^TP_l)\right)W_k$$
$$- W_k\left(\sum_{l=1,l\neq k}^{n}(B_{il}R_l^{-1}B_{il}^TP_l)\right)^T$$

for $i = 1, 2, \ldots, L$, and $k = 1, 2, \ldots, n$.

Note that $\text{tr}(L_iD_jD_j^TL_i^TW_k^{-1}) = \text{tr}(D_j^TL_i^TW_k^{-1}L_iD_j)$. Consider a new matrix variable $\Psi_k > D_j^TL_i^TW_k^{-1}L_iD_j$, then $\text{tr}(\Psi_k) > \text{tr}(D_j^TL_i^TW_k^{-1}L_iD_j)$. Also, $\Psi_k > D_j^TL_i^TW_k^{-1}L_iD_j$ is equivalent to

$$\begin{bmatrix} \Psi_k & D_j^TL_i^T \\ L_iD_j & W_k \end{bmatrix} > 0. \tag{7.58}$$

The suboptimal solution W_k can be solved by minimizing the upper bound $\text{tr}(\Psi_k)$ and can be found by solving the following minimization problem:

$$\begin{aligned} &\min_{W_k} \text{tr}(\Psi_k) \\ &\text{subject to } W_k = W_k^T > 0, (7.58) \text{ and } (7.57) \end{aligned} \tag{7.59}$$

for $i = 1, 2, \ldots, L$, and $k = 1, 2, \ldots, n$.

Although the Nash equilibrium solution is a natural solution concept for the non-cooperative stochastic H_2 game problem, its computational solution might yet require more effort. Thus, it is natural to investigate an iterative scheme for the determination of Nash equilibrium solution for (7.59). Consider the following updating algorithm [1]:

$$\begin{aligned} W_1^{(q+1)} &= \min_{W_1} J_1^{NC1}(W_1, P_2^{(q)}, \ldots, P_k^{(q)}, \ldots, P_N^{(q)}) \\ &\vdots \\ W_k^{(q+1)} &= \min_{W_k} J_k^{NC1}(P_1^{(q)}, P_2^{(q)}, \ldots, W_k, \ldots, P_N^{(q)}) \\ &\vdots \\ W_N^{(q+1)} &= \min_{W_N} J_N^{NC1}(P_1^{(q)}, P_2^{(q)}, \ldots, P_k^{(q)}, \ldots, W_N) \end{aligned} \tag{7.60}$$

where $W_k^{(q+1)} = (P_k^{(q+1)})^{-1}$.

To realize the above iterative updating algorithm, we can solve the following minimization problem iteratively:

$$\begin{aligned} &\min_{W_k} \mathrm{tr}(\Psi_k^{(q)}) \\ &\text{subject to } W_k^{(q)} = (W_k^{(q)})^T > 0, (7.58) \text{ and } (7.57) \end{aligned} \tag{7.61}$$

where $q = 1, 2, \ldots, q_f$ (q is increased by one after each iteration) and P_l in (7.57) is replaced by $P_l^{(q-1)}$ and $P_l^{(0)}$ is a starting choice for player l ($l = 1, \ldots, n$ and $l \neq k$). The procedure is repeated until all $\left|\mathrm{tr}(\Psi_k^{(q_f)}) - \mathrm{tr}(\Psi_k^{(q_f-1)})\right| < \varepsilon$ (for $k = 1, \ldots, n$) is satisfied where ε is a given small value. Therefore, the suboptimal $W_k = W_k^{(q_f)}(k = 1, \ldots, n)$. And the initial $P_l^{(0)}(l = 1, \ldots, n)$ can be obtained as follows. Note that, with $Y_{lk} = P_l W_k$ (change of variables), (7.57) is equivalent to

$$\begin{bmatrix} H_{ki} & W_k \\ W_k & -Q^{-1} \end{bmatrix} < 0 \tag{7.62}$$

where $H_{ki} = A_i W_k + W_k A_i^T - \frac{1}{2}(B_{ik} R_k^{-1} B_{jk}^T + B_{jk} R_k^{-1} B_{ik}^T) - \left(\sum_{l=1, l\neq k}^{n} (B_{il} R_l^{-1} B_{il}^T Y_{lk})\right) - \left(\sum_{l=1, l\neq k}^{n} (B_{il} R_l^{-1} B_{il}^T Y_{lk})\right)^T$. We can solve the initial $P_k^*(k = 1, \ldots, n)$ from the following minimization problem, denoted as $P_k^{(0)}$

$$\begin{aligned} &\min_{\{W_k^{(0)}, Y_{lk}\}} \mathrm{tr}(\Psi_k) \\ &\text{subject to } W_k^{(0)} = (W_k^{(0)})^T > 0, (7.58) \text{ and } (7.62) \end{aligned} \tag{7.63}$$

and the initial $\mathrm{tr}(\Psi_k)$ obtained from (7.63) is denoted as $\mathrm{tr}(\Psi_k^{(0)})$ and $P_k^{(0)} = \left(W_k^{(0)}\right)^{-1}$ for $(k = 1, \ldots, n)$. Obviously, the initial solutions $P_k^{(0)}$ for $(k = 1, \ldots, n)$ are not Nash equilibrium solution since they are solutions obtained for that the player k uses his best policy when for each $j \neq k$, player j does not use their best policy, i.e.,

$$J_k^{NC1}([u_k^*(t)|u_{-k}(t)]) = \min_{u_k} J_k^{NC1}([u_k(t)|u_{-k}(t)]).$$

Therefore, from (7.54), we get

$$J_k^{NC1}(u^*) < \mathrm{tr}(\Psi_k^{(q_f)}). \tag{7.64}$$

Therefore, by solving the iterative EVP in (7.61), a suboptimal solution can be obtained finally. In this situation, the solution approaches its optimal values of $\mathrm{tr}(\Psi_k^{(q_f)})$ and $J_k^{NC1}(u^*)$.

Based on the above analysis, we obtain the following result.

Theorem 7.1
In the noncooperative fuzzy stochastic H_2 game with fuzzy observer of (7.19), if the observer parameters is chosen as

$$L_i^{\text{sub}} = \hat{W}^{-1} C_i^T \tag{7.65}$$

for $i = 1, 2, \ldots, L$ where $\hat{W} = \hat{W}^T > 0$ is the common solution of the EVP in (7.53) and if the fuzzy control law

$$u_k^{\text{sub}}(t) = -\sum\nolimits_{i=1}^{L} h_i(z(t))(K_{ik}^{\text{sub}} \hat{x}(t)) \tag{7.66}$$

is employed with

$$K_{ik}^{\text{sub}} = R_k^{-1} B_{ik}^T (W_k)^{-1} \tag{7.67}$$

for $i = 1, 2, \ldots, L$, and $k = 1, 2, \ldots, n$ where $R_k = R_k^T > 0$ is a weighting matrix and $W_k = (W_k)^T > 0$ can be obtained by solving the EVP in (7.61), then the fuzzy observer (7.19) is suboptimal and $u_k^{\text{sub}}(t)$ is suboptimal fuzzy control action of the kth player for the noncooperative control performance in (7.2).

Proof: Based on the previous analysis, the proof is immediately followed.

B. Fuzzy Cooperative Stochastic H_2 Game Design

The design purpose of the cooperative stochastic H_2 control is to specify the control gain K_i and the estimator gain L_i (for $i = 1, 2, \ldots, L$) such that the common cost function in (7.5) is minimized for the cooperative fuzzy stochastic H_2 game strategy problem. By the same argument as above, (7.5) can be rewritten as follows:

$$\begin{aligned} J^C(u) &= \lim_{t_f \to \infty} \frac{1}{t_f} E\left\{ \hat{x}^T(t_f) S \hat{x}(t_f) + \int_0^{t_f} (\hat{x}^T(t) Q \hat{x}(t) + u^T(t) R u(t)) dt \right\} \\ &+ \lim_{t_f \to \infty} \frac{1}{t_f} \left\{ \text{tr}(S \Sigma_{\tilde{x}}(t_f)) + \int_0^{t_f} \text{tr}(Q \Sigma_{\tilde{x}}(t)) dt \right\} \\ &= J^{C1}(u) + J^{C2}(\tilde{x}) \end{aligned} \tag{7.68}$$

where

$$J^{C1}(u) = \lim_{t_f \to \infty} \frac{1}{t_f} E\left\{ \hat{x}^T(t_f) S \hat{x}(t_f) + \int_0^{t_f} (\hat{x}^T(t) Q \hat{x}(t) + u^T(t) R u(t)) dt \right\}$$

is related to the design of fuzzy controller and

$$J^{C2}(\tilde{x}) = \lim_{t_f \to \infty} \frac{1}{t_f} \left\{ \text{tr}(S \Sigma_{\tilde{x}}(t_f)) + \int_0^{t_f} \text{tr}(Q \Sigma_{\tilde{x}}(t)) dt \right\}$$

is related to the design of fuzzy observer.

For the observer design part, it is the same as that in noncooperative stochastic H_2 game case. For the control design part, similar to the stochastic Hamilton–Jacobi–Bellman equation, we define

$$V(\hat{x}(t),t) = E\left\{ \hat{x}^T(t_f)S\hat{x}(t_f) + \int_t^{t_f} [\hat{x}(\tau)^T Q\hat{x}(\tau) + u^T(\tau)Ru(\tau)]d\tau | \hat{x}(t) \right\}. \quad (7.69)$$

The stochastic Hamilton–Jacobi–Bellman equation then implies that

$$\begin{aligned} \min_{u(t)} \Bigg\{ & \frac{\partial V}{\partial t} + \hat{x}^T Q\hat{x} + u^T Ru + \left(\frac{\partial V}{\partial \hat{x}}\right)^T \left[\sum_{i=1}^{L} h_i(z(t))(A_i\hat{x}(t) + B_i u(t))\right] \\ & + \frac{1}{2}\mathrm{tr}\left(\sum_{i=1}^{L} h_i(z(t)) \sum_{j=1}^{L} h_j(z(\tau))\left(L_i D_j D_j^T L_i^T \frac{\partial^2 V}{\partial \hat{x}^2}\right)\right)\Bigg\} = 0 \end{aligned} \quad (7.70)$$

with endpoint condition $V(\hat{x}(t_f), t_f) = \hat{x}^T(t_f)S\hat{x}(t_f)$.

By the same argument as that in noncooperative stochastic H_2 game case, a solution of the above equation is of the following form:

$$V(\hat{x}(t),t) = \hat{x}^T(t)P^*\hat{x}(t) + \eta(t) \quad (7.71)$$

By substituting (7.71) into (7.70), at the steady state, we get

$$\begin{aligned} & \sum_{i=1}^{L} h_i(z(t)) \sum_{j=1}^{L} h_j(z(t))\{\hat{x}^T[Q + P^*A_i + A_i^T P^* - \frac{1}{2}(P^*B_i R^{-1}B_j^T P^* \\ & + P^*B_j R^{-1}B_i^T P^*)]\hat{x}\} + u^T Ru + \hat{x}^T P^*(\sum_{i=1}^{L} h_i(z(t))(B_i u)) \\ & + (\sum_{i=1}^{L} h_i(z(t))(B_i u))^T P^*\hat{x} + \frac{1}{2}\sum_{i=1}^{L} h_i(z(t)) \sum_{j=1}^{L} h_j(z(t))[\hat{x}^T \\ & \times (P^*B_i R^{-1}B_j^T P^* + P^*B_j R^{-1}B_i^T P^*)\hat{x}] = 0 \end{aligned} \quad (7.72)$$

By the fact that $\sum_{i=1}^{L} h_i(z(t)) = \sum_{j=1}^{L} h_j(z(t))$, (7.72) can be rewritten as

$$\begin{aligned} & \sum_{i=1}^{L} h_i(z(t)) \sum_{j=1}^{L} h_j(z(t))\{\hat{x}^T[Q + P^*A_i + A_i^T P^* - \frac{1}{2}(P^*B_i R^{-1}B_j^T P^* \\ & + P^*B_j R^{-1}B_i^T P^*)]\hat{x}\} + (Ru + \sum_{i=1}^{L} h_i(z(t))(B_i P^*\hat{x}))^T \\ & \times \mathrm{R}^{-1}(Ru + \sum_{j=1}^{L} h_j(z(t))(B_j P^*\hat{x})) = 0. \end{aligned} \quad (7.73)$$

It is found that if we let

$$u(t) = -\sum_{i=1}^{L} h_i(z(t))(K_i^* \hat{x}(t)) \tag{7.74}$$

be denoted as $u^*(t)$, where $K_i^* = R^{-1}B_iP^*$ for $i = 1, 2, \ldots, L$, then by substituting (7.74) into (7.73), we obtain

$$\begin{aligned}\sum_{i=1}^{L}\sum_{j=1}^{L} h_i(z(t))h_j(z(t))\{\hat{x}^T[Q - \frac{1}{2}(P^*B_iR^{-1}B_j^TP^* \\ + P^*B_jR^{-1}B_i^TP^*) + P^*A_i + A_i^TP^*]\hat{x}\} = 0.\end{aligned} \tag{7.75}$$

A sufficient condition for (7.75) implies that

$$P^*A_i + A_i^TP^* - \frac{1}{2}(P^*B_iR^{-1}B_j^TP^* + P^*B_jR^{-1}B_i^TP^*) + Q = 0. \tag{7.76}$$

Therefore, the cooperative stochastic H_2 optimal control performance is obtained as

$$J^{C1}(u^*) = \lim_{t_f \to \infty} \frac{1}{t_f}\left\{\int_0^{t_f} \mathrm{tr}\left(\sum_{i=1}^{L} h_i(z(\tau))\sum_{j=1}^{L} h_j(z(\tau))[L_iD_jD_j^TL_i^TP_k]d\tau\right)\right\}. \tag{7.77}$$

Furthermore, the cooperative H_2 optimal control performance

$$J^C(u^*) = \mathrm{tr}(Q\tilde{P}) + \lim_{t_f \to \infty} \frac{1}{t_f}\left\{\int_0^{t_f} \mathrm{tr}\left(\sum_{i=1}^{L} h_i(z(\tau))\sum_{j=1}^{L} h_j(z(\tau))(L_iD_jD_j^TL_i^TP_k)\right)d\tau\right\}.$$

Similarly, it is difficult to solve P^* from the Riccati-like equations in (7.76). The following suboptimal solution is employed to deal with this problem. Recall that

$$J^{C1}(u^*) < \lim_{t_f \to \infty} \frac{1}{t_f}\left\{\int_0^{t_f} \mathrm{tr}\left(\sum_{i=1}^{L} h_i(z(\tau))\sum_{j=1}^{L} h_j(z(\tau))(L_iD_jD_j^TL_i^TP_k)\right)d\tau\right\} \tag{7.78}$$

for any $P = P^T > 0$ such that

$$PA_i + A_i^TP - \frac{1}{2}(PB_iR^{-1}B_j^TP + PB_jR^{-1}B_i^TP) + Q < 0. \tag{7.79}$$

With $W = P^{-1}$, (7.79) is equivalent to

$$A_iW + WA_i^T - \frac{1}{2}(B_iR^{-1}B_j^T + B_jR^{-1}B_i^T) + WQW < 0. \tag{7.80}$$

By the Schur complements [23], (7.80) is equivalent to the following LMIs:

$$\begin{bmatrix} A_iW + WA_i^T - \frac{1}{2}(B_iR^{-1}B_j^T + B_jR^{-1}B_i^T) & W \\ W & -Q^{-1} \end{bmatrix} < 0 \tag{7.81}$$

for $i, j = 1, 2, \ldots, L$. Therefore, the upper bound of $J^{C1}(u^*)$ can be found by solving the following EVP:

$$\begin{aligned} &\min_W \operatorname{tr}(\Psi) \\ &\text{subject to } W = W^T > 0 \\ &\begin{bmatrix} \Psi & D_j^T L_i^T \\ L_i D_j & W \end{bmatrix} > 0 \text{ and (7.81)} \end{aligned} \tag{7.82}$$

for $i, j = 1, 2, \ldots, L$. Therefore, from (7.78), we get

$$J^{C1}(u^*) < \operatorname{tr}(\Psi). \tag{7.83}$$

Based on the analysis above, we obtain the following result.

Theorem 7.2
In the cooperative stochastic H_2 game with the fuzzy observer of (7.19), if the observer parameters are chosen as

$$L_i^{sub} = \hat{W}^{-1} C_i^T \tag{7.84}$$

where $\hat{W} = \hat{W}^T > 0$ is a common solution of the EVP in (7.53) and suppose the fuzzy control law

$$u^{sub}(t) = -\sum_{i=1}^{L} h_i(z(t))[K_i^{sub}\hat{x}(t)] \tag{7.85}$$

is employed with

$$K_i^{sub} = R^{-1} B_i^T W^{-1} \tag{7.86}$$

for $i = 1, 2, \ldots, L$, where $R = R^T > 0$ is a weighting matrix and $W = W^T > 0$ is obtained by solving the EVP in (7.82), then the fuzzy estimator (7.19) is suboptimal and $u^{sub}(t)$ in (7.85) is the suboptimal fuzzy control for the cooperative H_2 game strategy performance in (7.5).

Proof: Based on the analysis of suboptimal approach, the proof is immediately followed. Based on the above analysis, the robust control design for the suboptimal noncooperative and cooperative stochastic H_2 game problems with fuzzy observer are summarized as the following design procedure.
Design Procedure:
Step 1) Select membership function and construct fuzzy model to approximate the nonlinear stochastic system.
Step 2) Select weighting matrices Q and R_k for the noncooperative stochastic H_2 game (or Q and R for the cooperative stochastic H_2 game) according to the design objective.

Step 3) Solve the EVP in (7.53) for the noncooperative stochastic H_2 game suboptimal fuzzy observer (or solve the EVP in (7.53) for the cooperative H_2 game suboptimal fuzzy observer) to obtain $\hat{W}$.

Step 4) Solve the minimization problem in (7.59) for noncooperative stochastic H_2 game to obtain W_k (or solve the EVP in (7.82) for cooperative stochastic H_2 game to obtain W).

Step 5) Obtain fuzzy observer parameters L_i^{sub} from (7.65) for noncooperative stochastic H_2 game case (or from (7.84) for cooperative stochastic H_2 game case) and then construct the fuzzy observer in (7.19).

Step 6) Obtain control parameters K_{ik}^{sub} from (7.67) for noncooperative stochastic H_2 game case (or K_i^{sub} from (7.86) for cooperative stochastic H_2 game case) and then obtain the fuzzy control rule of (7.20).

Remark 7.1 The fuzzy observer for the noncooperative and cooperative fuzzy stochastic H_2 game problems is the same. ■

7.4 Simulation Design Examples

A three-machine interconnected power system is described as follows [112]:

$$\begin{aligned}
\dot{x}_1(t) &= x_2(t) + w_1(t) \\
\dot{x}_2(t) &= f_2(t) + w_2(t) \\
\dot{x}_3(t) &= x_4(t) + w_3(t) \\
\dot{x}_4(t) &= f_4(t) + w_4(t) \\
\dot{x}_5(t) &= x_6(t) + w_5(t) \\
\dot{x}_6(t) &= f_6(t) + w_6(t) \\
y(t) &= \begin{bmatrix} x_1(t) \\ x_3(t) \\ x_5(t) \end{bmatrix} + \begin{bmatrix} v_1(t) \\ v_3(t) \\ v_5(t) \end{bmatrix}
\end{aligned} \tag{7.87}$$

where

$$\begin{aligned}
f_2(t) &= -\frac{D_1}{M_1}x_2(t) + \frac{1}{M_1}u_1(t) + \frac{E_1E_2Y_{12}}{M_1}\left[\cos(\delta_{12}^0 - \theta_{12}) - \cos(\theta_{12} - x_1(t) + x_3(t))\right] \\
&\quad + \frac{E_1E_3Y_{13}}{M_1}\left[\cos(\delta_{13}^0 - \theta_{13}) - \cos(\theta_{13} - x_1(t) + x_5(t))\right] \\
f_4(t) &= -\frac{D_2}{M_2}x_4(t) + \frac{1}{M_2}u_2(t) + \frac{E_2E_1Y_{21}}{M_2}\left[\cos(\delta_{21}^0 - \theta_{21}) - \cos(\theta_{21} - x_3(t) + x_1(t))\right] \\
&\quad + \frac{E_2E_3Y_{23}}{M_2}\left[\cos(\delta_{23}^0 - \theta_{23}) - \cos(\theta_{23} - x_3(t) + x_5(t))\right] \\
f_6(t) &= -\frac{D_3}{M_3}x_6(t) + \frac{1}{M_3}u_3(t) + \frac{E_3E_1Y_{31}}{M_3}\left[\cos(\delta_{31}^0 - \theta_{31}) - \cos(\theta_{31} - x_5(t) + x_1(t))\right] \\
&\quad + \frac{E_3E_2Y_{32}}{M_3}\left[\cos(\delta_{32}^0 - \theta_{32}) - \cos(\theta_{32} - x_5(t) + x_3(t))\right]
\end{aligned}$$

where $x_1(t) = \delta_1(t)$, $x_3(t) = \delta_2(t)$, and $x_5(t) = \delta_3(t)$ are the absolute rotor angle of the 1st, 2nd, and 3rd machines, respectively, and assume that $x_i(t) \in [-\pi/2, \pi/2]$ for $i = 1,3,5$; $x_2(t) = \dot{\delta}_1(t)$, $x_4(t) = \dot{\delta}_2(t)$, and $x_6(t) = \dot{\delta}_3(t)$ are the absolute angular velocity of the 1st, 2nd, and 3rd machines, respectively; M_i is the inertia coefficient; D_i is the damping coefficient; E_i is the internal voltage; Y_{ij} is the modulus of the transfer admittance between the ith and jth machines; θ_{ij} is the phase angle of the transfer admittance between the$i$th and jth machines, for $i,j = 1,2,3$.

At the steady state of three-machine systems, the mechanical power delivered to the ith machine is equal to electrical power delivered to the network and the synchronization is achieved. In this situation, $u_i(t) = 0$. However, some initial conditions and disturbances due to short circuit and sudden increment of power load may occur in the interconnected three-machine power system. The control $u_i(t)$ must be employed to eliminate the transient phenomenon of three-machine system or the synchronization will be destroyed.

We assume the three-machine interconnected systems' parameters as follows [112]:

$$\begin{aligned}
&E_1 = 1.017\ E_2 = 1.005 \text{ and } E_3 = 1.033\\
&M_1 = 1.03\ M_2 = 1.25 \text{ and } M_3 = 1.4\\
&D_1 = 0.8\ D_2 = 1.2 \text{ and } D_3 = 1.1\\
&Y_{12} = 1.98\ Y_{13} = 1.14 \text{ and } Y_{23} = 1.06\\
&\theta_{12} = 1.5\ \theta_{13} = 1.55 \text{ and } \theta_{23} = 1.56\\
&\delta_{12}^0 = 0.5\ \delta_{13}^0 = 0.18 \text{ and } \delta_{23}^0 = -0.32
\end{aligned}$$

and $w(t) = [w_1(t), w_2(t), w_3(t), w_4(t), w_5(t), w_6(t)]^T$ and $v(t) = [v_1(t), v_3(t), v_5(t)]^T$ are external disturbance and measurement noise, respectively, with $E\{w(t), w(\tau)\} = I\delta(t-\tau)$ and $E\{v(t), v(\tau)\} = I\delta(t-\tau)$.

Example 7.1: In the above three-machine interconnected power system, in order to achieve synchronization, each machine designs a fuzzy controller to minimize its individual performance in (7.2) to eliminate the transient behavior due to short circuit and sudden changes of power load. This is a noncooperative stochastic H_2 game control design problem. Now, following the **Design Procedure** in the previous section, the suboptimal control policy for the noncooperative stochastic H_2 game using suboptimal fuzzy observer is determined by the following steps: Step 1): To use the fuzzy control approach, we must have a fuzzy model which represents the dynamics of the nonlinear plant. In these examples, we specify three fuzzy sets for x_1, x_3 and x_5, respectively, to construct the fuzzy model. This makes twenty-seven $(3 \times 3 \times 3)$ fuzzy rules for the example where membership functions and fuzzy sets are shown in Fig. 7.1. Step 2): Select $Q = I$ and $R_k = 0.001$ for $k = 1,2,3$. Step 3): Solve the EVP in (7.53) for the suboptimal fuzzy observer to obtain W. Step 4): Solve the iterative EVP in (7.61) for noncooperative stochastic H_2 game to obtain W_k (for $k = 1,2,3$). The updating process stops after four iterations with $\varepsilon = 0.001$ (refer to Fig. 7.2). Step 5): Construct the suboptimal fuzzy observer. Step 6): Construct the noncooperative stochastic H_2 game fuzzy control law. Figures 7.3–7.6 present the simulation results for the noncooperative stochastic H_2 game fuzzy control. The initial condition is assumed to be $(x_1(0), x_2(0), x_3(0), x_4(0), x_5(0), x_6(0), \hat{x}_1(0), \hat{x}_2(0), \hat{x}_3(0), \hat{x}_4(0), \hat{x}_5(0), \hat{x}_6(0))^T = (1,0,1,0,1,0,0,0,0,0,0,0,)^T$. The external disturbance $w(t)$ and measurement noise $v(t)$ are assumed to be white noise with identity power spectrum. Figure 7.3 shows the trajectories

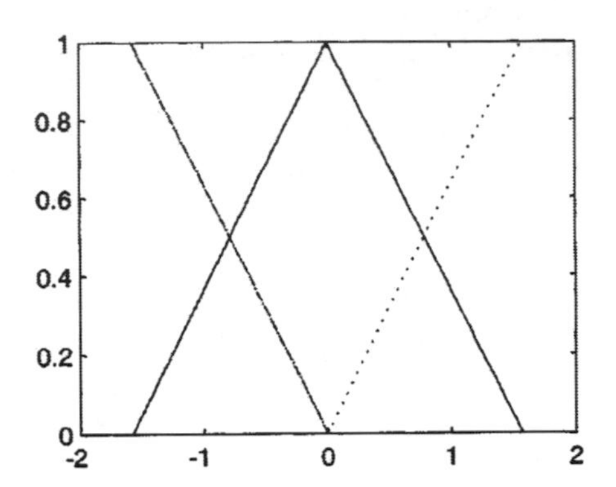

FIGURE 7.1
Membership functions and fuzzy sets for $x_i \in [-\pi/2, \pi/2] (i = 1, 3, 5)$.

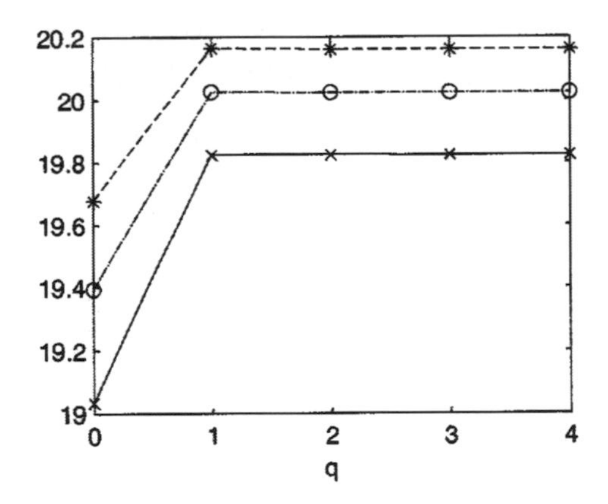

FIGURE 7.2
Iterations of $\mathrm{tr}(\Psi^{(q)}{}_1)$ denoted by "×", $\mathrm{tr}(\Psi^{(q)}{}_2)$ denoted by "○", and $\mathrm{tr}(\Psi^{(q)}{}_3)$ denoted by "∗".

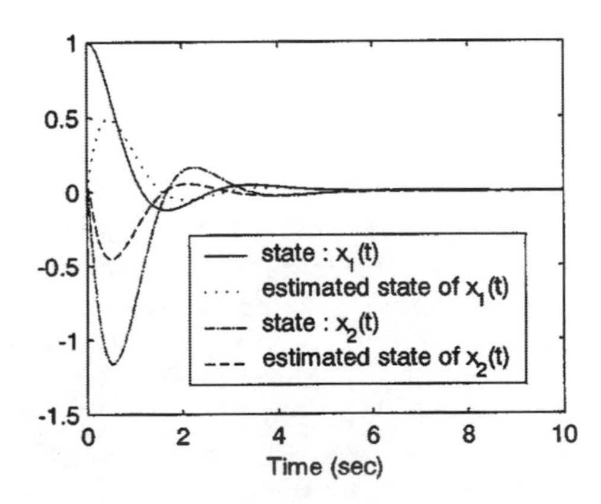

FIGURE 7.3
The trajectories of the states x_1 and x_2 including estimated states $\hat{x}_1$ and $\hat{x}_2$ (noncooperative stochastic H_2 game case).

of the states $x_1(t)$, and $x_2(t)$ (including the estimated states $\hat{x}_1(t)$, and $\hat{x}_2(t)$). Figure 7.4 shows the trajectories of the states $x_3(t)$ and $x_4(t)$ (including the estimated states $\hat{x}_3(t)$ and $\hat{x}_4(t)$). Figure 7.5 shows the trajectories of the states $x_5(t)$ and $x_6(t)$ (including the estimated states $\hat{x}_5(t)$ and $\hat{x}_6(t)$). The control inputs are presented in Fig. 7.6.

Example 7.2: In the above three-machine interconnected power system, suppose all three machines cooperate to design their fuzzy controller to compensate its transient behavior to achieve synchronization by minimizing the common control performance (7.5). This is a cooperative stochastic H_2 game design problem. The suboptimal control policy for the cooperative stochastic H_2 game using suboptimal fuzzy observer can be determined by the same procedure as Example 7.1 with $Q = I$ and $R = 0.001 \times I$. Figures 7.7–7.11 present the simulation results for the suboptimal fuzzy observer-based cooperative H_2 game fuzzy control. Figure 7.7 shows the trajectories of the states $x_1(t)$ and $x_2(t)$

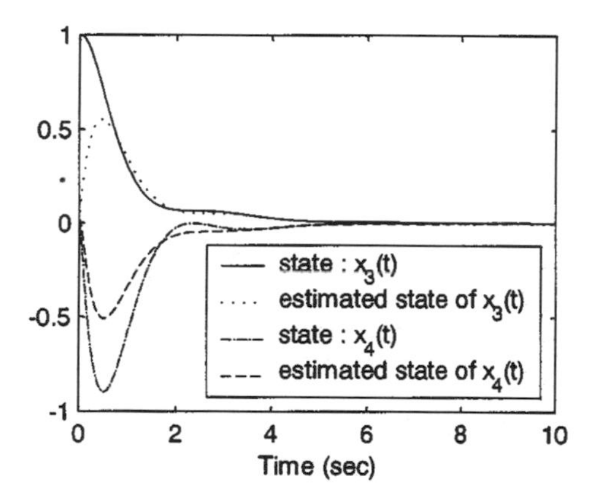

FIGURE 7.4
The trajectories of the states x_3 and x_4 including estimated states $\hat{x}_3$ and $\hat{x}_4$ (noncooperative stochastic H_2 game case).

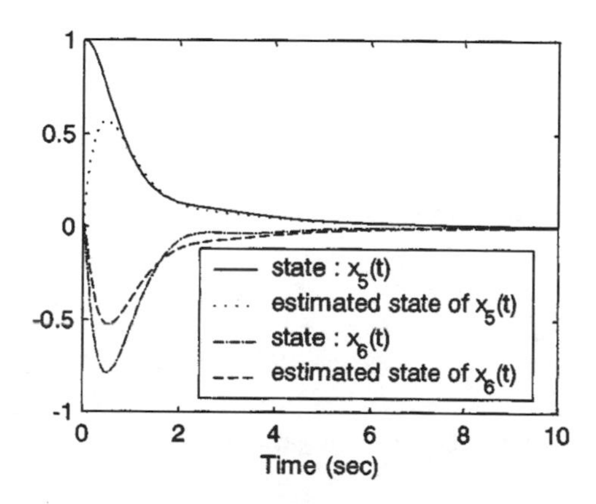

FIGURE 7.5
The trajectories of the states x_5 and x_6 including estimated states $\hat{x}_5$ and $\hat{x}_6$ (noncooperative stochastic H_2 game case).

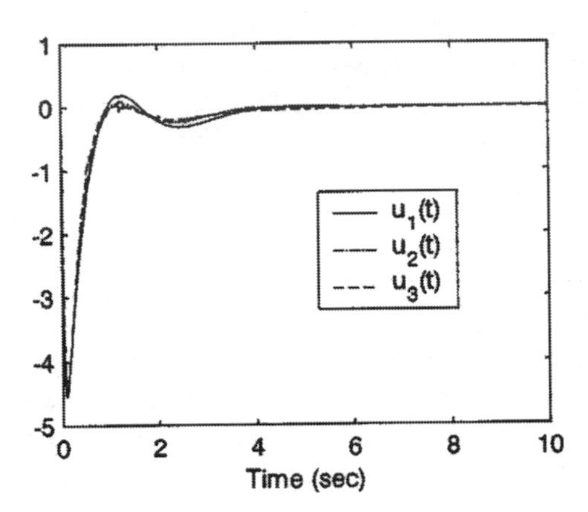

FIGURE 7.6
The noncooperative stochastic H_2 game control inputs.

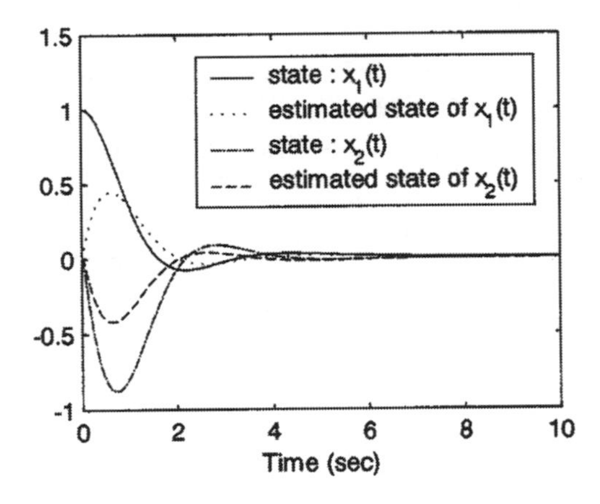

FIGURE 7.7
The trajectories of the states x_1 and x_2 including estimated states $\hat{x}_1$ and $\hat{x}_2$ (cooperative stochastic H_2 game case).

(including the estimated states $\hat{x}_1(t)$ and $\hat{x}_2(t)$). Figure 7.8 shows the trajectories of the states $x_3(t)$ and $x_4(t)$ (including the estimated states $\hat{x}_3(t)$ and $\hat{x}_4(t)$). Figure 7.9 shows the trajectories of the states $x_5(t)$ and $x_6(t)$ (including the estimated states $\hat{x}_5(t)$ and $\hat{x}_6(t)$). The control inputs are presented in Fig. 7.10. The simulation results show that the stochastic H_2 game cooperative fuzzy controller yields better performance as shown in Fig. 7.11 since it features the property that no other joint decision of the players can improve the H_2 game performance of at least one of them, without degrading the H_2 game performance of others.

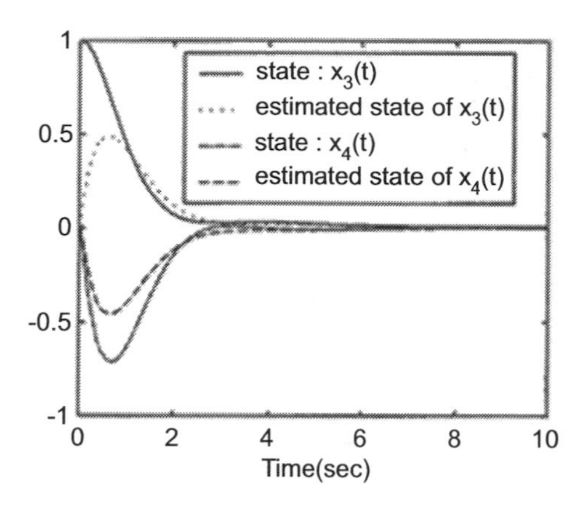

FIGURE 7.8
The trajectories of the states x_3 and x_4 including estimated states $\hat{x}_3$ and $\hat{x}_4$ (cooperative stochastic H_2 game case).

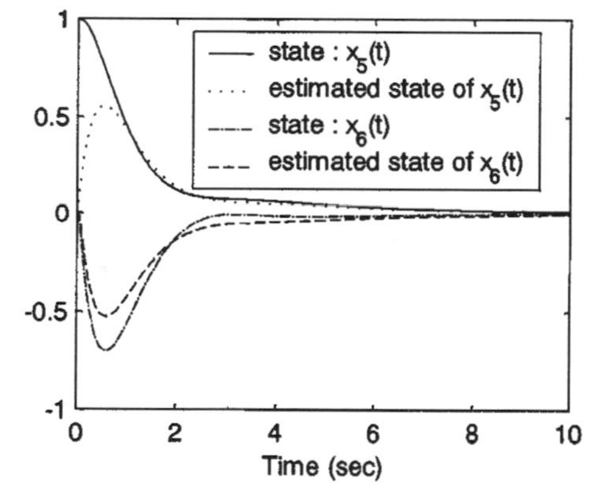

FIGURE 7.9
The trajectories of the states x_5 and x_6 including estimated states $\hat{x}_5$ and $\hat{x}_6$ (cooperative stochastic H_2 game case).

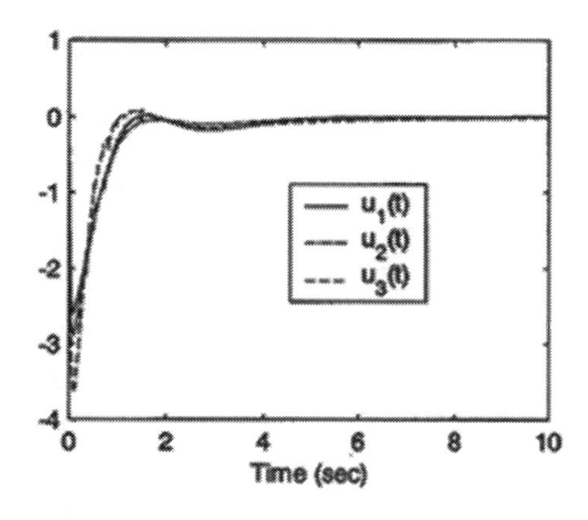

FIGURE 7.10
The cooperative stochastic H_2 game control inputs.

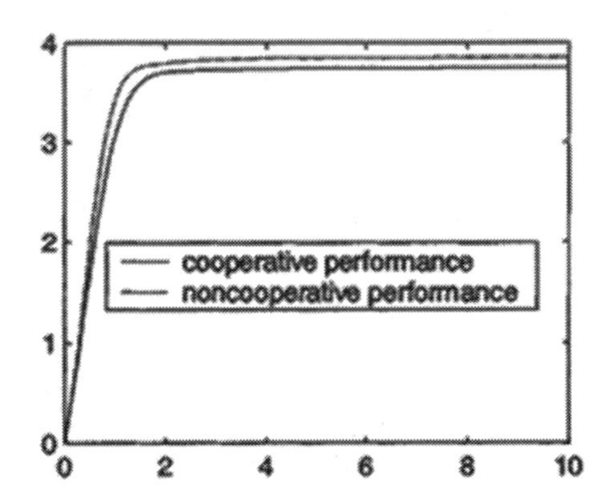

FIGURE 7.11
The plots of $\int_0^{t_f} (x^T Qx + u^T Ru)dt$ for cooperative H_2 game case and noncooperative H_2 game case.

7.5 Conclusion

In this chapter, both stochastic noncooperative and cooperative fuzzy H_2 game problems of nonlinear stochastic systems are solved using the suboptimal approach. Based on the fuzzy model and the suboptimal approach, the outcome of the noncooperative and cooperative fuzzy stochastic H_2 game problems is parameterized in terms of an iterative update EVP. A suboptimal fuzzy observer has also been introduced in the case that the state variables are unavailable. Based on the separation principal method, the solution of the observer-based fuzzy stochastic H_2 game problems is also parameterized in terms of an iterative update EVP. The proposed design methods are very simple and more efficient than other control methods to deal with the general n-person nonlinear stochastic H_2 game problems. Simulation examples indicate that the desired performance of noncooperative and cooperative stochastic H_2 game control designs for nonlinear interconnected power systems can be achieved using the proposed stochastic H_2 game methods. Hence, the proposed methods are suitable for solving the practical stochastic H_2 game problems in real applications.

8

Stochastic Noncooperative and Cooperative Multi-player H_∞ Game Strategy Design for Nonlinear Stochastic Jump Diffusion Control Systems with External Disturbance

8.1 Introduction

The differential game has been extensively researched in the last few decades [1,9,113,114] and the concept has been applied to many fields such as financial market investment [20,115], control system [116,117], and communication system [118]. In the cooperative differential game, the game strategy design for each player must be compromised with the common desired objective. At the same time, each player's game strategy depends on the information from other players. On the other hand, in the case of noncooperative game, the goal of each player is different from others and the information of each player is not exchanged. Hence, the strategy of each player is chosen to optimize his/her own desired goals. Under the concept of Pareto solution [21,119,120], the Pareto strategy is that in all the goals of using this strategy, at least one goal cannot be dominated by other strategies. Since there exist many Pareto optimal solutions to the multi-objective optimization problem (MOP) for the noncooperative game strategy design, designers are free to choose a preferable strategy from the Pareto front to achieve the desired goals. The non-cooperative differential games have been in-depth studied in many fields such as communications, signal processing, and financial markets [121–124].

In recent decades, stochastic dynamic systems have been widely studied and applied. In the field of control, the stochastic systems with intrinsic random fluctuation due to Wiener process and Poisson process have been concerned in recent years [8,22,125–128]. In general, the Wiener process is treated as a continuous random fluctuation and Poisson process can be regarded as discontinuous random fluctuation. For example, in the physical phenomena, the system state jumps caused by failure of power supply, short circuit, or some unexpected events can be regarded as a Poisson process. Due to the fact that the system dynamics will be disturbed by these intrinsic random fluctuations, it is more realistic to consider these internal fluctuations into the system model. By introducing the stochastic properties into stochastic game model, stochastic H_2 cooperative game strategy and noncooperative strategy are widely studied and applied to several fields in literature [3,19,129–131].

Since the effect of the external disturbance is unpredictable, the conventional stochastic game strategies based on the H_2 quadratic performance could not solve the stochastic noncooperative game and cooperative game problem of the nonlinear stochastic system with external disturbance [1,132]. Further, the present iterative schemes with updating algorithm have been employed to solve the Nash equilibrium for noncooperative game strategy indirectly [1,9]. In general, it is not easy to solve the design problem of multi-player stochastic noncooperative game strategies of nonlinear stochastic systems by these iterative schemes. As a result, the two difficult design problems mentioned above motivate us to address the design of stochastic H_∞ cooperative game strategy and stochastic H_∞ noncooperative game strategy for nonlinear stochastic jump diffusion system with external disturbance.

In this chapter, we consider the multi-player stochastic H_∞ noncooperative and cooperative game strategy design for nonlinear stochastic jump diffusion system with external disturbance. In general, designers have used H_∞ robust control method [8,133] to reduce the effect of external disturbances in the system on the desired target tracking and maintain stability under external disturbances. Hence, by taking the worst-case effect of external disturbance and other strategies of competitive players into account, a novel stochastic robust H_∞ utility function is proposed to replace the conventional stochastic H_2 utility function for each player.

In the case of n-player H_∞ noncooperative game, without knowing the strategies of other players, each player has his/her own desired target and considers the effect of the strategies from other players as unpredictable disturbances. Thus, the strategy of each player is chosen to minimize the worst-case effect from the external disturbance and strategies of competitive players on the desired target tracking from the H_∞ game perspective. Therefore, we need to solve n H_∞ noncooperative game strategies for all players simultaneously. In general, it is very difficult to solve the n-player H_∞ noncooperative game strategy directly. By using the proposed indirect method, we transformed the n-player H_∞ noncooperative game strategy design problem to an equivalent MOP under n HJII constraints. We find that the stochastic noncooperative H_∞ game strategy solved by a HJIIs-constrained MOP is a Nash equilibrium solution. In the case of the n-player stochastic H_∞ cooperative game, the information of all players is available for each player and the common desired target is compromised beforehand. Thus, the strategies of every player are chosen together to achieve the tracking of common desired target with an efficient attenuation of the worst-case effect of the external disturbance. Further, the n-player stochastic H_∞ cooperative game strategies design problem is transformed to an equivalent HJII-constrained single-objective optimization problem (SOP).

Due to the difficulty in solving HJIIs for multi-player stochastic noncooperative H_∞ game and HJII for stochastic cooperative H_∞ game, we use the Takagi–Sugeno (T-S) fuzzy model to interpolate the nonlinear stochastic system with a set of local linearized stochastic systems to facilitate the design procedure of noncooperative H_∞ game strategy design and cooperative H_∞ game strategy design [4,6,105,134,135]. Based on the fuzzy interpolation method, the HJIIs-constrained MOP for the noncooperative stochastic H_∞ game strategy design could be transformed to an equivalent LMIs-constrained MOP which can be easily solved with the help of Matlab LMI toolbox to find the best multiobjective strategy [5,23,100], i.e., the multi-player stochastic H_∞ noncooperative game strategy problem of the nonlinear stochastic system with external disturbance could be transformed into an LMIs-constrained MOP. Also, based on the fuzzy interpolation method, the HJII-constrained SOP for cooperative stochastic H_∞ game strategy design is transformed to an equivalent LMI-constrained SOP which could be easily solved with the help of LMI toolbox in Matlab.

On the other hand, instead of using the indirect iterative schemes to search the Nash equilibrium solution in previous works [119–124], we developed an LMIs-constrained multi-objective evolution algorithm (MOEA) based on NSGA-II MOEA in [136] to efficiently solve the LMIs-constrained MOP for multi-player noncooperative H_∞ game strategy for nonlinear stochastic jump diffusion system. The LMIs-constrained MOEA could optimize the multi-objective goals of all players in a single round, i.e., find the corresponding Pareto optimal solution at the Pareto front by a single round. Normally, we will use the strategy on the knee point of the Pareto front in the case of non-cooperative H_∞ game because of its compromised characteristic of the Pareto front. The knee solution will achieve the best solution for each player at a less sacrifice of the others' goals.

The main aims of this chapter are summarized as follows: Unlike the conventional multi-player stochastic noncooperative H_2 game strategy design with a common desired target $x_d = 0$ and zero external disturbance $v(t)$, i.e., each player considers stabilization problem only with different weighting matrices Q_k and R_k in the H_2 utility function; for $k = 1, \ldots, n$, the proposed multi-player H_∞ noncooperative game strategy has different desired targets x_d^k for each player and considers the effect of external disturbance and the effect of competitive players strategies on the desired target of each player k, for $k = 1, \ldots, n$, i.e., each player considers the tracking problem with different desired targets x_d^k, external disturbance $v(t)$, other players' strategies, and different weighting matrices Q_k and R_k in the H_∞ utility function, for $k = 1, \ldots, n$. Besides, by using the proposed indirect method, the noncooperative stochastic H_∞ game strategy design problem and cooperative stochastic H_∞ game strategy design problem are transformed to an equivalent HJIIs-constrained MOP with Nash equilibrium solution and an equivalent HJIIs-constrained SOP, respectively. Further, the HJIIs-constrained MOP for noncooperative stochastic H_∞ game and HJII-constrained SOP for the cooperative stochastic H_∞ game can be transformed to an LMIs-constrained MOP and an LMI-constrained SOP to simplify the design procedure of noncooperative stochastic H_∞ game strategy and cooperative stochastic H_∞ game strategy for nonlinear stochastic systems with the help of T-S fuzzy interpolation method, respectively. Finally, instead of using the conventional indirect iterative algorithm to search the Nash equilibrium solution, an LMIs-constrained MOEA algorithm is developed to efficiently solve the LMIs-constrained MOP in a single run for the multi-player noncooperative stochastic H_∞ game strategy design of the nonlinear stochastic system.

The chapter is organized as follows: In Section 8.2, we propose the n-player nonlinear stochastic jump diffusion system model and construct n-player stochastic H_∞ noncooperative game strategy problem. In Section 8.3, we transform the n-player stochastic H_∞ noncooperative game strategy problem to a HJIIs-constrained MOP. In Section 8.4, we construct the n-player stochastic H_∞ cooperative game and transform the problem to a HJII-constrained SOP. In Section 8.5, based on fuzzy interpolation method, the HJIIs-constrained MOP for n-player stochastic H_∞ noncooperative game is transformed to an equivalent LMIs-constrained MOP which could be efficiently solved by the proposed LMIs-constrained MOEA. In Section 8.6, the HJII-constrained SOP for n-player stochastic H_∞ cooperative game is transformed to an equivalent LMI-constrained SOP by using fuzzy interpolation method. In Section 8.7, a nonlinear financial market with multi-investors is given as an example to demonstrate the effectiveness of the proposed noncooperative and cooperative stochastic H_∞ game strategy design in the nonlinear stochastic jump diffusion system with external disturbance. A brief summary of this chapter and the directions of future research are given in Section 8.8.

Notation:

A^T: The transpose of matrix A; $A \geq 0 (A \leq 0)$: Symmetric positive semi-definite (symmetric positive definite) matrix A; I_n: n-dimensional identity matrix; C^n: The class of functions which have continuous n derivatives with respect to x; $\frac{\partial V(x)}{\partial x}$: The gradient column vector of n_x-dimensional continuously differentiable function $V(x)$; $\frac{\partial^2 V(x)}{\partial x^2}$: The Hessian matrix with elements of second partial derivatives of n_x-dimensional twice continuously differentiable function $V(x)$; E: The expectation operator.

8.2 Description of N-Player NonCooperative Stochastic H$_\infty$ Game in Nonlinear Stochastic Jump Diffusion System

Let us consider the following n-player nonlinear stochastic jump diffusion system:

$$\begin{aligned} dx(t) =& [F(x(t)) + \sum_{i=1}^{n} G_i(x(t))u_i(t) + v(t)]dt \\ &+ H(x(t))dw(t) + L(x(t))dp(t) \end{aligned} \tag{8.1}$$

where $x(t) \in \mathbb{R}^m$ is system state, $u_i(t) \in \mathbb{R}^m, i = 1, 2, \ldots, n$, are noncooperative game strategies, $v(t) \in \mathbb{R}^m$ is finite-energy external disturbance, $F(x), G_i(x)$, $i = 1, 2, \ldots, n$, $L(x)$, and $H(x)$ are Borel measurable $\mathbb{C}^1$ functions with Lipschitz continuous property. $w(t)$ and $p(t)$ are the standard one-dimensional (1-D) Wiener process with zero mean and unit variance and Poisson counting process with jump intensity defined on the probability space (Ω, F, F_t, P) relative to an increasing family $\{F_t\}_{t \in R^+}$ of σ-algebras $F_t \subset F$ where Ω denotes the sample space, F is the sigma algebra generated from $w(t)$ and $p(t)$, and P denotes the probability measure on Ω. $H(x(t))dw(t)$ denotes intrinsic continuous random fluctuation and $L(x(t))dp(t)$ is regarded as intrinsic discontinuous random fluctuation (jumping process). It is assumed that the Wiener process $w(t)$ and Poisson counting process $p(t)$ are independent.

In the noncooperative game problem of nonlinear stochastic system, the information of other competitive control strategies $[u_1(t), \ldots, u_{k-1}(t), u_{k+1}(t), \ldots u_n(t)]$ and external disturbance $v(t)$ are unavailable for the kth player. Since their effects are unpredictable, each player should consider them from the worst-case perspective. Also, the kth player has his own desired steady-state trajectory x_d^k. Hence, by augmenting the external disturbance and other control strategies as a competitive strategy $\bar{v}_k(t)$, the utility function of stochastic n-player noncooperative H_∞ game for the kth player with a desired target x_d^k is formulated as follows [8]:

$$\begin{aligned} \rho_k^* =& \min_{u_k(t)} \max_{\bar{v}_k(t)} \frac{E\{\int_0^T (x(t) - x_d^k)^T(t) Q_k (x(t) - x_d^k) + u_k^T(t) R_k u_k(t)) dt\}}{E\{(x(0) - x_d^k)^T (x(0) - x_d^k) + \int_0^T \bar{v}_k^T(t) \bar{v}_k(t) dt\}} \\ &\text{for} k = 1, 2, ..., n \end{aligned} \tag{8.2}$$

where $x(0)$ denotes the initial states, $\bar{v}_k(t) = [v^T(t), u_1^T(t), \ldots, u_{k-1}^T(t), u_{k+1}^T(t), \ldots, u_n^T(t)]^T, Q_k \geq 0$, and $R_k < 0$ with an appropriate dimension denote the corresponding weighting matrices for each player to tradeoff between the state tracking error $x(t) - x_d^k$ and the strategy effort of $u_k(t)$ from the perspective of player k. The term ρ_k^* in (8.2) denotes the H_∞ tracking performance of the stochastic H_∞ noncooperative game with the corresponding optimal game strategy $u_k^*(t)$ of player k. The term $E\{(x(0) - x_d^k)^T(x(0) - x_d^k)\}$ in (8.2) is used to consider the effect of uncertain initial condition $x(0) - x_d^k$ on the stochastic noncooperative H_∞ game strategy.

The main purpose of the noncooperative stochastic H_∞ game strategy $u_k(t)$ in (8.2) is to most attenuate the worst-case influence of the unpredictable external disturbance and the worst-case effect of other unavailable strategies on the individual goals to optimally achieve their individual targets, i.e., the noncooperative game strategy $u_k(t)$ of the kth player in (8.2) is chosen to minimize the worst-case effect of unpredictable augmented competitive strategy and external disturbance $\bar{v}_k(t)$ on the desired state tracking error $x(t) - x_d^k$ with a parsimonious control effort $u_k(t)$ simultaneously for all players from the H_∞ performance perspective.

Since every player has his/her own desired target in the n-player noncooperative H_∞ game (8.2), for the simplicity of design, the shifted nonlinear stochastic jump diffusion system with respect to the kth player is constructed as follows:

$$\begin{aligned} d\tilde{x}_k(t) =& [F_k(\tilde{x}_k(t)) + G_{k,k}(\tilde{x}_k(t))u_k(t) + \bar{B}_k(\tilde{x}_k(t))\bar{v}_k(t)]dt \\ &+ H_k(\tilde{x}_k(t))dw(t) + L_k(\tilde{x}_k(t))dp(t) \end{aligned} \tag{8.3}$$

where $\tilde{x}_k(t) = x(t) - x_d^k$ denotes the desired state tracking error, x_d^k denotes the desired state of player k. For the convenience of notation,
$F_k(\tilde{x}_k(t)) \triangleq F(\tilde{x}(t) - x_d^k), G_{k,k}(\tilde{x}_k(t)) \triangleq G_k(x(t) + x_d^k), H_k(\tilde{x}_k(t)) \triangleq H_k(x(t) + x_d^k), L_k(\tilde{x}_k(t)) \triangleq L_k(x(t) + x_d^k), \bar{v}_k(t) \triangleq [v^T(t), u_1^T(t), \ldots, u_{k-1}^T(t), u_{k+1}^T(t), \ldots, u_n^T(t)]^T, \bar{B}_k(\tilde{x}_k(t)) \triangleq [I, G_{k,1}(\tilde{x}_k(t)), \ldots, G_{k,k-1}(\tilde{x}_k(t)), G_{k,k+1}(\tilde{x}_k(t)), \ldots, G_{k,n}(\tilde{x}_k(t))], G_{k,i}(\tilde{x}_k(t)) \triangleq G_i(x(t) + x_d^k), \quad \forall i \in \{1, \ldots, k-1, k+1, \ldots, n\}$

Then, the utility function of stochastic H_∞ noncooperative game in (8.2) can be reformulated as

$$\rho_k^* = \min_{u_k(t)} \max_{\bar{v}_k(t)} \frac{E\{\int_0^T \tilde{x}_k(t)^T(t)Q_k\tilde{x}_k(t) + u_k^T(t)R_ku_k(t))dt\}}{E\{\tilde{x}_k(0)^T\tilde{x}_k(0) + \int_0^T \bar{v}_k^T(t)\bar{v}_k(t)dt\}} \tag{8.4}$$
$$\text{for} k = 1, 2, \ldots, n$$

i.e., we need to solve the above 2-tuple n-player noncooperative stochastic H_∞ games simultaneously.

8.3 N-Player Noncooperative Stochastic H_∞ Game Strategy Design of N Players

In general, it is very difficult to solve the n-player stochastic noncooperative H_∞ game problem (8.4) directly. Therefore, the following indirect approach is proposed to solve the stochastic n-player noncooperative H_∞ game problem in (8.4) by minimizing their corresponding upper bounds simultaneously

$$\rho_k^* = \min_{u_k(t)} \max_{\bar{v}_k(t)} \frac{E\{\int_0^T \tilde{x}_k^T(t)Q_k\tilde{x}_k(t) + u_k^T(t)R_ku_k(t))dt\}}{E\{\tilde{x}_k(0)^T\tilde{x}_k(0) + \int_0^T \bar{v}_k^T(t)\bar{v}_k(t)dt\}} \leq \rho_k \quad \text{for} k = 1, 2, \ldots, n \tag{8.5}$$

where the positive value ρ_k denotes the upper bound of the kth player's H_∞ tracking performance ρ_k^*. The upper bound ρ_k of ρ_k^* in (8.5) will be given beforehand and then be decreased as small as possible to approach ρ_k^* for all players simultaneously to attain the real n minimax solutions of n-player noncooperative stochastic H_∞ game in (8.4), which is the so-called suboptimal approach to the n-player noncooperative stochastic H_∞ game problem in (8.4), i.e., we need to solve the following MOP by minimizing their corresponding upper bounds ρ_k in (8.5) simultaneously

$$(\rho_1^*, \ldots, \rho_k^*, \ldots, \rho_n^*) = \min_{u_1, \ldots, u_n} (\rho_1, \ldots, \rho_k, \ldots, \rho_n) \tag{8.6}$$

subject to

$$\rho_k^* = \min_{u_k(t)} \max_{\bar{v}_k(t)} \frac{E\{\int_0^T \tilde{x}_k^T(t)Q_k\tilde{x}_k(t) + u_k^T(t)R_ku_k(t))dt\}}{E\{\tilde{x}_k^T(0)\tilde{x}_k(0) + \int_0^T \bar{v}_k^T(t)\bar{v}_k(t)dt\}} \leq \rho_k \quad \text{for} k = 1, 2, \ldots, n \tag{8.7}$$

where $\min(\rho_1, \ldots, \rho_k, \ldots, \rho_n)$ denotes the minimization of $\rho_1, \ldots, \rho_k, \ldots, \rho_n$ simultaneously, i.e., an n-tuple MOP.

Remark 8.1 By using the concept of Pareto Dominance in the MOP of (8.6) [21,22,115], a vector $(\rho_1^*, \ldots, \rho_k^*, \ldots, \rho_n^*)$ dominates the vector $(\rho_1, \ldots, \rho_k, \ldots, \rho_n)$ if $\rho_k^* \leq \rho_k$, $\forall k \in \{1, \ldots, n\}$ and at least one of the inequalities is strict inequality. Hence, the MOP in (8.6) is well defined to find the Pareto optimal solution $(\rho_1^*, \ldots, \rho_k^*, \ldots, \rho_n^*)$ which cannot be dominated by other solutions. The formal definition of Pareto Dominance can be referred to Definition 8.1 in the sequel.

Theorem 8.1

The indirect multi-objective problem in (8.6) and (8.7) is equivalent to the noncooperative game problem in (8.4) if the multi-objective optimization solution in (8.6) is achieved.

Proof: Please refer to Appendix A.

Since $\bar{v}_k(t)$ is independent of $u_k(t)$, the constraints of stochastic noncooperative H_∞ games in (8.7) are equivalent to the following n-constrained stochastic Nash quadratic games [1,9]:

$$\begin{aligned} &\min_{u_k(t)} \max_{\bar{v}_k(t)} E\{\int_0^T \tilde{x}_k^T(t)Q_k\tilde{x}_k(t) + u_k^T(t)R_ku_k(t) - \rho_k\bar{v}_k^T(t)\bar{v}_k(t)dt\} \\ &\leq \rho_k E\{\tilde{x}_k^T(0)\tilde{x}_k(0)\} \\ &\text{for} k = 1, 2, \ldots, n \end{aligned} \tag{8.8}$$

Let us denote

$$J_k = E\{\int_0^T \tilde{x}_k^T(t)Q_k\tilde{x}_k(t) + u_k^T(t)R_ku_k(t) - \rho_k\bar{v}_k^T(t)\bar{v}_k(t)dt\} \quad \text{for} k = 1, 2, \ldots, n \tag{8.9}$$

We then need two steps to solve the constrained stochastic Nash quadratic game problem in (8.8). The first step is to solve the following noncooperative stochastic Nash quadratic game problem

$$J_k^* = \min_{u_k(t)} \max_{\bar{v}_k(t)} J_k, \quad k = 1, 2, \ldots, n \tag{8.10}$$

and the second step is to solve the following inequality constraint problem:

$$J_k^* \leq \rho_k E\{\tilde{x}_k^T(0)\tilde{x}_k(0)\}, \text{for } \ k = 1, 2, \ldots, n. \tag{8.11}$$

By solving (8.10) and (8.11) for the constrained stochastic Nash quadratic game in (8.8), the n-player noncooperative Nash quadratic game strategies can be constructed by solving n corresponding Halmition-Jacobi-Issac inequalities (HJIIs) in Theorem 8.2. Before we state our main theorem, two lemmas are given in the following to help us solve the constrained stochastic Nash quadratic games in (8.8):

Lemma 8.1 [23] For any matrix X and Y with appropriate dimensions, we have

$$X^TY + Y^TX \leq X^TP^{-1}X + Y^TPY \tag{8.12}$$

where P is any positive-definite symmetric matrix.

Lemma 8.2 [125,126] Let $V : \mathbb{R}^m \to \mathbb{R}, V(\cdot) \in \mathbb{C}^2(\mathbb{R}^m)$ and $V(\cdot) \geq 0$. For the shifted nonlinear stochastic jump diffusion system with respect to the kth player in (8.3), the Itô–Lévy formula of $V(\tilde{x}_k(t))$ is given as follows:

$$\begin{aligned} dV(\tilde{x}_k(t)) = & \left(\frac{\partial V(\tilde{x}_k(t))}{\tilde{x}_k(t)}\right)^T [F_k(\tilde{x}_k(t)) + G_{k,k}(\tilde{x}_k(t))u_k(t) + \bar{B}_k(\tilde{x}_k(t))\bar{v}_k(t)] \\ & + \frac{1}{2}H_k^T(\tilde{x}_k(t))\left(\frac{\partial^2 V(\tilde{x}_k(t))}{\tilde{x}_k^2(t)}\right)H_k(\tilde{x}_k(t))dt \\ & + \left(\frac{\partial V(\tilde{x}_k(t))}{\tilde{x}_k(t)}\right)^T H_k^T(\tilde{x}_k(t))dw(t) \\ & + [V(\tilde{x}_k(t) + L_k(\tilde{x}_k(t))) - V(\tilde{x}_k(t))]dp(t) \end{aligned} \tag{8.13}$$

Theorem 8.2

The constrained stochastic Nash quadratic game problem in (8.7) or (8.8) can be solved by the following game strategy $u_k^*(t)$ and the worst-case competitive strategy $\bar{v}_k^*(t), \forall k = 1, 2, \ldots, n$

$$\begin{aligned} u_k^*(t) &= -\frac{1}{2}R_k^{-1}G_{k,k}^T(\tilde{x}_k(t)\left(\frac{\partial V_1(\tilde{x}_k(t))}{\tilde{x}_k(t)}\right) \\ \bar{v}_k^*(t) &= \frac{1}{2\rho_k}\bar{B}_k^T(\tilde{x}_k(t)\left(\frac{\partial V_1(\tilde{x}_k(t))}{\tilde{x}_k(t)}\right) \end{aligned} \tag{8.14}$$

where the positive Lyapunov function $V_1 : \mathbb{R}^m \to \mathbb{R}$, which satisfies $V_1(\cdot) \in \mathbb{C}^2(\mathbb{R}^m)$, $V_1(\cdot) \geq 0$, is the solution of the following Hamilton–Jacobi–Issac inequalities (HJIIs):

$$\begin{aligned}
&\left(\frac{\partial V_1(\tilde{x}_k(t))}{\partial \tilde{x}_k(t)}\right)^T F_k(\tilde{x}_k(t)) + \tilde{x}_k^T Q_k \tilde{x}_k(t) + \frac{1}{2} H_k^T(\tilde{x}_k(t)) \left(\frac{\partial^2 V_1(\tilde{x}_k(t))}{\partial \tilde{x}_k^2(t)}\right) H_k(\tilde{x}_k(t)) \\
&+ \lambda[V_1(\tilde{x}_k(t) + L_k(\tilde{x}_k(t))) - V_1(\tilde{x}_k(t))] \\
&- \frac{1}{4}\left(\frac{\partial V_1(\tilde{x}_k(t))}{\partial \tilde{x}_k(t)}\right)^T G_{k,k}(\tilde{x}_k(t)) R_k^{-1} G_{k,k}^T(\tilde{x}_k(t) \left(\frac{\partial V_1(\tilde{x}_k(t))}{\partial \tilde{x}_k(t)}\right) \\
&+ \frac{1}{4\rho_k}\left(\frac{\partial V_1(\tilde{x}_k(t))}{\partial \tilde{x}_k(t)}\right)^T \bar{B}_k(\tilde{x}_k(t)) \bar{B}_k^T(\tilde{x}_k(t) \left(\frac{\partial V_1(\tilde{x}_k(t))}{\partial \tilde{x}_k(t)}\right) \leq 0 \\
&\forall k = 1, 2, \ldots, n
\end{aligned} \tag{8.15}$$

with the following initial constraints:

$$V_1(\tilde{x}_k(0)) \leq \rho_k \tilde{x}_k^T(0) \tilde{x}_k(0), \quad \forall k = 1, 2, \ldots, n \tag{8.16}$$

Proof: Please refer to Appendix B.

$u_k^*(t)$ and $\bar{v}_k^*(t)$ in (8.14) are the solution of constrained stochastic Nash quadratic games problem in (8.7) or (8.8), which are the suboptimal solution of noncooperative H_∞ game in (8.4). Therefore, the MOP in (8.6) and (8.7) for the noncooperative stochastic H_∞ game in (8.4) of the nonlinear stochastic system in (8.3) could be transformed into the following MOP to simultaneously minimize their corresponding upper bounds under HJIIs-constrained in (8.15) and (8.16):

$$\begin{aligned}
&(\rho_1^*, \ldots, \rho_k^*, \ldots, \rho_n^*) = \min_{V_1(\tilde{x}_k)<0} (\rho_1, \ldots, \rho_k, \ldots, \rho_n) \\
&\text{subject to}(8.15)\text{and}(8.16).
\end{aligned} \tag{8.17}$$

After solving the MOP in (8.17) for the optimal solution $V_1^*(\tilde{x}_k)$, the noncooperative stochastic H_∞ game strategies for each player could be constructed as $u_k^*(t) = -\frac{1}{2} R_k^{-1} G_{k,k}^T(\tilde{x}_k(t)) \times \left(\frac{\partial V_1^*(\tilde{x}_k(t))}{\partial \tilde{x}_k(t)}\right)$ from (8.14) for $k = 1, 2, \ldots, n$.

Definition 8.1 [138,139] (Pareto Dominance) For the MOP in (8.17), suppose there are two multi-objective feasible solutions $(u_1^1, \ldots, u_n^1)$ and $(u_1^2, \ldots, u_n^2)$ with the corresponding objective values $(\rho_1^1, \ldots, \rho_n^1)$ and $(\rho_1^2, \ldots, \rho_n^2)$, respectively, the solution $(u_1^1, \ldots, u_n^1)$ is said to dominate $(u_1^2, \ldots, u_n^2)$ if $\rho_1^1 \leq \rho_1^2, \ldots, \rho_n^1 \leq \rho_n^2$ and at least one of the inequalities is a strict inequality.

Remark 8.2 Based on Definition 8.1, the multi-objective optimal solution $(u_1^*, \ldots, u_n^*)$ is nondominated by objective vectors of other solutions. In general, the multi-objective optimal solution $(u_1^*, \ldots, u_n^*)$ with the corresponding objective value $(\rho_1^*, \ldots, \rho_n^*)$ in (8.17) is not unique and there may exist many solutions in the Pareto optimal sense (also called Pareto optimal solutions).

Definition 8.2 [138,139] (Pareto Optimality) The feasible solution $(u_1^*, \ldots, u_n^*)$ with the corresponding objective value $(\rho_1^*, \ldots, \rho_n^*)$ of MOP in (17) is said to be Pareto optimality with respect to the feasible solution set if and only if there does not exist another feasible solution that dominates it.

Remark 8.3 The optimal solution $(u_1^*, \ldots, u_n^*)$ with the corresponding objective values $(\rho_1^*, \ldots, \rho_n^*)$ of n-person noncooperative game in (8.6) and (8.7) or in (8.17) constitutes a Nash equilibrium solution if and only if [1,9]

$$\begin{aligned}
&(\rho_1^*, \ldots, \rho_k^*, \ldots, \rho_n^*) \leq (\rho_1, \rho_2^*, \ldots, \rho_k^*, \ldots, \rho_n^*) \\
&\vdots \\
&(\rho_1^*, \ldots, \rho_k^*, \ldots, \rho_n^*) \leq (\rho_1^*, \ldots, \rho_{k-1}^*, \rho_k, \rho_{k+1}^*, \ldots, \rho_n^*) \\
&\vdots \\
&(\rho_1^*, \ldots, \rho_k^*, \ldots, \rho_n^*) \leq (\rho_1^*, \ldots, \rho_k^*, \ldots, \rho_{n-1}^*, \rho_n)
\end{aligned} \tag{8.18}$$

i.e., no player has anything to gain by changing their own strategy.

In the end of this section, the following theorem shows the optimal solution in (8.6) and (8.7) or in (8.17) is a Nash equilibrium solution of n-player H_∞ noncooperative game in (8.1).

Theorem 8.3

The multi-objective optimal solution $(u_1^*, \ldots, u_n^*)$ with the corresponding objective values $(\rho_1^*, \ldots, \rho_n^*)$ in (8.17) is the Nash equilibrium solution of n-player noncooperative game for nonlinear stochastic system in (8.3).

Proof: Please refer to Appendix C.

8.4 N-Player Cooperative Stochastic H_∞ Game Strategy Design

If the n players in (8.1) have compromised a common desired state x_d (common target) with each other at the beginning, then the n-player cooperative stochastic H_∞ game strategy of nonlinear stochastic jump diffusion system in (8.1) becomes how to specify $u_1, \ldots, u_n$ together to achieve the following n-player cooperative stochastic H_∞ game tracking performance:

$$\rho^0 = \min_{u(t)} \max_{v(t)} \frac{E\{\int_0^T \tilde{x}^T(t)Q\tilde{x}(t) + u^T(t)Ru(t))dt\}}{E\{\tilde{x}^T(0)\tilde{x}(0) + \int_0^T v^T(t)v(t)dt\}} \tag{8.19}$$

where $\tilde{x}(t) = x(t) - x_d$ denotes the tracking error between the state $x(t)$ and desired state x_d, $u(t) = [u_1(t)^T, \ldots, u_n(t)^T]^T$ denotes the augmented matrix combining the strategies of all players, $Q \geq 0$ denotes the tracking weighting matrix, and $R<0$ denotes the strategy effort weighting matrix.

In the n-player cooperative stochastic H_∞ game, for the convenience of design, the shifted nonlinear stochastic jump diffusion system can be represented as follows:

$$d\tilde{x}(t) = [F_c(\tilde{x}(t)) + G_c(\tilde{x}(t))u(t) + v(t)]dt + H_c(\tilde{x}(t))dw(t) + L_c(\tilde{x}(t))dp(t) \tag{8.20}$$

where $F_c(\tilde{x}(t)) \triangleq F_c(\tilde{x}(t)+x_d), G_c(\tilde{x}(t)) \triangleq [\bar{G}_1(\tilde{x}(t)),\ldots,\bar{G}_n(\tilde{x}(t))], G_i(\tilde{x}(t)) \triangleq G_i(\tilde{x}(t)$ $+x_d), \forall i=1,\ldots,n, H_c(\tilde{x}_k(t)) \triangleq H(\tilde{x}(t)+x_d), L_c(\tilde{x}_k(t)) \triangleq L(\tilde{x}(t)+x_d)$. Also, the cooperative stochastic H_∞ game strategy design for nonlinear stochastic jump diffusion system in (8.1) becomes how to solve the cooperative game strategy in (8.19) for nonlinear stochastic jump diffusion system in (8.20).

By using the similar technique in (8.5), the indirect approach is proposed to solve the stochastic n-player H_∞ cooperative game in (8.19) by minimizing its upper bound ρ, i.e.,

$$\rho^0 = \min_{u(t)}\max_{v(t)} \frac{E\{\int_0^T \tilde{x}^T(t)Q\tilde{x}(t)+u^T(t)Ru(t))dt\}}{E\{\tilde{x}^T(0)\tilde{x}(0)+\int_0^T v^T(t)v(t)dt\}} \leq \rho. \tag{8.21}$$

The above suboptimal stochastic n-player H_∞ cooperative game is equivalent to the following constrained minmax stochastic Nash quadratic game because of the independence of $v(t)$ on $u(t)$ [1,9]:

$$\min_{u(t)}\max_{v(t)} E\int_0^T \tilde{x}^T(t)Q\tilde{x}(t)+u^T(t)Ru(t)-\rho v^T(t)v(t)dt \leq \rho E\{\tilde{x}^T(0)\tilde{x}(0)\}. \tag{8.22}$$

We define

$$J(u,v) = E\int_0^T (\tilde{x}^T(t)Q\tilde{x}(t)+u^T(t)Ru(t)-\rho v^T(t)v(t)dt \tag{8.23}$$

Then, the following two-step procedure is also proposed to solve the constrained stochastic Nash quadratic game problem in (8.22). The first step is to solve the following stochastic Nash quadratic game problem:

$$J^o = \min_{u(t)}\max_{v(t)} E\int_0^T (\tilde{x}^T(t)Q\tilde{x}(t)+u^T(t)Ru(t)-\rho v^T(t)v(t))dt \tag{8.24}$$

and the second step is to solve the following constrained inequality problem:

$$J^o \leq \rho E\{\tilde{x}^T(0)\tilde{x}(0)\}$$

The main results of stochastic n-player H_∞ cooperative game is stated as follows:

Theorem 8.4

The suboptimal cooperative stochastic H_∞ game problem in (8.22) could be solved by the following strategy:

$$\begin{aligned} u^o(t) &= -\frac{1}{2}R^{-1}G_c^T(\tilde{x}(t))(\frac{\partial V_2(\tilde{x}(t))}{\partial \tilde{x}(t)}) \\ v^o(t) &= \frac{1}{2\rho}(\frac{\partial V_2(\tilde{x}(t))}{\partial \tilde{x}(t)}) \end{aligned} \tag{8.25}$$

where the positive Lyapunov function $V_2:\mathbb{R}^m \rightarrow \mathbb{R}$, which satisfies $V_2(\cdot) \in \mathbb{C}^2(\mathbb{R}^m)$, $V_2(\cdot) \geq 0$, is the solution of the following HJII:

$$\begin{aligned}&\left(\frac{\partial V_2(\tilde{x}(t))}{\partial \tilde{x}(t)}\right)^T F_c(\tilde{x}(t)) + \tilde{x}^T Q\tilde{x}(t) + \frac{1}{2}H_c^T(\tilde{x}(t))\left(\frac{\partial^2 V_2(\tilde{x}_k(t))}{\partial \tilde{x}^2(t)}\right)H_c(\tilde{x}(t))\\&+\lambda[V_2(\tilde{x}(t)+L_c(\tilde{x}(t)))-V_2(\tilde{x}(t))]\\&-\frac{1}{4}\left(\frac{\partial V_2(\tilde{x}(t))}{\partial \tilde{x}(t)}\right)^T G_c(\tilde{x}_k(t))R^{-1}G_c^T(\tilde{x}(t)\left(\frac{\partial V_2(\tilde{x}(t))}{\partial \tilde{x}(t)}\right)\\&+\frac{1}{4\rho}\left(\frac{\partial V_2(\tilde{x}(t))}{\partial \tilde{x}(t)}\right)^T\left(\frac{\partial V_2(\tilde{x}(t))}{\partial \tilde{x}(t)}\right)\leq 0\end{aligned} \tag{8.26}$$

with the following constraint on the initial condition

$$V_2(\tilde{x}(0)) \leq \rho\tilde{x}^T(0)\tilde{x}(0). \tag{8.27}$$

Proof: The proof is similar to Theorem 8.2. Hence, we omit the proof here.

Since the solution $u^o(t)$ and $v^o(t)$ in (8.25) is the suboptimal stochastic H_∞ cooperative game solution with the upper bound ρ in (8.21) or (8.22), the optimal stochastic cooperative H_∞ game solution for ρ^o in (8.21) could be solved through minimizing the upper bound ρ by the following HJII-constrained optimization problem:

$$\begin{aligned}&\rho^o = \min_{V_2(\tilde{x})<0} \rho\\&\text{subject to HJII in (8.26 and (8.27)}\end{aligned} \tag{8.28}$$

After solving the SOP in (8.28) for the optimal solution $V_2^o(\tilde{x})$, the stochastic cooperative H_∞ game strategy could be constructed as $u^o(t) = -\frac{1}{2}R^{-1}G_c^T(\tilde{x}(t))\times(\frac{\partial V_2^o(\tilde{x}(t))}{\partial \tilde{x}(t)})$ from (8.25).

8.5 N-Player Noncooperative Stochastic H_∞ Game Strategy Design Based on Fuzzy Interpolation Methods

In general, it is very difficult to solve the HJIIs-constrained MOP in (8.17) for the noncooperative stochastic H_∞ game strategy of the n-player nonlinear stochastic jump diffusion system in (8.1). At present, there does not exist good method to solve the HJIIs problem either analytically or numerically. In this section, in order to simplify the HJIIs-constrained MOP for the n-player noncooperative stochastic H_∞ game problem in (8.17), the fuzzy interpolation technique is used to interpolate a nonlinear stochastic jump diffusion system with a set of local linearized stochastic systems.

The plant rule i with the corresponding ith local system of nonlinear stochastic jump diffusion system in (8.3) is defined as [5]

Plant Rule i:

$$\begin{aligned}&\text{If } z_1(t) \text{ is } F_{i1},\ldots, \text{ and } z_g(t) \text{ is } F_{ig}\\&\text{Then } d\tilde{x}_k(t) = [A_{ki}\tilde{x}_k(t) + B_{ki}u_k(t) + \bar{B}_{ki}\bar{v}_k(t)]dt\\&\qquad + C_{ki}\tilde{x}_k(t)dw(t) + D_{ki}\tilde{x}_k(t)dp(t)\\&\text{for} i = 1, 2, \ldots, l.\end{aligned} \tag{8.29}$$

where $F_{i1},\ldots,F_{ig}$ are the fuzzy set, l is the number of If-Then rules and $z_1(t),\ldots,z_g(t)$ are the premise variable; $A_{ki}, B_{ki}, \bar{B}_{ki}, C_{ki}$ and D_{ki} are constant matrices with appropriate dimensions for $i = 1, 2, \ldots, l$.

The fuzzy system of the kth player in (8.29) can be represented as follows [5]:

$$\begin{aligned}d\tilde{x}_k(t) = &\sum_{i=1}^{l} h_i(z(t))([A_{ki}\tilde{x}_k(t) + B_{ki}u_k(t) + \bar{B}_{ki}\bar{v}_k(t)]dt\\&+ C_{ki}\tilde{x}_k(t)dw(t) + D_{ki}\tilde{x}_k(t)dp(t)\end{aligned} \tag{8.30}$$

where

$$\mu_i(z(t)) = \prod_{j=1}^{g} F_{ij}(z_j(t)), h_i(z(t)) = \frac{\mu_i(z(t))}{\sum_{i=1}^{l}\mu_i(z(t))}$$

$$z(t) = [z_1(t), z_2(t), \ldots, z_g(t)]$$

and $F_{ij}(z_j(t))$ is the grade of membership of $z_j(t)$ in F_{ij}.
It is assumed that

$$\mu_i(z(t)) \geq 0 \text{ and } \sum_{i=1}^{l}\mu_i(z(t)) > 0, \quad \text{for } i = 1, 2, \ldots, l. \tag{8.31}$$

Therefore, we get the fuzzy interpolation functions as follows:

$$h_i(z(t)) \geq 0 \text{ and } \sum_{i=1}^{l} h_i(z(t)) = 1, \text{for } i = 1, 2, \ldots, l. \tag{8.32}$$

Hence, the shifted nonlinear stochastic system in (8.3) with respect to the kth player can be rearranged as the following fuzzy stochastic system:

$$\begin{aligned}d\tilde{x}_k(t) = &\sum_{i=1}^{l} h_i(z(t))\{[A_{ki}\tilde{x}_k(t) + B_{ki}u_k(t) + \bar{B}_{ki}\bar{v}_k(t)]dt\\&+ \Delta F_k(\tilde{x}_k(t)) + \Delta G_k(\tilde{x}_k(t))u_k + \Delta\bar{B}_k(\tilde{x}_k(t))\bar{v}_k(t)dt\\&+ (C_{ki}\tilde{x}_k(t) + \Delta H_k(\tilde{x}_k(t)))dw(t) + (D_{ki}\tilde{x}_k(t) + \Delta L_k(\tilde{x}_k(t)))\}dp(t)\end{aligned} \tag{8.33}$$

where

$$\Delta F_k(\tilde{x}_k(t)) = F_k(\tilde{x}_k(t)) - \sum_{i=1}^{l} h_i(z(t))A_{ki}\tilde{x}_k(t)$$

$$\Delta G_k(\tilde{x}_k(t)) = G_{k,k}(\tilde{x}_k(t)) - \sum_{i=1}^{l} h_i(z(t))B_{ki}$$

$$\Delta \bar{B}_k(\tilde{x}_k(t)) = \bar{B}_k(\tilde{x}_k(t)) - \sum_{i=1}^{l} h_i(z(t))\bar{B}_{ki}$$

$$\Delta H_k(\tilde{x}_k(t)) = H_k(\tilde{x}_k(t)) - \sum_{i=1}^{l} h_i(z(t))C_{ki}\tilde{x}_k(t)$$

$$\Delta L_k(\tilde{x}_k(t)) = L_k(\tilde{x}_k(t)) - \sum_{i=1}^{l} h_i(z(t))D_{ki}\tilde{x}_k(t)$$

To deal with the fuzzy approximation error, the following assumption is given:

Assumption 8.1 There exist some bounding matrices $\Delta A_k, \Delta C_k, \Delta D_k$ with appropriate dimensions and scalars $\varepsilon_k, \bar{\varepsilon}_k$ such that

$$\|\Delta F_k(\tilde{x}_k(t))\|_2 \leq \|\Delta A_k(\tilde{x}_k(t))\|_2, \|\Delta G_k(\tilde{x}_k(t))\|_2 \leq \varepsilon_k, \|\Delta H_k(\tilde{x}_k(t))\|_2 \leq \|\Delta C_k(\tilde{x}_k(t))\|_2,$$

$$\left\|\Delta \bar{B}_k(\tilde{x}_k(t))\right\|_2 \leq \bar{\varepsilon}_k, \|\Delta L_k(\tilde{x}_k(t))\|_2 \leq \|\Delta D_k(\tilde{x}_k(t))\|_2, \forall k \in \{1, \ldots, n\}.$$

Remark 8.4 Since (i) the functions $F_k(\tilde{x}_k(t)), H_k(\tilde{x}_k(t)), L_k(\tilde{x}_k(t)), G_{k,k}(\tilde{x}_k(t))$, in the shifted nonlinear stochastic system (8.3) are continuous on $\mathbb{R}^m$ and (ii) the universe of course is compact, the Fuzzy Universal Approximation Theorem [137] could guarantee the estimation of each approximation error can be as small as possible if we increase the plant rule number.

Define the Lyapunov function of the stochastic fuzzy system of (8.33) as

$$V_1(\tilde{x}_k(t)) = \tilde{x}_k^T(t)P_1\tilde{x}_k(t) \tag{8.34}$$

where $P_1 = P_1^T > 0$, then we can get the following theorem.

Theorem 8.5

Based on the stochastic fuzzy system in (8.33), the indirect suboptimal approach to the n-player noncooperative stochastic H_∞ game problem in (8.7) or (8.8)–(8.11) could be solved by the following strategies:

$$\begin{aligned} u_k^*(t) &= -R_k^{-1}G_{k,k}^T(\tilde{x}_k(t))P_1\tilde{x}_k(t) \\ v_k^*(t) &= \frac{1}{\rho_k}\bar{B}_k^T(\tilde{x}_k(t))P_1\tilde{x}_k(t) \\ &\text{for } k = 1, 2, \ldots, n \end{aligned} \tag{8.35}$$

where the positive matrix P_1 is the solution of the following Riccati-like inequalities:

$$
\begin{aligned}
&P_1A_{ki} + A_{ki}^TP_1 + P_1P_1 + \Delta A_k^T\Delta A_k + Q_k + C_{ki}^TP_1C_{ki} + C_{ki}^TP_1P_1C_{ki} \\
&+ (1+\rho_k)\Delta C_k^T\Delta C_k + \lambda(P_1D_{ki} + (2+\rho_k)\Delta D_k^T\Delta D_k + P_1P_1 \\
&+ D_{ki}^TP_1 + D_{ki}^TP_1D + D_{ki}^TP_1P_1D) + \frac{2}{\rho_k}P_1\bar{B}_{ki}\bar{B}_{ki}{}^TP_1 + \frac{2\bar{\varepsilon}_k}{\rho_k}P_1P_1 < 0 \\
&\text{for } k = 1,2,\ldots,n, \ i = 1,2,\ldots,l
\end{aligned}
\tag{8.36}
$$

with the following eigenvalue constraints on P_1

$$
0 < P_1 \leq \rho_k I, \text{for } k = 1,2,\ldots,n \tag{8.37}
$$

Proof: Please refer to Appendix D.

Using the Schur complement transformation method in Lemma (2.3) [23], the Riccati-like inequalities in (8.36) are equivalent to the following LMIs:

$$
\begin{bmatrix}
\Xi_{ki} & P_1 & C_{ki}^TP_1 & C_{ki}^TP_1 & \sqrt{\lambda}D_{ki}^TP_1 & \sqrt{\lambda}D_{ki}^TP_1 & \sqrt{\frac{2}{\rho_k}}P_1\bar{B}_{ki} \\
* & -\left(\frac{2\bar{\varepsilon}_k}{\rho_k}+1+\lambda\right)^{-1}I & 0 & 0 & 0 & 0 & 0 \\
* & * & -P_1 & 0 & 0 & 0 & 0 \\
* & * & * & -I & 0 & 0 & 0 \\
* & * & * & * & -P_1 & 0 & 0 \\
* & * & * & * & * & -I & 0 \\
* & * & * & * & * & * & -I
\end{bmatrix} < 0 \tag{8.38}
$$

for $k = 1,2,\ldots,n$, for $i = 1,2,\ldots,l$

where $\Xi_{ki} = P_1A_{ki} + A_{ki}^TP_1 + \Delta A_k^T\Delta A_k + Q_k + (1+\rho_k)\Delta C_k^T\Delta C_k + \lambda(P_1D_{ki} + (2+\rho_k)\Delta D_k^T \Delta D_k + D_{ki}^TP_1)$.Therefore, by transforming (8.36) into (8.38), the LMI constraints in (8.38) can be easily solved with the help of Matlab LMI toolbox.

Remark 8.5 The multi-objective optimization problem for n-player noncooperative stochastic H_∞ game strategies in (8.6) becomes the following LMIs-constrained MOP:

$$
\begin{aligned}
&(\rho_1^* \ \cdots \ \rho_k^* \ \cdots \ \rho_n^*) = \min_{P_1<0}(\rho_1 \ \cdots \ \rho_k \ \cdots \ \rho_n) \\
&\text{subject to LMIs in (8.37) and LMIs in (8.38)}
\end{aligned}
\tag{8.39}
$$

From the above analysis, based on fuzzy stochastic system in (8.33), the HJIIs-constrained MOP in (8.17) of the stochastic n-player H_∞ noncooperative control strategy design problem of the nonlinear stochastic jump diffusion system in (8.1) is transformed to an LMIs-constrained MOP in (8.39). To solve the MOP for n-player stochastic H_∞ noncooperative game in (8.39), we always employ the MOEA algorithm [35]. Since the MOEA can parallelly search the Pareto optimal solutions and escape from local optimal solutions [25,138–141], it is particularly suitable for solving MOP in (8.39) with some modifications. The MOEAs could solve an MOP via a stochastic search method based on a "survival of the fittest" law to approach a set of compromised solutions called Pareto optimal solutions through the evolutionary algorithm (EA), including crossover, mutation, and elitist selection operator. Since the constraints in (8.37) and (8.38) of MOP in (8.39) are a set of LMIs, an LMIs-based MOEA is developed to efficiently solve the MOP in (8.39). Some important definitions about Pareto optimality of the LMIs-constrained MOP in (8.39) are given as follows:

Definition 8.3 [138,139] (Pareto Optimal Solution Set) For the given LMIs-constrained MOP in (8.39), the Pareto optimal solution set P_s is defined as $P_s \triangleq \{(u_1^*, \dots, u_n^*)|$. There does not exist another feasible solution $(u_1^o, \dots, u_n^o)$ such that $(\rho_1^o, \dots, \rho_n^o)$ dominates $(\rho_1^*, \dots, \rho_n^*)\}$.

Definition 8.4 [138,139] (Pareto Front) For the given LMIs-constrained MOP in (39), the Pareto front P_F is defined as $P_F \triangleq \{(\rho_1^*, \dots, \rho_n^*)|(u_1^*, \dots, u_n^*) \in P_s\}$.

For the n-player noncooperative stochastic H_∞ game strategy design of nonlinear stochastic jump diffusion system in (8.1), based on the LMIs-constrained MOP in (8.39), a design procedure based on MOEA is developed and proposed as follows:

Design Procedure of LMIs-Constrained MOEA for N-Player NonCooperative stochastic H_∞ Game Strategy of Nonlinear Stochastic Jump Diffusion Systems

Step 1 Construct the fuzzy rules and obtain the fuzzy system in (8.33) of nonlinear stochastic system in (8.1).

Step 2 Initialization

Step 2.1 Set the searching region $[\rho_{\min,1}, \rho_{\max,1}] \times \dots \times [\rho_{\min,n}, \rho_{\max,n}] \triangleq S$, the maximum number N_g of individuals (population), the crossover rate M_c, the mutation rate M_r, and the iteration number N_i in the LMIs-constrained MOEA, where $\rho_{\min,k}$ and $\rho_{\max,k}$ denote the lower and upper bounds of ρ_k, for $k = 1, \dots, n$, respectively.

Step 2.2 Set algorithm counter $t = 1$. Randomly generate a feasible initial population $a_1, a_2, \dots, a_{N_g} \in S$. Define $\{a_1, a_2, \dots, a_{N_g}\}$ as parent population and $\bar{P}_t = \{a_1, a_2, \dots, a_{N_g}\}$, i.e., the set $\bar{P}_t$ collects the parent population.

Step 3 Differential evolution

Step 3.1 Perform the crossover operation on crossover rate M_c.

Step 3.2 Perform the mutation operation $\bar{P}_t$ on mutation rate M_r.

Step 3.3 After crossover and mutation operation, N_g feasible populations are generated and we denote these population as the child population. Let the set P_t^C collect the child population by examining whether their corresponding objective vectors $(\rho_1, \dots, \rho_n)$ are feasible objective vectors satisfying the LMIs in (8.37) and (8.38), i.e., P_t^C collects the feasible child populations generated by crossover operation and mutation operation.

Step 4 Update

Let the set $\bar{P}$ select N_g populations from the set $P_t^C \cup \bar{P}_t$ by using nondominated sorting method and crowded-comparison operator. Set the algorithm counters $t = t + 1$ and $\bar{P}_t = \bar{P}$, i.e., the set $\bar{P}$ collects the next-generation parent populations from the current parent populations and child population.

Step 5 Stopping criterion

Repeat steps 3 and 4 until the iteration number N_i is satisfied, then set the final population as the Pareto front P_F.

Step 6 Solution selection

Select a preferable feasible objective individual $(\rho_1^*, \ldots, \rho_n^*)$ and the corresponding solution P_1^* according to designer own preference. In this chapter, the knee point in the Pareto front is preferred in simulation example.

Step 7 Obtain the Pareto optimal solution of knee point and construct the kth player's strategy $u_k^*(t) = -R_k^{-1} G_{k,k}^T(\tilde{x}_k(t)) P_1^* \tilde{x}_k(t)$, for $k = 1, 2, \ldots, n$.

8.6 N-Player Cooperative Stochastic H_∞ Game Strategy Design Based on Fuzzy Interpolation Methods

In the n-player cooperative stochastic H_∞ game control design, we need to solve HJII-constrained SOP in Theorem 8.4. At present, there is no any efficient way to solve HJII in (8.26) and (8.27). In this section, the T-S Fuzzy interpolation methods is proposed to approximate the nonlinear stochastic jump diffusion system with a set of local linearized stochastic systems. Sequently, by using the Shur complement, the complex HJII in Theorem 8.4 is transformed to a set of LMIs which can be easily solved with the help of Matlab toolbox.

The plant rule i of T-S fuzzy system with the corresponding ith local system of nonlinear stochastic system in (8.20) is defined as [5]

Plant Rule i:

$$\begin{aligned}
&\text{If } z_1(t) \text{ is } F_{i1}, \ldots, \text{ and } z_g(t) \text{ is } F_{ig} \\
&\text{Then } d\tilde{x}(t) = [A_i \tilde{x}(t) + B_i u(t) + v(t)]dt \\
&\qquad\qquad + C_i \tilde{x}(t) dw(t) + D_i \tilde{x}(t) dp(t) \\
&\text{for } i = 1, 2, \ldots, l.
\end{aligned}$$

where $F_{i1}, \ldots, F_{ig}$ is the fuzzy set, l is the number of If–Then rules and $z_1(t), \ldots, z_g(t)$ are the premise variables; $A_i, B_i, \bar{B}_i, C_i$ and D_i are constant matrices with appropriate dimensions for $i = 1, 2, \ldots, l$.

In this situation, the fuzzy system of nonlinear stochastic system in (8.20) can be constructed as

$$\begin{aligned}
d\tilde{x}(t) = &\sum_{i=1}^{l} h_i(z(t))([A_i \tilde{x}(t) + B_i u(t) + v(t)]dt \\
&+ C_i \tilde{x}(t) dw(t) + D_i \tilde{x}(t) dp(t)
\end{aligned} \tag{8.40}$$

where

$$\begin{aligned}
&\mu_i(z(t)) = \prod_{j=1}^{g} F_{ij}(z_j(t)), h_i(z(t)) = \frac{\mu_i(z(t))}{\sum_{i=1}^{l} \mu_i(z(t))} \\
&z(t) = [z_1(t), z_2(t), \ldots, z_g(t)]
\end{aligned}$$

Similar to previous definitions, we assume

$$\mu_i(z(t)) \geq 0 \text{ and } \sum_{i=1}^{l} \mu_i(z(t)) > 0, \text{ for } i = 1, 2, \ldots, l. \tag{8.41}$$

Hence, we get the fuzzy interpolation functions as follows:

$$h_i(z(t)) \geq 0 \text{ and } \sum_{i=1}^{l} h_i(z(t)) = 1, \text{ for } i = 1, 2, \ldots, l. \tag{8.42}$$

If the above T-S fuzzy rules with l local linearized stochastic systems are employed to approximate nonlinear stochastic jump diffusion system (8.20), the nonlinear stochastic system in (8.20) can be represented as

$$\begin{aligned} d\tilde{x}(t) = \sum_{i=1}^{l} h_i(z(t))\{[A_i\tilde{x}(t) + B_i u(t) + v(t)]dt + C_i\tilde{x}(t)dw(t) \\ + D_i\tilde{x}(t)dp(t) + (\Delta F_c(\tilde{x}_k(t)) + \Delta G_c(\tilde{x}(t))u)dt \\ + \Delta H_c(\tilde{x}(t)))dw(t) + \Delta L_c(\tilde{x}(t))dp(t) \end{aligned} \tag{8.43}$$

where

$$\Delta F_c(\tilde{x}(t)) = F_c(\tilde{x}(t)) - \sum_{i=1}^{l} h_i(z(t))A_i\tilde{x}(t), \Delta G_c(\tilde{x}(t)) = G_c(\tilde{x}(t)) - \sum_{i=1}^{l} h_i(z(t))B_i$$
$$\Delta H_c(\tilde{x}(t)) = H_c(\tilde{x}(t)) - \sum_{i=1}^{l} h_i(z(t))C_i\tilde{x}(t), \Delta L_c(\tilde{x}(t)) = L_c(\tilde{x}_k(t)) - \sum_{i=1}^{l} h_i(z(t))D_i\tilde{x}(t)$$

are the fuzzy estimation errors between the fuzzy systems in (8.40) and the shifted system in (8.20).

Assumption 8.2 There exist some bounding matrices $\Delta A, \Delta C, \Delta D$ with appropriate dimension and scalar $\varepsilon > 0$ such that

$$\begin{aligned} &\|\Delta F_c(x(t))\|_2 \leq \|\Delta A x(t)\|_2, \|\Delta G_c(x(t))\|_2 \leq \varepsilon \\ &\|\Delta H_c(x(t))\|_2 \leq \|\Delta C x(t)\|_2, \|\Delta L_c(x(t))\| \leq \|\Delta D x(t)\|_2 \end{aligned}$$

Define the Lyapunov function of the system in (8.43) as

$$V_2(x) = x^T(t)P_2x(t) \tag{8.44}$$

where $P_2 = P_2^T > 0$. Our main result for the cooperative H_∞ game strategy design is stated as follows:

Theorem 8.6

Based on the stochastic fuzzy system (8.43), the suboptimal n-player cooperative stochastic H_∞ game in (8.21) or (8.22) could be solved by the following strategy:

$$\begin{aligned} u(t) &= -R^{-1}G_c^T(\tilde{x}(t))P_2\tilde{x}(t) \\ v(t) &= \frac{1}{\rho}P_2\tilde{x}(t) \end{aligned} \tag{8.45}$$

where matrix P_2 is the solution of the following Riccati-like inequalities:

$$\begin{aligned} &Q + P_2A_i + A_i^TP_2 + (1+\lambda+\frac{1}{\rho})P_2P_2 + \Delta A^T\Delta A + C_i^TP_2P_2C_i + C_i^TP_2C_i \\ &+ (1+\rho)\Delta C^T\Delta C + \lambda(D_i^TP_2 + P_2D_i + D_i^TP_2D_i + D_i^TP_2P_2D_i \\ &+ (2+\rho)\Delta D^T\Delta D) < 0 \\ &\forall i = 1, \ldots, l. \end{aligned} \tag{8.46}$$

with the following eigenvalue constraint on P_2

$$0 < P_2 \leq \rho I \tag{8.47}$$

Proof: The proof of Theorem 8.6 is similar to Theorem 8.5. Hence, we omit the proof here.

By using the Schur complement to (8.46), the Riccati-like inequalities in (8.46) can be rewritten as the following linear matrix inequalities (LMIs):

$$\begin{bmatrix} \Xi_i & P_2 & C_i^TP_2 & C_i^TP_2 & D_i^TP_2 & D_i^TP_2 \\ * & -(1+\frac{1}{\rho}+\lambda)^{-1} & 0 & 0 & 0 & 0 \\ * & * & -I & 0 & 0 & 0 \\ * & * & * & -P_2 & 0 & 0 \\ * & * & * & * & -I & 0 \\ * & * & * & * & * & -P_2 \end{bmatrix} < 0 \tag{8.48}$$
$$\forall i = 1, 2, \ldots, l$$

where
$\Xi_i = Q + P_2A_i + A_i^TP_2 + \Delta A^T\Delta A + (1+\rho)\Delta C^T\Delta C + \lambda(D_i^TP_2 + P_2D_i + (2+\rho)\Delta D^T\Delta D)$.

Remark 8.6 Based on T-S fuzzy interpolation method, the HJII-constrained SOP in (8.28) for n-player cooperative stochastic H_∞ game control strategy design becomes the following LMIs-constrained SOP:

$$\rho^o = \min_{P_2>0} \rho$$
$$\text{subjecttoLMIsin (8.47) andLMIsin (8.48)} \tag{8.49}$$

which could be easily solved with the help of LMI Toolbox in Matlab by decreasing ρ until no positive definite solution P_2 exists in LMI constraints (8.47), (8.48).

8.7 Simulation Example

To illustrate the design procedure and confirm the performance of the proposed n-player noncooperative stochastic H_∞ game strategy and cooperative stochastic H_∞ game strategy for the nonlinear stochastic jump diffusion system, an example of finding 3-player H_∞ noncooperative and cooperative game investment strategies for a nonlinear stochastic jump diffusion financial system is given to mimic an emerging market. In the following, a financial dynamic model is given to illustrate the interaction between interesting rate $x(t)$, investment demand $y(t)$, and the price index $z(t)$ [142]:

$$\begin{cases} \dot{x} = z(t) + (y(t) - a)x(t) \\ \dot{y} = 1 - by(t) - (x(t))^2 \\ \dot{z} = -x(t) - cz(t) \end{cases} \tag{8.50}$$

where the parameters $a = 1.5$ is the saving amount, $b = 0.2$ is the per-investment cost, and $c = 0.25$ is the elasticity of demands of commercials.

In the realistic financial system, however, the system may suffer from random intrinsic fluctuation and external disturbance due to the variation of oil price, the occurrence of war, the change of government policy, the surplus between investment and saving, etc. Hence, the dynamic financial system in (8.50) should be modified by continuous and discontinuous intrinsic random fluctuation as well as environmental disturbance and the investment strategies of investors as follows [22]:

$$\begin{aligned} dx(t) =& (z(t) + (y(t) - a)x(t) + \sum_{k=1}^{n} u_{k1}(t) + v_1(t))dt + h_1(x(t), y(t), z(t))dw(t) \\ & + l_1(x(t), y(t), z(t))dp(t) \\ dy(t) =& 1 - by(t) - (x(t))^2 + \sum_{k=1}^{n} u_{k2}(t) + v_2(t)dt) + h_2(x(t), y(t), z(t))dw(t) \\ & + l_2(x(t), y(t), z(t))dp(t) \\ dz(t) =& (-x(t) - cz(t) + \sum_{k=1}^{n} u_{k3}(t) + v_3(t)dt) + h_3(x(t), y(t), z(t))dw(t) \\ & + l_3(x(t), y(t), z(t))dp(t) \end{aligned} \tag{8.51}$$

where $h_1(x(t), y(t), z(t)) = 0.01(z(t) + (y(t) - a)x(t)), h_2(x(t), y(t), z(t)) = 0.01(1 - by(t) - (x(t))^2), h_3(x(t), y(t), z(t)) = 0.02(-x(t) - cz(t)), l_1(x(t), y(t), z(t)) = 0.1x(t), l_2(x(t), y(t), z(t)) = 0.4y(t), l_3(x(t), y(t), z(t)) = 0.3z(t)$. The u_{ki} denotes the strategy of the kth investor applied to the ith state, for $k = 1, \ldots, n$ and $i = 1, 2, 3$. The Wiener process $w(t)$ and the Poisson counting process $p(t)$ are introduced to model the continuous and discontinuous random fluctuation, $v_1(t), v_2(t),$ and $v_3(t)$ denote the external disturbances. Nonlinear stochastic system (8.51) may describe a real financial market, which refers to the buying and selling behaviors or the price changes of the objects of transaction like stocks, bounds, futures, and oil or economic index changes. Suppose there are three investors, for example, government, bank consortium, and the public, in the nonlinear stochastic financial model in (8.51), then (8.51) can be written as follows:

$$d\bar{x}(t) = (F(\bar{x}(t)) + \sum_{k=1}^{3} B_k u_k(t) + v(t))dt + H(\bar{x}(t)dw(t) + L(\bar{x}(t))dp(t) \tag{8.52}$$

with $\bar{x}(t) = [x(t), y(t), z(t)]^T, u_k(t) = [u_{k1}(t), u_{k2}(t), u_{k3}(t)]^T, v(t) = [v_1(t), v_2(t), v_3(t)]^T,$ $F(\bar{x}(t)) = [z(t) + (y(t) - a)x(t)), 1 - by(t) - (x(t))^2, (-x(t) - cz(t))]^T, H(\bar{x}(t)) = [h_1(\bar{x}(t)), h_2(\bar{x}(t)), h_3(\bar{x}(t))]^T, L(\bar{x}(t)) = [l_1(\bar{x}(t)), l_2(\bar{x}(t)), l_3(\bar{x}(t))]^T$ and $B_k = I_{3\times3}$, for $k = 1, 2, 3$.

Suppose the initial states and the external disturbance of stochastic nonlinear financial system in (8.52) are given as

$$\bar{x}_0 = [0.37, -3.06, 0.71]^T.$$
$$v(t) = [0.02\sin(3t), -0.01\sin(2t), -0.02\sin(3t)]^T.$$

Figure 8.1 is used to describe the dynamical behaviors of the nonlinear stochastic jump diffusion financial system in (8.52) with continuous and discontinuous fluctuations in real situation. It is seen that the trajectories of three states $x(t)$, $y(t)$, and $z(t)$ of the nonlinear stochastic financial system are with random fluctuations and jumps.

Suppose three investors of government $u_1(t)$, bank consortium $u_2(t)$, and the public $u_3(t)$ want to invest the nonlinear stochastic financial market through noncooperative game strategy. Therefore, based on the T-S fuzzy interpolation method, the 3-player noncooperative H_∞ game investment policy of the MOP in (8.39) is employed by three investors to regulate the stochastic financial system and achieve their individual desired steady states:

$$\begin{aligned} x_d^1 &= [0.9, 0.51, 1]^T \\ x_d^2 &= [1.5, 1.5, 2]^T \\ x_d^3 &= [0, 3, -1]^T \end{aligned} \tag{8.53}$$

Define $\tilde{x}_k(t) = \bar{x}(t) - x_d^k$ where x_d^k denotes the desired state (target) of inventor k for $k = 1, 2, 3$, then we get the following shifted nonlinear stochastic financial system in (8.52) as follows:

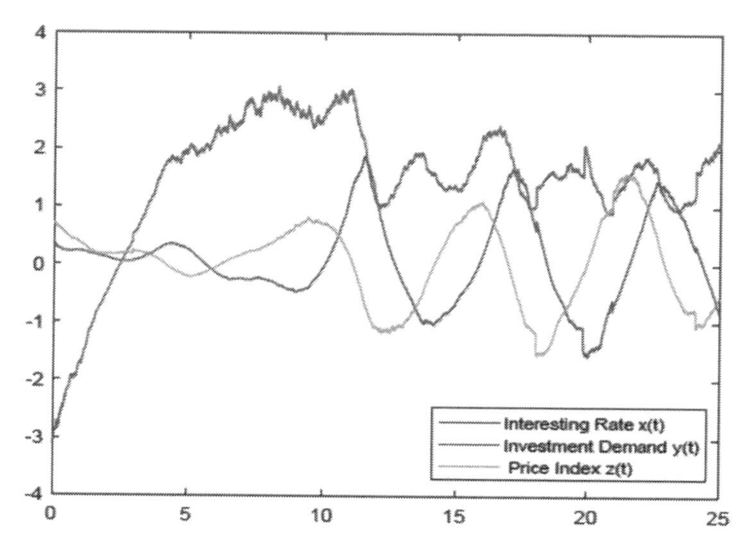

FIGURE 8.1
State trajectories of stochastic nonlinear jump diffusion financial system in (8.52) without investment strategies u_k, for k=1,2,3.

$$d\tilde{x}_k(t) = (F_k(\tilde{x}_k(t)) + B_k u_k(t) + \bar{B}_k \bar{v}_k(t))dt + H_k(\tilde{x}_k(t))dw(t) + L_k(\tilde{x}_k(t))dp(t) \tag{8.54}$$

where $\bar{v}_k(t) =: [v(t)^T, u_1^T(t), \ldots, u_{k-1}^T(t), u_{k+1}^T(t), \ldots, u_3^T(t)]^T, \bar{B}_k =: [I_{3\times3}, \ldots, I_{3\times3}], F_k(\tilde{x}_k(t)) =$

$F(\tilde{x}_k(t) + x_d^k), H_k(\tilde{x}_k(t)) = H(\tilde{x}_k(t) + x_d^k)$ and $L_k(\tilde{x}_k(t)) = L(\tilde{x}_k(t) + x_d^k)$, for $k = 1, 2, 3$. In this simulation example, we assume the vector of the fuzzy premise variables $z = [x(t), y(t)]^T = [z_1(t), z_2(t)]^T$ is available in the investment strategy design. Since there are eight fuzzy sets associated with premise variable $z_1(t)$ and eight fuzzy sets associated with premise variable $z_2(t)$, there are a total of 64 fuzzy If–Then rules in the T-S fuzzy financial system. Suppose the weighting matrices in the noncooperative investment policy in (8.2) are used to invest the stochastic financial system with the following weighting matrices:

$$R_k = 0.01I_{3\times3}, Q_1 = \begin{bmatrix} 0.9 & 0 & 0 \\ 0 & 0.1 & 0 \\ 0 & 0 & 0.1 \end{bmatrix}, Q_2 = \begin{bmatrix} 0.1 & 0 & 0 \\ 0 & 0.9 & 0 \\ 0 & 0 & 0.1 \end{bmatrix}, Q_1 = \begin{bmatrix} 0.1 & 0 & 0 \\ 0 & 0.1 & 0 \\ 0 & 0 & 0.9 \end{bmatrix}.$$

for $k = 1, 2, 3$

(8.55)

and the parameters used in LMIs-constrained MOEA are given as follows:

Iteration Number N_i: 100
Searching Region S: [0, 10] × [0, 10] × [0, 10]
Population Number N_g: 1000
Crossover Rate M_c: 0.8
Mutation Rate : 0.2

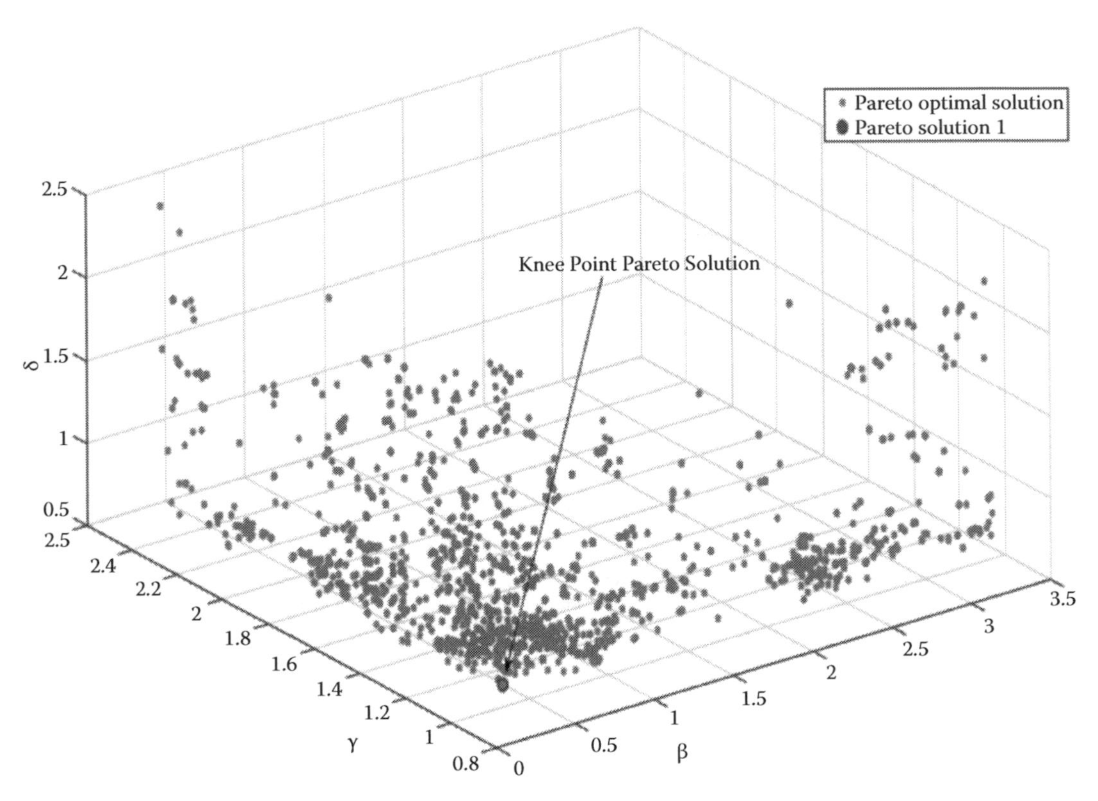

FIGURE 8.2
The Pareto front of $(\rho_1^*, \rho_2^*, \rho_3^*)$. By the proposed MOEA in Section 8.4, the Pareto optimal solutions of 3-player noncooperative H_∞ game strategies are found. In this figure, each red point represents a Pareto solution with the corresponding investment strategy. The knee point is at (0.412, 1.115, 0.613).

Once the iteration number $N_i = 100$ is achieved, we get the Pareto front in Fig. 8.2 for the Pareto optimal solutions of 3-player noncooperative H_∞ game investment strategy from the LMIs-constrained MOP in (8.39). In order to illustrate the performance of proposed 3-player noncooperative H_∞ game investment strategy in the stochastic financial system, the knee solution $(\rho_1^*, \rho_2^*, \rho_3^*) = (0.412, 1.115, 0.613)$ with the corresponding positive definite matrix

$$P_1^* = \begin{bmatrix} 1.816 & -0.025 & 0.306 \\ -0.025 & 1.189 & -0.014 \\ 0.306 & -0.014 & 1.849 \end{bmatrix}. \tag{8.56}$$

is chosen as the optimal multiobjective solution at the knee point of the Pareto front in Fig. 8.2 and investment strategy can be constructed as $u_k^*(t) = -R_k^{-1} B_k P_1^* \tilde{x}_k(t)$ for three investors $k = 1, 2, 3$.

The simulation results based on the knee solution in Fig. 8.3 are given to demonstrate the performance of 3-player noncooperative stochastic H_∞ game strategies of three investors. From the simulation in Fig. 8.3, it is seen that even the random fluctuations of three state trajectories have been smoothed, the steady state of these trajectories could

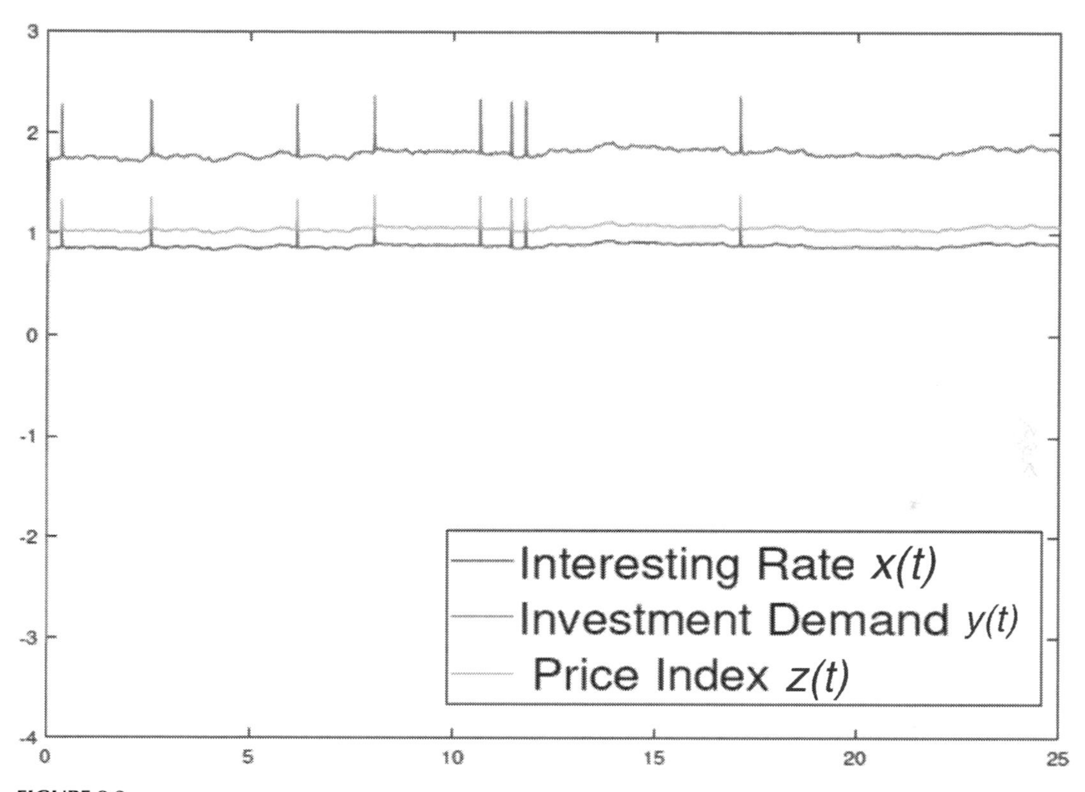

FIGURE 8.3
System state trajectories with the 3-player noncooperative H_∞ game investment strategies $u_k^*(t) = -R_k^{-1}B_kP_1^*\tilde{x}_k(t)$, for three investors $k = 1,2,3$. In the case of 3-player noncooperative H_∞ game, each state trajectory fluctuates around a point. Since each player has its own desired target, these trajectories show the conflict result among these players and the simulation shows the compromised result among three players' strategies.

not achieve the desired values of three investors but the compromised results of desired targets of three investors according to their weighting matrices in (8.55); i.e., $x(t)$ more approaches to the desired value of the first investor with more weighting 0.9 than 0.1 of others; $y(t)$ more approaches to the desired value of the second investor with more weighting 0.9 than 0.1 of others; $z(t)$ more approaches to the desired value of the third investor with more weighting 0.9 than 0.1 of others.

To compare the effect of 3-player noncooperative and cooperative stochastic H_∞ game investment strategies on the financial system, the SOP in (8.49) is carried out to demonstrate the effectiveness of our proposed stochastic 3-player cooperative H_∞ game investment strategy. In the 3-player cooperative H_∞ game investment strategy design, we need to compromise the common desired target $\bar{x}_d$ and weighting matrices Q,R in (8.19) beforehand. Suppose the compromised weighting matrices in (8.19) are given as

$$Q = \begin{bmatrix} 0.01 & 0 & 0 \\ 0 & 0.01 & 0 \\ 0 & 0 & 0.01 \end{bmatrix}, R = 0.5I_{9\times 9}$$

with the compromised desired target $\bar{x}_d = [0.097, 0.557, 0.086]^T$. By solving the LMIs-constrained SOP in (8.49), we have $\rho_c^* = 1.11$,

$$P_2^* = \begin{bmatrix} 0.02 & -0.0303 & 0.0062 \\ -0.303 & 0.0725 & -0.0134 \\ 0.0062 & -0.0134 & 0.0069 \end{bmatrix}$$

and the corresponding 3-player cooperative H_∞ game strategy $u(t) = -R^{-1}[B_1 \quad B_2 \quad B_3]^T P_2^* \tilde{x}(t)$. Figure 8.4 shows the state trajectories controlled by the proposed 3-player cooperative H_∞ investment strategy. In Fig. 8.4, since all players are compromised with a common desired objective beforehand and share with investment information, the system trajectories quickly converge to a compromised desired steady state. Clearly, the 3-player cooperative H_∞ investment strategy could quickly achieve their common objective which is compromised beforehand while the 3-player noncooperative H_∞ investment strategy could achieve the results which are compromised based on the conflict multi-objective strategies and different weighting matrices. However, how to achieve their common objective is always a complicated and difficult compromised problem because some investors need to sacrifice for the common target.

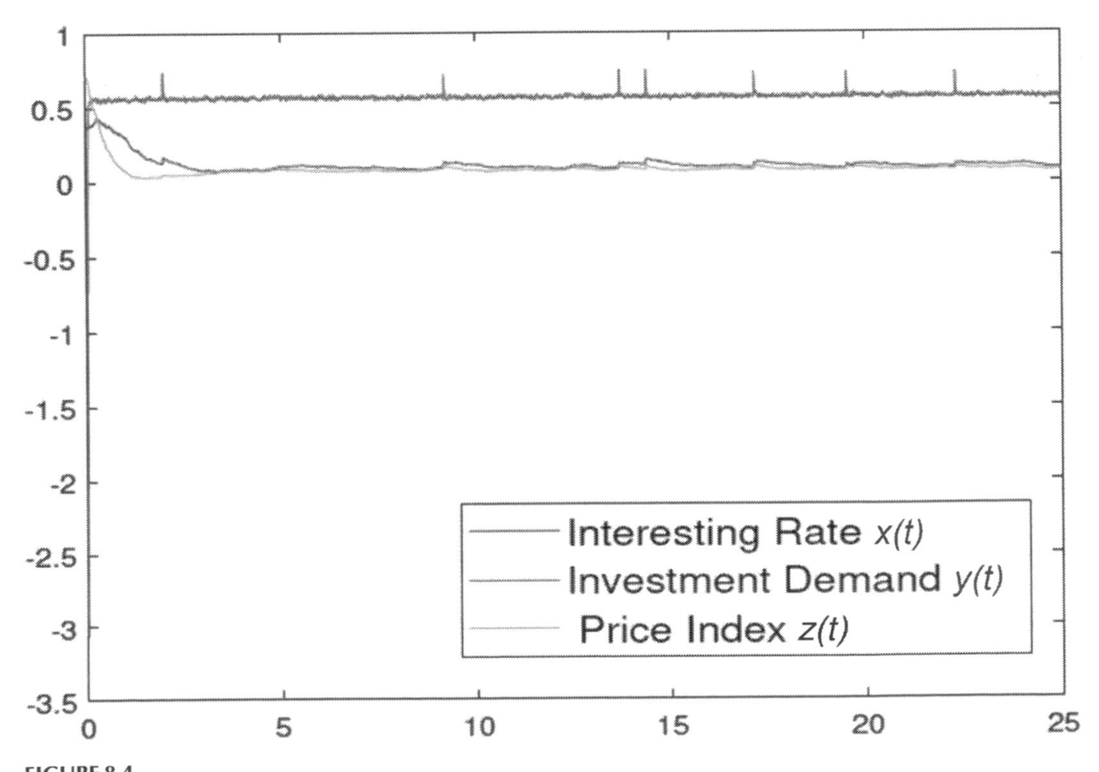

FIGURE 8.4
System start trajectories with the 3-player cooperative H_∞ investment strategies. Since the investment strategy of each player is compromised to a common desired objective beforehand, the simulation results illustrate that the state trajectories tend to converge to the common target.

8.8 Conclusion

Stochastic H_∞ noncooperative and cooperative game strategy for multi-player nonlinear stochastic jump diffusion system with external disturbance are proposed for target tracking strategy design through the multi-objective and single-objective optimal methods in this chapter. First, the stochastic H_∞ n-player noncooperative game strategy problem is constructed and then transformed to a corresponding HJIIs-constrained MOP with a Nash equilibrium solution to the n-player noncooperative stochastic H_∞ game by the proposed suboptimal method while the stochastic H_∞ n-player cooperative game strategy problem is transformed to a corresponding HJII-constrained SOP. Due to the difficulty in solving HJIIs for noncooperative stochastic H_∞ game and HJII for stochastic H_∞ cooperative game, the T-S fuzzy interpolation method is proposed to approximate the nonlinear stochastic system so that the HJIIs-constrained MOP and HJII-constrained SOP could be transformed into an LMIs-constrained MOP and LMIs-constrained SOP, respectively. Next, an LMIs-based MOEA algorithm is also developed to efficiently solve the LMIs-constrained MOP for the nonlinear multi-player noncooperative stochastic game strategy. The effectiveness of stochastic proposed noncooperative and cooperative H_∞ game strategies is illustrated by the simulation example of the nonlinear stochastic financial market with noncooperative and cooperative H_∞ game investment strategy. From the simulation results of 3-player noncooperative H_∞ investment strategy, we can find that the system state trajectories could not achieve the desired targets of three investors but the compromised results by the noncooperative stochastic H_∞ game strategies of each investor in the financial market. In the case of stochastic H_∞ cooperative game, the simulation result shows that the system state trajectories could quickly approach to the common desired target which was compromised by each player beforehand. In the future, we will focus on the extension and application of noncooperative and cooperative stochastic H_∞ game strategy to different kinds of control and filtering design problems in different nonlinear stochastic systems.

8.9 Appendix

8.9.1 Appendix A: Proof of Theorem 8.1

Let $(\rho_1^*, \ldots, \rho_k^*, \ldots, \rho_n^*)$ be a solution of (8.6). If $(\rho_1^*, \ldots, \rho_k^*, \ldots, \rho_n^*)$ is not a solution of (8.4), then there at least exist a $\rho_k^{'}$ be the solution of (8.4) such that

$$\min_{u_k(t)} \max_{\bar{v}_k(t)} \frac{E\{\int_0^T \tilde{x}_k^T(t) Q_k \tilde{x}_k(t) + u_k^T(t) R_k u_k(t))dt\}}{E\{\tilde{x}_k^T(0)\tilde{x}_k(0) + \int_0^T \bar{v}_k^T(t)\bar{v}_k(t)dt\}} = \rho_k^{'}$$

and $\rho_k^{'} > \rho_k^*$ for some k. As a result, we immediately find the vector $(\rho_1^*, \ldots, \rho_k^{'}, \ldots, \rho_n^*)$ dominates the vector $(\rho_1^*, \ldots, \rho_k^*, \ldots, \rho_n^*)$ and this implies $(\rho_1^*, \ldots, \rho_k^*, \ldots, \rho_n^*)$ is not a solution of (8.6). Indeed, the fact that $(\rho_1^*, \ldots, \rho_k^{'}, \ldots, \rho_n^*)$ dominates the vector $(\rho_1^*, \ldots, \rho_k^*, \ldots, \rho_n^*)$ shows $(\rho_1^*, \ldots, \rho_k^{'}, \ldots, \rho_n^*)$ is the solution of (8.6). Contradiction.

8.9.2 Appendix B: Proof of Theorem 8.2

Let $V_1 : \mathbb{R}^m \to \mathbb{R}^+$ be some Lyapunov function that satisfies $V_1(\cdot) \in \mathbb{C}^2(\mathbb{R}^m)$ and $V_1(\cdot) \geq 0$, then (8.9) can be rewritten as follows:

$$\begin{aligned} J_k = EV_1(\tilde{x}_k(0)) - EV_1(dt\tilde{x}_k(T)) + E\int_0^T \tilde{x}_k^T(t)Q_k\tilde{x}_k(t) + u_k^T(t)R_ku_k(t) \\ - \rho_k\bar{v}_k^T(t)\bar{v}_k(t) + dV_1(\tilde{x}_k(t)). \end{aligned} \tag{8.57}$$

By applying Lemma 8.2 and using the fact that $E[dw(t)] = 0, E[dp(t)] = \lambda dt$ (λ is the Poisson jump intensity), (8.57) can be rewritten as

$$\begin{aligned} J_k =& EV_1(\tilde{x}_k(0)) - EV_1(\tilde{x}_k(T)) + E\int_0^T \tilde{x}_k^T(t)Q_k\tilde{x}_k(t) + u_k^T(t)R_ku_k(t) \\ &- \rho_k\bar{v}_k^T(t)\bar{v}_k(t) + \left(\frac{\partial V_1(\tilde{x}_k(t))}{\tilde{x}_k(t)}\right)^T F_k(\tilde{x}_k(t)) + \left(\frac{\partial V_1(\tilde{x}_k(t))}{\tilde{x}_k(t)}\right)^T G_{k,k}(\tilde{x}_k(t))u_k(t) \\ &+ \left(\frac{\partial V_1(\tilde{x}_k(t))}{\tilde{x}_k(t)}\right)^T \bar{B}_k(\tilde{x}_k(t))\bar{v}_k(t) + \frac{1}{2}H_k^T(\tilde{x}_k(t))\left(\frac{\partial^2 V_1(\tilde{x}_k(t))}{\tilde{x}_k^2(t)}\right)H_k(\tilde{x}_k(t)) \\ &+ \lambda[V_1(\tilde{x}_k(t) + L_k(\tilde{x}_k(t))) - V_1(\tilde{x}_k(t))]dt \end{aligned} \tag{8.58}$$

By completing the square for u_k and $\bar{v}_k$, we have following equalities:

$$\begin{aligned} & u_k^T(t)R_ku_k(t) + \left(\frac{\partial V_1(\tilde{x}_k(t))}{\tilde{x}_k(t)}\right)^T G_{k,k}(\tilde{x}_k(t))u_k(t) \\ &= \left(R_ku_k(t) + \frac{1}{2}G_{k,k}^T(\tilde{x}_k(t))\left(\frac{\partial V_1(\tilde{x}_k(t))}{\tilde{x}_k(t)}\right)\right)^T R_k^{-1} \\ &\times \left(R_ku_k(t) + \frac{1}{2}G_{k,k}^T(\tilde{x}_k(t))\left(\frac{\partial V_1(\tilde{x}_k(t))}{\tilde{x}_k(t)}\right)\right) \\ &- \frac{1}{4}\left(\frac{\partial V_1(\tilde{x}_k(t))}{\tilde{x}_k(t)}\right)^T G_{k,k}(\tilde{x}_k(t))R_k^{-1}G_{k,k}^T(\tilde{x}_k(t))\left(\frac{\partial V_1(\tilde{x}_k(t))}{\tilde{x}_k(t)}\right) \end{aligned} \tag{8.59}$$

$$\begin{aligned} & - \rho_k\bar{v}_k^T(t)\bar{v}_k(t) + \left(\frac{\partial V_1(\tilde{x}_k(t))}{\tilde{x}_k(t)}\right)^T \bar{B}_k(\tilde{x}_k(t))\bar{v}_k(t) \\ &= -\left[\sqrt{\rho_k}\bar{v}_k(t) - \frac{1}{2\sqrt{\rho_k}}\bar{B}_k^T(\tilde{x}_k(t))\left(\frac{\partial V_1(\tilde{x}_k(t))}{\tilde{x}_k(t)}\right)\right]^T \\ &\times \left[\sqrt{\rho_k}\bar{v}_k(t) - \frac{1}{2\sqrt{\rho_k}}\bar{B}_k^T(\tilde{x}_k(t))\left(\frac{\partial V_1(\tilde{x}_k(t))}{\tilde{x}_k(t)}\right)\right] \\ &+ \frac{1}{4\rho_k}\left(\frac{\partial V_1(\tilde{x}_k(t))}{\tilde{x}_k(t)}\right)^T \bar{B}_k(\tilde{x}_k(t))\bar{B}_k^T(\tilde{x}_k(t))\left(\frac{\partial V_1(\tilde{x}_k(t))}{\tilde{x}_k(t)}\right). \end{aligned} \tag{8.60}$$

Next, substituting (8.59) and (8.60) into (8.58), we get

$$
\begin{aligned}
J_k = {} & EV_1(\tilde{x}_k(0)) - EV_1(\tilde{x}_k(T)) + E\int_0^T \left(\frac{\partial V_1(\tilde{x}_k(t))}{\tilde{x}_k(t)}\right)^T F_k(\tilde{x}_k(t)) + \tilde{x}_k^T(t)Q_k\tilde{x}_k(t) \\
& + \frac{1}{2}H_k^T(\tilde{x}_k(t))\left(\frac{\partial^2 V_1(\tilde{x}_k(t))}{\tilde{x}_k^2(t)}\right)H_k(\tilde{x}_k(t)) + \lambda[V_1(\tilde{x}_k(t) + L_k(\tilde{x}_k(t))) - V_1(\tilde{x}_k(t))] \\
& - \frac{1}{4}\left(\frac{\partial V_1(\tilde{x}_k(t))}{\tilde{x}_k(t)}\right)^T G_{k,k}(\tilde{x}_k(t))R_k^{-1}G_{k,k}^T(\tilde{x}_k(t))\left(\frac{\partial V_1(\tilde{x}_k(t))}{\tilde{x}_k(t)}\right) \\
& + \frac{1}{4\rho_k}\left(\frac{\partial V_1(\tilde{x}_k(t))}{\tilde{x}_k(t)}\right)^T \bar{B}_k(\tilde{x}_k(t))\bar{B}_k^T(\tilde{x}_k(t))\left(\frac{\partial V_1(\tilde{x}_k(t))}{\tilde{x}_k(t)}\right) \\
& + \left(R_k u_k(t) + \frac{1}{2}G_{k,k}^T(\tilde{x}_k(t))\frac{\partial V_1(\tilde{x}_k(t))}{\tilde{x}_k(t)}\right)^T R_k^{-1}\left(R_k u_k(t) + \frac{1}{2}G_{k,k}^T(\tilde{x}_k(t))\frac{\partial V_1(\tilde{x}_k(t))}{\tilde{x}_k(t)}\right) \\
& - \left[\sqrt{\rho_k}\bar{v}_k(t) - \frac{1}{2\sqrt{\rho_k}}\bar{B}_k^T(\tilde{x}_k(t))\left(\frac{\partial V_1(\tilde{x}_k(t))}{\tilde{x}_k(t)}\right)\right]^T \\
& \times \left[\sqrt{\rho_k}\bar{v}_k(t) - \frac{1}{2\sqrt{\rho_k}}\bar{B}_k^T(\tilde{x}_k(t))\left(\frac{\partial V_1(\tilde{x}_k(t))}{\tilde{x}_k(t)}\right)\right]dt.
\end{aligned} \tag{8.61}
$$

Then,

$$
\begin{aligned}
J_k^* = {} & \min_{u_k(t)}\max_{\bar{v}_k(t)}\{EV_1(\tilde{x}_k(0)) - EV_1(\tilde{x}_k(T)) + E\int_0^T \left(\frac{\partial V_1(\tilde{x}_k(t))}{\tilde{x}_k(t)}\right)^T F_k(\tilde{x}_k(t)) \\
& + \tilde{x}_k^T(t)Q_k\tilde{x}_k(t) + u_k^T(t)R_k u_k(t) + \frac{1}{2}H_k^T(\tilde{x}_k(t))\left(\frac{\partial^2 V_1(\tilde{x}_k(t))}{\tilde{x}_k^2(t)}\right)H_k(\tilde{x}_k(t)) \\
& + \lambda[V_1(\tilde{x}_k(t) + L_k(\tilde{x}_k(t))) - V_1(\tilde{x}_k(t))] - \frac{1}{4}\left(\frac{\partial V_1(\tilde{x}_k(t))}{\tilde{x}_k(t)}\right)^T G_{k,k}(\tilde{x}_k(t)) \\
& \times R_k^{-1}G_{k,k}^T(\tilde{x}_k(t))\left(\frac{\partial V_1(\tilde{x}_k(t))}{\tilde{x}_k(t)}\right) + \frac{1}{4\rho_k}\left(\frac{\partial V_1(\tilde{x}_k(t))}{\tilde{x}_k(t)}\right)^T \bar{B}_k(\tilde{x}_k(t)) \\
& \times \bar{B}_k^T(\tilde{x}_k(t))\left(\frac{\partial V_1(\tilde{x}_k(t))}{\tilde{x}_k(t)}\right) + \left(R_k u_k(t) + \frac{1}{2}G_{k,k}^T(\tilde{x}_k(t))\frac{\partial V_1(\tilde{x}_k(t))}{\tilde{x}_k(t)}\right)^T R_k^{-1} \\
& \times \left(R_k u_k(t) + \frac{1}{2}G_{k,k}^T(\tilde{x}_k(t))\frac{\partial V_1(\tilde{x}_k(t))}{\tilde{x}_k(t)}\right) - \left[\sqrt{\rho_k}\bar{v}_k(t) - \frac{1}{2\sqrt{\rho_k}}\bar{B}_k^T(\tilde{x}_k(t)) \times \right. \\
& \left.\left(\frac{\partial V_1(\tilde{x}_k(t))}{\tilde{x}_k(t)}\right)\right]^T \left[\sqrt{\rho_k}\bar{v}_k(t) - \frac{1}{2\sqrt{\rho_k}}\bar{B}_k^T(\tilde{x}_k(t))\left(\frac{\partial V_1(\tilde{x}_k(t))}{\tilde{x}_k(t)}\right)\right]dt\}
\end{aligned} \tag{8.62}
$$

Obviously, we can choose the following strategy as the minimax solution to the above Nash game problem:

$$u_k^*(t) = -\frac{1}{2}R_k^{-1}G_{k,k}^T(\tilde{x}_k(t)\left(\frac{\partial V_1(\tilde{x}_k(t))}{\tilde{x}_k(t)}\right)$$
$$\bar{v}_k^*(t) = \frac{1}{2\rho_k}\bar{B}_k^T(\tilde{x}_k(t)\left(\frac{\partial V_1(\tilde{x}_k(t))}{\tilde{x}_k(t)}\right). \tag{8.63}$$

Substituting (8.63) into (8.62), we have

$$J_k^* = EV_1(\tilde{x}_k(0)) - EV_1(\tilde{x}_k(T)) + E\int_0^T \left(\frac{\partial V_1(\tilde{x}_k(t))}{\tilde{x}_k(t)}\right)^T F_k(\tilde{x}_k(t)) + \tilde{x}_k^T(t)Q_k\tilde{x}_k(t)$$
$$+\frac{1}{2}H_k^T(\tilde{x}_k(t))\left(\frac{\partial^2 V_1(\tilde{x}_k(t))}{\tilde{x}_k^2(t)}\right)H_k(\tilde{x}_k(t)) + \lambda[V_1(\tilde{x}_k(t) + L_k(\tilde{x}_k(t)))$$
$$- V_1(\tilde{x}_k(t))] - \frac{1}{4}\left(\frac{\partial V_1(\tilde{x}_k(t))}{\tilde{x}_k(t)}\right)^T G_{k,k}(\tilde{x}_k(t))R_k^{-1}G_{k,k}^T(\tilde{x}_k(t))\left(\frac{\partial V_1(\tilde{x}_k(t))}{\tilde{x}_k(t)}\right) \tag{8.64}$$
$$+\frac{1}{4\rho_k}\left(\frac{\partial V_1(\tilde{x}_k(t))}{\tilde{x}_k(t)}\right)^T \bar{B}_k(\tilde{x}_k(t))\bar{B}_k^T(\tilde{x}_k(t))\left(\frac{\partial V_1(\tilde{x}_k(t))}{\tilde{x}_k(t)}\right)dt\}$$

Next, by using the HJIIs constraint in (8.15) and the initial condition constraints in (8.16), we immediately have the following results:

$$J_k^* = EV_1(\tilde{x}_k(0)) - EV_1(\tilde{x}_k(T)).$$
$$\leq V(\tilde{x}_k(0)) \leq \rho_k\tilde{x}_k^T(0)\tilde{x}_k(0)$$

The proof is completed.

8.9.3 Appendix C: Proof of Theorem 8.3

It could be proven by contradiction. If Nash equilibrium solution in (8.18) is violated, for example,

$$(\rho_1^*, \ldots, \rho_k^*, \ldots, \rho_n^*) > (\rho_1^*, \ldots, \rho_{k-1}^*, \rho_k, \rho_{k+1}^*, \ldots, \rho_n^*)$$

for some k, it will violate the domination of $(\rho_1^*, \ldots, \rho_k^*, \ldots, \rho_n^*)$ of the multi-objective optimization in (8.17). Therefore, when MOP in (8.17) is solved, the inequalities in (8.18) will all hold and multi-objective optimal solution $(u_1^*, \ldots, u_n^*)$ with the corresponding objective values $(\rho_1^*, \ldots, \rho_n^*)$ is the Nash equilibrium solution. The proof is completed.

8.9.4 Appendix D: Proof of Theorem 8.5

By substituting the Lyapunov function in (8.34) and the stochastic fuzzy system in (8.33) into (8.61), J_k can be rewritten as

$$
\begin{aligned}
J_k =& E\{\tilde{x}_k(0)^T P_1 \tilde{x}_k(0)\} - E\{\tilde{x}_k(T)^T P_1 \tilde{x}_k(T)\} + E\{\sum_{i=1}^{l} h_i(z(t)) \int_0^T \tilde{x}_k(t)^T P_1 A_{ki} \tilde{x}_k(t)^T \\
&+ \tilde{x}_k(t)^T A_{ki}^T P_1 \tilde{x}_k(t)^T + \tilde{x}_k(t)^T P_1 \Delta F_k \tilde{x}_k(t) + \Delta F_k \tilde{x}_k(t)^T P_1 \tilde{x}_k(t) + \tilde{x}_k^T(t) Q_k \tilde{x}_k(t) \\
&+ [\tilde{x}_k(t)^T C_{ki}^T P_1 C_{ki} \tilde{x}_k(t) + \Delta H_k(\tilde{x}_k(t))^T P_1 C_{ki} \tilde{x}_k(t) + (\tilde{x}_k(t))^T C_{ki}^T P_1 \Delta H_k(\tilde{x}_k(t)) \\
&+ \Delta H_k(\tilde{x}_k(t))^T P_1 \Delta H_k(\tilde{x}_k(t))] + \lambda[\tilde{x}_k(t)^T P_1 D_{ki} \tilde{x}_k(t) + \tilde{x}_k(t)^T P_1 \Delta L_k(\tilde{x}_k(t)) \\
&+ \tilde{x}_k(t)^T D_{ki}^T P_1 \tilde{x}_k(t) + \tilde{x}_k(t)^T D_{ki}^T P_1 D_{ki} \tilde{x}_k(t) + \tilde{x}_k(t)^T D_{ki}^T P_1 \Delta L_k(\tilde{x}_k(t)) \\
&+ \Delta L_k(\tilde{x}_k(t))^T P_1 \tilde{x}_k(t) + \Delta L_k(\tilde{x}_k(t))^T P_1 D_{ki} \tilde{x}_k(t) + \Delta L_k(\tilde{x}_k(t))^T P_1 \Delta L_k(\tilde{x}_k(t))] \\
&- \tilde{x}_k(t)^T P_1 (B_{ki} + \Delta G_k(\tilde{x}_k(t)) R_k^{-1} (B_{ki} + \Delta G_k(\tilde{x}_k(t)))^T P_1 \tilde{x}_k(t) \\
&+ \frac{1}{\rho_k} \tilde{x}_k(t)^T P_1 (\bar{B}_{ki} + \Delta \bar{B}_k(\tilde{x}_k(t)))(\bar{B}_{ki} + \Delta \bar{B}_k(\tilde{x}_k(t)))^T P_1 \tilde{x}_k(t) \\
&+ \left(R_k u_k(t) + G_{k,k}^T(\tilde{x}_k(t)) P_1 \tilde{x}_k(t)\right)^T R_k^{-1} \left(R_k u_k(t) + G_{k,k}^T(\tilde{x}_k(t)) P_1 \tilde{x}_k(t)\right) \\
&- \left[\sqrt{\rho_k}\bar{v}_k(t) - \frac{1}{\sqrt{\rho_k}} \bar{B}_k^T(\tilde{x}_k(t)) P_1 \tilde{x}_k(t)\right]^T \left[\sqrt{\rho_k}\bar{v}_k(t) - \frac{1}{\sqrt{\rho_k}} \bar{B}_k^T(\tilde{x}_k(t)) P_1 \tilde{x}_k(t)\right] dt.\}
\end{aligned}
$$

Thus, similar to the deviation in Theorem 8.2, we have

$$
\begin{aligned}
J_k^* &= \min_{u_k(t)} \max_{\bar{v}_k(t)} J_k \\
&= \min_{u_k(t)} \max_{\bar{v}_k(t)} E\{\tilde{x}_k(0)^T P_1 \tilde{x}_k(0)\} - E\{\tilde{x}_k(T)^T P_1 \tilde{x}_k(T)\} + E\{\sum_{i=1}^{l} h_i(z(t)) \int_0^T \tilde{x}_k(t)^T P_1 A_{ki} \tilde{x}_k(t)^T \\
&+ \tilde{x}_k(t)^T A_{ki}^T P_1 \tilde{x}_k(t)^T + \tilde{x}_k(t)^T P_1 \Delta F_k \tilde{x}_k(t) + \Delta F_k \tilde{x}_k(t)^T P_1 \tilde{x}_k(t) + \tilde{x}_k^T(t) Q_k \tilde{x}_k(t) \\
&+ [\tilde{x}_k(t)^T C_{ki}^T P_1 C_{ki} \tilde{x}_k(t) + \Delta H_k(\tilde{x}_k(t))^T P_1 C_{ki} \tilde{x}_k(t) + (\tilde{x}_k(t))^T C_{ki}^T P_1 \Delta H_k(\tilde{x}_k(t)) \\
&+ \Delta H_k(\tilde{x}_k(t))^T P_1 \Delta H_k(\tilde{x}_k(t))] + \lambda[\tilde{x}_k(t)^T P_1 D_{ki} \tilde{x}_k(t) + \tilde{x}_k(t)^T P_1 \Delta L_k(\tilde{x}_k(t)) \\
&+ \tilde{x}_k(t)^T D_{ki}^T P_1 \tilde{x}_k(t) + \tilde{x}_k(t)^T D_{ki}^T P_1 D_{ki} \tilde{x}_k(t) + \tilde{x}_k(t)^T D_{ki}^T P_1 \Delta L_k(\tilde{x}_k(t)) \\
&+ \Delta L_k(\tilde{x}_k(t))^T P_1 \tilde{x}_k(t) + \Delta L_k(\tilde{x}_k(t))^T P_1 D_{ki} \tilde{x}_k(t) + \Delta L_k(\tilde{x}_k(t))^T P_1 \Delta L_k(\tilde{x}_k(t))] \\
&- \tilde{x}_k(t)^T P_1 (B_{ki} + \Delta G_k(\tilde{x}_k(t)) R_k^{-1} (B_{ki} + \Delta G_k(\tilde{x}_k(t)))^T P_1 \tilde{x}_k(t) \\
&+ \frac{1}{\rho_k} \tilde{x}_k(t)^T P_1 (\bar{B}_{ki} + \Delta \bar{B}_k(\tilde{x}_k(t)))(\bar{B}_{ki} + \Delta \bar{B}_k(\tilde{x}_k(t)))^T P_1 \tilde{x}_k(t) dt\}
\end{aligned}
\tag{8.65}
$$

with the following strategies as the minmax solution to the above Nash quadratic game problem

$$u_k^*(t) = -R_k^{-1} G_{k,k}^T (\tilde{x}_k(t) P_1 \tilde{x}_k(t)$$
$$\bar{v}_k^*(t) = \frac{1}{\rho_k} \bar{B}_k^T (\tilde{x}_k(t) P_1 \tilde{x}_k(t)$$

By Lemma 8.1 and Assumption 8.1, we have the following inequalities of the fuzzy estimation error:

$$\begin{aligned}
&\tilde{x}_k^T(t) P_1 \Delta F_k(\tilde{x}_k(t)) + \Delta F_k^T(\tilde{x}_k(t)) P_1 \tilde{x}_k(t) \leq \tilde{x}_k^T(t)(P_1 P_1 + \Delta A_k^T \Delta A_k) \tilde{x}_k(t) \\
&\tilde{x}_k(t)^T (P_1 \bar{B}_{ki} \Delta \bar{B}_k(\tilde{x}_k(t))^T P_1 + P_1 \Delta \bar{B}_k(\tilde{x}_k(t)) \bar{B}_{ki}^T P_1) \tilde{x}_k(t) \\
&\qquad \leq \tilde{x}_k(t)^T P_1 \bar{B}_{ki} \bar{B}_{ki}^T P_1 \tilde{x}_k(t) + \bar{\varepsilon}_k \tilde{x}_k(t)^T P_1 P_1 \tilde{x}_k(t) \\
&\tilde{x}_k^T(t) P_1 \Delta L_k(\tilde{x}_k(t)) + \Delta L_k^T(\tilde{x}_k(t)) P_1 \tilde{x}_k(t) \\
&\qquad \leq \tilde{x}_k^T(t)(P_1 P_1 + \Delta D_k^T \Delta D_k) \tilde{x}_k(t) \qquad (8.66) \\
&\tilde{x}_k^T(t) D_{ki}^T P_1 \Delta L_k(\tilde{x}_k(t)) + \Delta L_k^T(\tilde{x}_k(t)) P_1 D_{ki} \tilde{x}_k(t) \\
&\qquad \leq \tilde{x}_k^T(t)(D_{ki}^T P_1 P_1 D_{ki} + \Delta D_k^T \Delta D_k) \tilde{x}_k(t) \\
&\tilde{x}_k^T(t) C_{ki}^T P_1 \Delta H_k(\tilde{x}_k(t)) + \Delta H_k^T(\tilde{x}_k(t)) P_1 C_{ki} \tilde{x}_k(t) \\
&\qquad \leq \tilde{x}_k^T(t)(C_{ki}^T P_1 P_1 C_{ki} + \Delta C_k^T \Delta C_k) \tilde{x}_k(t)
\end{aligned}$$

Sequently, by using the matrix inequalities in (8.66), we have the following matrix inequalities of the integral part in (8.65):

$$\begin{aligned}
&E\{\sum_{i=1}^{l} h_i(z(t)) \int_0^T \tilde{x}_k(t)^T P_1 A_{ki} \tilde{x}_k(t)^T + \tilde{x}_k(t)^T A_{ki}^T P_1 \tilde{x}_k(t)^T + \tilde{x}_k(t)^T P_1 \Delta F_k(\tilde{x}_k(t)) \\
&+ \Delta F_k \tilde{x}_k(t)^T P_1 \tilde{x}_k(t) + \tilde{x}_k^T(t) Q_k \tilde{x}_k(t) + [\tilde{x}_k(t)^T C_{ki}^T P_1 C_{ki} \tilde{x}_k(t) \\
&+ \Delta H_k(\tilde{x}_k(t))^T P_1 C_{ki} \tilde{x}_k(t) + (\tilde{x}_k(t))^T C_{ki}^T P_1 \Delta H_k(\tilde{x}_k(t)) \\
&+ \Delta H_k(\tilde{x}_k(t))^T P_1 \Delta H_k(\tilde{x}_k(t))] + \lambda[\tilde{x}_k(t)^T P_1 D_{ki} \tilde{x}_k(t) + \tilde{x}_k(t)^T P_1 \Delta L_k(\tilde{x}_k(t)) \\
&+ \tilde{x}_k(t)^T D_{ki}^T P_1 \tilde{x}_k(t) + \tilde{x}_k(t)^T D_{ki}^T P_1 D_{ki} \tilde{x}_k(t) + \tilde{x}_k(t)^T D_{ki}^T P_1 \Delta L_k(\tilde{x}_k(t)) \qquad (8.67) \\
&+ \Delta L_k(\tilde{x}_k(t))^T P_1 \tilde{x}_k(t) + \Delta L_k(\tilde{x}_k(t))^T P_1 D_{ki} \tilde{x}_k(t) + \Delta L_k(\tilde{x}_k(t))^T P_1 \Delta L_k(\tilde{x}_k(t))] \\
&- \tilde{x}_k(t)^T P_1 (B_{ki} + \Delta G_k(\tilde{x}_k(t)) R_k^{-1} (B_{ki} + \Delta G_k(\tilde{x}_k(t)))^T P_1 \tilde{x}_k(t) \\
&+ \frac{1}{\rho_k} \tilde{x}_k(t)^T P_1 (\bar{B}_{ki} + \Delta \bar{B}_k(\tilde{x}_k(t)))(\bar{B}_{ki} + \Delta \bar{B}_k(\tilde{x}_k(t)))^T P_1 \tilde{x}_k(t) dt\} \\
&\leq E\{\sum_{i=1}^{l} h_i(z(t)) \int_0^T \tilde{x}_k(t)^T [P_1 A_{ki} + A_{ki}^T P_1 + P_1 P_1 + \Delta A_k^T \Delta A_k + Q_k + C_{ki}^T P C_{ki}
\end{aligned}$$

$$+ (1+\rho_k)\Delta C_k^T \Delta C_k + \lambda(P_1 D_{ki} + (2+\rho_k)\Delta D_k^T \Delta D_k + P_1 P_1 + D_{ki}^T P_1 + D_{ki}^T P_1 D$$
$$+ D_{ki}^T P_1 P_1 D) + \frac{2}{\rho_k} P_1 \bar{B}_{ki} \bar{B}_{ki}^T P_1 + \frac{2\bar{\varepsilon}_k}{\rho_k} P_1 P_1] \tilde{x}_k(t) dt\}$$

Thus, by the applying Riccati-like inequalities in (8.36), eigenvalue constraints on P_1 in (8.37) and matrix inequality in (8.67) to (8.65), we immediately have

$$J_k^* \leq E\{\tilde{x}_k(0)^T P_1 \tilde{x}_k(0)\} - E\{\tilde{x}_k(T)^T P_1 \tilde{x}_k(T)\}$$
$$\leq E\{\tilde{x}_k(0)^T P_1 \tilde{x}_k(0)\} \leq \rho_k \tilde{x}_k^T(0) \tilde{x}_k(0)$$

The proof is completed.

Part III

Stochastic Game Strategies in Signal Processing and Communication

9

Robust Design of Deconvolution Filters in Signal Transmission Systems under Stochastic Parametric and Noise Uncertainties: Stochastic Game Approach

9.1 Introduction

The deconvolution filter design problem is widely encountered in the signal processing literature. There are wide range of applications, including seismology [143], equalization [144], acoustics [145], etc. Many of the papers [146–148] have supplied with analysis or improvement for the optimal deconvolution filtering design method. However, in most cases, the transmission characteristics are not ideal. Often, they are also not precisely known and may be time-varying. The usual causes of these varying phenomena are due to the perturbation of transmission medium, linearization, and model reduction. Generally, it may be deterministic in some cases; and stochastic in other cases. Another case which does usually occur is the uncertainty in the second-order statistical properties of the noises. This is due to the variation of transmission environment. When any one of the above conditions occurs, the performance of the optimal deconvolution filter associated with some nominal transmission medium will be degraded.

One approach for deconvolution filter design under such uncertain conditions is to use an adaptive filtering scheme [149,150]. There do arise situations, however, in which adaptive deconvolution filter may not be practical because of cost, complexity, and the requirement to adapt rapidly to the changing transmission conditions. It may be desirable to use a fixed deconvolution filter which gives acceptable performance over the whole range of uncertain conditions in signal transmission system. One of the major techniques for designing fixed deconvolution filter designs that are robust with respect to modeling uncertainties is the minimax stochastic game approach, in which to be on the safe side the design goal is to optimize the worst-case performance for some criterion measure over the all uncertain class.

Early applications of minimax game approach to a great deal of robust filter design problems about modeling uncertainties in the signal transmission can be found in the literature. For example, the design works of [151–153] are about matched filtering, [154,155] about Wiener filtering, [156–159] about Kalman filtering, and [160] about channel equalization. However, the application of minimax stochastic game approach to the robust deconvolution filter design problems cannot be found in the literature yet. Although the design work of [160] was closely related to the robust deconvolution filter design problem, it could only treat the signal reconstruction of *time-invariant* uncertain transmission channel with known statistics of *stationary noises*. This is a severe restriction to the deconvolution filtering practical aspects of the design work since the actual transmission channel is often time-varying and the noises are usually nonstationary and not precisely known.

Furthermore, the resultant robust deconvolution filter of the design work [160] is not in causal form, and some filter realization problems are still needed to be solved.

In this chapter, we focus our attention on the design of minimax robust deconvolution filters based on stochastic game strategy under the more *general* uncertain *cases*. The linear discrete-time systems with continuous-valued stochastic parameters and noise uncertainties are considered. It is assumed that we have known the exact first moments (i.e. mean values) of random noises, initial conditions, and stochastic channel parameters. And we want to design a robust deconvolution filter for linear transmission systems with large but bounded uncertainties in second moments of these stochastic variations of channel parameters.

A recursive linear state estimator, which has nominal system matrix and unspecified gain vector, is chosen a *priori* as a robust deconvolution filter. Some useful sensitivity measures for this deconvolution filter are its mean square estimation error (S_1) and the standard deviation of this error from the minimum estimation error in an absolute (S_2) or relative (S_3) sense. These sensitivity measures are based on the functions of both the uncertain statistics and the unspecified filter gain. The robust deconvolution filters are related with S_1, S_2, and S_3 minimax estimation criteria based on MSDF, ADDF, and RDDF respectively. The search for minimax stochastic game solution is a difficult work if there does not exist any amenable analytical solution. The cases in which there exists an amenable analytical solution for finding the minimax filter gain are those of Nash equilibrium solution for which saddle-points exist. Stochastic game theory will be used to verify the existence of saddle-point solutions and to simplify the search procedure for minimax filter gain.

This chapter consists of eight sections. In Section 9.2, we introduce the adopted mathematical notation and the linear stochastic uncertain signal transmission system under consideration. In Section 9.3, the original signal transmission system is transformed to a state-space system model with nominal system matrix and state-dependent process noise, then the optimal deconvolution filter design under exactly known statistics is derived. In Section 9.4, we formulate the minimax robust deconvolution filter design problems under the case of uncertain statistics and derive the fixed-point minimax deconvolution filter based on stochastic game-theoretic strategy. Then, the more practical time-interval minimax deconvolution filter design is addressed in more detail in Section 9.5. In Section 9.6, we first consider the existence of steady-state deconvolution filters, and then solutions for the steady-state S_1, S_2, and S_3 minimax stochastic game criteria are derived. A numerical example is presented in Section 9.7 to illustrate the robust properties of the resultant minimax deconvolution filters and the utility of the minimax stochastic game design approaches. Finally, we conclude this chapter with remarks in Section 9.8.

9.2 System Description of Deconvolution Filter

This section provides the preliminary notation and essential structural information of the uncertain linear stochastic signal transmission system.

9.2.1 Mathematical Notation

The following notation applies unless otherwise noted. Lower case symbols denote scalar quantities. Upper case symbols denote vectors, matrices, or subsets of a Euclidean space. Furthermore, if we emphasize some quantity as a function of time, we would manipulate the quantity in the time interval associated with it rather than at a fixed time point. We adopt the following specific conventions:

I	identity matrix of appropriate dimension
O	zero matrix of appropriate dimension
A^T	transpose of the matrix A
E^n	n dimensional Euclidean space
V_R	bounding set for the time function of uncertain quantity $R(n)$
$\delta(n)$	Kronecker delta function
$\rho(A)$	spectral radius of the matrix A [161]
$E\lambda(v)$	expectation for random quantity v with probability distribution λ
$\mathrm{tr}(A)$	trace of the matrix A
$A \otimes B$	Kronecker product of the matrices A and B [162]
$\mathrm{con}(\Omega)$	convex hull of the arbitrary Euclidean subset Ω [163]
$\mathrm{cov}[w(n), m(k)]$	covariance of the random signals $w(n)$ and $m(k)$
$\mathrm{vec}(A)$	vector of the matrix A; i.e., $\mathrm{vec}(A) = [a_1^T \ \ a_2^T \ \ \cdots]^T$, where a_i is the ith column of the matrix A
$\bar{A}, \bar{P}$	mean of A if A is a stochastic quantity and the steady-state value of P if P is a deterministic, time-varying quantity
$\nabla_K J(K_1, v)$	gradient of the scalar function $J(K, v)$ with respect to argument K at K_1 point

9.2.2 ARMA Signal Transmission System Model Description

Figure 9.1 shows the robust deconvolution system under consideration. Signal $u(n)$, which is generated by the signal generator G with uncertain driving noise $d(n)$ as input, is the quantity to be estimated. It is passed through a uncertain linear stochastic transmission system (channel) L with ARMA model description, and corrupted by measurement noise $m(n)$ which has uncertain second-order statistical property. Then, we want to construct a robust deconvolution filter H, which is identical in form to a recursive linear estimator, to estimate $u(n)$.

The signal generator G has the following model description:

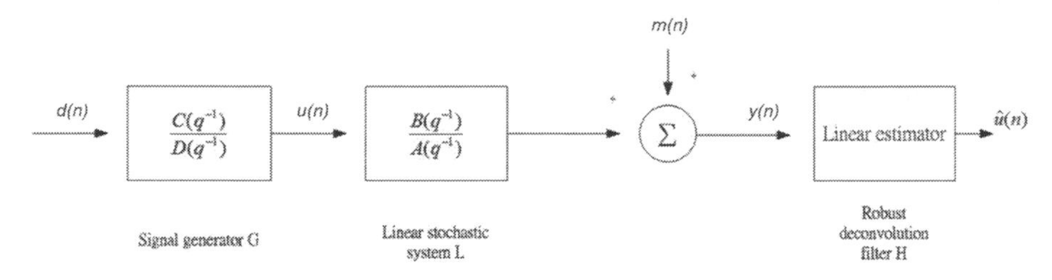

FIGURE 9.1
Robust deconvolution system

$$u(n) = \frac{C(q^{-1})}{D(q^{-1})} d(n) \tag{9.1}$$

where q^{-1} denotes unit-delay operator and both denominator and numerator polynomials are composed of deterministic constant coefficients such that

$$\begin{aligned} C(q^{-1}) &= c_1 q^{-1} + \cdots + c_{n_c} q^{-n_c} \\ D(q^{-1}) &= 1 + d_1 q^{-1} + \cdots + d_{n_d} q^{-n_d} \end{aligned} \tag{9.2}$$

where $n_d \geq n_c$. And the uncertain linear stochastic signal transmission system (channel) L has the following model description:

$$y(n) = \frac{B(q^{-1})}{A(q^{-1})} u(n) + m(n) \tag{9.3}$$

where denominator and numerator polynomials are composed of real random coefficients such that

$$\begin{aligned} A(q^{-1}) &= 1 + a_1(n) q^{-1} + \cdots + a_{n_a}(n) q^{-n_a} \\ B(q^{-1}) &= b_1(n) q^{-1} + \cdots + b_{n_b}(n) q^{-n_b} \end{aligned} \tag{9.4}$$

where $n_a \geq n_b$. It is assumed that $d(n)$ and $m(n)$ are uncorrelated, zero mean, white noise processes such that

$$\begin{aligned} \text{cov}[d(n), d(k)] &= \Sigma(n)\delta(n-k) \\ \text{cov}[m(n), m(k)] &= R(n)\delta(n-k). \end{aligned} \tag{9.5}$$

Furthermore, let us define the stochastic vector $\alpha(n)$ by

$$\alpha(n) = [a_1(n) \cdots a_{n_a}(n)\ b_1(n) \cdots b_{n_b}(n)], n \geq 0. \tag{9.6}$$

And we assume that $\{\{\alpha(n), n \geq 0\}\}$ is a sequence of independent stochastic vectors with constant statistics and the first moment of $\alpha(n)$ is exactly known a *priori*. Moreover, we assume that all stochastic quantities, such as initial conditions of both signals $u(n)$

and $y(n)$, parametric vector $\alpha(n)$ as well as random noises $d(n)$ and $m(n)$ for each $n>0$ are mutually uncorrelated with each other.

It is assumed that the uncertainties of stochastic characteristics will exhibit in the second moments of initial conditions, noise processes and random parameters in the linear stochastic transmission system (channel). However, we know that they will belong to previously specified compact uncertain classes.

9.3 The Optimal Deconvolution Filter Design under Exactly Known Statistics

In this section, we derive an optimal deconvolution filter under the condition of exactly known statistics in transmission system. Before we can design an optimal deconvolution filter, we must transform the transmission system to a special system description, called the state-dependent noise model. The processes of the transformation are described in the following.

9.3.1 State-Space Model Description of Transmission System

It is well known that the ARMA model description can be easily transformed to the state-space model description, which has the Kalman filter as the optimal linear minimum variance state estimator under the criterion of minimizing the state estimation error covariance. So we can rewrite the uncertain linear stochastic transmission system L as the following observable state space model [164]:

$$\begin{aligned} x_2(n+1) &= A_2(n)x_2(n) + B_2(n)u(n) \\ y(n) &= C_2x_2(n) + m(n) \end{aligned} \tag{9.7}$$

where $x_2(n) \in R^{n_a}$ for each $n \geq 0$ and $C_2 = [0 \cdots 01]$. $A_2(n)$ and $B_2(n)$ can further be decomposed as follows:

$$\begin{aligned} A_2(n) &= \begin{bmatrix} 0 & \cdots & \cdots & 0 & -a_{n_a}(n) \\ 1 & \ddots & & \vdots & \vdots \\ 0 & \ddots & \ddots & \vdots & \vdots \\ \vdots & \ddots & \ddots & 0 & -a_2(n) \\ 0 & \cdots & 0 & 1 & -a_1(n) \end{bmatrix} = \bar{A}_2 + \tilde{A}_2(n), \\ B_2(n) &= \begin{bmatrix} b_{n_a}(n) \\ \vdots \\ b_2(n) \\ b_1(n) \end{bmatrix} = \bar{B}_2 + \tilde{B}_2(n) \end{aligned} \tag{9.8}$$

where the bar denotes expectation and index n is deleted when a variable is constant. In the above equation, $\{b_i; i = n_b + 1, \ldots, n_b\}$ are all set to be zero and $\tilde{A}_2(n)$ and $\tilde{B}_2(n)$ are zero mean random matrices to represent the random fluctuation of transmission channel.

Similarly, we can rewrite the signal generator G as the following observable state-space model:

$$\begin{aligned} x_1(n+1) &= A_1 x_1(n) + B_1 d(n) \\ u(n) &= C_1 x_1(n) \end{aligned} \tag{9.9}$$

where $x_1(n) \in R^{n_d}$ for each $n \geq 0$. A_1, B_1 and C_1 are identical in form to $A_2(n), B_2(n)$ and C_2, respectively, except having different dimensions and being deterministic constant quantities.

Then, let us combine the signal generating system with the linear stochastic transmission system and consider the measurement noise, which results in the following combined signal system model:

$$\begin{aligned} x(n+1) &= A(n)x(n) + w(n) \\ y(n) &= Cx(n) + m(n) \end{aligned} \tag{9.10}$$

where

$$\begin{aligned} A(n) &= \begin{bmatrix} A_1 & O \\ B_2(n)C_1 & A_2(n) \end{bmatrix} = \begin{bmatrix} A_1 & O \\ \bar{B}_2 C_1 & \bar{A}_2 \end{bmatrix} + \begin{bmatrix} O & O \\ \tilde{B}_2(n)C_1 & \tilde{A}_2(n) \end{bmatrix} = \bar{A} + \tilde{A}(n), \\ C &= [O \quad C_2],\ w(n) = \begin{bmatrix} B_1 \\ O \end{bmatrix} d(n),\ x(n) = \begin{bmatrix} x_1(n) \\ x_2(n) \end{bmatrix}. \end{aligned} \tag{9.11}$$

We shall call the zero mean random process $w(n)$ the process noise and the zero mean random matrix $\tilde{A}(n)$ the perturbative system matrix. It is known from previous assumptions that $w(n)$ and $m(n)$ have the following second-order statistical properties:

$$\begin{aligned} \text{cov}[w(n), w(k)] &= \begin{bmatrix} B_1 B_1^T & O \\ O & O \end{bmatrix} \Sigma(n)\delta(n-k) \triangleq Q(n)\delta(n-k) \\ \text{cov}[w(n), m(k)] &= O, \text{ for all } n, k \end{aligned} \tag{9.12}$$

Also, from the previous assumptions on $\alpha(n)$, we know that the second moment of perturbative system matrix $\tilde{A}(n)$ can be written as follows:

$$\text{cov}[\text{vec}\tilde{A}(n), \text{vec}\tilde{A}(k)] = \Phi\delta(n-k) \tag{9.13}$$

where Φ is a constant positive semidefinite covariance matrix.

The initial state $x(0)$ can be shown to be a linear transformation of initial conditions of $u(n)$ and $y(n)$ [164], so it can be assumed to have known mean $\bar{x}_0$ and uncertain initial state covariance P_0. Additionally, $x(0)$ is not correlated with $\tilde{A}(n)$, $w(n)$ and $m(n)$, respectively.

9.3.2 State-Dependent Noise Model

We know that the system matrix $A(n)$ is a random quantity, which leads the dynamic behavior of linear signal transmission system (9.10) to be unexpected. So it will be reasonable to decompose $A(n)$ into $\bar{A}$ and $\tilde{A}(n)$, then view the signal transmission system behaving according to the nominal system matrix $\bar{A}$ and having an additional noise $\tilde{A}(n)x(n)$. Then, the so-called state-dependent noise model is derived as follows:

$$\begin{aligned} x(n+1) &= \bar{A}x(n) + \tilde{w}(n) \\ y(n) &= Cx(n) + m(n) \end{aligned} \tag{9.14}$$

where

$$\tilde{w}(n) = \tilde{A}(n)x(n) + w(n) \tag{9.15}$$

Using the statistical properties of linear transmission system (9.10), it is easy to see that $\{\tilde{w}(n), n \geq 0\}$ is a sequence of uncorrelated stochastic vectors and $x(0)$ is uncorrelated with $\{\tilde{w}(n), n \geq 0\}$. This implies that $x(n_1)$ is uncorrelated with $\tilde{w}(n_2)$ for $n_1 \leq n_2$. Moreover, the state-dependent process noise $\tilde{w}(n)$ has zero mean and covariance

$$\tilde{Q}(n) = \overline{\tilde{A}X(n)A^T} + \wp(n) \tag{9.16}$$

where the bar denotes expectation and $X(n) \triangleq E\{x(n)x^T(n)\}$. It can be shown that $X(n)$ has a recursive form as follows:

$$X(n+1) = \overline{AX(n)A^T} + Q(n) = \bar{A}X(n)\bar{A}^T + \tilde{Q}(n) \tag{9.17}$$

with the initial condition

$$X(0) = \bar{x}_0\bar{x}_0^T + P_0. \tag{9.18}$$

From the above description, we know that the matrices $\tilde{Q}(n)$ and $X(n)$ are real symmetric, nonnegative definite, and $\tilde{Q}(n)$ is a transformation of $Q(n)$ as well as second moment Φ of perturbative system matrix $\tilde{A}(n)$.

9.3.3 Optimal Deconvolution Filter

If the second-order statistics $P_0, \Phi, R(n)$ and $Q(n)$ are known previously, then the covariance $Q(n)$ will be known and can be evaluated in advance. In this case, the Kalman filter [164] can be applied as a linear minimum variance estimator of system state $x(n)$. Then, the optimal recursive deconvolution filter of signal $u(n)$ for the state-dependent noise model in (9.14) is given by

$$\begin{aligned} &\hat{u}(n) = [C_1 \quad O]\hat{x}(n), \hat{x}(0) = \bar{x}_0 \\ &\hat{x}(n+1) = \bar{A}\hat{x}(n) + K(n)[y(n) - C\hat{x}(n)] \\ &K(n) = \bar{A}P(n)C^T[CP(n)C^T + R(n)]^{-1} \\ &P(n+1) = \bar{A}P(n)A^T + \tilde{Q}(n) - \bar{A}P(n)C^T[R(n) + CP(n)C^T]^{-1}CP^T(n)\bar{A}^T, \\ &P(0) = P_0. \end{aligned} \tag{9.19}$$

However, in fact, we often have no exact knowledge of the second-order stochastic characteristics. So the optimal deconvolution filter (9.19) for some nominal signal transmission system cannot be applied to the uncertain systems directly. Further sensitivity analysis to the above Kalman deconvolution filter must be taken.

9.4 Fixed-Point Minimax Deconvolution Filter Design Based on Stochastic Game

In this section, we first formulate the minimax deconvolution filter design problems; then further developments and simplification based on stochastic game theories are followed to treat these robust deconvolution filter design problems.

9.4.1 The Minimax Robust Deconvolution Filter Design Problems

From the above description, we know that the uncertain quantity $\tilde{Q}(n)$ in the state-dependent noise model (9.14) is a combination of Φ and $Q(n)$. So, we have simplified the complicated uncertain parametric and noise second moments to a simpler uncertain noise covariance.

Lacking exact knowledge of the initial state covariance P_0 as well as time functions $\tilde{Q}(n)$ and $R(n)$, the form of the recursive linear estimator (9.19) is chosen *a priori* as a robust deconvolution filter. However, the time function of filter gain $K(n)$ now will be adjusted to satisfy an appropriate sensitivity criterion. Specifically, the deconvolution filter takes the form

$$\begin{aligned} \hat{u}(n) &= [C_1 \quad O]\hat{x}(n), \hat{x}(0) = \bar{x}_0 \\ \hat{x}(n+1) &= A\hat{x}(n) + K(n)[y(n) - C\hat{x}(n)] \end{aligned} \tag{9.20}$$

where $K(n)$, the filter gain, is an independent time function to be selected by the filter designer. Since the structure of the robust deconvolution filter has been decided as (9.20), the sole nontrivial unknown quantity which must be solved by further work is the minimax filter gain $K(n)$. In the rest of this chapter, we emphasize our efforts on the search of the filter gain solutions for several minimax filtering criteria based on stochastic game strategy.

The state estimation error covariance denoted by $M(n)$ is expressed as follows:

$$M(n) = E\{[x(n) - \hat{x}(n)][x(n) - \hat{x}(n)]^T\}. \tag{9.21}$$

Then, for some given initial state covariance P_0 as well as time functions $\tilde{Q}(n)$ and $R(n), M(n)$ will have a recursive form

$$M(n+1) = [\bar{A} - K(n)C]M(n)[\bar{A} - K(n)C]^T + K(n)R(n)K^T(n) + \tilde{Q}(n); \ M(0) = P_0 \tag{9.22}$$

The mean square signal estimation error $E\{[u(n) - \hat{u}(n)]^2\}$ of the deconvolution filter (9.20) will be adopted as a sensitivity measure, and can be expressed at some fixed time N as

$$J_M(N) = E\{[u(N) - \hat{u}(N)]^2 = \mathrm{tr}[WM(N)] \tag{9.23}$$

where

$$W = \begin{bmatrix} C_1^T C_1 & O \\ O & O \end{bmatrix} \tag{9.24}$$

For some given initial state covariance P_0 as well as time functions $\tilde{Q}(n)$ and $R(n)$ in the time interval, the minimum value of mean square error $J_M(N)$ is

$$J_o(N) = \mathrm{tr}[WP(N)] \tag{9.25}$$

where $P(N)$, the state error covariance of the optimal estimate, can be solved recursively by the discrete matrix Ricatti equation

$$\begin{aligned} &P(n+1) = \bar{A}P(n)\bar{A}^T - \bar{A}P(n)C^T[R(n) + CP(n)C^T]^{-1}CP^T(n)\bar{A}^T + \tilde{Q}(n), \\ &P(0) = P_0;\ 0 \le n \le N-1 \end{aligned} \tag{9.26}$$

and the optimal deconvolution filter gain associated with these given covariances is

$$K_o(n) = \bar{A}P(n)C^T[R(n) + CP(n)C^T]^{-1}, 0 \le n \le N-1. \tag{9.27}$$

From (9.22) and (9.23), it is known that $J_M(N)$ is convolved with not only the uncertain covariance matrices Φ and P_o as well as time functions of the unknown quantities $Q(n)$ and $R(n)$, but also the filter gain $K(n)$ in the time interval $[0, N-1]$. For the purpose of robust deconvolution filter design, the time function $K(n)$ may belong to any compact convex set $\tilde{V}_K$, which covers the set of time functions of all optimal filter gains corresponding to uncertain noise covariances. Furthermore, the uncertain $V = V_Q \times V_R \times V_\Phi \times V_{P_0}$, which are composed of time functions of uncertain noise covariances as well as uncertain second moments of perturbative system matrix and initial state vector, are defined for convenience. Hence, any time function $v \in V$ will be convolved with the Cartesian product $Q(n) \times R(n) \times \Phi \times P_o$. For the purpose of robust deconvolution filter design, it seems appropriate to select time function $K(n)$ in the time interval $[0, N-1]$ to satisfy the following fixed-point minimax mean-square-error deconvolution game problem:

$$\min_{K \in \tilde{V}_K} \max_{v \in V} J_M(K, v, N). \tag{9.28}$$

Then, the fixed-point minimax deconvolution game places a least upper bound on J_M at time N under the presence of uncertain second moments in the time interval $[0, N-1]$. And it may be considered as a recursive realization of the optimal "worst-case" deconvolution filter design for the mean square estimation error criterion J_M at time N.

However, the time-interval minimax mean-square-error deconvolution game may be more practical in actual applications because we are always concerned with the robust deconvolution filter design problem in a time interval rather than at some fixed time N.

For the time-interval minimax deconvolution filter design problem, the desirable object is to design a robust recursive estimator, which places the least upper bounds on the objective sensitivity measure J_M for all time $N \in [N_o, N_f]$ under the presence of uncertain statistical characteristics in the time interval $[0, N_f - 1]$. Hence, in the next section, we will, by a little modification to fixed-point minimax deconvolution filter analysis, develop a design algorithm for the recursive realization of time-interval minimax deconvolution filter based on the fundamental structure in (9.20).

9.4.2 Fixed-Point Minimax Game MSDF

In this subsection, we shall be concerned with other reformulated fixed-point minimax deconvolution game problems at fixed time N, which are more suitable to recursive realization of analytical solutions. The cases in which there exists an amenable analytical solution for finding the minimax deconvolution filter gain are those for which saddle points exist. We can use game theory to verify the existence of saddle-point (Nash equilibrium) solutions and to simplify the search for minimax deconvolution filter gain. However, the sufficient conditions for a game to have saddle point solutions require some topological properties on the sets of the stochastic game and continuity and quasi concave-convexity of the sensitivity measure on the maximizing and minimizing sets, respectively.

For the fixed-point minimax mean-square-error deconvolution game possessed in (9.28), the sensitivity measure J_M cannot be guaranteed to be concave in the time function $v \in V$. Then, this minimax stochastic game problem in general has no saddle-point solutions, and it will be difficult to find the minimax stochastic game solutions. However, the minimax stochastic game problem can be reformulated by allowing stochastic game strategies, i.e. the original maximizing set V is replaced by a set of probability distributions defined on them, and the sensitivity measure J_M is replaced by its expected value associated with those probability distributions. The reformulated results can be summarized as follows.

First, we define the reformulated minimizing set V_K by

$$V_K = \text{con}(K_o(v, n), v \in \text{con}(V)) \tag{9.29}$$

where time function $K_o(v, n)$ is expressed in (9.27), and the inclusion of argument v into K_o emphasizes the fact that K_o is the optimal deconvolution filter gain associated with the uncertain element v. Furthermore, let M_V denote the class of all probability measures on V. If λ denotes any probability distribution for V, then a new randomized fixed-point minimax game criterion can be reformulated as follows:

$$S_1'(N) = \min_{K \in V_K} \max_{\lambda \in M_V} E_\lambda[J_M(K, v, n)]. \tag{9.30}$$

Then, from game theory, the triple (J_M, V_K, M_V) will form a stochastic game with randomized strategies in the set M_V of probability distributions. Now, let us show that the stocahstic game (J_M, V_K, M_V) satisfies all sufficient conditions in Lemma 9.A.l. in Appendix A. Hence, by Lemma 9.A.l, it has a saddle-point solution (K^*, λ^*).

Theorem 9.1

The game (J_M, V_K, M_V) has a saddle-point solution (K^*, λ^*) such that the min-max game criterion S_1' in (9.30) will be equivalent to the max-min game criterion and can be simplified to a maximization problem as

$$S_1'(N) = \max_{\lambda \in M_V} \min_{K \in V_K} E_\lambda [J_M(K, v, n)] \\ = \max_{\lambda \in M_V} J_o(\bar{v}, N) = J_o(\bar{v}_N^*, N) \tag{9.31}$$

where $\bar{v} = E_\lambda(v)$, and the time function $\bar{v}_N^*$ is the mean value of v under a least favorable priori distribution λ^* for the stocahstic game (J_M, V_K, M_V). In addition, the solution to minimax deconvolution filter gain can be solved by

$$K_1^*(n) = K_o(\bar{v}_N^*, N), 0 \leq n \leq N - 1. \tag{9.32}$$

Proof: The set V_K is compact convex by definition, and V is a compact set by assumption. Continuity of J_M in V_K and V for all finite n is easily demonstrated by applying the properties of matrix norms to (9.22) and (9.23). It remains to be shown that J_M is convex in V_K for each $v \in V$. From (9.22) and (9.23), one has the second partial of $J_M(n+1)$ with respect to $K(n)$ as

$$\frac{\partial^2 J_M(n+1)}{\partial K^2(n)} = 2[CM(n)C^T + \mathrm{R(n)}]\mathrm{W} > 0, \; n \geq 0 \tag{9.33}$$

then $J_M(n+1)$ is strictly convex in $V_K(n)$ for each $v \in V$ and $\forall n \geq 0$, where $V_K(n)$ denotes the bounding set of filter gain K at time point n only. In addition to the above convex property, we know that $M(n+1)$ is a linear transformation of $M(n)$ for $\forall n \geq 0$, so $J_M(n)$ is strictly convex in $V_K(n)$ for each $v \in V$ and $\forall n \geq 0$. Since the convex property of $J_M(n)$ in $V_K(n)$ holds for any $n \geq 0$, $J_M(n)$ must be strictly convex in $V_K(n)$ for each $v \in V$. Thus, all conditions of Lemma 9.A. 1 (see Appendix A) are satisfied and the stochastic game (J_M, V_K, M_V) has a saddle-point solution (K^*, λ^*). Furthermore, the min-max criterion, S_1' will be equivalent to the max-min criterion and can be further simplified to a maximization problem as (9.31). If λ denotes any probability distribution for V, then the filter gain which minimizes $E_\lambda[J_M(K, v, n)]$ is $K_o(E_\lambda(v), n), 0 \leq n \leq N-1$. Therefore, (9.32) follows immediately by the existence of the saddle-point solution. Q.E.D.

From the above discussion, we know that the minimizing set V_K has an optimal pure strategy K^* for time function K, and the maximizing set V has an optimal mixed strategy λ^* for time function v. And the solution of minimax deconvolution filter gain is simply the optimal filter gain for the mean value of time function v under a least favorable priori distribution λ^*. However, the maximizing of J_o on the probability measure set M_V in (9.31) remains to be difficult to realization. Fortunately, the following theorem provides us further simplification of fixed-point minimax filter design deconvolution problem. Lemma 9.A.2 (see Appendix A) motivates us to reformulate a maximizing set V to form concave property of sensitivity measure J_M in the reformulated maximizing set. So, we define another compact convex maximizing set $\tilde{V}$ as

$$\tilde{V} = \mathrm{con}(V_R) \times \mathrm{con}(V_{\tilde{Q}}) \times \mathrm{con}(V_{P_o}) \tag{9.34}$$

and the associated reformulated minimizing set $\tilde{V}_K$ is then defined by

$$\tilde{V}_K = \mathrm{con}(K_o(\tilde{v}, n), \tilde{v} \in \tilde{V}). \tag{9.35}$$

With the above preparation for further simplification, then a new fixed-point minimax mean-square-error deconvolution game problem is reformulated as follows:

$$S_1(N) = \min_{K \in \tilde{V}_K} \max_{\tilde{v} \in \tilde{V}} J_M(K, \tilde{v}, n). \tag{9.36}$$

It is noted that any element $\tilde{v} \in \tilde{V}$ is merely convolved with the Cartesian product $\tilde{Q}(n) \times R(n) \times P_o$.

Theorem 9.2
For the reformulated maximizing set $\tilde{V}$ and the associated reformulated minimizing one $\tilde{V}_K$, not only the stochastic game $(J_M, \tilde{V}_K, \tilde{V})$ has a saddle-point solution, but also each set in it has an optimal game strategy. Then we have the following simplifications:

$$\begin{aligned} S_1(N) &= \max_{\tilde{v} \in \tilde{V}} \min_{K \in \tilde{V}_K} J_M = (K, \tilde{v}, n) \\ &= \max_{\tilde{v} \in \tilde{V}} J_o(\tilde{v}, n) = J_o(\tilde{v}_N^{\#}, n) \end{aligned} \tag{9.37}$$

where $\tilde{v}_N^{\#}$ is the maximizing element of time function $\tilde{v}$ for the stochastic game $(J_M, \tilde{V}_K, \tilde{V})$.
Proof: After the above definition of maximizing set $\tilde{V}$, we know that $\tilde{V}$ is a compact convex set. Since J_M is linear in con(V_R) and con$(V_{\tilde{Q}})$, it is concave in $\tilde{V}$. The continuity of J_M in $\tilde{V}$ for all finite time n is easily demonstrated by applying the properties of matrix norms to (9.22) and (9.23). Furthermore, all necessary continuity and compact convex properties for $\tilde{V}_K$ can be verified as the proof of Theorem 9.1 for that of the minimizing set V_K. So, the reformulated stochastic game $(J_M, \tilde{V}_K, \tilde{V})$ satisfies the sufficient conditions of Lemma 9.A.l and 9.A.2. Not only the stochastic game $(J_M, \tilde{V}_K, \tilde{V})$ has a saddle-point solution, but also each set in it has an optimal game strategy. Then, (9.37) results from the characteristics of the saddle-point. Q.E.D.

The existence of the saddle-point solution and the uniqueness of the Kalman filter immediately implies that the solution to fixed-point minimax deconvolution filter gain can be solved by

$$K^{\#}(N) = K_o(\tilde{v}_N^{\#}, n), 0 \le n \le N-1 \tag{9.38}$$

where $K^{\#}$ is the minimax solution of the filter gain K in stochastic game $(J_M, \tilde{V}_K, \tilde{V})$. Thus, the relatively difficult problem of minimax mean-square-error J_M can be replaced with the simpler problem of maximizing J_o over $\tilde{V}$. And hence (9.20) together with (9.38) will synthesize the fixed-point minimax mean-square-error deconvolution filter (MSDF).

The stochastic games (J_M, V_K, M_V) and $(J_M, \tilde{V}_K, \tilde{V})$ are closely related by their structure since the reformulated maximizing set $\tilde{V}$ is the convex hull of a transformation of the original maximizing set $\tilde{V}$. We should expect a correspondence between the saddle-point solutions (K^*, λ^*) and $(K^{\#}, \tilde{v}_N^{\#})$.

Such a relation is obtained by noting that any point $\tilde{v} \in V$ can be expressed by a convex combination of at most $s+1$ points in $V_R \times V_{\tilde{Q}} \times V_{P_o}$, where s denotes the Euclidean dimension of $V_R \times V_{\tilde{Q}} \times V_{P_o}$. Let p denote $(p_1, p_2, \ldots, p_{s+1})$, an appropriate choice of weights for expressing $\tilde{v}_N^{\#}$ as a convex combination of points in $V_R \times V_{\tilde{Q}} \times V_{P_o}$. It is observed that we can interpret p (and hence $\tilde{v}_N^{\#}$) as a mixed game strategy on $V_R \times V_{\tilde{Q}} \times V_{P_o}$. Further, due to the relationship between $V_R \times V_{\tilde{Q}} \times V_{P_o}$ and V, we can select a mixture of points in V. Thus, a solution $\tilde{v}_N^{\#}$ to stochastic game

$(J_M, \tilde{V}_K, \tilde{V})$ induces a corresponding discrete probability distribution $\lambda^{\#}$ on V. Conversely, a least favorable prior distribution λ^* for stochastic game (J_M, V_K, M_V) induces a corresponding pure solution $\tilde{v}_N^*$ on $\tilde{V}$. So, the associated least favorable prior distribution must be discrete and assign probability to at most $\min(r+1, s+1)$ points, where r denotes the Euclidean dimension of V_K and s denotes the Euclidean dimension of $V_R \times V_{\tilde{Q}} \times V_{P_o}$. Hence, the value [165] of S_1 criterion is equal to that of S_1' criterion and the minimax deconvolution filter gain K^* in stochastic game(J_M, V_K, M_V) will be equivalent to minimax deconvolution filter gain $K^{\#}$ in stochastic game $(J_M, \tilde{V}_K, \tilde{V})$. Hence, if the minimax deconvolution filter gain $K^{\#}$ was used to serve as the solution of deconvolution filter gain in (9.20), it will result in a recursive realization of fixed-point minimax deconvolution filter at time N under the minimax game criterion (9.36) [and hence the randomized minimax game criterion (9.30)].

The stochastic games (J_M, V_K, M_V) and $(J_M, \tilde{V}_K, \tilde{V})$ are also closely related since the set M_V of all probability measures on V obviously contain the uncertain set V. Sometimes the least favorable priori distribution λ^* in stochastic game (J_M, V_K, M_V) supplies the sole element $v \in V$ with probability one. In this special case, the values for both games are equal and the minimax deconvolution filter gains are the same. However, in general, λ^* supplies probability measure on many elements in V. In this general case, the values for stochastic game (J_M, V_K, M_V) will be greater than that of stochastic game $(J_M, \breve{V}_K, V)$ and the minimax filter gains for both games are different. The utility function of the stochastic game $(J_M, \breve{V}_K, V)$ is that it guarantees a least upper bound (i.e. the value of this game) on J_M for each element $v \in V$. And it may be considered as an optimal "worst-case" design in the "absolute sense." On the other hand, the stochastic game (J_M, V_K, M_V) places a least upper bound on the mean values of J_M for the set M_V of all probability measures on V. And it may be considered as an optimal "worst-case" design in the "mean-square sense." After the above discussion, it is concluded that the relative merit between the two games is decided by the desire of filter designer. However, for the case of robust deconvolution filter design problem possessed in this chapter, the stochastic game (J_M, V_K, M_V) will be the favorite since it supplies an analytical solution to find the minimax deconvolution filter gain.

9.5 Time-interval minimax deconvolution filter based on stochastic game under general uncertain case

From the above discussion, a recursive realization of fixed-point minimax MSDF has been analytically solved by stochastic game approach. It will obtain a fixed-point minimax deconvolution filter gain $K^{\#}$ in the time interval $[0, N-1]$. However, for the more practical consideration, the recursive filtering problem of time-interval minimax mean-square-error deconvolution will be addressed in more detail in the following design procedure.

9.5.1 Design Procedure

By a little modification to fixed-point minimax MSDF, the time-interval minimax MSDF in the time interval $[N_o, N_f]$ can be constructed as follows: Given signal polynomials

$C(q^{-1})$ and $D(q^{-1})$, the uncertain ranges of second moments of the system polynomials $A(q^{-1})$ and $B(q^{-1})$ as well as initial conditions of signals $u(n)$ and $y(n)$, and bounds of noise covariances $\Sigma(n)$ and $R(n)$, a design procedure of minimax filter is proposed for signal deconvolution step by step as follows:

1) According to (9.7)–(9.11), we first construct the state-dependent noise model (9.14).
2) According to (9.12), (9.13), (9.16), and (9.17), we then obtain the convex maximizing set $\tilde{V} = \text{con}(V_R) \times \text{con}(V_{\tilde{Q}}) \times \text{con}(V_{P_o})$ in (9.34).
3) Solve the following maximization problems:

$$\tilde{v}_N^{\#} = (\tilde{Q}_N^{\#} \times R_N^{\#} \times P_{oN}^{\#} = \arg\max_{\tilde{v}\in\tilde{V}} J_o(\tilde{v}, N), \text{for all } N_o \leq N \leq N_f \tag{9.39}$$

where $0 \leq N_o \leq N_f < \infty$ and J_o is expressed in (9.25) and (9.26).
4) Find $\tilde{v}_{N_1}^{\#}, \tilde{v}_{N_2}^{\#}, \ldots, \tilde{v}_{N_L}^{\#}$ such that $\tilde{v}_N^{\#} = \tilde{v}_{N_i}^{\#}$ for all $0 \leq n \leq N-1$ and $N_{i-1} \leq N \leq N_i$, where $N_0 = N_o - 1$ and $N_L = N_f$. It should be noted that L is a positive integer less than $N_f - N_o + 1$ and its value illustrates how heavily $\tilde{v}_N^{\#}$'s depend on N in the time interval $[N_o, N_f]$.
5) Define the time function of the minimax deconvolution filter gain in stochastic game $(J_M, \tilde{V}_K, \tilde{V})$ for maximizing element $\tilde{v}_N^{\#}$, as

$$K_i^{\#}(n) = K_o(\tilde{v}_{N_i}^{\#} n), 0 \leq n \leq N_i - 1 \tag{9.40}$$

where $i = 1, 2, \ldots, L$ and K_o us expressed in (9.27)

9.5.2 Time-Interval Minimax MSDF

In the above design procedure, we have obtained the time function of minimax deconvolution filter gain $K_i^{\#}$'s for $i = 1, 2, \ldots, L$. Let each $K_i^{\#}$ serve as the filter gain solution of the recursive linear estimator (9.20) in the time interval $[0, N_i - 1]$, then each of the L resultant recursive estimate in (9.20) yields a minimax deconvolution of signal $u(n)$ in the segmented time interval $(N_i - 1, N_i]$. In conclusion, the recursive realization of time interval minimax MSDF is constructed as follows: The signal estimator must be segmented in different time intervals as

$$\begin{cases} \hat{u}(n) = [C_1 \quad O]\hat{x}_i(n), 0 \leq n \leq N_0; \text{any } i \in [1, L] \\ \hat{u}(n) = [C_1 \quad O]\hat{x}_i(n), N_{i-1} + 1 \leq n \leq N_i;\ \forall i \in [1, L] \end{cases} \tag{9.41}$$

where $\hat{x}_i$ for $i = 1, 2, \ldots, L$ are iteratively updated by the following recursive state estimator:

$$\begin{aligned} &\hat{x}_i(n+1) = \bar{A}\hat{x}_i(n) + K_i^{\#}(n)[y(n) - C\hat{x}_i(n)],\ 0 \leq n \leq N_i - 1 \\ &\hat{x}_i(0) = \bar{x}_0, \forall i \in [1, L]. \end{aligned} \tag{9.42}$$

Then, the resultant signal estimator $\hat{u}(n)$ will satisfy the minimax deconvolution game criterion (9.36) [and hence the randomized criterion (9.30)] in the time interval $[N_o, N_f]$.

It will be noted that in the time interval $[0, N_o)$, any of the L state estimates in the state estimator (9.42) can be used to obtain the estimate $\hat{u}(n)$ of desirable signal in the signal estimator (9.41). This is owing to the reason that the original formulated time-interval minimax game deconvolution filter is concerned with the robustness of signal estimator in the time interval $[N_o, N_f]$ only. Furthermore, at any time $n \in [N_{i-1}, N_i - 1]$, there will be $L - i + 1$ iterative state updates for $\hat{x}_i(n), \ldots, \hat{x}_L(n)$ respectively, but only $\hat{x}_i(n)$ will be adopted to update the signal estimator $\hat{u}(n)$ in the time interval $[N_{i-1} + 1, N_i]$.

9.5.3 Further Remarks

The properties of the time-interval minimax MSDF game possessed above will be emphasized in more detail in the following Remarks:

Remark 9.1 The sets of uncertain statistics which apply to the minimax deconvolution filter design problems have been defined quite arbitrarily with only compact constraint so the minimax game deconvolution filter is adequate to the general uncertain classes. Furthermore, for the further problem of random signal generator parameters whose mean values are known but with uncertain second moments, all the above developments are followed without any change to solve the further problem. It merely increases the computational complexity of steps 2) and 3) in the above minimax game deconvolution filter algorithm. Hence, from the above discussion, the possessed algorithm has addressed the stochastic game strategy of minimax deconvolution filters under fairly general uncertain conditions.

Remark 9.2 In fact, the deconvolution filter adopted in this chapter is in a one-step-ahead prediction form. It estimates $u(n)$ only by using the measurements up to $y(n-1)$. For the "deconvolution filtering" problem, i.e. estimating $u(n)$ by the given measurements up to $y(n)$, the recursive filtering form of Kalman filter [166] can be used in (9.42) as a state estimator. For the smoothing problem, we can introduce additional states [164] which contain time delay vision of the signal $u(n)$ into the original state-space model (9.10), and construct the canonical form of the deconvolution filter (9.20) according to the augmented state-space model. From the above description, both filtering and smoothing design would not alter the organization of the uncertainty problem. Hence, in addition to an one-step-ahead prediction, the above minimax game strategy can also be applied to treat the design problem of minimax deconvolution filtering as well as smoothing. Furthermore, the usage of observable canonical form of state-space model is not necessary, and it is merely one of many simple and fast transformations from ARMA model to state space model. Any other special canonical forms which further simplify the minimax deconvolution filter design game problems will be adopted to replace the observable canonical form. Hence, the algorithm has supplied a much more flexibility in the design procedure.

Remark 9.3 We have constructed a canonical realization to the robust deconvolution filter in the form of recursive linear estimator. The nominal system matrix $\bar{A}$ of the deconvolution filter is merely a simple transformation of $C(q^{-1})$ and $D(q^{-1})$ as well as mean values of $A(q^{-1})$ and $B(q^{-1})$. Then the sole nontrivial unknown quantity is the minimax filter gain. Even the least favorable prior distribution assigns the probability to a large number of points in the original stochastic game (J_M, V_K, M_V). The reformulated stochastic game $(J_M, \tilde{V}_K, \tilde{V})$ needs only to search a pure maximizing element $\tilde{v}_N^{\#}$. and the minimax filter gain is simply the optimal filter gain associated with $\tilde{v}_N^{\#}$. Hence, although the determination of minimax filter gain generally requires some search techniques, the stochastic game theory however has simplified the minimax deconvolution filter design problem remarkably.

Remark 9.4 At the first glance, the end-product (9.41) and (9.42) of the possessed deconvolution filter design algorithm looks like a filter bank and seems horrendously complex. However, this is just an illusion due to its complex profile. The structure of the minimax deconvolution filter (9.41) and (9.42) is truly a single filter and its computational complexity is certainly moderate. The reasons of these claims are as follows. Once the maximizing elements $\tilde{v}_N^{\#}$'s $i = 1, 2, \ldots L$ have been found in step 4), each time function of minimax deconvolution filter gain $K_i^{\#}(n)$'s in step 5) can be calculated for $0 \le n \le N_{i-1}$ prior to any processing of real data since its computation does not involve the input data $y(n)$ of the minimax deconvolution filter (9.41) and (9.42). Hence, in the operation duration of the minimax deconvolution filter, each update of estimated signal $\hat{u}(n)$ for $n \in [N_{i-1}, N_i - 1]$ involves only $L - i + 1$ iterative updates $\hat{x}_i(n), \ldots, \hat{x}_L(n)$ of state estimator (9.42). Since it involves only an iterative calculation of the same computational algorithm (9.42), the minimax game deconvolution filter (9.41) and (9.42) is truly a single filter as the previous claim. Furthermore, in view of the sparse structure of coefficient matrix/vectors $\bar{A}$, C, and C_1 in (9.41) and (9.42), we can show that each update of the estimated signal $\hat{u}(n)$ requires only $(L - i + 1)$ $(3n_a + 2n_d)$ multiplications and $(L - i + 1) \times (3n_a + 2n_d - 1)$ additions for each $n \in [N_{i-1}, N_i - 1]$ and $1 \le i \le L$. Hence, the computational complexity of the minimax deconvolution filter (9.41) and (9.42) is quite moderate. Furthermore, the computational complexity will decease with increased time, which results in the proposed minimax game deconvolution filter more attractive.

Remark 9.5 From Remark 9.4, we know that the computational complexity of each update of estimated signal $\hat{u}(n)$ heavily depends on the quantity L. Hence, if $\tilde{v}_N^{\#}$ is either independent of N (i.e. $L = 1$) or only weakly dependent upon N (i.e. L is small), then the computational complexity of minimax deconvolution filter (9.41) and (9.42) will be certainly very simple. A frequent situation in which $\tilde{v}_N^{\#}$ is independent of N will be as follows [157]: The uncertain second moments $P_o, \tilde{Q}(n)$ and $R(n)$ are confined to the classes characterized by $0 < P_o \le P_{om}$, $0 < \tilde{Q}(n) \le \tilde{Q}_m(n)$, and $0 < R(n) \le R_m(n)$, respectively, for all $n \ge 0$. In this case, the unique worst-case covariance matrices, in the sense of maximizing the cost (9.39), are the over-bounding covariance matrices $P_{om}, \tilde{Q}_m(n)$ and $R_m(n)$, $0 \le n \le N - 1$. Another special condition in which $\tilde{v}_N^{\#}$ is weakly dependent upon N will be the stationary case, i.e., the uncertain quantities Q and R are not of time-varying but constant. In this case, as described in the next section, the linear stochastic system will approach the steady-state fast and hence there will be an unique maximizing element $\bar{\tilde{v}}^{\#}$ after the steady-state is achieved. This results in $\tilde{v}_N^{\#}$ is dependent on N only during the transition period, and hence L will be small.

9.6 Steady-state minimax deconvolution filters based on stochastic game under stationary uncertain case

In the previous two sections, we have derived the finite-stage deconvolution filters based on stochastic game strategy for both fixed-point and time-interval minimax mean-square-error criteria. However, under fairly general conditions, the solution of the state estimation error covariance equation (9.26) starting at a symmetric non-negative matrix P_o

converges to a unique constant non-negative matrix $\bar{P}$ as $n \to \infty$. In this case, J_o is independent of P_o as $N \to \infty$. So, in this section, we will focus on the realization of the steady-state minimax deconvolution filters based on stochastic game strategy.

9.6.1 Existence of the Steady-State Minimax Game Deconvolution Filter

Assume the uncertain quantities Q and R are not of time-varying but constant. In this stationary case, (9.17) becomes $X(n+1) = \overline{AX(n)A^T} + Q$. Moreover, we assume that the uncertain system is mean-square stable, i.e., $\rho(\overline{A \oplus A}) < 1$ for all possible second moments of system matrix $A(n)$. The mean-square stable assumption can be verified to be the sufficient and almost necessary condition for $\bar{X} = \lim_{n\to\infty} X(n)$ existing [167], no matter what initial state covariance P_o was. Furthermore, under the mean-square stable assumption, $\bar{X}$ is the unique solution of the generalized Lyapunov equation

$$\bar{X} = \overline{A\bar{X}A^T} + Q. \tag{9.43}$$

Therefore, $\bar{\tilde{Q}} = \lim_{n\to\infty} \tilde{Q}(n)$ exists and it satisfies the following equation:

$$\bar{\tilde{Q}} = Q + \overline{\tilde{A}\bar{X}\tilde{A}^T}. \tag{9.44}$$

In addition to Q and R being constant by assumption and $\bar{\tilde{Q}}$ existing as $n \to \infty$, if the filter gain K of deconvolution filter (9.20) is selected adequately such that it is constant and all eigenvalues of the matrix $\bar{A} - KC$ are inside the unit circle, then $M(n)$ in (9.22) will approach to a constant one as $n \to \infty$ and hence the linear estimator (9.20) is a steady-state deconvolution filter.

9.6.2 The Steady-State Minimax Game MSDF

In the above subsection, we have proposed the sufficient condition for the existence of the steady-state deconvolution filter. Now, let us find the steady-state solution to the minimax filter gain for the infinite time mean-square-error criterion S_1 and discuss the maximization of $\bar{J}_o(\tilde{v})$ with respect to the set $\tilde{V}$.

We know that the mean-square stable condition (i.e., $\rho(\overline{A \oplus A}) < 1$) also implies that the nominal system matrix $\bar{A}$ is stable (i.e., $\rho(\bar{A}) < 1$). Therefore, by application of the standard linear estimation theory to (9.26) and (9.27), $\bar{P} = \lim_{n\to\infty} P(n)$ and $\bar{K}_o = \lim_{n\to\infty} K_o(n)$ exist where $\bar{P}$ is the unique nonnegative definite solution of the equation

$$P = \bar{A}\bar{P}\bar{A}^T + \bar{\tilde{Q}} - \bar{A}\bar{P}C^T(R + C\bar{P}C^T)^{-1}C\bar{P}^T\bar{A}^T \tag{9.45}$$

and the steady-state optimal filter gain associated with the given R and $\bar{\tilde{Q}}$ is

$$\bar{K}_o = \bar{A}\bar{P}C^T(C\bar{P}C^T + R)^{-1} \tag{9.46}$$

Clearly, the resultant steady-state deconvolution filter will be stable (i.e. $\rho(\bar{A} - \bar{K}_o C) < 1$) and optimal (minimum) mean square estimation error J_o will approach a constant value $\bar{J}_o = \text{tr}(W\bar{P})$.

Theorem 9.3
Under the above assumptions for the existence of the steady-state minimax game deconvolution filter, it is shown that in the limit as $N \to \infty$

$$\bar{K}^{\#} = \bar{K}_o(\tilde{v}_{\infty}^{\#}) \tag{9.47}$$

where the bar denotes the value at steady-state. That is, the infinite-time S_1 minimax filter gain is a fixed gain equal to the steady-state optimal filter gain for the uncertain quantities that maximize $\bar{J}_o$.

The proof of Theorem 9.3 can be found in Appendix B.

Some of the important properties of the infinite time maximization are now developed. From saddle-point theorem, $\tilde{v}_{\infty}^{\#}$ is unique. Furthermore, $\bar{J}_o(\tilde{v})$ is continuous and concave in $\tilde{V}$ because $\bar{J}_o$ is linear in con$(V_{\tilde{Q}})$ and

$$\frac{\partial^2 \bar{J}_o}{\partial R^2} = -2(R + C\bar{P}C^T)^{-3}(\text{vec}\bar{P}\bar{A}^T)^T W(\text{vec}\bar{P}\bar{A}^T) < 0. \tag{9.48}$$

The above properties are sufficient to guarantee that $\bar{J}_o$ has only a global maximum.

Next, the maximization of $\bar{J}_o(\tilde{v})$ with respect to the set $\tilde{V}$ will be discussed. $\bar{P}$ for some nominal $\bar{\tilde{Q}}_N$ and R_N is obtained by

$$\bar{P}_N = \bar{A}\bar{P}_N\bar{A}^T + \bar{\tilde{Q}}_N - \bar{A}\bar{P}_N C^T(R_N + C\bar{P}_N C^T)^{-1}C\bar{P}_N^T\bar{A}^T \tag{9.49}$$

Consider an uncertain element q_{ij} in $\bar{\tilde{Q}}$ and let

$$Q_{ij} = \frac{\partial \bar{\tilde{Q}}_N}{\partial q_{ij}} \text{ and } \bar{P}_{ij} = \frac{\partial \bar{P}_N}{\partial q_{ij}} \tag{9.50}$$

Observe that Q_{ij} is symmetric with two forms:

$$Q_{ii} = [1_{ii}], i = j; \; Q_{ij} = [1_{ij}] + [1_{ji}], i \neq j \tag{9.51}$$

where $[1_{ij}]$ denotes a matrix whose entries are all zero except for a 1 in the ijth position. Differentiating (9.49) with respect to q_{ij}, one obtains

$$\bar{P}_{ij} = (\bar{A} - \bar{K}_N C)\bar{P}_{ij}(\bar{A} - \bar{K}_N C)^T + \bar{K}_N R_{ij}\bar{K}_N^T \tag{9.52}$$

where $\bar{K}_N = \overline{AP}_N C^T(R_N + C\bar{P}_N C^T)^{-1}$. Similarly, defining

$$R_{ij} = \frac{\partial R_N}{\partial r_{ij}} \text{ and } \tilde{r}P_{ij} = \frac{\partial \bar{P}_N}{\partial r_{ij}} \tag{9.53}$$

and differentiating (9.49) with respect to r_{ij}, one obtains

$$\tilde{P}_{ij} = (\bar{A} - \bar{K}_N C)\tilde{P}_{ij}(\bar{A} - \bar{K}_N C)^T + \bar{K}_N R_{ij}\bar{K}_N^T. \tag{9.54}$$

Equations in (9.52) and (9.54) are the discrete Lyapunov equations for $\bar{P}_{ij}$ and $\tilde{P}_{ij}$ which have solutions whenever all eigenvalues of $(\bar{A} - \bar{K}_N C)$ are inside the unit circle. And this condition has been verified in the previous discussion of this subsection; therefore, $\bar{P}_{ij}$ and $\tilde{P}_{ij}$ always exist. It follows immediately that

$$\frac{\partial \bar{J}_o}{\partial q_{ij}} = \mathrm{tr}(W\bar{P}_{ij}) \text{ and } \frac{\partial \bar{J}_o}{\partial r_{ij}} = \mathrm{tr}(W\tilde{P}_{ij}) \tag{9.55}$$

always exist. Q_{ii} and R_{ii} are positive semi-definite and therefore so are $\bar{P}_{ii}$ and $\tilde{P}_{ii}$. Then,

$$\frac{\partial \bar{J}_o}{\partial q_{ii}} = \mathrm{tr}(W\bar{P}_{ii}) > 0 \text{ and } \frac{\partial \bar{J}_o}{\partial r_{ii}} = \mathrm{tr}(W\tilde{P}_{ii}) \tag{9.56}$$

The above results indicate that $\bar{J}_o(\tilde{v}_\infty^{\#})$ can be easily found by the following method. $\bar{J}_o$ is maximized with respect to the diagonal elements of $\bar{Q}$ or R by setting those elements to their maximum assumed values. Furthermore, because $\bar{J}_o(\tilde{v})$ is continuous and concave in $\tilde{V}$, the simple steepest accent techniques are appropriate for the search of the off-diagonal elements. One needs to only ensure that the positive definiteness requirements on $\bar{\bar{Q}}$ and R are met at each step in the search.

The construction of the steady-state minimax game MSDF is the same as the canonical form (9.20) possessed in Section 9.4. However, the search of maximizing element $\tilde{v}_\infty^{\#}$ is much simpler than that of maximizing element $\tilde{v}_N^{\#}$ in (9.39) for the time-varying case. Through the above analysis, $\tilde{v}_\infty^{\#}$ can be solved by the steepest accent technique and the S_1 minimax filter gain is a fixed gain equal to the steady-state optimal filter gain for maximizing element $\tilde{v}_\infty^{\#}$. Of course, the uncertain system must satisfy the mean-square stable condition $\rho(\overline{A \otimes A}) < 1$ and the uncertain quantities Q and R must be constant for the existence of the steady-state filter.

9.6.3 Other Steady-State Minimax Game Deconvolution Filters

The mean square error J_M is one of sensitivity measures for deconvolution filter in (9.20). On the other hand, it is also appropriate to measure the performance of deconvolution filter in (9.20) in terms of its absolute or relative departure from the optimality. These sensitivity measures take the form

$$S^A(K, v', N) = J_M(K, v', N) - J_o(v', N) \tag{9.57}$$

and

$$S^R(K, v', N) = \frac{J_M(K, v', N) - J_o(v', N)}{J_o(v', N)} \tag{9.58}$$

where v' will belong to uncertain set V' defined by $V' = V_R \times V_{\bar{\bar{Q}}} \times V_{P_o}$. These sensitivity measures result in the following minimax deconvolution filter games respectively:

$$S_2(N) = \min_{K \in V'_K} \max_{v' \in V'} S^A(K, v', N) \tag{9.59}$$

and

$$S_3(N) = \min_{K \in V'_K} \max_{v' \in V'} S^R(K, v', N) \tag{9.60}$$

where the minimizing set V'_K is defined by the compact collection of all filter gains with dimension $(n_a + n_d) \times 1$ since we have not yet known what region the minimax filter gain for S_2 or S_3 criterion will belong to. The S_2 and S_3 criteria seek to control filter sensitivity directly by minimizing the maximum absolute or relative deviation of J_M from its optimal value at time N over the uncertain quantity set V'. And hence these two kinds of robust deconvolution filters will be named as minimax game ADDF and RDDF, respectively.

It can be known from Theorem 9.4 at the end of this subsection that even in the steady-state case, both of the minimax game criteria S_2 and S_3 have no amenable analytical procedures for the implementation of the relative deconvolution filters ADDF and RDDF. Therefore, the recursive realization of these minimax game deconvolution filters under the time-varying uncertain systems will be much more formidable than the steady-state case and would not be discussed in this subsection. However, certain useful properties of the steady-state minimax game ADDF and RDDF will be presented in the following. In the steady-state case, S^A and S^R are both the utility functions of K and v' only, and we shall denote by $\hat{K}$ the value of $K \in V'_K$ for which

$$\max_{v' \in V'} S(\hat{K}, v') = \min_{K \in V'_K} \max_{v' \in V'} S(K, v') \tag{9.61}$$

where S may be the steady-state value of S^R or S^A. It is known that $\hat{K}$ is the steady-state solution of minimax deconvolution filter gain for S_2 or S_3 criterion. Since $\min_{K \in V'_K} S(K, v') = 0$, it is clear that min-max does not equal max-min for the S^A and S^R sensitivity measures. However, it will be shown that the steady-state minimax game ADDF and RDDF exist and are unique. Also, the smaller regions than V' and V'_K in which the minimax game criterion S_2 or S_3 will be found can be determined.

Lemma 9.1 For the minimax deconvolution filter games in (9.59) and (9.60), the steady-state minimax ADDF and RDDF exist and are unique.

Proof: The set V'_K is compact by definition. From proof of Theorem 9.1, we know that J_M is also continuous and strictly convex in V'_K. In addition to J_o being independent of K, we know that $S(K, v')$ and $\max_{v' \in V'} S(K, v')$ are continuous and strictly convex in V'_K. Since the sufficient conditions of Lemma 9.A.3 are all satisfied, $\max_{v' \in V'} S(K, v')$ must attain the minimum value on the set V'_K and $\hat{K}$ is the unique minimum point for $\max_{v' \in V'} S(K, v')$. Thus, the steady-state minimax ADDF and RDDP exist and are unique. Q.E.D.

After we have verified the existence and uniqueness of the steady-state minimax ADDF and RDDF, the following lemma shows the smaller region to which the minimax deconvolution filter gain $\hat{K}$ will belong.

Lemma 9.2 The steady-state minimax ADDF and RDDF are optimal for some $v' \in V'$, that is

$$\hat{K} = \Xi \tag{9.62}$$

where $\Xi = \{K | K = \bar{K}_o(v'), v' \in V'\}$.

The proof of Lemma 9.2 can be found in Appendix C. Next, we show that the search for steady-state S_2 or S_3 minimax filter gain may be restricted to the set of the extreme points of V' only.

Lemma 9.3 For the minimax deconvolution filter criteria in (9.59) and (9.60) in the steady-state case, we have

$$\min_{K \in V'_K} \max_{v' \in V'} S(K, v') = \min_{K \in V'_K} \max_{v' \in V'_E} S(K, v') \tag{9.63}$$

where the set V'_E is composed of the extreme points of set V'.

Proof: From (9.48), we know that $\bar{J}_o$ is also concave in V'. This, in addition to $\bar{J}_M$ being linear in V', may be used to show that S^A and S^R are convex in V'. Since a convex scalar function of a vector defined on a compact set attains its maximum on the extreme points of that set, (9.63) follows immediately. Q.E.D.

Combining the above three lemmas, we have immediately the following conclusion for the design of steady-state minimax ADDF and RDDF.

Theorem 9.4

For the minimax deconvolution filter games in (9.59) and (9.60), the steady-state minimax ADDF and RDDF exist and are unique. And the search for minimax deconvolution filter gain of the steady-state minimax ADDF or RDDF may be restricted to the set of all optimal gains corresponding to V' and the set of extreme points of V', i.e.,

$$\min_{K \in V'_K} \max_{v' \in V'} S(K, v') = \min_{K \in \Xi} \max_{v' \in V'_E} S(K, v') \tag{9.64}$$

Based on the above theorem, the design procedure of the steady-state minimax ADDF or RDDF can be constructed as follows: First form the state-dependent noise model (9.14) and construct the canonical form of robust deconvolution filter (9.20). Then, the unspecified filter gain must be decided by solving the min-max deconvolution filter design problem in (9.64). In general, the set V' is the convex polyhedron, the search for the maximizing element v' can then be reduced to the finite set of the vertices of V'. Based on the above observation, the maximization part of the deconvolution filter design min-max game problem in (9.64) will be solved by simply making use of an exhaustive search over a set of finite vertices, and the minimizing element $\hat{K}$ can be obtained by an *ad hoc* computer search method. The required *ad hoc* search depends on the number of set V'_E and dimension of Ξ. However, sensitivity analysis to Kalman filter in (9.22)–(9.27) simplifies the computational complexity of the *ad hoc* search remarkably.

9.7 Numerical Example

Now, we study a particular numerical example among those treated in the previous section. Consider the following uncertain linear stochastic signal transmission system:

$$
\begin{aligned}
u(n) &= \frac{q^{-1}}{1-0.8q^{-1}}d(n) \\
y(n) &= \frac{bq^{-1}}{1+aq^{-1}}u(n)+m(n)
\end{aligned}
\tag{9.65}
$$

where stochastic parameters a and b are uncorrelated and have mean values $\bar{a}=-0.2, \bar{b}=0.4$. The bounding sets of uncertain quantities are $\overline{a^2}\in[0.1,0.3]$, $\overline{b^2}\in[0.2,0.4], \Sigma\in[0.5,1]$, and $R\in[0.5,1]$, respectively. Our signal deconvolution design problem is how to design the minimax steady-state deconvolution filters under these parameter and noise uncertainties.

Solution: From (9.11) and (9.16), we know that the relative nominal system matrix and the covariance matrix of state-dependent process noise are

$$
\bar{A}=\begin{bmatrix}0.8 & 0\\ 0.4 & 0.2\end{bmatrix} \text{ and } \tilde{Q}(n)=\begin{bmatrix}\Sigma & 0\\ 0 & \overline{\tilde{b}^2}X_{11}(n)+\overline{\tilde{a}^2}X_{22}(n)\end{bmatrix} \tag{9.66}
$$

respectively, where $\overline{\tilde{a}^2}\in[0.06,0.26]$ and $\overline{\tilde{b}^2}\in[0.04,0.24]$. The quantities $X_{11}(n)$ and $X_{22}(n)$ are the diagonal elements of state covariance $X(n)$.

First, let us verify the existence of steady-state by examining the mean-square stable condition. The spectral radius $\rho(\overline{A\otimes A})$ can be easily shown to equal 0.64, no matter what values $\overline{a^2}$ and $\overline{b^2}$ are. Since $\rho(\overline{A\otimes A})<1$ satisfies the requirement of the mean-square stable condition, $\bar{\tilde{Q}}=\lim_{n\to\infty}\tilde{Q}(n)$, we can construct steady-state robust deconvolution filters according to the minimax stochastic game strategy.

The minimax value of $\bar{J}_M$ occurs at the maximum points of noise covariances R and $\bar{\tilde{Q}}$ which can be numerically solved as

$$
R=1 \text{ and } \bar{\tilde{Q}}=\begin{bmatrix}1 & 0\\ 0 & 1.1423\end{bmatrix} \tag{9.67}
$$

respectively. So, from (9.45) and (9.46), the minimax filter gain of steady-state minimax MSDF is $\bar{K}^{\#}=[0.2322 \quad 0.2380]^T$. After solving (9.64) by an *ad hoc* computer search method, we know that the minimax value of S^A is attained at the extreme points

$$
R=1, \bar{\tilde{Q}}=\begin{bmatrix}0.5 & 0\\ 0 & 0.5711\end{bmatrix} \text{ and } R=0.5, \bar{\tilde{Q}}=\begin{bmatrix}1 & 0\\ 0 & 0.1594\end{bmatrix} \tag{9.68}
$$

and the minimax deconvolution filter gain of steady-state minimax ADDF is $\hat{K}=[0.3091 \quad 0.2453]^T$. By the similar method to the minimax ADDF, we know that the minimax value of S^R is attained at the same extreme points as (9.68) and the minimax filter gain of steady-state minimax RDDF is $\hat{K}=[0.2893 \quad 0.2364]^T$.

For the purpose of manifesting the low sensitive properties of minimax deconvolution filters, let the nominal values of uncertain quantities $\overline{a^2}, \overline{b^2}, \Sigma$ and R be 0.2, 0.3, 0.75, and 0.75 respectively, then the relative nominal filter gain is $K_N=[0.2874 \quad 0.2435]^T$. A numerical comparison between the steady-state minimax MSDF, ADDF, RDDF, and the nominal deconvolution filter (NDF) is contained in Table 9.1. Note that the minimax MSDF provides a least upper bound on $\bar{J}_M$ at the expense of a greater maximum

TABLE 9.1

The comparison of the steady-state minimax game and nominal deconvolution filters at nominal and extreme points.

$\bar{\bar{Q}}$	R	$\bar{J}_o$	$\bar{J}_M$				$\bar{S}^A$				$\bar{S}^R$			
			NDF	MSDF	ADDF	RDDF	NDF	MSDF	ADDF	RDDF	NDF	MSDF	ADDF	RDDF
$\begin{bmatrix} 0.5 & 0 \\ 0 & 0.0797 \end{bmatrix}$	0.5	1.1176	1.1248	1.1423	1.1208	1.1240	0.0071	0.0247	0.0032	0.0064	0.0064	0.0221	0.0029	0.0058
	1	1.1830	1.1867	1.1843	1.1916	1.1873	0.0037	0.0013	0.0086	0.0043	0.0032	0.0011	0.0073	0.0036
$\begin{bmatrix} 0.5 & 0 \\ 0 & 0.5711 \end{bmatrix}$	0.5	1.1971	1.2049	1.1971	1.2121	1.2059	0.0079	0.0000	0.0150	0.0088	0.0066	0.0000	0.0126	0.0074
	1	1.2308	1.2669	1.2390	1.2829	1.2691	0.0361	0.0082	0.0521	0.0383	0.0294	0.0067	0.0423	0.0312
$\begin{bmatrix} 1 & 0 \\ 0 & 0.1594 \end{bmatrix}$	0.5	2.1188	2.1875	2.2426	2.1709	2.1849	0.0687	0.1238	0.0521	0.0661	0.0324	0.0584	0.0246	0.0312
	1	2.2352	2.2495	2.2846	2.2417	2.2481	0.0143	0.0494	0.0065	0.0129	0.0064	0.0221	0.0029	0.0058
$\begin{bmatrix} 1 & 0 \\ 0 & 1.1423 \end{bmatrix}$	0.5	2.3460	2.3479	2.3522	2.3534	2.3486	0.0019	0.0062	0.0075	0.0026	0.0008	0.0026	0.0032	0.0011
	1	2.3941	2.4099	2.3941	2.4242	2.4118	0.0157	0.0000	0.0301	0.0177	0.0066	0.0000	0.0126	0.0074
Nominal point		1.7398	1.7398	1.7494	1.7412	1.7398	0.0000	0.0096	0.0014	0.0000	0.0000	0.0055	0.0008	0.0000
Maximum values		2.3941	2.4099	2.3941	2.4242	2.4118	0.0687	0.1238	0.0521	0.0661	0.0324	0.0584	0.0423	0.0312

deviation from the optimality, whereas the opposite is true for the minimax ADDF and RDDF which provide the least upper bounds on $\bar{S}^A$ and $\bar{S}^R$, respectively, at the expense of a greater maximum mean square estimation error.

Of the three minimax deconvolution filters, the minimax MSDF is the simplest to compute. Its maximum departure from the optimality is also easily determined since it will occur at one of the extreme points of $\tilde{V}$. It is therefore recommended that the minimax MSDF is evaluated first in any design effort. If a closer tracking of the optimal error than that provided by the minimax MSDF is then deemed necessary, one may proceed with a design with the evaluation of the minimax ADDF or RDDF.

9.8 Conclusion

Several minimax stochastic game approaches have been introduced to the design of robust deconvolution filters for signal transmission systems. The linear stochastic signal transmission systems under consideration are assumed to have uncertain second moments in the noise covariances, initial conditions, and stochastic parameters. The sets of uncertain statistics have been defined quite arbitrarily with only compact constraint. The possessed minimax game deconvolution filter is identical in form to a recursive linear estimator. It would be derived by transforming the signal transmission system to one with nominal parameters and state-dependent process noise.

Three kinds of sensitivity measures are used, namely, the mean square error (S_1) and the deviation of this error from the minimum estimation error in an absolute (S_2) or relative (S_3) sense. Our stochastic game approach yields three minimax game deconvolution filters (MSDF, ADDF, and RDDF, respectively), which place a least upper bound on the relative sensitivity measure over the assumed range of uncertain quantities.

For the S_1 minimax deconvolution filter game, by applying two standard saddle-point theorems, we have examined the existence and behavior of stochastic game-theoretic solutions for finitestate minimax MSDF. The resulting MSDF design algorithm is of recursive filtering form for both fixed-point and time-interval minimax deconvolution filter games. Furthermore, it can be applied to the further problems of minimax robust smoothing and the case of uncertain random signal generator parameters as discussed in Remarks 9.1 and 9.2. From Remark 9.3, we conclude that the minimax deconvolution filter design problem has been adequately simplified by stochastic game strategy. Furthermore, by Remarks 9.4 and 9.5, the resultant minimax game deconvolution filters possess the moderate computational complexity and are no more difficult than that of the Kalman filter. And therefore it could represent an attractive alternative to adaptive filters which are necessarily more complex in structure. Finally, several properties of the required maximization for the infinite-time (steady-state) case are also developed and discussed.

For the S_2 and S_3 minimax deconvolution filter games, only the steady-state case was discussed. The relative steady-state minimax ADDF and RDDF are shown to be unique and optimal for at least one point in the set of uncertain quantities. Furthermore, the maximum of these sensitivity measures is attained over a finite set of extreme points. So, the maximization part of steady-state S_2 and S_3 criteria will be solved by an exhaustive search over a set of finite vertices, and the minimax filter gains can be obtained by an *ad hoc* computer search method.

In conclusion, we have solved the robust deconvolution problem from the well-known minimax stochastic game strategy for several sensitivity measures. The proposed stochastic game strategies address the design of minimax deconvolution filters under fairly general uncertain conditions and have supplied much more flexibility in the robust deconvolution filter design procedure. The computational complexity of the proposed deconvolution filter is quite moderate in general and will be very simple for some special uncertain cases. All of these excellent robust deconvolution properties result in the proposed minimax deconvolution algorithms based on stochastic game being attractive.

9.9 Appendix

9.9.1 Appendix A

Lemma 9.A.1 Let the triple (G, X, Y) be a stochastic game such that X is a compact convex subset of E^m, Y is a compact subset of E^n, and $G(X,Y)$ is a real-valued mapping defined on the Cartesian product of X and Y. If for each $y \in Y, G(x, y)$ is convex and continuous in $x \in X$ for each $x \in X, G(x, y)$ is continuous in $y \in Y$, then, the game (G, X, Y) has a saddle-point solution (x^*, λ^*) such that

$$\begin{aligned} &\lim_{x\in X} \max_{\lambda\in M_Y} E_\lambda(G(x,y)) = \max_{\lambda\in M_Y} \min_{x\in Y} E_\lambda(G(x,y)) \\ &= E_{\lambda^*}(G(x^*,y)) = \int_Y G(x^*,y)d\lambda^* \end{aligned} \tag{9.69}$$

where $x^* \in X, \lambda^* \in M_Y$, M_Y denotes the class of all probability measures on Y, and λ^* is a discrete probability distribution which assigns mass to at most $m+1$ points in Y.

Proof: See [168].

Lemma 9.A.2 In addition to the hypothesis of Lemma 9.A. l, which requires that Y is convex; for each $x \in X, G(x, y)$ is concave in $y \in Y$. Then, each player has an optimal pure strategy such that

$$\min_{x\in X} \max_{y\in Y} E_\lambda(G(x,y)) = \max_{y\in Y} \min_{x\in X} E_\lambda(G(x,y)) \tag{9.70}$$

Proof: See [165].

Lemma 9.A.3 If a function f is continuous on a compact set U, it must attain both a maximum and a minimum value on set U. If, in addition, f is strictly convex (concave) on U, then the minimum (maximum) point is unique.

Proof: See [163].

9.9.2 Appendix B

The Proof of Theorem 9.3: Let us examine (9.37) in the limit as $N \to \infty$. Since $\tilde{V}$ is compact, the convergence of $P(N)$ to $\bar{P}$ and therefore $J_o(N)$ to $\bar{J}_o$ is uniform in $\tilde{V}$. That is, there exists a $N(\varepsilon)$ such that for each $N > N(\varepsilon)$ and any $\tilde{v} \in \tilde{V}$

$$\left|J_o(\tilde{v},N) - \bar{J}_o(\tilde{v})\right| < \varepsilon. \tag{9.71}$$

$\bar{J}_o$ is always nonnegative, so (9.71) implies that

$$J_o(\tilde{v},N) - \varepsilon < \bar{J}_o(\tilde{v}) \leq \bar{J}_o(\tilde{v}^{\#}_{\infty}) \tag{9.72}$$

where the rightmost inequality follows from the definition of $\tilde{v}^{\#}_{\infty}$ Since (9.72) is true for each $\tilde{v} \in \tilde{V}$, it is true for $\tilde{v}^{\#}$ and one has

$$J_o(\tilde{v}^{\#},N) - \bar{J}_o(\tilde{v}^{\#}_{\infty}) < \varepsilon \tag{9.73}$$

Again from (9.71) and the definition of $\tilde{v}^{\#}$

$$\bar{J}_o(\tilde{v}^{\#}_{\infty}) < \varepsilon + J_o(\tilde{v}^{\#}_{\infty},N) < \varepsilon + J_o(\tilde{v}^{\#},N) \tag{9.74}$$

which is equivalent to

$$\bar{J}_o(\tilde{v}^{\#}_{\infty}) - J_o(\tilde{v}^{\#},N) < \varepsilon. \tag{9.75}$$

Equations in (9.73) and (9.75) together imply that

$$\left|J_o(\tilde{v}^{\#}_{N},N) - \bar{J}_o(\tilde{v}^{\#}_{\infty})\right| < \varepsilon;\ N > N(\varepsilon) \tag{9.76}$$

Substituting (9.37) into (9.76) now yields the desired result

$$\lim_{N\to\infty} S_1(N) = \max_{\tilde{v}\in\tilde{V}} \bar{J}_o(\tilde{v}) = \bar{J}_o(\tilde{v}^{\#}_{\infty}) \tag{9.77}$$

It follows that

$$\bar{K}^{\#} = \bar{K}_o(\tilde{v}^{\#}_{\infty}) \tag{9.78}$$

where the bar denotes the value at steady-state. Q.E.D.

9.9.3 Appendix C

The proof of Lemma 9.2: Any point $v' \in V'$ can be written as a convex combination of the extreme points of V' denoted here by V'_E. That is,

$$\bar{J}_M(K,v') = \bar{J}_M(K,\sum_{i=1}^{r} a_i v_i)\ v_i \in V'_E, 0 \leq a_i \leq 1, \sum_{i=1}^{r} a_i = 1 \tag{9.79}$$

where r denotes the number of extreme points. Since $\bar{J}_M$ is linear in V', (9.79) may be written as

$$\bar{J}_M(K,v') = \sum_{i=1}^{r} a_i \bar{J}^i{}_M(K,v_i). \tag{9.80}$$

where $\bar{J}^i{}_M(K, v_i), i = 1, \ldots, r$ constitute a set of basis functionals for the functional $\bar{J}_M(K, v')$. By Lemma 9.C.1, we know that for $K \notin \Xi$, a direction of motion in V'_K, say ΔK, always exists such that

$$\bar{J}^i{}_M(K + \varepsilon\Delta K, v_i) < \bar{J}^i{}_M(K, v_i) \tag{9.81}$$

where ε is an appropriate small positive constant. Since the a_i's are all greater than or equal to zero, so far any $K \notin \Xi$, a ΔK always exists such that

$$\bar{J}_M(K, v') > \bar{J}_M(K + \varepsilon\Delta K, v'), \ \forall v' \in V' \tag{9.82}$$

However, (9.82) implies that

$$S(K, v') > S(K + \varepsilon\Delta K, v') \tag{9.83}$$

and thus

$$\max_{v' \in V'} S(K, v') = \max_{v' \in V'_E} S(K + \varepsilon\Delta K, v'). \tag{9.84}$$

Therefore, any $K \notin \Xi$ cannot be the minimax deconvolution filter gain for the S_2 and S_3 criteria. This, in addition to the existence and uniqueness of $\hat{K}$, results that $\hat{K}$ is optimal for some $v' \in V'$. Q.E.D.

Lemma 9.C.1: Let $\bar{J}^i{}_M(K, v_i); v_i \in V'_E, i = 1, \ldots, r$ represent the set of basis functionals for the performance index $\bar{J}_M(K, v')$, and let K_1 be such that $K_1 \notin \Xi$. Then, there exists a direction ΔK in V'_K such that

$$\bar{J}^i{}_M(K_1 + \varepsilon\Delta K, v_i) < \bar{J}^i{}_M(K_1, v_i), \ \forall i \tag{9.85}$$

where ε is an appropriate small positive constant.

Proof: An inductive proof is given as follows:

When $K_1 \notin \Xi$ is the gradient, $\nabla_K \bar{J}_M(K, v')|_{K=K_1} \neq 0$ for any $v' \in V'$. Now, denote by a_i the gradient $\nabla_K \bar{J}^i{}_M(K, v_i)|_{K=K_1} \neq 0$ and select $\Delta K \in \{\Delta K | \langle a_1, K \rangle < 0\}$, then (9.85) holds for $i = 1$.

Assume (9.85) is true for $i = 1, 2, \ldots, n$, then ΔK will belong to a convex polyhedral cone C_n, defined as follows:

$$C_n = \{\Delta K | \langle a_i, \Delta K \rangle < 0, i = 1, 2, \ldots, n\}. \tag{9.86}$$

To establish the next step in the induction, we must show that the intersection of negative half space

$$S = \{\Delta K_{n+1} | \langle a_{n+1}, \Delta K_{n+1} \rangle < 0\} \tag{9.87}$$

with C_n being nonvoid. If it is true, then we have that (9.85) is also true for $i = n + 1$ and the proof is completed by induction.

Assuming the contrary, i.e., $C_n \cap S = \emptyset$. Observe that $C_n \cap S = \emptyset$ if and only if

$$a_{n+1} = -t\left(\sum_{i=1}^{n} \alpha_i a_i\right);\ 0 \le \alpha_i \le 1, \sum_{i=1}^{n} \alpha_i = 1, t>0 \tag{9.88}$$

i.e., a_{n+1} lies in the negative convex span of $a_1, \ldots, a_n$. Now consider a point $v' \in V'$ such that

$$v' = \sum_{i=1}^{n+1} \gamma_i v_i;\ v_i \in V'_E, 0 \le \gamma_i \le 1, \sum_{i=1}^{n+1} \gamma_i = 1 \tag{9.89}$$

and the performance functional

$$\bar{J}_M(K_1, v') = \sum_{i=1}^{n+1} \gamma_i \bar{J}^i{}_M(K_1, v_i). \tag{9.90}$$

Then,

$$\nabla_K \bar{J}_M(K_1, v') = \sum_{i=1}^{n+1} \gamma_i \nabla_K \bar{J}^i{}_M(K_1, v_i) \tag{9.91}$$

and using (9.88), one obtains

$$\nabla_K \bar{J}_M(K_1, v') = \sum_{i=1}^{n} (\gamma_i - \alpha_i t \gamma_{n+1}) \nabla_K \bar{J}^i{}_M(K_1, v_i). \tag{9.92}$$

This gradient (9.92) can be made zero at K_1 by equating all coefficients to zero and invoking the constraint on the γ_i's.

This leads to a set of γ_i's linear equations in $n+1$ unknowns. The have the solution

$$\gamma_i = \frac{\alpha_i t}{1+t} < 1,\ i = 1, 2, \ldots, n;\ \gamma_{n+1} = \frac{1}{1+t} < 1. \tag{9.93}$$

Thus, a set of γ_i's satisfying (9.89) exist. Therefore, there exists a $v' \in V'$ such that (9.92) is zero,

$$\nabla_K \bar{J}_M(K, v')|_{K=K_1} = 0. \tag{9.94}$$

This implies $K_1 \in \Xi$ and is contrary to the assumption. Therefore, the intersection of S with C_n is nonvoid.

Q.E.D.

10

Robust Equalizer Designs via Minimax Stochastic Quadratic Game Approach for MIMO Wireless Communications with Time-Varying Uncertain Channel

10.1 Introduction

Recently, a huge interest has been generated in the design of wireless communication systems with multi-input–multi-output (MIMO) antennas since such MIMO systems are able to exploit antenna diversity for increasing capacity and spectral efficiency in fading environments [169–171]. For the equalizer design of such wireless communication systems, several methods have been developed, e.g., [172–177], under the assumption that channel state information (CSI) is known precisely at the receiver. In a real-world wireless communication environment, CSI is generally acquired via channel estimation. As estimation errors in the estimated CSI may be inevitable, they can lead to high bit error rates (BERs) in the channel equalization if not being handled properly. This together with the rising demand of reliable recovery of information in modern wireless communications has led to the design of MIMO equalizers that can extract symbols at the receiver robustly and effectively from noisy channel subject to model uncertainties.

At present, various approaches for robust MIMO equalization have been proposed, e.g., [178,179]. Conventional MIMO equalizers like the MMSE [180] and the Kalman-based [181] equalizers may still be applicable subject to channel uncertainties, but their performance may be degraded noticeably. The minimax mean-square error (MSE) design in [178] takes variations in both channel and noise covariance into consideration, and yields an equalizer by minimizing the MSE of error symbol vectors for the least favorable model in a neighborhood of the nominal model. In [179], robust equalization against model uncertainties and unknown statistics of exogenous signals is developed based on the H_∞ criterion in [182]. Although a minimax H_∞ scheme may not result in satisfactory average performance, it can ensure performance at a fixed level under a worst-case situation. Aside from the foregoing results, the minimax estimation schemes in [183–185] that construct estimators robust to modeling errors may be employed for developing robust equalizers. However, those estimation methods can only be applied under the assumption that all signals are affected by uncertainties. They need to be modified in order to be applicable to communication problems in which the statistics of transmitted symbols is not influenced by channel uncertainties.

In this chapter, we are mainly concerned with the design of robust MIMO equalizers that minimize the worst-case MSE under channel uncertainties within a neighborhood of the estimated channel. Unlike [178] using a bound on the Kullback–Leibler divergence between the actual and estimated channels, we consider a bound on the spectral matrix norm of channel estimation errors. The problem formulation and proposed solutions herein are different from those in [178,179]. Despite those differences, it seems that the integration of the proposed weighted combination of multiple model-based equalizers approach herein into the equalizer designs in [178,179] may result in some improvements to MSE and BER performance of equalizers in [178,179].

The development of minimax stochastic quadratic game solutions to the robust equalization design problem is given as follows. First, we show that a robust minimax MSE equalizer can be obtained by a linear matrix inequality (LMI) approach [23]. The LMI-based equalizer is solvable using the toolbox in [100], but its computational cost may grow rapidly as the MIMO dimension increases. Hence, we next derive a guaranteed cost-based equalizer, which may save computations, based on the application of the minimax stochastic quadratic game scheme to a modified MSE utility function of the robust equalization design problem. In practice, the foregoing two equalizers are not realizable unless time-varying norm bounds of channel uncertainty are known. To overcome the shortcoming, the least upper bound of the set of the tightest upper bounds on the spectral matrix norm of the channel uncertainty is utilized since it is accessible and is dominated by the highest allowable relative speed of mobile stations. This facilitates us to partition channel uncertainty into some channel uncertain states whose state transitions are characterized by a Markov process akin to that in [186,187]. As a result, a multiple model-based minimax stochastic quadratic game approach is proposed. Based on minimax stochastic quadratic game, we derive a feasible robust equalizer from a weighted combination of multiple over-guaranteed cost-based equalizers, each of which is designed with respect to a channel uncertain state. Simulation results show that this weighting combined robust equalizer is more effective than an LMI-based equalizer, and yields superior MSE and BER performance in comparison with an over-guaranteed robust equalizer. In addition, it can outperform conventional equalizers like the MMSE and Kalman-based equalizers based on the tradeoff between robustness and effectiveness. Some preliminary results of this work have been presented in [188].

Notations: Vectors and matrices are denoted by boldface lowercase letters and by boldface capital letters, respectively. The notation $(\cdot)^T$ denotes the transpose of a matrix or a vector; $(\cdot)^H$ denotes the Hermitian transpose of a matrix or a vector; $\mathrm{Tr}\{\cdot\}$ stands for taking trace of a squared matrix; $E\{\cdot\}$ means taking expectation. The notation $A \geq B$ for arbitrary matrices A and B with appropriate dimension means that the matrix $A - B$ is positive semidefinite; $\lambda_{\max}(A)$ stands for the maximum eigenvalue of the matrix A; $\|A\|_2$ denotes the spectral norm of the matrix A[189], i.e., $\|A\|_2 = \sqrt{\lambda_{\max}(AA^H)}$; $\|A\|_F$ denotes the Fobenious norm of the matrix A, i.e., $\|A\|_F = (\mathrm{Tr}(AA^H))^{1/2}$. The notation $N(a; b, R_{aa})$ denotes the complex Gaussian probability density function of a random vector a with mean b and the associated covariance R_{aa}. Finally, I_q and 0 represent an $q \times q$ identity matrix and a zero matrix with appropriate dimension.

10.2 System Model of Wireless Communication

We consider MIMO channels with N_t transmit and N_r receive antennas where $N_r \geq N_t$. The received signal vector $y(k) \in C^{N_r \times 1}$ can be expressed as

$$y(k) = H(k)Gx(k) + n(k) \tag{10.1}$$

where $x(k) \in C^{Nx \times 1}$ is the transmitted symbol vector; $G \in C^{N_t \times N_x}$ is the precoding matrix; $H(k) \in C^{N_r \times N_t}$ is the channel gain matrix; and $n(k) \in C^{N_r \times 1}$ is the additive noise at the receiving end. The selection for the number of parallel transmitted symbol streams N_x needs to satisfy $N_x \leq \min(N_r, N_t)$ [174] where $\min(N_r, N_t)$ stands for the number N_r or N_t whichever is smaller. Without loss of generality, we assume that the transmitted symbol vector $x(k)$ is zero-mean and spatially white with unit variance (i.e., identically independent distributed); moreover, $x(k)$ and $n(k)$ are statistically independent. Accordingly, the correlation matrices are

$$\begin{aligned} R_{xx} &= E\{x(k)x(k)^H\} = I_{N_x} \\ R_{nn} &= E\{n(k)n(k)^H\} \\ R_{xn} &= R_{nx} = 0. \end{aligned} \tag{10.2}$$

It is sensible to assume the precoder matrix G has a finite total power budget, i.e.,

$$\|G\|_F^2 = \text{Tr}\{GG^H\} = p_0 \tag{10.3}$$

where p_0 is a positive value. This results in the total transmitted power $E[\text{Tr}\{Gx(Gx)^H\}] = \text{Tr}\{GR_{xx}G^H\} = p_0$. The transmitted symbol vector $x(k)$ is estimated as

$$\hat{x}(k) = F(k)y(k) = F(k)H(k)Gx(k) + F(k)n(k) \tag{10.4}$$

where $F(k)$ is an equalizer.

Remark 10.1 Examples of precoder matrices satisfying (10.3) are $G = \sqrt{p_0/N_t} I_{N_t}$ in simple spatial multiplexing systems (direct transmission), and $G = \sqrt{p_0/N_t} D_{N_t}$ in discrete multi-tone (DMT) systems where D_{N_t} is the discrete Fourier transform (DFT) matrix [190]. Other examples of the fixed precoder matrix combined with block transmission can be found, e.g., in [191].

10.3 Problem Formulation of Robust Equalizer Based on Minimax Stochastic Quadratic Game

In [192–194], robust precoder design methods via Lagrangian optimization are proposed in the case that channel state information is imperfect. Unlike those studies, we mainly focus on robust equalizer design by assuming a precoder is given.

The minimax stochastic quadratic game approach has been selected to include the imperfect channel information in the design of robust equalizers. Specifically, with the received signal vector $y(k)$ and a given precoder matrix G satisfying the constraint in (10.3), we design an equalizer $F(k)$ based on minimizing the MSE of the error symbol vector $x(k) - \hat{x}(k)$, i.e., the cost function $\mathrm{Tr}\{\mathrm{E}[(x(k) - \hat{x}(k))(x(k)-\hat{x}(k))^H]\}$, under the worst channel uncertainty. Although the design approach may not guarantee a minimum BER quality for each single symbol stream, it can lead to a good BER performance on average under the worst-case channel uncertainties. In practice, channel uncertainty arises owing to measurement noise, estimation error, quantization error, and the time-varying nature. Thus, we define the channel uncertainty matrix $H_\Delta(k)$ as

$$H_\Delta(k) = H(k) - \hat{H}(k) \tag{10.5}$$

where the nominal channel gain matrix $\hat{H}(k)$ is known, and may not equal the actual channel gain matrix $H(k)$. Then, the channel gain uncertainty matrix $H_\Delta(k)$ forms a ball centering at the zero matrix with radius $r(k)$ at each time k, i.e.,

$$B_{H_\Delta}(r(k)) = \{H_\Delta(k) : \|H_\Delta(k)\|_2 \leq r(k)\} \tag{10.6}$$

where $r(k)$ is the tightest upper bound of $\|H_\Delta(k)\|_2$, i.e., $H_\Delta(k)H_\Delta^H(k) \leq r(k)^2 \cdot I_{N_r}$. The spectral norm of channel uncertainty $H_\Delta(k)$ is adopted herein mainly because it is widely used in the literature (e.g., [182,185]) on linear estimation.

Accordingly, the robust equalization design problem is formulated mathematically as the following minimax stochastic quadratic game problem:

$$\begin{aligned} &\min_{F(k)}\left\{\max_{H_\Delta(k)\in B_{H_\Delta}(r(k))} \mathrm{Tr}\{E[(x(k) - \hat{x}(k))(x(k) - \hat{x}(k))^H]\}\right\} \\ &= \min_{F(k)}\left\{\max_{H_\Delta(k)\in B_{H_\Delta}(r(k))} \mathrm{Tr}\{L(F(k), H_\Delta(k))\}\right\} \end{aligned} \tag{10.7}$$

where

$$\begin{aligned} &L(F(k), H_\Delta(k)) \\ &= (I_{N_x} - F(k)(\hat{H}(k) + H_\Delta(k))G)(I_{N_x} - F(k)(\hat{H}(k) + H_\Delta(k))G)^H \\ &\quad + F(k)R_{nn}(k)F^H(k). \end{aligned} \tag{10.8}$$

10.4 Minimax Stochastic Quadratic (H_2) Game Equalizer Design

In this section, solutions for the robust equalization design problem are derived by using various approaches, and their feasibility and computational complexities are analyzed. Like [178], the solutions derived for flat-fading channels can also be adapted to frequency-selective channels.

10.4.1 LMI-based Optimization Equalizer

To derive a solution to the robust equalization design problem, we first introduce the following lemma to facilitate us to transform the original minmax stochastic quadratic game problem (10.7) into a convex optimization problem. The lemma states that a symmetric matrix inequality (10.9) containing a norm-bounded uncertainty matrix M can be transformed into the LMI (10.10) which is primarily related to the bound on the spectral norm of the matrix M.

Lemma 10.1 [185]: Given matrices $A = A^H$, B, and C with appropriate dimensions

$$A > B^H MC + C^H M^H B, \quad \forall M : \|M\|_2 \leq \rho \tag{10.9}$$

if and only if there exists some $\sigma > 0$ such that

$$A \geq \sigma \cdot C^H C + \rho^2 \sigma^{-1} \cdot B^H B$$

or equivalently, the LMI

$$\begin{bmatrix} A - \sigma \cdot C^H C & -\rho \cdot B^H \\ -\rho \cdot B & \sigma I \end{bmatrix} \geq 0. \tag{10.10}$$

Based on Lemma 10.1, the minmax stochastic quadratic game problem (10.7) can be converted into an LMI-based optimization design problem given by (10.11)–(10.13). The solution is referred to as the **LMI-based optimization equalizer** for the minimax stochastic quadratic game problem formulated in (10.7).

Theorem 10.1

For the received signal vector $y(k) = (\hat{H}(k) + H_\Delta(k))Gx(k) + n(k)$ in (10.1) where $\hat{H}(k)$ is the nominal channel and $H_\Delta(k)$ is the channel uncertainty matrix satisfying (10.6), a solution to the minimax stochastic quadratic game problem (10.7) is obtained by solving the following:

$$\min_{t,\tau,\sigma,F(k)} t \tag{10.11}$$

subject to

$$t > 0, \tau > 0, \sigma > 0, N_x \cdot \tau \leq t \tag{10.12}$$

and

$$\begin{bmatrix} \tau \cdot I_{N_x} & I_{N_x} - F(k)\hat{H}(k)G & F(k)R_{nn}^{1/2}(k) & -r(k) \cdot F(k) \\ * & I_{N_x} - \sigma \cdot G^H G & 0 & 0 \\ * & * & I_{N_r} & 0 \\ * & * & * & \sigma \cdot I_{N_r} \end{bmatrix} \geq 0. \tag{10.13}$$

Proof: See Appendix A.

The computational complexity of the LMI-based optimization equalizer is analyzed as follows.

Proposition 10.1 Suppose each multiplication–addition is a single operation. The complexity of the LMI-based equalizer is

$$O(2N_x^3N_r^4 + (2N_x^4 + 4N_x^3 + 18N_x^2)N_r^3 + (18N_x^3 + 36N_x^2 + 54N_x)N_r^2$$
$$+ (54N_x^2 + 108N_x + 54)N_r + (54N_x + 108)).$$

Proof: See Appendix B.

10.4.2 Guaranteed Minimax Stochastic Quadratic Game Equalizer

The amount of computations for solving (10.11)–(10.13) may grow rapidly as the MIMO dimension increases. Thus, a simple and efficient solution to the minimax stochastic quadratic game problem formulated in (10.7) is derived.

In the derivation process, we will replace the cost function $\mathrm{Tr}[L(F(k), H_\Delta(k))]$ in (10.7) with a modified cost function $J(F(k), H_\Delta(k))$, which bounds the cost function $\mathrm{Tr}[L(F(k), H_\Delta(k))]$ tightly from above and whose maximization under the worst-case channel uncertainty leads to an explicit expression. By applying the following lemma, we can obtain the modified game cost function $J(F(k), H_\Delta(k))$ expressed in (10.39) in Appendix.

Lemma 10.2 $\forall A, B \in V, V \in C^{N_r \times N_t}$ is a real or complex inner product space, then the following inequality holds:

$$\mathrm{Tr}\{AB^H\} + \mathrm{Tr}\{BA^H\} \leq \mathrm{Tr}\{AA^H\} + \mathrm{Tr}\{BB^H\}. \tag{10.14}$$

Proof: Since $\|A - B\|_F^2 \geq 0$, (10.14) follows.

Because the minimax value of the stochastic quadratic game problem (10.7) cannot exceed the minimax value of the modified utility function $J(F(k), H_\Delta(k))$ (given in (10.39)) under channel uncertainties, the resultant equalizer that minimizes the modified utility function $J(F(k), H_\Delta(k))$ under the worst channel uncertainty is referred to as **a guaranteed minimax quadratic game equalizer** for the problem (10.7). We have the following result.

Theorem 10.2
For the received signal vector $y(k) = (\hat{H}(k) + H_\Delta(k))Gx(k)+n(k)$ in (10.1) where $\hat{H}(k)$ is the nominal channel and $H_\Delta(k)$ is the channel uncertainty matrix satisfying (10.6), a guaranteed minimax robust equalization solution to the minimax stochastic quadratic game problem (10.7) is given by

$$F(k) = (\hat{H}(k)G)^H(\hat{H}(k)GG^H\hat{H}^H(k) + R_\Delta(k))^{-1} \tag{10.15}$$

where

$$R_\Delta(k) = \lambda_{\max}(GG^H)r^2(k) \cdot I_{N_r} + \frac{1}{2}R_{nn}(k). \tag{10.16}$$

Proof: See Appendix C.

The computational complexity of the guaranteed minimax stochastic quadratic game equalizer is analyzed in the following.

Proposition 10.2: Suppose each multiplication–addition is a single operation. The complexity of the guaranteed robust equalizer is $O(N_r N_t N_x + 2N_r^2 N_x + N_r^2 + N_x N_t^2)$ lower than the complexity of the LMI-based optimization equalizer.

Proof: See Appendix D.

10.4.3 Over-Guaranteed Minimax Stochastic Quadratic Game Equalizer

To obtain the LMI-based optimization equalizer or the guaranteed minimax robust equalizer requires the tightest upper bound $r(k)$ of $\|H_\Delta(k)\|_2$ at each time k. However, they are not implementable since $r(k)$ is time-varying and unknown beforehand.

For conventional design in wireless communications, the least upper bound $\bar{r}$ of the set of the tightest upper bounds $r(k)$ of $\|H_\Delta(k)\|_2$ for all time k, i.e.,

$$\|H_\Delta(k)\|_2 \leq \bar{r}, \bar{r} = \sup_k r(k) \tag{10.17}$$

can be obtained in advance since the value $\bar{r}$ is generally dominated by the highest allowable relative velocity of mobile stations. Hence, we can get a quasi-LMI-based optimization equalizer by solving (10.11)–(10.13) with $r(k)$ replaced by $\bar{r}$, and an over-guaranteed minimax robust equalizer (10.15) with $r(k)$ in (10.16) replaced by $\bar{r}$. From Propositions 10.1 and 10.2, the guaranteed minimax stochastic quadratic game solution is more computationally efficient than the LMI-based optimization solution. Thus, an over-guaranteed cost-based equalizer is preferred to a quasi-LMI-based optimization equalizer.

10.4.4 Weighted Combination of Over-Guaranteed Minimax Stochastic Quadratic Game Equalizers

When $r(k)$ in (10.15) is replaced by $\bar{r}$ to yield an over-guaranteed cost-based equalizer, this may lead to conservatism such that the equalizer may not perform satisfactorily. To overcome this shortcoming, we characterize the time-varying behavior of channel uncertainty using a Markov-transitioned process, and then integrate the characterization of channel uncertainty together with the result of (10.15) into a robust equalizer design.

Based on the least upper bound of the set of the tightest upper bounds on the matrix spectral norm of channel uncertainty, we partition channel uncertainty into M distinct Markov-transitioned channel uncertain states. Each channel uncertain state i, $i = 1, \ldots, M$, represents that $H_\Delta(k)$ belongs to the difference set between two balls $B_{H_\Delta}(r_i)$ and $B_{H_\Delta}(r_i - 1)$ i.e., $H_\Delta(k) \subset B_{H_\Delta}(r_i) \quad B_{H_\Delta}(r_{i-1})$ where

$$B_{H_\Delta}(r_i) - B_{H_\Delta}(r_{i-1}) = \{H_\Delta(k) : r_{i-1} < \|H_\Delta(k)\|_2 \leq r_i\}, r_{i-1} < r_i. \tag{10.18}$$

In fact, $H_\Delta(k) \in B_{H_\Delta}(r_i) - B_{H_\Delta}(r_{i-1})$ implies that $H_\Delta(k) \in B_{H_\Delta}(r_i)$ holds, i.e., $H_\Delta(k)$ satisfies (10.6). So, according to Theorem 10.2 and the description in Section 10.4.C, an over-guaranteed minimax stochastic quadratic game equalizer designed with respect to $H_\Delta(k) \in B_{H_\Delta}(r_i)$ is the same as that designed with respect to $H_\Delta(k) \in B_{H_\Delta}(r_i) - B_{H_\Delta}(r_{i-1})$.

Though there are many schemes to choose values for r_i, a plausible way is assuming that the distance between any two neighboring channel uncertain states i and $i-1$ is equally spaced, i.e., $r_i - r_{i-1} = \bar{r}/(M-1)$ if $M>1$. This implies

$$r_i \approx \begin{cases} \bar{r} \times \frac{(i-1)}{(M-1)}, & \text{if } M>1 \\ \bar{r}, & \text{if } M=1 \end{cases} \tag{10.19}$$

and $r_0 = 0$. As soon as $\bar{r}$ and M are assigned, r_i are obtained off-line based on (10.19). The state transitions of M channel uncertain states follow a Markov process akin to that in [186,187] where for each state i, $\sum_{j=1}^{M} p_{ij} = 1$,

$$p_{ii} > p_{ij} \text{ for } i \neq j \tag{10.20}$$

and

$$p_{ij} > p_{im} \text{ for } |i-j| \leq 1 \text{ and } |i-m|>1. \tag{10.21}$$

Equation (10.20) means that each state is more likely to stay than to move to other states. Equation (10.21) means that each state is more probable to visit its adjacent states than to move to other far separated states. The M Markov-transitioned channel uncertain states are displayed in Figure. 10.1. Accordingly, we have M received signal models satisfying (10.18)–(10.21), $i = 1, \ldots, M$

$$y(k) = (\hat{H}(k) + H_\Delta(k))Gx(k) + n(k) \text{with} H_\Delta(k) \in B_{H_\Delta}(r_i) - B_{H_\Delta}(r_{i-1}) \tag{10.22}$$

Next, we propose a multiple model-based minimax MSE game approach using the foregoing development together with the concept of the maximal ratio combining technique in [195]. We derive a weighted combination of M over-guaranteed cost-based equalizers $F_i(k)$ shown in Figure. 10.2, each of which is designed based on one of the M

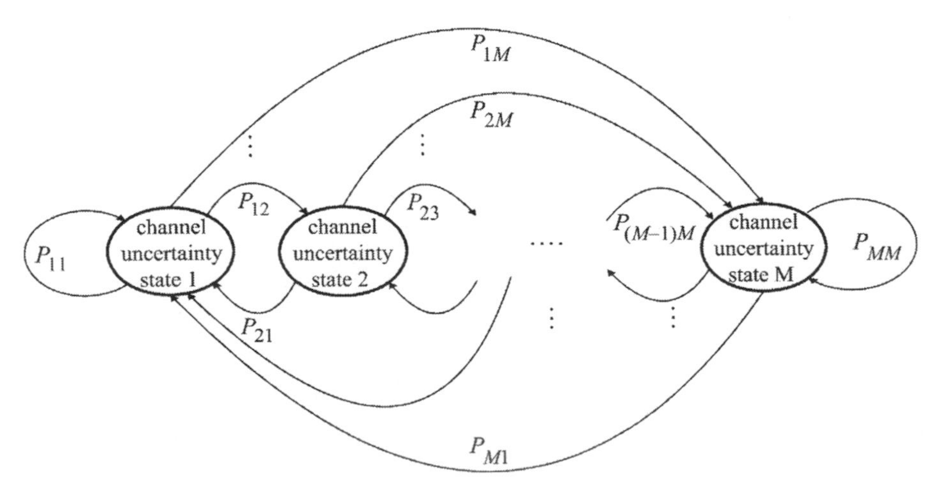

FIGURE 10.1
Markov-transitioned model for time-varying channel uncertainties in MIMO systems.

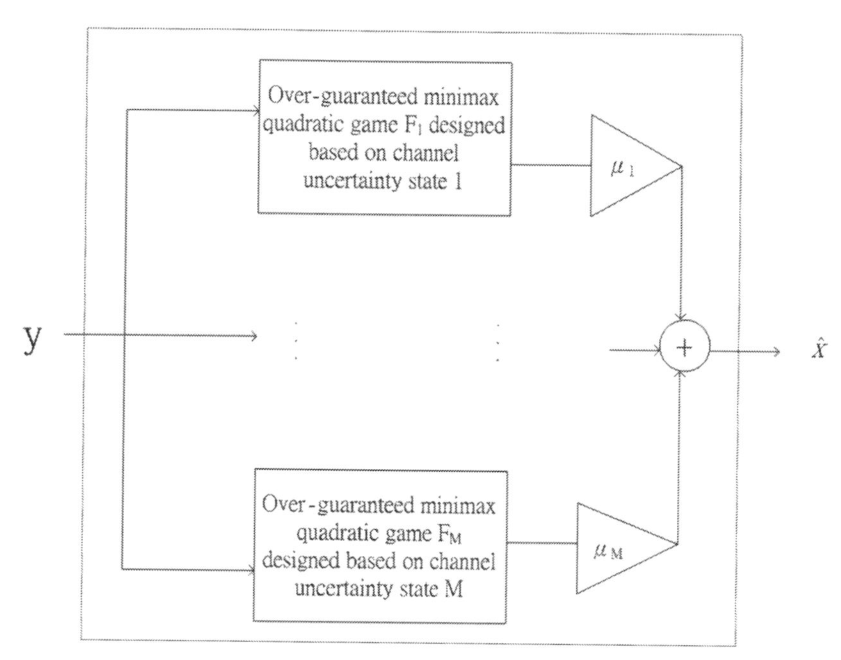

FIGURE 10.2
Weighted combination of over-guaranteed minimax quadratic game equalizers.

received signal models (10.22). The derived equalizer is referred to as a **combined over-guaranteed minimax quadratic game equalizer**. A weighted combination of quasi-LMI-based equalizers is not recommended because an LMI-based optimization equalizer generally requires immense computational effort.

Theorem 10.3
For the received signal vector $y(k) = (\hat{H}(k) + H_{\Delta}(k))Gx(k) + n(k)$ in (10.1), where $\hat{H}(k)$ is the nominal channel and the overall range of $H_{\Delta}(k)$ is covered by M distinct Markov-transitioned channel uncertain states satisfying (10.18)–(10.21), a guaranteed minimax robust equalization solution to the minimax stochastic quadratic game problem (10.7) is a combined over-guaranteed minimax stochastic quadratic game equalizer given by

$$\sum_{i=1}^{M} \mu_i(k) F_i(k) \tag{10.23}$$

where each $F_i(k)$ is an over-guaranteed minimax stochastic quadratic game equalizer given in (10.15)–(10.16) with $r(k)$ in (10.16) replaced by r_i given in (10.19); $\mu_i(k)$ denoting the mode probability of the ith equalizer $F_i(k)$ is given by

$$\mu_i(k) = \frac{1}{c} \Lambda_i(k) \sum_{j=1}^{M} p_{ji} \mu_j(k-1) \tag{10.24}$$

where c denotes the normalization constant as

$$c = \sum_{i=1}^{M} \Lambda_i(k) \sum_{j=1}^{M} p_{ji}\mu_j(k-1) \tag{10.25}$$

and

$$\Lambda_i(k) = \frac{1}{\pi^{N_r} \left| R_{\tilde{y}_i\tilde{y}_i}(k) \right|} \exp(-\tilde{y}_i^H(k) R_{\tilde{y}_i\tilde{y}_i}^{-1}(k) \tilde{y}_i(k)) \tag{10.26}$$

with

$$\tilde{y}_i(k) = y(k) - \hat{H}(k) G F_i(k) y(k) \tag{10.27}$$

$$\begin{aligned} R_{\tilde{y}_i\tilde{y}_i}(k) \approx & \, (I_{N_r} - \hat{H}(k)GF_i(k))(\hat{H}(k)GG^H\hat{H}^H(k) + \lambda_{\max}(GG^H) \cdot r_i^2 \cdot I_{N_r} \\ & + R_{nn}(k)) \times (I_{N_r} - \hat{H}(k)GF_i(k))^H. \end{aligned} \tag{10.28}$$

Proof: See Appendix E.

A combined over-guaranteed minimax quadratic game equalizer can outperform other equalizers derived herein, the MMSE and the Kalman-based equalizers because it utilizes the probability distributions (mode probabilities) of the multiple models. Its complexity is analyzed as follows.

Proposition 10.3 Suppose each multiplication–addition is a single operation. The complexity of a combined over-guaranteed cost-based equalizer, i.e., (10.23), is $O(M(N_rN_tN_x + 2N_r^2N_x + N_t^2N_x + 2N_r^3 + N_r^2))$, lower than that of an LMI-based optimization equalizer for any moderately small positive integer M.

Proof: See Appendix F.

Accordingly, it is plausible to choose a moderately small positive integer M with $M>1$ based on the tradeoff between performance and computation cost. In mobile communications, since the spectral norm of channel uncertainty is related to mobile speed, the number of channel uncertain states M means that mobile speeds are partitioned roughly into M velocity regions. For a quasi-static channel (i.e., the channel remains almost constant throughout the transmission of data), the channel uncertainty only resides at a channel uncertain state j for most of the time. This can yield a small absolute value of the error in $\tilde{y}_j(k)$ (10.27) and thereby lead to a large value $\Lambda_j(k)$ in (10.26), i.e., weighting is likely to be given to the jth equalizer $F_j(k)$ according to (10.24)–(10.26). In this case, the equalizer $F_j(k)$ and a combined over-guaranteed minimax robust equalizer with the inclusion of $F_j(k)$ perform similarly.

10.5 Performance Evolution of Minimax Quadratic Game Equalizer

In this section, we illustrate the performance and effectiveness of equalizers derived in Section 10.4. Moreover, we compare the robustness capabilities of the combined over-guaranteed minimax robust equalizer with that of other conventional solutions such as the MMSE equalizer [180] and the Kalman-based equalizer [181].

10.5.1 Simulation Example

We consider MIMO mobile communications operating with the carrier frequency 2.4 GHz and the transmission data rate of 5 Mb/s. The precoder of the direct transmission scheme (i.e., $G = \sqrt{p_0/N_t I_{N_t}}$) with the power constraint $p_0 = 1$ is adopted. Suppose the symbol vector in (10.1) is generated using an uncoded quadrature amplitude modulation scheme and N_x is chosen as $N_x = \min(N_r, N_t)$. The input signal-to-noise ratio (SNR) is defined as

$$\begin{aligned} \text{SNR(dB)} &= 10\log_{10}\frac{\text{Tr}\{E\{H(k)Gx(k)x^H(k)G^H H\}\}}{\text{Tr}\{E\{n(k)n^H(k)\}\}} \\ &= 10\log_{10}\frac{\text{Tr}\{H(k)GG^H H^H(k)\}}{\text{Tr}\{R_{nn}(k)\}}. \end{aligned} \tag{10.29}$$

For each fixed SNR (dB), we get

$$R_{nn}(k) = \frac{\text{Tr}\{H(k)GG^H H^H(k)\}}{\text{Tr}\{N_r \cdot 10^{\text{SNR}/10}\}} \cdot I_{N_r}. \tag{10.30}$$

It is indicated in [196,197] that relative velocities of mobiles (e.g., cellphones operated by people in high-speed trains) are allowed to be quite high (e.g., ≥400 km/h). Thus, we assume the allowable highest relative velocity is 400 (km/h), and consider the channel gain matrices $H(k)$ due to time-varying relative velocities taking place in the following order: 50, 100, 150, 200, 250, 300, 350, and 400 (km/h). For each velocity, 25000 symbols are sent as the information data. The channel gain matrices $H(k)$ are generated by the Jakes' model in [195]. The nominal channel gain matrices $\hat{H}(k)$ are obtained by both sampling the channel gain matrices $H(k)$ every 250 time steps and keeping the sampled results remaining unaffected between sampling occurrences. For simplicity, channel uncertainties arise from the channel gain errors during sampling and holding operations. Hence, the nominal least upper bounds $\bar{r}$ of the sets of the tightest upper bounds $r(k)$ of $\|H_\Delta(k)\|_2$ for 2×3 and 2×4 (i.e., $N_t = 2$inputs, $N_r = 3, 4$outputs) flat-fading channels corresponding to the relative velocity 400 km/h are obtained empirically as 0.58 and 0.69, respectively. On the other hand, if channel estimation error is assumed to be the main cause for channel uncertainty, then the relation between the spectral norm of channel uncertainty and mobile velocity can be established approximately based on the channel model and the model-based channel estimator (e.g., [177,186]).

To yield minimax stochastic quadratic game equalizers presented in Section 10.4, we utilize computational tools and select relevant parameters as follows. The LMI toolbox of Matlab [100] is employed for computing an LMI-based optimal equalization solution to (10.11)–(10.13). To yield a combined over-guaranteed minimax game equalizer, we suggest the number of channel uncertain states $M = 3$, which means that mobile speeds are partitioned roughly into low-, medium-, and high-velocity regions. The radius of each uncertain state is given in accordance with (10.19). The transition process of channel uncertain states is assumed to be a discrete-time, first-order Markov chain whose transition probabilities satisfy (10.20) and (10.21). In practice, a relative velocity between mobiles seems more likely to remain in the current relative velocity subrange. It is noted that the simulation scenario matches this situation. Accordingly, a channel uncertain state is more probable to stay as it is. Based on this and (10.20), we let $p_{ii} = \alpha$, where α is a large probability value, e.g., $0.8 < \alpha < 1$. It is improbable that a relative velocity subrange i can directly change to a relative velocity subrange j with $i, j = 1, 2, 3$ and $|i - j| > 1$ without going through some intermediate relative velocity subranges. Based on this and (10.21), we assign

$$p_{ij} = \begin{cases} 0.9(1-\alpha), & \text{if } i = 1,3 \text{ and } |i-j| = 1 \\ 0.1(1-\alpha), & \text{if } i = 1,3 \text{ and } |i-j| > 1 \\ 0.5(1-\alpha), & \text{if } i = 2 \text{ and } |i-j| = 1. \end{cases} \tag{10.31}$$

Let $\alpha = 0.999$. We get the transition probability matrix of the three channel uncertain states as

$$[p_{ij}] = \begin{bmatrix} 0.999 & 9\times 10^{-4} & 1\times 10^{-4} \\ 5\times 10^{-4} & 0.999 & 5\times 10^{-4} \\ 1\times 10^{-4} & 9\times 10^{-4} & 0.999 \end{bmatrix}.$$

Assuming each of the three equalizers is equally probable in the beginning, we set the mode probability of the ith equalizer as $\mu_i(0) = 1/3$. As time evolves, $\mu_i(k)$ are computed using (10.24)–(10.28).

10.5.2 Simulation Results of Minimax Stochastic Quadratic Game Equalizer

In Figs. 10.3–10.6, each performance curve is obtained based on the result of 100 Monte–Carlo runs. In Figs. 10.3 and 10.4, the LMI-based optimization and the guaranteed minimax robust equalizers are denoted as "genie-aided LMI" and "genie-aided guaranteed cost," respectively, since their implementations require the genie-aided knowledge of $r(k)$ of $\|H_\Delta(k)\|_2$ at each time k. The over-guaranteed minimax stochastic quadratic game equalizer is denoted as "over-guaranteed." The combined over-guaranteed minimax stochastic quadratic game equalizer is marked as "combined." In Figures. 10.5 and 10.6, the MMSE equalizer1 is denoted as "MMSE [180]." Based on (10.1), the results of MMSE-decision feedback equalizers in [180] can be reduced to the MMSE equalizer given by $F_{MMSE} = (\hat{H}(k)G)^H(\hat{H}(k)GG^H\hat{H}^H(k) + R_{nn}(k))^{-1}$. The Kalman-based equalizer2 is denoted as "Kalman-based [181]." It is developed based on the state equation

$$\begin{bmatrix} x(k) \\ x(k-1) \end{bmatrix} = \begin{bmatrix} 0_{N_x} & 0_{N_x} \\ I_{N_x} & 0_{N_x} \end{bmatrix} \begin{bmatrix} x(k-1) \\ x(k-2) \end{bmatrix} + \begin{bmatrix} x(k) \\ 0_{N_x \times 1} \end{bmatrix}$$

and the measurement equation

$$\begin{bmatrix} y(k) \\ y(k-1) \end{bmatrix} = \begin{bmatrix} \hat{H}(k)G & 0_{N_x} \\ 0_{N_x} & \hat{H}(k-1)G \end{bmatrix} \begin{bmatrix} x(k) \\ x(k-1) \end{bmatrix} + \begin{bmatrix} H_\Delta(k)Gx(k) + n(k) \\ H_\Delta(k-1)Gx(k-1) + n(k-1) \end{bmatrix}$$

with the noise covariance matrix

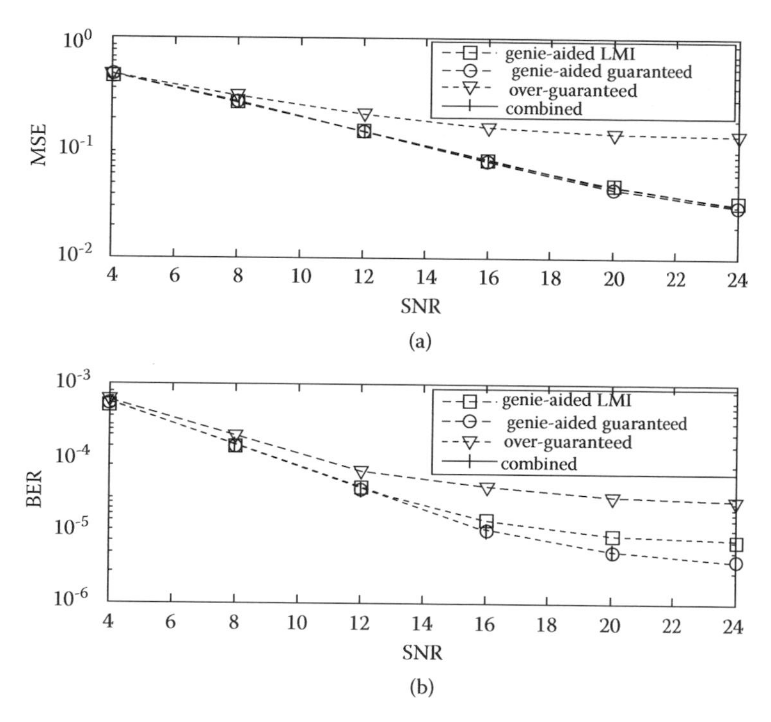

FIGURE 10.3
(a) MSE comparison and (b) BER comparison of derived equalizers under a 2 × 3 flat-fading channel.

$$\begin{bmatrix} \lambda_{\max}(GG^H) \cdot r_i^2 \cdot I_{N_r} + R_{nn}(k) & 0_{N_r} \\ 0_{N_r} & \lambda_{\max}(GG^H) \cdot r_i^2 \cdot I_{N_r} + R_{nn}(k) \end{bmatrix}.$$

It is noted in Figs. 10.3–10.6 that the equalization performance of the communication system is improved when the number of antennas increases. Figures 10.3(a) and 10.4(a) show the MSE performance of "combined" and that of "genie-aided guaranteed" are similar; moreover, they outperform that of "over-guaranteed" noticeably. The "over-guaranteed" does not perform reasonably well perhaps because it uses the fixed quantity $\bar{r}$. Consistent with this, Figs. 10.3(b) and 10.4(b) show that the BER equalization performance of "combined" and that of "genie-aided guaranteed" are similar; in addition, they are superior to that of "over-guaranteed." Figures 10.3(a) and 10.4(a) also show that "genie-aided guaranteed" and "combined" can achieve the same MSE equalization performance as "genie-aided LMI." Figure 10.3(b) indicates that the BER performance of "genie-aided LMI" from moderate to high SNRs is not superior to that of "genie-aided guaranteed" and to that of "combined." This situation arises perhaps because the utility function of the minimax stochastic H_2 game in (10.7) is the MSE of symbol vectors rather than the BER so that the minimax stochastic quadratic game equalizer for (10.7), i.e., "genie-aided LMI," may not guarantee a minimum BER equalization for each single stream. But even so, "genie-aided LMI" achieves a good BER equalization performance on average under the worst-case channel uncertainties. The ratio of execution times of "genie-aided guaranteed," "over-guaranteed," "combined," and "genie-aided LMI" is approximately 1:1:1.7:126. Although "genie-aided LMI" and

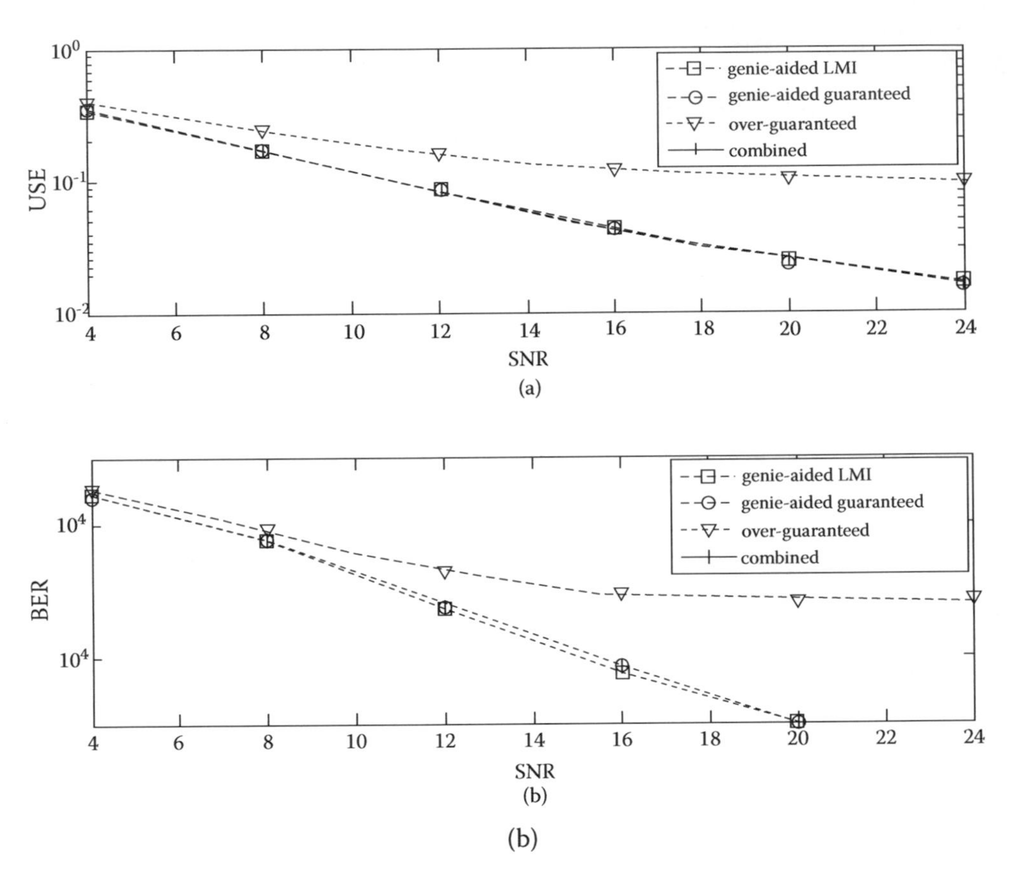

FIGURE 10.4
(a) MSE comparison and (b) BER comparison of derived equalizers under a 2 × 4 flat-fading channel.

"genie-aided guaranteed" have good MSE and BER equalization performance, they are not feasible. Based on the tradeoff between equalization performance and efficiency, "combined" among equalizers presented in Section 10.4 best suits the need in applications.

Figures 10.3–10.6 show that "Kalman-based [181]" and "over-guaranteed" perform quite similarly. Perhaps, it is because they similarly use the fixed quantity $\bar{r}$ which overestimates time-varying channel uncertainty gain. Figures 10.5(a) and 10.6(a) show that "combined" is slightly better than "MMSE [180]" which does not use $\bar{r}$, and significantly outperforms "Kalman-based [181]." Consistent with this, Figs. 10.5(b) and 10.6(b) show that the BER equalization performance of "combined" is superior to that of "MMSE" and to that of "Kalman-based." The ratio of execution times of "MMSE", "combined," and "Kalman-based" is approximately 1:1.7:22. Accordingly, it is suggested that the combined over-guaranteed minimax quadratic game equalizer is a viable choice for minimax robust equalization due to its superior performance and effectiveness.

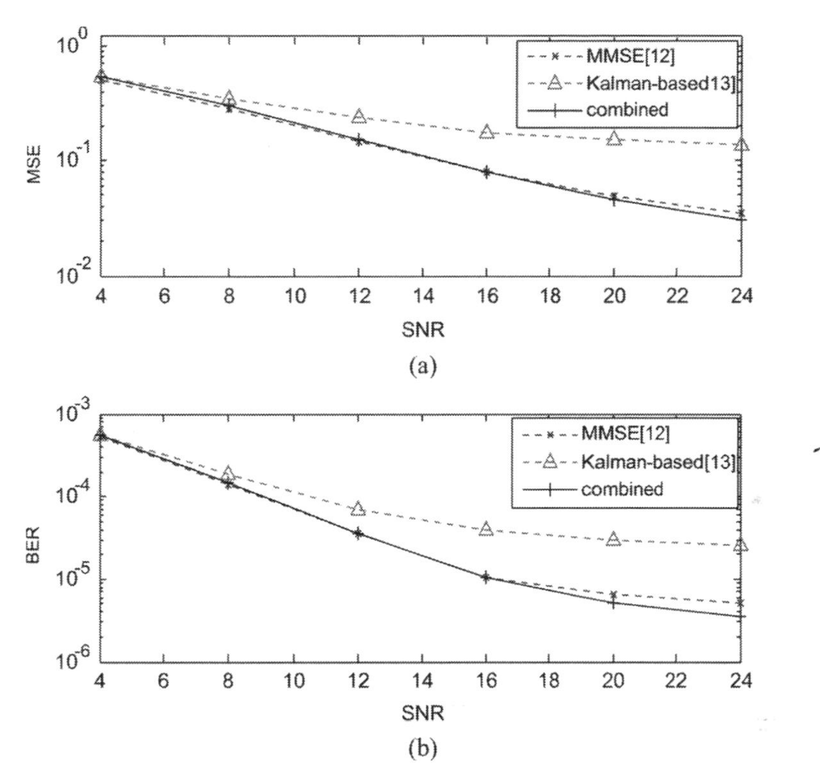

FIGURE 10.5
(a) MSE comparison and (b) BER comparison of various equalizers under a 2×3 flat-fading channel.

10.6 Conclusion

In this chapter, we have dealt with the MIMO robust equalization problem formulated as the design of a minimax stochastic quadratic game equalizer that minimizes the MSE of error symbol vectors under the worst-case channel uncertainty in a neighborhood of the estimated channel. In this respect, we have derived LMI-based optimization and guaranteed minimax stochastic quadratic game equalizers; however, they are not implementable owing to their dependence on time-varying norm bounds of channel uncertainty. To characterize time-varying channel uncertainty, we suggest that channel uncertainty should be partitioned into some Markov-transitioned channel uncertain states based on the least upper bound of the set of the tightest upper bounds on the matrix spectral norm of channel uncertainty. This leads to a feasible robust equalizer based on a weighted combination of multiple over-guaranteed minimax stochastic quadratic game equalizers, each of which is designed with respect to a channel uncertain state. It has been demonstrated that a weighted combination of a moderate number of over-guaranteed minimax stochastic quadratic game equalizers can achieve better MSE and BER performance effectively than other equalizers derived herein. Moreover, it is superior to conventional equalizers in terms of both equalization performance and computational cost.

10.7 Appendix

10.7.1 Proof of Theorem 10.1

The inner maximization term in (10.7), $\max_{H_\Delta(k)\in B_{H_\Delta(r(k))}} \mathrm{Tr}[L(F(k), H_\Delta(k))]$, can be expressed as the following form:

$$\min_{\tau} N_x \cdot \tau \tag{10.32}$$

subject to $\forall H_\Delta(k)$, $\|H_\Delta(k)\|_2 \leq r(k)$

$$\begin{aligned}&(I_{N_x} - F(k)(\hat{H}(k) + H_\Delta(k))G)(I_{N_x} - F(k)(\hat{H}(k) + H_\Delta(k))G)^H\\ &+ F(k)R_{nn}(k)F^H(k) \leq \tau \cdot I_{N_x}\end{aligned} \tag{10.33}$$

We can rewrite (10.33) as

$$\begin{aligned}&\tau \cdot I_{N_x} - [I_{N_x} - F(k)(\hat{H}(k) + H_\Delta(k))GF(k)R_{nn}^{1/2}]\\ &\begin{bmatrix} I_{N_x} - F(k)(\hat{H}(k) + H_\Delta(k))G \\ F(k)R_{nn}^{1/2}(k) \end{bmatrix}^H \geq 0.\end{aligned} \tag{10.34}$$

After applying the Schur's complement [184] to (10.34), we obtain

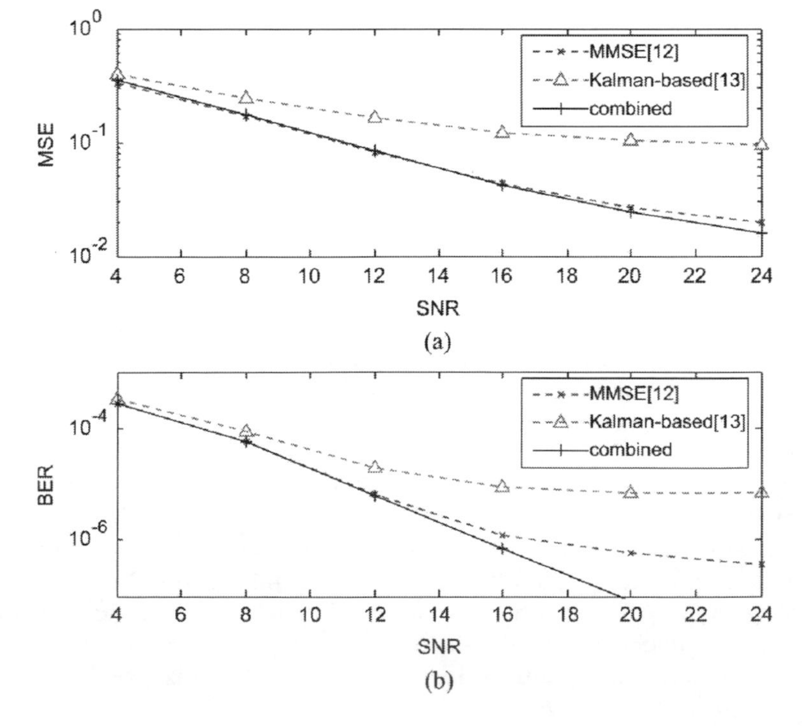

FIGURE 10.6
(a) MSE comparison and (b) BER comparison of various equalizers under a 2×4 flat-fading channel.

$$A(\tau, F(k)) \geq B(F(k))^H H_\Delta(k) C + C^H H_\Delta^H(k) B(F(k)),$$
$$\forall H_\Delta(k), \|H_\Delta(k)\|_2 \leq r(k) \tag{10.35}$$

where

$$A(\tau, F(k)) = \begin{bmatrix} \tau \cdot I_{N_x} & I_{N_x} - F(k)\hat{H}(k)G & F(k)R_{nn}^{1/2}(k) \\ * & I_{N_x} & 0 \\ * & * & I_{N_r} \end{bmatrix} \tag{10.36}$$
$$B(F(k)) = [F^H(k) \quad 0 \quad 0]$$
$$C = [0 \quad G \quad 0]$$

with * denoting a symmetric entry. Applying Lemma 10.1 to (10.35) yields (10.13).

10.7.2 Proof of Proposition 10.1

The LMI optimization problem (10.11)–(10.13) can be solved by efficient interior-point algorithms and the number of operations to compute an ε-accurate solution is bounded by the value $a_1 a_2^3 \log(a_3/\varepsilon)$ [100], where a_1 is the row size of the LMI system, a_2 is the total number of scalar decision variables, and a_3 is a data-dependent scaling factor. Hence, the complexity of an LMI-based solution to (10.11)–(10.13) is $O(a_1 a_2^3)$, where $a_1 = 2N_x + 2N_r + 4$ and $a_2 = N_x N_r + 3$. This yields the result of Proposition 10.1.

10.7.3 Proof of Theorem 10.2

Using (10.14), (10.7) is bounded from the upper as shown in (10.37)

$$\begin{aligned}
&\min_{F(k)} \left\{ \max_{H_\Delta(k) \in B_{H_\Delta(r(k))}} \mathrm{Tr}[L(F(k), H_\Delta(k))] \right\} \\
&= \min_{F(k)} \{ \max_{H_\Delta(k) \in B_{H_\Delta(r(k))}} \mathrm{Tr}[F(k) H_\Delta(k) G (G^H \hat{H}^H(k) F^H(k) - I_{N_t}) \\
&+ (F(k)\hat{H}(k)G - I_{N_t}) G^H H_\Delta^H(k) F^H(k) + (G^H \hat{H}^H(k) F^H(k) \\
&- I_{N_t})(F(k)\hat{H}(k)G - I_{N_t}) + F(k) H(k) G G^H H_\Delta^H(k) F^H(k) \\
&+ F(k) R_{nn}(k) F^H(k)]\} \\
&\leq \min_{F(k)} \{ \max_{H_\Delta(k) \in B_{H_\Delta(r(k))}} \mathrm{Tr}[F(k) H_\Delta(k) G G^H H_\Delta^H(k) F^H(k) \\
&+ 2 \cdot (G^H \hat{H}^H(k) F^H(k) - I_{N_t})(F(k)\hat{H}(k)G - I_{N_t}) \\
&+ F(k) H_\Delta(k) G G^H H_\Delta^H(k) F^H(k) + F(k) R_{nn}(k) F^H(k)]\}. \\
&\leq \min_{F(k)} \{ \max_{H_\Delta(k) \in B_{H_\Delta(r(k))}} J(F(k), H_\Delta(k))\}
\end{aligned} \tag{10.37}$$

where

$$\begin{aligned} &J(F(k), H_\Delta(k)) \\ &= 2 \cdot \text{Tr}[(G^H \hat{H}^H(k)F^H(k) - I_{N_t})(F(k)\hat{H}(k)G - I_{N_t}) \\ &\quad + F(k)(\lambda_{\max}(GG^H)H_\Delta(k)H_\Delta^H(k) + \frac{1}{2}R_{nn}(k)F^H(k)] \end{aligned} \tag{10.38}$$

It is apparent that $J(F(k), H_\Delta(k))$ is a convex function of $H_\Delta(k)$. Therefore, the maximum of $J(F(k), H_\Delta(k))$ over the set $B_{H_\Delta(r(k))}$ in (10.6) is achieved at the boundary of $B_{H_\Delta(r(k))}$, i.e.,

$$\tilde{J}(F(k)) = \max_{H_\Delta(k) \in B_{H_\Delta(r(k))}} J(F(k), H_\Delta(k)) \tag{10.39}$$

where

$$\begin{aligned} \tilde{J}(F(k)) &= 2 \cdot \text{Tr}[(G^H \hat{H}^H(k)F^H(k) - I_{N_t})(F(k)\hat{H}(k)G - I_{N_t}) \\ &\quad + F(k)(\lambda_{\max}(GG^H)r^2(k) \cdot I_{N_r} + \frac{1}{2}R_{nn}(k)F^H(k)]. \end{aligned} \tag{10.40}$$

Hence,

$$\min_{F(k)} \tilde{J}(F(k)) = \min_{F(k)}\{\max_{H_\Delta(k) \in B_{H_\Delta(r(k))}} J(F(k), H_\Delta(k))\} \tag{10.41}$$

Next, $\min_{F(k)}\tilde{J}(F(k))$ can be solved based on the necessary and sufficient condition for the optimal solution of $\tilde{J}(F(k))$ given by

$$\nabla_{F(k)}\tilde{J}(F(k)) = 0 \tag{10.42}$$

This leads to

$$[-(\hat{H}(k)G)^H + F(k)(\hat{H}(k)GG^H\hat{H}^H(k) + R_\Delta(k))] = 0 \tag{10.43}$$

where $R_\Delta(k)$ is defined as (10.16). Since the matrix $(\hat{H}(k)GG^H\hat{H}^H(k) + R_\Delta(k))$ is invertible, (10.43) yields (10.15).

10.7.4 Proof of Proposition 10.2

The complexity of the guaranteed robust equalizer, i.e., (10.15), hinges on the terms $\hat{H}(k)G(\hat{H}(k)$, $R_\Delta(k)$, and $(\hat{H}(k)GG^H\hat{H}^H(k) + R_\Delta(k))^{-1}$ whose complexities are $O(N_rN_tN_x + N_r^2N_x)$, $O(N_xN_r^2)$ and $O(N_r^2)$, respectively, as well as the multiplication of both $(\hat{H}(k)G)^H$ and $(\hat{H}(k)GG^H\hat{H}^H(k) + R_\Delta(k))^{-1}$, i.e., $O(N_xN_r^2)$. Thus, the complexity of (10.15) is around $O(N_rN_tN_x + 2N_r^2N_x + N_r^2 + N_xN_t^2)$. This together with the result of Proposition 10.1 leads to the result of Proposition 10.2.

10.7.5 Proof of Theorem 10.3

Since the estimates of a symbol vector $x(k)$ yielded by M over-guaranteed cost-based equalizers $F_i(k)$ are combined probabilistically in accordance with the M-state Markov transitions (10.18)–(10.21), we have the following result using the total probability theorem:

$$p\{x(k)|Y(k)\} = \sum_{i=1}^{M} p\{F_i(k)|Y(k)\} \cdot p\{x(k)|F_i(k), Y(k)\} \tag{10.44}$$

with $Y(k)$ standing for a collection of received measurements as $Y(k) = \{y(1), y(2), \ldots, y(k)\}$. Using the Bayes' rule and Markov chain properties, we get

$$\begin{aligned}
\mu_i(k) &= p\{F_i(k)|Y(k)\} \\
&= \frac{1}{c} p\{y(k)|F_i(k), Y(k-1)\} \cdot p\{F_i(k)|Y(k-1)\} \\
&= \frac{1}{c} p\{y(k)|F_i(k), Y(k-1)\} \cdot \sum_{j=1}^{M} [p\{F_i(k)|F_i(k-1), Y(k-1)\} \\
&\quad \cdot p\{F_i(k-1)|Y(k-1))] \\
&= \frac{1}{c} \Lambda_i(k) \cdot \sum_{j=1}^{M} [p_{ji} \cdot \mu_j(k-1)]
\end{aligned} \tag{10.45}$$

where c is expressed as (10.25) and the ith likelihood function $\Lambda_i(k)$ is expressed as (10.26) since

$$\Lambda_i(k) = p\{y(k)|F_i(k), Y(k-1)\} = N(\tilde{y}_i(k); 0, R_{\tilde{y}_i\tilde{y}_i}(k)) \tag{10.46}$$

$$\begin{aligned}
R_{\tilde{y}_i\tilde{y}_i}(k) &= E\{\tilde{y}_i(k)\tilde{y}_i^H(k)\} \\
&= (I_{N_r} - \hat{H}(k)GF_i(k))(\hat{H}(k)GR_{xx}G^H\hat{H}^H(k) \\
&\quad + H_i(k)GR_{xx}G^H H_i^H(k) + R_{nn}(k)(I_{N_r} - \hat{H}(k)GF_i(k))^H
\end{aligned} \tag{10.47}$$

with $\tilde{y}_i(k) = y(k) - \hat{H}(k)GF_i(k)y(k)$ and the covariance matrix $R_{\tilde{y}_i\tilde{y}_i}(k)$ satisfying (10.47). Moreover, $R_{\tilde{y}_i\tilde{y}_i}(k)$ is expressed as (10.28) because of both (10.47) and the relation $H_i(k)GR_{xx}G^H H_i^H(k) \leq \lambda_{\max}(GG^H) \cdot r_i^2 \cdot I_{N_r}$. In accordance with (10.44), the weighted sum of state estimates can be expressed as

$$\hat{x}(k) = \sum_{i=1}^{M} \mu_i(k) \cdot F_i(k)y(k)) = \left(\sum_{i=1}^{M} \mu_i(k) \cdot F_i(k)\right) \cdot y(k). \tag{10.48}$$

Thus, we get the equalizer expressed as (10.23).

From (10.37) and (10.45), we have the second equation

$$\min_{F(k)}\left\{\max_{H_\Delta(k)\in B_{H_\Delta(r(k))}} \mathrm{Tr}[L(F(k),H_\Delta(k))]\right\}$$

$$\leq \sum_{i=1}^{M}\mu_i(k)\times \min_{F_i(k)}\{\max_{H_\Delta(k)\in B_{H_\Delta(r_i)}-B_{H_\Delta(r_{i-1})}} J(F_i(k),H_\Delta(k))\}.$$

This implies that (10.23) is a guaranteed cost-based solution to the problem (10.7).

10.7.6 Proof of Proposition 10.3

The complexity of a combined over-guaranteed stochastic quadratic game equalizer hinges on the computation of model probabilities (10.24), i.e., (10.26). The complexity of (10.26) due to a matrix inversion is $O(N_r^2)$. The complexity of both $\hat{H}(k)GG^H\hat{H}^H(k)+\lambda_{\max}(GG^H)r_i^2\cdot I_{N_r}+R_{nn}(k)$ and $(\hat{H}(k)G)F(k)$ is $O(N_rN_tN_x+2N_r^2N_x+N_t^2N_x)$. Moreover, the multiplications in (10.28) take $O(2N_r^3)$. Hence, the complexity of (10.26) is $O(N_rN_tN_x+2N_r^2N_x+N_t^2N_x+2N_r^3+N_r^2)$, so is the complexity of (10.24). The complexity of (10.24) is equivalent to the complexity of each equalizer $F_i(k)$ in (10.23) which has similar mathematical manipulations as (10.15) plus $O(N_r^2)$. Hence, the complexity of $\mu_i(k)F_i(k)$ is $O(N_rN_tN_x+2N_r^2N_x+N_t^2N_x+2N_r^3+N_r^2)$. This further yields the result of Proposition 10.3.

11

Min-max l_2 Game Approach to Robust Sensor Location Estimation Problem in Wireless Sensor Networks

11.1 Introduction

The information of relative locations of sensors is very useful applications of wireless sensor networks (WSNs). An accurate sensor location estimation is a key requirement of various practical applications in WSNs [198–200]. In the literature on location estimation in WSNs, location-aware nodes are also referred to as anchors, beacons, landmarks, or reference devices, whereas location-unaware nodes are simply termed sensors, or blindfolded devices [201–205]. The relative sensor location estimation process can be divided into three stages [204]. The first stage relates to obtaining distances or angles between neighboring sensor nodes. In this stage, either of three popular sensor position techniques could be applied, namely time (difference) of arrival (T(D)OA), angle of arrival (AOA), and received signal strength (RSS). The RSS method is adopted in this chapter due to its simplicity and low cost. We could refer to [202,206–209] for more details about the RSS-based localization of sensors. The second stage involves estimating the positions of location-unaware nodes according to the measurements obtained in the first stage. The third stage is optional, and involves in refining the sensor location estimation [210,211]. This chapter mainly focuses on the second stage of relative location estimation process of sensors.

Some studies on sensor location estimation problems assume that the exact positions of location-aware nodes are known. However, in reality, such position information has errors [212] and the sensor location estimation performance will deteriorate if these position errors are neglected. The reasons for these sensor position errors can include inaccurate range measurements, algorithm artifacts, or a combination of both. For related studies on location errors for location-aware nodes, the reader can refer to [212–214] for geographical routing problems, and [215–217] for target localization problems.

For solving a sensor location estimation problem, convex optimization methods have been used in previous literature. For example, Doherty et al. first studied convex position estimation in WSNs [218], and several convex approaches have since then been developed. Biswas et al. [203] proposed a semidefinite programming (SDP) relaxation method. In this approach, the refinement after positioning was also addressed. Tseng [205] studied the second-order cone programming (SOCP) relaxation technique; however, SOCP relaxation causes the estimated sensor locations to lie in the convex hull of the location-aware sensor nodes [217]. In other words, it fails if there are a large number of location-unaware sensors positioned outside the convex hull.

Except the convex positioning methods, many studies employed conventional maximum likelihood (ML) estimation methods for sensor location estimation in WSNs. For instance, Chang and Liao [219] proposed a probability-based ML estimation approach if some prior information is available. Moses et al. [199] used ML estimation for AOA and TOA measurements, whereas Patwari et al. [201,202] employed the same estimation method for TOA and RSS measurements and studied the Cramer-Rao lower bound (CRLB). The performance of ML estimation for the location estimation of sensor in WSNs is highly accurate if the corresponding likelihood optimization problem can be solved, for example, by grid search. However, the solving process often takes a great deal of time. Although other iterative search techniques may be applied for the likelihood optimization problem, the accuracy usually depends heavily on the starting point of the solving process. Therefore, a different formulation of the sensor location estimation problem that does not rely on a good starting point and can be solved by existing numerical techniques is needed for robust sensor location estimation in WSNs.

In this chapter, we assume the location errors of location-aware sensors in WSNs to be bounded [217,220] with an unknown distribution. A robust sensor location estimation method is proposed to deal with the inexact position problem. Using the distance measurements obtained from the RSS technique, a min-max quadratic game method is proposed to minimize the worst-case estimation error (the maximum error) from the two-person min-max l_2 game perspective. Therefore, it is the worst-case scenario design. Owing to the nonconvexity of the proposed optimization formulation, it is then transformed into a convex optimization formulation, which is solvable by existing numerical techniques [221].

The aims of this chapter are stated as follows: Unlike existing studies that focus on a Gaussian model for the inexact position problem [222,223], this chapter examines the robust sensor location estimation problem from the worst-case perspective. In other words, exact knowledge of the location error distribution is not needed in the proposed min-max l_2 game approach. To the best of our knowledge, the worst-case scenario for the sensor location estimation in WSNs has not been fully investigated. Furthermore, the estimation performance is conventionally evaluated for a particular deployment of sensors, with a theoretical lower bound. Conversely, in this paper, for a fair comparison of different approaches, the estimation performance has been evaluated by the average of random deployments, which means that different topologies of sensors have been considered. Although only the RSS measurement technique is employed, the proposed min-max l_2 game approach is very flexible in the sense that other techniques such as TOA and TDOA can be easily integrated into the sensor location estimation scheme in WSNs.

In contrast to the ML estimators for sensor localization, whose accuracy depends heavily on the starting point for the solving process, the proposed min-max l_2 game method is relatively robust in sensor location estimation in WSNs. In comparison to the SOCP methods, which result in a poor performance if location-unaware nodes lie outside the convex hull of the location-aware nodes, a more accurate location estimation performance is achieved by the proposed min-max game method, as evidenced by our simulation results. Convex positioning techniques have been applied to target localization problems in the presence of location errors of location-aware sensors in WSNs [217]. This is the first study (that we are aware of) that has considered convex optimization methods for sensor location estimation with the inexact position problem in WSNs.

The rest of this chapter is organized as follows: The related sensor location estimation works are examined in Section 11.2. Section 11.3 describes the notations used and the estimation schemes concerned. Motivated by [201,203,217], a robust relative sensor location estimation scheme based on min-max l_2 game is proposed for sensors of WSNs in Section 11.4. Section 11.5 presents the numerical results for the proposed min-max l_2 game approach and other sensor location estimation methods. Finally, some concluding remarks are given in Section 11.6.

11.2 Related Sensor Location Estimation Works

In the context of the relative location estimation problem, a hybrid Cramer-Rao bound (HCRB) was examined for passive localization with uncertain anchor positions in [222,223]. However, the presented bounds in [222,223] may not necessarily be tight in certain circumstances [224]. It is noted that all bounds for the estimation error depend on the distribution of measurements and hence could vary over the different distributions. Conventionally, an estimation problem involves a comparison between the proposed method and the theoretical Cramer-Rao bound (CRB) [225]. The bound can be derived for a particular distribution of measurements. For most cases, a Gaussian model is adopted because of its simplicity, thus resulting in a closed-form bound. However, since the distribution of location estimation errors is assumed to be bounded with unknown statistical properties in this chapter, the CRB cannot be derived. Even if a particular distribution is assumed, the proposed min-max l_2 game method cannot be proven to be unbiased and hence, the comparison between the proposed min-max l_2 game estimator and the theoretical bound could be misleading. Therefore, the CRB analysis is not included in this chapter.

In [201,203,204,217,226,227], we could find for more related works of sensor location estimation. In [204,226,227], mobile sensor localization was considered. The inclusion of mobile sensors could be a future research direction. In [201], ML estimators were used for sensor location estimation and will be compared to the proposed min-max l_2 game approach in the simulation. Besides, the distance measurement model in this chapter is based on [201]. In [203], SDP localization methods were proposed. Some SDP relaxation and problem formulation techniques are also used in this chapter. In [217], a target localization problem was dealt with. The consideration of anchor location errors motivates our study of the inexact position problem of robust sensor location estimation in this chapter.

Lastly, it should be mentioned that the non-line-of-sight (NLOS) measurement propagation is a very important issue that still needs to be addressed, as most existing studies on sensor location estimation problems in WSNs only focus on the line-of-sight (LOS) environment. This issue is mostly considered in the mobile location estimation problem [228–231]. For those studies on NLOS mitigation in WSNs, such as [232,233], exact positions of location-aware sensors are assumed. For WSNs, the NLOS problem is difficult to address because each range estimate could be LOS or NLOS, and cannot be easily identified. The WSN relative sensor location estimation in the presence of both the inexact position and the NLOS problems could be an interesting research topic in robust sensor location estimation of WSNs in the future.

Remark 11.1 For other WSN localization studies, the reader can refer to [210,211,234–239]. A fundamental upper bound for sensor location estimation accuracy using TOA was derived in [234]. A parking management system for sensor location estimation of vehicles was examined in [237]. In [238], a segmentation-aided and density-aware hop-count localization algorithm was proposed. Vision-based bearing measurements were employed for cooperative localization in [239]. Sensor localization problems were handled by successive refinement in [210,211]. Distributed sensor location algorithms with constrained sensor resources were discussed in [235]. A GPS-free positioning scheme was studied in [236]. More approaches to obtain the topology of a WSN can be found in [240–245].

11.3 Problem Description

Suppose that there are N location-unware sensors and M location-aware sensors. For $n = 1, 2, \ldots, N, x_n \in R^2$ represent the unknown positions of location-unaware sensors to be estimated. For $m = 1, 2, \ldots, M$, let $z_m \in R^2$ denote the true positions of location-aware sensors. In practice, the known positions of location-aware sensors are corrupted with location errors [216,217]. The relationship between the true position z_m and the inexact position $\hat{z}_m$ can be defined by

$$z_m = \hat{z}_m + \Delta_m, \quad \text{for } m = 1, 2, \ldots, M, \tag{11.1}$$

where

$$\|\Delta_m\| \leq \zeta. \tag{11.2}$$

In (11.2), $\|\|$ represents the Euclidean norm. For example, if $\Delta_m = (\Delta_{mx}\Delta_{my})^T \in R^2$, then $\|\Delta_m\| = \sqrt{\Delta_{mx}^2 + \Delta_{my}^2}$. Equation (11.1) indicates that the true sensor position z_m, which is unavailable, can be obtained by perturbing the inexact position $\hat{z}_m$, which is available, with the bounded error ζ. The existence of the location error Δ_m for $m = 1, 2, \ldots, M$ in (11.1) is called the "inexact position problem" in this chapter. By using the RSS technique for range estimation, the relative distance measurement $\hat{d}_{nm}$ between sensors can be obtained as [201,202]

$$\hat{d}_{nm} = d_0 \cdot 10^{\frac{p_0 - \hat{p}_{nm}}{10n_p}} = d_{nm} \cdot 10^{\frac{\xi_{nm}}{10n_p}}, \tag{11.3}$$

where

$$\begin{aligned} \hat{p}_{nm} &= p_{nm} + \xi_{nm}, \\ p_{nm} &= p_0 - 10n_p \log_{10}\left(\frac{d_{nm}}{d_0}\right). \end{aligned} \tag{11.4}$$

In (11.3), p_0 represents the received power at a reference distance d_0, n_p is the path loss exponent [246], d_{nm} is the Euclidean distance between two sensors, $\hat{p}_{nm}$ is the measured power, and ξ_{nm} models the fading channel, which is a zero-mean Gaussian with variance σ_p^2.

Due to the ranging limitation, pairwise estimation $\hat{d}_{nm}$ may not be available. Let $\tilde{d}$ be the threshold for a possible connection. For $n = 1, 2, ..., N$, define

$$\mathbb{N}(n) = \mathbb{N}_1(n) \cup \mathbb{N}_2(n) \tag{11.5}$$

as the index set that indicates neighbors of the location-unaware sensor, where

$$\begin{aligned} \mathbb{N}_1(n) &= \{n' : n < n' \leq N, \|x_n - x_{n'}\| \leq \tilde{d}\}, \\ \mathbb{N}_2(n) &= \{m : 1 \leq m \leq M, \|x_n - z_m\| \leq \tilde{d}\}. \end{aligned} \tag{11.6}$$

Therefore, $\hat{d}_{nm}$ is available if and only if $m \in \mathbb{N}(n)$. Note that $\mathbb{N}_1(n)$ represents the connection between location-unaware sensors, whereas $\mathbb{N}_2(n)$ represents the connection between location-unaware and location-aware sensors. Based on (11.6), the relative distance d_{nm} in (11.3) can be formulated as

$$d_{nm} = \begin{cases} \|x_n - z_m\|, \text{if } m \text{ comes from } \mathbb{N}_2(n) \\ \|x_n - x_m\|, \text{if } m \text{ comes from } \mathbb{N}_1(n). \end{cases} \tag{11.7}$$

The sensor location estimation problem can be described as follows. Given the measured relative distances $\hat{d}_{nm}$ in (11.3) for $n = 1, 2, ..., N$ and $m \in \mathbb{N}(n)$, and the inexact positions of location-aware sensors $\hat{z}_m \in R^2$ in (11.1) for $m = 1, 2, ..., M$, it is desirable to estimate the positions of location-unaware sensors $x_n \in R^2$ for $n = 1, 2, ..., N$.

11.4 Formulation of Robust Sensor Location Estimation Problem in WSNs as a Minimax L$_2$ Game Problem

In this section, the sensor location estimation problem in WSNs is formulated as a minimization problem that minimizes the distance errors among location-unaware sensors, which is relevant to $\mathbb{N}_1(n)$, and the distance errors between location-unaware and location-aware sensors, which is relevant to $\mathbb{N}_2(n)$. To facilitate the understanding, the proposed robust sensor location estimation scheme in WSNs is divided into two parts. In the first part, only the set $\mathbb{N}_1(n)$ is considered. A semidefinite program relaxation approach is employed to form a convex optimization problem for robust relative location estimation; in the second part, only the set $\mathbb{N}_2(n)$ is considered. Due to the inexact knowledge of the positions of location-aware sensors in WSNs, the worst-case error is considered to be minimized and a min-max l_2 game approach is proposed to treat the robust sensor location estimation problem in WSNs. By employing the "S-procedure" [221], the min-max problem will then be transformed into an SDP problem to simplify the design procedure of robust sensor location problem in WSNs. Finally, a robust sensor location estimation scheme will be developed by combining the two parts into a convex optimization scheme.

In the first part, based on the information in the set $\mathbb{N}_1(n)$, the unknown positions of location-unaware sensors can be obtained by solving

$$\min_{X} \sum_{n=1}^{N-1} \sum_{n' \in \mathbb{N}_1(n)} \left| \|x_n - x_{n'}\|^2 - \hat{d}_{nn'}^2 \right|. \tag{11.8}$$

It should be noted that since n' comes from the set $\mathbb{N}_1(n)$ as shown in (11.8), $\hat{d}_{nn'}$ relates to $\|x_n - x_{n'}\|$ according to (11.7). Let e_n denote the nth unit vector in R^N. For convenience, we define $e_{nn'} = e_n - e_{n'}$ and $X = (x_1\ x_2\ \cdots\ x_N) \in R^{2\times N}$. The optimization problem in (11.8) is equivalent to

$$\min_{X} \sum_{n=1}^{N-1} \sum_{n' \in \mathbb{N}_1(n)} \left| \begin{pmatrix} e_{nn'} \\ 0 \end{pmatrix}^T \begin{pmatrix} Y & X^T \\ X & I_2 \end{pmatrix} \begin{pmatrix} e_{nn'} \\ 0 \end{pmatrix} - \hat{d}_{nn'}^2 \right| s.t.\ Y = X^T X. \tag{11.9}$$

However, (11.9) is not convex. The SDP relaxation method [203] is then employed to relax $Y = X^T X$ into $Y \succeq X^T X$, which is equivalent to

$$\begin{pmatrix} Y & X^T \\ X & I_2 \end{pmatrix} \succeq 0 \tag{11.10}$$

(see *Schur complement* in [221]).

Utilizing (11.9) and (11.10), (11.8) can be transformed into the following SDP:

$$\min_{X,Y} \sum_{n=1}^{N-1} \sum_{n' \in \mathbb{N}_1(n)} \left| \begin{pmatrix} e_{nn'} \\ 0 \end{pmatrix}^T \begin{pmatrix} Y & X^T \\ X & I_2 \end{pmatrix} \begin{pmatrix} e_{nn'} \\ 0 \end{pmatrix} - \hat{d}_{nn'}^2 \right| \tag{11.11}$$

$$s.t. \begin{pmatrix} Y & X^T \\ X & I_2 \end{pmatrix} \succeq 0.$$

In this part, the relative sensor location estimation is done by solving the above constrained optimization problem based on the information in the set $\mathbb{N}_1(n)$.

In the second part, consider the set $\mathbb{N}_2(n)$, the information between location-unaware and location-aware sensors. A min-max optimization approach is proposed for the location estimation of location-unaware sensors in the following min-max l_2 game problem:

$$\begin{aligned} &\min_{X} \sum_{n=1}^{N} \max_{\|\Delta_m\| \le \zeta} \sum_{m \in \mathbb{N}_2(n)} \left(\|x_n - z_m\| - \hat{d}_{nm} \right)^2 \\ &s.t.\ z_m = \hat{z}_m + \Delta_m, m \in \mathbb{N}_2(n). \end{aligned} \tag{11.12}$$

It should be noted that since m comes from $\mathbb{N}_2(n)$ as shown in (11.12), $\hat{d}_{nm}$ relates to $\|x_n - z_m\|$ according to (11.7). Equation (11.12) is the worst-case design, with ζ as the error bound of $\|\Delta_m\|$ in (11.2). In this part, based on the information $\mathbb{N}_2(n)$, the relative sensor location estimation is done by minimizing the estimation error in the presence of location errors of location-aware sensors from the worst-case perspective.

By applying the Taylor expansion, the term $\|x_n - z_m\|$ in (11.12) can be expanded as

$$\|x_n - z_m\| = \|x_n - \hat{z}_m\| - \frac{\Delta_m^T (x_n - \hat{z}_m)}{\|x_n - \hat{z}_m\|} + o(\|\Delta_m\|). \tag{11.13}$$

Let $\delta_{nm} = \frac{\Delta_m^T (x_n - \hat{z}_m)}{\|x_n - \hat{z}_m\|}$. We have

$$|\delta_{nm}| \leq \zeta. \tag{11.14}$$

Using (11.13) and (11.14), the min-max l_2 game problem in (11.12) can be transformed into

$$\min_X \sum\nolimits_{n=1}^{N} \max_{|\delta_{nm}|\leq\zeta} \sum\nolimits_{m\in\mathbb{N}_2(n)} \left(\|x_n - \hat{z}_m\| - \delta_{nm} - \hat{d}_{nm}\right)^2. \tag{11.15}$$

To facilitate the ensuing discussion, let us denote $\mathbb{N}_2(n)=\{\mu(n,1),\mu(n,2),\ldots,\mu(n,|\mathbb{N}_2(n)|)\}$, where $|\mathbb{N}_2(n)|$ represents the size of the set $\mathbb{N}_2(n)$ and $\mu(n,m)>\mu(n,m')$ if m'. Therefore, $\mu(n,m)$ is strictly increasing in m. For example, if $N=10, M=5$ and $\mathbb{N}_2(1)=\{1,3,5\}$, then $|\mathbb{N}_2(1)|=3$, $\mu(1,1)=1$, $\mu(1,2)=3$, and $\mu(1,3)=\mu(1,|\mathbb{N}_2(1)|)=5$. It can be noted that, by the definitions in (11.6), $1\leq|\mathbb{N}_2(n)|\leq M$ and $1\leq\mu(n,1)\leq\mu(n,|\mathbb{N}_2(n)|)\leq M$ if $\mathbb{N}_2(n)\neq\phi$. Denote

$$\delta_n = \begin{pmatrix} \delta_{n\mu(n,1)} \\ \delta_{n\mu(n,1)} \\ \vdots \\ \delta_{n\mu(n,|\mathbb{N}_2(n)|)} \end{pmatrix}, \text{and } \hat{d}_n = \begin{pmatrix} \hat{d}_{n\mu(n,1)} \\ \hat{d}_{n\mu(n,1)} \\ \vdots \\ \hat{d}_{n\mu(n,|\mathbb{N}_2(n)|)} \end{pmatrix}. \tag{11.16}$$

Adopting the notations in (11.16), (11.15) can be rewritten as

$$\min_X \sum\nolimits_{n=1}^{N} \max_{|\delta_{nm}|\leq\zeta} \left\|l_n - \delta_n - \hat{d}_n\right\|^2$$
$$\text{s.t. } l_n = \begin{pmatrix} \|x_n - \hat{z}_{\mu(n,1)}\| \\ \|x_n - \hat{z}_{\mu(n,2)}\| \\ \vdots \\ \|x_n - \hat{z}_{\mu(n,|\mathbb{N}_2(n)|)}\| \end{pmatrix}, n = 1,2,\ldots,N. \tag{11.17}$$

Note that, from (11.16),

$$|\delta_{nm}| \leq \zeta, m\in\mathbb{N}_2(n) \Rightarrow \|\delta_n\| \leq \zeta\cdot\sqrt{|\mathbb{N}_2(n)|} \triangleq \zeta_n. \tag{11.18}$$

Using (11.18), (11.17) can be relaxed into

$$\min_X \sum\nolimits_{n=1}^{N} \max_{|\delta_n|\leq\zeta_n} \left\|l_n - \delta_n - \hat{d}_n\right\|^2$$
$$\text{s.t. } l_n = \begin{pmatrix} \|x_n - \hat{z}_{\mu(n,1)}\| \\ \|x_n - \hat{z}_{\mu(n,2)}\| \\ \vdots \\ \|x_n - \hat{z}_{\mu(n,|\mathbb{N}_2(n)|)}\| \end{pmatrix}, n = 1,2,\ldots,N. \tag{11.19}$$

In (11.19), $\min_{X}\sum_{n=1}^{N}\max_{|\delta_n|\le\zeta_n}\left\|l_n-\delta_n-\hat{d}_n\right\|^2$ can be transformed into the following constrained optimization problem:

$$\begin{aligned}&\min_{X}\sum_{n=1}^{N}\rho_n\\ &s.t.\ \max_{|\delta_n|\le\zeta_n}\left\|l_n-\delta_n-\hat{d}_n\right\|^2\le\rho_n, n=1,2,\ldots,N.\end{aligned}\tag{11.20}$$

Unfortunately, (11.20) is nonconvex. To obtain a convex formulation so that the existing numerical algorithms can be employed to derive a solution, some conditions are transformed or relaxed as shown in the following four steps (S1)–(S4):

(S1) First step: Letting

$$Y=X^TX \text{ and } L_n=l_nl_n^T\tag{11.21}$$

with $[L_n]_{ij}$ as the element in the ith row and jth column of the matrix L_n, it can be readily shown that

$$[L_n]_{ij}=\begin{pmatrix}e_n\\-\hat{z}_{\mu(n,i)}\end{pmatrix}^T\begin{pmatrix}Y & X^T\\X & I_2\end{pmatrix}\begin{pmatrix}e_{nn'}\\-\hat{z}_{\mu(n,i)}\end{pmatrix}.\tag{11.22}$$

Meanwhile, by the Cauchy–Schwartz inequality, we have

$$[L_n]_{ij}\ge\left|\begin{pmatrix}e_n\\-\hat{z}_{\mu(n,i)}\end{pmatrix}^T\begin{pmatrix}Y & X^T\\X & I_2\end{pmatrix}\begin{pmatrix}e_{nn'}\\-\hat{z}_{\mu(n,i)}\end{pmatrix}\right|.\tag{11.23}$$

(S2) Second step: The constraint of the optimization problem in (11.20) is transformed into a matrix inequality constraint, as shown in the following.

Theorem 11.1
The constraint on the optimization problem in (11.20)

$$\max_{|\delta_n|\le\zeta_n}\left\|l_n-\delta_n-\hat{d}_n\right\|^2\le\rho_n,\tag{11.24}$$

is equivalent to

$$\exists\ \lambda_n\ge0\ s.t.$$

$$\begin{pmatrix}I_{|\mathbb{N}_2(n)|} & -(l_n-\hat{d}_n)\\-(l_n-\hat{d}_n)^T & s_n-\rho_n\end{pmatrix}\preceq\lambda_n\begin{pmatrix}I_{|\mathbb{N}_2(n)|} & 0\\0 & -\zeta_n^2\end{pmatrix},\tag{11.25}$$

where

$$s_n=trace\left\{\begin{pmatrix}L_n & l_n\\l_n^T & 1\end{pmatrix}\begin{pmatrix}I_{|\mathbb{N}_2(n)|} & -\hat{d}_n\\-\hat{d}_n^T & \hat{d}_n^T\hat{d}_n\end{pmatrix}\right\}.$$

Proof: It can be noted that (11.24) holds true if and only if the statement

$$\|\delta_n\|^2 \leq \zeta_n^2 \Rightarrow \left\|l_n - \delta_n - \hat{d}_n\right\|^2 \leq \rho_n \tag{11.26}$$

holds true. By using the notations in (11.16), (11.26) can be further expressed as the statement in the following matrix form:

$$\begin{aligned} &\begin{pmatrix} \delta_n \\ 1 \end{pmatrix}^T \begin{pmatrix} I_{|\mathbb{N}_2(n)|} & 0 \\ 0 & -\zeta_n^2 \end{pmatrix} \begin{pmatrix} \delta_n \\ 1 \end{pmatrix} \leq 0 \\ \Rightarrow &\begin{pmatrix} \delta_n \\ 1 \end{pmatrix}^T \begin{pmatrix} I_{|\mathbb{N}_2(n)|} & -(l_n - \hat{d}_n) \\ -(l_n - \hat{d}_n)^T & s_n - \rho_n \end{pmatrix} \begin{pmatrix} \delta_n \\ 1 \end{pmatrix} \leq 0. \end{aligned} \tag{11.27}$$

Finally, employing the S-procedure [221], (11.27) holds true if and only if (11.25) holds true, which completes the proof.

(S3) Third step: To transform (11.20) into a convex optimization problem, $Y = X^T X$ and $L_n = l_n l_n^T$ in (11.21) are relaxed into $Y \succeq X^T X$ and $L_n \succeq l_n l_n^T$, which are equivalent to (see *Schur complement* in [221])

$$\begin{pmatrix} Y & X^T \\ X & I_2 \end{pmatrix} \succeq 0, \text{ and} \begin{pmatrix} L_n & l_n \\ l_n^T & 1 \end{pmatrix} \succeq 0, \tag{11.28}$$

respectively.

(S4) Fourth step: Incorporating (11.22), (11.23), (11.25), and (11.28) in steps (S1)–(S3) into (11.20), a convex formulation modified from (11.20) is obtained as

$$\begin{aligned} &\min_{X,Y,L_n,l_n,\rho_n,\lambda_n} \sum_{n=1}^{N} \rho_n + c \sum_{n=1}^{N} \sum_{j>i}^{|\mathbb{N}_2(n)|} [L_n]_{ij} \\ &\text{s.t.} \begin{pmatrix} (\lambda_n - 1) I_{|\mathbb{N}_2(n)|} & l_n - \hat{d}_n \\ (l_n - \hat{d}_n)^T & \rho_n - s_n - \lambda_n \zeta_n^2 \end{pmatrix} \geq 0 \end{aligned}$$

where

$$\begin{aligned} &s_n = trace\left\{ \begin{pmatrix} L_n & l_n \\ l_n^T & 1 \end{pmatrix} \begin{pmatrix} I_{|\mathbb{N}_2(n)|} & -\hat{d}_n \\ -\hat{d}_n^T & \hat{d}_n^T \hat{d}_n \end{pmatrix} \right\}, \\ &[L_n]_{ij} = \begin{pmatrix} e_n \\ -\hat{z}_{\mu(n,i)} \end{pmatrix}^T \begin{pmatrix} Y & X^T \\ X & I_2 \end{pmatrix} \begin{pmatrix} e_{nn'} \\ -\hat{z}_{\mu(n,i)} \end{pmatrix}, \\ &[L_n]_{ij} \geq \left| \begin{pmatrix} e_n \\ -\hat{z}_{\mu(n,i)} \end{pmatrix}^T \begin{pmatrix} Y & X^T \\ X & I_2 \end{pmatrix} \begin{pmatrix} e_{nn'} \\ -\hat{z}_{\mu(n,j)} \end{pmatrix} \right|, \\ &\begin{pmatrix} L_n & l_n \\ l_n^T & 1 \end{pmatrix} \succeq 0, \begin{pmatrix} Y & X^T \\ X & I_2 \end{pmatrix} \succeq 0, \ \lambda_n \geq 0, \rho_n \geq 0, \\ &n = 1, 2, \ldots, N; i, j = 1, 2, \ldots, |\mathbb{N}_2(n)|, j > i. \end{aligned} \tag{11.29}$$

Remark 11.2 For convex relaxation techniques, such as those presented in this section, a penalty function is usually needed so that the original problem can be approximated accurately by the relaxed problem. In (11.29), the term $c\sum_{n=1}^{N}\sum_{j>i}^{|\mathbb{N}_2(n)|}[L_n]_{ij}$ with the positive number c in the cost function serves this purpose. The value of c is typically very small, for example, $c \in [10^{-8}, 10^{-4}]$. Yang et al. [217] confirmed that the performance of the corresponding relaxation is not sensitive to the value of c if c is within the interval $[10^{-8}, 10^{-4}]$.

Until now, the robust relative location estimations for location-unaware sensors were performed separately according to two sets $\mathbb{N}_1(n)$ and $\mathbb{N}_2(n)$. The corresponding convex optimization approaches based on the information of two sets $\mathbb{N}_1(n)$ and $\mathbb{N}_2(n)$ are presented in (11.11) and (11.29), respectively. In order to utilize all the information set $\mathbb{N}(n)$, which is the union of two sets $\mathbb{N}_1(n)$ and $\mathbb{N}_2(n)$, the proposed sensor relative location estimation $\hat{X}_P$ combines (11.11) and (11.29) as the following convex optimization problem:

$$
\begin{aligned}
\hat{X}_P = \arg_X & \min_{X,Y,L_n,l_n,\rho_n,\lambda_n} \sum_{n=1}^{N}\rho_n + c\sum_{n=1}^{N}\sum_{j>i}^{|\mathbb{N}_2(n)|}[L_n]_{ij} \\
& +\min_{X,Y}\sum_{n=1}^{N}\sum_{n'\in\mathbb{N}_1(n)}\left|\begin{pmatrix} e_{nn'} \\ 0\end{pmatrix}^T\begin{pmatrix} Y & X^T \\ X & I_2\end{pmatrix}\begin{pmatrix} e_{nn'} \\ 0\end{pmatrix} - \hat{d}_{nn'}^2\right| \\
& s.t. \begin{pmatrix} (\lambda_n - 1)I_{|\mathbb{N}_2(n)|} & l_n - \hat{d}_n \\ (l_n - \hat{d}_n)^T & \rho_n - s_n - \lambda_n\zeta_n^2 \end{pmatrix} \geq 0
\end{aligned}
$$

where

$$
\begin{aligned}
& s_n = trace\left\{\begin{pmatrix} L_n & l_n \\ l_n^T & 1\end{pmatrix}\begin{pmatrix} I_{|\mathbb{N}_2(n)|} & -\hat{d}_n \\ -\hat{d}_n^T & \hat{d}_n^T\hat{d}_n\end{pmatrix}\right\}, \\
& [L_n]_{ij} = \begin{pmatrix} e_n \\ -\hat{z}_{\mu(n,i)}\end{pmatrix}^T\begin{pmatrix} Y & X^T \\ X & I_2\end{pmatrix}\begin{pmatrix} e_{nn'} \\ -\hat{z}_{\mu(n,i)}\end{pmatrix}, \\
& [L_n]_{ij} \geq \left|\begin{pmatrix} e_n \\ -\hat{z}_{\mu(n,i)}\end{pmatrix}^T\begin{pmatrix} Y & X^T \\ X & I_2\end{pmatrix}\begin{pmatrix} e_{nn'} \\ -\hat{z}_{\mu(n,j)}\end{pmatrix}\right|, \\
& \begin{pmatrix} L_n & l_n \\ l_n^T & 1\end{pmatrix} \succeq 0, \begin{pmatrix} Y & X^T \\ X & I_2\end{pmatrix} \succeq 0,\ \lambda_n \geq 0, \rho_n \geq 0, \\
& n = 1,2,\ldots,N; i,j = 1,2,\ldots,|\mathbb{N}_2(n)|, j>i.
\end{aligned}
\tag{11.30}
$$

In (11.30), $X=(x_1x_2\cdots x_N)\in R^{2\times N}$ represents the position matrix of interest, and $\hat{X}_P$ represents the corresponding estimator of the decision variable X. The remaining variables are $Y\in R^{N\times N}$, $L_n\in R^{|\mathbb{N}_2(n)|\times|\mathbb{N}_2(n)|}$, $l_n\in R^{|\mathbb{N}_2(n)|\times 1}$, $\rho_n\in R$, and $\lambda_n\in R$ for $n=1,2,\ldots,N$.

It can be noted that all constraints on the optimization problem in (11.30) are expressed by matrix inequalities. The constrained optimization in (11.30) is termed an SDP problem, which is convex. Hence, the global optimum can be achieved by applying existing numerical techniques, for example, the interior-point methods [221].

Remark 11.3 (Difference from the estimation problem of target localization). Generally speaking, a target localization problem may be regarded as a special case of relative location estimation problems when the number of location-unaware sensor nodes is equal to one. However, the corresponding situations are quite different in two ways. First, collaborative sensor nodes are involved for self-configuration, whereas the concerned target is uncooperative [217]. In such a case, the TOA technique can be employed for relative location estimation, but not for target localization. Instead, the TDOA technique is often used [216] in a target localization problem. Second, for target localization, the position of interest is estimated according to the information between it and the sensors of known locations. However, due to the ranging limitation, some sensors of interest in a sensor location estimation problem need to be located by the information between sensors of unknown locations. Therefore, target localization methods cannot be directly applied to sensor location estimation problems.

In contrast to target localization problems, the proposed formulation in (11.30) employs not only the information between location-aware and location-unaware sensors, i.e., $\mathbb{N}_2(n)$, but also the information between location-unaware sensors, i.e., $\mathbb{N}_1(n)$. The use of $\mathbb{N}_1(n)$ differentiates the relative location estimation problem from a target localization problem.

Remark 11.4 The first constraint in (11.30) corresponds to the worst-case scenario design, i.e., the maximum error introduced by the inexact positions of location-aware sensors is minimized. This differentiates the proposed approach from existing studies. Instead of considering normal distributions of location errors, the proposed approach utilizes the knowledge of the error bound ζ in (11.2). In particular, the inexact position problem for sensor relative location estimation in WSNs has been dealt with from the worst-case perspective in this chapter.

The proposed approach may be applicable to a monitoring system for biological groups. For example, assume that we are interested in the influence of humidity on a certain biological group in a certain area. Sensors with humidity detectors are then deployed to the region of interest (ROI). However, it may not be cost-effective to carefully position all sensors such that their corresponding locations are known immediately after the deployment. A basic approach is to intentionally place some sensors as location-aware nodes. The remaining sensors would then be randomly deployed by robots and hence, are regarded as location-unaware nodes. The topology of the deployed sensors could be obtained by using existing localization approaches [201], [203], [205], [218]. Once the topology has been obtained, all location-unaware sensors are then regarded as location-aware nodes with a certain degree of location error, which corresponds to the error bound ζ in this chapter. The bound can be estimated by using prior simulation results.

Extending the aforementioned example, let us now suppose that another factor, which can have a great impact on the biological group, has been found, e.g., temperature. Hence, either the deployed sensors need to be equipped with temperature-sensing functionality or new sensors with this functionality are to be added to the existing topology. The latter is chosen due to the lower costs involved. Once again, the random deployment is adopted. The final localization problem is the inexact position problem considered in this chapter, and, hence, can be dealt with by the proposed approach.

This chapter provides a framework for related problems, which require the localization process more than twice. In such a process, blindfolded nodes will become reference nodes with a certain degree of location error after the estimation. The inexact problem occurs when new sensors are added into the ROI and the reference nodes resulting from the estimation are used as the location-aware sensors for another localization process. It can be noted that, for future work, as this chapter has primarily focused on mathematical formulations, several practical issues are yet to be addressed, such as the update of the value of ζ, the effect of changes in the propagation environment, and the interference. More experiments and tests should also be carried out.

11.5 Simulation Examples

Numerical simulations were performed in this section to verify the robustness of the proposed sensor location estimation approach in (11.30). For comparison with our method, the maximum likelihood estimation modified from [201] and the SOCP relaxation method modified from [205] are employed to position the coordinates of location-unaware sensors in Section 11.5.1. Section 11.5.2 describes the figure of merit for all simulations. To illustrate the pros and cons of different sensor location estimation methods, a scenario with different parameters is considered in Section 11.5.3.

11.5.1 Review of Other Sensor Location Estimation Methods

Modified ML estimation [201] and SOCP methods [205] were used in the simulation. Originally, these were only designed in the absence of the inexact position problem. To enable comparison with the proposed method, these relative location estimation methods are modified in this chapter by replacing the true location z_m with the inexact location $\hat{z}_m$ for $m = 1, 2, .., M$. The corresponding estimators are presented as follows.

The ML estimation modified from [201] is obtained as

$$\hat{X}_{ML} = \arg_X \min_X \sum_{n=1}^{N-1} \sum_{m \in \mathbb{N}_1(n)} (\ln \frac{\hat{d}_{nm}^2}{\|x_n - x_m\|^2})^2 + \sum_{n=1}^{N} \sum_{m \in \mathbb{N}_2(n)} (\ln \frac{\hat{d}_{nm}^2}{\|x_n - \hat{z}_m\|^2})^2, \quad (11.31)$$

which can be dealt with by nonlinear least squares algorithms [247]. The main drawback of using the ML method in (11.31) is the requirement of a starting point for the solving process. The optimality of the obtained solution often depends on the choice of starting points. In the simulation, two starting points (X), true positions of location-unaware sensors, and (C), the centroid of the neighboring location-aware sensors, were used. For example, if a location-unaware sensor x_n connects with location-aware sensors z_1, z_2, and z_3, then the nth column of the starting point (X) is set by

$$\frac{\hat{z}_1 + \hat{z}_2 + \hat{z}_3}{3}.$$

If x_n does not connect with any location-aware sensors, then the nth column of the starting point (X) is set by the centroid of the region monitored. To distinguish the resulting estimates obtained by solving (11.31), let us denote $\hat{X}_{ML(X)}$ and $\hat{X}_{ML(C)}$ as the use of the starting points (X) and (C), respectively. It can be noted that the starting point (X) is an ideal choice and not achievable in practice, whereas (C) is a natural choice and realistic.

The SOCP method modified from [201] is obtained as

$$\begin{aligned}\hat{X}_{SOCP} &= \arg_X \min_{X,y_{nm}} \sum_{n=1}^{N-1} \sum_{m\in\mathbb{N}_1(n)} \left|y_{nm} - \hat{d}^2_{nm}\right| + \sum_{n=1}^{N} \sum_{m\in\mathbb{N}_2(n)} \left|y_{nm} - \hat{d}^2_{nm}\right| \\ s.t.\ & y_{nm} \geq \|x_n - x_m\|^2, n = 1,2,\ldots,N-1, m \in \mathbb{N}_1(n) \\ & y_{nm} \geq \|x_n - \hat{z}_m\|^2, n = 1,2,\ldots,N, m \in \mathbb{N}_2(n),\end{aligned} \tag{11.32}$$

which is convex and can be solved by interior-point methods [221]. The physical meaning of SOCP method in (11.32) can be readily understood by replacing all inequality constraints with equality constraints. The relaxation of feasible solutions, i.e., using inequality constraints rather than equality constraints, transforms the problem into a convex SOCP, which is solvable with existing solvers. However, it is known that this relaxation can result in a "convex-hull" problem [217]. In particular, the performance can severely degrade if location-unaware sensors mostly lie outside the convex hull of $\hat{z}_m$, $m = 1,2,\ldots,M$. Therefore, a good deployment of sensors is needed for a good performance of SOCP method, for example, location-aware nodes positioned on the edges or vertices of the region monitored.

It is constructive to examine the iteration complexity for producing results of the relative location estimation approaches mentioned previously. It should be noted that, for one particular optimization problem, the iteration complexity can vary with different algorithms used. In the ensuing discussion, for simplicity, only one particular algorithm is considered for each optimization problem. Hopefully, the reader can have an idea of the order of magnitude.

For the ML estimator in (11.31), the Levenberg-Marquardt method in [248] is concerned. Let us denote the objective function of (11.31) by ϕ. The corresponding iteration complexity is $O(\varepsilon^{-2})$, where ε satisfies $\|\nabla_\phi\| \leq \varepsilon$. For the SOCP and the proposed estimators in (11.32) and (11.30), primal-dual algorithms in [249,250] are considered, respectively. The main computation effort for solving the SOCP in (11.32) comes from the number of second-order cones, denoted by K. The iteration complexity can be expressed as $O(\sqrt{K}\log\varepsilon'^{-1})$, where ε' represents the reduced duality gap [249]. Referring to (11.32), we have

$$K = \sum_{k=1}^{N-1} |\mathbb{N}_1(k)| + \sum_{k=1}^{N-1} |\mathbb{N}_2(k)|.$$

For the proposed sensor location estimator in (11.30), the dimension of the variables, denoted by K', mainly contributes to the iteration complexity, which is $O(K'^{\frac{3}{2}}\log\varepsilon'^{-1})$ [250]. In order to use the algorithms in [250], the optimization problem in (11.30) is transformed into a standard form by introducing extra variables to eliminate the absolute function in the objective space and to replace the inequality constraints with equality constraints. By doing so, we have the following estimation of K':

$$K' > 4N + 2\sum_{k=1}^{N-1} |\mathbb{N}_1(k)| + \sum_{k=1}^{N-1} |\mathbb{N}_2(k)|.$$

11.5.2 Performance Evaluation of Sensor Location Estimators by Simulations

All simulations have been carried out in MATLAB and K = 50 Monte–Carlo (MC) runs were performed. Solvers, such as *SeDuMi* [251] and the MATLAB function "lsqnonlin," are readily available, and, hence, are used to produce results of the mentioned estimators. The root mean square error (RMSE) of an estimator is defined as

$$\sqrt{\frac{1}{K}\sum_{k=1}^{K}\left\|\hat{X}^{(k)}-X^{(k)}\right\|^{2}}=\sqrt{\frac{1}{K}\sum_{k=1}^{K}\left(\left(\hat{x}_{n}^{(k)}-x_{n}^{(k)}\right)^{2}+\left(\hat{y}_{n}^{(k)}-y_{n}^{(k)}\right)^{2}\right)}, \tag{11.33}$$

where

$$X^{(k)}=\left(\mathbf{x}_1^{(k)}\ \mathbf{x}_2^{(k)}\ \ldots\ \mathbf{x}_N^{(k)}\right)=\begin{pmatrix} x_1^{(k)} & x_2^{(k)} & \cdots & x_N^{(k)} \\ y_1^{(k)} & y_2^{(k)} & \cdots & t_N^{(k)} \end{pmatrix}\in R^{2\times N}$$

represents the true positions of location-unaware sensors at the kth MC run and $\hat{X}^{(k)}$ represents the corresponding location estimation. For example, if the ML method using (C) as the starting point is considered, then $\hat{X}^{(k)}$ stands for the kth estimates resulting from the use of $\hat{X}_{ML(C)}$ in (31). It can be noted that the value of RMSE defined in (11.33) often increases upon increasing N, the number of location-unaware sensors. The RMSE is not normalized by dividing it by N because the whole X is regarded as the unknown parameter to be estimated.

Apart from the RMSE, a "boxplot" [252] of $\{\left\|\hat{X}^{(1)}-X^{(1)}\right\|,\left\|\hat{X}^{(2)}-X^{(2)}\right\|,\ldots,\left\|\hat{X}^{(K)}-X^{(K)}\right\|\}$ was employed as well. The function "boxplot" in MATLAB transforms the input data into a box with several marks. The central mark stands for the median, and the edge marks stand for the 25th and 75th percentiles. The reader can refer to MATLAB "help" for a detailed description of the "boxplot" function. Generally speaking, a boxplot diagram illustrates the distribution of the input data. In our case, the boxplot illustrates the distribution of the location estimation error. For an estimator to be robust, it is desirable to have the corresponding box with a shorter length and a lower median mark.

11.5.3 Case Study

To compare the performance of the proposed estimator $\hat{X}_p$ with $\hat{X}_{ML(C)}$, $\hat{X}_{ML(X)}$, and $\hat{X}_{SOCP}$, a case with different parameters has been considered. Suppose that there are 10 location-unaware sensors (i.e., $N = 10$) to be estimated. All sensors were randomly deployed in a normalized 1×1 m^2 area. In particular, for each MC run (for each k in (11.33)), the deployment

$$\hat{X}^{(k)}=\left(z_1^{(k)}\ z_2^{(k)}\ \ldots\ z_M^{(k)}\right)=rand(2,N+M)$$

has been used, where $rand(2, N+M)$ represents a $2\times(N+M)$ matrix with all elements randomly chosen from [0, 1]. The inexact positions $\hat{z}_m$'s have been generated according to (11.1) with Δ_m randomly chosen from the ball

TABLE 11.1

Typical values of parameters for scenario

Parameter	Scenario
M	4
$\tilde{d}$ in (6)	0.5 m
σ_p (variance of ξ_{nm} in (4))	3.9 db
ζ in (3)	0.06 m
n_p in (3)	2.8

$$\{\Delta_m \in R^2 : \|\Delta_m\| \leq \zeta\}, \tag{11.34}$$

for $m = 1, 2, ..., M$. Equation (11.34) can be readily realized by setting

$$\Delta_m = (r\cos\theta r\sin\theta)^T,$$

where $\hat{X}^{(k)} = (z_1^{(k)}\, z_2^{(k)} \ldots z_M^{(k)}) = rand\,(2,N+M)$ and $\hat{X}^{(k)} = (z_1^{(k)}\, z_2^{(k)} \ldots z_M^{(k)}) = rand\,(2,N+M)$. The typical values of the parameters for the scenario are listed in Table 11.1. Performance of the proposed method is compared with other methods by adjusting these in turn and the simulation results are discussed as follows:

1. *Number of location-aware sensors M*: The estimation error will generally be reduced if the number of location-aware sensors is increased. Figure 11.1a illustrates the RMSEs of different sensor location estimation methods versus the number of location-aware sensors. As can be observed in Fig. 11a, the estimator $\hat{X}_{ML(X)}$ with X as the starting point achieves a high level of accuracy, whereas $\hat{X}_{ML(C)}$ with a starting point selected by the centroid of the neighboring location-aware sensors has the worst performance of the presented methods. With respect to the performance of $\hat{X}_{SOCP}$, the possible probability of all the estimation locations lying in the convex hull of the location-aware sensor nodes increases if the number of location-aware sensors becomes greater. Overall, the proposed estimator $\hat{X}_P$ is relatively robust. Consider the boxplot in Fig. 11.1b. The y-axis denotes the estimation error, whereas the x-axis denotes different estimators as detailed in the caption. An estimator is more robust in the boxplot if the corresponding box has a shorter length, i.e., a narrower error distribution, and a lower central mark (median mark), i.e., a lower average estimation error. As can be observed, the error distribution of $\hat{X}_{SOCP}$ is the widest among the estimators. Note that the performance of $\hat{X}_{ML}$ depends heavily on the starting point. Although $\hat{X}_{ML(X)}$ can achieve a high level of accuracy, it is impractical due to the use of the unknown parameter X. With the exception of $\hat{X}_{ML(X)}$, $\hat{X}_P$ has a superior performance in terms of a lower value of RMSE and a narrower error distribution with lower values of estimation error.
2. *Threshold for a possible connection $\tilde{d}$*: When $\tilde{d}$ becomes large, it implies that the probability of the connection between two sensors would increase and that the estimation algorithm would obtain more information. Consequently, the performance would improve. The simulation results obtained are illustrated in Fig. 11.2. With the exception of $\hat{X}_{ML(X)}$, $\hat{X}_P$ has the best performance among the presented estimators in terms of a lower value of RMSE and a narrower error distribution

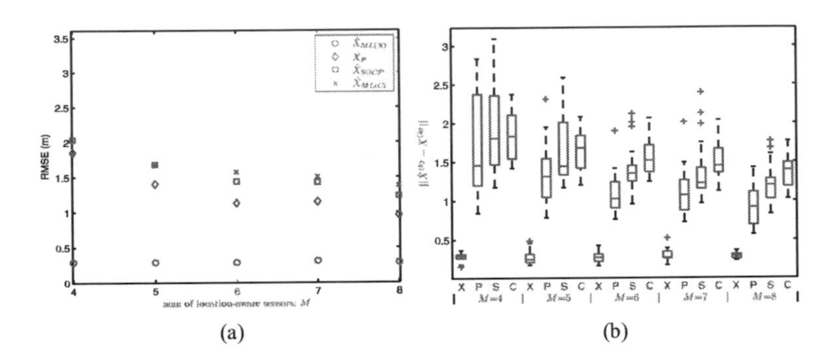

(a) (b)

FIGURE 11.1
Numerical simulation results of different location estimation methods with various number of the location-aware sensors (ROI = 1×1 m^2, sensor range = 0.5m, $N = 10$). (a) "RMSE versus M" plot. $\hat{X}_P$ denotes the proposed min-max l_2 game method, $\hat{X}_{SOCP}$ denotes the SOCP method, $\hat{X}_{ML(R)}$ denotes the ML method with a starting point selected by the centroid of the neighboring location-aware sensors, and $\hat{X}_{ML(X)}$ denotes the ML method with the starting point X. (b) is the boxplot of estimation errors $\left\|\hat{X}^{(K)} - X\right\|$. In x-axis label, "P," "S," "C," and "X" stand for $\hat{X}_P$, $\hat{X}_{SOCP}$, $\hat{X}_{ML(R)}$, and $\hat{X}_{ML(X)}$, respectively. All of the following figures have the same notations, except for the x-axis in (a).

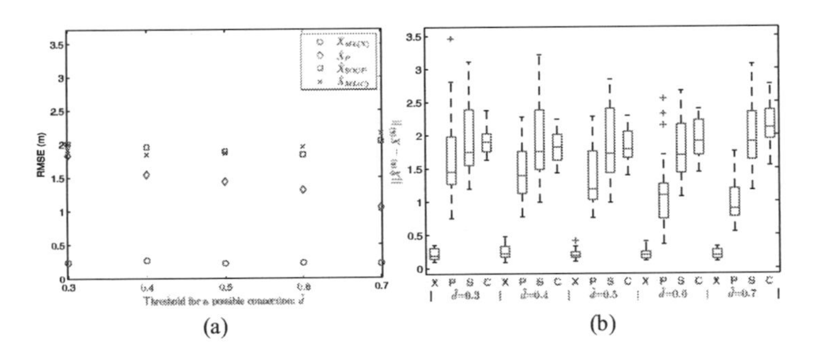

(a) (b)

FIGURE 11.2
Numerical simulation results of different methods with various values of $\tilde{d}$(ROI = 1×1 m^2, $M = 4$, $N = 10$).

with lower values of estimation error. It should be noted that although the sensing range $\tilde{d}$, varying from 0.3 to 0.7 m, seems relatively large with respect to the normalized 1×1 m^2 region, it is still possible that a large proportion of location-unaware nodes do not connect with the location-aware nodes due to the random deployment. This phenomenon is related to the "k-coverage problem," which is often discussed in studies relating to the relationship between the sensing range and the coverage. The reader can refer to [253–257], for related works.

3. *Variance of measured power error* σ_p: From (11.3) and (11.4), we know that the variance of the relative distance measurement $\hat{d}_{nm}$ is exponentially proportional to σ_p. The RMSE becomes large if the variance of $\hat{d}_{nm}$ is increased. The trend can be found in Fig. 11.3. Estimator $\hat{X}_p$ has the best performance among the realizable estimation methods.

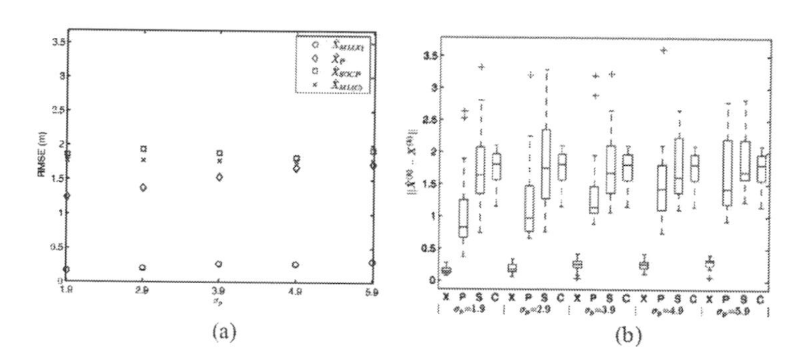

FIGURE 11.3
Numerical simulation results of different methods with various values of σ_p (ROI $= 1 \times 1$ m^2, sensor range $= 0.5$m, $M = 4$, $N = 10$).

4. Error bound ζ: The large bounding error ζ implies that the positions of the location-aware sensors are more inaccurate. Therefore, the RMSE is proportional to ζ. The simulation results for different values of ζ are illustrated in Fig. 11.4. It should be noted that ζ is the worst possible error for all location-aware nodes. The proposed sensor location estimation approach uses the knowledge of ζ and hence, it is a worst-case design. Similarly, using $\hat{X}_p$ as the relative location estimator results in the smallest estimation error, excluding the impractical estimator $\hat{X}_{ML(X)}$.
5. Path loss exponent n_p: The variance of $\hat{d}_{nm}$ is inversely exponentially proportional to n_p, according to (11.3). This implies that the RMSE will decrease if the value of np increases. The simulation results are illustrated in Fig. 11.5. Estimator $\hat{X}_p$ still has the best performance among the feasible estimation methods.

From (11.30) to (11.32), it is evident that the proposed min-max l_2 game method has more variables and constraints than the other methods. This means that the proposed min-max l_2 game method would take more time to derive the relative location estimation. However, from the above simulations, it can be observed that the proposed min-

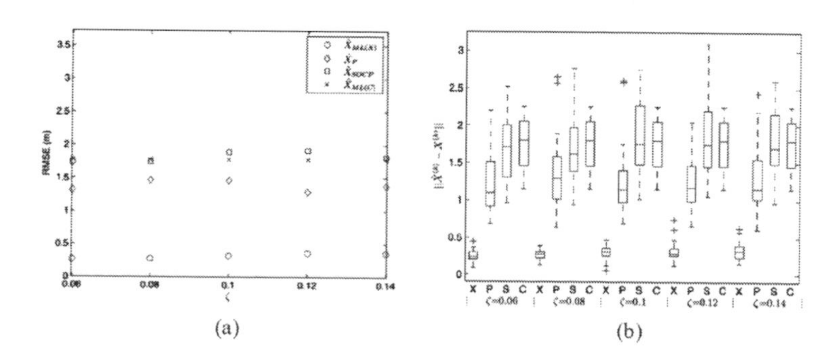

FIGURE 11.4
Numerical simulation results of different methods with various values of ζ (ROI $= 1 \times 1$ m^2, sensor range $= 0.5$m, $M = 4$, $N = 10$).

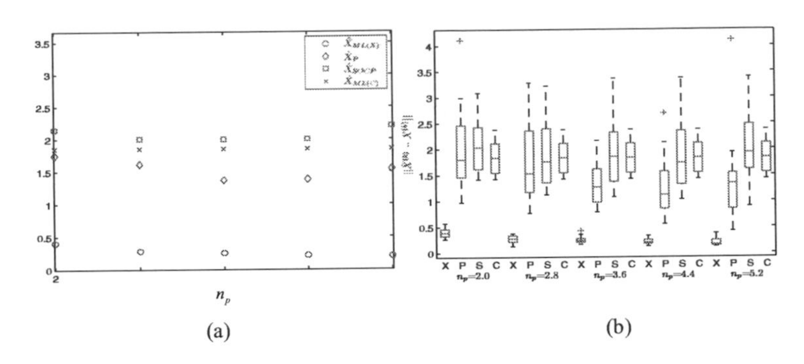

FIGURE 11.5
Numerical simulation results of different location estimation methods with various values of n_p (ROI $= 1 \times 1$ m^2, sensor range $= 0.5$m, $M = 4$, $N = 10$).

max l_2 game method $\hat{X}_p$ outperforms $\hat{X}_{ML(C)}$ and $\hat{X}_{SOCP}$. Moreover, unlike the ML and SOCP location estimation methods, a good starting point for the solving process and a good deployment of sensors are not required in the proposed scheme. Therefore, to minimize the worst estimation error, the proposed min-max l_2 game method has better performance without exact location knowledge of location-aware sensors. It also has greater potential for robust sensor location estimation in WSNs with location errors in the future.

11.6 Conclusion

The inexact position problem of location-aware sensors in WSNs has been considered. A robust sensor location estimation method based on a min-max l_2 game method is proposed in the presence of location errors. To the best of our knowledge, the existing approaches for sensor location estimation in WSNs only consider the Gaussian distribution model for the inexact positions of location-aware sensors. In this chapter, exact knowledge of the location error distribution is not needed because the problem is examined from a worst-case perspective, which has not been fully investigated in the literature.

Our simulations have evidenced that the proposed min-max l_2 game method outperforms the modified ML and SOCP methods. Rather than focusing on some particular topologies of WSNs, the random deployment of sensor nodes has been considered. The influence of different environmental parameters on the localization performance has been examined. There are three benefits from the proposed min-max l_2 game approach. First, the sensor location estimation problem is formulated as a convex optimization problem, which can be solved by existing numerical techniques. The feasibility of the proposed approach is then validated. Second, instead of knowing the distribution of the location error, only an error bound is required. In this way, the proposed min-max l_2 game approach can be applied to a more general situation. Third, unlike the conventional ML estimation and SOCP methods, a good starting point for the solving process and a good deployment of sensors are not needed in our robust sensor estimation scheme.

Finally, it should be noted that this chapter aims to provide a framework for the inexact sensor position problem with a feasible solution. Many mathematical formulations and models have been investigated as well. Any future work should include detailed experiments in practical scenarios.

Part IV

Stochastic Game Strategies in Management and Financial Systems

12

Stochastic H_∞ Game Approach to Robust Resource Management Control for CO_2 Emission and Reduction of Greenhouse Effect

12.1 Introduction

In recent decades, the world has attracted much attention to environmental protection issues such as atmospheric pollution, conservation of water reserves, and the reduction of tropical forests cover. For example, people feel concern about global warming, caused by greenhouse gases (GHG) such as carbon dioxide (CO_2), methane, nitrous oxide, sulfur hexafluoride, hydrofluorocarbons, and perfluorocarbons, which is leading to ecological destruction, climatic anomalies, and sea level rise [258,259]. However, despite the increasing environmental awareness, the global economic success heavily relies on the industrial throughput. People have gained a better life following the expansion of industrial sector and the number of job positions. This better life target has been achieved following the expressions of urban environmental quality, significant increase in pollution, and loss of natural habitats [260]. In order to reduce the emissions of GHG, especially CO_2, without limiting economic growth, more substantial investments should target the development of clean technology, expansion of forested areas, and some political actions [261–263].

A major problem associated with economic growth is the need for the energy, for which fossil fuel is the primary source. Such economic growth resulted in an increase of atmospheric emission of CO_2, as shown in Table 12.1 [264]. From Fig. 12.1 [265], it is seen that from 1900, the global CO_2 emission increased year by year except in the European Union (EU) that CO_2 emission had been decreased by 2% in the later period (1990–1996), but it is still very high elsewhere. According to United Nations Environment Programme (UNEP) in 2007 [264], this decrease was possible due to many initiatives taken by Germany such as the investings in renewable energy, solar power, new technology for car production, reforestation, and political actions creating laws requiring 5% reduction of carbon emission. Recently, an indicator called the Ecological Footprint (EF) was concerned by UNEP to relate the "pressure" exerted by human pollutions on the global ecosystems (Table 12.2) [264]. The EF is expressed in terms of area, and according to the definition provided by WWF [266], it representshow much productive land and sea is needed to provide the resources such as energy, water and raw materials used everyday. It also calculates the emission generated from the oil, coal and gas burnt, and determines how much land is required to absorb the waste.

This indicator is very useful in establishing how far the present situation is from the ideal condition in terms of the emission of CO_2. The worst EF indicators are found in North America and Western Europe. In order to mitigate the threat of an escalating greenhouse effect, it is necessary to establish a rigorous management control process of the available resources to reduce CO_2 emission. These management controls should include direct government incentive to promote pro-environment actions by the private sector and the establishment of stricter pollution regulations. To meet the CO_2 emission limitations and combat the global warming, 193 parties (192 states and the EU) have signed and ratified the Kyoto Protocol to the United Nations Framework Convention on Climate Change (UNFCCC) [267]. The cost estimated for the industrialized countries to implement the Kyoto Protocol ranges from 0.1 to 0.2% of their gross domestic product (GDP) [260]. Based on mathematical dynamic models, these costs can be efficiently optimized through control theory methods.

In order to manage the resources commitment to achieve the desired model reference tracking control of CO_2 concentration, mathematical models are required. In previous studies, Nordbous (1991) [268,269] presented a mathematical model to describe the effect of GHG in the economy and to maximize a social welfare function, subject to some dynamic constraints for the global temperature and atmospheric concentration of CO_2. He carried out a study considering low, medium, and high level of damages as a function of the concentration of CO_2. In another study, Nordhous (1993) [270] used the same mathematical model to evaluate some optimal taxation policies to stabilize climate and carbon emissions, i.e. enforcing political actions about taxes on the CO_2 emissions from burning coal, petroleum-based products, and natural gas. Poterba (1993) [271] has discussed the relationship between the global warming and GDP growth and considered the influence of certain macroeconomic initiatives on the decrease of the atmospheric emission of CO_2. For example, a consumption-linked carbon tax to reduce CO_2 emissions by 50% would reduce GDP by 4% in North America, 1% in Europe, and by 19% in some oil exporting countries. In the study of Stollery (1998) [272], an optimal CO_2 emission tax could be initially high, but it would eventually be lowered as the emission declines due to energy resource depletion. He also showed that to sustain consumption in the face of both energy resource depletion and economic damage from the global warming, it suffices to reinvest the sum of carbon tax revenues and the net energy rents. Caetano et al. (2008) [260] follows the ideas in Stollery [272] and offers a quantitative tool for the efficient allocation of resources to reduce the greenhouse effect caused by CO_2 emission. Their approach was developed by a mathematical model to describe the dynamic relation of CO_2 emission with the investment in reforestation and clean technology, and propose a method

TABLE 12.1

Carbon dioxide emission by region (1998)

Region	Million tones carbon/year
Asia and Pacific	2167
Europe	1677
North America	1614
Latin America & Caribbean	365
West Asia	187

Source: UNEP(2007).

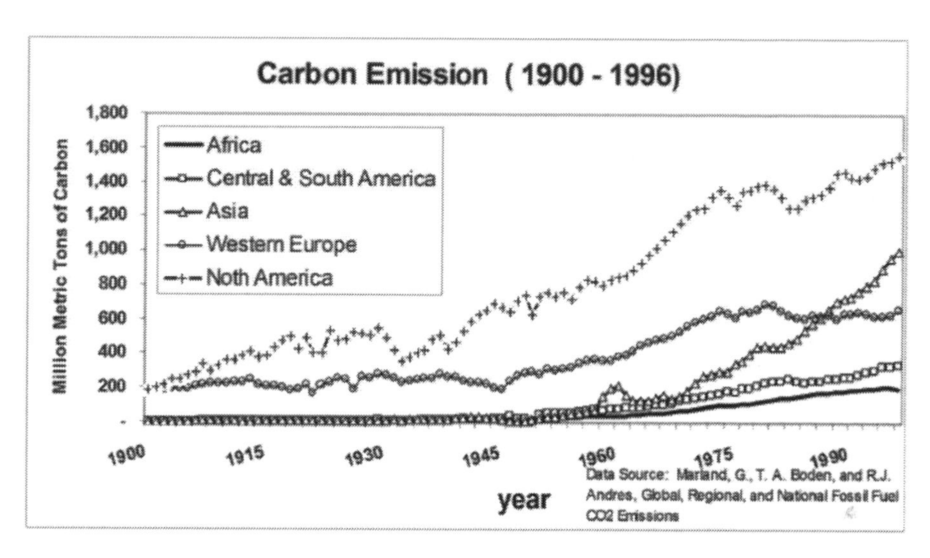

FIGURE. 12.1
Global carbon emission. The CO_2 emission has an increasing trend in the world, except for an about 2% decreasing in recent period (1990–1996) in EU.

TABLE 12.2
Ecological footprints.

Region	Million tones carbon/year
Africa	<2
Asia and Pacific	<2
Latin America & Caribbean	Between 2 and 3
West Europe	Between 4 and 5
North America	>11

Source: UNEP(2007).

to efficiently manage the available resources by casting an optimal control problem. Also, an optimal tracking control of CO_2 emission was addressed to achieve the emission targets proposed in the Kyoto Protocol for European countries by numerically solving a nonlinear partial Hamiltonian differential function [273].

In the above CO_2 control methods, ordinary nonlinear differential equations with time-invariant parameters are used to describe the systematic dynamics among CO_2 emissions, forest area expansion, and GDP growth. However, these intrinsic parameters may fluctuate area to area and time to time for different regional development or unpredictable situations, like sub-prime crash that was initiated in 2007, which may lead to the necessity of estimating new parameters as time changes or let the control strategies be limited in some specific area. Further, the external disturbances, due to modeling error and environmental noise, should also be considered in order to mimic the real dynamics of CO_2 emission system. Therefore, the more real dynamical system model of CO_2 emission system should be described by stochastic differential equations (SDEs). A differential equation containing a deterministic part and an additional random fluctuation term is called a SDE, which has been frequently used to model diverse phenomena in physics, biology, and finance [274].

In this chapter, a nonlinear stochastic model is proposed to describe the dynamic system with model uncertainties from intrinsic parametric fluctuations, for the CO_2 emission with some investments in the reforestation, clean technology, and political action about carbon tax. In addition to the intrinsic parametric random fluctuations, external disturbances from modeling error and environmental noise are also included in the nonlinear stochastic dynamic model of CO_2 emission system; thus, the generalized stochastic dynamic model could be widely applied to different area and time. Then, a reference model is developed to generate the desired dynamics of CO_2 emission system for which people hope in future world. Finally, a robust model reference tracking control is proposed based on stochastic H_∞ game strategy to manage these available resources in the earth, so that the nonlinear stochastic CO_2 emission system can track the desired output of the reference model, in spite of parametric fluctuations and external disturbances [74,105]. Since the statistical knowledge of the parametric random fluctuation, external disturbance, and uncertain initial condition is always unavailable, based on robust stochastic minimax H_∞ game theory in Chapter 3, the worst-case effect of random parametric fluctuations, external disturbances, and uncertain initial conditions on the model reference tracking error should be minimized by the control efforts, so that all possible effects on the desired model reference tracking, due to these random fluctuations and external disturbances, could be attenuated as small as possible from the stochastic H_∞ game perspective.

The random parametric fluctuations, external disturbances, and uncertain initial conditions are considered as one player to maximize the tracking error, while the control of resource management is considered as another player to minimize the model reference tracking error, from the stochastic minimax H_∞ game theory perspective. This stochastic minimax H_∞ game problem could be equivalent to a robust minimax tracking problem, to achieve a prescribed reference output, in spite of the worst-case effect of parametric fluctuations, external disturbances, and uncertain initial condition. Thus, solving the stochastic minimax H_∞ game problem for the robust reference control of nonlinear stochastic CO_2 emission system will need to solve the Hamilton–Jacobi–Issac inequality (HJII). At present, there is no analytic or numerical solution for the HJII except simple cases. To avoid solving the HJII for the nonlinear stochastic minimax H_∞ game problem, a Takagi–Sugeno (T-S) fuzzy model [4] is proposed to interpolate several linearized stochastic systems at different operation points, to approximate the nonlinear dynamics of CO_2 emission system. With the help of fuzzy approximation method to simplify the nonlinear stochastic game problem, it can be easily solved by the proposed fuzzy stochastic H_∞ game approach via linear matrix inequality (LMI) technique with the help of Robust Control Toolbox in Matlab. Finally, some simulation results are given to confirm the robust model reference tracking performance of the proposed stochastic minimax H_∞ game approach for reducing the CO_2 emissions and greenhouse effect.

Mathematical Preliminaries:

Before the further analysis of the stochastic CO_2 emission dynamic system, some definition and lemma of SDE are given in the following for the convenience of stochastic H_∞ game problem description and robust resource management control design:

Definition 12.1 (*Ito* SDE) [274]:

For a given SDE,

$$\dot{x}(t) = a(x(t)) + b(x(t))n(t)$$

where $x(t)$ is a stochastic process; $a(x(t))$ and $b(x(t))$ are nonlinear functions of $x(t)$; $n(t)$ is the standard white noise with zero mean and unit variance to denote the random fluctuation, which can be considered as the derivative of Wiener process (or the Brownian motion). The Ito-type SDE (*Ito SDE*) of $x(t)$ is represented by

$$dx(t) = a(x(t)) + b(x(t))dw(t)$$

where $w(t)$ denotes the standard Wiener process, i.e. $dx(t) = n(t)dt$.

Lemma 12.1 (Ito's formula) [275,276]:
Let $x(t)$ be an Ito stochastic process in the above equation; if $f{:}\mathbb{R} \to \mathbb{R}$ is a twice continuous differentiable function of $x(t)$, then $f(x(t))$ is also a stochastic process satisfied with the following stochastic dynamic equation:

$$df(x(t)) = f'(x(t))dx(t) + \frac{1}{2}b^T(x(t))f''(x(t)b(x(t))dt$$

where $f'(x(t)) =: \frac{\partial f(x(t))}{\partial x(t)}$ and $f''(x(t)) =: \frac{\partial^2 f(x(t))}{\partial^2 x(t)}$.

12.2 Nonlinear Stochastic Dynamical Model of CO_2 Emission System under Random Parametric Fluctuation and External Disturbance

For the convenience of illustration, some ordinary nonlinear differential equations [260,273] have been proposed to model the dynamics of CO_2 emission system. Taking account of the political actions mentioned in the previous section, the modified dynamic equations by introducing a carbon tax control term are used [277,278]. The more general deterministic model deals only with a few parameters to represent the nonlinear dynamics of atmospheric CO_2 $\chi(t)$, forest area $z(t)$, and GDP $y(t)$ as follows [260,273]:

$$\begin{aligned}
\dot{\chi}(t) &= r_1\chi(t)\left(1 - \frac{\chi(t)}{s}\right) - \alpha_1 z(t) + (\alpha_2 - u_2)y(t) \\
\dot{z}(t) &= u_1 y(t) - hz(t) \\
\dot{y}(t) &= \gamma y(t) - u_3\chi(t)
\end{aligned} \tag{12.1}$$

The first equation in (12.1) is to model the CO_2 emission, in which r_1 is the emission rate, s is the carrying capacity of the atmosphere in terms of CO_2, $\alpha_1 z(t)$ denotes the removal of CO_2 from the atmosphere and is proportional to the total forest area $z(t)$, and $(\alpha_2 - u_2)y(t)$ denotes the production of CO_2 due to GDP. It is assumed that CO_2 emission increases with the economic activity term $\alpha_2 y(t)$ and decreases with the clean technology investment term $u_2 y(t)$, which are both proportional to the GDP $y(t)$. The second equation in (12.1) is to model the total area of forest, which depends on the reforestation term $u_1 y(t)$ and the forest depletion term

$hz(t)$. Consider the impact of economic activity is much larger than natural growth, the reforestation term is assumed be a fraction of GDP, i.e. with laws and incentive to promote reforestation, whereas the total area of forest decreasing is mainly due to forest logging or other economic activities. The coefficient u_1 represents the intensity of incentives directed to reforestation and the coefficient h represents the forest depletion rate, which amalgamates a variety of factors such as expansion of cattle ranching, fire, commercial logging, shifted cultivators, and colonization, among others. The third equation in (12.1) is to model the GDP. Usually, $y(t)$ is assumed to present an exponential growth with rate γ and u_3 is a tax rate which is proportional to the CO_2 emission, thus the CO_2 revenue term $u_3\chi(t)$including the effects of carbon tax and "virtual tax" – all the effects that are similar to carbon tax, i.e. energy cost rise, consumer prices rise, real wages fall and output, and employment fall [279] that can also be directly or indirectly controlled by government order. In (12.1), u_1, u_2 and u_3 are control variables to be specified, so that the state variables $\chi(t), z(t)$ and $y(t)$ can achieve their desired model reference outputs.

The nonlinear dynamic model of CO_2 emission in (12.1) has some limitations such as 1) the deterministic nature of economic growth (as expressed by GDP), 2) the difficulty in limiting the geographic area, as one country in political sense, can effect a neighboring state, 3) the absence of time-varying parameters to adapt the model to changing situation. Further, the model is too simple and some factors may be neglected, i.e. there exist some unmodeled dynamics. In order to mimic the real stochastic dynamics of CO_2 emission, the random parameter fluctuations and external disturbances should be considered in the following stochastic model:

$$\begin{aligned}\dot{\chi}(t) &= (r_1 + \Delta r_1)\chi(t)\left(1 - \frac{\chi(t)}{s}\right) - (\alpha_1 + \Delta\alpha_1)z(t)\\ &\quad + ((\alpha_2 + \Delta\alpha_2) - u_2(t))y(t) + v_1(t)\\ \dot{z}(t) &= u_1(t)y(t) - (h + \Delta h)z(t) + v_2(t)\\ \dot{y}(t) &= (\gamma + \Delta\gamma)y(t) - u_3(t)\chi(t) + v_3(t)\end{aligned} \tag{12.2}$$

where Δr_1, $\Delta\alpha_1$, $\Delta\alpha_2$, Δh and $\Delta\gamma$ denote the parametric fluctuations of the coefficients, from one country to one country and from time to time to adapt the dynamic system model to the changing situation. $v_1(t), v_2(t)$ and $v_3(t)$ denote the external disturbances, due to modeling error and environmental noise.

Suppose the random parametric fluctuations can be separated into a deterministic part and a random part as follows:

$$\Delta r_1 = \delta_1 n(t), \Delta\alpha_1 = \delta_2 n(t), \Delta\alpha_2 = \delta_3 n(t), \Delta h = \delta_4 n(t), \Delta\gamma = \delta_5 n(t) \tag{12.3}$$

where δ_i denotes the standard deviation of stochastic parametric fluctuation, and $n(t)$ denotes a standard white noise with unit variance, i.e. $\text{var}(\Delta r_1) = \delta_1^2$,$\text{var}(\Delta\alpha_1) = \delta_2^2$ and so on, i.e. the stochastic property of random parametric fluctuations is absorbed by a white noise $n(t)$, and the amplitudes of parametric fluctuations are determined by their standard deviations δ_i respectively. Then, the nonlinear stochastic dynamic model for system dynamics of CO_2 emission could be represented by

$$
\begin{bmatrix} \dot{\chi}(t) \\ \dot{z}(t) \\ \dot{y}(t) \end{bmatrix} = \begin{bmatrix} r_1\chi(t)\left(1-\frac{\chi(t)}{s}\right) - \alpha_1 z(t) + (\alpha_2 - u_2(t))y(t) \\ u_1(t)y(t) - hz(t) \\ \gamma y(t) - u_3(t)\chi(t) \end{bmatrix}
$$

$$
+ \begin{bmatrix} \delta_1\chi(t)(1-\frac{\chi}{s}) - \delta_2 z(t) + \delta_3 y(t) \\ -\delta_4 z(t) \\ \delta_5 y(t) \end{bmatrix} n(t) + \begin{bmatrix} v_1(t) \\ v_2(t) \\ v_3(t) \end{bmatrix} \tag{12.4}
$$

For the convenience of analysis and design, the above stochastic CO_2 emission dynamic system can be represented by the following Ito stochastic system [126,274]:

$$
\begin{bmatrix} d\chi(t) \\ dz(t) \\ dy(t) \end{bmatrix} = \begin{bmatrix} r_1\chi(t)\left(1-\frac{\chi(t)}{s}\right) - \alpha_1 z(t) + \alpha_2 y(t) \\ -hz(t) \\ \gamma y(t) \end{bmatrix} dt
$$

$$
+ \begin{bmatrix} 0 & -y(t) & 0 \\ y(t) & 0 & 0 \\ 0 & 0 & -\chi(t) \end{bmatrix} \begin{bmatrix} u_1(t) \\ u_2(t) \\ u_3(t) \end{bmatrix} dt + \begin{bmatrix} v_1(t) \\ v_2(t) \\ v_3(t) \end{bmatrix} dt \tag{12.5}
$$

$$
+ \begin{bmatrix} \delta_1\chi(t)(1-\frac{\chi}{s}) - \delta_2 z(t) + \delta_3 y(t) \\ -\delta_4 z(t) \\ \delta_5 y(t) \end{bmatrix} dw(t)
$$

where $w(t)$ with $w(t) = n(t)dt$ denotes a standard Wiener process or Brownian motion. Actually, the nonlinear stochastic dynamic system for CO_2 emission in (12.5) can be extended to a more general nonlinear stochastic CO_2 emission dynamic system as follows:

$$
\begin{aligned} &dx(t) = (f(x(t)) + g(x(t))u(t) + cv(t))dt + h(x(t))dw(t), \\ &x(0) = x_0 \end{aligned} \tag{12.6}
$$

where $x(t) = [x_1(t)\ldots x_n(t)]^T, u(t) = [u_1(t)\ldots u_m(t)]^T, v(t) = [v_1(t)\ldots v_n(t)]^T$ denote the state vector, control input vector, and external disturbance vector, respectively. $f(x(t)) \in R^{n\times 1}$ denotes the nonlinear interaction vector among the state variables of the CO_2 emission system. $g(x(t)) \in R^n$ denotes the control input matrix. $h(x(t)) \in R^{n\times 1}$ denotes the noise-dependent parameter fluctuation vector. In the more general model of (12.6), let $x_1(t) = \chi(t), x_2(t) = z(t), x_3(t) = y(t)$ and so on.

Consider a reference model of the stochastic CO_2 emission system in (12.6) with a desired state output as follows:

$$
\dot{x}_r(t) = A_r x_r(t) + r(t) \tag{12.7}
$$

where $x_r(t) \in R^{n\times 1}$ is the reference state vector, $A_r \in R^{n\times n}$ is a specific asymptotically stable matrix and $r(t)$ is a desired reference signal. Based on the model reference tracking control, A_r and $r(t)$ are specified beforehand by designer, so that $x_r(t)$ can represent a desired system's state output for the stochastic system of CO_2 emission in (12.6) to follow. At the steady state, (12.7) becomes $x_r(t) = A_r^{-1} r(t)$. If we specify $A_r = -I$, then $x_r(t) = r(t)$. Then, the robust model reference tracking control is to design control strategy $u(t)$ to make $x(t)$ in (12.6) track the desired $x_r(t)$, such that the tracking error $\tilde{x}(t) = x(t) - x_r(t)$ must be as small as possible, in spite of the influence of stochastic parametric fluctuations, external disturbances, and the uncertain initial condition $x(0)$ in (12.6). Since the parametric fluctuations are stochastic, external disturbance $v(t)$ and initial state $x(0)$ are uncertain, and reference signal $r(t)$ could be arbitrarily assigned by control manager, the robust model reference tracking control design should be specified, so that the worst-case effect of three uncertainties $v(t)$, $x(0)$, and $r(t)$ on the model reference tracking error $\tilde{x}(t)$ and control effort $u(t)$ should be minimized to an optimal value ρ_0^2, i.e. both the minimax H_∞ tracking and robustness against uncertainties $v(t)$, $x(0)$, and $r(t)$ should be achieved simultaneously as the following stochastic minmax H_∞ game problem:

$$\rho_0^2 = \min_{u(t)} \max_{v(t),r(t)} \frac{E\int_0^{t_f} (\tilde{x}^T(t)Q\tilde{x}(t) + u^T(t)Ru(t))dt}{E\int_0^{t_f} (v^T(t)v(t) + r^T(t)r(t))dt + \tilde{x}^T(0)\tilde{x}(0)} \le \rho^2, \forall \tilde{x}(0) \tag{12.8}$$

where E denotes the expectation, ρ^2 denotes the upper bound of ρ_0^2 in the stochastic H_∞ game, and the weighting matrices Q and R are assumed diagonal as follows:

$$Q = \begin{bmatrix} q_{11} & 0 & \cdots & 0 \\ 0 & q_{22} & & \vdots \\ \vdots & & \ddots & 0 \\ 0 & \cdots & 0 & q_{nn} \end{bmatrix}, R = \begin{bmatrix} r_{11} & 0 & \cdots & 0 \\ 0 & r_{22} & & \vdots \\ \vdots & & \ddots & 0 \\ 0 & \cdots & 0 & r_{mm} \end{bmatrix}$$

The diagonal element q_{ii} of Q denotes the punishment on the corresponding tracking error and the diagonal r_{ii} of R denotes the relative control cost. ρ^2 denotes the upper bound of the stochastic game problem in (12.8). Since the worst-case effect of $v(t)$, $x(0)$ and $r(t)$ on the tracking error $\tilde{x}(t)$ is minimized by control effort $u(t)$, from the energy point of view, the stochastic minimax H_∞ game problem in (12.8) is suitable for a robust H_∞ tracking problem in which the statistics of $v(t)$, $\tilde{x}(t)$ and $r(t)$ are unknown or uncertain, that are always met in practical control design case, for example, in the stochastic CO_2 emission system.

Remark 12.1 If $v(t)$ and $r(t)$ are all deterministic, then the expectation E in the denominator of (12.8) can be neglected.

Because it is not easy to solve the minimax stochastic H_∞ game problem in (12.8) subject to (12.6) and (12.7) directly, an upper bound ρ^2 of the stochastic minimax H_∞ game problem is proposed to formulate a sub-optimal minimax stochastic H_∞ game problem. After that, the sub-optimal minimax stochastic H_∞ game problem is solved first, then the upper bound ρ^2 is decreased as small as possible to approximate the real minimax stochastic H_∞ game ρ_0^2 of the stochastic CO_2 emission system indirectly.

Since the denominator in (12.8) is independent of $u(t)$ and is not zero, the stochastic minmax H_∞ game problem in equation (12.8) is equivalent to the following constrained stocahstic Nash quadratic game problem [23]:

$$\begin{aligned}&\min_{u(t)}\max_{v(t),r(t)} E\int_0^{t_f}(\tilde{x}^T(t)Q\tilde{x}(t)+u^T(t)Ru(t)-\rho^2 v^T(t)v(t)-\rho^2 r^T(t)r(t))dt\\&\leq \rho^2 E\{\tilde{x}^T(0)\tilde{x}(0)\},\ \forall \tilde{x}(0)\end{aligned} \tag{12.9}$$

Let us denote

$$\begin{aligned}&\min_{u(t)}\max_{v(t),r(t)} J(u(t),v(t),r(t))\\&=\min_{u(t)}\max_{v(t),r(t)} E\int_0^{t_f}(\tilde{x}^T(t)Q\tilde{x}(t)+u^T(t)Ru(t)-\rho^2 v^T(t)v(t)-\rho^2 r^T(t)r(t))dt\end{aligned} \tag{12.10}$$

From the above analysis, the stochastic minimax Nash game problem in (12.9) or (12.10) is equivalent to finding the worst-case disturbance $v^*(t)$ and reference signal $r^*(t)$ which maximize $J(u(t),v(t),r(t))$, and a minimax stochastic Nash quadratic game tracking control $u^*(t)$ which minimizes $J(u(t),v^*(t),r^*(t))$, such that the minimax Nash game value $J(u^*(t),v^*(t),r^*(t))$ is less than $\rho^2E\{\tilde{x}^T(0)\tilde{x}(0)\}$, i.e.

$$\begin{aligned}J(u^*(t),v^*(t),r^*(t))&=\min_{u(t)} J(u(t),v^*(t),r^*(t))\\&=\min_{u(t)}\max_{v(t),r(t)} J(u(t),v(t),r(t))\\&\leq \rho^2 E\{\tilde{x}^T(0)\tilde{x}(0)\},\ \forall \tilde{x}(0)\end{aligned} \tag{12.11}$$

If there exist $u^*(t)$, $v^*(t)$ and $r^*(t)$, such that the minimax stochastic Nash quadratic game problem in (12.11) is solved, then they can satisfy the suboptimal stochastic minimax H_∞ game problem in (12.8) as well. Therefore, the first step of robust minimax H_∞ game tracking control design of stochastic CO_2 emission system is to solve the following minimax stochastic Nash quadratic game problem:

$$\min_{u(t)}\max_{v(t),r(t)} J(u(t),v(t),r(t)) \tag{12.12}$$

subject to the CO_2 emission system (12.6) and the desired reference model (12.7).

After that, the next step is to check whether the condition $J(u^*(t),v^*(t),r^*(t))\leq \rho^2E\{\tilde{x}^T(0)\tilde{x}(0)\}$ is satisfied or not for any $\tilde{x}(0)$.

To solve the minimax stochastic Nash tracking game problem in (12.12), it is convenience to transform the minimax H_2 quadratic tracking game problem into an equivalent minimax stochastic H_2 quadratic regulation game problem.

Let us denote

$$F(\bar{x}(t)) = \begin{bmatrix} f(x(t)) \\ A_r x_r(t) \end{bmatrix}, \bar{x}(t) = \begin{bmatrix} x(t) \\ x_r(t) \end{bmatrix}, \bar{v}(t) = \begin{bmatrix} v(t) \\ r(t) \end{bmatrix},$$
$$G(\bar{x}(t)) = \begin{bmatrix} g(x(t)) \\ 0 \end{bmatrix}, H(\bar{x}(t)) = \begin{bmatrix} h(x(t)) \\ 0 \end{bmatrix} \tag{12.13}$$

Thus, an augmented stochastic system of (12.6) and (12.7) is obtained as follows:

$$d\bar{x}(t) = (F(\bar{x}(t)) + G(\bar{x}(t))u(t) + \bar{C}\bar{v}(t))dt + H(\bar{x}(t))dw(t),$$
$$\bar{x}(0) = \bar{x}_0 \tag{12.14}$$

where $\bar{C} = \begin{bmatrix} C & 0 \\ 0 & I \end{bmatrix}$.

Then, the minimax stochastic Nash tracking game problem in (12.12) can be rewritten as the following minimax stocahstic Nash regulation game problem:

$$\min_{u(t)} \max_{\bar{v}(t)} J(u(t), \bar{v}(t))$$
$$= \min_{u(t)} \max_{\bar{v}(t)} E \int_0^{t_f} (\bar{x}^T(t)\bar{Q}\bar{x}(t) + u^T(t)Ru(t) - \rho^2\bar{v}^T(t)\bar{v}(t))dt, \forall \bar{x}(0) \tag{12.15}$$

subject to (12.12)
where

$$\bar{Q} = \begin{bmatrix} Q & -Q \\ -Q & Q \end{bmatrix}.$$

Then, the robust minimax stochastic Nash tracking game problem in (12.11) is equivalent to the following constrained minimax stocahstic Nash regulation game problem:

$$\min_{u(t)} \max_{\bar{v}(t)} E \int_0^{t_f} (\bar{x}^T(t)\bar{Q}\bar{x}(t) + u^T(t)Ru(t) - \rho^2\bar{v}^T(t)\bar{v}(t))dt$$
$$\leq \rho^2 E\{\bar{x}^T(0)\bar{I}\bar{x}(0)\} \tag{12.16}$$

subject to (12.14)
where

$$\bar{I} = \begin{bmatrix} I & -I \\ -I & I \end{bmatrix}.$$

Theorem 12.1

The minimax stochastic Nash quadratic regulation game problem in (12.16) for robust tracking control of stochastic CO_2 emission system could be solved by the following minimax Nash tracking control u^* and the worst-case disturbance $\bar{v}^*$

$$u^*(t) = -\frac{1}{2}R^{-1}G^T(\bar{x}(t))\frac{\partial V(\bar{x}(t))}{\partial \bar{x}(t)} \tag{12.17}$$

$$\bar{v}^*(t) = \frac{1}{2\rho^2}\bar{C}^T\frac{\partial V(\bar{x}(t))}{\partial \bar{x}(t)} \tag{12.18}$$

where $V(\bar{x}(t))>0$ is the positive solution of the following HJII:

$$\begin{aligned}
&\left(\frac{\partial V(\bar{x}(t))}{\partial \bar{x}(t)}\right)^T F(\bar{x}(t)) + \bar{x}^T(t)\bar{Q}\bar{x}(t) \\
&-\frac{1}{4}\left(\frac{\partial V(\bar{x}(t))}{\partial \bar{x}(t)}\right)^T G(\bar{x}(t))R^{-1}G^T(\bar{x}(t))\frac{\partial V(\bar{x}(t))}{\partial \bar{x}(t)} \\
&+\frac{1}{4\rho^2}\left(\frac{\partial V(\bar{x}(t))}{\partial \bar{x}(t)}\right)^T \bar{C}\bar{C}^T\frac{\partial V(\bar{x}(t))}{\partial \bar{x}(t)} \\
&+\frac{1}{2}H^T(\bar{x}(t))\frac{\partial^2 V(\bar{x}(t))}{\partial \bar{x}^2(t)}H(\bar{x}(t))<0
\end{aligned} \tag{12.19}$$

with

$$E(V(\bar{x}(0))) \leq \rho^2 E\{\bar{x}^T(0)\bar{x}(0)\} \tag{12.20}$$

Proof: See Appendix A.

Since ρ^2 is the upper bound of minimax tracking problem in (12.8), based on the analysis above, the minimax stocahstic Nash tracking game control $u^*(t)$ and the worst-case disturbance $\bar{v}^*(t)$ still need to minimize the upper bound ρ^2 as the following constrained optimization problem:

$$\rho_0^2 = \min_{V(\bar{x}(t))>0} \rho^2 \tag{12.21}$$

subject to (12.19) and (12.20)

After solving a $V_0(\bar{x})$ and ρ_0^2 from the constrained optimization problem in (12.21), the solution $V_0(\bar{x})$ is substituted into (12.17) to obtain the stochastic minimax H_∞ game tracking control $u^*(t)$, for the stochastic CO_2 emission system in (12.6), to achieve the stochastic minimax H_∞ tracking of the desired reference model in (12.7), in spite of stochastic intrinsic parametric fluctuation and external disturbance.

Remark 12.2 If $\rho \to \infty$ in (12.8), i.e. the effect of $v(t)$, $r(t)$, and $\tilde{x}(0)$ on tracking error $\tilde{x}(t)$ and control effort $u(t)$ is neglected in the tracking control design procedure, then the minimax stochastic H_∞ game tracking control problem in (12.8) is reduced to the following stochastic H_2 optimal quadratic tracking control problem [8,275,280]:

$$\min_{u(t)} E\int_0^{t_f} (\tilde{x}^T(t)Q\tilde{x}(t) + u^T(t)Ru(t))dt \tag{12.22}$$

In this case, the stochastic H_2 optimal tracking control $u^*(t)$ is also given by (12.17), i.e. $u^*(t) = -\frac{1}{2}R^{-1}G^T(\bar{x}(t))\frac{\partial V(\bar{x}(t))}{\partial \bar{x}(t)}$. However, $V(\bar{x})$ of HJII in (12.19) should be replaced via solving the following HJI:

$$\begin{aligned}&\left(\frac{\partial V(\bar{x}(t))}{\partial \bar{x}(t)}\right)^T F(\bar{x}(t)) + \bar{x}^T(t)\bar{Q}\bar{x}(t)\\&-\frac{1}{4}\left(\frac{\partial V(\bar{x}(t))}{\partial \bar{x}(t)}\right)^T G(\bar{x}(t))R^{-1}G^T(\bar{x}(t))\frac{\partial V(\bar{x}(t))}{\partial \bar{x}(t)}\\&+\frac{1}{2}H^T(\bar{x}(t))\frac{\partial^2 V(\bar{x}(t))}{\partial \bar{x}^2(t)}H(\bar{x}(t)) \leq 0\end{aligned} \tag{12.23}$$

which is obtained from (12.19) but with $\rho \to \infty$. Therefore, if $\rho \to \infty$ in the stochastic minimax H_∞ game tracking problem, it is equivalent to the H_2 optimal tracking control problem. Because the effect of external disturbance $\bar{v}(t)$ is neglected by the stochastic H_2 optimal tracking control, its tracking performance will be deteriorated by the external disturbance $\bar{v}(t)$ and uncertain initial condition $\tilde{x}(0)$. Therefore, stochastic H_2 optimal tracking control design in (12.22) is not suitable for stochastic minimax H_∞ game tracking control design with external disturbance.

12.3 Robust Minimax H_∞ Game Tracking Control of the Stochastic CO_2 Emission System via Fuzzy Interpolation Method

In general, there is no analytic or numerical solution for the HJII in (12.19) to solve the constrained optimization problem in (12.21), for robust minimax H_∞ game tracking control of the stochastic CO_2 emission system in (12.6). Recently, the T-S fuzzy model has been widely applied to approximate the nonlinear system via interpolating several linearized systems at different operation points [4,74,105]. By the fuzzy interpolation method, the nonlinear minimax stochastic H_∞ game problem could be transformed to a fuzzy stochastic game problem so that the HJII in (12.19) could be replaced by a set of LMIs. In this situation, the nonlinear stochastic minimax H_∞ game problem in (12.8) could be easily solved by fuzzy stochastic H_∞ game approach for the design of minimax H_∞ game tracking control of stochastic CO_2 emission system.

Suppose the nonlinear stochastic dynamic CO_2 emission system in (12.6) can be represented by T-S fuzzy model [4]. The T-S model is a piecewise interpolation of several local linearized models through membership functions. The fuzzy model is described by fuzzy if–then rules and will be employed to deal with the nonlinear stochastic minimax H_∞ game problem for robust minimax H_∞ tracking control to achieve a desired CO_2 emission, under stochastic fluctuations, external disturbances, and uncertain initial conditions. The ith rule of fuzzy model for nonlinear stochastic system in (12.6) is of the following form [74,105]:

$$\begin{aligned}&\textit{If } z_1(t) \textit{ is } F_{i1} \text{ and}\dots \text{ and } z_g(t) \text{ is } F_{ig}, \textit{ then}\\&dx(t) = (A_i x(t) + B_i u(t) + Cv(t))dt + D_i x(t)dw(t), i = 1, \dots, L\end{aligned} \tag{12.24}$$

where F_{ij} is the fuzzy set, A_i, B_i and D_i are linearized system matrices, g is the number of premise variables, and $z_1(t), \ldots, z_g(t)$ are the fuzzy premise variables.

The fuzzy system in (12.24) is inferred as follows [4,74,105]:

$$\begin{aligned} dx(t) &= \frac{\sum_{i=1}^{L} \mu_i(z(t))\{(A_i x(t) + B_i u(t) + Cv(t))dt + D_i x(t)dw(t)\}}{\sum_{i=1}^{L} \mu_i(z(t))} \\ &= \sum_{i=1}^{L} h_i(z(t))\{(A_i x(t) + B_i u(t) + Cv(t))dt + D_i x(t)dw(t)\} \end{aligned} \tag{12.25}$$

where

$$\mu_i(z(t)) = \prod_{j=1}^{g} F_{ij}(z(t)), h_i(z(t)) = \frac{\mu_i(z(t))}{\sum_{i=1}^{L} \mu_i(z(t))}, z(t) = (z_1(t), \ldots, z_g(t))^T$$

and $F_{ij}(z_j(t))$ is the grade of membership of $z_j(t)$ in F_{ij}. We assume $\mu_i(z(t)) \geq 0$ and $\sum_{i=1}^{L} \mu_i(z(t)) > 0$. Therefore, we get

$$h_i(z(t)) \geq 0 \text{ and} \sum_{i=1}^{L} h_i(z(t)) = 1. \tag{12.26}$$

The T-S fuzzy model in (12.25) is to interpolate L local linear stochastic systems in (12.24) to approximate the nonlinear stochastic system in (12.6) via the fuzzy basis function $h_i(z(t))$. We could specify parameters A_i, B_i and D_i easily in (12.25), so that $\sum_{i=1}^{L} h_i(z(t))A_i x(t), \sum_{i=1}^{L} h_i(z(t))B_i$ and $\sum_{i=1}^{L} h_i(z(t))D_i x(t)$ can approximate the nonlinear functions $f(x(t)), g(x(t))$ and $h(x(t))$ in (12.6), respectively, by Fuzzy identification method [4]

Remark 12.3 Actually, in (12.25), other interpolation methods such as cubic spline method, can be also employed to interpolate several local linear stochastic systems to approximate the nonlinear stochastic system in (12.6), i.e. the smoothing bases $h_i(z(t))$ could be replaced by other interpolation bases of other interpolation methods.

After the nonlinear stochastic system in (12.6) is approximated by the T-S fuzzy system in (12.25), the augmented system in (12.14) can be also approximated by the following fuzzy system:

$$d\bar{x}(t) = \sum_{i=1}^{L} h_i(z(t))\{(\bar{A}_i x(t) + \bar{B}_i u(t) + \bar{C}v(t))dt + \bar{D}_i x(t)dw(t)\} \tag{12.27}$$

where $\bar{A}_i = \begin{bmatrix} A_i & 0 \\ 0 & A_r \end{bmatrix}, \bar{B}_i = \begin{bmatrix} B_i \\ 0 \end{bmatrix}, \bar{C} = \begin{bmatrix} C & 0 \\ 0 & I \end{bmatrix}, \bar{D}_i = \begin{bmatrix} D_i & 0 \\ 0 & 0 \end{bmatrix}$.

After the nonlinear augmented stochastic system in (12.14) is approximated by the T-S fuzzy system in (12.27), the nonlinear stochastic minimax H_∞ game problem in

(12.14) and (12.16) is replaced by solving the fuzzy stochastic minimax H_∞ game problem in (12.27) and (12.16).

Theorem 12.2

The stochastic minimax H_∞ game tracking control and worst-case disturbance for fuzzy stochastic Nash game problem in (12.16) subject to (12.27) are solved, respectively, as follows:

$$u^*(t) = -\sum_{i=1}^{L} h_i(z(t))R^{-1}\bar{B}_j^T P\bar{x}(t), \bar{v}^*(t) = \frac{1}{\rho^2}\bar{C}^T P\bar{x}(t) \tag{12.28}$$

where P is the positive definite symmetric solution of the following Riccati-like inequalities:

$$\begin{gathered} P\bar{A}_i + \bar{A}_i^T P + \bar{Q} - P\bar{B}_i R^{-1}\bar{B}_j^T P + \frac{1}{\rho^2}P^T\bar{C}\bar{C}^T P + D_i^T P D_j < 0;\ i,j = 1\ldots L \\ P \le \rho^2 \bar{I} \end{gathered} \tag{12.29}$$

Proof: See Appendix B

By fuzzy approximation, obviously, the HJII in (12.19) is approximated by a set of local algebraic inequalities in (12.29) and the inequality in (12.20) is also equivalent to the second inequality in (12.29). Since ρ^2 is the upper bound of minimax H_∞ game tracking control problem in (12.8), the robust minimax stochastic H_∞ game tracking control problem still needs to minimize ρ^2 as follows

$$\rho_0^2 = \min_{P>0} \rho^2 \tag{12.30}$$

subject to (12.29)

In order to solve the above constrained optimization problem in (12.30) by the conventional LMI method, the Riccati-like inequalities in (12.29) can be rewritten as the following relaxed conditions [5]:

$$\begin{gathered} P\bar{A}_i + \bar{A}_i^T P + \bar{Q} - P\bar{B}_i R^{-1}\bar{B}_i^T P + \frac{1}{\rho^2}P^T\bar{C}\bar{C}^T P + D_i^T P D_i < 0; i = j \\ P\bar{A}_i + \bar{A}_i^T P + \bar{Q} - P\bar{B}_i R^{-1}\bar{B}_j^T P + \frac{1}{\rho^2}P^T\bar{C}\bar{C}^T P + \left(\frac{D_i + D_j}{2}\right)^T P\left(\frac{D_i + D_j}{2}\right) < 0; i \ne j \\ P \le \rho^2 \bar{I} \end{gathered} \tag{12.31}$$

Then, we multiply $W = P^{-1} > 0$ to both sides of (12.31), and the Riccati-like inequalities in (12.31) can be equivalent to

$$\begin{gathered} \bar{A}_i W + W\bar{A}_i^T + W\bar{Q}W - \bar{B}_i R^{-1}\bar{B}_i^T + \frac{1}{\rho^2}\bar{C}\bar{C}^T + WD_i^T W^{-1} D_i W \le 0; i = j \\ \bar{A}_i W + W\bar{A}_i^T + W\bar{Q}W - \bar{B}_i R^{-1}\bar{B}_j^T + \frac{1}{\rho^2}\bar{C}\bar{C}^T + W\left(\frac{D_i + D_j}{2}\right)^T W^{-1}\left(\frac{D_i + D_j}{2}\right)W \le 0; i \ne j \\ \rho^2 W > \bar{I} \end{gathered}$$

or

$$\bar{A}_i W + W\bar{A}_i^T + W\bar{\bar{Q}}^T\bar{\bar{Q}}W - \bar{B}_i R^{-1}\bar{B}_i^T + \frac{1}{\rho^2}\bar{C}\bar{C}^T + WD_i^T W^{-1} D_i W \le 0; i = j$$

$$\bar{A}_i W + W\bar{A}_i^T + W\bar{\bar{Q}}^T\bar{\bar{Q}}W - \bar{B}_i R^{-1}\bar{B}_j^T + \frac{1}{\rho^2}\bar{C}\bar{C}^T + W\left(\frac{D_i + D_j}{2}\right)^T W^{-1}\left(\frac{D_i + D_j}{2}\right)W \le 0; i \ne j$$

$$\rho^2 W > \bar{I}$$

where $\bar{\bar{Q}} = \left[Q^{\frac{1}{2}} - Q^{\frac{1}{2}}\right]$.

By applying the Schur complements several times [23], the constrained optimization problem in (12.30) is equivalent to the following LMI-constrained optimization problem:

$$\rho_0^2 = \min_{W>0} \rho^2 \tag{12.32}$$

subject to

$$\begin{bmatrix} \bar{A}_i W + W\bar{A}_i^T - \bar{B}_i R^{-1}\bar{B}_i^T + \frac{1}{\rho^2}\bar{C}\bar{C}^T & WD_i^T & W\bar{\bar{Q}}^T \\ D_i W & -W & 0 \\ \bar{\bar{Q}}W & 0 & -I \end{bmatrix} < 0; i = j$$

$$\begin{bmatrix} \bar{A}_i W + W\bar{A}_i^T - \bar{B}_i R^{-1}\bar{B}_i^T + \frac{1}{\rho^2}\bar{C}\bar{C}^T & W\left(\frac{D_i+D_j}{2}\right)^T & W\bar{\bar{Q}}^T \\ \left(\frac{D_i+D_j}{2}\right)W & -W & 0 \\ \bar{\bar{Q}}W & 0 & -I \end{bmatrix} < 0; i \ne j \tag{12.33}$$

$$\rho^2 W > \bar{I}$$

Remark 12.4

1. The fuzzy basis function $h_i(z(t))$ in (12.25) can be replaced by other interpolation function, for example, cubic spline function.
2. By fuzzy approximation, the HJII in (12.19) of nonlinear stochastic game problem for the minimax stochastic H_∞ game tracking of nonlinear stochastic CO_2 emission system is replaced by a set of Riccati-like inequalities in (12.29), which can be easily solved by the LMI-constrained optimization in (12.33).
3. The constrained optimization to solve ρ_0 and W_0 in (12.32), (12.33), can be easily solved by decreasing ρ^2 until there exists no $W>0$ solution in (12.32), (12.33).
4. After solving W_0 and then $P_0 = W_0^{-1}$ from the constrained optimization problem in (12.32), (12.33), ρ_0 can be solved by Robust Control Toolbox in Matlab efficiently.
5. If the conventional H_2 optimal quadratic tracking control in (12.22) is considered, i.e. the effect of disturbance $v(t)$ is not considered in the management control design problem, then the H_2 optimal quadratic 1 tracking control problem is

equivalent to letting $\rho^2 = \infty$ in (12.8). The H_2 optimal fuzzy quadratic tracking control design $u^*(t) = -\sum_{i=1}^{L} h_i(z(t))R^{-1}\bar{B}_j^T P_0 \bar{x}(t)$ can be solved by a common positive definite symmetric matrix P_0 from the inequalities in (12.29) with $\rho^2 = \infty$, i.e. solving a common positive definite symmetric matrix $P_0 > 0$ from the following Riccati-like inequalities [23]

$$P\bar{A}_i + \bar{A}_i^T P + \bar{Q} - P\bar{B}_i R^{-1}\bar{B}_j^T P + D_i^T P D_j < 0; i, j = 1 \ldots L \tag{12.34}$$

or the following relaxed conditions [5]:

$$\begin{aligned} &P\bar{A}_i + \bar{A}_i^T P + \bar{Q} - P\bar{B}_i R^{-1}\bar{B}_i^T P + D_i^T P D_i < 0; i = j \\ &\bar{A}_i + \bar{A}_i^T P + \bar{Q} - P\bar{B}_i R^{-1}\bar{B}_j^T P + \left(\frac{D_i + D_j}{2}\right)^T P \left(\frac{D_i + D_j}{2}\right) < 0; i \neq j \end{aligned}$$

In order to solve the stochastic H_2 optimal quadratic tracking problem by LMI technique, the stochastic H_2 optimal quadratic tracking control is equivalent to solving a common $W_0 = P_0^{-1}$ from the following inequalities:

$$\begin{aligned} &\bar{A}_i W + W\bar{A}_i^T + W\bar{\bar{Q}}^T I \bar{\bar{Q}} W - \bar{B}_i R^{-1}\bar{B}_i^T + W D_i^T W^{-1} D_i W < 0; i = j \\ &\bar{A}_i W + W\bar{A}_i^T + W\bar{\bar{Q}}^T I \bar{\bar{Q}} W - \bar{B}_i R^{-1}\bar{B}_j^T + W\left(\frac{D_i + D_j}{2}\right)^T W^{-1} \left(\frac{D_i + D_j}{2}\right) W < 0; i \neq j \end{aligned}$$

or the following LMIs:

$$\begin{aligned} &\begin{bmatrix} \bar{A}_i W + W\bar{A}_i^T - \bar{B}_i R^{-1}\bar{B}_i^T & W D_i^T & W\bar{\bar{Q}}^T \\ D_i W & -W & 0 \\ \bar{\bar{Q}} W & 0 & -I \end{bmatrix} < 0; i = j \\ &\begin{bmatrix} \bar{A}_i W + W\bar{A}_i^T - \bar{B}_i R^{-1}\bar{B}_i^T & W\left(\frac{D_i + D_j}{2}\right)^T & W\bar{\bar{Q}}^T \\ \left(\frac{D_i + D_j}{2}\right) W & -W & 0 \\ \bar{\bar{Q}} W & 0 & -I \end{bmatrix} < 0; i \neq j \end{aligned} \tag{12.36}$$

i.e., if $W_0 = P_0^{-1}$ is solved from (12.36), then the stochastic H_2 optimal fuzzy quadratic tracking control can be obtained as

$$u^*(t) = -\sum_{j=1}^{L} h_j(z(t))R^{-1}\bar{B}_j^T P_0 \bar{x}(t)$$

According to the analysis above, the minimax stochastic H_∞ game tracking control of CO_2 emission system via fuzzy interpolation method is summarized as follows.

Design Procedure:

Step 1. Give a desired reference model in (12.7) for the stochastic CO_2 emission system in (12.6).

Step 2. Select membership functions and construct fuzzy plant rules in (12.24) and (12.25).

Step 3. Give the weighting matrices Q and R of minimax stochastic H_∞ game tracking problem in (12.8).

Step 4. Solve the LMI-constrained optimization in (12.32), (12.33) to obtain W (thus $P_0 = W_0^{-1}$) and ρ_0^2.

Step 5. Construct the minimax stochastic H_∞ game tracking control $u^*(t)$ in (12.28).

Remark 12.5 The software package such as Robust Control Toolbox in Matlab can be employed to solve the LMI-constrained optimization problem in (12.32), (12.33) easily.

12.4 Computational Simulation Example

Consider the stochastic CO_2 emission system in (12.5). The values of system parameters are given in Table 12.3 to fit the actual CO_2 emission in Western Europe [260]. In order to emphasize the influence of disturbances on the CO_2 emission system, the bounded standard deviations are assumed that $\delta_1 = r_1, \delta_2 = \alpha_1, \delta_3 = \alpha_2, \ \delta_4 = h, \delta_5 = \gamma$, i.e. the standard deviations of parametric fluctuations are equal to the original system parameters; $w(t)$ is a standard Wiener process with unit variance. The environmental disturbances $v_1(t) \sim v_3(t)$ are unknown but bounded signals. For the convenience of simulation, $v_1(t) \sim v_3(t)$ are assumed to be zero mean white noise with variances equal to $10^2, 10^3$ and 10^4, respectively. To simulate the dynamic CO_2 emission in Western Europe, the initial values in 1960 are given as $\chi(t) = 398$ million tones of CO_2 emission [264], $z(t) = 43$ million m^3 of conifer forest area [281,282], and $y(t) = 2787$ billion international dollars of GDP [283–285]. The control efforts were assumed to be invariant from 1960 to 2010, i.e. $u_1 = 0.00012, u_2 = 0.0008, u_3 = 0$, to fit the actual data [260] (Fig. 12.2). But these constant control efforts would limit the system behavior too rigid for actual performance demand, which could not guarantee the control ability of CO_2 emission under disturbances (Fig. 12.3). In order to attenuate the effect of stochastic disturbance on CO_2 emission system and make a flexible control design for actual demand immediately, the minimax stochastic H_∞ game tracking control strategy will be applied after 2010.

TABLE 12.3

Model parameters for Western Europe [260].

Parameters	Value	Parameters	Value
r_1	0.15	u_3	0
s	700	γ	0.035
h	0.0001	α_1	0.0006
u_1	0.00012	α_2	0.00005
u_2	0.0008		

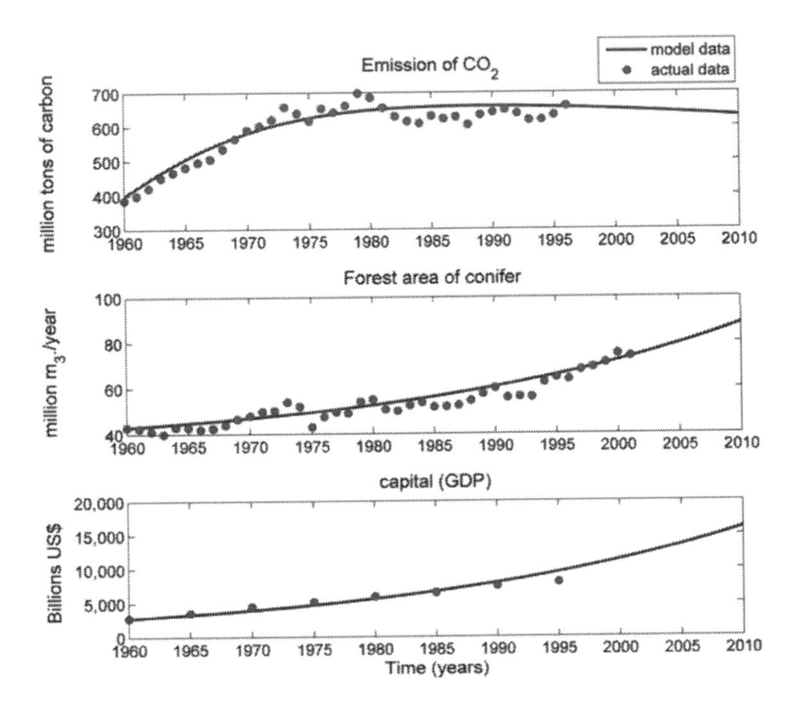

FIGURE. 12.2
Simulation and comparison between model and actual data. To fit the actual data, the invariant control efforts, i.e. reforestation u1, clean technology u_2, and CO_2 tax u_3, are assumed to be 0.012%, 0.08%, and 0%, respectively [260].

For the robust minimax stochastic H_∞ game tracking control purpose, the reference model design requests a prescribed trajectory behavior for CO_2 emission system. Thus, the system matrix A_r and reference signal $r(t)$ should be specified, based on some standards in prior, to determine the transient response and steady state of the reference model, so that the desired reference signal can perform as a guideline for the model reference tracking control system, for example, if the real parts of eigenvalues of A_r are more negative, the model reference tracking control system follows a trajectory prescribed by $r(t)$ sooner. In Europe, consider the historical data starting at 1960, it is reasonable to assume an average growth rate of GDP around 3.5%, and the present growth rate of GDP is around 4% for Europe. Moreover, the change in total forest cover from 1990 to 2000 was positive due to reforestation, but corresponding to only 0.3% per year [260]. Thus, for the purpose of robust resource management control for CO_2 emission and efficient reduction of greenhouse effect, the reference model is set via the following

$$A_r \begin{bmatrix} -1 & 0 & 0 \\ 0 & -0.1 & 0 \\ 0 & 0 & -0.1 \end{bmatrix}, \; r(t) = \begin{bmatrix} 398 \\ 19.31 \\ 3514.14 \end{bmatrix}$$

and the initial state value in 2010 as $x_r(0) = x(0)$ to simulate the desired progressive process of clean technology improvement, forest expansion, and GDP increase after

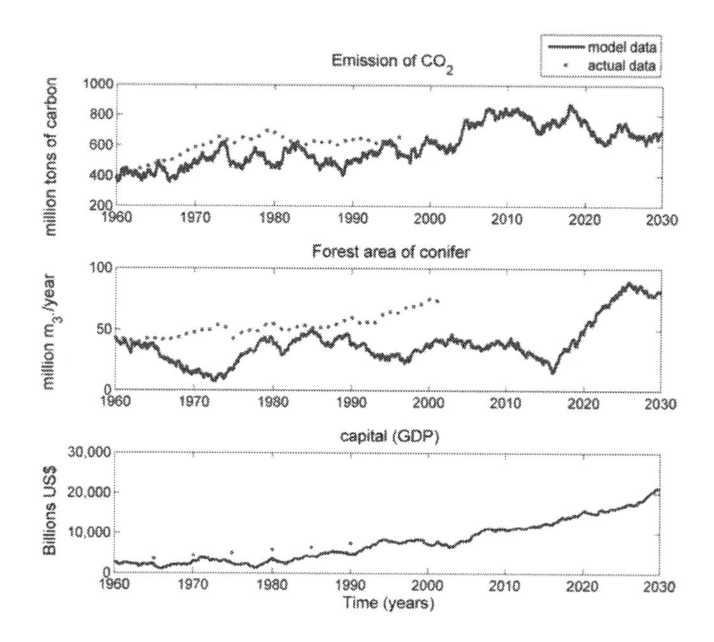

FIGURE. 12.3
The CO_2 emission system with invariant control efforts under stochastic disturbance. It is seen that the control ability would not be guaranteed under parametric fluctuations and environmental noises.

2010. Therefore, based on the reference model, the CO_2 concentration could be decreased to the value in 1960, and GDP could reach a desired steady state that is prescribed without limiting the growth of GDP (Fig. 12.4), i.e. the GDP growth cannot be less than the original GDP growth rate 4% in Europe. And the expansion rate of forested area can also be higher than 0.3% until reaching an appropriate value.

To avoid solving the HJII in Theorem 12.1, the T-S fuzzy model is employed to approximate the nonlinear stochastic CO_2 emission system described in the above section. For the convenience of management control design, each state is taken with 3 operation points respectively, and triangle-type membership functions are taken for the T-S fuzzy model 27 Rules (Fig. 12.5). In order to accomplish the robust minimax stochastic H_∞ game tracking control performance of the desired reference signal, in spite of the worst influence of stochastic parametric fluctuation, environmental noise, and with a parsimonious control efforts, a set of weighting matrices Q and R are tuned up as follows:

$$Q = \begin{bmatrix} 10^{-4} & 0 & 0 \\ 0 & 10^{-4} & 0 \\ 0 & 0 & 10^{-4} \end{bmatrix}, \quad R = \begin{bmatrix} 10^{4} & 0 & 0 \\ 0 & 10^{4} & 0 \\ 0 & 0 & 10^{4} \end{bmatrix}$$

i.e. with a heavy penalty on the control effort and a light penalty on the tracking error in (12.8).

After that, the LMI-constrained optimization problem in (12.32) and (12.33) for the robust minimax stochastic H_∞ game tracking control can be solved by using Matlab

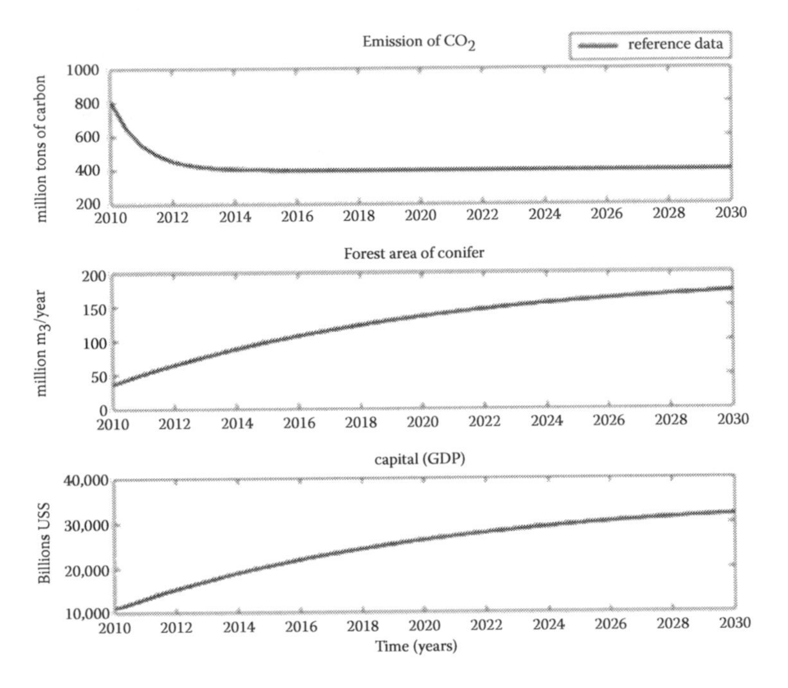

FIGURE. 12.4
The desired trajectory of reference model for CO_2 emission system, from 2010 to 2030. The tracking control goal is to decrease CO_2 to the value in 1960, i.e. $\chi(t) = 398$ million tones and make sure the increasing rate of forest area and GDP are both higher than 4% in each year.

Robust Control Toolbox. Finally, a minimum $\rho_0 = 1.39$ and the associated common positive definite symmetric matrix P_0 can be obtained as follows:

$$P_0 = \begin{bmatrix} 0.0009 & 0.0007 & 0.0000 & -0.0009 & -0.0007 & -0.0000 \\ 0.0007 & 0.00015 & 0.0000 & -0.0007 & -0.00015 & -0.0000 \\ 0.0000 & 0.0000 & 0.0161 & -0.0000 & -0.0000 & -0.0161 \\ -0.0009 & -0.0007 & -0.0000 & 0.0022 & 0.0010 & 0.0000 \\ -0.0007 & -0.00015 & -0.0000 & 0.0010 & 0.0030 & 0.0000 \\ -0.0000 & -0.0000 & -0.0161 & 0.0000 & 0.0000 & 0.0252 \end{bmatrix}$$

Thus, the minimax stochastic H_∞ game tracking management control is designed according to these imperative parameters and matrix, i.e. the management control $u^*(t) = -\sum_{j=1}^{27} h_j(z(t))R^{-1}\bar{B}_j^T P_0\bar{x}(t)$ is employed for the stochastic CO_2 emission system in (12.5) to track the desired reference signal to the end (Fig. 12.6). Fig. 12.7 shows the responses of the controlled CO_2 emission system with the stochastic minimax H_∞ game tracking management control. As the CO_2 emission target is approached, both investments in reforestation and clean technology tend to decrease, and a positive carbon tax revenue could be achieved in the end. From the simulation results, it is seen that the effect of intrinsic parametric fluctuations and external disturbances on the reference model tracking of CO_2 emission system can be overcome efficiently by the proposed

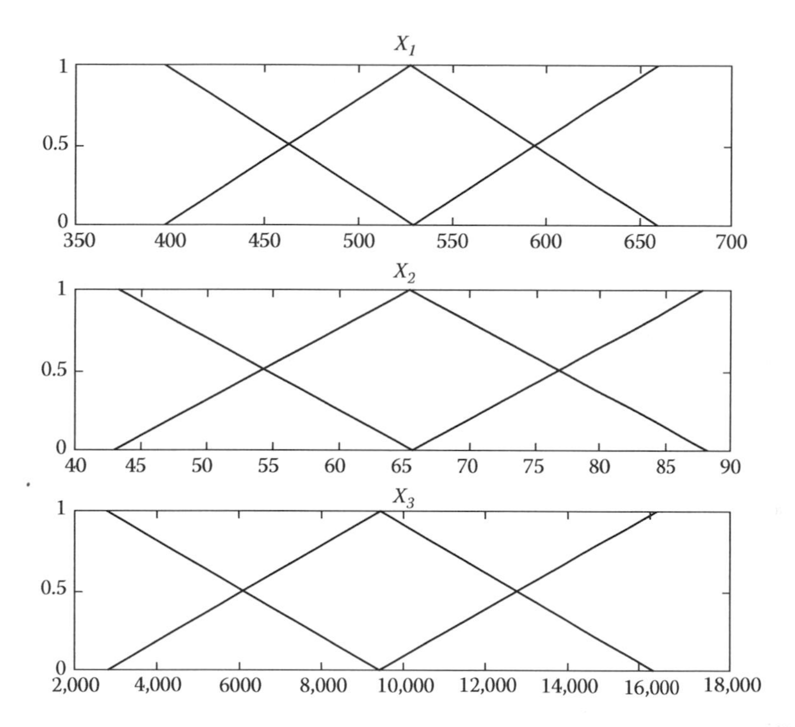

FIGURE. 12.5
Membership functions for the three states x_1, x_2, and x_3 in the CO_2 emission system.

stochastic minimax H_∞ game tracking management control design. Thus, the tracking performance of the stochastic minimax H_∞ game tracking management control via T-S fuzzy interpolation is quite satisfactory.

12.5 Discussion

From the computer simulation, it is shown that the CO_2 emission system with invariant control efforts can fit the actual data perfectly from 1960 to 2010, but could not guarantee its performance under intrinsic or external disturbances (Fig. 12.3). To achieve actual demands, i.e. the system can track an appropriate reference model as soon as possible without limiting GDP growth and forest area increase to guarantee CO_2 emission could decrease under disturbances or modeling errors, the robust model reference tracking control is proposed from a stochastic dynamic minimax H_∞ game theory perspective, which can be efficiently solved by fuzzy stochastic game approach.

By employing the stochastic minimax H_∞ game tracking management controls $u^*(t)$ (Fig. 12.6) instead of using the invariant controls from 2010 to 2030, the stochastic minimax H_∞ game tracking performance is guaranteed, no matter what stochastic property of external disturbance $v(t)$ and what value of the uncertain initial condition $x(0)$. Within the controlled period (2010–2030) (Fig. 12.7), the government and

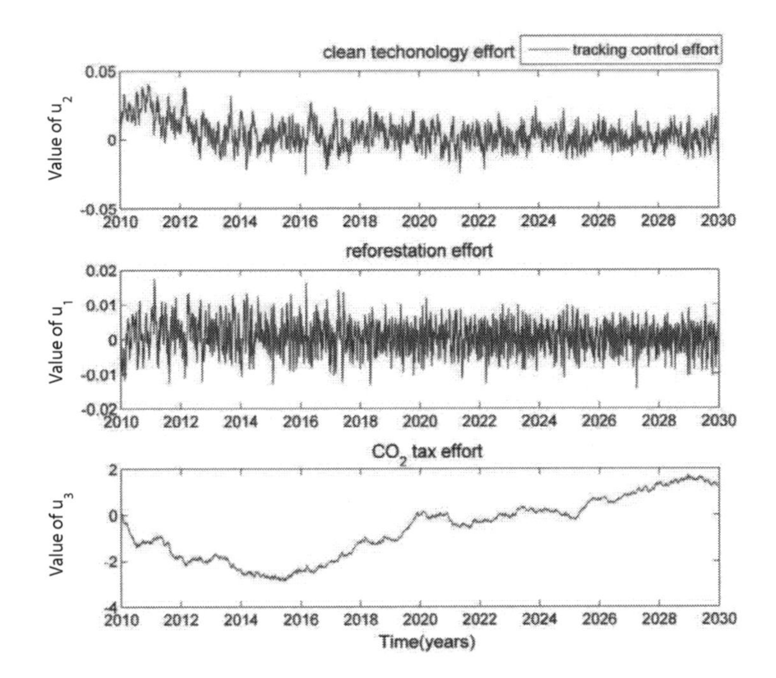

FIGURE. 12.6
The stochastic minimax H_∞ game tracking control in the simulation example.

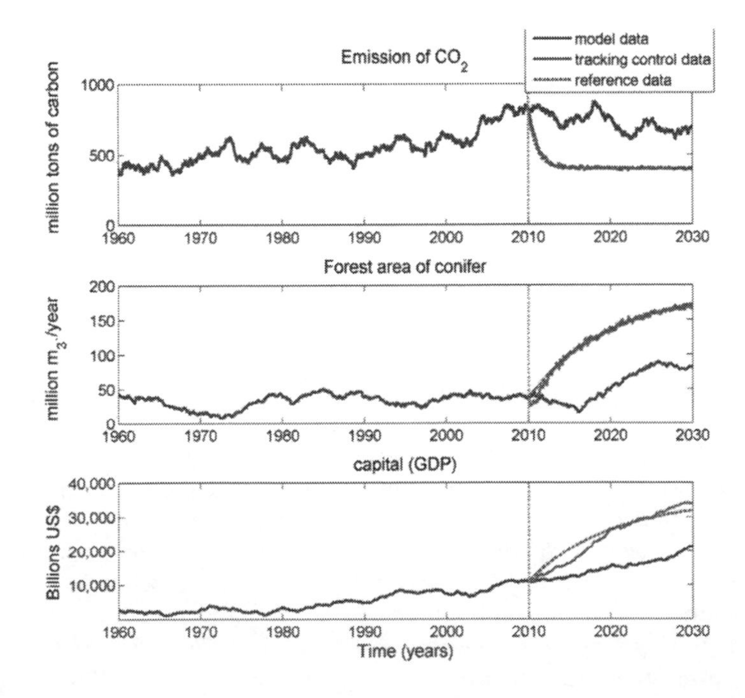

FIGURE. 12.7
The tracking performance of CO_2 emission system to a desired reference model by the stochastic minimax H_∞ game tracking management control, under the influence of parametric fluctuations and environmental noises.

companies following the stochastic minimax H_∞ game tracking management control $u^*(t)$ can reduce CO_2 emission to a desired value without limiting forest area and GDP increasing by managing expenditure in clean technology $u_2(t)$, reforestation $u_1(t)$, and CO_2 revenue $u_3(t)$.

From Fig. 12.6, the management control efforts of reforestation $u_1(t)$, and the investment in innovating or improving clean technology $u_2(t)$, in simulation increase acutely during the early stage and then converge to the stable values, which clearly point out the urgent need of a concerted effort of reforestation and clean technology to change the CO_2 emission dynamics into a desired reference trajectory. From the perspective of desired economic development, at the beginning, companies produced more pollutions (which means they emit more CO_2) may also have more economic contributions to GDP growth [286] and then GDP deteriorates because of the cost of global warming and greenhouse effect until a balance. The control effort of CO_2 revenue $u_3(t)$ in our simulation also interprets this situation reasonably by an acute descent to negative and then a gentle climbing to positive, which means to ensure the unlimited GDP growth. By following a desired reference model, the government should provide a financial subsidy to improve industrial throughput in early years, even if it creates more pollution, until the scale enterprises can bear the loss of carbon tax.

By tuning the weighting matrices of error punishment Q and management control cost R, more situation in reality can be taken into account, because the cost of management control may not be invariant when the stochastic minimax H_∞ game tracking management control starts. In this chapter, it is shown that the tracking error is punished by a low Q and a high control cost R, which means to guarantee the stochastic minimax H_∞ game tracking control performance, the management control strategy can endure more tracking error by using less control, thus making the management control strategy efficiently and viably.

If the CO_2 emission model is free of external disturbance, i.e. $c = 0$, robust tracking control performance has a lower suboptimal upper bound $\rho_0 = 0.41$. It implies that if the measurements of states are more accurate, and the controlled CO_2 emission system could track the desired reference trajectory more sophistically, then the management control design for the CO_2 emission system is more precise.

Although international cooperation from tradable quotas and permits can reduce CO_2 emission efficiently, uncertainties about compliance costs have caused countries to withdraw from negotiations. Without tuning any system parameters, these time-invariant control efforts could make the CO_2 emission system too rigid to respond for an immediate need or lead the CO_2 emission system toward an uncontrollable circumstance under intrinsic random fluctuation and external disturbance, which may finally lose its control ability for actual dynamics of CO_2 emission system. Optimal control method without taking account of the effect of intrinsic uncertainties and external disturbances in the design procedure could even not guarantee the tracking control performance. If the more flexible CO_2 emission targets can be made to incorporate an optimum choice of investments with a minimum impact on the GDP growth, i.e. if taking account of the stochastic disturbances with respect to stochastic minimax H_∞ game management control problem, then climate agreements for reducing greenhouse effect may become more attractive and efficient [279,287].

In this chapter, the fuzzy interpolation technique is employed to approximate the nonlinear stochastic CO_2 emission system, so that LMIs technique is used to efficiently solve the nonlinear minimax stochastic H_∞ game tracking problem to simplify our robust minimax management control design procedure of nonlinear stochastic CO_2

emission. Since the proposed stochastic minimax H_∞ game tracking management control design can efficiently control the CO_2 emission in real time to protect environment from the global warming and reduce greenhouse effect, in the future, the applications of robust stochastic minimax H_∞ game tracking management control design for environmental resource conservation and pollution control under stochastic intrinsic random fluctuation and external disturbance would be potential in the robust management control strategies of ecological and economic field.

12.6 Conclusions

If current GHG concentrations remain constant, the world would be committed to several centuries of increasing global mean temperatures and sea level rise. Slowing such climate change requires overcoming inertia in political, technological, and geophysical systems. To efficiently manage the resources commitment for decreasing the atmospheric CO_2, mathematical methods have been proposed to help people make decision. However, the important thing of how to ensure the desired CO_2 emission performance under stochastic disturbances is still in infancy.

In this chapter, based on robust H_∞ tracking control theory and stochastic game theory, a nonlinear stochastic minimax H_∞ game tracking control problem is suboptimally solved by indirectly solving a constrained nonlinear robust minimax Nash quadratic regulation game problem, for controlling the CO_2 emission system to achieve a desired time response under the influence of random parametric fluctuations, environmental noises, and unknown initial conditions.

To solve the nonlinear HJI-constrained problem for the stochastic minimax H_∞ game tracking control design is generally difficult. Instead of solving the HJI-constrained problem, a fuzzy stochastic game approach is proposed to transform this nonlinear stochastic minimax H_∞ game tracking control problem into a set of equivalent local linear robust minimax Nash game problems. Such transformation allows us to solve an equivalent LMI-constrained optimization problem for this stochastic minimax H_∞ game tracking control design in an easier way with the help of Robust Control Toolbox in Matlab.

This stochastic minimax H_∞ game tracking control strategy not only considers the random parametric fluctuation and environmental noise but also guarantees the robust tracking performance in a suboptimal condition. And the unknown initial condition of the CO_2 emission system is also considered as a random factor; thus, this method can be used to control the CO_2 emission system tracking around any feasible reference model whenever the management control of this CO_2 emission system starts. The fact that the theoretical stochastic minimax H_∞ game management strategy rests on the conservative suboptimal method doesn't frustrate its potential as a government policy guideline and the power of prediction in the public decision-making. Once these obstacles have been surmounted, i.e. a more rapid response by real-time monitor via e-government implementation, this stochastic minimax H_∞ game strategy would be powerful to control and manage the economic and ecological resource. What is more is that for its convenient and efficient control design for stochastic nonlinear systems with parametric fluctuation and stochastic uncertainties, this stochastic minimax H_∞ game strategy can be applied in other fields with similar demands.

12.7 Appendix

12.7.1 Appendix A: Proof of Theorem 12.1

For the augmented stochastic system in (12.14), let us denote a Lyapunov energy function $V(\bar{x}(t))>0$ for $\bar{x}(0) \neq 0$ with $V(0) = 0$. Then, the Nash regulation game problem in (12.15) is equivalent to the following minimax Nash quadratic game problem:

$$\begin{aligned} &\min_{u(t)} \max_{\bar{v}(t)} J(u(t), \bar{v}(t)) \\ &= \min_{u(t)} \max_{\bar{v}(t)} E\{V(\bar{x}(0)) - V(\bar{x}(t_f)) \\ &+ \int_0^{t_f} (\bar{x}^T(t)\bar{Q}\bar{x}(t) + u^T(t)Ru(t) - \rho^2\bar{v}^T(t)\bar{v}(t) + \frac{dV(\bar{x}(t))}{dt})dt\}, \forall\bar{x}(0) \end{aligned} \tag{12.A1}$$

By Ito's formula in Lemma 12.1 [275,276], we get

$$\begin{aligned} \frac{dV(\bar{x}(t))}{dt} &= \left(\frac{\partial V(\bar{x}(t))}{\partial \bar{x}(t)}\right)^T \frac{d\bar{x}(t)}{dt} + \frac{1}{2}H^T(\bar{x}(t))\frac{\partial^2 V(\bar{x}(t))}{\partial \bar{x}^2(t)}H(\bar{x}(t)) \\ &= \left(\frac{\partial V(\bar{x}(t))}{\partial \bar{x}(t)}\right)^T (F(\bar{x}(t)) + G(\bar{x}(t))u(t) + \bar{C}\bar{v}(t))dt + H(\bar{x}(t))\frac{dw(t)}{dt}) \\ &+ \frac{1}{2}H^T(\bar{x}(t))\frac{\partial^2 V(\bar{x}(t))}{\partial \bar{x}^2(t)}H(\bar{x}(t)) \end{aligned} \tag{12.A2}$$

Subsitituting (12.A2) to (12.A1) and by the fact that $E\left\{\frac{dw(t)}{dt}\right\} = 0$, *we get*

$$\begin{aligned} &\min_{u(t)} \max_{\bar{v}(t)} J(u(t), \bar{v}(t)) \\ &= \min_{u(t)} \max_{\bar{v}(t)} E\{V(\bar{x}(0)) - V(\bar{x}(t_f)) \\ &+ \int_0^{t_f} \left(\bar{x}^T(t)\bar{Q}\bar{x}(t) + u^T(t)Ru(t) - \rho^2\bar{v}^T(t)\bar{v}(t) + \left(\frac{\partial V(\bar{x}(t))}{\partial \bar{x}(t)}\right)^T F(\bar{x}(t)) \right. \\ &+ \left(\frac{\partial V(\bar{x}(t))}{\partial \bar{x}(t)}\right)^T G(\bar{x}(t))u(t) + \left(\frac{\partial V(\bar{x}(t))}{\partial \bar{x}(t)}\right)^T \bar{C}\bar{v}(t)) \\ &\left. + \frac{1}{2}H^T(\bar{x}(t))\frac{\partial^2 V(\bar{x}(t))}{\partial \bar{x}^2(t)}H(\bar{x}(t))\right)dt\}, \forall\bar{x}(0) \\ &= \min_{u(t)} \max_{\bar{v}(t)} E\{V(\bar{x}(0)) - V(\bar{x}(t_f)) \\ &+ \int_0^{t_f} \left(\left(\frac{\partial V(\bar{x}(t))}{\partial \bar{x}(t)}\right)^T F(\bar{x}(t)) + \bar{x}^T(t)\bar{Q}\bar{x}(t) - \frac{1}{4}\left(\frac{\partial V(\bar{x}(t))}{\partial \bar{x}(t)}\right)^T G(\bar{x}(t))R^{-1}G^T(\bar{x}(t))\frac{\partial V(\bar{x}(t))}{\partial \bar{x}(t)}\right. \end{aligned}$$

$$+\frac{1}{4\rho^2}\left(\frac{\partial V(\bar{x}(t))}{\partial \bar{x}(t)}\right)^T \bar{C}\bar{C}^T \frac{\partial V(\bar{x}(t))}{\partial \bar{x}(t)} + \frac{1}{2}H^T(\bar{x}(t))\frac{\partial^2 V(\bar{x}(t))}{\partial \bar{x}^2(t)}H(\bar{x}(t))$$
$$+\left(u(t)+\frac{1}{2}R^{-1}G^T(\bar{x}(t))\frac{\partial V(\bar{x}(t))}{\partial \bar{x}(t)}\right)^T R\left(u(t)+\frac{1}{2}R^{-1}G^T(\bar{x}(t))\frac{\partial V(\bar{x}(t))}{\partial \bar{x}(t)}\right)$$
$$-\left(\rho\bar{v}(t)-\frac{1}{2\rho}\bar{C}^T\frac{\partial V(\bar{x}(t))}{\partial \bar{x}(t)}\right)^T\left(\rho\bar{v}(t)-\frac{1}{2\rho}\bar{C}^T\frac{\partial V(\bar{x}(t))}{\partial \bar{x}(t)}\right)\Bigg)dt\},\forall \bar{x}(0)$$

Then, the minimax Nash quadratic game solution is given as follows:

$$J(u^*(t),\bar{v}^*(t)) = E\{V(\bar{x}(0)) - V(\bar{x}(t_f))$$
$$+\int_0^{t_f}\Bigg(\left(\frac{\partial V(\bar{x}(t))}{\partial \bar{x}(t)}\right)^T F(\bar{x}(t)) + \bar{x}^T(t)\bar{Q}\bar{x}(t) - \frac{1}{4}\left(\frac{\partial V(\bar{x}(t))}{\partial \bar{x}(t)}\right)^T G(\bar{x}(t))R^{-1}G^T(\bar{x}(t))\frac{\partial V(\bar{x}(t))}{\partial \bar{x}(t)}$$
$$+\frac{1}{4\rho^2}\left(\frac{\partial V(\bar{x}(t))}{\partial \bar{x}(t)}\right)^T \bar{C}\bar{C}^T \frac{\partial V(\bar{x}(t))}{\partial \bar{x}(t)} + \frac{1}{2}H^T(\bar{x}(t))\frac{\partial^2 V(\bar{x}(t))}{\partial \bar{x}^2(t)}H(\bar{x}(t))\},\forall \bar{x}(0)$$

with

$$u^*(t) = \frac{1}{2}R^{-1}G^T(\bar{x}(t))\frac{\partial V(\bar{x}(t))}{\partial \bar{x}(t)}, \bar{v}^*(t) = \frac{1}{2\rho^2}\bar{C}^T\frac{\partial V(\bar{x}(t))}{\partial \bar{x}(t)}.$$

If equation (12.19) holds, then

$$J(u^*(t),\bar{v}^*(t)) \leq EV\{(\bar{x}(0)) - V(\bar{x}(t_f))\}.$$

From the inequality in (12.16), the minimax Nash regulation game solution should be less than $\rho^2 E\{\bar{x}^T(0)\bar{I}\bar{x}(0)\}$.

After that, the equality in (12.20) is obtained as follows:

$$J(u^*(t),\bar{v}^*(t)) \leq EV\{(\bar{x}(0)) - V(\bar{x}(t_f))\} \leq EV\{(\bar{x}(0))\} \leq \rho^2 E\{\bar{x}^T(0)\bar{I}\bar{x}(0)\},\forall \bar{x}(0).$$

Q.E.D.

12.7.2 Appendix B: Proof of Theorem 12.2

For the fuzzy system in (12.27), let us denote a Lyapunov function $V(\bar{x}(t)) = \bar{x}^T(t)P\bar{x}(t) > 0$, for $\bar{x}(0) \neq 0$ with $V(0) = 0$. Then, the minimax Nash regulation game problem in (12.16) is equivalent to the following:

$$\min_{u(t)}\max_{\bar{v}(t)} J(u(t),\bar{v}(t)) = \min_{u(t)}\max_{\bar{v}(t)} E\{\bar{x}^T(0)P\bar{x}(0) - \bar{x}^T(t_f)P\bar{x}(t_f)$$
$$+\int_0^{t_f}(\bar{x}^T(t)\bar{Q}\bar{x}(t) + u^T(t)Ru(t) - \rho^2\bar{v}^T(t)\bar{v}(t) + \frac{dV(\bar{x}(t))}{dt})dt\},\forall \bar{x}(0) \tag{12.B1}$$

By Ito's formula in Lemma 12.1 [275,276], we get

$$\frac{dV(\bar{x}(t))}{dt} = 2\bar{x}^T(t)P\sum_{i=1}^{L} h_i(z(t))\left(\bar{A}_i\bar{x}(t) + \bar{B}_i u(t) + \bar{C}\bar{v}(t) + \bar{D}_i\frac{dw(t)}{dt}\right)$$
$$+\sum_{i=1}^{L}\sum_{j=1}^{L} h_i(z(t))h_j(z(t))\bar{x}^T(t)\bar{D}_i^T P\bar{D}_j\bar{x}(t) \tag{12.B2}$$

Substituting (12.B2) with (12.B1) and by the fact that $E\left\{\frac{dw(t)}{dt}\right\} = 0, weget$

$$\min_{u(t)}\max_{\bar{v}(t)} E\{\bar{x}^T(0)P\bar{x}(0) - \bar{x}^T(t_f)P\bar{x}(t_f) + \int_0^{t_f}(\bar{x}^T(t)\bar{Q}\bar{x}(t) + u^T(t)Ru(t)$$
$$-\rho^2\bar{v}^T(t)\bar{v}(t) + 2\bar{x}^T(t)P\sum_{i=1}^{L} h_i(z(t))(\bar{A}_i\bar{x}(t) + \bar{B}_i u(t) + \bar{C}\bar{v}(t))$$
$$+\sum_{i=1}^{L}\sum_{j=1}^{L} h_i(z(t))h_j(z(t))\bar{x}^T(t)\bar{D}_j^T P\bar{D}_j\bar{x}(t))dt\}$$
$$= \min_{u(t)}\max_{\bar{v}(t)} E\{\bar{x}^T(0)P\bar{x}(0) - \bar{x}^T(t_f)P\bar{x}(t_f) + \int_0^{t_f}(\bar{x}^T(t)\bar{Q}\bar{x}(t)$$
$$+\sum_{i=1}^{L} h_i(z(t))(\bar{x}^T(t)P\bar{A}_i\bar{x}(t) + \bar{x}^T(t)\bar{A}_i^T P\bar{x}(t)) + \frac{1}{\rho^2}\bar{x}^T(t)P\bar{C}\bar{C}^T P\bar{x}(t)$$
$$-\sum_{i=1}^{L}\sum_{j=1}^{L} h_i(z(t))h_j(z(t))\bar{x}^T(t)P\bar{B}_i R^{-1}\bar{B}_j^T P\bar{x}(t)$$
$$+\sum_{i=1}^{L}\sum_{j=1}^{L} h_i(z(t))h_j(z(t))\bar{x}^T(t)\bar{D}_i^T P\bar{D}_j\bar{x}(t)$$
$$+\left(Ru(t) + \sum_{i=1}^{L} h_i(z(t))\bar{B}_i^T P\bar{x}(t)\right)^T R^{-1}\left(Ru(t) + \sum_{i=1}^{L} h_i(z(t))\bar{B}_i^T P\bar{x}(t)\right)$$
$$-\left(\rho\bar{v}(t) - \frac{1}{\rho}\bar{C}^T P\bar{x}(t)\right)^T\left(\rho\bar{v}(t) - \frac{1}{\rho}\bar{C}^T P\bar{x}(t)\right)\Bigg)dt\}$$
$$= \min_{u(t)}\ \max_{\bar{v}(t)} E\{\bar{x}^T(0)P\bar{x}(0) - \bar{x}^T(t_f)P\bar{x}(t_f)$$
$$+\int_0^{t_f}\sum_{i=1}^{L}\sum_{j=1}^{L} h_i(z(t))h_j(z(t))\bar{x}^T(t)(P\bar{A}_i + \bar{A}_i^T P + Q - P\bar{B}_i R^{-1}\bar{B}_j^T P + \frac{1}{\rho^2}P\bar{C}\bar{C}^T P$$
$$+\bar{D}_i^T P\bar{D}_j)\bar{x}(t) + \left(Ru(t) + \sum_{i=1}^{L} h_i(z(t))\bar{B}_i^T P\bar{x}(t)\right)^T R^{-1}\left(Ru(t) + \sum_{i=1}^{L} h_i(z(t))\bar{B}_i^T P\bar{x}(t)\right)$$
$$-\left(\rho\bar{v}(t) - \frac{1}{\rho}\bar{C}^T P\bar{x}(t)\right)^T\left(\rho\bar{v}(t) - \frac{1}{\rho}\bar{C}^T P\bar{x}(t)\right)\Bigg)dt\}$$

The minimax Nash regulation game solution is given as follows:

$$J(u^*(t), \bar{v}^*(t)) = E\{\bar{x}^T(0)P\bar{x}(0) - \bar{x}^T(t_f)P\bar{x}(t_f) + \int_0^{t_f} \sum_{i=1}^{L}\sum_{j=1}^{L} h_i(z(t))h_j(z(t))\bar{x}^T(t)$$

$$\cdot (P\bar{A}_i + \bar{A}_i^T P + Q - P\bar{B}_i R^{-1}\bar{B}_j^T P + \frac{1}{\rho^2}P\bar{C}\bar{C}^T P + \bar{D}_i^T P\bar{D}_j)\bar{x}(t))dt\}$$

with

$$u^*(t) = -\sum_{j=1}^{L} h_j(z(t))R^{-1}\bar{B}_j^T P\bar{x}(t), \bar{v}^*(t) = \frac{1}{\rho^2}\bar{C}^T P\bar{x}(t).$$

In order to simplify the above equation, suppose the inequalities in (12.29) hold, then

$$\begin{aligned}
&\min_{u(t)} \max_{\bar{v}(t)} J(u(t), \bar{v}(t)) = J(u^*(t), \bar{v}^*(t)) \leq E\{\bar{x}^T(0)P\bar{x}(0) - \bar{x}^T(t_f)P\bar{x}(t_f)\} \\
&\leq E\bar{x}^T(0)P\bar{x}(0) \leq \rho^2 E\{\bar{x}^T(0)\bar{I}\bar{x}(0)\}, \\
&i.e.P \leq \rho^2\bar{I}.
\end{aligned}$$

Q.E.D.

13

Noncooperative and Cooperative Energy Management Strategy of Smart Power Network

13.1 Introduction

Due to environmental concerns and the climate change, the use of renewable energy sources (RESs) such as solar panels and wind turbines is getting more and more attention recently. Since the smart grid often associates with RESs, it has become a relevant research field of the future generation in power systems. A smart grid contains many types of energy generation options such as central, distributed, and intermittent energy sources. The managers can interact with the utility or company to manage their energy consumption and reduce the cost of energy [288]. With the development of the smart grid technologies and the coverage of the smart meter [289–292], the interactions between managers do not need to be manual, but can be automatic through digital communication and intelligent structures called microgrids. Each microgrid will be linked with each other to exchange information, power, and control signal through a particular method in a smart grid network [293–294].

Conceptually, a microgrid can operate in either grid-connected mode or separated mode from the grid, i.e., the islanded mode [295,296]. Taking advantage of the ability for self-managing, the microgrids in a neighborhood can cooperate by exchanging the information and the electric power. With the cooperative characteristics of a smart grid system, each microgrid can compensate the deviations through buying and selling surplus energy of other microgrids. This mechanism will allow them to use the energy more efficiently and also reduce the burden on the environment. Considering above advantages of smart grid, a power flow management strategy for a power grid which consists of cooperating microgrids is formulated as a management problem in a model predictive control (MPC) framework to minimize the power needed from the micro gas turbines [297]. Unlike the traditional power flow management problem, some researchers have thought that the power flow management would need to be distributed [298]. Therefore, the optimal power flow (OPF) problem was dealt as the minimum cost power flow problem in the corresponding graph, and a cost-scaling push-relabel algorithm was applied to solve the OPF problem in a distributed agent system [298].

Although the RESs are good options to provide clean energy to a smart grid, the power supplied from RESs is usually fluctuating/intermittent over time because of the time-varying and unpredictable weather conditions. The power system equipped with RESs may be unstable and unpredictable due to these fluctuations and uncertainties. Recently, many kinds of research have focused on energy storage technologies to alleviate the problems arising from fluctuating power sources [299,300]. In [301], the authors formulated the power flow control problem regarding a state-space model with energy storage equipment. Further, the instantaneous flows of power in the grid network could be obtained by solving a linear quadratic Gaussian problem. Motivated by [301], the authors by [302] considered H_∞ control design to maintain the system robustness from the worst-case perspective by cooperatively sharing energy with one another. With the demand-side management (DSM), an economical way to manage demand-side energy storage systems in the smart grid was used in [303] to tolerate a certain degree of system uncertainty with some physical constraints on the stored energy level. For other related researches on the power control in smart grids, the reader can refer to [304–307].

Different from the previous studies considering the power flow management problem in a cooperative sense of the smart grid, some of the researches have focused on the situation that different participants only care about their own targets in the power flow management problem for the smart grids. An energy consumption scheduling game is formulated by [308], where the player's strategy is denoted as the daily power schedules of household appliances to minimize their cost function. Moreover, the authors also presented a distributed algorithm to find Nash equilibrium of this noncooperative game. In [309], the primary goal becomes to maximize the net value from providing power to the power grid. By contrast to a power grid, a microgrid is desirable to maximize the net gain from consuming the received power. Therefore, a market operator (MO) and a distribution network operator (DNO) are designed for a microgrids network in consideration of multiple objectives. To this end, the study in [309] proposed a Multiobjective Immune Algorithm (MOIA) to solve this multiobjective optimization problem (MOP). Considering the interactions and energy trading decisions of some geographically distributed storage units, the study in [310] formulated a noncooperative game where each storage unit's owner could decide the maximum amount of energy to sell in a local market.

In this chapter, we model a dynamic energy transfer system to describe the behavior of the power flow among the microgrids which are equipped with the storage device and the RESs in a smart grid network. There are many managers (users) such as owners of microgrids, government or energy investors involved in the dynamic energy transfer system. Different managers may expect to use the mechanism of power flow in the grid to achieve the desired energy storage levels they want and determine management strategies according to their own cost consideration. Taking into account whether the managers can communicate and would compromise with each other, we then classify management strategies of the smart grid network into noncooperative management strategy and cooperative management strategy with the proposed dynamic energy transfer system in a smart grid network.

In the noncooperative management, a manager cannot immediately get information from other managers who have different targets. This may lead to conflicting strategies in the smart grid network. Different from the noncooperative model in [310], the microgrids in our model are equipped with RESs and energy storage devices. Also, each microgrid can not only buy electricity from the main grid but also sell the surplus power to other microgrids. Since the energy generated by RESs is unpredictable and other strategies are unavailable, we assume that the noncooperative H_∞ management strategy of each manager is designed to use the power more effectively to minimize the target regulation error over all possible effects of random fluctuation and other strategies in the smart grid network. Furthermore, an indirect method is proposed to transform the noncooperative H_∞ management strategy design problem into a linear matrix inequalities (LMIs)-constrained MOP. Considering NSGA-II algorithm in [311] as the theoretical foundation [25], we further developed an LMIs-constrained multiobjective evolution algorithm (MOEA) to efficiently solve the LMIs-constrained MOP for noncooperative H_∞ management strategy of dynamic energy transfer system in the smart grid network. Note that although there exist lots of studies developed for different MOEAs to search the Pareto optimal solutions of MOP, most of them focus on the MOP with algebraic constraints [17,25,35,138,311]. Few of these studies have discussed the LMIs-constrained MOP to deal with the management strategy problem in smart grid, especially about the multiperson noncooperative H_∞ management strategy. For the cooperative H_∞ management strategy, we extend the research of [302] by assuming that the information of all managers are available for each manager and the common desired target is compromised by all managers beforehand. Thus, the designer should provide the cooperative H_∞ management strategy of all managers to minimize the effect of the unpredictable electric power supply from RESs on the regulation error robustly with parsimonious strategy effort. Further, this chapter transforms the cooperative H_∞ management strategy design problem to an LMI-constrained single-objective problem (SOP).

To illustrate the design procedure and confirm the performance of two H_∞ management strategies, two design examples of noncooperative and cooperative H_∞ management strategy are given for the dynamic energy transfer system with RESs in the smart grid network. Since managers with the cooperative H_∞ management strategy have compromised a common desired target with each other at the beginning, the management strategy in this situation will be more effective and robust than the noncooperative H_∞ management strategy with conflicting targets. But the managers would pay more cost on exchanging the information and longer compromising process for common target at the beginning. Also, they may sacrifice some of their own individual interests for the common target inevitably in the cooperative management.

The main aims of this chapter are summarized as follows:

- First, a dynamic energy transfer system is introduced for a smart grid network so that the power flow among microgrids in the smart grid network could be controlled by many managers through the storage equipment and power supply from the RESs. To achieve the desired energy storage level of each microgrid, two management strategies, i.e., noncooperative management strategy and cooperative management strategy, are considered by the managers of the smart grid in this chapter.
- Second, for the noncooperative management strategy, a novel utility function is proposed for each noncooperative manager to achieve his/her desired energy storage level and attenuate the effect of conflicting strategies of other managers and external disturbance simultaneously.
- Third, an indirect method is introduced to solve the noncooperative and cooperative management strategy design problem in the dynamic energy transfer system in a smart grid network by transforming them into an equivalent LMIs-constrained MOP and SOP, respectively.
- Fourth, an LMIs-constrained MOEA algorithm is developed to efficiently solve the LMIs-constrained MOP for multi-person noncooperative game strategy of dynamic energy transfer system in a smart grid network with the help of Matlab Toolbox.

This chapter is organized as follows: Section 13.2 introduces the energy transfer dynamic system model of each microgrid in a smart grid network. In Section 13.3, a novel utility function is proposed for the noncoopertive management strategy design problem in a smart grid network. Then, we transform the above noncooperative management problem into an LMIs-constrained MOP. An LMIs-constrained MOEA to solve the MOP for noncooperative management strategy is proposed in Section 13.4. In Section 13.5, we introduce the cooperative management strategy design problem for a smart grid network, which is transformed to an LMIs-constrained SOP. Two simulation examples for noncooperative and cooperative management strategy design in a smart grid network are given in Section 13.6. Finally, we provide the conclusion in Section 13.7.

Notation: $P>0$ ($P \geq 0$) means that P is a symmetric positive definite (symmetric positive semi-definite) matrix. $P^T(P^{-1})$ is the transpose (inverse) of matrix P. I_n denotes the n-dimensional identity matrix. $[A]_{ij}$ denotes the (i,j)-entry of the matrix A. For a diagonal matrix $B \geq 0$, $B^{1/2}$ is also a diagonal matrix with the element $[B^{1/2}]_{ii} = \sqrt{[B]_{ii}}$. Similarly, for a column vector x, $[x]_i$ represents the ith element of the vector x. The notation $\|x\|_2$ denotes the Euclidean norm for the given vector $x \in \mathbb{R}^n$, i.e., $\|x\|_2 = x^T x$ and the $|\cdot|$ denotes the cardinality of a set.

13.2 System Description of Dynamic Energy Transfer System in Smart Grid Network

In this section, we consider a smart grid network that consists of a power grid and n microgrids. The power grid is a conventional power provider to microgrid, and a microgrid may be a smart structure, such as a smart building, an industrial area, or just a charging station, that can manage and integrate energy resources by itself within its operating area. Generally, a microgrid will equip with an energy storage equipment, which can acquire extra power input from local RESs in the future. In this chapter, we assume that each microgrid is connected to the power grid. These microgrids may have connections with all other users so that power can be transferred among microgrids in the smart grid network. At the same time, there are m managers involved in this noncooperative management strategy which can individually determine the power flow among these microgrids in the smart grid network. The dynamic energy transfer system in a smart grid network with n microgrids and m managers is introduced as follows:

$$s(k+1) = As(k) + \sum_{i=1}^{m} B_i u_i(k) + d(k) + r(k), \tag{13.1}$$

where the energy storage vector of n microgrids

$$s(k) = [s_1(k) \quad s_2(k) \quad \cdots \quad s_n(k)]^T \in \mathbb{R}^n \tag{13.2}$$

denotes the energies stored in each microgrid energy storage equipment at time slot k, $d(k)$ denotes the power demand and $r(k)$ is the green supplied energy from the connected RES. $A \in \mathbb{R}^{n\times n}$ is a diagonal matrix which is termed as a storage efficiency matrix and its diagonal values depends on the physical characteristics of storage media used by the microgrids. For the basic characteristics of energy storage battery [302], we assume that $0<[A]_{ii}<1$, for $i=1\ldots n$, and $[A]_{ij}=0$, for $i\neq j$.

For the smart grid network, there are many camps or managers which can manage the power flow among the microgrids they are interested, such as the owners of microgrids, the energy investors, or the government. Similar to [302], the ith management strategy

$$u_i(k)=\left[\,u_{i,1}(k)\quad u_{i,2}(k)\quad\cdots\quad u_{i,n_i}(k)\,\right]^T\in\mathbb{R}^{n_i}\tag{13.3}$$

is designed by the ith manager with n_i interested power transfer links. A directed graph $G_i(V_i,E_i)$ is employed to describe the management strategy of the ith manager. For the convenience of illustration, an example of management model in the dynamic energy system with $n=3$ and $m=3$ in a smart grid network is given in Fig. 13.1. A graphical example G_1 of such a topology is depicted in Fig. 13.1(a) to describe the management strategy $\left[\,u_{1,1}(k)\quad u_{1,2}(k)\quad u_{1,3}(k)\,\right]^T$ of manager 1, where $V_1=\{1,2,3\}$ and $E_1=\{1,2,3\}$ with $|V_1|=3$ and $|E_1|=3$, respectively. Each vertex in V_i represents a microgrid that is connected to a local RES with a stored energy state $s_i(k)$. The ith manager will decide each edge in E_i, which represents a link that allows the power to flow between the connected grids. In Fig. 13.1, the edge is directed by an arrow to indicate the flow from start to end. The bold edges indicate the connection with a power grid, i.e., a local energy provider that could produce a stable and controlled energy. For the convenience of mathematical manipulation, a topology matrix B_i can alternatively represent the directed graph $G_i(V_i,E_i)$. Let $[B_i]_{pq}$ denote the (p,q)-entry of B_i. The matrix B_i is defined by

$$[B_i]_{pq}=\begin{cases}1, & \text{if the } q \text{ th edge ends at the } p\text{th microgrid.}\\ -1, & \text{if the } q \text{ th edge starts from the } p\text{th microgrid.}\\ 0, & \text{if the } q \text{ th edge is not to the } p\text{th microgrid.}\end{cases}\tag{13.4}$$

where $p=1,2,\ldots,|V_i|$ and $q=1,2,\ldots,|E_i|$. Although we have determined the direction of the power flow in a directed graph, the real direction can still be decided by the manager. In our scheme, the element $u_{i,n_i}(k)>0$ implies that the power flow follows the direction indicated by the n_i arrowed edge. Conversely, if $u_{i,n_i}(k)<0$, then the transmitting power will be controlled to flow with the opposite direction at time k. Finally, we denote the power demand of the smart grid network at time k by $d(k)\in\mathbb{R}^{n\times1}$, and the $r(k)\in\mathbb{R}^{n\times1}$ denotes the extra energy supplied from the connected RES.

A realistic and complete example is given in the following: Let us consider a smart grid network consists of a power grid and three microgrids with three managers involved in the dynamic energy transfer system (13.1) as shown in Fig. 13.1. We assume that manager 1 is the 1*st* microgrid owner in the residential area, and he/she expects his/her storage energy can be regulated to a particular level s_{d1} to prevent emergencies. Then he/she will just care about microgrid 1 with the power flow link he/she is interested such as the directed graph G_1 in Fig. 13.1(a), so that the management strategy $u_1(k)$ can be denoted $\left[\,u_{1,1}(k)\quad u_{1,2}(k)\quad u_{1,3}(k)\,\right]^T\in\mathbb{R}^{3\times1}$. Manager 2 is the 2*nd* microgrid owner in the industrial area and he/she expects the storage level can be maintained at a particular level s_{d2} to work in safety through the power flow mechanism of smart grid network. The power link he/she is interested in is presented as the directed graph G_2 in Fig. 13.1(b), and the management strategy $u_2(k)=\left[\,u_{2,1}(k)\quad u_{2,2}(k)\quad u_{2,3}(k)\,\right]^T\in\mathbb{R}^{3\times1}$. Manager 3 becomes the 3*rd* microgrid owner who is an energy investor. Manager 3 wishes the produced energy can be sold as much as possible. Thus, he/she may be interested in the power link such as the directed graph G_3 in Fig. 13.1(c) with a storage level s_{d3}, then the management strategy $u_3(k)$ can be denoted by $[u_{3,1}(k) \quad u_{3,2}(k) \quad u_{3,3}(k)]^T \in \mathbb{R}^{3\times1}$. Based on the directed graphs G_1, G_2, and G_3 of the smart grid network in Fig.13.1, the corresponding topology matrices $B_1 \in \mathbb{R}^{3\times3}$, $B_2 \in \mathbb{R}^{3\times3}$, and $B_3 \in \mathbb{R}^{3\times3}$ in the dynamic energy transfer system (13.1) of the smart grid network can be respectively denoted as [302]

$$B_1 = \begin{bmatrix} 1 & 1 & 1 \\ 0 & -1 & 0 \\ 0 & 0 & -1 \end{bmatrix}, B_2 = \begin{bmatrix} -1 & 0 & 0 \\ 1 & 1 & 1 \\ 0 & 0 & -1 \end{bmatrix}, B_3 = \begin{bmatrix} 1 & 0 & 0 \\ 0 & 1 & 0 \\ -1 & -1 & -1 \end{bmatrix}. \quad (13.5)$$

Finally, the whole smart grid network can be regarded as a 3-manager management model as shown in Fig. 13.1(d). As for the management strategies, noncooperative and cooperative H_∞ management strategies will be considered in this chapter for the smart grid network according to whether these managers can communicate and would compromise with each other.

13.3 The Noncooperative H_∞ Management Strategy Design for Dynamic Energy Transfer System in the Smart Grid Network

After the example of dynamic energy transfer system (13.1) in a smart grid network is given, we then want to design the noncooperative management strategies for each manager to achieve their desired targets as possible despite the effects of unpredictable supplied electric power, the uncertain power demand, and the unavailable strategies of other managers. For each manager's strategy $u_i(k)$ in the noncooperative management, all other management strategies $u_1(k), \ldots, u_{i-1}(k), u_{i+1}(k), \ldots, u_m(k)$, the unpredictable extra energy vector $w(k) = r(k) + d(k)$, where power demand $d(k)$ and the green supplied energy $r(k)$ are all unavailable for the ith manager due to the time-varying and unpredictable weather conditions. For the simplicity of analysis, all other management strategies $u_1(k), \ldots, u_{i-1}(k), u_{i+1}(k), \ldots, u_m(k)$ can be bundled together as an augmented conflicting strategy $u_{-i}(k)$ to $u_i(k)$. Since the power demand $d(k)$ and the green supplied energy $r(k)$ play the important roles in the management of the smart grid network, their effect on the target tracking control of each manager should be considered. In this situation, the role of unpredictable $w(k)$ can be seen as an extra player of the noncooperative strategy in the smart grid network. Therefore, we can represent the dynamic energy transfer system (13.1) in a smart grid network as

$$s(k+1) = As(k) + B_i u_i(k) + \bar{B}_i u_{-i}(k), \quad (13.6)$$

where

$$\begin{aligned} u_{-i}(k) &= \left[(w(k))^T \quad (u_1(k))^T \quad \cdots \quad (u_{i-1}(k))^T \quad (u_{i+1}(k))^T \quad \cdots \quad (u_m(k))^T \right], \\ \bar{B}_i &= \left[I_{n\times n} \quad B_1 \quad \cdots \quad B_{i-1} \quad B_{i+1} \quad \cdots \quad B_m \right]. \end{aligned} \quad (13.7)$$

For each noncooperative manager in (13.6) of a smart grid network, they may expect to regulate the energy storage state $s(k)$ to their desired working levels. For instance, the managers of different microgrids in Fig. 13.1 may expect the storage energy levels could achieve their desired levels; the energy investor may expect the power he/she generated can be sold to these microgrids as much as possible. The government would wish these microgrids not to be too selfish, i.e., they can work through the power flow among the microgrids to reduce the burden on the power grid. As a result, the government may want the storage energy state to be regulated to a lower storage state (or just the emergence level). Without loss of generality, we assume that the strategy of the ith manager is expected to make storage state $s(k)$ track to his/her desired working level (target) s_{di}.

When a manager applies the power flow mechanism among the microgrids in a smart grid network, the noncooperative H_∞ management strategy is employed because a manager cannot immediately know the information about how other managers transfer the power in the smart grid network. Even if he/she can get the partial information about other managers, they may also have the competing goals and different cost considerations. On the other hand, although the electric energy from RESs can effectively reduce the burden on the power grid and the environment, it is often fluctuating over time due to the weather condition. Because of the reasons mentioned above, the noncooperative H_∞ management strategy in the dynamic energy transfer system (13.1) should be more effective in the smart grid network. Hence, the following noncooperative H_∞ utility function in [2,8,275,312] is proposed as the design objective for each manager to specify noncooperative H_∞ management strategy $u_i(k)$:

$$J_\infty^i(u_i(k)) = \frac{\sum_{k=0}^{t_f} (s(k) - s_{di})^T Q_i (s(k) - s_{di}) + u_i^T(k) R_i u_i(k)}{(s(0) - s_{di})^T (s(0) - s_{di}) + \sum_{k=0}^{t_f} u_{-i}^T(k) u_{-i}(k)}, \tag{13.8}$$
$$\text{for all } i = 1, 2, \ldots, m,$$

where $Q_i > 0, R_i > 0$ are diagonal matrix selected by the manager i to denote the corresponding weighting matrices for each manager to trade off between the regulation error $(s(k) - s_{di})$ and the strategy effort of $u_i(k)$ from the view of the manager i. The physical meaning of the H_∞ utility function for noncooperative management strategy in (13.18) considers of the effect of the unpredictable factor and other conflicting strategies (i.e., $u_{-i}(k)$) and the unknown initial condition $s(0) \neq s_{di}$ on the regulation error $s(k) - s_{di}$ and control effort $u_i(k)$ (i.e. the target regulation performance of the ith manager).

The diagonal elements of Q_i denote the different degree of penalty on the corresponding regulation error, and its value can be interpreted as the penalty on manager i for not achieving his/her desired energy storage level s_{di}. By contrast, the R_i is concerned with the relative consideration of the strategy effort for the manager i, and its diagonal values can be interpreted as the strategy cost consideration on the corresponding power transfer links among the microgrids he/she is interested in. The t_f denotes the terminal time with the energy transfer system (13.1) in the smart grid network.

Since all other strategies and the unpredictable factors are unavailable in the dynamic energy transfer system (13.1) of the smart grid network, each noncooperative management strategy $u_i(k)$ of manager i attempts to simultaneously minimize the effect of the combined competitive strategy and the unpredictable vector $u_{-i}(k)$ on the target regulation error of each manager to achieve his/her desired target with parsimonious effort, i.e., to simultaneously minimize the noncooperative H_∞ utility functions in (13.8).

$$\min_{u_i(k)} J_\infty^i(u_i(k)) = \min_{u_i(k)} \frac{\sum_{k=0}^{t_f} (s(k)-s_{di})^T Q_i (s(k)-s_{di}) + u_i^T(k) R_i u_i(k)}{(s(0)-s_{di})^T (s(0)-s_{di}) + \sum_{k=0}^{t_f} u_{-i}^T(k) u_{-i}(k)}, \tag{13.9}$$

for all $i = 1, 2, \ldots, m$,

For the convenience of the noncooperative H_∞ management strategy design, the origin of the dynamic energy transfer system (13.1) should be shifted to s_{di} at first. In such situation, if the shifted energy transfer system is robustly stabilized at the origin, then the robust regulation of $s(k)$ to the desired state s_{di} will be achieved equivalently. This will simplify the noncooperative H_∞ management strategy design procedure of the smart grid system. By redefining the unpredictable vector $w_i(k) = r(k) - d(k) + (As_{di} - s_{di})$ and denoting the $\tilde{s}_i(k) = s(k) - s_{di}$ as the regulation error between the storage level and the desired level of the ith manager at time slot k, then we get the following shifted energy transfer system from the perspective of the i manager:

$$\tilde{s}_i(k+1) = A\tilde{s}_i(k) + B_i u_i(k) + \bar{B}_i \bar{u}_{-i}(k), \tag{13.10}$$

where

$$\begin{aligned} \bar{u}_{-i}(k) &= \begin{bmatrix} w_i^T(k) & (u_1(k))^T & \cdots & (u_{i-1}(k))^T & (u_{i+1}(k))^T & \cdots & (u_m(k))^T \end{bmatrix}, \\ \bar{B}_i &= [\, I_{n\times n} \quad B_1 \quad \cdots \quad B_{i-1} \quad B_{i+1} \quad \cdots \quad B_m \,]. \end{aligned} \tag{13.11}$$

Then, the noncooperative H_∞ management strategy problem in (13.9) at the shifted energy transfer system in (13.10) could be formulated as the following MOP:

$$\min_{(u_1(k), u_2(k), \ldots, u_m(k)) \in U} (J_\infty^1(u_1(k)), \ldots, J_\infty^m(u_m(k))) \tag{13.12}$$

where

$$J_\infty^i(u_i(k)) = \frac{\sum_{k=0}^{t_f} \tilde{s}_i^T(k) Q_i \tilde{s}_i(k) + u_i^T(k) R_i u_i(k)}{\tilde{s}_i^T(0)\tilde{s}_i(0) + \sum_{k=0}^{t_f} \bar{u}_{-i}^T(k) \bar{u}_{-i}(k)}, \quad i = 1, 2, \ldots, m,$$

subject to

$$\tilde{s}_i(k+1) = A\tilde{s}_i(k) + B_i u_i(k) + \bar{B}_i \bar{u}_{-i}(k), \ \textit{for all } i = 1, 2, \ldots, m,$$

and the U is the set of all the admissible strategies for the given m-manager noncooperative H_∞ strategy problem; the vector of the objective functions $(J^1_\infty(u_1(k)), J^2_\infty(u_2(k)), \ldots, J^m_\infty(u_m(k)))$ is called objective vector of noncooperative H_∞ management strategy $(u_1(k), u_2(k), \ldots, u_m(k))$.

Remark 13.1 (i) In the conventional H_∞ robust control problem [2,8,275,312], only the stabilization problem by one control strategy with the attenuation of external disturbance is considered, i.e., $s_{di} = 0, u_{-i}(k) = w(k)$ and $i = 1$ in (13.8). (ii) For the noncooperative H_∞ utility function we proposed in (13.8), the selection of different targets s_{di} with different weighting matrices Q_i and R_i in (13.8) could lead to different strategies with different levels of noncooperative behavior within these managers in the noncooperative H_∞ management strategy design problem in (13.12). However, there is still a certain level of cooperation present in these noncooperative strategies, because the individual goals of these noncooperative H_∞ management strategies in (13.8) are not in completely conflict with one another. All managers need to select their target s_{di} and specify weighting matrices Q_i and R_i according to their own consideration so that the noncooperative H_∞ management strategies have some levels of cooperative behavior to achieve their goals within the dynamic energy transfer system (13.1) in a smart grid network.

Remark 13.2 The physical meaning of the noncooperative H_∞ management strategy design problem in (13.12) with the dynamic energy transfer system (13.1) is twofold: (i) the primary goal of each manager is to minimize the energy state deviation $\tilde{s}_i(k)$ from the desired energy storage level s_{di} due to daily power consumption and storage efficiency by a parsimonious effort $u_i(k)$; (ii) since the extra power generated by RESs is intermittent over time and the information of other strategies is unavailable, the effect of both renewable energy production and the unpredictable strategies of other managers on the noncooperative H_∞ management strategy of each manager should be simultaneously minimized as possible. The aim of noncooperative H_∞ management strategy design problem for each manager in (13.12) is to design management strategies of all managers simultaneously under the effects of the unpredictable combined competitive strategies and disturbances, to robustly achieve their desired energy storage level by using more efficient power flow among the microgrids from the multiobjective H_∞ robust tracking perspective.

Clearly, it is still very difficult to solve the noncooperative H_∞ management strategy design problem in (13.12) directly. Therefore, the following indirect method is employed to simultaneously minimize the corresponding upper bounds from the suboptimal perspective. Suppose e_i is the upper bound of the utility function of the ith manager in (13.12), i.e., $J^i_\infty(u_i(k)) \leq e_i$, for $i = 1, 2, \ldots, m$, then the indirect method to solve the noncoopertive H_∞ management strategy design problem in (13.12) is formulated by the suboptimal method as the following MOP:

$$(e_1^*, e_2^*, \ldots, e_m^*) = \min_{(u_1(k), u_2(k), \ldots, u_m(k)) \in U} (e_1, e_2, \ldots, e_m) \tag{13.13}$$

$$s.t.\ J^i_\infty(u_i(k)) \leq e_i, \text{for all } i = 1, 2, \ldots, m \tag{13.14}$$

Before further discussion of how to solve the MOP in (13.13) and (13.14) for noncooperative H_∞ management strategy, some fundamental concepts of MOP are needed as follows:

Definition 13.1 *(Pareto dominance)* ([17,25,35,138,311]). For two feasible solutions $(u_1^1(k),...,u_m^1(k))$ and $(u_1^2(k),...,u_m^2(k))$ corresponding to objective values $(e_1^1,...,e_m^1)$ and $(e_1^2,...,e_m^2)$ of MOP in (13), respectively, the solution $(u_1^1(k),...,u_m^1(k))$ is said to dominate $(u_1^2(k),...,u_m^2(k))$ if $e_1^1 \le e_1^2,...,e_m^1 \le e_m^2$ and at least one of the inequalities is a strict inequality.

Definition 13.2 (Pareto optimality) ([17,25,35,138,311]). The feasible solution $(u_1(k),...,u_m(k))$ of MOP in (13.13) is said to be Pareto optimality with respect to the feasible set U if and only if there does not exist another feasible solution that dominates it.

Definition 13.3 (Pareto optimal solution set) ([17,25,35,138,311]). For the MOP in (13.13) and (13.14), Pareto optimal solution set σ^* is defined as

$$\sigma^* = \{(u_1^*(k),...,u_m^*(k))|(u_1^*(k),...,u_m^*(k)) \text{ is Pareto optimality}\}$$

where $(u_1^*(k),...,u_m^*(k))$ denotes the Pareto optimal solution of the objective values $(e_1^*,...,e_m^*)$.

Definition 13.4 (Pareto front) ([17,25,35,138,311]). For the MOP in (13.13) and (13.14) with the Pareto optimal solution set σ^*, the Pareto front P_F is defined as

$$P_F = \{(e_1^*,...,e_m^*)|(u_1^*(k),...,u_m^*(k)) \ \in \sigma^*\}.$$

Theorem 13.1

The MOP in (13.13) is equivalent to the MOP in (13.12) for noncooperative H_∞ management strategy of dynamic energy transfer system (13.1) in the smart grid network.

Proof: The proof of this theorem is straightforward. One only needs to prove that m inequalities contained on the MOP in (13.14) become equalities for Pareto optimal solutions. We show this by contradiction. Given a Pareto optimal solution $(e_1^*,e_2^*,...,e_m^*)$ of (13.13), we assume that any one of the inequalities in (13.14) remains a strict inequality at the optimal solution.

Without loss of generality, suppose that $J_\infty^i(u_i'(k)) \le e_i'$, such that $e_i' \le e_i^*$. As a result, $(e_1^*,...,e_i',...,e_m^*)$ dominates the Pareto optimal solution $(e_1^*,...,e_i^*,...,e_m^*)$, leading to a contradiction by the above definitions. This implies that m inequalities in (13.14) indeed become m equalities for Pareto optimal solutions. The MOP in (13.13) is hence equivalent to the MOP in (13.12).

Q.E.D.

For the proof of next theorem, we still need a lemma:

Lemma 13.1 (Schur complements, [23]). Suppose that $P = P^T > 0$ and $A > 0$. For any matrix B (or vectors) with an appropriate dimension, the following inequality

$$A - B^T P^{-1} B > 0, \tag{13.15}$$

is equivalent to the following LMI:

$$\begin{bmatrix} A & B^T \\ B & P \end{bmatrix} > 0, \tag{13.16}$$

Then, the following theorem will provide the sufficient condition for the noncooperative H_∞ management strategy $u_i(k) = K_i\tilde{s}_i$ for all $i = 1,...,m$, to solve the MOP in (13.13) and (13.14) for the noncooperative H_∞ management strategy design problem of the smart grid network in (13.12).

Theorem 13.2
(i) The MOP for noncooperative H_∞ management strategies in (13.13) and (13.14) could be transformed to the following LMIs-constrained MOP:

$$(e_1^*, e_2^*, \ldots, e_m^*) = \min_{\{P, K_1, K_2, \ldots, K_m\}} (e_1, e_2, \ldots, e_m) \tag{13.17}$$

subject to the following LMIs, for all $i = 1, 2, \ldots, m$

$$\begin{bmatrix} e_i I_n & I_n \\ I_n & W \end{bmatrix} \geq 0, \tag{13.18}$$

$$\begin{bmatrix} W & W & Y_i^T & 0 & \Theta_i^T \\ W & Q_i^{-1} & 0 & 0 & 0 \\ Y_i & 0 & R_i^{-1} & 0 & 0 \\ 0 & 0 & 0 & e_i I_{(n+m\times n_i)} & \bar{B}_i^T \\ \Theta_i & 0 & 0 & \bar{B}_i & W \end{bmatrix} \geq 0, \tag{13.19}$$

where $W = P^{-1} > 0, Y_i = K_i P^{-1} = K_i W$ and the matrix $\Theta_i = AW + B_i Y_i$.

(ii) If the above LMIs-constrained MOP is solved by $P^* = (W^*)^{-1}$ and $K_1^*, K_2^*, \ldots, K_m^*$ with $K_i^* = Y_i^* P^*$, then the corresponding noncoopreative H_∞ management strategies $u_i^*(k) = K_i^* \tilde{s}_i(k) = Y_i^* P^* \tilde{s}_i(k)$, for all $i = 1, 2, \ldots, m$.

Proof: See Appendix A.

In the following section, the proposed LMIs-constrained MOEA searching algorithm will be employed to efficiently solve the Pareto optimal solution $(e_1^*, e_2^*, \ldots, e_m^*)$ with the P^* and K_i^* of MOP in (13.17) for the noncooprative H_∞ management strategy design problem of dynamic energy transfer system (13.10) in the smart grid network.

13.4 LMIs-constrained MOEA Algorithm for Noncooperative H_∞ Management Strategy in Smart Grid Network

Now, the feasible solutions of LMIs-constrained MOP in (13.17) can be obtained for noncooperative H_∞ management strategies with the help of the convex optimization toolbox in MATLAB for the smart grid network. The MOEA is a stochastic algorithm inspired by biological evolution and the survival of the better fitness [25,35]. Since the algorithm can avoid local optimality with the parallel search of the Pareto optimality with multiple conflicting objectives, MOEA is suitable for solving the MOP for noncooperative H_∞ management strategy. Although many MOEAs have been proposed to solve the algebra-constrained MOPs [17,25,35,138,311], few of the researches have discussed the LMIs-constrained MOP in (13.17) for the noncooperative H_∞ management strategy of the dynamic energy transfer system in the smart grid network. Further, the conventional MOEAs search for $(K_1^*, \ldots, K_m^*)$ in MOP to achieve $(e_1^*, e_2^*, \ldots, e_m^*)$ directly. Therefore, it could not apply to the MOP in (13.17)–(13.19) because the complexity of control matrices $(K_1^*, \ldots, K_m^*)$ is not suitable for searching. After solving $(e_1^*, e_2^*, \ldots, e_m^*)$ with the help of LMI toolbox in Matlab, we could solve $W^*, Y_1^*, \ldots, Y_m^*$ to obtain control matrices $(K_1^*, \ldots, K_m^*)$ for $(u_1^*, \ldots, u_m^*)$. Thus, some modifications of the conventional MOEAs are necessary in the sense of LMIs. In this chapter, we choose the NSGA-II in [25–311] to search $(e_1^*, e_2^*, \ldots, e_m^*)$ as the foundation to develop the proposed LMIs-constrained MOEA algorithm to solve MOP in (13.17).

For the LMIs-constrained MOP in (13.17), Pareto optimal solutions $(u_1^*(k) \ldots u_m^*(k))$ can be guaranteed obviously on a given feasible set. Different from the traditional MOEA in [17,25,35,138,311], the searching region of the LMIs-constrained MOEA is the feasible set of objective values $(e_1, e_2, \ldots, e_m)$ under LMIs constraints in (13.18) and (13.19). First, the MOEA encodes the individuals $(e_1^n, e_2^n, \ldots, e_m^n)$ for all $n = 1, 2, \ldots, N_p$ as the so-called initial parent population P_1 in the feasible set where P_r is the population at the rth iteration. In other words, the nth individual $(e_1^n, e_2^n, \ldots, e_m^n)$ must satisfy the LMIs in (13.18) and (13.19) with $e_i^L \leq e_i^n \leq e_i^U, i = 1, \ldots, m$ for all $n = 1, 2, \ldots, N_p$, where N_p is the population size. e_i^L and e_i^U denote the upper and lower bounds of each e_i^n, respectively. Second, the candidate Pareto domination solutions can be obtained by the crowded-comparison operator [25,35]. Then, the child population can also be generated by crossover and mutation. Note that if some individuals are not feasible, i.e., they cannot satisfy with the requirement of LMIs in (13.18) and (13.19), these individuals need to be deleted from the candidate set. The algorithm is implemented iteratively until the Pareto front of the final population is obtained. Finally, once the set of Pareto front P_F is achieved, the Pareto optimal solutions $(u_1^*(k), u_2^*(k), \ldots, u_m^*(k))$ of the MOP in (13.17) can be obtained for the noncooperative H_∞ management strategy of the dynamic energy transfer system (13.1) in the smart grid network. According to previous analysis, the algorithm of the LMIs-constrained MOEA to solve the MOP in (13.17) for the noncooperative H_∞ management strategy of dynamic energy transfer system (13.1) is proposed as follows.

13.4.1 LMIs-constrained MOEA algorithm for noncooperative H_∞ management strategy of dynamic energy transfer system in smart grid network

Step 1: Select the searching range $(e_1^L, e_2^L, \ldots, e_m^L) \times (e_1^U, e_2^U, \ldots, e_m^U)$ for the feasible objective vector $(e_1, e_2, \ldots, e_m)$ and set the iteration number N_i, the population number N_p, the crossover rate c_r, and the mutation ratio m_r, in the LMIs-constrained MOEA. Set iteration number $i = 1$.

Step 2: Select N_p feasible individuals (chromosomes) from the feasible chromosome set randomly to be the initial population P_1.

Step 3: Operate the EA with the crossover rate c_r, the mutation ratio m_r, and generate $2N_p$ feasible chromosomes by examining whether their corresponding objective vectors $(e_1, e_2, \ldots, e_m)$ are feasible objective vectors for the LMIs in (13.18) and (13.19).

Step 4: Set the iteration index $i = i + 1$, and select the N_p chromosomes from the $2N_p$ feasible chromosomes in **Step 3** through non-dominated sorting method to be the population P_{i+1}.

Step 5: Repeat **Steps 3 and 4** until the iteration number N_i is reached. If the iteration number N_i is satisfied, then we set $P_{N_i} = P_F$ as Pareto front.

Step 6: Select a "preferable" feasible objective individual $(e_1^*, e_2^*, \ldots, e_m^*) \in P_F$ according to designer's own preference with the optimal $W^* = (P^*)^{-1}$, Y_i^*, and $K_i^* = Y_i^* W^{*^{-1}}$, where the W^* and K_i^* are the solution of MOP in (13.17). Once the "preferable" feasible objective individual is selected as the noncooperative management strategy

$(u_1^*(k), u_2^*(k), \ldots, u_m^*(k))$ with the $u_i^*(k) = K_i^* \tilde{s}_i(k), i = 1, 2, \ldots m$, then the noncooperative H_∞ management strategy design problem in (13.12) of dynamic energy transfer system (13.1) in smart grid network can be solved from (13.18) and (13.19).

Remark 13.3 The computational complexity of the proposed LMIs-constrained MOEA algorithm is approximately estimated as $O(n(n+1)mN_p^2 N_i)$, including $O(\frac{n(n+1)m}{2})$ for solving the LMIs, and $O(2N_p^2 N_i)$ for the MOEA algorithm where n is the dimension of the system state $s(k)$, N_i is the iteration number of MOEA, and N_p is the population number of the LMIs-constrained MOEA algorithm.

13.5 The Cooperative H_∞ Management Strategy Design for the Dynamic Energy Transfer System in Smart Grid Network

Motivated by [302], if the m managers in dynamic energy transfer system (13.1) have compromised a common desired state s_d (target) with each other at the beginning, then the cooperative H_∞ management strategy of the dynamic energy transfer system (13.1) in the smart grid network becomes how to specify $u_1(k), u_2(k), \ldots, u_m(k)$ to minimize the following cooperative H_∞ utility function:

$$J_\infty(u(k)) = \frac{\sum_{k=0}^{t_f} (s(k) - s_d)^T Q (s(k) - s_d) + u^T(k) R u(k)}{(s(0) - s_d)(s(0) - s_d) + \sum_{k=0}^{t_f} w^T(k) w(k)} \tag{13.20}$$

where $u(k) = [u_1(k), \ldots, u_m(k)]^T$ and $w(k) = r(k) - d(k)$. The common desired state s_d and weighting matrices Q and R are decided together by all managers in the smart grid network at the beginning. Similar to the noncooperative management case, although each manager cooperates with other managers to use the power flow mechanism of the smart grid to achieve the compromised target, the unpredictable energy $r(k)$ generated from the RESs and the unavailable power demand $d(k)$ may affect the regulation error so that each manager tries to design a cooperative H_∞ management strategy to minimize the effect of the unavailable factor $w(k)$ on the tracking of common desired target with parsimonious effort.

Let us denote $\tilde{s}(k) = s(k) - s_d$, then the cooperative H_∞ management strategy of dynamic energy transfer system (13.1) in smart grid network can be reformulated as the following SOP:

$$\min_{u(k)} J_\infty(u(k) \tag{13.21}$$

where

$$J_\infty(u(k) = \frac{\sum_{k=0}^{t_f} \tilde{s}^T(k)\ Q \tilde{s}(k) + u^T(k)\ R u(k)}{\tilde{s}^T(0)\ \tilde{s}(0) + \sum_{k=0}^{t_f} \bar{w}^T(k) \bar{w}(k)},$$

s.t.

$$\tilde{s}(k+1) = A\tilde{s}(k) + Bu(k) + \bar{w}(k), \tag{13.22}$$

where $u(k) = [u_1(k), \ldots, u_m(k)]^T$ and the $B = [B_1, \ldots, B_m]$, and the unpredictable vector $\bar{w}(k) = r(k) - d(k) + (As_d - s_d)$.

Remark 13.4 The cooperative H_∞ utility function has some differences with the noncooperative management case. In the noncooperative H_∞ management strategy, the information of other managers are unavailable so that the noncooperative H_∞ utility function is specified by each manager according to his/her own consideration to deal with the unpredictable factor including other managers' strategies. Here, the desired state s_d and weighting matrices Q and R are decided together with each other at the beginning to design the effective strategy through the cooperative H_∞ utility function $J_\infty(u(k)$ in (3.20) against the effect of unpredictable factors.

By using the similar technique in (13.12), an indirect approach is employed to solve the cooperative H_∞ management strategy of design problem in (13.21) by minimizing its upper bound, i.e.,

$$e^* = \min_{u(k)} e \tag{13.23}$$

s.t.

$$J_\infty(u(k)) = \frac{\sum_{k=0}^{t_f} \tilde{s}^T(k)Q\tilde{s}(k) + u^T(k)Ru(k)}{\tilde{s}^T(0)\tilde{s}(0) + \sum_{k=0}^{t_f} \bar{w}^T(k)\bar{w}(k)} \leq e \tag{13.24}$$

which can be modified as

$$\sum_{k=0}^{t_f} \tilde{s}^T(k)Q\tilde{s}(k) + u^T(k)Ru(k) \leq e\tilde{s}^T(0)\tilde{s}(0) + e\sum_{k=0}^{t_f} \bar{w}^T(k)\bar{w}(k) \tag{13.25}$$

The physical meaning of (13.23) is that the effects of $\bar{w}(k)$ and $\tilde{s}(0)$ on the tracking error $\tilde{s}(k)$ can be attenuated below a desired level e with a parsimonious effort $u(k) = K\tilde{s}(k)$ from the viewpoint of energy, no matter what $\bar{w}(k)$ is, i.e., the l_2 gain from $\bar{w}(k)$ to $\tilde{s}(k)$ must be equal to or less than the prescribed level e. Hence, the cooperative H_∞ management strategy of dynamic energy transfer system (13.1) in the smart grid network is to design the most effective common strategy to achieve the desired common target with a parsimonious effort despite the effect of unpredictable factors as possible. Then, we get the following theorem for the H_∞ cooperative management strategy design problem.

Theorem 13.3

(i) For the dynamic energy transfer system (13.1) in a smart grid network, the cooperative H_∞ management strategy design problem in (13.23) is solved if the following LMIs-constrained SOP can be solved:

$$e^* = \min_{\{P,K\}} e \tag{13.26}$$

s.t.

$$\begin{bmatrix} e_i I_n & I_n \\ I_n & W \end{bmatrix} \geq 0, \tag{13.27}$$

$$\begin{bmatrix} W & W & Y^T & 0 & \Gamma^T \\ W & Q^{-1} & 0 & 0 & 0 \\ Y & 0 & R^{-1} & 0 & 0 \\ 0 & 0 & 0 & eI_n & I_n \\ \Gamma & 0 & 0 & I_n & W \end{bmatrix} \geq 0 \tag{13.28}$$

where $W = P^{-1} > 0, Y = KP^{-1} = KW$, and $\Gamma = AW + BY$.

(ii) If the LMIs-constrained SOP in (13.26) is solved by $P^* = (W^*)^{-1}$ and Y^*, then the cooperative H_∞ management strategy is given by $u^*(k) = K^*\tilde{s}(k) = Y^*P^*\tilde{s}(k)$.

Proof: See Appendix B.

The SOP in (13.26) could be solved by decreasing e until no positive definite solution W exists with the help of LMI toolbox in Matlab.

13.6 Simulation Results

To illustrate the design procedure and confirm the performance of the proposed H_∞ noncooperative and cooperative management strategies for the dynamic energy transfer system in smart grid network, a numerical design example of the management strategy of the smart grid network in Fig. 13.1 is introduced as follows:

The smart grid consists of a power grid and three microgrids, and there are only three managers involved in the dynamic energy transfer system (13.1) of the smart grid network as shown in Fig. 13.1, that is, $n = 3$ and $m = 3$. We assumed that manager 1 is the owner of the 1st microgrid in a residential area; manager 2 is the owner of the 2nd microgrid in an industrial area, and manager 3 is the owner of the 3rd microgrid of an energy investor. The strategies of each manager are shown in Fig. 13.1(a)–(c) and the whole smart grid network is shown in Fig. 13.1(d).

In the noncooperative H_∞ management strategy, different managers may have different requirements of the energy storage levels to determine different H_∞ utility functions with themselves. For example, the microgrid in the industrial and residential area will expect the energy storage level to be maintained at an emergency level to work in safety. In particular, the manager in industrial area will pay more attention to this issue. On the other hand, the energy investor will expect the generated energy can be sold as much as possible. So, the storage efficiency matrices and the strategies of each manager with the corresponding topology matrices are given as follows:

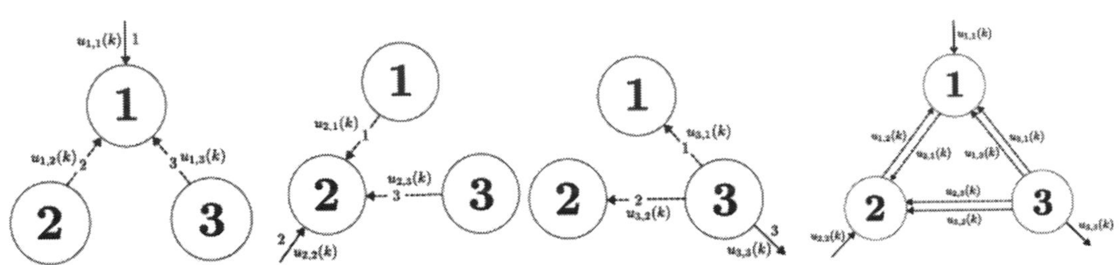

(a) The directed graph G_1 to represent the management strategy of manager 1.

(b) The directed graph G_2 to represent the strategy of manager 2.

(c) The directed graph G_3 to represent the strategy of manager 3.

(d) The whole smart grid network contains 3 managers with their management strategies.

FIGURE 13.1
An example of the management model in the dynamic energy transfer system with $n = 3$ and $m = 3$ in a smart grid network.

$$A = \begin{bmatrix} 0.98 & 0 & 0 \\ 0 & 0.99 & 0 \\ 0 & 0 & 0.99 \end{bmatrix} \in \mathbb{R}^{3\times 3}, u_1(k) = [\,u_{1,1}(k) \quad u_{1,2}(k) \quad u_{1,3}(k)\,]^T \in \mathbb{R}^{3\times 1},$$
$$u_2(k) = [\,u_{2,1}(k) \quad u_{2,2}(k) \quad u_{2,3}(k)\,]^T \in \mathbb{R}^{3\times 1}, u_3(k) = [\,u_{3,1}(k) \quad u_{3,2}(k) \quad u_{3,3}(k)\,]^T \in \mathbb{R}^{3\times 1}, \tag{13.29}$$

$$B_1 = \begin{bmatrix} 1 & 1 & 1 \\ 0 & -1 & 0 \\ 0 & 0 & -1 \end{bmatrix}, B_2 = \begin{bmatrix} -1 & 0 & 0 \\ 1 & 1 & 1 \\ 0 & 0 & -1 \end{bmatrix}, B_3 = \begin{bmatrix} 1 & 0 & 0 \\ 0 & 1 & 0 \\ -1 & -1 & -1 \end{bmatrix}. \tag{13.30}$$

Since the manager at the industrial area should pay more attention to the problem of electricity storage [314], this microgrid may use more efficient electricity storage equipment so that the storage efficiency value will be larger than the other two microgrids in our assumption. The strategies of manager 1 and manager 2 can be interpreted as they want to use the power flow to regulate the energy storage level they want and manager 3 wants to sell the energy to other microgrids. Without loss of generality, we suppose that each manager will take the more attention on his/her energy storage level than the others so that the weighting matrices of each H_∞ utility function are assumed as follows:

$$\begin{aligned} &Q_1 = 0.01diag[10,1,1], Q_2 = 0.01diag[1,10,1], Q_3 = 0.01diag[1,1,10], \\ &R_1 = 1.3diag[1,1,1], R_1 = 1.3diag[1,1,1], R_1 = 1.3diag[1,1,1], \end{aligned} \tag{13.31}$$

In this example, whether the microgrid is in the industrial area, residential area, or an energy investor, they will more care about if their targets can be achieved. Suppose the initial state and the desired energy storage levels s_{di} are given as

$$s(0) = [9 \quad 13 \quad 10]^T, s_{d1} = [8.5 \quad 13 \quad 8]^T, s_{d2} = [5 \quad 10 \quad 10]^T, s_{d3} = [10 \quad 15 \quad 3.8]^T. \tag{13.32}$$

We can see that each manager may not have the same desired storage level. For the energy investor, he/she may wish the other microgrids buy the energy he/she produced as much as possible so that his/her storage equipment does not need to contain too much power. The manager in industrial and residential area then expect the storage level can achieve his/her target to work in safety. Because of the time-varying and unpredictable weather conditions, the extra power from the RESs and the power demand are usually fluctuating/intermittent over time. We assume that the renewable energy $r(k)$ is getting from the solar panels and the data can be got from [315]. Reasonably, there is no extra energy produced by the solar panels in the night and the energy will be disturbed due to the unpredictable cloudiness. Therefore, the unpredictable power demand $d(k)$ is assumed as follows [302]:

$$d(k) = \begin{cases} \begin{bmatrix} \psi_{[0.1,0.4]}(k) \\ \psi_{[0.1,0.4]}(k) \\ \psi_{[0.1,0.4]}(k) \end{bmatrix} \text{if } r(k) = 0, \\ \begin{bmatrix} \psi_{[-0.4,0.6]}(k) \\ \psi_{[-0.4,0.6]}(k) \\ \psi_{[-0.4,0.6]}(k) \end{bmatrix} + r(k) \text{ otherwise.} \end{cases}$$

where $\psi_{[q_1,q_2]}(k)$ represents a random variable that is uniformly distributed over the range $[q_1, q_2]$. Actually, the power demands of these microgrids are not the same; the power demand of the microgrid in the industrial area will be larger than in the residential area and the power demand of the energy investor will be the lowest because he/she just produce energy to sell. Though the daily power demand is uncertain, the power demand will be lower in the night or the early morning because less people are engaged in production at that time, that is $r(k) = 0$. On the contrary, the power demand will be at the highest level at the operating hours and the extra power generated from the RESs will be more at this time. These quantities are chosen to be independent from time to time. Figures 13.2–13.4 show the energy provided by RESs $r(k)$ in [315] and the energy demand $d(k)$. To obtain a more clear view of system trajectory, the energy storage level was examined from time $k = 0$ to $k = 500$. Based on the proposed LMIs-constrained MOEA algorithm, we could solve the LMIs-constrained MOP in (13.17)–(13.19) for the noncooperative H_∞ management strategies of the example, with the searching range $0 \leq J_\infty^i(u_i(k)) \leq 1$, the population number $N_p = 1500$, iteration number $N_i = 30$, crossover rate $c_r = 0.9$, and the mutation ratio $m_r = 0.3$. Once the iteration number $N_i = 30$ is achieved, we get the (e_1^*, e_2^*, e_3^*) and $P^* = W^{*^{-1}}$. The Pareto front for the Pareto optimal solutions of MOP of the noncooperative H_∞ management strategy of this example can be obtained as shown in Fig. 13.5.

Figure 13.5 shows that the Pareto optimal solutions indeed approach to the Pareto front, which can guarantee the effectiveness of the proposed LMIs-constrained

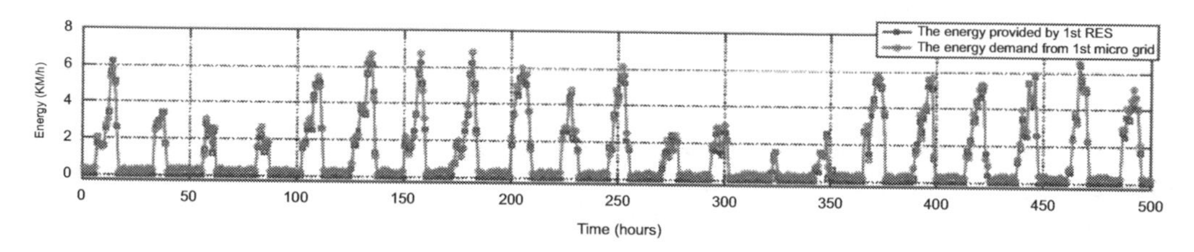

FIGURE 13.2
The energy demand $d_1(k)$ of the first microgrid and the energy $r_1(k)$ provided by the corresponding RES.

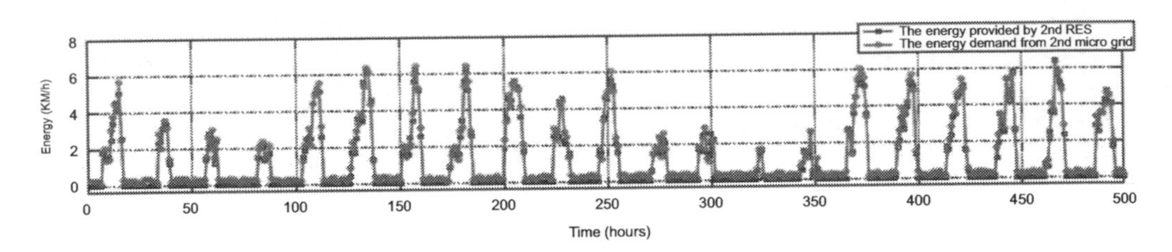

FIGURE 13.3
The energy demand $d_2(k)$ of the second microgrid and the energy $r_2(k)$ provided by the corresponding RES.

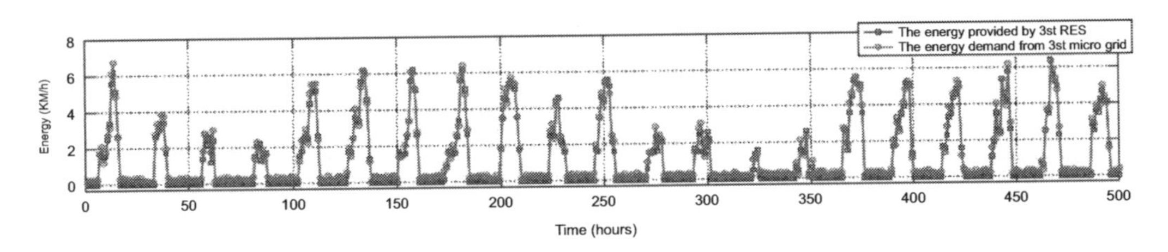

FIGURE 13.4
The energy demand $d_3(k)$ of the third microgrid and the energy $r_3(k)$ provided by the corresponding RES.

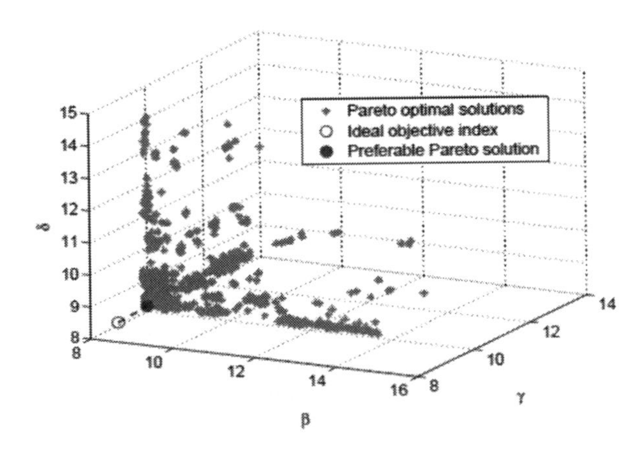

FIGURE 13.5
Pareto front obtained by the proposed LMIs-constrained MOEA for the three managers noncooperative H_∞ management strategy design problem in dynamic energy transfer system in the smart grid network.

MOEA algorithm for the proposed noncooperative H_∞ management strategy of dynamic energy transfer system (13.1) in the smart grid network. Since the knee point solution does not bias toward each manager, the knee point solution, i.e., the preferable Pareto solution in Fig. 13.5, is selected as the design strategy in the Pareto optimal set. In this case,

$$(e_1^*, e_2^*, e_3^*) = (9.0081, 8.5284, 8.9534) \tag{13.33}$$

and

$$K_1^* = \begin{bmatrix} -0.2880 & -0.2489 & -0.1587 \\ 0.0245 & 0.4467 & -0.0232 \\ -0.1155 & -0.1204 & 0.2759 \end{bmatrix},$$
$$K_2^* = \begin{bmatrix} 0.3369 & -0.0417 & -0.0982 \\ -0.2022 & -0.3770 & -0.1849 \\ -0.1286 & -0.1035 & 0.2958 \end{bmatrix}, \tag{13.34}$$
$$K_3^* = \begin{bmatrix} -0.2775 & 0.1150 & 0.0354 \\ 0.0389 & 0.4267 & -0.0740 \\ -0.0294 & -0.1278 & -0.4575 \end{bmatrix},$$

In Fig. 13.6, the dynamical trajectories of the energy storage level in each microgrid of the smart grid network without noncoopertive H_∞ management strategies are always with the random fluctuations driven by unpredictable power demand and the extra power supplied from the RESs in Fig. 13.2–13.4. It can be seen that the states of the dynamic energy transfer systems (13.1) suffer from the unavailable power supply at each time slot. Legitimately, the unpredictable power supplied from the RESs $r(k)$ is from the solar panels and may suffer from the effect of the night and the overcast sky so that the energy storage level will be perturbed. From a long time point of view, we can see that there is a decreasing trend in the energy storage level and this phenomenon is rational because the extra power from the RESs is intermittent. Then, the three managers decide to design a more efficient noncooperative H_∞ management strategies to regulate the states of dynamic energy transfer system (13.1) to achieve their desired targets s_{d1}, s_{d2} and s_{d3} in (13.32), respectively.

In Fig. 13.7, since each player of noncooperative power management strategy design problem does not know other players' information, the strategies of three managers need larger control efforts in order to reach their own targets as possible. To attenuate

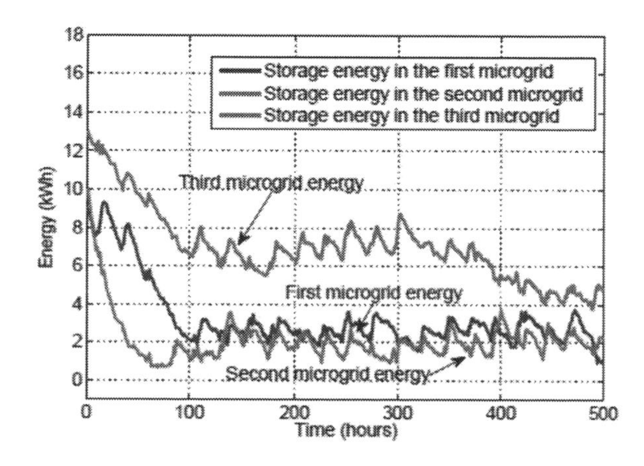

FIGURE 13.6
The energy storage levels of microgrids in a smart grid network without management strategies.

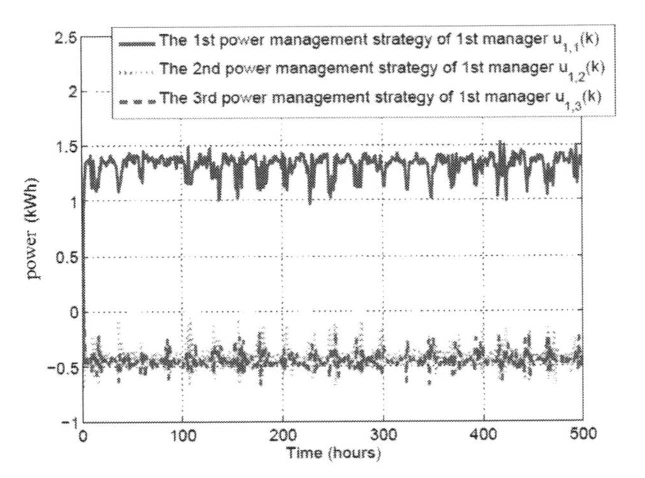

(a) The noncooperative H_∞ power management strategy of the first manager.

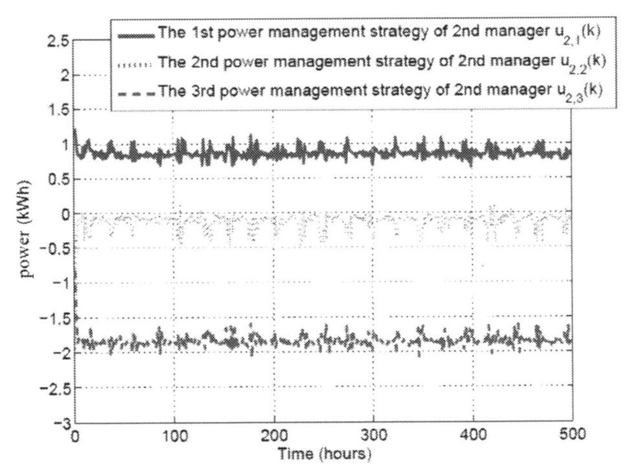

(b) The noncooperative H_∞ power management strategy of the second manager.

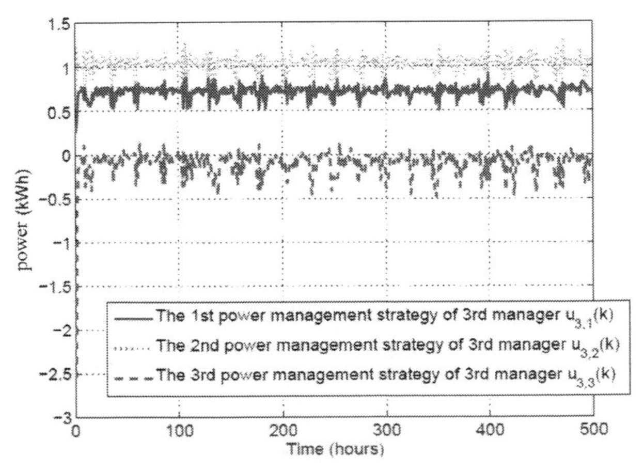

(c) The noncooperative H_∞ power management strategy of the third manager.

FIGURE 13.7
The noncooperative H_∞ power management strategy of all managers.

the effect from the power energy of RESs and power demand from the microgrid, the noncooperative strategies also reveal some regulation corresponding to the external fluctuation. In Fig. 13.8, the energy storage levels $s_1(k), s_2(k)$ and $s_3(k)$ of the dynamical energy transfer system in smart grid network in (13.1) achieve some compromised steady state among their designed targets s_{d1}, s_{d2} and s_{d3} based on the tradeoff between weighting matrices Q_i and R_i for the proposed noncooperative management strategies of these three managers. Obviously, the energy storage in these three microgrids can follow the compromised target of three desired targets s_{d1}, s_{d2} and s_{d3} in the dynamic energy transfer system (13.1). Here we can find that the compromised result will have a certain level of relationship with Q_i. The larger diagonal value of Q_i means that the manager pays more attention and penalty on this microgrid's storage level so that the $s_1(k)$ more approaches to the desired value of the first manager with a more weighting of 0.1 than 0.01 of others; $s_2(k)$ more approaches to the desired value of the second manager with a more weighting of 0.1 than 0.01 of others; $s_3(k)$ more approaches to the desired value of the third manager with a more weighting of 0.1 than 0.01 of others as shown in Fig. 13.8. Further, the proposed noncooperative H_∞ management strategy in (13.9) could eliminate the random power supply from the RESs due to the weather condition and attenuate the effect from other competitive strategies to compromise their desired targets for the dynamic energy transfer system of the smart grid network by a parsimonious effort. The reason why the noncooperative management strategy in (13.9) is arobust management strategy is that it could efficiently minimize the effect of other competitive strategies, random power supply from the RESs, and the unavailable power demand on the desired target regulation.

To compare the noncooperative H_∞ management strategy and cooperative H_∞ management strategy in the smart grid network, the SOP in (13.26) is carried out for the same example to demonstrate the effectiveness of the proposed cooperative H_∞ management strategy. In the cooperative H_∞ management strategy design problem, we need to compromise the common desired target s_d and weighting matrices Q, R in (13.20)

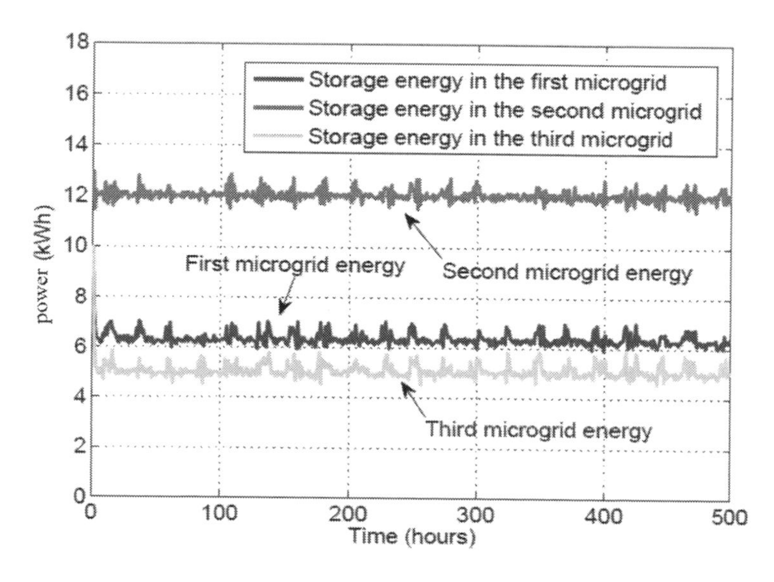

FIGURE 13.8
The power storage levels of microgrids in the smart grid network with the proposed noncooperative H_∞ management strategies.

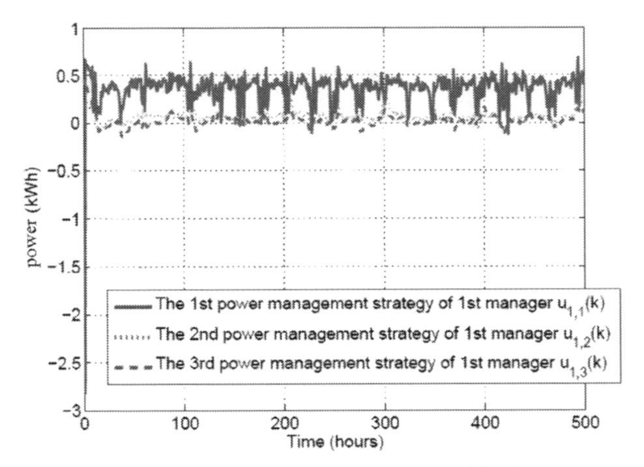

(a)The cooperative H_∞ power management strategy of the first manager.

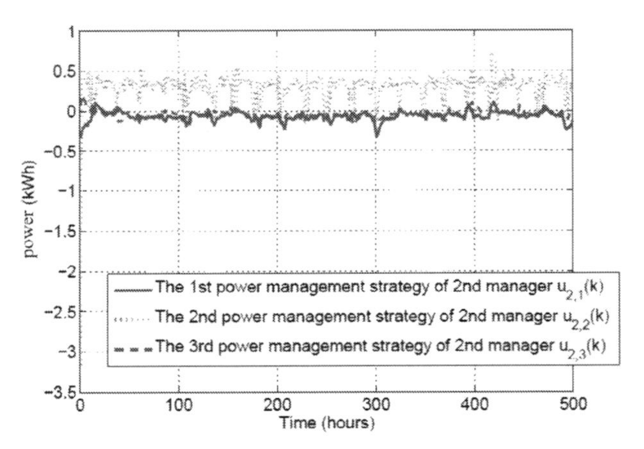

(b)The cooperative H_∞ power management strategy of the second manager.

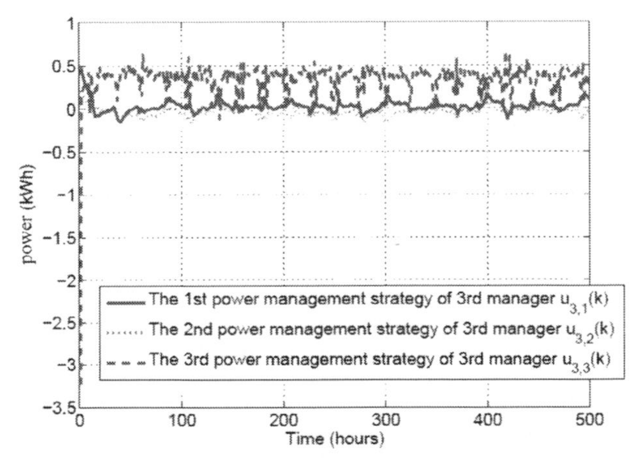

(c) The cooperative H_∞ power management strategy of the third manager.

FIGURE 13.9
The cooperative H_∞ power management strategy of all managers.

beforehand. Suppose the compromised desired target $s_d = [8.5 \quad 10 \quad 3.8]^T$ and weighting matrices are respectively given as follows:

$$Q = 0.01 \times diag[1, 1, 1],$$
$$R = 1.3 \times diag[1, 1, 1, 1, 1, 1, 1, 1, 1], \tag{13.35}$$

By solving the LMIs-constrained SOP in (13.25), we have $e^* = 1.3481$, and

$$K^* = \begin{bmatrix} -0.3979 & -0.2973 & -0.2811 \\ -0.1036 & 0.1037 & 0.0110 \\ -0.1079 & 0.0070 & 0.676 \\ 0.1036 & -0.1037 & -0.0110 \\ -0.2943 & -0.4010 & -0.2920 \\ -0.0043 & -0.0967 & 0.0566 \\ -0.1079 & 0.0070 & 0.0676 \\ -0.0043 & -0.0967 & 0.0566 \\ -0.2900 & -0.3043 & -0.3486 \end{bmatrix} \tag{13.36}$$

and the cooperative H_∞ management strategy $u(k) = [u_1(k), \ldots, u_3(k)]^T = K^*\tilde{s}(k)$. Figure 13.10 shows the state trajectories of dynamic energy transfer system in the smart grid network by the proposed cooperative H_∞ management strategy. In Fig. 13.10, since all managers are compromised with a common desired objective at the beginning and share with the information of power transfer management, the system trajectories quickly converge to the compromised steady state. Obviously, cooperative H_∞ management strategy could achieve their common objective quickly. However, manager should spend much time to compromise for the common target s_d and weighting matrix Q and R beforehand while the noncooperative

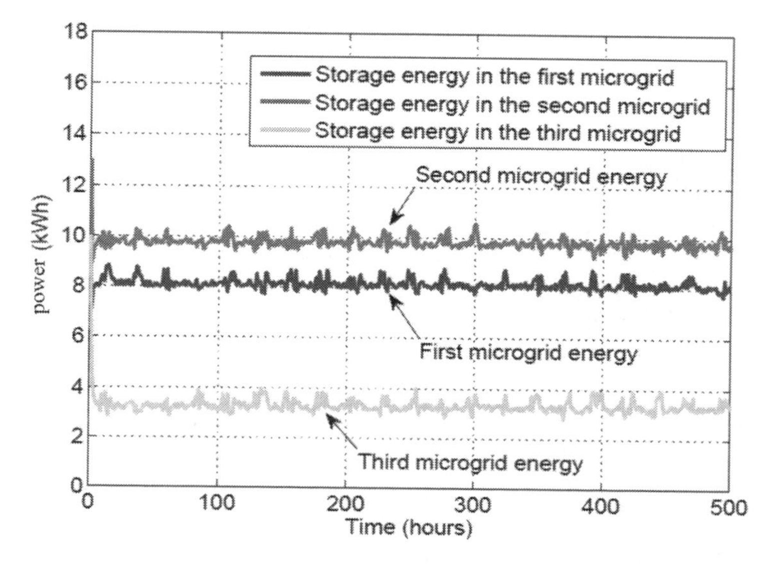

FIGURE 13.10
The power storage levels of microgrids in the smart grid network with the proposed cooperative H_∞ management strategies.

H_∞ management strategy could achieve the results which are compromised based on the conflict multiobjective strategies and different weighting matrices. Furthermore, the cooperative H_∞ management strategy is more robust and effective to reduce the effect of the random power supply from the RESs and the unavailable power demand on the compromised desired target. However, how to achieve their common objective at the beginning is always a complicated and difficult compromised procedure.

13.7 Conclusions

In this chapter, the dynamic energy transfer system in smart grid network can be seen as a multiperson power transfer system with power supply from RESs as a fluctuating/intermittent process over time. In noncooperative H_∞ management strategy, each manager could determine his/her utility based on his/her individual concern to minimize the effect from other conflicting strategies and fluctuation to achieve their desired targets, respectively. This chapter introduced an indirect method to transform the noncooperative H_∞ management strategy design problem into an equivalent LMIs-constrained MOP, which could be solved simultaneously by all managers. Further, the noncooperative H_∞ management strategy design problem for the smart grid network could be solved efficiently through the proposed LMIs-constrained MOEA with the help of MATLAB via LMI toolbox. When the Pareto optimal solutions of the MOP are obtained, each manager can determine a preferable strategy from the set of Pareto optimal solutions according to his/her own preference to obtain a noncooperative H_∞ management strategy. Furthermore, if all managers in the smart grid network have compromised a common desired target with each other at the beginning, then cooperative H_∞ management strategy could be designed. The cooperative H_∞ management strategy design problem of a smart grid network could be transformed to an LMIs-constrained SOP to guarantee a H_∞ robust regulation of common target in the dynamic energy transfer system of a smart grid network. Finally, a design example of these two management strategies is also given to illustrate the design procedure and to confirm the performance of the H_∞ noncooperative and cooperative management strategy of dynamic energy transfer system in a smart grid network with the unpredictable power demand and extra power supply from the RESs through the computer simulation. We found that the proposed H_∞ noncooperative and cooperative management strategies have enough system robustness to reduce the effect of these unpredictable factors in the smart grid network. Further research is still needed to address with a more complicated and practical system structure; for example, the electric price may also be discussed in the later research.

13.8 Appendix

13.8.1 Appendix A: Proof of Theorem 13.2

Let $V(\tilde{s}_i(k)) = \tilde{s}_i^T(k)P\tilde{s}_i(k)$ be the Lyapunov function for the shifted dynamic energy transfer system in (13.10), where the $P = P^T > 0$ is a positive definite matrix. For the ith manager, we have

$$\begin{aligned}&\sum_{k=0}^{t_f}\tilde{s}_i^T(k)Q_i\tilde{s}_i(k)+u_i^T(k)R_iu_i(k)\\&\le \tilde{s}_i^T(0)P\tilde{s}_i(0)-e_i\sum_{k=0}^{t_f}\bar{u}_{-i}^T(k)\bar{u}_{-i}(k)+\sum_{k=0}^{t_f}[\tilde{s}_i^T(k)Q_i\tilde{s}_i(k)+u_i^T(k)R_iu_i(k)\\&+\tilde{s}_i^T(k+1)P\tilde{s}_i(k+1)-\tilde{s}_i^T(k)P\tilde{s}_i(k)-e_i\bar{u}_{-i}^T(k)\bar{u}_{-i}(k)],\end{aligned} \tag{13.37}$$

where the inequality comes from the neglecting term $\tilde{s}_i^T(t_f+1)P\tilde{s}_i(t_f+1)$. Then if the following both inequalities hold for all possible $\bar{u}_{-i}(k)$ simultaneously,

$$\tilde{s}_i^T(0)P\tilde{s}_i(0)\le e_i\tilde{s}_i^T(0)\tilde{s}_i(0) \tag{13.38}$$

and

$$\begin{aligned}&\tilde{s}_i^T(k)Q_i\tilde{s}_i(k)+u_i^T(k)R_iu_i(k)+\tilde{s}_i^T(k+1)P\tilde{s}_i(k+1)\\&-\tilde{s}_i^T(k)P\tilde{s}_i(k)-e_i\bar{u}_{-i}^T(k)\bar{u}_{-i}(k)\le 0,\end{aligned} \tag{13.39}$$

then we have

$$\sum_{k=0}^{t_f}\tilde{s}_i^T(k)Q_i\tilde{s}_i(k)+u_i^T(k)R_iu_i(k)\le e_i\tilde{s}_i^T(0)\tilde{s}_i(0)+e_i\sum_{k=0}^{t_f}\bar{u}_{-i}^T(k)\bar{u}_{-i}(k) \tag{13.40}$$

i.e.,

$$\frac{\sum_{k=0}^{t_f}\tilde{s}_i^T(k)Q_i\tilde{s}_i(k)+u_i^T(k)R_iu_i(k)}{\tilde{s}_i^T(0)\tilde{s}_i(0)+e_i\sum_{k=0}^{t_f}\bar{u}_{-i}^T(k)\bar{u}_{-i}(k)}\le e_i \tag{13.41}$$

for all possible $\bar{u}_{-i}(k)$. For the LMI in (13.18), let the variable matrix $W=P^{-1}$, then we have the following inequality from (13.38), i.e.,

$$e_iI-P=e_iI-W^{-1}\ge 0 \tag{13.42}$$

by Schur complement in lemma 13.1. For obtaining the LMI in (13.19), from (13.39) and (13.10) with $u_i(k)=K_i\tilde{s}_i(k)$, we get

$$\begin{aligned}&\tilde{s}_i^T(k)Q_i\tilde{s}_i(k)+\tilde{s}_i^T(k)K_i^TR_iK_i\tilde{s}_i(k)+\tilde{s}_i^T(k)A^TPA\tilde{s}_i(k)+\tilde{s}_i^T(k)A^TPB_iK_i\tilde{s}_i(k)\\&+\tilde{s}_i^T(k)A^TP\bar{B}_iK_i\bar{u}_{-i}(k)+\tilde{s}_i^T(k)K_i^TB_i^TPA\tilde{s}_i(k)+\tilde{s}_i^T(k)K_i^TB_i^TPB_iK_i\tilde{s}_i(k)\\&+\tilde{s}_i^T(k)K_i^TB_i^TP\bar{B}_i\bar{u}_{-i}(k)+\bar{u}_{-i}^T(k)\bar{B}_i^TPA\tilde{s}_i(k)+\bar{u}_{-i}^T(k)\bar{B}_i^TPB_iK_i\tilde{s}_i(k)\\&+\bar{u}_{-i}^T(k)\bar{B}_i^TP\bar{B}_i\bar{u}_{-i}(k)-\tilde{s}_i^T(k)P\tilde{s}_i(k)-e_i\bar{u}_{-i}^T(k)\bar{u}_{-i}(k)\le 0\end{aligned} \tag{13.43}$$

which can be rearranged in a matrix form as

$$\begin{bmatrix} \tilde{s}_i(k) \\ \bar{u}_{-i}(k) \end{bmatrix}^T \left\{ \begin{bmatrix} (A+B_iK_i)^T \\ \bar{B}_i^T \end{bmatrix} P \begin{bmatrix} (A+B_iK_i)^T \\ \bar{B}_i^T \end{bmatrix}^T - \begin{bmatrix} P - Q_i - K_i^T R_i K_i & 0 \\ 0 & e_i I \end{bmatrix} \right\} \begin{bmatrix} \tilde{s}_i(k) \\ \bar{u}_{-i}(k) \end{bmatrix} \leq 0, \tag{13.44}$$

The matrix inequality in (13.44) holds for all possible $\tilde{s}_i(k)$ and $\bar{u}_{-i}(k)$ if the following inequality holds:

$$\begin{bmatrix} (A+B_iK_i)^T \\ \bar{B}_i^T \end{bmatrix} PP^{-1}P \begin{bmatrix} (A+B_iK_i)^T \\ \bar{B}_i^T \end{bmatrix}^T - \begin{bmatrix} P - Q_i - K_i^T R_i K_i & 0 \\ 0 & e_i I \end{bmatrix} \leq 0. \tag{13.45}$$

Hence, by using the Schur complement in Lemma 13.1, we obtain the following equivalent bilinear matrix inequality (BMI):

$$\begin{bmatrix} P - Q_i - K_i^T R_i K_i & 0 & (A+B_iK_i)^T P \\ 0 & e_i I & \bar{B}_i^T P \\ P(A+B_iK_i) & P\bar{B}_i & P \end{bmatrix} \geq 0, \tag{13.46}$$

Now pre-multiplying and post-multiplying the diagonal matrix $diag[P^{-1} \quad I \quad P^{-1}]$ to the above matrix inequality, with and $Y_i = K_i P^{-1} = K_i W$, it follows that

$$\begin{bmatrix} W - WQ_iW - Y_i^T R_i Y_i & 0 & WA^T + Y_i^T B_i^T \\ 0 & e_i I & \bar{B}_i^T \\ AW + B_i Y_i & \bar{B}_i & W \end{bmatrix} \geq 0, \tag{13.47}$$

then, by using the Schur complement in Lemma 13.1 again, we obtain the equivalent LMIs in (13.19).

13.8.2 Appendix B: Proof of Theorem 13.3

Let $V(\tilde{s}(k)) = \tilde{s}^T(k)P\tilde{s}(k)$ be the Lyapunov function for the shifted dynamic energy transfer system in (13.22), where the $P = P^T > 0$ is a positive definite matrix. From the left side of the (13.25),

$$\begin{aligned} &\sum_{k=0}^{t_f} \tilde{s}^T(k)Q\tilde{s}(k) + u^T(k)Ru(k) \\ &= \tilde{s}^T(0)P\tilde{s}(0) - \tilde{s}^T(t_f+1)P\tilde{s}(t_f+1) + \sum_{k=0}^{t_f} \tilde{s}^T(k)Q\tilde{s}(k) \\ &\quad + u^T(k)Ru(k) + \tilde{s}^T(k+1)P\tilde{s}(k+1) - \tilde{s}^T(k)P\tilde{s}(k) \\ &\leq \tilde{s}^T(0)P\tilde{s}(0) + e\sum_{k=0}^{t_f} \bar{w}^T(k)\bar{w}(k) + \sum_{k=0}^{t_f} [\tilde{s}^T(k)Q\tilde{s}(k) \\ &\quad + u^T(k)Ru(k) + \tilde{s}^T(k+1)P\tilde{s}(k+1) - \tilde{s}^T(k)P\tilde{s}(k) - e\bar{w}^T(k)\bar{w}(k)] \end{aligned} \tag{13.48}$$

where the inequality comes from neglecting the term $\tilde{s}^T(t_f+1)P\tilde{s}(t_f+1)$.

Then if the following both inequalities hold for all possible $\bar{w}(k)$

$$\tilde{s}^T(0)P\tilde{s}(0) \leq e\,\tilde{s}^T(0)\tilde{s}(0) \tag{13.49}$$

and

$$\tilde{s}^T(k)Q\tilde{s}(k) + u^T(k)Ru(k) + \tilde{s}^T(k+1)P\tilde{s}(k+1) - \tilde{s}^T(k)P\tilde{s}(k) - e\bar{w}^T(k)\bar{w}(k) \leq 0, \tag{13.50}$$

then we have

$$\sum_{k=0}^{t_f} \tilde{s}^T(k)Q\tilde{s}(k) + u^T(k)Ru(k) \leq e\tilde{s}^T(0)\tilde{s}(0) + e\sum_{k=0}^{t_f} \bar{w}^T(k)\bar{w}(k) \tag{13.51}$$

which is equivalent to

$$\frac{\sum_{k=0}^{t_f} \tilde{s}^T(k)Q\tilde{s}(k) + u^T(k)Ru(k)}{\tilde{s}^T(0)\tilde{s}(0) + e\sum_{k=0}^{t_f} \bar{w}^T(k)\bar{w}(k)} \leq e \tag{13.52}$$

for all possible $\bar{w}(k)$.
For the LMI in (13.27), let the $W = P^{-1}$, then it can be obtained from (13.49), i.e.,

$$eI - P = eI - W^{-1} \geq 0, \tag{13.53}$$

by Schur complement in Lemma 13.1.
For obtaining the LMI in (13.28), from (13.50) and (13.22) with $u(k) = K\tilde{s}(k)$, we get

$$\begin{aligned}
&\tilde{s}^T(k)Q\tilde{s}(k) + \tilde{s}^T(k)K^TRK\tilde{s}(k) + \tilde{s}^T(k)A^TPA\tilde{s}(k) + \tilde{s}^T(k)A^TPBK\tilde{s}(k)\\
&+ \tilde{s}^T(k)A^TP\bar{w}(k) + \tilde{s}^T(k)K^TB^TPA\tilde{s}(k) + \tilde{s}^T(k)K^TB^TPBK\tilde{s}(k)\\
&+ \tilde{s}^T(k)K^TB^TP\bar{w}(k) + \bar{w}^T(k)PA\tilde{s}(k) + \bar{w}^T(k)PBK\tilde{s}(k)\\
&+ \bar{w}^T(k)P\bar{B}\bar{w}(k) - \tilde{s}^T(k)P\tilde{s}(k) - e\bar{w}^T(k)\bar{w}(k) \leq 0,
\end{aligned} \tag{13.54}$$

which can be rearranged in a matrix for as

$$\begin{bmatrix} \tilde{s}(k) \\ \bar{w}(k) \end{bmatrix}^T \left\{ \begin{bmatrix} (A+BK)^T \\ I_n \end{bmatrix} P \begin{bmatrix} (A+BK)^T \\ I_n \end{bmatrix}^T - \begin{bmatrix} P - Q - K^TRK & 0 \\ 0 & eI_n \end{bmatrix} \right\} \begin{bmatrix} \tilde{s}(k) \\ \bar{w}(k) \end{bmatrix} \leq 0, \tag{13.55}$$

The above inequality in (13.55) holds for all possible $\tilde{s}(k)$ and $\bar{w}(k)$ if the following inequality holds:

$$\begin{bmatrix} (A+BK)^T \\ I_n \end{bmatrix} PP^{-1}P \begin{bmatrix} (A+BK)^T \\ I_n \end{bmatrix}^T - \begin{bmatrix} P - Q - K^TRK & 0 \\ 0 & eI_n \end{bmatrix} \leq 0, \tag{13.56}$$

Hence, by using the Schur complement in Lemma 13.1, we obtain the following equivalent BMI:

$$\begin{bmatrix} P - Q - K^T RK & 0 & (A + BK)^T P \\ 0 & e_i I_n & P \\ P(A + BK) & P & P \end{bmatrix} \geq 0, \tag{13.57}$$

By pre-multiplying and post-multiplying the diagonal matrix $diag[P^{-1}, I_n, P^{-1}]$ to the above matrix inequality, with $W = P^{-1}$ and $Y = KP^{-1} = KW$, it follows that

$$\begin{bmatrix} W - WQW - Y^T RY & 0 & WA^T + Y^T B^T \\ 0 & eI_n & I_n \\ AW + BY & I_n & W \end{bmatrix} \geq 0, \tag{13.58}$$

then, by using the Schur complements in Lemma 13.1 again, we obtain the equivalent LMI in (13.28).

14

Robust Model Reference Investment Strategy of Nonlinear Stochastic Financial Systems: Stochastic H_∞ Game Approach

14.1 Introduction

In the last decades, all kinds of economic and financial systems have become more and more complicated because of nonlinear interactions among finance, stock, and social economic. Therefore, different mathematical models for economic and financial systems have been developed [316–324]. In recent years, researchers focus on system dynamical model to describe the real economic and financial systems [319–325]. Therefore, systematic computation analysis and control have attracted a significant and increasing interest from the financial engineering community, and an emerging interest from analytic economics researchers [125,126,326]. They are instrumental in the design of market mechanisms, contagion mechanisms, and are contributing to the analysis, simulation, and regulation of micro- and macroeconomic processes.

In practical cases, the economic and financial systems are always nonlinear stochastic systems with continuous and discontinuous intrinsic random fluctuations and external disturbances due to national and international situation change, oil price change, the surplus between investment and savings, the variable of interest rate, false economy policy, etc. [322–324]. Since nonlinear stochastic system with Wiener process and Poisson counting process can efficiently describe the nonlinear interaction of economic factors and agents and the effects of continuous and discontinuous intrinsic random fluctuations as well as external disturbances, using a nonlinear stochastic dynamic model with Wiener and Poisson counting process to describe a real economic or financial system could be more appealing [22,125,126,325,326].

In this chapter, the desired trajectories and target of economic or financial system are prescribed by a reference dynamic model (i.e. the desired transient and steady states for target are given based on a reference model by policy-makers) and the design purpose is how to specify an investment strategy to let the economic or financial system follow the trajectories of the reference dynamic model to achieve the desired trajectory and steady state target as possible despite the intrinsic random fluctuations and external disturbance [61], [327]. Because the intrinsic random fluctuation and external disturbance are unavailable beforehand, the stochastic fractional Nash game theory is employed to achieve the robust optimal model reference investment strategy design so that the worst-case effect of external disturbance and random intrinsic fluctuation on the reference tracking error could be as small as possible from the minimax perspective [1,2,9,17,18]. In general, the robust model reference investment strategy of nonlinear stochastic economic and financial systems needs to solve a nonlinear Hamilton–Jacobic–Issac inequality (HJII)-constrained optimization problem. Because HJII is a partial differential inequality, it is very difficult to solve except a very simple case [61,327]. In this situation, in order to simplify the design procedure of robust model reference investment strategy, the global linearization technique is employed to interpolate several local linearized stochastic systems to approximate the nonlinear stochastic economic or financial system so that the nonlinear HJII could be replaced by a set of LMIs [23,32] and therefore the HJII-constrained optimization problem for robust model reference control of economic or financial system could be transformed to a convex LMIs-constrained optimization problem, which could be efficiently solved with the help of LMI toolbox in Matlab [23].

In the investment strategy for a financial system, managers and investors expect their investment policies could let the trajectories of a financial system robustly follow a desired reference model despite intrinsic continuous and jumping random fluctuations and external disturbance. The proposed robust model reference investment strategy based on stochastic H_∞ Nash game is used to design a robust investment strategy for a nonlinear stochastic financial system to follow a prescribed reference tracking to achieve the desired target. Further, an economic system reflects a dynamic interaction of a large number of economic entities. The current economic crisis could illustrate a critical need for a new and fundamental understanding of the structure and dynamics of economic systems. In order to facilitate model reference investment strategy designs to reduce the conflict between individual interest and global efficiency in the economic system, the nonlinear stochastic dynamics of financial contagion is one of important problems to address the systemic complexity, nonlinearity, and volatility in the capital flow of international economic systems under a liberal economy [324,325]. Therefore, the proposed robust model reference investment strategy could provide policy-makers with a valuable financial policy-making for macroeconomic control to establish effective policies.

The outline of this chapter is as follows. Section 14.2 gives the system description and problem formulation of robust model reference investment strategy of nonlinear stochastic financial systems through stochastic H_∞ Nash game to efficiently track the desired trajectories of reference system and eliminate the effect of intrinsic random fluctuation and external disturbance. In Section 14.3, the H_∞ Nash game problem for robust model reference investment strategy of nonlinear stochastic financial systems is discussed by an indirect method of an HJII-constrained optimization through a two-step scheme. In Section 14.4, the global linearization method is suggested to solve the fractional Nash game problem. Section 14.5 gives simulations of two economic or financial systems to illustrate and confirm the results of the proposed robust model reference investment strategy.

14.2 System Description of Stochastic Economic and Financial Systems

In practice, most economic and financial systems are stochastic systems. They may suffer from random parametric fluctuation and external disturbance due to variations of oil price, war, government policy changes, surplus between investment and saving, etc.

Consider the following economic or financial system:

$$dx(t) = [f(x(t)) + g(x(t))u(t) + v(t)]dt + M(x(t))dw(t) + N(x(t))dp(t) \tag{14.1}$$

where $x(t) = [x_1(t), \dots, x_i(t), \dots, x_n(t)]^T$ denotes the state vector of an economic or a financial system, $u(t) = [u_1(t), \dots, u_m(t)]^T$ denotes the control strategy or policy, $v(t) = [v_1(t), \dots, v_n(t)]^T$ denotes the stochastic external disturbance, $w(t)$ is a standard 1-dimensional Wiener process. $M(x(t))dw(t)$ denotes a continuous state dependent random fluctuation, $p(t)$ denotes Poisson counting process with mean λ in a unit time, and $N(x(t))dp(t)$ is regarded as an intrinsic discontinuous random turbulence (jumping process). The nonlinear interaction vector $f(x)$ of the economic or financial system satisfies with Lipschitz continuity, and $g(x)$ denotes the input matrix of control strategy $u(t)$.

In order to illustrate the nonlinear stochastic financial system (14.1), the following two financial systems are given as two examples in this chapter. A stochastic financial dynamical model is given to illustrate the interaction between interesting rate $x(t)$, investment demand $y(t)$, and the price index $z(t)$ of a financial system in which continuous and discontinuous intrinsic random fluctuations and environmental disturbance are also considered as follows [22,142,331]:

$$\begin{cases} dx(t) = [z(t) + (y(t) - a)x(t) + u_1(t) + v_1(t)]dt \\ \qquad + m_1(x(t), y(t), z(t))dw(t) + n_1(x(t), y(t), z(t))dp(t) \\ dy(t) = [1 - by(t) - x^2(t) + u_2(t) + v_2(t)]dt \\ \qquad + m_2(x(t), y(t), z(t))dw(t) + n_2(x(t), y(t), z(t))dp(t) \\ dz(t) = [-x(t) - cz(t) + u_3(t) + v_3(t)]dt \\ \qquad + m_3(x(t), y(t), z(t))dw(t) + n_3(x(t), y(t), z(t))dp(t) \end{cases} \tag{14.2}$$

where $W(t)$ is Wiener process and $p(t)$ is Poisson counting process which are introduced to model the continuous and discontinuous random turbulence and fluctuation, respectively. $v(t) = [v_1(t), v_2(t), v_3(t)]^T$ denotes the corresponding external disturbance, and $u(t) = [u_1(t), u_2(t), u_3(t)]^T$ denotes the control strategy or policy to be designed. The parameter $a \geq 0$ is the saving amount, $b \geq 0$ is the per-investment cost, and $c \geq 0$ is the elasticity of commercial demands. In the nonlinear financial model (14.2), the interest rate $x(t)$ can be influenced by the surplus between investment and saving as well as adjustments of the prices. The investment demand $y(t)$ is proportional to the rate of investment and inversely proportional to the cost of investment and the interest rate. The price index $z(t)$ depends on the difference between supply and demand which determines the price in the market, and it is also negatively influenced by the interesting rate [22,142,331]. According to the nonlinear stochastic dynamic system in (14.2), it can be employed to describe a real financial system such as a financial market, which refers to the buy-and-sell behavior, or to describe the price changes of the objects of transaction (like stocks, bonds, futures, and oil) or economic index changes (investment rate, price index, and investment demand).

Due to the global impact of financial crisis, in recent years, the stochastic dynamic system theory has become a frontier issue in financial analysis to address financial contagion problem. When the impact of the financial shock is felt, the net capital inflow in source country of financial turbulence sharply decreases, adding to driving force [325]. On the other hand, the net capital inflow for the volatility-affected country also deviates in response to the contagion effects from the source country of financial turbulence [320,325]. Considering that, in the international capital flow volatility system, the two financial coupling strengths are always uncertain and stochastic, the stochastic nonlinear dynamic financial system of international capital flow volatility between the source country of financial turbulence and the volatility-affected country, in which continuous and discontinuous random turbulence as well as external disturbance are included, is described as the following [325]:

$$\begin{cases} dx_v(t) = [a - c_{sv}x_v(t)x_s^2(t) + u_1(t) + v_1(t)]dt \\ \qquad + m_1(x_v(t), x_s(t))dw(t) + n_1(x_v(t), x_s(t))dp(t) \\ dx_s(t) = [-bx_s(t) + c_{sv}x_v(t)x_s^2(t) + u_2(t) + v_2(t)]dt \\ \qquad + m_2(x_v(t), x_s(t))dw(t) + n_2(x_v(t), x_s(t))dp(t) \end{cases} \tag{14.3}$$

where $w(t)$ is the Wiener process and $p(t)$ is the Poisson counting process which are introduced to model the continuous and discontinuous random turbulence and fluctuation, $x_s(t)$ denotes the historical trend of net capital inflow for source country of financial turbulence, and $x_v(t)$, for the volatility-affected country. The notation $v(t) = [v_1(t), v_2(t)]^T$ denotes the corresponding external disturbance, and $u(t) = [u_1(t), u_2(t)]^T$ denotes the control strategy or policy to be designed; $a \neq 0$ is the inertial coefficient of the volatility-affected country; $b>0$ denotes the inertial coefficient of the source country of financial turbulence; c_{sv} denotes the coupling coefficient for the impact from the source country of the financial turbulence to the volatility-affected country; c_{vs} is the coupling coefficient for the impact from the volatility-affected country to the source country of financial turbulence.

For the nonlinear stochastic financial system in (14.1), suppose we want to design an investment strategy $u(t)$ to let the system state $x(t)$ follow the following desired reference model:

$$\dot{x}_r(t) = A_r x_r(t) + r(t) \tag{14.4}$$

where $x_r(t) = [x_{r1}(t) \cdots x_{rn}(t)]^T$ is the desired reference state vector to be tracked by the financial state vector $x(t)$in (14.1), A_r is a specified asymptotically stable matrix, and $r(t)$ is a desired reference input. At the steady state, i.e. $\dot{x}_r = 0$, $x_r(t) = -A_r^{-1}r(t)$. For example, if we specify $A_r = -I$, i.e. all the eigenvalues are all at –1 in the complex s-domain, then $x_r(t) = r(t)$ at the steady state. In the reference model, we specify an adequate A_r to fit a desired transient behavior of the desired $x_r(t)$ and select a reference signal input $r(t)$ as a prescribed target at the steady state.

After we specify a desired reference model (14.4) of the nonlinear stochastic financial system in (14.1), we then want to design an investment strategy or policy so that the state vector $x(t)$ in (14.1) could track the reference state vector $x_r(t)$ as possible in spite of the effect of intrinsic continuous and jumping random fluctuations and external disturbance. Since the external disturbance $v(t)$ is unavailable and difficult to predict, the following stochastic H_∞ Nash game is employed for designing a robust model reference investment strategy of nonlinear stochastic financial system to follow the desired reference model in (14.4):

$$\min_{u(t)} \max_{v(t)} \frac{E\{\int_0^{t_p} [(x(t) - x_r(t))^T Q(x(t) - x_r(t)) + u^T(t)Ru(t)]dt\}}{E\{\int_0^{t_p} [v^T(t)v(t)dt\}} \tag{14.5}$$

where Q and R are the weighting matrices between the tracking error and the strategy effort, $E\{\cdot\}$ denotes the expectation and t_p denotes the terminal time. The numerator of (14.5) denotes the model reference tracking performance with the consideration of the cost of control strategy which could efficiently eliminate intrinsic random continuous and discontinuous fluctuation to achieve the model reference tracking with a parsimonious control effort. The physical meaning of stochastic H_∞ Nash game problem in (14.5) for stochastic financial system in (14.2) is that the worst-case effect of stochastic environmental disturbance at the denominator on the reference tracking error and control strategy cost at the numerator must be minimized from the energy perspective. This minimax model reference tracking game in (14.5) is the so-called stochastic H_∞ Nash game problem of the robust model reference investment strategy in the nonlinear stochastic financial system in (14.2).

In order to simplify the design procedure, let us augment the financial system and its reference model as follows:

$$d\begin{bmatrix} x(t) \\ x_r(t) \end{bmatrix} = \{\begin{bmatrix} f(x(t)) \\ A_r x_r(t) \end{bmatrix} + \begin{bmatrix} g(x(t)) \\ 0 \end{bmatrix} u(t) + \begin{bmatrix} v(t) \\ r(t) \end{bmatrix}\}dt + \begin{bmatrix} M(x(t)) \\ 0 \end{bmatrix} dw(t) + \begin{bmatrix} N(x(t)) \\ 0 \end{bmatrix} dp(t), \tag{14.6}$$

which could be denoted by

$$d\bar{X}(t) = [F(\bar{X}(t)) + G(\bar{X}(t))u(t) + \bar{v}(t)]dt + (\bar{X}(t))dw(t) + \bar{N}(\bar{X}(t))dp(t) \tag{14.7}$$

where $\bar{X}(t) = \begin{bmatrix} x(t) \\ x_r(t) \end{bmatrix}, F(\bar{X}) = \begin{bmatrix} f(x) \\ A_r x_r \end{bmatrix}, G(\bar{X}) = \begin{bmatrix} g(x) \\ 0 \end{bmatrix}, \bar{v}(t) = \begin{bmatrix} v(t) \\ r(t) \end{bmatrix}, (\bar{X}) = \begin{bmatrix} M(x) \\ 0 \end{bmatrix},$

$\bar{N}(\bar{X}) = \begin{bmatrix} N(x) \\ 0 \end{bmatrix}.$

Then let us denote the model reference tracking error as

$$\tilde{x}(t) := x(t) - x_r(t) = [I - I]\begin{bmatrix} x(t) \\ x_r(t) \end{bmatrix} := \bar{I}\bar{X}(t), \tag{14.8}$$

where I is the identity matrix.

In this situation, the stochastic H_∞ game problem for robust model reference investment design of nonlinear stochastic financial system in (14.5) should be modified as

$$\min_{u(t)} \max_{\bar{v}(t)} \frac{E\{\int_0^{t_p} [(\bar{X}^T(t)\bar{Q}\bar{X}(t) + u^T(t)Ru(t)]dt\}}{E\{\int_0^{t_p} \bar{v}^T(t)\bar{v}(t)dt\}} \tag{14.9}$$

where $\bar{Q} = \bar{I}^T \bar{Q} \bar{I}$.

Remark 14.1 In the stochastic H_∞ Nash game problem (14.9) for the robust model reference investment design of nonlinear stochastic financial systems, we assume the initial conditions $x(0)$ in (14.1) and $x_r(0)$ in (14.4) are 0. If this is not the case, the effect of these initial conditions should be considered in the stochastic Nash game so that the stochastic H_∞ Nash game problem in (14.9) should be modified as

$$\min_{u(t)} \max_{v(t)} \frac{E\{\int_0^{t_p} [(\bar{X}^T(t)\bar{Q}\bar{X}(t) + u^T(t)Ru(t)]dt\}}{E\{\int_0^{t_p} \bar{v}^T(t)\bar{v}(t)dt + V(\bar{X}(0))\}} \tag{14.10}$$

for some positive Lyapunov function $V(\bar{X}(t)) > 0(\bar{X}(t) \neq 0)$ and $V(0) = 0$.

14.3 Robust Model Reference H_∞ Investment Strategy for Nonlinear Stochastic Financial Systems

Based on the analysis of the above section, the robust model reference investment strategy for the nonlinear stochastic financial system in (14.1) to track the desired reference model in (14.5) becomes how to solve the stochastic H_∞ Nash game problem in (14.10) subject to the augmented stochastic system in (14.7). Because of its difficulty to solve the stochastic H_∞ game in (14.10) directly in [1,2,9,17,18], it is always solved indirectly by a sub-minimax method. First, let the upper bound ρ of (14.10) be

$$\min_{u(t)} \max_{\bar{v}(t)} \frac{E\{\int_0^{t_p} [(\bar{X}^T(t)\bar{Q}\bar{X}(t) + u^T(t)Ru(t)]dt\}}{E\{\int_0^{t_p} \bar{v}^T(t)\bar{v}(t)dt + V(\bar{X}(0))\}} \leq \rho \tag{14.11}$$

We will first solve the sub-minimax problem in (14.11) and then minimize the upper bound ρ to approach the stochastic H_∞ Nash game of investment strategy. In general, because $\bar{v}(t)$ is independent of $u(t)$, the constrained minimax H_∞ game problem in (14.11) is equivalent to the following constrained minimax quadratic game problem [1,2,17,18]:

$$\min_{u(t)} \max_{\bar{v}(t)} E\{\int_0^{t_p} [(\bar{X}^T(t)\bar{Q}\bar{X}(t) + u^T(t)Ru(t) - \rho\bar{v}^T(t)\bar{v}(t)]dt\} \leq \rho V(\bar{X}(0)), \tag{14.12}$$

where ρ is to be minimized.

Let us denote

$$J(u(t), \bar{v}(t)) = E\{\int_0^{t_p} [(\bar{X}^T(t)\bar{Q}\bar{X}(t) + u^T(t)Ru(t) - \rho\bar{v}^T(t)\bar{v}(t)]dt\}. \tag{14.13}$$

Then, the constrained stochastic H_∞ game problem in (14.11) and (14.12) is equivalent to finding the worst-case disturbance $\bar{v}^*(t)$ which maximizes $J(u(t), \bar{v}(t))$ and then finding optimal strategy $u^*(t)$ which minimizes $J(u(t), \bar{v}^*(t))$ such that $J(u^*(t), \bar{v}^*(t))$ is less than $\rho E\{V(\bar{X}(0))\}$, i.e.

$$J(u^*(t), \bar{v}^*(t)) = \min_{u(t)} J(u(t), \bar{v}^*(t)) = \min_{u(t)} \max_{\bar{v}(t)} J(u(t), \bar{v}(t)) \leq \rho E\{V(\bar{X}(0))\} \tag{14.14}$$

Hence, if there exist $u^*(t)$ and $\bar{v}^*(t)$ such that the minimax quadratic game strategy design problem in (14.14) is solved, then we can satisfy the sub-minimax performance of robust model reference control in (14.11) as well. Therefore, the first step of robust H_∞ model reference investment strategy is to solve the following conventional Nash minimax quadratic game problem:

$$\begin{aligned}&\min_{u(t)}\max_{\bar{v}(t)} J(u(t),\bar{v}(t))\\&\text{subject to (14.7) and } J(u^*(t),\bar{v}^*(t))\leq \rho E\{V(\bar{X}(0))\}\end{aligned}\tag{14.15}$$

Since $J(u^*(t),\bar{v}^*(t))\leq \rho E\{V(\bar{X}(0))\}$, according to (14.14), in the second step, after solving Nash minimax quadratic game problem in (14.15), the sub-minimax method has to make the upper bound ρ in (14.11) as small as possible.

From the above analysis, we obtain the following Nash minimax quadratic game result as the first step of solving the sub-minimax problem in (14.11) for the robust H_∞ model reference investment strategy design of nonlinear stochastic financial system. Since the Wiener process $w(t)$ and Poisson counting process $p(t)$ are non-differentiable, the following lemma is crucial for the solution of stochastic fractional H_∞ game problem for the model reference investment strategy of nonlinear stochastic financial systems.

Lemma 14.1 For the augmented nonlinear stochastic financial systems in (14.7), the Itô–Lévy formula is given as follows [125,126]:

$$\begin{aligned}dV(\bar{X}(t)) =& \frac{\partial V(\bar{X}(t))}{\partial \bar{X}}[F(\bar{X}(t))+G(\bar{X}(t))+\bar{v}(t)]dt+\frac{\partial V(\bar{X}(t))}{\partial \bar{X}}\bar{M}(\bar{X}(t))dw(t)\\&+\frac{1}{2}\bar{M}(\bar{X})^T\frac{\partial^2 V(\bar{X}(t))}{\partial \bar{X}^2}\bar{M}(\bar{X})+[V(\bar{X}(t)+\bar{N}(\bar{X}))-V(\bar{X}(t))]dp(t)\end{aligned}\tag{14.16}$$

Proposition 14.1 The stochastic Nash minimax quadratic game problem in (14.15) is solved by

$$u^*(t)=-\frac{\rho}{2}R^{-1}G(\bar{X}(t))\frac{\partial V(\bar{X}(t))}{\partial \bar{X}},\bar{v}^*(t)=\frac{1}{2}\frac{\partial V(\bar{X}(t))}{\partial \bar{X}}\tag{14.17}$$

where $V(\bar{X}(t))$ is the solution of the following HJII:

$$\begin{aligned}H(V(\bar{X})):=&\left(\frac{\partial V(\bar{X})}{\partial \bar{X}}\right)^T F(\bar{X})+\frac{1}{\rho}\bar{X}^T\bar{Q}\bar{X}+\frac{1}{4}\left(\frac{\partial V(\bar{X})}{\partial \bar{X}}\right)^T\frac{\partial V(\bar{X})}{\partial \bar{X}}\\&-\frac{1}{4}\left(\frac{\partial V(\bar{X})}{\partial \bar{X}}\right)^T G(\bar{X})R^{-1}G^T(\bar{X})\frac{\partial V(\bar{X})}{\partial \bar{X}}\\&+\frac{1}{2}\bar{M}(\bar{X})^T\frac{\partial^2 V(\bar{X}(t))}{\partial \bar{X}^2}\bar{M}(\bar{X})+\lambda[V(\bar{X}+\bar{N}(\bar{X}))-V(\bar{X})]\leq 0\end{aligned}\tag{14.18}$$

with $V(\bar{X})>0(\bar{X}\neq 0)$.

Proof: See Appendix.

Since ρ is the upper bound of stochastic H_∞ Nash game in (14.11), after solving the above Nash minimax quadratic game problem of (14.15), the second step to solve the stochastic H_∞ game problem in (14.10) still needs to minimize the upper bound ρ as follows

$$\rho_0 = \min_{V(\bar{X})>0} \rho \quad \text{subject to HJII in (14.18)} \tag{14.19}$$

After solving $V^*(\bar{X}(t))$ and ρ_0 from the HJII-constrained optimization problem in (14.19), we substitute this solution $V^*(\bar{X}(t))$ to obtain the robust model reference investment strategy

$$u^*(t) = -\frac{\rho_0}{2} R^{-1} G(\bar{X}(t)) \frac{\partial V^*(\bar{X}(t))}{\partial \bar{X}}$$

and the worst-case external disturbance

$$\bar{v}^*(t) = \frac{1}{2} \frac{\partial V^*(\bar{X}(t))}{\partial \bar{X}}$$

from (14.17) as the solution of the stochastic H_∞ game problem in (14.9).

14.4 Robust H_∞ Model Reference Investment Game Strategy through Global Linearization Scheme

Because it is very difficult to solve the HJII in (14.18), no simple approach is available to solve the HJII-constrained optimization problem in (14.19) for the robust reference investment strategy based on stochastic fractional game in (14.10) of the nonlinear stochastic financial system in (14.7). In this chapter, the global linearization technique in [23,32] will be employed to interpolate several local linearized stochastic financial systems to approximate the nonlinear stochastic financial system in (14.7) so that the HJII in (14.18) could be replaced by a set of linear matrix inequalities (LMIs). In this situation, the HJII-constrained optimization problem in (14.19) will be replaced by a simple LMIs-constrained optimization problem.

Using the global linearization method [23,32], we suppose all the global linearization of the nonlinear stochastic financial system (14.1) are bounded by a polytope, consisting of L vertices as follows:

$$\begin{bmatrix} \frac{\partial f(x)}{\partial x} \\ \frac{\partial g(x)}{\partial x} \\ \frac{\partial M(x)}{\partial x} \\ \frac{\partial N(x)}{\partial x} \end{bmatrix} \in C_0 \left\{ \begin{bmatrix} A_1 \\ B_1 \\ M_1 \\ N_1 \end{bmatrix}, \cdots, \begin{bmatrix} A_i \\ B_i \\ M_i \\ N_i \end{bmatrix}, \cdots, \begin{bmatrix} A_L \\ B_L \\ M_L \\ N_L \end{bmatrix}, \forall x(t) \right\} \tag{14.20}$$

where C_0 denotes the convex hull of the polytope with L vertices. That is, if the local linearized systems at all $x(t)$ are inside the convex hull C_0, then the evolutionary trajectory $x(t)$ of the nonlinear stochastic financial system in (14.1) will belong to the convex combination of the trajectories of the following L locally linearized stochastic financial systems at L vertices of the polytope [23,32]:

$$dx(t) = [A_i x(t) + B_i u(t) + v(t)]dt + M_i x(t)dw(t) + N_i x(t)dp(t), i = 1, 2, \ldots, L \tag{14.21}$$

Based on the global linearization theory [23,32], if (14.20) holds, then every evolutionary trajectory of nonlinear stochastic financial system in (14.1) can be represented by the following convex combination of L local linearized stochastic financial systems:

$$dx(t) = \sum_{i=1}^{L} \alpha_i(x)\{[A_i x(t) + B_i u(t) + v(t)]dt + M_i x(t)dw(t) + N_i x(t)dp(t)\}, \tag{14.22}$$

where the interpolation function $\alpha_i(x)$ satisfies $0 \leq \alpha_i(x) \leq 1$ and $\sum_{i=1}^{L} \alpha_i(x) = 1$, i.e. if (14.20) holds, then the evolutionary trajectory of the nonlinear stochastic financial system in (14.1) can be represented by the evolutionary trajectory of the interpolated financial system in (14.22).

By the global linearization system in (14.22), the augmented financial system in (14.7) can be represented by

$$d\bar{X}(t) = \sum_{i=1}^{L} \alpha_i(x)\{[\bar{A}_i \bar{X}(t) + \bar{B}_i u(t) + \bar{v}(t)]dt + \bar{M}_i \bar{X}(t)dw(t) + \bar{N}_i \bar{X}(t)dp(t)\}, \tag{14.23}$$

where

$$\bar{A}_i = \begin{bmatrix} A_i & 0 \\ 0 & A_r \end{bmatrix}, \bar{B}_i = \begin{bmatrix} B_i \\ 0 \end{bmatrix}, \bar{M}_i = \begin{bmatrix} M_i & 0 \\ 0 & 0 \end{bmatrix}, \bar{N}_i = \begin{bmatrix} N_i & 0 \\ 0 & 0 \end{bmatrix}.$$

Based on the above global linearization scheme, the result in Proposition 14.1 can be replaced by Proposition 14.2.

Proposition 14.2 The Nash minimax quadratic game problem in (14.15) can be solved based on the global linearization system in (14.23) as follows:

$$u^*(t) = -\sum_{i=1}^{L} \alpha_i(\bar{X}) R^{-1} \bar{B}_i \bar{P} \bar{X}(t), \bar{v}^*(t) = \frac{1}{\rho} \bar{P} \bar{X}(t) \tag{14.24}$$

where $\bar{P} > 0$ is the solution of the following Riccati-like inequalities

$$\begin{aligned} \bar{H}_i(\bar{P}) := \bar{A}_i^T \bar{P} + \bar{P}\bar{A}_i + \frac{1}{\rho}\bar{Q} + \bar{P}\bar{P} - \rho \bar{P}\bar{B}_i R^{-1} \bar{B}_i^T \bar{P} + \bar{M}_i^T \bar{P}_i \\ + \lambda(\bar{N}_i^T \bar{P} \bar{N}_i + \bar{N}_i^T \bar{P} + \bar{P}\bar{N}_i) \leq 0, i = 1, 2, \ldots, L \end{aligned} \tag{14.25}$$

Proof: See Appendix.

In the proof of Proposition 14.2, the following lemma is necessary [22,32].

Lemma 14.2 For any matrix S_i with approximate dimension and interpolation function α_i with $0 \leq \alpha_i(x) \leq 1$ and $\sum_{i=1}^{L} \alpha_i(x) = 1$, then for $P>0$, we get

$$(\sum_{i=1}^{L} \alpha_i S_i)^T P (\sum_{i=1}^{L} \alpha_i S_i) \leq \sum_{i=1}^{L} \alpha_i S_i^T P S_i \tag{14.26}$$

Since ρ is the upper bound of stochastic minimax H_∞ game in (14.11), the stochastic minimax H_∞ game problem in (14.10) for the global linearization stochastic financial system in (14.22) still needs to solve the following Riccati-like inequalities-constrained optimization:

$$\rho_0 = \min_{\bar{P}>0} \rho \quad \text{subject to (14.25)} \tag{14.27}$$

In order to use the LMI's method to efficiently solve the Riccati-like inequalities (14.25), we let $W = \bar{P}^{-1}$, then, by Schur's Lemma [23], (14.25) is equivalent to the following LMIs:

$$\begin{bmatrix} \bar{H}_{11} & W\bar{M}_i^T & W\bar{N}_i^T & W\bar{I}^T \\ * & -W & 0 & 0 \\ * & * & -\frac{1}{\lambda}W & 0 \\ * & * & * & -\rho Q^{-1} \end{bmatrix} \leq 0, \tag{14.28}$$

where

$$\bar{H}_{11} = W\bar{A}_i^T + \bar{A}_i W + \lambda W\bar{N}_i^T + \lambda \bar{N}_i W + I - \rho \bar{B}_i R^{-1} \bar{B}_i^T, i = 1, 2, \ldots, L.$$

Therefore, the robust H_∞ model reference strategy based on stochastic H_∞ Nash game in (14.5) for the interpolated linearization financial system in (14.23) becomes how to solve the following LMIs-constrained optimization problem:

$$\rho_0 = \min_{W>0} \rho \quad \text{subject to (14.28)} \tag{14.29}$$

The above constrained optimization problem is called as an eigenvalue problem (EVP). This EVP can also be solved very efficiently by decreasing ρ until no positive $W>0$ exists. Software packages, such as the LMI optimization toolbox of Matlab, have been developed for this purpose and can be utilized to solve the problem. After solving ρ_0 and W^* from the LMIs-constrained optimization problem in (29). From (17), we get the robust model reference investment strategy $u^*(t)$ and the worst-case external disturbance $v^*(t)$ as follows:

$$u^*(t) = -\sum_{i=1}^{L} \alpha_i(\bar{X}) R^{-1} \bar{B}_i W^{*^{-1}} \bar{X}(t), \bar{v}^*(t) = \frac{1}{\rho_0} W^{*^{-1}} \bar{X}(t) \tag{14.30}$$

After we finish the robust model reference strategy of stochastic financial systems based on stochastic minimax H_∞ game strategy, two simulation examples of stochastic financial system are given to illustrate the design procedure and to confirm the performance of the proposed robust model reference strategy in the following two cases.

14.5 Simulation Examples

14.5.1 Simulation Results of the Stochastic Financial System (14.2)

To illustrate the design procedure and to compare the performance of the proposed robust H_∞ model reference investment policy for nonlinear stochastic jump diffusion financial system (14.2) to that of Pareto multi-objective strategy suggested in [22], the related parameters in system (14.2) are postulated as follows [22]:

$$
\begin{aligned}
&a = 1.5, b = 0.2, c = 0.25, \lambda = 0.5;\\
&m_1(x(t), y(t), z(t)) = 0.03 \times [z(t) + (y(t) - a)x(t)]\\
&m_2(x(t), y(t), z(t)) = 0.01 \times [1 - by(t) - (x(t))^2]\\
&m_3(x(t), y(t), z(t)) = 0.02 \times [-x(t) - cz(t)]\\
&n_1(x(t), y(t), z(t)) = 0.3 \times [z(t) + (y(t) - a)x(t)]\\
&n_2(x(t), y(t), z(t)) = 0.1 \times [1 - by(t) - (x(t))^2]\\
&n_3(x(t), y(t), z(t)) = 0.2 \times [-x(t) - cz(t)]
\end{aligned}
$$

Suppose the initial states are given as

$$x(0) = 0.37, y(0) = -1.06, z(0) = 0.71$$

Figure 14.1 describes the dynamical behaviors of the nonlinear stochastic jump diffusion financial system in (14.2) with random fluctuations in real situation including Wiener process and Poisson jumps illustrated in Fig. 14.2. It is seen that the three states $x(t)$, $y(t)$, and $z(t)$ of the nonlinear stochastic financial system fluctuate with random form to reflect the impact of financial turbulence and shock in the real world. The interest rate $x(t)$, investment demand $y(t)$, and the price $z(t)$ fluctuate wildly with high volatility, and the economics are not in the equilibrium state.

Therefore, the government's or investor's policy of the proposed model reference strategy of (14.4) can be employed by the government or decision-makers to regulate the stochastic financial system to follow along with the desired processes $(x_r(t), y_r(t), z_r(t))^T$ given by (14.4), see Fig. 14.3. Obviously, the desired trajectory of the reference $(x_r(t), y_r(t), z_r(t))^T$ is relatively smooth and mild, and the model references have a property of trending to a desired steady state gradually; see Fig. 14.3. For the desired model reference references prescribed in Fig. 14.3, the matrix A_r and the desired steady state process $r(t)$ in model reference system (14.4) are taken as

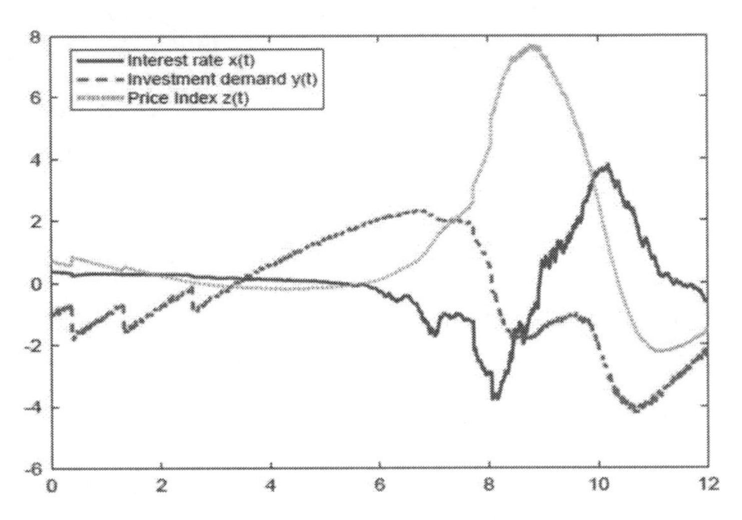

FIGURE 14.1
The system dynamic behaviors for nonlinear stochastic jump diffusion financial system in (14.2) with $u(t) = 0$.

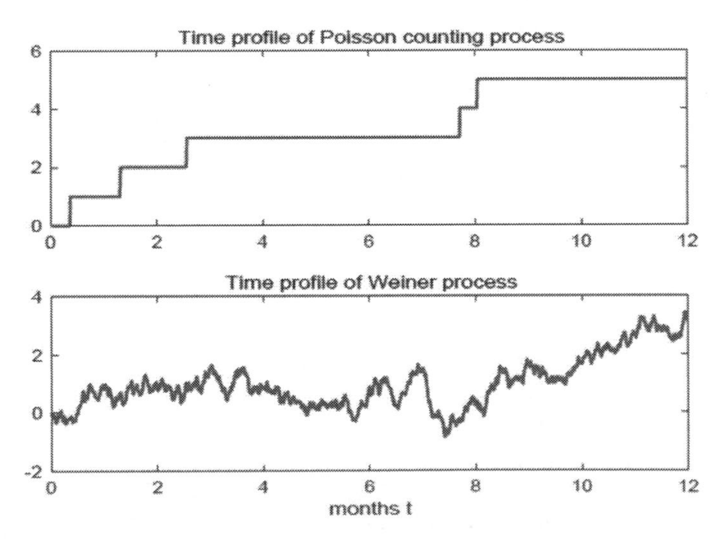

FIGURE 14.2
Time profiles of Brownian motion and Poisson counting process in system (14.2).

$$A_r = I_3, r(t) = \begin{bmatrix} 0.16 \\ 5 \\ 0.02 \end{bmatrix}$$

The weighting matrices Q and R in performance (14.5) are given as $Q = I_3$, and $R = I_3$. In order to design the robust model reference strategy based on the stochastic H_∞ Nash game, for the convenience of design, we first shift the origin of the coordinate to the

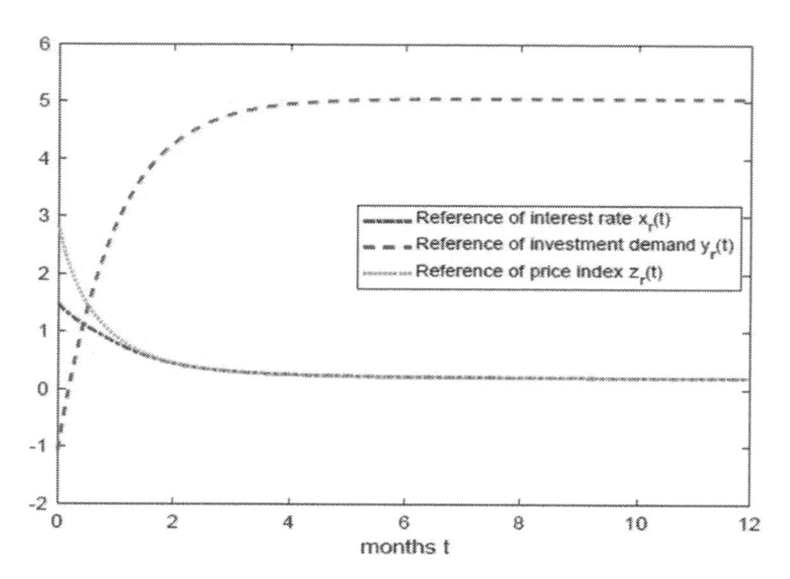

FIGURE 14.3
Trajectories of the prescribed model reference $x_r(t)$, $y_r(t)$, and $z_r(t)$. (Because the values of $yr(t)$ is too larger than that of $x_r(t)$ and $z_r(t)$, the scales of $x_r(t)$ and $z_r(t)$ are amplified by 4 times in this figure.)

desired target $X_e = (0.16, 5, 0.02)$ at the desired steady state, then every evolutionary trajectory of nonlinear stochastic financial system in (14.1) can be represented by the following convex combination of L local linearized stochastic financial system:

$$dX(t) = \sum_{i=1}^{12} \alpha_i(X(t))\{[A_i X(t) + B_i u(t) + v(t)]dt + M_i X(t)dw(t) + N_i X(t)dp(t)\}, \qquad (14.31)$$

where $X = (x - 0.16, y - 5, z - 0.02)^T$, and the interpolation function

$$\alpha_i(X(t)) = (1/\|X_i - X(t)\|_2^2)/[\sum_{i=1}^{M} (1/\|X_i - X(t)\|_2^2)]$$

satisfies $0 \le \alpha_i(x) \le 1$ and $\sum_{i=1}^{L} \alpha_i(x) = 1$ and A_i, M_i, N_i, and X_i are given in Appendix and

$$B = \begin{bmatrix} 1 & 0 & 0 \\ 0 & 1 & 0 \\ 0 & 0 & 1 \end{bmatrix}$$

Considering the above model reference system, by (14.23), the augmented financial system in (14.6) can be represented by

$$d\bar{X}(t) = \sum_{i=1}^{12} \alpha_i(x)\{[\bar{A}_i \bar{X}(t) + \bar{B}_i u(t) + \bar{v}(t)]dt + \bar{M}_i \bar{X}(t)dw(t) + \bar{N}_i \bar{X}(t)dp(t)\}. \qquad (14.32)$$

By solving the LMI-constrained optimization problem in (14.29) with

$$\bar{P}^* = W^{*^{-1}} = \begin{bmatrix} 4.7441 & -0.2882 & -0.5416 & -0.1995 & 0.0590 & 0.4357 \\ -0.2882 & 18.7425 & -2.3033 & 0.0418 & -1.8161 & -0.0114 \\ -0.5416 & -2.3033 & 4.8962 & -0.4462 & 0.1282 & -1.0259 \\ -0.1995 & 0.0418 & -0.4462 & 0.8143 & -0.0022 & -0.0224 \\ 0.0590 & -1.8161 & 0.1282 & -0.0022 & 0.7311 & -0.0018 \\ 0.4357 & -0.0114 & -1.0259 & -0.0224 & -0.0018 & 0.8242 \end{bmatrix},$$

we could obtain the optimal $\rho_0 = 0.9940$. Furthermore, by (14.24), the optimal model reference investment strategy $u^*(t)$ and the worst-case disturbance $v^*(t)$ are obtained

$$u^*(t) = -\rho_0 \bar{B}^T \bar{P}^* \bar{X}(t), \bar{v}^*(t) = \bar{P}^* \bar{X}(t) \tag{14.33}$$

where

$$\bar{B} = \begin{bmatrix} B \\ 0 \end{bmatrix} = \begin{bmatrix} 1 & 0 & 0 & 0 & 0 & 0 \\ 0 & 1 & 0 & 0 & 0 & 0 \\ 0 & 0 & 1 & 0 & 0 & 0 \end{bmatrix}^T.$$

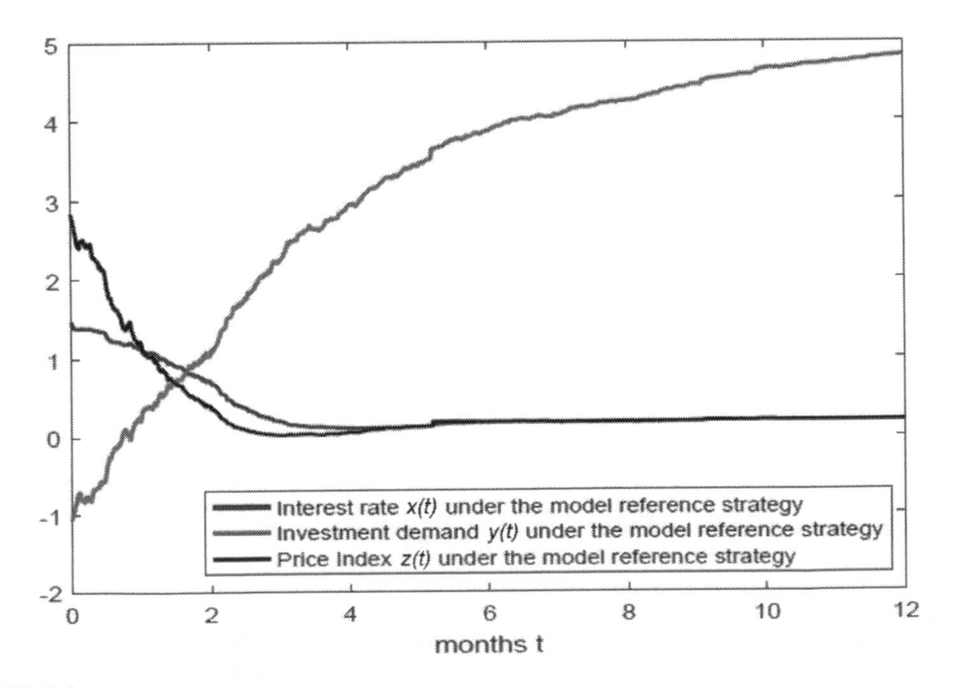

FIGURE 14.4
Trajectories of interest $x(t)$, investment demand $y(t)$, and the price index $z(t)$ under the proposed model reference strategy based on the stochastic H_∞ game scheme. (Since the scale of $y(t)$ is too larger than that of $x(t)$ and $z(t)$, in order to compare the trends of them, the values of $x(t)$ and $z(t)$ are amplified by 4 times in this figure.)

Figure 14.4 illustrates the trajectories of $x(t)$, $y(t)$, and $z(t)$ under the proposed model reference strategy based on stochastic minimax H_∞ game scheme under the worst-case disturbance $v^*(t)$. It shows that under the guidance of reference model, the government could apply the interest rate policy and the price policy to promote the demand growth in investment. At the beginning of the financial system under the proposed model reference investment strategy, the interest rate and the price index are too high, which leads to a low investment demand. When the interest rate and the price go down, the investment demand rises up. At the 4th month, the interest rate $x(t)$ and the price index $z(t)$ are near the desired state $x_d = 0.016$ and $z_d = 0.02$, then the country's or social economics has entered a stable development phase. In this financial system, the investment demand $y(t)$ has also entered a step of steady growth. From Fig. 14.4, we also found that, under the robust H_∞ model reference investment strategy, $x(t)$ and $z(t)$ need about 4 months to achieve their desired state (or near the steady state), but $y(t)$ needs about 10 months to achieve. This shows that, comparing to the trend of investment demand $y(t)$, the trends of interest $x(t)$ and price index $z(t)$ are more sharply to achieve the desired steady state. This is the reason why the interest and price are always employed by the government to adjust the rate of economic development, playing an early and better role in the national economical system.

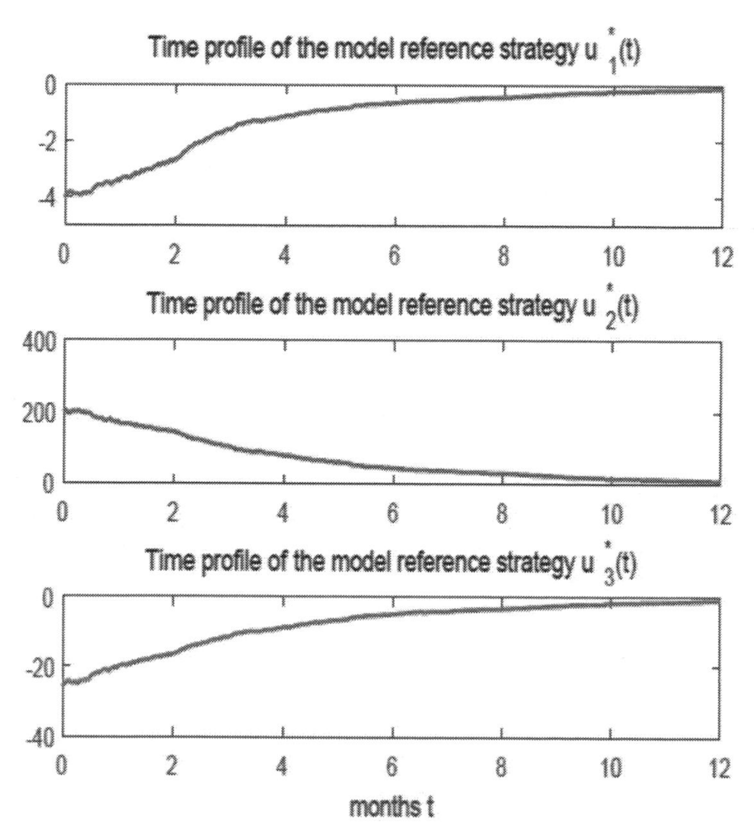

FIGURE 14.5
Time profiles of the proposed model reference control strategy $u^*(t) = (u_1^*(t), u_2^*(t), u_3^*(t))^T$

Figure 14.5 illustrates the financial system under the proposed robust H_∞ model reference investment strategy $u^*(t)$ based on the stochastic minimax H_∞ game scheme. It shows that the time profiles $u^*(t)$ are going smaller and smaller, i.e. their control values tend to zeros. This means that, along with the effect of applying economic leverage $x(t)$ and $z(t)$, the investment demand of the social economics has increased to a desired steady state, and the interest rate and price index have been decreased to the desired state, i.e. the social economic could enter to a new steady state. After entering the desired steady stage, the government's investment on the proposed model reference investment policy is also reduced.

Figure 14.6 illustrates the time profiles of worst-case disturbance. It shows that the worst-case disturbance $v^*(t)$ is trending to zero as the interest $x(t)$, investment demand $y(t)$, and price $z(t)$ are going to the new desired steady state. This implies that the capacity of eliminating random fluctuations and resisting disturbance of the financial system has been enhanced by the robust model reference investment strategy under the stochastic minimax H_∞ Nash game scheme.

Figures 14.7–14.9 illustrate the trajectories of $(x(t), y(t), z(t))^T$ under the proposed model reference investment strategy based on H_∞ game scheme, the desired trajectories $(x_r(t), y_r(t), z_r(t))^T$, and the trajectories of $(x(t), y(t), z(t))^T$ under the multi-objective control of Pareto strategy suggested by [22], respectively. The three figures show that under the proposed model reference investment strategy, $(x(t), y(t), z(t))^T$ can

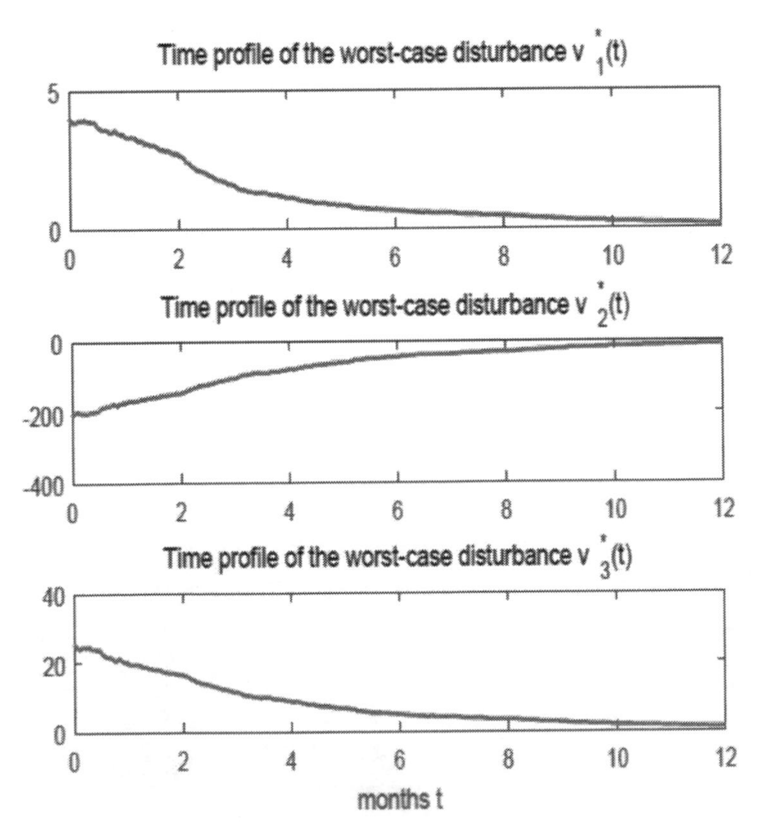

FIGURE 14.6
The time profiles of the worst-case disturbance $v^*(t) = (v_1^*(t), v_2^*(t), v_3^*(t))^T$.

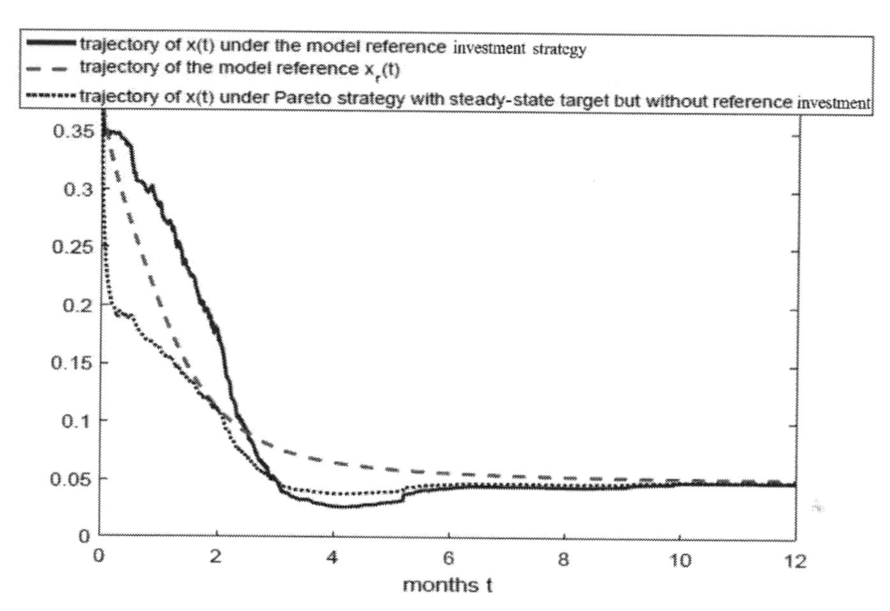

FIGURE 14.7
Comparing the interest rate trajectories $x(t)$ of the proposed model reference investment strategy with reference $x_r(t)$ and $x(t)$ of Pareto optimal strategy in [22].

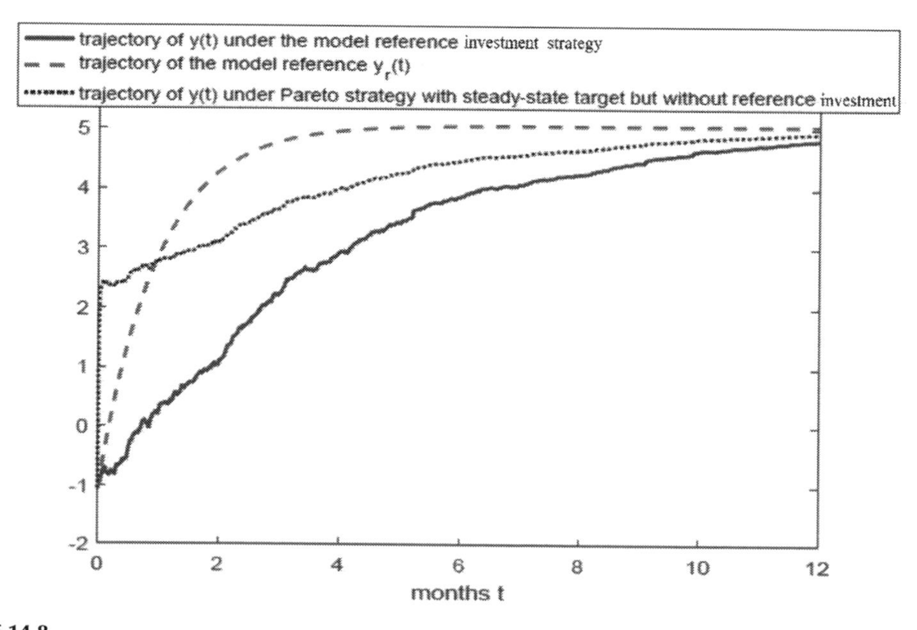

FIGURE 14.8
Comparing the investment demand trajectory $y(t)$ of the proposed model reference investment strategy with model reference $y_r(t)$ and $y(t)$ of Pareto optimal strategy in [22].

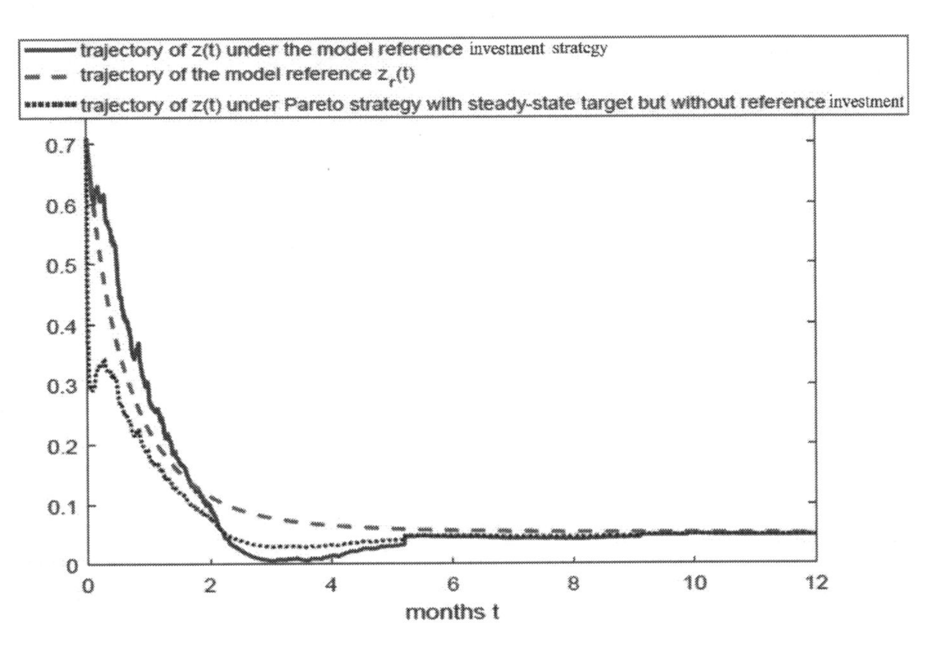

FIGURE 14.9
Comparing the price index trajectory $z(t)$ of the proposed model reference investment strategy with model reference $z_r(t)$ and $z(t)$ of Pareto optimal strategy in [22].

follow the trend of the model reference $(x_r(t), y_r(t), z_r(t))^T$ to more gradually achieve the desired target. Comparing to the results under the multi-objective control of Pareto optimal strategy suggested by [22], the trajectories of $(x(t), y(t), z(t))^T$ of the financial system are more gentle and gradual.

This shows that the proposed H_∞ model reference strategy can lead to a smoother and gradual change in the economic development than the robust multi-objective investment strategy based on Pareto optimal strategy suggested by [22], which has a sharp jump at the beginning of control strategy.

14.5.2 Simulation Results of the Financial System in (14.3)

To illustrate the design procedure and the performance of the proposed minimax H_∞ model reference investment strategy for nonlinear stochastic international capital flow volatility system in (14.3), some related parameters in system (14.3) are given as follows [325]:

$$
\begin{aligned}
&a = 0.223, b = 0.106, c_{sv} = 0.98, c_{vs} = 1.03, \lambda = 2;\\
&m_1(x(t), y(t), z(t)) = 0.1 \times [a - c_{sv}x_v(t)x_s^2(t)]\\
&m_2(x(t), y(t), z(t)) = 0.2 \times [-bx_s(t) + c_{vs}x_v(t)x_s^2(t)]\\
&n_1(x(t), y(t), z(t)) = 0.0.1 \times [a - c_{sv}x_v(t)x_s^2(t)]\\
&n_2(x(t), y(t), z(t)) = 0.03 \times [-bx_s(t) + c_{vs}x_v(t)x_s^2(t)]
\end{aligned}
$$

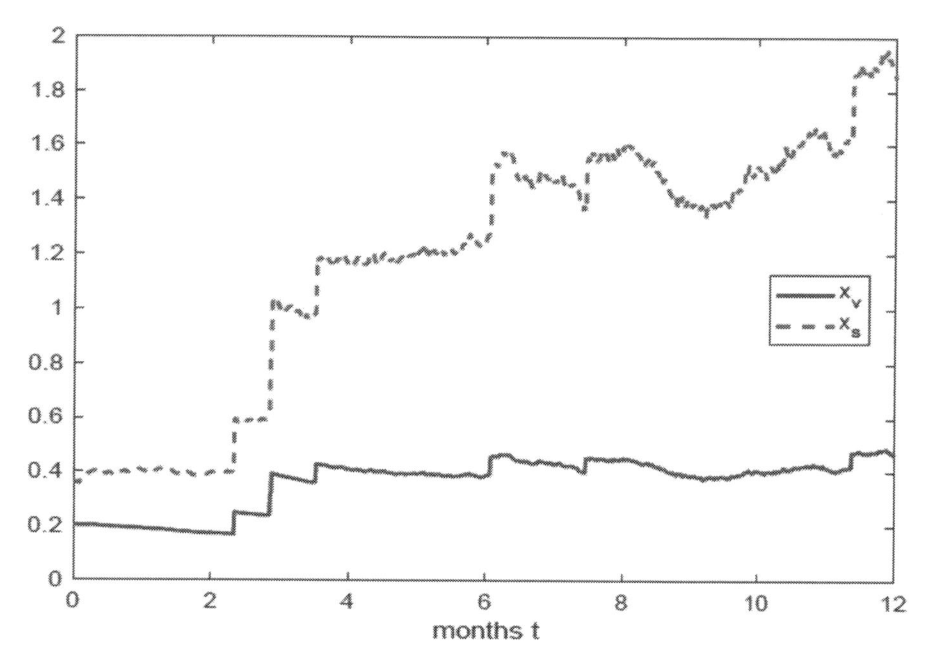

FIGURE 14.10
The system dynamic behaviors for nonlinear stochastic jump diffusion financial system in (14.3) with $u(t) = 0$. (Because the scale of $x_s(t)$ is too larger than that of $x_v(t)$, in order to compare the trends of them, the value of $x_v(t)$ is amplified by 10 times in this figure.)

Suppose the initial states are given as

$$x_v(0) = 0.092, x_s(0) = 1.61$$

Figure 14.10 describes the dynamical behaviors of the nonlinear stochastic jump diffusion financial system in (14.3) with much random fluctuations in real situation including Wiener process and Poisson jumps illustrated in Fig. 14.11. It is seen that the two states $x_v(t)$ and $x_s(t)$ of the nonlinear stochastic financial system fluctuate with random turbulence and jumps. From Fig. 14.10, we see that the net capital inflows of the source country maintains increasing level. This causes the net capital inflow of the volatility-affected country decreasing. Of course, those trends are not volatility around their equilibriums. Our objective is to design a proper control strategy to maintain the capital account balance, and we also hope this process goes smoothly and not so sharply, especially at the beginning stage. So the proposed H_∞ model reference tracking control strategy of (14.14) is employed to follow along with the model reference $x_{vr}(t)$and $x_{sr}(t)$; see Fig. 14.12.

Obviously, the variation tendency of $x_{vr}(t)$and $x_{sr}(t)$ given by (14.4) is relatively mild, and the references model also has a property of trending to a steady state gradually; see Fig. 14.12. The matrix A_r and process $r(t)$ in reference system (14.4) are taken as

$$A_r = -I_2, r(t) = \begin{bmatrix} x_{ve} \\ x_{se} \end{bmatrix}$$

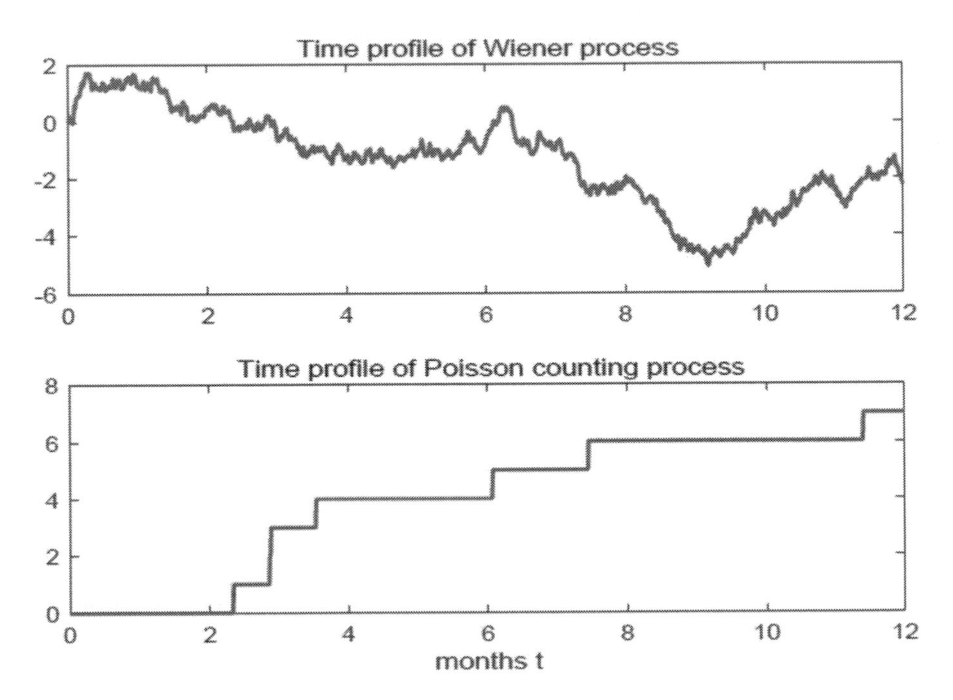

FIGURE 14.11
Time profiles of Brownian motion and Poisson counting process in system (14.3).

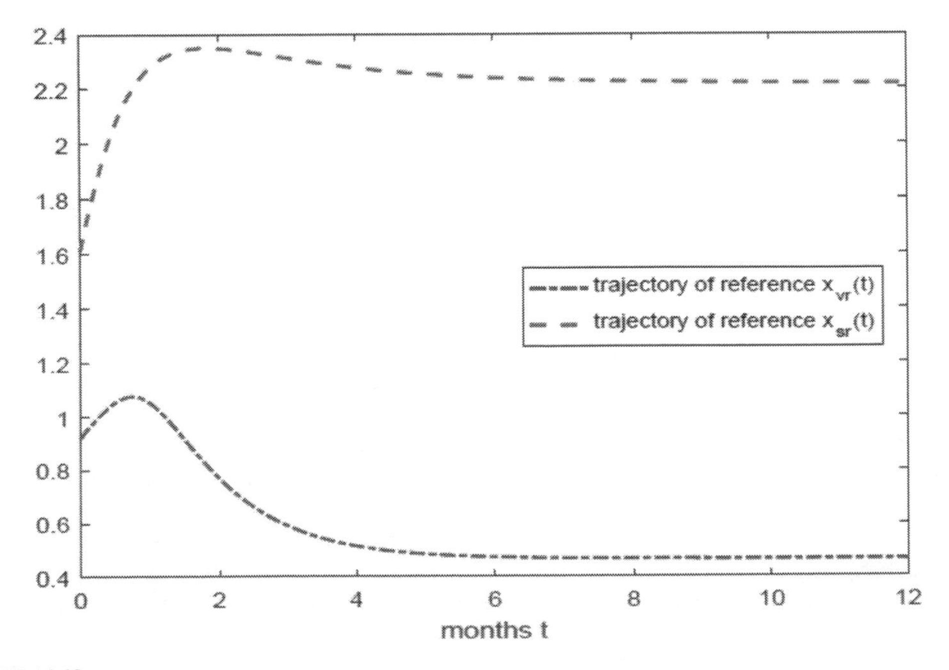

FIGURE 14.12
Trajectories of $x_{vr}(t)$ and $x_{sr}(t)$. (Since the scale of $x_{sr}(t)$ is too larger than that of $x_{vr}(t)$, in order to compare the trends of them, the time profile of $x_{vr}(t)$ is amplified by 10 times in this figure.)

The weighting matrices Q and R in performance (14.5) are given as $Q = I_2$ and $R = I_2$. In order to design the H_∞ Nash game tracking strategy by the proposed global linearization method, we first shift the origin coordinates to the desired steady state $X_e = (x_{ve}, x_{se})^T = (0.045, 2.21)^T$ of (14.3), then every evolutionary trajectory of nonlinear stochastic financial system in (14.3) can be represented by the following convex combination of 4 local linearized stochastic financial system:

$$dX(t) = \sum_{i=1}^{4} \alpha_i(X(t))\{[A_iX(t) + Bu(t) + v(t)]dt + M_iX(t)dw(t) + N_ix(t)dp(t)\} \quad (14.34)$$

where $X = (x_v - x_{ve}, x_s - x_{se})^T$, and the interpolation function

$$\alpha_i(X(t)) = (1/\|X_i - X(t)\|_2^2)/[\sum_{i=1}^{M} (1/\|X_i - X(t)\|_2^2)]$$

satisfies $0 \le \alpha_i(x) \le 1$ and $\sum_{i=1}^{L} \alpha_i(x) = 1$ and A_i, M_i, N_i, and X_i are given in Appendix and

$$B = \begin{bmatrix} 1 & 0 \\ 0 & 1 \end{bmatrix}$$

Considering the reference system, by (14.23), the augmented financial system in (14.6) can be represented by

$$d\bar{X}(t) = \sum_{i=1}^{4} \alpha_i(x)\{[\bar{A}_i\bar{X}(t) + \bar{B}u(t) + \bar{v}(t)]dt + \bar{M}_i\bar{X}(t)dw(t) + \bar{N}_i\bar{X}(t)dp(t)\} \quad (14.35)$$

By solving the LMI-constrained optimization problem in (14.27) with

$$\bar{P}^* = \begin{bmatrix} 0.8282 & 0.3899 & -0.1586 & -0.2327 \\ 0.3899 & 0.6253 & 0.1067 & -0.4201 \\ -0.1586 & 0.1067 & 1.0100 & 0.04020 \\ -0.2327 & -0.4201 & 0.04020 & 0.3604 \end{bmatrix},$$

we could obtain the optimal $\rho_0 = 0.9726$. Furthermore, by (14.24), the robust H_∞ model reference control strategy u^* and the worst-case disturbance v^* are obtained

$$u^*(t) = -\rho_0\bar{B}^T\bar{P}\bar{X}(t), \bar{v}^*(t) = \bar{P}\bar{X}(t) \quad (14.36)$$

where

$$\bar{B} = \begin{bmatrix} B \\ 0 \end{bmatrix} = \begin{bmatrix} 1 & 0 & 0 & 0 \\ 0 & 1 & 0 & 0 \end{bmatrix}^T$$

Figure 14.13 illustrates the trajectories of $x_v(t)$ and $x_s(t)$ under the proposed model reference tracking control strategy $u^*(t)$ based on the minimax H_∞ Nash game scheme (see Fig. 14.14) and the worst-case disturbance $v^*(t)$ (see Fig. 14.15). We see that, under

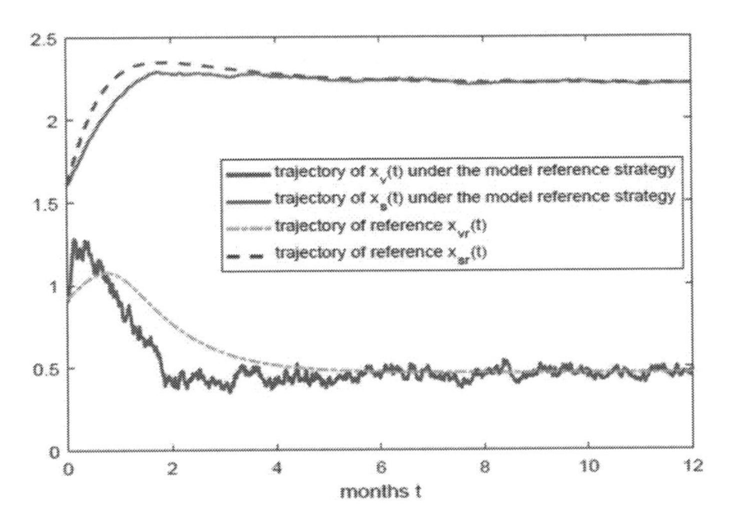

FIGURE 14.13
Trajectories of $x_v(t)$ and $x_s(t)$ under the proposed model reference tracking strategy. (Since the scale of $x_v(t)$ is too larger than that of $x_s(t)$, in order to compare the trends of them, the value of $x_v(t)$ is amplified by 10 times in this figure.)

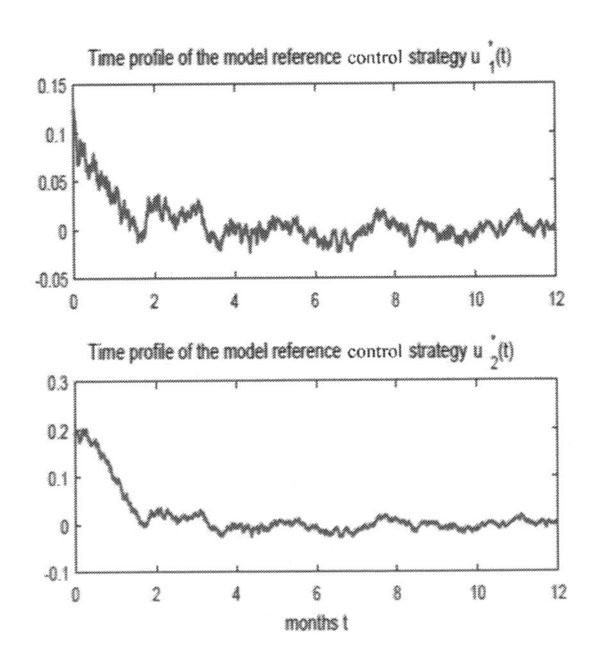

FIGURE 14.14
Time profiles of the proposed robust model reference tracking strategy $u^*(t) = (u_1^*(t), u_2^*(t))^T$.

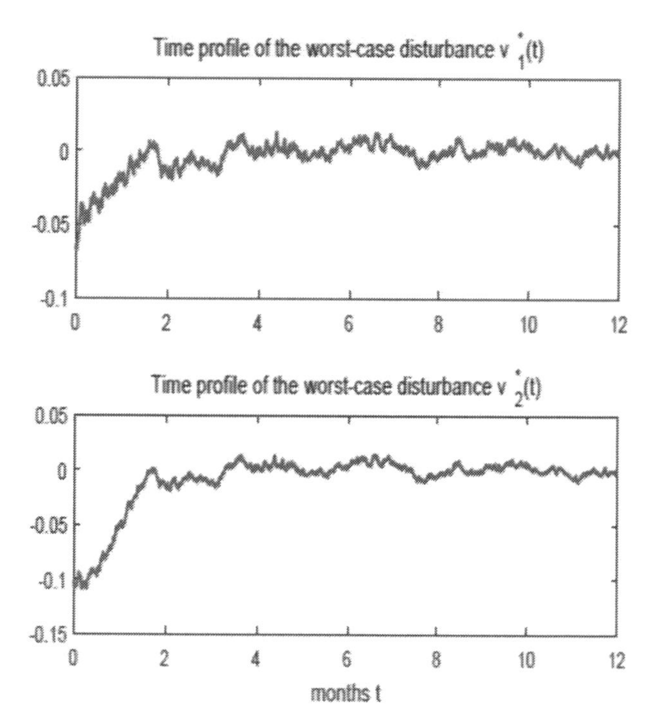

FIGURE 14.15
Time profiles of worst-case disturbance $v^*(t) = (v_1^*(t), v_2^*(t))^T$.

the effect of the proposed model reference tracking strategy, the net capital inflow x_s of the source country increases gradually to the desired steady state. Because the source country has a large net capital inflow, the increase of x_s leads to the decrease of x_v, until they arrive the states around the desired steady state.

14.6 Conclusion

In this chapter, a robust H_∞ model reference investment strategy for nonlinear stochastic system based on the minimax H_∞ game scheme is introduced to solve the desired reference tracking control problems in the stochastic financial and economic systems with continuous random fluctuation and discontinuous random jumps. For the nonlinear stochastic financial systems, based on the global linearization technique, we obtain the robust H_∞ model reference investment strategy by solving a LMI-constrained optimization problem with the help of LMI-toolbox. From the two simulation results, based on the proposed robust H_∞ model reference investment strategy, the managers or policy-maker could achieve the desired dynamic behavior and target of the financial system because its trajectories could be efficiently guided by the prescribed reference model despite the random turbulence and jumping fluctuation.

14.7 Appendix

14.7.1 Proof of proposition 14.1

Proof. Applying Itô's formula (14.16) to $\rho V(\bar{X}(t))$, we have

$$E\{\rho V(\bar{X}(t_p))\} - E\{\rho V(\bar{X}(0))\} = \rho E\{\int_0^{t_p} [\frac{\partial V}{\partial \bar{X}}[F(\bar{X}) + G(\bar{X})u + \bar{v}(t)] + \frac{1}{2}\bar{M}^T(\bar{X})\frac{\partial^2 V}{\partial \bar{X}^2}\bar{M}(\bar{X}) + \lambda[V(\bar{X} + \bar{N}(\bar{X})) - V(\bar{X})]]dt\}$$

So

$$\begin{aligned}
&E\{\int_0^{t_p} [\bar{X}^T\bar{Q}\bar{X} + u^T(t)Ru(t) - \rho\bar{v}^T(t)\bar{v}(t)]dt\} \\
&= \rho E\{V(\bar{X}(0))\} - \rho E\{V(\bar{X}(t_p))\} + \rho E\{\int_0^{t_p} [\frac{\partial V}{\partial \bar{X}}[F(\bar{X}) + G(\bar{X})u + \bar{v}(t)] \\
&+ \frac{1}{2}\bar{M}^T(\bar{X})\frac{\partial^2 V}{\partial \bar{X}^2}\bar{M}(\bar{X}) + \lambda[V(\bar{X} + \bar{N}(\bar{X})) - V(\bar{X})] + \frac{1}{\rho}\bar{X}^T(t)\bar{Q}\bar{X}(t) \\
&+ \frac{1}{\rho}u^T(t)Ru(t) - \bar{v}^T(t)\bar{v}(t)]dt\}
\end{aligned}$$

Completing the square for u and $\bar{v}$ on the right-hand side, we obtain

$$\begin{aligned}
&E\{\int_0^{t_p} [\bar{X}^T\bar{Q}\bar{X} + u^T(t)Ru(t) - \rho\bar{v}^T(t)\bar{v}(t)]dt\} \\
&= \rho E\{V(\bar{X}(0))\} - \rho E\{V(\bar{X}(t_p))\} + \rho E\{\int_0^{t_p} H(V(\bar{X}) \\
&+ \frac{1}{\rho}(u(t) - u^*(t))^T R(u(t) - u^*(t)) - (\bar{v}(t) - \bar{v}^*(t))^T(\bar{v}(t) - \bar{v}^*(t))]dt\}
\end{aligned}$$

Let $u(t) = u^*(t)$, by (14.18), and reminding $E\{V(\bar{X}(t_p))\} \geq 0$, we have

$$\max_u \min_{\bar{v}} E\{\int_0^{t_p} [\bar{X}^T\bar{Q}\bar{X} + u^T(t)Ru(t) - \rho\bar{v}^T(t)\bar{v}(t)]dt\} \leq \rho E\{V(\bar{X}(0))\}$$

i.e. the minimax problem in (14.12) or (14.14) is achieved, which implies that $u^*(t)$ and $\bar{v}^*(t)$ are solutions of the Nash quadratic game problem (14.15). This ends the proof.

14.7.2 Proof of Proposition 14.2

Proof. Let $V(\bar{X}(t)) = \bar{X}^T(t)P\bar{X}(t)$, and then applying Itô's formula to $\rho V(\bar{X}(t)) = \rho\bar{X}^T(t)P\bar{X}(t)$, there exists

$$E\{\int_0^{t_p} [\bar{X}^T(t)\bar{Q}\bar{X}(t) + u^T(t)Ru(t) - \rho\bar{v}^T(t)\bar{v}(t)]dt\}$$
$$= \rho E\{V(\bar{X}(0))\} - \rho E\{V(\bar{X}(t_p))\} + \rho \int_0^{t_p} [\bar{X}^T(t)\bar{H}_i(\bar{P})\bar{X}(t)$$
$$+ \frac{1}{\rho}(u(t) - u_i^*(t))^T R(u(t) - u_i^*(t)) - (\bar{v}(t) - \bar{v}^*(t))^T R(\bar{v}(t) - \bar{v}^*(t))]dt$$

where

$$u_i^*(t) = -R^{-1}\bar{B}_i\bar{P}\bar{X}(t).$$

Keeping (14.23) in mind, we get

$$E\{\int_0^{t_p} [\bar{X}^T(t)\bar{Q}\bar{X}(t) + u^T(t)Ru(t) - \rho\bar{v}^T(t)\bar{v}(t)]dt\}$$
$$= \rho E\{V(\bar{X}(0))\} - \rho E\{V(\bar{X}(t_p))\} + \rho \int_0^{t_p} \sum_{i=1}^{L} \alpha_i(\bar{X})[\bar{X}^T(t)\bar{H}_i(\bar{P})\bar{X}(t)$$
$$+ \frac{1}{\rho}(u(t) - u_i^*(t))^T R(u(t) - u_i^*(t)) - (\bar{v}(t) - \bar{v}^*(t))^T R(\bar{v}(t) - \bar{v}^*(t))]dt$$

By (14.25) and (14.26), we have

$$E\{\int_0^{t_p} [\bar{X}^T(t)\bar{Q}\bar{X}(t) + u^T(t)Ru(t) - \rho\bar{v}^T(t)\bar{v}(t)]dt\}$$
$$\leq \rho E\{V(\bar{X}(0))\} - \rho E\{V(\bar{X}(t_p))\} + \rho\{\int_0^{t_p} \{\sum_{i=1}^{L} \alpha_i(\bar{X})[\bar{X}^T(t)\bar{H}_i(\bar{P})\bar{X}(t)$$
$$+ \frac{1}{\rho}[u(t) - \sum_{i=1}^{L} \alpha_i(\bar{X})u_i^*(t)]^T R[u(t) - \sum_{i=1}^{L} \alpha_i(\bar{X})u_i^*(t)] - (\bar{v}(t) - \bar{v}^*(t))^T R(\bar{v}(t) - \bar{v}^*(t))]\}dt\}$$

Let $u = u^*$, and then applying (14.25) and reminding $E\{V(\bar{X}(t_p))\} \geq 0$, we have

$$\max_u \min_{\bar{v}} E\{\int_0^{t_p} [\bar{X}^T(t)\bar{Q}\bar{X}(t) + u^T(t)Ru(t) - \rho\bar{v}^T(t)\bar{v}(t)]dt\} \leq \rho E\{V(\bar{X}(0))\}$$

i.e. the minimax problem in (14.12) or (14.14) is achieved, which implies that $u^*(t)$ and $\bar{v}^*(t)$ are solutions of the Nash quadratic game problem (14.15). This ends the proof.

Matrices A_i, M_i, N_i and X_i in (14.31)

$$A_1=\begin{bmatrix}-1.5&0&1\\-0.05&-0.2&0\\-1&0&-0.25\end{bmatrix},A_2=\begin{bmatrix}-1.5&0&1\\-0.1&-0.2&0\\-1&0&-0.25\end{bmatrix},A_3=\begin{bmatrix}-1.5&0&1\\-0.25&-0.2&0\\-1&0&-0.25\end{bmatrix},$$

$$A_4=\begin{bmatrix}-2&0&1\\-0.05&-0.2&0\\-1&0&-0.25\end{bmatrix},A_5=\begin{bmatrix}-2&0&1\\-0.1&-0.2&0\\-1&0&-0.25\end{bmatrix},A_6=\begin{bmatrix}-2&0&1\\-0.25&-0.2&0\\-1&0&-0.25\end{bmatrix},$$

$$A_7=\begin{bmatrix}-3&0&1\\-0.05&-0.2&0\\-1&0&-0.25\end{bmatrix},A_8=\begin{bmatrix}-3&0&1\\-0.1&-0.2&0\\-1&0&-0.25\end{bmatrix},A_9=\begin{bmatrix}-3&0&1\\-0.25&-0.2&0\\-1&0&-0.25\end{bmatrix},$$

$$A_{10}=\begin{bmatrix}-4&0&1\\-0.05&-0.2&0\\-1&0&-0.25\end{bmatrix},A_{11}=\begin{bmatrix}-4&0&1\\-0.1&-0.2&0\\-1&0&-0.25\end{bmatrix},A_{12}=\begin{bmatrix}-4&0&1\\-0.25&-0.2&0\\-1&0&-0.25\end{bmatrix},$$

$$M_1=\begin{bmatrix}-0.045&0&0.03\\-0.0005&-0.002&0\\-0.02&0&-0.005\end{bmatrix},M_2=\begin{bmatrix}-0.045&0&0.03\\-0.001&-0.002&0\\-0.02&0&-0.005\end{bmatrix},M_3=\begin{bmatrix}-0.045&0&0.03\\-0.0025&-0.002&0\\-0.02&0&-0.005\end{bmatrix},$$

$$M_4=\begin{bmatrix}-0.06&0&0.03\\-0.0005&-0.002&0\\-0.02&0&-0.005\end{bmatrix},M_5=\begin{bmatrix}-0.06&0&0.03\\-0.001&-0.002&0\\-0.02&0&-0.005\end{bmatrix},M_6=\begin{bmatrix}-0.06&0&0.03\\-0.0025&-0.002&0\\-0.02&0&-0.005\end{bmatrix},$$

$$M_7=\begin{bmatrix}-0.09&0&0.03\\-0.0005&-0.002&0\\-0.02&0&-0.005\end{bmatrix},M_8=\begin{bmatrix}-0.09&0&0.03\\-0.001&-0.002&0\\-0.02&0&-0.005\end{bmatrix},M_9=\begin{bmatrix}-0.09&0&0.03\\-0.0025&-0.002&0\\-0.02&0&-0.005\end{bmatrix},$$

$$M_{10}=\begin{bmatrix}-0.12&0&0.03\\-0.0005&-0.002&0\\-0.02&0&-0.005\end{bmatrix},M_{11}=\begin{bmatrix}-0.12&0&0.03\\-0.001&-0.002&0\\-0.02&0&-0.005\end{bmatrix},M_{12}=\begin{bmatrix}-0.012&0&0.03\\-0.0025&-0.002&0\\-0.02&0&-0.005\end{bmatrix},$$

$$N_1=\begin{bmatrix}-0.45&0&0.3\\-0.005&-0.02&0\\-0.2&0&-0.05\end{bmatrix},N_2=\begin{bmatrix}-0.45&0&0.3\\-0.01&-0.02&0\\-0.2&0&-0.05\end{bmatrix},N_3=\begin{bmatrix}-0.45&0&0.3\\-0.025&-0.02&0\\-0.2&0&-0.05\end{bmatrix},$$

$$N_4=\begin{bmatrix}-0.6&0&0.3\\-0.005&-0.02&0\\-0.2&0&-0.05\end{bmatrix},N_5=\begin{bmatrix}-0.6&0&0.3\\-0.01&-0.02&0\\-0.2&0&-0.05\end{bmatrix},N_6=\begin{bmatrix}-0.6&0&0.3\\-0.025&-0.02&0\\-0.2&0&-0.05\end{bmatrix},$$

$$N_7=\begin{bmatrix}-0.9&0&0.3\\-0.005&-0.02&0\\-0.2&0&-0.05\end{bmatrix},N_8=\begin{bmatrix}-0.9&0&0.3\\-0.01&-0.02&0\\-0.2&0&-0.05\end{bmatrix},N_9=\begin{bmatrix}-0.9&0&0.3\\-0.025&-0.02&0\\-0.2&0&-0.05\end{bmatrix},$$

$$N_{10}=\begin{bmatrix}-1.2&0&0.3\\-0.005&-0.02&0\\-0.2&0&-0.05\end{bmatrix},N_{11}=\begin{bmatrix}-0.6&0&0.3\\-0.005&-0.02&0\\-0.2&0&-0.05\end{bmatrix},N_{12}=\begin{bmatrix}-1.2&0&0.3\\-0.025&-0.02&0\\-0.2&0&-0.05\end{bmatrix},$$

$X_1=(0.05,-5)^T, X_2=(0.1,-5)^T, X_3=(0.25,-5)^T, X_4=(0.05,-5.5)^T, X_5=(0.1,-5.5)^T, X_6=(0.25,-5.5)^T,$
$X_7=(0.05,-6.5)^T, X_8=(0.1,-6.5)^T, X_9=(0.25,-6.5)^T, X_{10}=(0.05,-7.5)^T, X_{11}=(0.1,-7.5)^T,$
$X_{12}=(0.25,-7.5)^T.$

Matrices A_i, M_i, N_i and X_i in (14.34)

$$A_1 = \begin{bmatrix} -2.5438 & -0.17434 \\ 2.6735 & 0.077236 \end{bmatrix}, A_2 = \begin{bmatrix} -3.0395 & -0.18118 \\ 3.1945 & 0.084427 \end{bmatrix}, A_3 = \begin{bmatrix} -3.3945 & -0.18574 \\ 3.5676 & 0.089221 \end{bmatrix},$$
$$A_4 = \begin{bmatrix} -3.769 & -0.19031 \\ 3.9613 & 0.094015 \end{bmatrix}, M_1 = \begin{bmatrix} -0.25438 & -0.017434 \\ 0.53471 & 0.015447 \end{bmatrix}, M_2 = \begin{bmatrix} -0.30395 & -0.018118 \\ 0.63891 & 0.016885 \end{bmatrix},$$
$$M_3 = \begin{bmatrix} -0.33945 & -0.018574 \\ 0.71353 & 0.017844 \end{bmatrix}, M_4 = \begin{bmatrix} -0.3769 & -0.019031 \\ 0.79227 & 0.018803 \end{bmatrix}, N_1 = \begin{bmatrix} -0.025438 & 0.0017434 \\ 0.080206 & 0.0023171 \end{bmatrix},$$
$$N_2 = \begin{bmatrix} -0.30395 & -0.018118 \\ 0.095836 & 0.0025328 \end{bmatrix}, N_3 = \begin{bmatrix} -0.033945 & -0.0018574 \\ 0.10703 & 0.0026766 \end{bmatrix}, N_4 = \begin{bmatrix} -0.03769 & -0.0019031 \\ 0.11884 & 0.0028205 \end{bmatrix},$$
$$X_1 = (0, 1.6111)^T, X_2 = (0, 1.7611)^T, X_3 = (0, 1.8611)^T, X_4 = (0, 1.9611)^T.$$

15

Multi-Person Noncooperative H_∞ Stochastic Game Strategy of Stochastic Financial Systems with Random Fluctuations: Multiobjective Optimization Approach

15.1 Introduction

In the financial markets, stocks, and social economics, all kinds of economic problems due to the interaction between nonlinear factors become more and more complicated today [318,332–336]. Recently, nonlinear systematic dynamic models have been widely applied to represent the financial systems and practical economies [142,318,320,332,333,337]. Actually, however, financial systems are nonlinear stochastic systems, which may suffer from stochastic fluctuations because of the variations of domestic and international situation, varying interest rate, investment strategy, and economic environment [22,126,128,142,322,323]. Also, a large amount of external disturbances like the unpredictable investment changes or a world event can influence the random fluctuations of financial market, such as national disaster, war, fatal epidemic disease, etc. [22,126,128,322,323,329,330,334–336,338]. We can classify these stochastic parameter fluctuations as continuous state-dependent random fluctuation and discontinuous (jump) state-dependent random fluctuation. They will affect the stability and controllability of the stochastic financial system. Thus, a nonlinear stochastic system with continuous Wiener fluctuation and discontinuous Poisson fluctuation is employed to describe continuous and discontinuous fluctuations of a nonlinear stochastic financial system [32,34,329,330,338] in this chapter.

Inherently, game theory involves multi-person decision-making [1,2,9]. It is noncooperative if each person involved pursues his/her own interests which are partly conflicting with others [1,17,19]. Recently, extensive researches have been carried out on the new features of multi-person noncooperative game problems in different research fields [339–342], such as the study of the convergence for Nash equilibria [339], an adaptive algorithm for noncooperative game with unknown quadratic objectives [340], the Stackelberg game with leader and multiple noncooperative followers [341], and the solutions to the noncooperative consensus problems via mean field theory [342]. In the financial markets, managers and investors involve in pursuing their own interests partly conflicting with each other. Therefore, noncooperative game strategies have been widely applied to the multi-person noncooperative decision-making problem of financial market. However, at present, there still exists no effective way to solve the nonlinear stochastic multi-person noncooperative game strategy problem for nonlinear stochastic financial systems [1,2,9]. In general, iterative schemes with updating algorithm are employed to solve the Nash equilibrium points for multi-person noncooperative game strategy directly [1,9,339], which is not easily tackled by conventional methods, especially for nonlinear stochastic financial systems.

Compared with the aforementioned iterative schemes by updating other player's information [1,9,339] to solve the Nash equilibrium solution iteratively for multi-person noncooperative game strategy, the proposed multi-person noncooperative stochastic H_∞ game strategy design problem is transformed to an equivalent multiobjective optimization problem (MOP) to effectively obtain the Nash equilibrium solution for the strategies of all players simultaneously in a single run with the help of multiobjective evolutionary algorithm (MOEA). Therefore, the proposed noncooperative stochastic H_∞ game strategy is more realistic and applicable to the investment strategy design in the nonlinear stochastic financial systems.

At present, the existing noncooperative games are always based on quadratic utility functions. In general, these noncooperative games based on the H_2 (quadratic) utility function are not easy to treat the effect of unavailable competitive strategies of other players. In order to overcome the effect of unavailable competitive strategies of other players, the utility function of each player in the proposed multi-person noncooperative H_∞ game is based on the worst-case effect of combined strategies of other players on their strategy from the H_∞ regulation perspective to robustly achieve their desired targets. Further, the external disturbance, intrinsic Wiener process, and Poisson process have few appeared in the conventional noncooperative game problem. In this chapter, these external disturbance, intrinsic continuous Wiener process, and discontinuous Poisson process are modeled in our nonlinear stochastic financial system for realistic fluctuation of financial market so that we could obtain more practical m-person noncooperative stochastic H_∞ investment strategy for a more realistic financial market.

On the other hand, the cooperative game strategy design problem has attracted significant interest in both academia and industry, which has abundant applications in different research topics [343–346]. Generally, the cooperative game strategy design is to find the corresponding strategy of each player so that all of the interests of players can reach the same agreement. Specifically, the cooperative containment control problems in non-identical networks and discrete-time multi-agent systems have been investigated in [343,344], respectively. A linear quadratic regulator (LQR)-based optimal distributed cooperative game design has been proposed for synchronization control of linear discrete-time multi-agent systems [345]. To achieve the global optimality and acceptable convergence speed in the distributed cooperative consensus protocols, an optimal partial stabilization method has been developed for the distributed cooperative control of general linear multi-agent systems by using the inverse optimal approach [346].

In this chapter, the multi-person noncooperative nonlinear stochastic H_∞ game problem of the investment strategy of nonlinear stochastic financial systems is firstly transformed to an equivalent Halmition-Jacobi-Isacc inequalities (HJIIs)-constrained MOP. We could also prove the solution of HJIIs-constrained MOP is the Nash equilibrium solution of multi-person noncooperative H_∞ game strategy of nonlinear stochastic financial systems. Since there is no analytic or numerical good method yet to solve the HJIIs [8,34,134], the global linearization method for nonlinear stochastic financial systems is applied to approach a set of local linearized stochastic financial systems to close in the nonlinear stochastic financial system and simplify the multi-person noncooperative H_∞ stochastic game problem. According to the global linearization technique [23,32], the HJIIs-constrained MOP is transformed to an equivalent linear matrix inequalities (LMIs)-constrained MOP, i.e., for the nonlinear stochastic financial systems, the multi-person noncooperative H_∞ game for the investment strategy design problem could be transformed to an LMIs-constrained MOP.

At present, a number of MOEAs for MOP have been developed for their excellent searching capability for the Pareto optimal solutions by a single execution run [22]. However, to the best of our knowledge, current researchers only investigate MOP from the algebraic perspective [25,35,136,138,139]. For the nonlinear stochastic financial systems, few of these papers have discussed the LMIs-constrained MOP, especially about the multi-person noncooperative stochastic H_∞ game strategy. In this chapter, based on NSGA-II MOEA in [25,136], we have developed an LMIs-constrained MOEA algorithm for effectively solving the LMIs-constrained MOP for the multi-person noncooperative stochastic H_∞ game of investment strategy design in a single run. Finally, two designs of noncooperative stochastic H_∞ game strategy of stochastic nonlinear financial systems, one noncooperative H_∞ game strategy about dynamic investment [22,142] and another noncooperative H_∞ game strategy about international capital flow volatility [325,347] are given to describe the systematic design procedure and to confirm the performance of multi-person noncooperative stochasticH_∞ game strategy of nonlinear stochastic financial systems with continuous and jumping fluctuations and external disturbances.

The main aims of this chapter are summarized as follows: (i) Nonlinear stochastic systems with intrinsic continuous Wiener process and discontinuous Poisson jumping process are introduced to model random fluctuations and external disturbances of nonlinear stochastic financial systems and a nonlinear stochastic noncooperative H_∞ game is also introduced for multi-person noncooperative investment strategies in conflict with one another. (ii) An indirect method is introduced to solve the multi-person noncooperative stochastic H_∞ game strategy problem by transforming to an equivalent MOP with a set of HJII constraints. We also proved that the solutions of HJIIs-constrained MOP are just the Nash equilibrium solution of multi-person noncooperative stochastic H_∞ game strategy of nonlinear stochastic financial solutions. (iii) In order to solve the HJIIs-constrained MOP for noncooperative stochastic H_∞ game strategy design of nonlinear stochastic financial systems, a global linearization method is presented to interpolate several local linearized stochastic financial systems to approximate the nonlinear stochastic financial system so that the HJIIs-constrained MOP can be converted to an LMIs-constrained MOP to simplify the design procedure of noncooperative stochastic H_∞ game strategy of nonlinear stochastic financial systems. (iv) We developed an LMIs-constrained MOEA algorithm to effectively solve the LMIs-constrained MOP for multi-person noncooperative stochastic H_∞ game in the investment strategy of nonlinear stochastic financial system in a single run with the help of the Matlab LMI Toolbox.

15.2 System Description of Nonlinear Stochastic Financial System and Multi-Person Noncooperative Stochastic H_∞ Game for Investment Strategy

In general, most economic financial systems are nonlinear stochastic systems because they may suffer from random intrinsic fluctuation and external disturbance due to the variation of oil price, the occurrence of war, the change of government policy, the surplus between investment and saving, etc. [22,126,128,142,318,320,322,323,329,330,332–338]. Let us consider the following nonlinear economic or financial stochastic system with m-person decision-makers or investors:

$$dx(t) = \left[f(x(t)) + \sum_{i=1}^{m} g_i(x(t))u_i(t) + h(x(t))v(t) \right] dt \\ + l(x(t))dw(t) + n(x(t))dp(t), \tag{15.1}$$

where $x(t) = [x_1(t), ..., x_i(t), ..., x_n(t)]^T$ denotes the state vector of nonlinear economic or financial stochastic system. $u_i(t) = [u_{i1} \cdots u_{ik}]^T$, $i = 1, ..., m$ denote the control strategies or policies of m decision-makers or investors (we will call them agents or players in the following). $v(t)$ denotes the external disturbance. $w(t)$ is the standard 1-dimensional Wiener process. $l(x(t))dw(t)$ denotes continuous random fluctuation. $p(t)$ denotes Poisson counting process with mean λ in a unit time and $n(x(t))dp(t)$ is regarded as intrinsic discontinuous random fluctuation (jumping process). The nonlinear interaction vector $f(x(t))$ of economic or financial stochastic system satisfies with Lipschitz continuity and $g_i(x(t))$ denotes the input matrix of the ith control strategy $u_i(t)$ of the ith player. The following two simple examples are given to illustrate the nonlinear stochastic financial system in (15.1):

Example 15.1: The following is a nonlinear financial dynamic model to demonstrate the interplay between each state of the market [128,142]:

$$\begin{aligned} \dot{x}(t) &= z(t) + (y(t) - a)x(t), \\ \dot{y}(t) &= 1 - by(t) - x^2(t), \\ \dot{z}(t) &= -x(t) - cz(t), \end{aligned} \tag{15.2}$$

where the interest rate, the investment demand, and the price index are denoted as $x(t)$, $y(t)$, and $z(t)$, respectively. The parameters $a \geq 0$, $b \geq 0$, and $c \geq 0$ are the saving amount, the per-investment cost, and the resilience of demands of commercials, respectively. The regulations of the prices, and balance between savings and investment can affect the interest rate $x(t)$ in the nonlinear financial model (15.2). It is found that the investment demand $y(t)$ is in direct proportion to the investment rate. In addition, the investment demand is inversely related to the investment expense and interest rate. Moreover, the discrepancy between demand and supplying influences the price index $z(t)$, which decides the price in the nonlinear financial system. Also, the price index is negatively influenced by the interest rate [142].

However, in the realistic financial system, it is crucial that the nonlinear dynamic financial model in (15.2) must be amended by continuous and discontinuous inherent random fluctuation as well as environmental disturbance and the investment strategies of investors as follows [22,126,128,322]:

$$\begin{aligned} dx(t) &= [z(t) + (y(t) - a)x(t) + u_{11}(t) + 0.5u_{21}(t) + 0.3u_{31}(t) \\ &\quad + v_1(t)]dt + x(t)dw(t) + 0.5z(t)dp(t), \\ dy(t) &= [1 - by(t) - x^2(t) + u_{12}(t) + 0.5u_{22}(t) + 0.3u_{32}(t) \\ &\quad + v_2(t)]dt + 0.5y(t)dw(t) - 0.1x^2(t)dp(t), \\ dz(t) &= [x(t) - cz(t) + u_{13}(t) + 0.2u_{23}(t) + 0.2u_{33}(t) \\ &\quad + v_3(t)]dt - 0.7z(t)dw(t) + 0.3z(t)dp(t), \end{aligned} \tag{15.3}$$

where the Wiener process $w(t)$ and Poisson counting process $p(t)$ are presented to model the continuous and discontinuous random fluctuation, and $v(t) = [v_1(t)\ v_2(t)\ v_3(t)]^T$ denotes the corresponding external disturbance. $u_1(t) = [u_{11}(t)u_{12}(t)u_{13}(t)]^T$ denotes the control strategy of government, $u_2(t) = [u_{21}(t)u_{22}(t)u_{23}(t)]^T$ denotes the investment strategy of bank consortium, and $u_3(t) = [u_{31}(t)u_{32}(t)u_{33}(t)]^T$ denotes the investment strategy of public. Nonlinear stochastic system (15.3) may describe a practical financial market system, to model the sell and buy characteristics, or the price variations of the objects of transaction like futures, stocks, bonds, and oil or changes in economic indicator, comprising investment rate, investment demand, and price index in the more real financial markets.

Example 15.2: Recently, the nonlinear stochastic system theory has become a pioneering issue in financial contagion problem due to the global impact of financial crisis. It is discovered that while the impact of financial shock is detected, the net capital flow in source country of financial turbulence intensely declines, joining to the driving force [325]. Also, for the volatility-affected country, the net capital flow diverges from the normal equilibrium value point to respond to the infectiousness effects from the source country of financial turbulence [325]. A nonlinear dynamic financial system of international capital flow volatility between the source country of financial turbulence and the volatility-affected country could be expressed by the nonlinear dynamic equation [325] as follows:

$$\begin{aligned} \frac{dx_v(t)}{dt} &= a - c_{sv}x_v(t)x_s^2(t), \\ \frac{dx_s(t)}{dt} &= -bx_s(t) + c_{vs}x_v(t)x_s^2(t), \end{aligned} \tag{15.4}$$

where $x_v(t)$ indicates the historical tendency of net capital flow for the volatility-affected country of financial turbulence; $x_s(t)$ for the source country, $a \neq 0$ denotes the inertial coefficient of the volatility-affected country, $b>0$ is the inertial coefficient of the source country of financial turbulence; c_{sv} is the coefficient of the impact from the source country of the financial turbulence to the volatility-affected country; c_{vs} is the coupling coefficient for the impact from the volatility country to the source country of financial turbulence.

Actually, in the real international capital flow volatility system, the two financial coupling strengths are always stochastically uncertain and should be modified as the following nonlinear stochastic continuous and discontinuous random fluctuations as well as external disturbance and control strategy:

$$\begin{aligned} dx_v(t) &= [a - c_{sv}x_v(t)x_s^2(t) + u_{11}(t) + 0.7u_{21}(t) + v_1(t)]dt \\ &\quad + 3.4[a - c_{sv}x_v(t)x_s^2(t)]dw(t) + 4.1[a - c_{sv}x_v(t)x_s^2(t)]dp(t), \\ dx_s(t) &= [-bx_s(t) + c_{vs}x_v(t)x_s^2(t) + 0.7u_{12}(t) + u_{22}(t) + v_2(t)]dt \\ &\quad + 3.4[-bx_s(t) + c_{vs}x_v(t)x_s^2(t)]dw(t) + 4.1[-bx_s(t) \\ &\quad + c_{vs}x_v(t)x_s^2(t)]dp(t), \end{aligned} \tag{15.5}$$

where $u_1(t) = [u_{11}(t)u_{12}(t)]^T$ denotes the control strategy of volatility-affected country and $u_2(t) = [u_{21}(t)u_{22}(t)]^T$ denotes the control strategy of source country of financial turbulence.

After two examples in (15.3) and (15.5) of nonlinear stochastic financial system are given to illustrate nonlinear economic or financial stochastic system in (15.1) with m players, we then want to design m-person noncooperative stochastic H_∞ game strategies for each player to achieve their desired targets as possible despite the effect of intrinsic continuous and jumping random fluctuation as well as external disturbance and competitive strategies of other players. For each game strategy $u_i(t)$, it assumes that all other competitive game strategies $u_1(t)\ldots u_{i-1}(t), u_{i+1}(t)\ldots u_m(t)$ and environmental disturbance $v(t)$ are unavailable for the ith player, which can be bunched together as a competitive game strategy vector to $u_i(t)$ to simplify the analysis. Since external disturbance $v(t)$ plays an important role and its knowledge becomes unavailable to each player in the nonlinear stochastic financial control system, its influence on the nonlinear stochastic financial system must be thought from the worst-case scenario and each player tends to minimize the worst-case effect of external disturbance. Therefore, external disturbance $v(t)$ should be considered as a player in the m-person noncooperative stochastic H_∞ game strategy of stochastic financial system. Thus, the nonlinear stochastic financial system with m players in (15.1) could be expressed by

$$\begin{aligned} dx(t) = {} & [f(x(t)) + g_i(x(t))u_i(t) + g_{-i}(x(t))u_{-i}(t)]dt \\ & + l(x(t))dw(t) + n(x(t))dp(t), \end{aligned} \tag{15.6}$$

where

$$\begin{aligned} g_{-i}(x(t)) &= [g_1(x(t)), \ldots, g_{i-1}(x(t)), g_{i+1}(x(t)), \ldots, g_m(x(t)), h(x(t))], \\ u_{-i}(t) &= [u_1^T(t), \ldots, u_{i-1}^T(t), u_{i+1}^T(t), \ldots, u_m^T(t), v^T(t)]^T \end{aligned}$$

In the noncooperative stochastic H_∞ game strategy of nonlinear financial stochastic system, it is hard to acquire all other competitive strategies and external disturbance. Thus, each game strategy $u_i(t)$ of player i attempts to simultaneously minimize the worst-case effect of the combined competitive strategies and external disturbance to achieve his or her desired target despite the continuous and discontinuous random fluctuation (i.e., to simultaneously minimize the deviation to desired target $\tilde{x}_i(t) = x(t) - x_{di}$ with a parsimonious control effort $u_i(t)$ under the maximal effect of other competitive strategies and external disturbance as the following stochastic m-person noncooperative stochastic H_∞ game problem):

$$\begin{aligned} \rho_i^* = \min_{u_i(t)} \max_{u_{-i}(t)} \frac{E\int_0^{t_p} [(x(t) - x_{di})^T Q_i (x(t) - x_{di}) + u_i^T(t) R_i u_i(t)]dt}{E(x(0) - x_{di})^T (x(0) - x_{di}) + E\int_0^{t_p} u_{-i}^T(t) u_{-i}(t)dt}, \\ \text{for } i = 1, 2, \ldots, m, \end{aligned} \tag{15.7}$$

where x_{di} denotes the desired target of player i; Q_i and R_i indicate the corresponding weighting matrices selected by each player to compromise between the deviation $\tilde{x}_i(t)$ to desired target and the strategy effort of $u_i(t)$ towards the goal of player i; t_p denotes the terminal time. By the fact $\tilde{x}_i(t) = x(t) - x_{di}$, for the convenience of the desired target tracking design, we get the following shifted financial system of (15.6):

$$\begin{aligned} d\tilde{x}_i(t) = {} & [f(\tilde{x}_i(t)) + g_i(\tilde{x}_i(t))u_i(t) + g_{-i}(\tilde{x}_i(t))u_{-i}(t)]dt \\ & + l(\tilde{x}_i(t))dw(t) + n(\tilde{x}_i(t))dp(t), \end{aligned} \tag{15.8}$$

where $f(\tilde{x}_i(t)) = f(\tilde{x}_i(t) + x_{di}), g_i(\tilde{x}_i(t)) = g_i(\tilde{x}_i(t) + x_{di}), h(\tilde{x}_i(t)) = h(\tilde{x}_i(t) + x_{di}),$

$$g_{-i}(\tilde{x}_i(t)) = g_{-i}(\tilde{x}_i(t) + x_{di}), l(\tilde{x}_i(t)) = l(\tilde{x}_i(t) + x_{di}), n(\tilde{x}_i(t)) = n(\tilde{x}_i(t) + x_{di}).$$

Then, the m-person noncooperative stochastic H_∞ game problem in (15.7) is modified as

$$\rho_i^* = \min_{u_i(t)} \max_{u_{-i}(t)} \frac{E\int_0^{t_p} [\tilde{x}_i(t)^T Q_i \tilde{x}_i(t) + u_i^T(t) R_i u_i(t)]dt}{E\tilde{x}_i(0)^T \tilde{x}_i(0) + E\int_0^{t_p} u_{-i}^T(t) u_{-i}(t)dt}, \quad \text{for } i = 1, 2, \ldots, m, \tag{15.9}$$

Remark 15.1 (i) The m-person noncooperative stochastic H_∞ game strategy $(u_1^*, u_2^*, \ldots, u_m^*)$ constitutes a Nash equilibrium solution if and only if [1]

$$\begin{aligned}
(\rho_1^*, \rho_2^*, \ldots, \rho_m^*) &\le (\rho_1, \rho_2^*, \ldots, \rho_m^*),\\
&\vdots\\
(\rho_1^*, \rho_2^*, \ldots, \rho_i^*, \ldots, \rho_m^*) &\le (\rho_1^*, \rho_2^*, \ldots \rho_{i-1}^*, \rho_i, \rho_{i+1}^* \ldots, \rho_m^*),\\
&\vdots\\
(\rho_1^*, \rho_2^*, \ldots, \rho_m^*) &\le (\rho_1^*, \ldots, \rho_{m-1}^*, \rho_m),\\
&\text{for } i = 1, 2, \ldots, m,
\end{aligned} \tag{15.10}$$

where $(\alpha_1, \beta_1) \le (\alpha_2, \beta_2)$ if and only if $\alpha_1 \le \alpha_2, \beta_1 \le \beta_2$ and at least one aforementioned inequality is a strict inequality; i.e., dominates (α_2, β_2). In other words, no player could gain any benefit if he/she changes his/her strategy from Nash equilibrium solution.

(ii) The solution vector of the m-person noncooperative stochastic H_∞ game strategy $(u_1^*, u_2^*, \ldots, u_m^*)$ is indicated as $(\rho_1^*, \ldots, \rho_i^*, \ldots, \rho_m^*)$. Choice of weighting matrices Q_i, R_i and different targets x_{di} in (15.9) could give rise to different game strategies with different degrees of noncooperative characteristic among these players. Since these players' purposes of these strategies in (15.9) are not fully in conflict with one another, a certain degree of cooperation still exists in the multi-person noncooperative stochastic H_∞ game strategies. All players have to assign their weighting matrices Q_i, R_i and target x_{di} so that the game strategies involve some grades of cooperative feature to achieve their purposes within a given game in the financial market.

(iii) The physical significance of the noncooperative stocahstic H_∞ game strategy in (15.9) is given as follows:

(iii-a) The primary purpose of each player is to minimize the state error $\tilde{x}(t)$ from his target x_{di} by parsimonious effort $u_i(t)$; $x_{di} = 0$ for all i in the traditional noncooperative game strategy in [1]. Namely, only the stabilization issue is considered in traditional noncooperative game strategies.

(iii-b) The worst-case influences of both the strategies of other players and external disturbance on the target of each player are to be simultaneously minimized, because the external disturbance serves as a player in the financial system. Since the external disturbance is not easily handled by the conventional noncooperative H_2 (quadratic) game problem in [1], external disturbance $v(t)$ is always ignored to simplify the problem.

(iv) In the linear stochastic financial systems, equation (15.1) should be modified as

$$dx(t) = \left[Ax(t) + \sum_{i=1}^{m} B_i u_i(t) + Hv(t)\right]dt + Lx(t)dw(t) + Nx(t)dp(t).$$

In this situation, equation (15.6) should be modified as the following form:

$$\begin{aligned} dx(t) &= [Ax(t) + B_i u_i(t) + B_{-i}u_{-i}(t)]dt \\ &\quad + Lx(t)dw(t) + Nx(t)dp(t). \end{aligned} \tag{15.11}$$

where $B_{-i} = [B_1, \ldots, B_{i-1}, B_{i+1}, \ldots, B_m]$.

Since the Wiener process $w(t)$ and Poisson counting process $p(t)$ are non-differentiable, the following lemma is crucial for the solution of stochastic noncooperative H_∞ game strategy of nonlinear stochastic financial systems.

Lemma 15.1 (Itô–Lévy Lemma): The Itô–Lévy formula can be expressed as the following for the nonlinear stochastic financial system in (15.8) with a Lyapunov function $V(\tilde{x}(t)) > 0$ [34,126].

$$\begin{aligned} dV(\tilde{x}(t)) &= \left(\frac{\partial V(\tilde{x}_i(t))}{\partial \tilde{x}_i(t)}\right)^T (f(\tilde{x}_i(t)) + g_i(\tilde{x}_i(t))u_i(t) + g_{-i}(\tilde{x}_i(t))u_{-i}(t))dt \\ &\quad + \left(\frac{\partial V(\tilde{x}_i(t))}{\partial \tilde{x}_i(t)}\right)^T l(\tilde{x}_i(t))dw(t) + \frac{1}{2} l^T(\tilde{x}_i(t)) \frac{\partial^2 V(\tilde{x}_i(t))}{\partial \tilde{x}_i^2(t)} l(\tilde{x}_i(t))dt \\ &\quad + [V(\tilde{x}_i(t) + n(\tilde{x}_i(t))) - V(\tilde{x}_i(t))]dp(t). \end{aligned} \tag{15.11}$$

15.3 Noncooperative Stochastic H∞ Game Strategies for Nonlinear Stochastic Financial Systems

Basically, it is hard to resolve the multi-person noncooperative stochastic H_∞ game problem in (15.9) for nonlinear stochastic financial systems directly. Consequently, this study adopts the indirect method to simultaneously minimize the corresponding upper bounds as follows:

$$\rho_i^* = \min_{u_i(t)} \max_{u_{-i}(t)} \frac{E\int_0^{t_p} [\tilde{x}_i(t)^T Q_i \tilde{x}_i(t) + u_i^T(t) R_i u_i(t)]dt}{E\tilde{x}_i^T(0)\tilde{x}_i(0) + E\int_0^{t_p} u_{-i}^T(t) u_{-i}(t)dt} \le \rho_i, \quad \text{for } i = 1, 2, \ldots, m, \tag{15.13}$$

where $\rho_i > 0$ for the ith strategy indicates the upper bound of ρ_i^*. The design procedure provides an initial upper bound ρ_i of ρ_i^*. To approach ρ_i^*, $\forall i$ at the same time, these initial upper bounds need to be minimized as small as possible simultaneously. This indirect approach finally realizes the real m minimax optimization problems of m-person noncooperative stochastic H_∞ game strategy in (15.13). Namely, we must solve the MOP by simultaneously minimizing the relevant upper bounds in (15.13) as follows:

$$(\rho_1^*, \ldots, \rho_i^*, \ldots, \rho_m^*) = \min_{(u_1(t), \ldots, u_i(t), \ldots, u_m(t))} (\rho_1, \ldots, \rho_i, \ldots, \rho_m), \tag{15.14}$$

subject to

$$\rho_i^* = \min_{u_i(t)} \max_{u_{-i}(t)} \frac{E \int_0^{t_p} [\tilde{x}_i(t)^T Q_i \tilde{x}_i(t) + u_i^T(t) R_i u_i(t)] dt}{E \tilde{x}_i^T(0) \tilde{x}_i(0) + E \int_0^{t_p} u_{-i}^T(t) u_{-i}(t) dt} \leq \rho_i, \tag{15.15}$$

for $i = 1, 2, \ldots, m$,

In this chapter, the multi-person noncooperative stocahstic H_∞ game problem in (15.9) or (15.13) is transformed to a MOP in (15.14) and (15.15) and we need to design noncooperative game strategies $(u_1, \ldots, u_i, \ldots, u_m)$ to simultaneously minimize the objective vector $(\rho_1, \ldots, \rho_i, \ldots, \rho_m)$ in the Pareto optimal sense [22,138,139]. In general, the solution of MOP in (15.14) and (15.15) is not unique and called Pareto optimal solution. Before the further investigation of the MOP in (15.14) and (15.15) for m-person noncooperative stochastic H_∞ game strategy of nonlinear stochastic financial systems, some fundamental concepts of MOPs need to be introduced.

Definition 15.1 (Pareto dominance) [25,35,136,138,139]: For two feasible solutions $(u_1^1, \ldots, u_m^1)$ and $(u_1^2, \ldots, u_m^2)$ corresponding to objective values $(\rho_1^1, \ldots, \rho_m^1)$ and $(\rho_1^2, \ldots, \rho_m^2)$ of MOP in (15.14) and (15.15), respectively, the solution $(u_1^1, \ldots, u_m^1)$ is said to dominate $(u_1^2, \ldots, u_m^2)$ if $\rho_1^1 \leq \rho_1^2, \ldots, \rho_m^1 \leq \rho_m^2$ and at least one of the inequalities is of strict inequality.

Definition 15.2 (Pareto optimality) [25,35,136,138,139]: The feasible solution $(u_1, \ldots, u_m)$ is said to be Pareto optimality with respect to the feasible set if and only if there does not exist another feasible solution that dominates it.

Definition 15.3 (Pareto optimal solution set) [25,35,136,138,139]: For a given MOP in (15.14) and (15.15) with feasible solution $(u_1, \ldots, u_m)$ and corresponding objective vector $(\rho_1, \ldots, \rho_i, \ldots, \rho_m)$, Pareto optimal solution set σ^* is defined as $\sigma^* \triangleq \{(u_1^*, \ldots, u_m^*) | (u_1^*, \ldots, u_m^*) \text{is Pareto optimality}\}$, where $(u_1^*, u_2^*, \ldots, u_m^*)$ denotes the Pareto optimal solution of the objective values $(\rho_1^*, \ldots, \rho_m^*)$ in (15.14) and (15.15).

Definition 15.4 (Pareto front) [25,35,136,138,139]: For a given MOP in (15.14) and (15.15) with feasible solution $(u_1, \ldots, u_m)$, the corresponding objective vector $(\rho_1, \ldots, \rho_i, \ldots, \rho_m)$, and the Pareto optimal solution set σ^*, the Pareto front is defined as $P_F \triangleq \{(\rho_1^*, \ldots, \rho_m^*) | (u_1^*, \ldots, u_m^*) \in \sigma^*\}$.

Lemma 15.2 The indirect multiobjective problem in (15.14) and (15.15) is equivalent to the noncooperative H_∞ game problem in (15.9) or (15.13) when the Pareto optimality is satisfied.

Proof: The proof of Lemma 15.2 only needs to intuitively derive that the inequalities in (15.13) or (15.15) turn into equalities for the Pareto optimal solutions by contradiction. First, the problem defines the Pareto optimal solution $(\rho_1^*, \ldots, \rho_i^*, \ldots, \rho_m^*)$ of MOP in (15.14) and (15.15). We suppose that a strict inequality at the optimality holds for any one of the inequalities in (15.13) or (15.15). Without loss of generality, let us suppose $\min_{u_i(t)} \max_{u_{-i}(t)} \frac{E \int_0^{t_p} [\tilde{x}_i(t)^T Q_i \tilde{x}_i(t) + u_i^T(t) R_i u_i(t)] dt}{E \tilde{x}_i^T(0) \tilde{x}_i(0) + E \int_0^{t_p} u_{-i}^T(t) u_{-i}(t) dt} = \rho_i'$ such that $\rho_i' < \rho_i^*$, which means the $(\rho_1^*, \ldots, \rho_i', \ldots, \rho_m^*)$ dominates Pareto optimal solution $(\rho_1^*, \ldots, \rho_i^*, \ldots, \rho_m^*)$. The above consequence results in a contradiction. Thus, when the Pareto optimal solution of MOP in (15.14) and (15.15) is reached, we can conclude that m inequalities in (15.15) definitely become equalities.

The MOP in (15.14) and (15.15) is assuredly equivalent to the noncooperative game problem in (15.9).

Q.E.D.

Since the minimization of numerator of (15.15) by $u_i(t)$ is independent on $u_{-i}(t)$, the m minimax constraints in (15.15) are equivalent to the m Nash quadratic game constraints as follows [1,9]:

$$\begin{aligned}\rho_i^* &= \min_{u_i(t)} \max_{u_{-i}(t)} E \int_0^{t_p} [\tilde{x}_i(t)^T Q_i \tilde{x}_i(t) + u_i^T(t) R_i u_i(t) - \rho_i u_{-i}^T(t) u_{-i}(t)] dt \\ &\leq \rho_i E \tilde{x}_i^T(0) \tilde{x}_i(0), \\ &\text{for } i = 1, 2, \ldots, m,\end{aligned} \tag{15.16}$$

Before further discussion, we denote

$$\begin{aligned}J_i &= E \int_0^{t_p} [\tilde{x}_i(t)^T Q_i \tilde{x}_i(t) + u_i^T(t) R_i u_i(t) - \rho_i u_{-i}^T(t) u_{-i}(t)] dt \\ &\leq \rho_i E \tilde{x}_i^T(0) \tilde{x}_i(0), \\ &\text{for } i = 1, 2, \ldots, m,\end{aligned} \tag{15.17}$$

It requires two steps to solve the m stochastic Nash game problems with upper constraints in (15.16). In the first step, we need to solve the following m Nash game problems

$$\begin{aligned}J_i^* &= \min_{u_i(t)} \max_{u_{-i}(t)} J_i \\ &= \min_{u_i(t)} \max_{u_{-i}(t)} E \int_0^{t_p} [\tilde{x}_i(t)^T Q_i \tilde{x}_i(t) + u_i^T(t) R_i u_i(t) - \rho_i u_{-i}^T(t) u_{-i}(t)] dt \\ &\text{for } i = 1, 2, \ldots, m,\end{aligned} \tag{15.18}$$

In the second step, we should solve the following upper constrained problems:

$$J_i^* \leq \rho_i E \tilde{x}_i^T(0) \tilde{x}_i(0), \ \text{for } i = 1, 2, \ldots, m, \tag{15.19}$$

By two-step procedure for the m constrained Nash game problem in (15.16), solving (15.18) and (15.19) brings the following result in preparation for the MOP in (15.14) and (15.15).

Theorem 15.1

The problem of Nash games in (15.18) and with upper constraints in (15.19) or the constrained Nash quadratic game in (15.16) is able to be solved by the m-person noncooperative game strategies as follows:

$$u_i^*(t) = \frac{1}{2} R_i^{-1} g_i^T(\tilde{x}_i(t)) \left(\frac{\partial V(\tilde{x}_i(t))}{\partial \tilde{x}_i(t)} \right), \ \text{for } i = 1, 2, \ldots, m, \tag{15.20}$$

$$u_{-i}^*(t) = \frac{1}{2\rho_i} g_{-i}^T(\tilde{x}_i(t)) \left(\frac{\partial V(\tilde{x}_i(t))}{\partial \tilde{x}_i(t)} \right), \ \text{for } i = 1, 2, \ldots, m, \tag{15.21}$$

where the positive Lyapunov function $V(\tilde{x}_i(t))>0$ satisfies the following Hamilton–Jacobi–Issac inequalities (HJIIs):

$$\begin{aligned}
&\tilde{x}_i(t)^T Q_i \tilde{x}_i(t) + \left(\frac{\partial V(\tilde{x}_i(t))}{\partial \tilde{x}_i(t)}\right)^T f(\tilde{x}_i(t)) \\
&-\frac{1}{4}\left(\frac{\partial V(\tilde{x}_i(t))}{\partial \tilde{x}_i(t)}\right)^T g_i(\tilde{x}_i(t)) R_i^{-1} g_i^T(\tilde{x}_i(t)) \left(\frac{\partial V(\tilde{x}_i(t))}{\partial \tilde{x}_i(t)}\right) \\
&+\frac{1}{4\rho_i}\left(\frac{\partial V(\tilde{x}_i(t))}{\partial \tilde{x}_i(t)}\right)^T g_{-i}(\tilde{x}_i(t)) g_{-i}^T(\tilde{x}_i(t)) \left(\frac{\partial V(\tilde{x}_i(t))}{\partial \tilde{x}_i(t)}\right) \\
&+\frac{1}{2} l^T(\tilde{x}_i(t)) \frac{\partial^2 V(\tilde{x}_i(t))}{\partial \tilde{x}_i^2(t)} l(\tilde{x}_i(t)) \\
&+\lambda[V(\tilde{x}_i(t)+n(\tilde{x}_i(t))) - V(\tilde{x}_i(t))] \leq 0, \text{ for } i=1,2,\ldots,m,
\end{aligned} \tag{15.22}$$

with

$$E\{V(\tilde{x}_i(0))\} \leq \rho_i E\{\tilde{x}_i^T(0)\tilde{x}_i(0)\}. \tag{15.23}$$

Proof: See Appendix A.

The last two terms $l^T(\tilde{x}_i(t))\frac{\partial^2 V(\tilde{x}_i(t))}{\partial \tilde{x}_i^2(t)} l(\tilde{x}_i(t))$ and $\lambda[V(\tilde{x}_i(t)+n(\tilde{x}_i(t)))-V(\tilde{x}_i(t))]$ in (15.22) are the effects of intrinsic fluctuation of Wiener process and Poisson process, respectively. $u_i^*(t)$ in (15.20) and $u_{-i}^*(t)$ in (15.21) are Nash quadratic game solutions in (15.16) or (15.15). Hence, for nonlinear stochastic financial system, we can transform the MOP in (15.14) and (15.15) for the noncooperative H_∞ game strategies in (15.9) into the following HJIIs-constrained MOP:

$$\begin{aligned}
&(\rho_1^*,\ldots,\rho_i^*,\ldots,\rho_m^*) = \min_{(u_1(t),\ldots,u_i(t),\ldots,u_m(t))} (\rho_1,\ldots,\rho_i,\ldots,\rho_m) \\
&\text{s.t. } V(\tilde{x}_i(t))>0, (15.22) \text{ and } (15.23).
\end{aligned} \tag{15.24}$$

Lemma 15.3 The solution $(\rho_1^*,\ldots,\rho_i^*,\ldots,\rho_m^*)$ in (15.24) is the Nash equilibrium solution of m-person noncooperative H_∞ game strategy for the shifted stochastic financial system in (15.8).

Proof: See Appendix B.

Remark 15.2 Unlike the conventional iterative method with the updating algorithm one player by one player [1,9], the noncooperative H_∞ game strategy problem by MOP in (15.24) is solved by all players $(u_1(t),\ldots,u_m(t))$ simultaneously.

15.4 Solution of Noncooperative Stochastic H∞ Game Strategies for Nonlinear Stochastic Financial Systems

Generally, solving the HJIIs-constrained MOP problem in (15.24) of nonlinear stochastic financial systems is extremely difficult for the multi-person noncooperative stochastic H_∞ game strategies. Currently, the research still lacks an efficient method for solving it.

In this chapter, the global linearization with a set of local linearized systems is applied to interpolate a nonlinear stochastic financial system. This leads to bounding all local linearized systems of the nonlinear stochastic financial system in (15.8) by a polytope C_o with J vertices as follows [23,32]:

$$\begin{bmatrix} \frac{\partial f(\tilde{x}_i)}{\partial \tilde{x}_i} \\ \frac{\partial g_i(\tilde{x}_i)}{\partial \tilde{x}_i} \\ \frac{\partial g_{-i}(\tilde{x}_i)}{\partial \tilde{x}_i} \\ \frac{\partial l(\tilde{x}_i)}{\partial \tilde{x}_i} \\ \frac{\partial n(\tilde{x}_i)}{\partial \tilde{x}_i} \end{bmatrix} \in C_o \left[\begin{bmatrix} A_{i1} \\ B_{i1} \\ B_{-i1} \\ L_{i1} \\ N_{i1} \end{bmatrix} \cdots \begin{bmatrix} A_{ij} \\ B_{ij} \\ B_{-ij} \\ L_{ij} \\ N_{ij} \end{bmatrix} \cdots \begin{bmatrix} A_{iJ} \\ B_{iJ} \\ B_{-iJ} \\ L_{iJ} \\ N_{iJ} \end{bmatrix} \right], \forall \tilde{x}_i(t) \tag{15.25}$$

Namely, we can say that the trajectory $\tilde{x}_i(t)$ of the nonlinear financial system in (15.8) will be represented by the convex combination of the trajectories of the following J local linearized financial systems of the polytope if the convex hull C_o consists of all the local linearized systems at all $\tilde{x}_i(t)$ [23]:

$$\begin{aligned} d\tilde{x}_i(t) &= [A_{ij}\tilde{x}_i(t) + B_{ij}u_i(t) + B_{-ij}u_{-i}(t)]dt + L_{ij}\tilde{x}_i(t)dw(t) \\ &\quad + N_{ij}\tilde{x}_i(t)dp(t), \; i = 1,2,\ldots,m, j = 1,2,\ldots,J, \end{aligned} \tag{15.26}$$

where $B_{-ij} = [B_{1j} \ldots B_{i-1j} \; B_{i+1j} \ldots B_{mj}]$ and $u_{-i}(t)$ are defined in (15.6).

According to the global linearization theory [23], the trajectory of the shifted financial system in (15.8) can be represented by a convex combination of the trajectories of J local linearized financial systems in (15.26) as follows [23]:

$$\begin{aligned} d\tilde{x}_i(t) &= \sum_{j=1}^{J} \alpha_j(\tilde{x}_i(t))[(A_{ij}\tilde{x}_i(t) + B_{ij}u_i(t) + B_{-ij}u_{-i}(t))dt \\ &\quad + L_{ij}\tilde{x}_i(t)dw(t) + N_{ij}\tilde{x}_i(t)dp(t)] \end{aligned} \tag{15.27}$$

where $\alpha_j(\tilde{x}_i)$ denotes the interpolation functions, $0 \leq \alpha_j(\tilde{x}_i) \leq 1$ and $\sum_{j=1}^{J} \alpha_j(\tilde{x}_i) = 1$; i.e., we can replace the trajectory of the nonlinear stochastic financial system in (15.8) by the trajectory of the interpolated financial system in (15.27). The following lemma is necessary for the proof of the Theorem 15.2.

Lemma 15.4 [5] For any matrix S_i with appropriate dimensions and interpolation function $\alpha_i(\tilde{x}_j)$ with $0 \leq \alpha_i(\tilde{x}_j) \leq 1$, $\sum_{i=1}^{J} \alpha_i(\tilde{x}_j) = 1$. Then, for the matrix $P>0$, we have

$$\left(\sum_{i=1}^{J} \alpha_i(\tilde{x}_j) S_i \right)^T P \left(\sum_{i=1}^{J} \alpha_i(\tilde{x}_j) S_i \right) \leq \sum_{i=1}^{J} \alpha_i^T(\tilde{x}_j) S_i^T P S_i. \tag{15.28}$$

Thus, we can obtain the following result.

Theorem 15.2
The noncooperative Nash quadratic game problem in (15.16)–(15.19) for the interpolated stochastic financial system in (15.27) can be solved by

$$u_i^*(t) = -\sum_{j=1}^{J} \alpha_j(\tilde{x}_i) R_i^{-1} B_{ij}^T P \tilde{x}_i(t), \tag{15.29}$$

$$u_{-i}^*(t) = \frac{1}{\rho_i} \sum_{j=1}^{J} \alpha_j(\tilde{x}_i) B_{-ij}^T P \tilde{x}_i(t), \text{ for } i = 1, 2, \dots, m, \tag{15.30}$$

where $P>0$ is the common solution satisfying the following Riccati-like inequalities:

$$\begin{aligned} &Q_i + A_{ij}^T P + P A_{ij} - P B_{ij}^T R_i^{-1} B_{ij} P + \frac{1}{\rho_i} P B_{-ij} B_{-ij}^T P \\ &+ L_{ij}^T P L_{ij} + \lambda (N_{ij}^T P N_{ij} + N_{ij}^T P + P N_{ij}) \le 0, \text{for } i = 1, 2, \dots, m, \text{for } j = 1, 2, \dots, J, \end{aligned} \tag{15.31}$$

with

$$0 < P \le \rho_i I, \text{ for } i = 1, 2, \dots, m. \tag{15.32}$$

Proof: See Appendix C.

Remark 15.3 (i) We can consider the Riccati-like inequalities in (15.31) as local linearized HJIIs in (15.22) of the polytope at J vertices in (15.25). It still needs much effort to solve Riccati-like inequalities in (15.31) and (15.32). Therefore, for the simplicity, the Riccati-like inequalities in (15.31) can be readily transformed into the following equivalent LMIs through applying Schur complement transformations [23] several times after multiplying $W = P^{-1}$ to both sides of (15.31).

$$\begin{bmatrix} \begin{array}{c} WA_{ij}^T + A_{ij} - B_{ij}^T R_i^{-1} + \frac{1}{\rho_i} B_{-ij} B_{-ij}^T \\ + \lambda (W N_{ij}^T + N_{ij} W) \end{array} & W L_{ij}^T & \lambda^{\frac{1}{2}} W N_{ij}^T & W \\ L_{ij} W & -W & 0 & 0 \\ \lambda^{\frac{1}{2}} N_{ij} W & 0 & -W & 0 \\ W & 0 & 0 & -Q_i^{-1} \end{bmatrix} \le 0, \tag{15.33}$$

for $i = 1, 2, \dots, m,$ for $j = 1, 2, \dots, J.$

(ii) Similarly, the inequalities in (15.32) are equivalent to

$$\begin{bmatrix} \rho_i I & I \\ I & W \end{bmatrix} \ge 0, \text{ for } i = 1, 2, \dots, m. \tag{15.34}$$

(iii) The LMIs in (15.33) and (15.34) could be readily solved by the LMI toolbox in MATLAB.

Based on Theorem 15.2 and Remark 15.3, interpolating nonlinear stochastic financial system with J local linear financial systems in (15.27) transforms the HJIIs-constrained MOP in (15.24) for noncooperative stochastic H_∞ game strategy of nonlinear financial systems into the following LMIs-constrained MOP

$$\begin{aligned}(\rho_1^*, \dots, \rho_i^*, \dots, \rho_m^*) &= \min_{W>0}(\rho_1, \dots, \rho_i, \dots, \rho_m) \\ \text{s.t. } &LMIs \text{ in (15.33) and (15.34).}\end{aligned} \tag{15.35}$$

After solving the MOP in (15.35) with the optimal W^*, the noncooperative H_∞ game strategies $u_i^*(t)$, for $i = 1, 2, \dots, m$ in (15.29) can be acquired with P being replaced by $P^* = (W^*)^{-1}$. Finally, the control strategy $u_i(t)$ in (15.1) is designed by noncooperative H_∞ game strategy as $u_i^*(t) = -\sum_{j=1}^{J} \alpha_j(\tilde{x}_i) R_i^{-1} B_{ij}^T P^* \tilde{x}_i(t)$ in (15.29).

In the next section, a modified LMIs-constrained MOEA appropriate algorithm will be proposed for the noncooperative stochastic H_∞ game strategy of nonlinear stochastic financial systems to efficiently solve the Pareto optimal solutions $(\rho_1^*, \dots, \rho_i^*, \dots, \rho_m^*)$ and W^* of MOP in (15.35).

15.5 LMIs-constrained MOEA Algorithm for Noncooperative Stochastic H_∞ Game Strategies of Nonlinear Stochastic Financial System

Now, the feasible solutions of LMIs-constrained MOP in (15.35) can be obtained with the help of the convex optimization toolbox in MATLAB for the given $(\rho_1, \dots, \rho_i, \dots, \rho_m)$ in (15.35) subject to the LMIs in (15.33) and (15.34). The MOEA is a stochastic algorithm inspired by biological evolution and the survival of the better fitness [35,140]. Since the algorithm can avoid local optima with the parallel search of the Pareto optimality with multiple conflicting objectives, MOEA is suitable for solving the MOP for noncooperative stochastic H_∞ game strategy. Although many MOEAs have been proposed to solve the algebra-constrained MOPs [25,35,136,138,139], few of the researches have discussed the LMIs-constrained MOP in (15.35) for noncooperative stochastic H_∞ game strategies of nonlinear stochastic dynamic financial systems. Thus, some modifications of the conventional MOEAs are necessary in the sense of LMIs. In this chapter, we choose the NSGA-II in [25,136] as the foundation to develop the proposed LMIs-constrained MOEA algorithm to solve MOP in (15.35).

For the LMIs-constrained MOP in (15.35), the Definitions 15.1–15.4 of Pareto optimal solutions $(u_1^*, \dots, u_m^*)$ can be guaranteed obviously on a given feasible set Ω. Different from the traditional MOEA to search $(w_{1,1}, \dots, w_{n,n})$ of all components of $W>0$ in (15.35) [25,35,136,138,139], the searching region of the LMIs-constrained MOEA is the set of objective values under LMIs constraints in (15.33) and (15.34). First, the MOEA encodes the individuals $(\rho_1^n, \dots, \rho_m^n)$ for all $n = 1, 2, \dots, N_p$ as the so-called initial parent population P_r is the feasible set with the initial number $r = 1$, where P_r is the population at the rth iteration. In other words, the nth individual $(\rho_1^n, \dots, \rho_m^n)$ must satisfy the LMIs in (15.33) and (15.34) with $\rho_i^L \le \rho_i^n \le \rho_i^U$, $i = 1, \dots, m$ for all $n = 1, 2, \dots, N_p$, where N_p is the population size. ρ_i^U and ρ_i^L denote the upper and lower bounds of each ρ_i^n, respectively. Second, the candidate Pareto dominance solutions can be obtained by the crowded comparison operators [25,35,136]. Then the child population can also be generated by crossover and mutation. Note that if some individuals are not feasible, i.e., if they cannot satisfy with the requirement of LMIs in (15.33) and (15.34), these individuals should be deleted from the candidate set. The proposed algorithm is executed iteratively while the final population is achieved. At last, once the set of Pareto front P_F is reached, the Pareto optimal solutions $(u_1^*, ..., u_i^*, ...u_m^*)$ of the MOP in (15.35) for the m-person noncooperative stochastic H_∞ game strategy of nonlinear stochastic financial systems can also be acquired. According to previous analysis, the algorithm of the LMIs-constrained MOEA algorithm to solve the MOP in (15.35) for the m-person noncooperative stochastic H_∞ game strategy of nonlinear stochastic financial systems is proposed as follows:

LMIs-constrained MOEA algorithm for noncooperative stochastic H_∞ game strategy of nonlinear stochastic financial system

Step 1: Choose the searching limits $(\rho_1^L, ..., \rho_m^L) \times (\rho_1^U, ..., \rho_m^U)$ for the feasible objective vector $(\rho_1, ..., \rho_m)$, and provide the population number N_p, the iteration number N_i, the crossover rate c_r and mutation ratio m_r, in the proposed MOEA. Set iteration number $i = 1$.

Step 2: Choose N_p feasible individuals (chromosomes) as the initial population P_1.

Step 3: Employ EA and produce $2N_p$ feasible individuals by checking if their corresponding $(\rho_1, ..., \rho_m)$ are feasible (i.e., satisfying the LMIs in (15.33) and (15.34)).

Step 4: Place the iterative index $i = i + 1$. Choose N_p elite individuals from $2N_p$ feasible individuals in step 3 via the non-dominated sorting scheme and the crowded comparison method to the population P_{i+1}.

Step 5: Repeat steps 3 and 4 until the iteration number N_i is reached, then set the final population $P_{N_i} = P_F$ as Pareto front.

Step 6: Choose a desirable feasible objective vector $(\rho_1^*, ..., \rho_m^*) \in P_F$ according to designer own preference with the optimal $W^* = (P^*)^{-1}$, where W^* is the solution of MOP in (15.35). Once the "desirable" solution is selected as the control strategy $(u_1^*, ..., u_m^*)$ with $u_i^*(t) = -\sum_{j=1}^{J} \alpha_j(\tilde{x}_i) R_i^{-1} B_{ij}^T P^* \tilde{x}_i(t) i = 1, ..., m$, the stochastic noncooperative H_∞ game strategy of nonlinear financial systems in (15.1) can be solved.

Remark 15.4 For the proposed LMIs-constrained MOEA algorithm, its computational complexity can be approximately estimated as $O(n(n+1)mJN_p^2N_i)$, which consists of $O(\frac{n(n+1)mJ}{2})$ for solving the LMIs, and $O(2N_p^2N_i)$ for MOEA, where N_i is the iteration number of MOEA, n is the dimension of system state $x(t)$, J is the number of local linearized systems in (15.25) or (15.27), and N_p is the population number of the proposed MOEA algorithm.

Remark 15.5 The conventional MOEA algorithms [25,35,136,139] to solve the MOP in (15.35) need to search $W>0$ directly instead of $(\rho_1, ..., \rho_m)$ in the proposed method to achieve multi-objective optimization, i.e., $(w_{1,1}, w_{1,2}, ..., w_{i,j}, ..., w_{n,n})$ is the individual instead of $(\rho_1, ..., \rho_m)$ in the conventional MOEA, where $w_{i,j}$ is the (i,j) element of W. Therefore, if the dimension of $x(t)$ becomes large, it is very difficult to solve the MOP in (15.35) by the conventional MOEA algorithm.

15.6 Simulation Examples

In order to illustrate the design procedure of the proposed LMIs-constrained MOEA algorithm for noncooperative stochastic H_∞ game investment strategy of nonlinear stochastic financial systems and to confirm the performance of the proposed n-person noncooperative stochastic H_∞ game investment strategy, two design examples of the proposed m-person noncooperative stochastic H_∞ game investment strategy in the stochastic nonlinear financial systems in (15.3) and (15.5) are introduced as follows.

15.6.1 Simulation Results of Example 15.1 in (15.3)

Consider the nonlinear stochastic financial system in (15.3). Suppose the saving amount $a = 1.5$, the per-investment cost $b = 0.2$ and the elasticity of demands of commercials $c = 0.25$. The mean λ of Poisson process in a unit time is 0.2. The initial states of the stochastic financial system in (15.3) are set as $x(0) = 0.36, y(0) = -3.05, z(0) = 0.7$. The external disturbance $v(t)$ is assumed to be $[0.01\sin(2t) - 0.02\sin(2t) - 0.01\sin(2t)]$. The Wiener process and Poisson process of this financial system are shown in Fig. 15.1 and the fluctuant trajectories of the stochastic financial system without noncooperative stochastic H_∞ game strategies are shown in Fig. 15.3. There are three players in this noncooperative investment game. $u_1(t) = [u_{11}(t) \quad u_{12}(t) \quad u_{13}(t)]^T$ is the investment

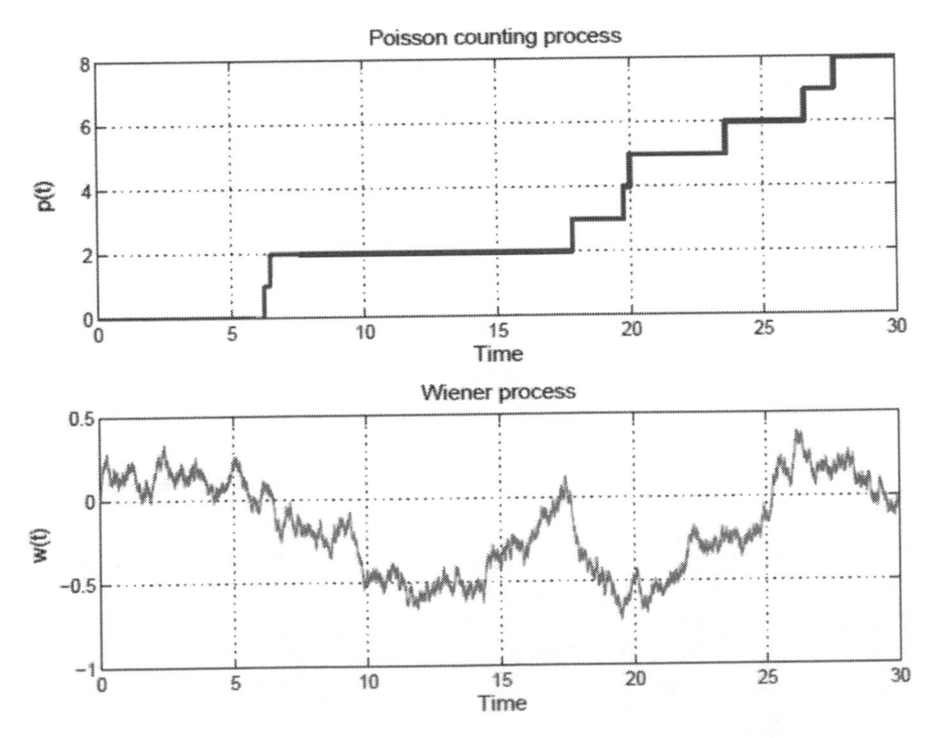

FIGURE 15.1
The time evolution of Wiener process w(t) and Poisson counting process $p(t)$ in stochastic financial system (15.3).

strategy of government with the desired target x_{d1} and weighting matrices Q_1 and R_1 as follows:

$$x_{d1} = \begin{bmatrix} 0.1 \\ 4.0 \\ -0.2 \end{bmatrix}, Q_1 = \begin{bmatrix} 1 & 0 & 0 \\ 0 & 1 & 0 \\ 0 & 0 & 8 \end{bmatrix}, R_1 = 0.01 I_{3\times 3}.$$

$u_2(t) = [u_{21}(t) \quad u_{22}(t) \quad u_{23}(t)]^T$ is the investment strategy of bank consortium with the desired target x_{d2} and weighting matrices Q_2 and R_2 as follows:

$$x_{d2} = \begin{bmatrix} 0.05 \\ 3.0 \\ -0.1 \end{bmatrix}, Q_2 = \begin{bmatrix} 1 & 0 & 0 \\ 0 & 8 & 0 \\ 0 & 0 & 1 \end{bmatrix}, R_2 = 0.01 I_{3\times 3}.$$

$u_3(t) = [u_{31}(t) \quad u_{32}(t) \quad u_{33}(t)]^T$ is the investment strategy of the public with the desired target x_{d3} and weighting matrices Q_3 and R_3 as follows:

$$x_{d3} = \begin{bmatrix} 0.05 \\ 4.5 \\ -0.15 \end{bmatrix}, Q_3 = \begin{bmatrix} 8 & 0 & 0 \\ 0 & 1 & 0 \\ 0 & 0 & 1 \end{bmatrix}, R_3 = 0.01 I_{3\times 3}.$$

Based on the global linearization technique with 10 local linearized financial systems, the financial system in (15.3) can be represented by (15.27) as the following interpolation of 10 local linearized financial systems with the origin being shifted to the desired target of each player:

$$\begin{aligned} d\tilde{x}_1(t) = {} & \sum_{j=1}^{10} \alpha_j(\tilde{x}_1(t))[(A_{1j}\tilde{x}_1(t) + B_{1j}u_1(t) + B_{-1j}u_{-1}(t))dt \\ & + L_{1j}\tilde{x}_1(t)dw(t) + N_{1j}\tilde{x}_1(t)dp(t)] \end{aligned} \tag{15.36}$$

where $\tilde{x}_1(t) = x(t) - x_{d1}$.

$$\begin{aligned} d\tilde{x}_2(t) = {} & \sum_{j=1}^{10} \alpha_j(\tilde{x}_2(t))[(A_{2j}\tilde{x}_2(t) + B_{2j}u_2(t) + B_{-2j}u_{-2}(t))dt \\ & + L_{2j}\tilde{x}_2(t)dw(t) + N_{2j}\tilde{x}_2(t)dp(t)] \end{aligned} \tag{15.37}$$

where $\tilde{x}_2(t) = x(t) - x_{d2}$.

$$\begin{aligned} d\tilde{x}_3(t) = {} & \sum_{j=1}^{10} \alpha_j(\tilde{x}_3(t))[(A_{3j}\tilde{x}_3(t) + B_{3j}u_3(t) + B_{-3j}u_{-3}(t))dt \\ & + L_{3j}\tilde{x}_3(t)dw(t) + N_{3j}\tilde{x}_3(t)dp(t)] \end{aligned} \tag{15.38}$$

where $\tilde{x}_3(t) = x(t) - x_{d3}$. In this chapter, the interpolation functions could be chosen as

$$\alpha_j(\tilde{x}_i(t)) = (1/\left\|x_j^c - \tilde{x}_i(t)\right\|_2^2)/\sum_{j=1}^{J}(1/\left\|x_j^c - \tilde{x}_i(t)\right\|_2^2), \text{ for } j = 1,\ldots,J, i = 1,\ldots,m.$$

where $\tilde{x}_i(t)$ denotes the shifted current state, and x_j^c denotes the operation point of the jth local linear system.

Based on the proposed LMIs-constrained MOEA algorithm, we solved the LMIs-constrained MOP in (15.35) for example 15.1, with the searching range $0 \le \rho_i \le 100$, the population number , iteration number $N_i = 30$, crossover rate $C_r = 0.9$, and the mutation rate $m_r = 0.1$. Once the iteration number $N_i = 30$ is achieved, we get $(\rho_1^*, \rho_2^*, \rho_3^*)$ and $P^* = (W^*)^{-1}$. The Pareto front for the Pareto optimal solution of MOP of the noncooperative stochastic H_∞ game investment strategy in (15.35) can be obtained as shown in Fig. 15.2.

In Fig. 15.2, according to the proposed design procedure, the Pareto optimal solutions consist of the Pareto front. Since the knee point solution does not bias toward each player, the knee point solution is selected as the preferable design strategy in the Pareto optimal solution set. In this case, $(\rho_1^*, \rho_2^*, \rho_3^*) = (13.9679, 13.6079, 13.9901)$ and

$$P^* = (W^*)^{-1} = \begin{bmatrix} 1.7124 & -0.1894 & 0.0575 \\ * & 13.6012 & 0.0015 \\ * & * & 0.7338 \end{bmatrix}.$$

Figure 15.2 shows that the Pareto optimal solutions indeed approach to the Pareto front, which can guarantee the effectiveness of the proposed LMIs-constrained MOEA algorithm for the proposed noncooperative stochastic H_∞ game strategy.

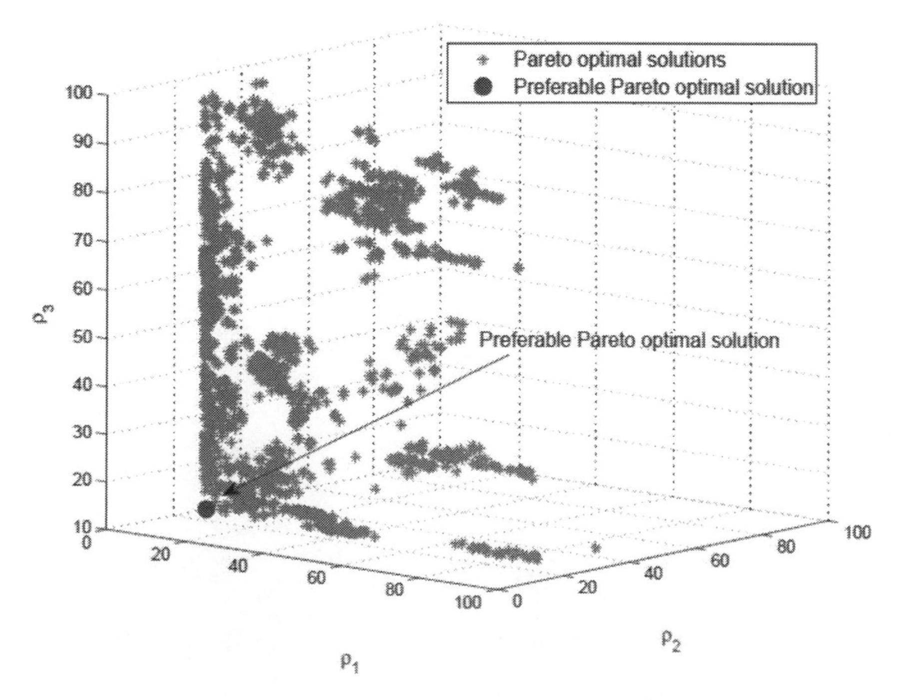

FIGURE 15.2
Pareto front obtained by the proposed LMIs-constrained MOEA for the three players noncooperative stochastic H_∞ game financial system in example 15.1.

Figure 15.3 describes the dynamical trajectories of the original nonlinear stochastic financial systems without noncooperative H_∞ stochastic game strategy in (15.3) with random fluctuations driven by Wiener process and Poisson counting process in Fig. 15.1. It can be seen that the dynamic states of the nonlinear stochastic financial systems suffer from stochastic Wiener turbulence and Poisson jumping. The objectives of three players are to design a noncooperative stochastic H_∞ game strategy to maintain the states in (15.3) and to achieve their desired target x_{d1}, x_{d2}, and x_{d3} in (15.36)–(15.38), respectively.

In Fig. 15.4, the trajectories of the stochastic financial systems $x_1(t)$, $x_2(t)$ and $x_3(t)$ achieve some compromised steady state among their designed targets x_{d1}, x_{d2}, and x_{d3} based on the tradeoff of weighting matrices Q_i and R_i for the proposed noncooperative stochastic H_∞ game strategies of three players. Obviously, the interest rate, investment demand, and the price index can follow the compromised target of three desired targets x_{d1}, x_{d2}, and x_{d3} in the nonlinear stochastic financial systems. However, since player 1 (government) addresses more weighting on the regulation of price index $z(t)$ to −0.2, the price index $z(t)$ is more regulated to −0.2 (the desired price index of player 1). Since the investment demand is more concerned by player 2 (bank consortium), the investment demand is more regulated to 3. Further, the interest rate is more concerned by player 3 (public), the interest rate is more regulated to 0.05. Moreover, the proposed noncooperative stochastic H_∞ game investment strategy in (15.9) could eliminate random fluctuations and attenuate external disturbance to compromise their desired targets for nonlinear stochastic financial systems. This result confirms that the minimax stochastic noncooperative H_∞ game strategy in (15.9) is a robust control strategy, which could efficiently minimize the worst-case effect of other strategies, random fluctuation and external disturbance on the desired target regulation.

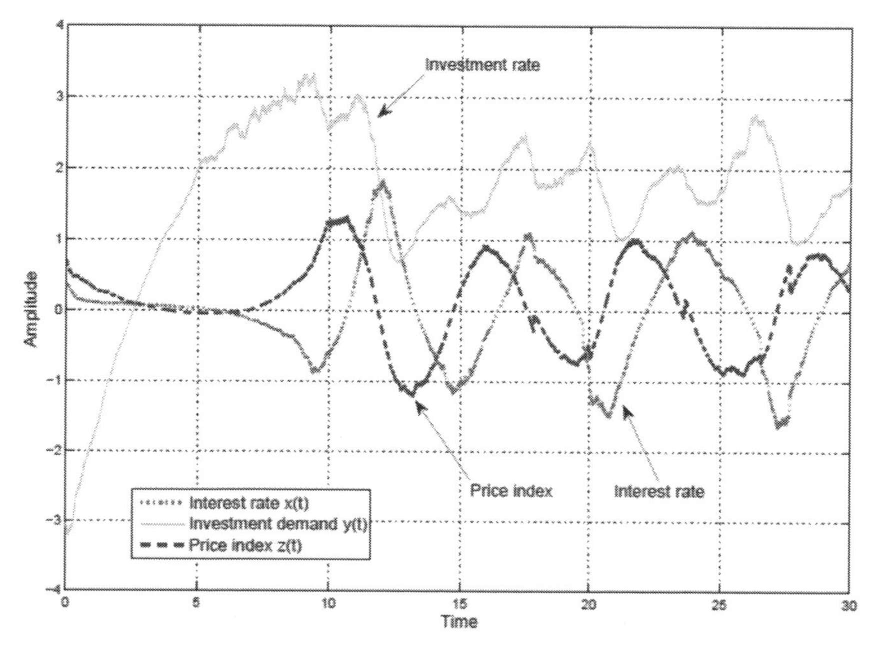

FIGURE 15.3
The fluctuant trajectories of stochastic financial system without noncooperative H_∞ control strategies in example 15.1.

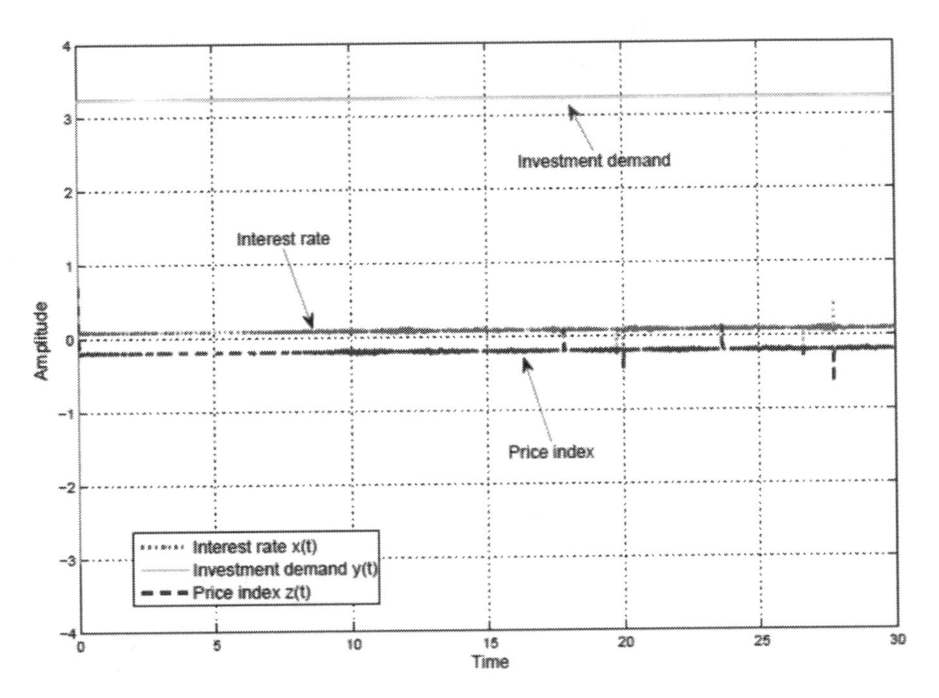

FIGURE 15.4
The compromised trajectories of the stochastic financial system by the proposed m-person noncooperative stochastic H_∞ game strategy in example 15.1.

15.6.2 Simulation Results of Example 15.2 in (15.5)

Consider the international capital flow volatility system in (15.5). Suppose the parameters in the financial system of (15.5) are given as follows [325]: $a = 0.223$, $b = 0.106$, $C_{vs} = 0.98$, $C_{sv} = 1.03$, $\lambda = 0.2$. The external disturbance $v(t)$ is assumed to be $[0.01\sin(2t) - 0.01\sin(2t)]$ and the initial states are given as $x_v(0) = 0.192, x_s(0) = 2.11$. Suppose the desired target and weighting matrices of player $u_1(t)$ of source country's government are given by

$$x_{d1} = \begin{bmatrix} 0.05 \\ 2 \end{bmatrix}, Q_1 = \begin{bmatrix} 0.08 & 0 \\ 0 & 0.01 \end{bmatrix}, R_1 = 0.01I_{2\times 2}.$$

Suppose the desired target and weighting matrices of player $u_2(t)$ of volatility country's government are given by

$$x_{d2} = \begin{bmatrix} 0.1 \\ 1.5 \end{bmatrix}, Q_2 = \begin{bmatrix} 0.01 & 0 \\ 0 & 0.08 \end{bmatrix}, R_2 = 0.01I_{2\times 2}.$$

Based on the global linearization technique, the international nonlinear financial system in (15.5) could be represented by (15.27) as the following interpolation of 10 local linearized financial systems with the origin being shifted to the desired targets of two players, respectively.

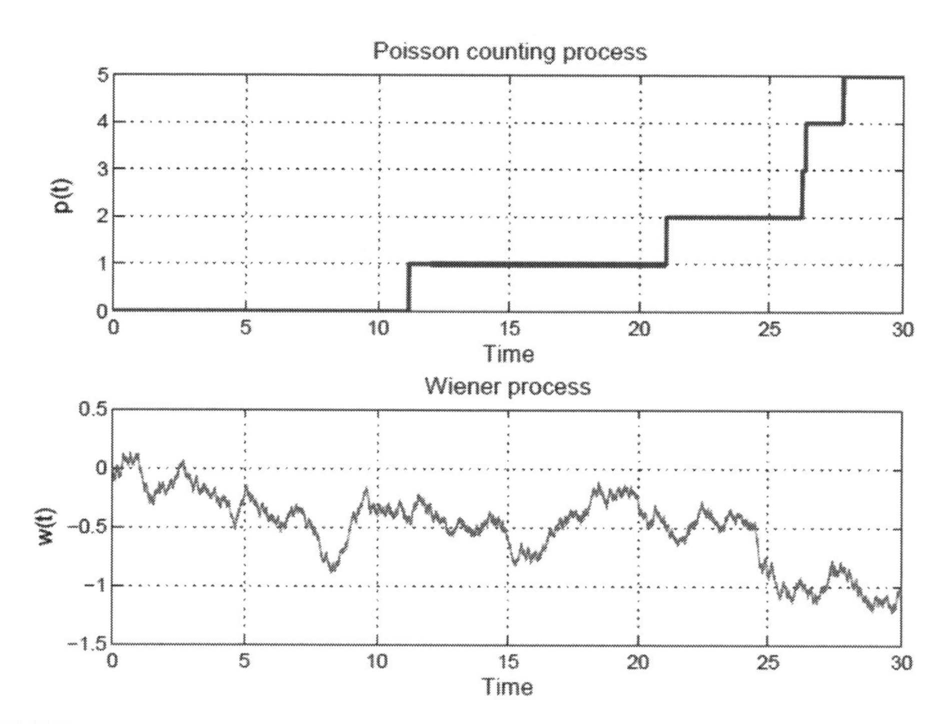

FIGURE 15.5
The time evolution of Wiener process $w(t)$ and Poisson counting process $p(t)$ in stochastic financial system (15.5).

$$d\tilde{x}_1(t) = \sum_{j=1}^{10} \alpha_j(\tilde{x}_1(t))[(A_{1j}\tilde{x}_1(t) + B_{1j}u_1(t) + B_{-1j}u_{-1}(t))dt + L_{1j}\tilde{x}_1(t)dw(t) + N_{1j}\tilde{x}_1(t)dp(t)], \tag{15.39}$$

where $\tilde{x}_1(t) = x(t) - x_{d1}$.

$$d\tilde{x}_2(t) = \sum_{j=1}^{10} \alpha_j(\tilde{x}_2(t))[(A_{2j}\tilde{x}_2(t) + B_{2j}u_2(t) + B_{-2j}u_{-2}(t))dt + L_{2j}\tilde{x}_2(t)dw(t) + N_{2j}\tilde{x}_2(t)dp(t)], \tag{15.40}$$

where $\tilde{x}_2(t) = x(t) - x_{d2}$. The interpolation functions are set as

$$\alpha_j(\tilde{x}_i(t)) = (1/\left\|x_j^c - \tilde{x}_i(t)\right\|_2^2)/\sum_{j=1}^{J}(1/\left\|x_j^c - \tilde{x}_i(t)\right\|_2^2),$$

where $j = 1, \ldots, J$, $i = 1, \ldots, m$ and x_j^c denote the operation point of the jth local linear system.

According to the proposed LMIs-constrained MOEA algorithm, the LMIs-constrained MOP in (15.35) for example 15.2 can be solved with the searching range $0 \le \rho_i \le 150$. In this example, we set the population number $N_p = 1000$, iteration number $N_i = 30$,

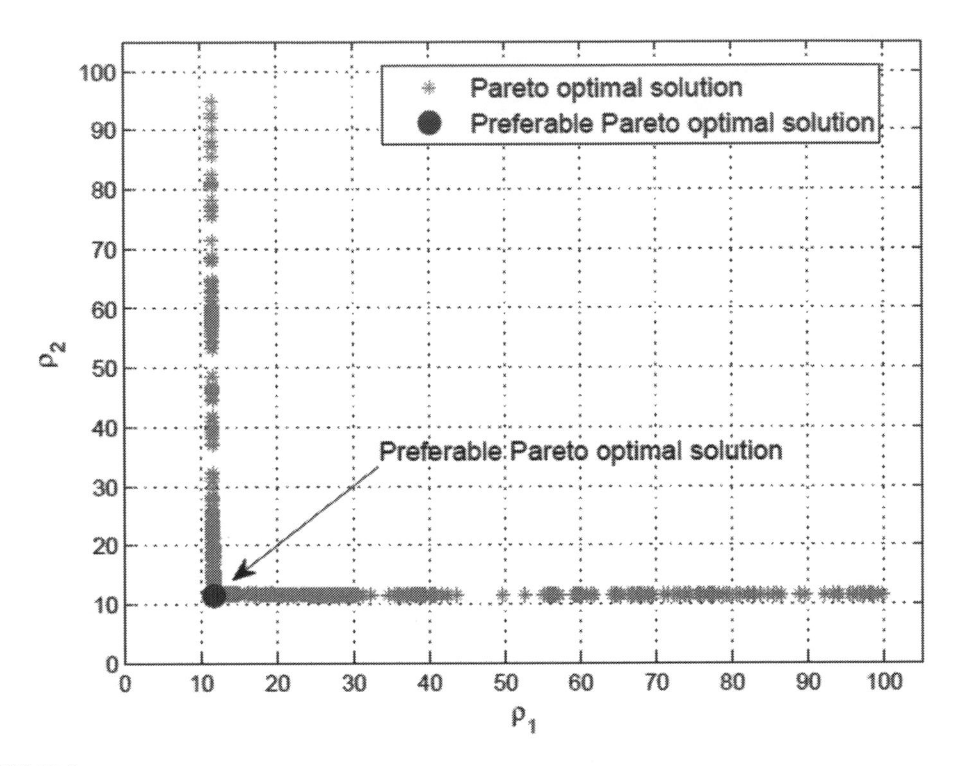

FIGURE 15.6
Pareto front obtained by the proposed LMIs-constrained MOEA for the two players noncooperative stochastic H_∞ game financial system in example 15.2.

crossover rate $C_r = 0.9$ and the mutation rate $m_r = 0.1$. We can obtain $(\rho_1^*, \rho_2^*, \rho_3^*)$ and $P^* = (W^*)^{-1}$ when the iteration number $N_i = 30$ is achieved. The Pareto front of MOP for the noncooperative stochastic H_∞ control game strategy for example 15.2 in (15.35) can be obtained as shown in Fig. 15.6.

The knee point of the Pareto optimal solutions in Fig. 15.6 is selected as the noncooperative H_∞ stochastic game strategy of example 15.2. In this preferable case, we obtain $(\rho_1^*, \rho_2^*) = (11.6760, 11.5275)$ and

$$P^* = (W^*)^{-1} = \begin{bmatrix} 11.4408 & 0.8759 \\ * & 2.5895 \end{bmatrix}.$$

Similar to the Pareto front of example 15.1 in Fig. 15.2, the Pareto optimal solutions can approach to the Pareto front efficiently, which ensure the applicability of the proposed LMIs-constrained MOEA algorithm for noncooperative stochastic H_∞ game control strategy in example 15.2.

Figure 15.7 describes the stochastic behaviors of nonlinear stochastic international volatility system in (15.5) with much random fluctuation due to Wiener process and Poisson jumps as shown in Fig. 15.5. Two countries use noncooperative H_∞ game strategy to solve the financial contagion problem. It also shows that the fluctuant trajectories of the nonlinear stochastic financial systems in (15.5) fluctuate with the driven Wiener noise and Poisson noise. The goals of two players are to design their

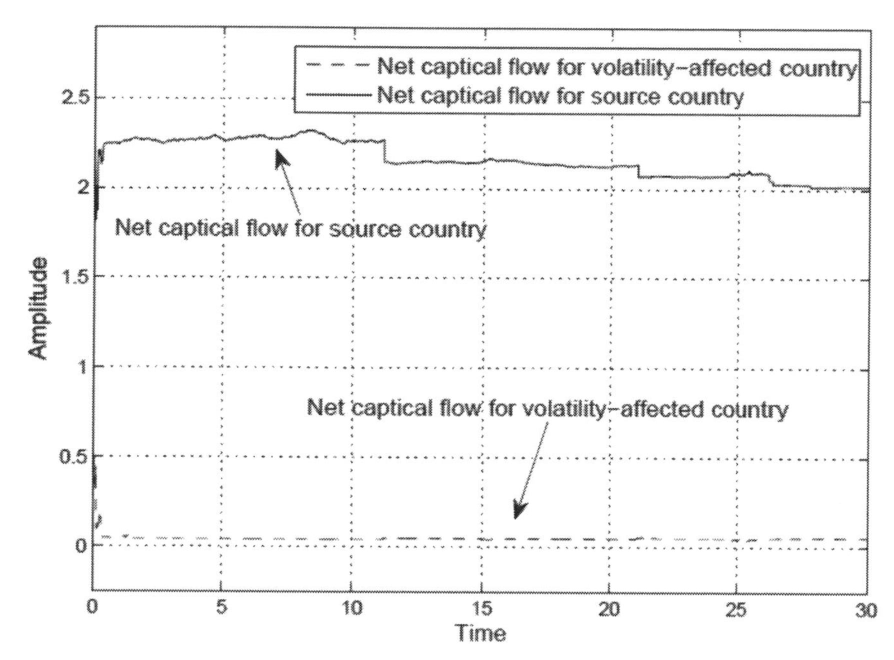

FIGURE 15.7
The fluctuant trajectories of stochastic financial system without noncooperative control strategy in example 15.2.

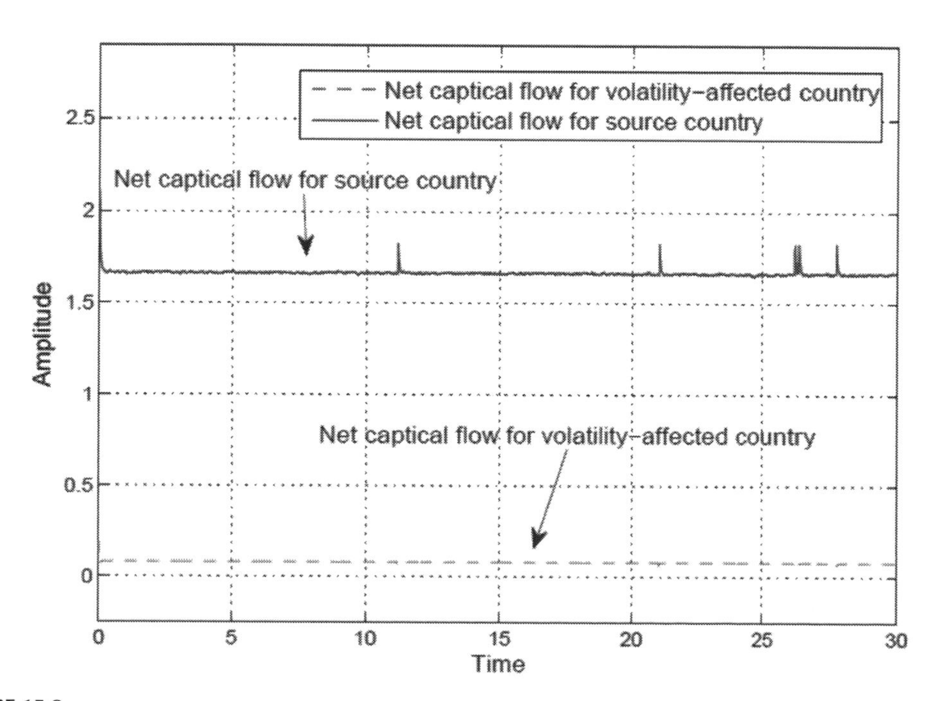

FIGURE 15.8
The compromised trajectories of the stochastic financial system by the proposed m-person noncooperative stochastic H_∞ game strategy in example 15.2.

appropriate noncooperative game strategy to hold the states in (15.5) and to achieve their desired target x_{d1}, x_{d2} in (15.39)–(15.40), respectively.

In Fig. 15.8, the dynamic states of the stochastic financial systems $x_1(t)$, $x_2(t)$ in (15.5) achieve some compromised steady state between their designed targets x_{d1}, x_{d2} based on the tradeoff of weighting matrices Q_i and R_i for the proposed noncooperative stochastic H_∞ game strategies of two players. Clearly, capital flows from source country of financial turbulence and volatility-affected country follows the compromised target of two desired targets x_{d1}, x_{d2} in the nonlinear stochastic financial systems. To sum up, the proposed noncooperative stochastic H_∞ game investment strategy has the robust ability to attenuate random fluctuations and external disturbance and to compromise their desired targets for nonlinear stochastic financial systems.

15.7 Conclusions

In this chapter, we have studied the m-person noncooperative stochastic H_∞ game strategy of nonlinear stochastic financial systems with continuous and discontinuous random fluctuations. Unlike conventional iterative methods, the noncooperative H_∞ stochastic game strategy design of nonlinear stochastic financial systems is converted into an equivalent HJIIs-constrained MOP, which could be solved simultaneously by all players and could achieve the Nash equilibrium solution. According to the global linearization method, we transformed the HJIIs-constrained MOP for stochastic financial systems to an LMIs-constrained MOP to obtain a simplified design procedure of m-person noncooperative stochastic H_∞ game investment strategy via the proposed LMIs-constrained MOEA algorithm. With the support of convex toolbox in MATLAB, the LMIs-constrained MOEA algorithm could be readily implemented. Once it achieves the Pareto front of MOP in (15.35) for the Pareto optimal solutions, the managers or investors, i.e., the players can choose a desirable investment policy for nonlinear stochastic financial systems from the set of Pareto optimal solutions based on his or her own favor. Finally, two simulation examples, one about investment strategy and another about robust control of international capital flow volatility, are given to confirm the performance of the proposed m-person noncooperative stochastic H_∞ game strategy design of nonlinear stochastic systems with continuous and discontinuous random fluctuations. We found that the proposed m-person noncooperative stochastic H_∞ game strategy has enough robustness to compromise the desired target of each player, and to filter the inherent continuous and intrinsic discontinuous fluctuation with the attenuation of the external disturbance.

15.8 Appendix

15.8.1 Appendix A: Proof of Theorem 15.1

From equation (15.17), the function J_i can be expressed as

$$
\begin{aligned}
J_i = \ & E\int_0^{t_p} [\tilde{x}_i(t)^T Q_i \tilde{x}_i(t) + u_i^T(t) R_i u_i(t) - \rho_i u_{-i}^T(t) u_{-i}(t)]dt \\
& = E\{V(\tilde{x}_i(0)) - V(\tilde{x}_i(t_p))\} + E\{V(\tilde{x}_i(t_p)) - V(\tilde{x}_i(0))\} \\
& \quad + E\int_0^{t_p} [\tilde{x}_i(t)^T Q_i \tilde{x}_i(t) + u_i^T(t) R_i u_i(t) - \rho_i u_{-i}^T(t) u_{-i}(t)]dt \\
& = E\{V(\tilde{x}_i(0)) - V(\tilde{x}_i(t_p))\} + E\{\int_0^{t_p} [(\tilde{x}_i(t)^T Q_i \tilde{x}_i(t) \\
& \quad + u_i^T(t) R_i u_i(t) - \rho_i u_{-i}^T(t) u_{-i}(t))dt + dV(\tilde{x}_i(t))]\}.
\end{aligned} \tag{15.41}
$$

By applying the Itô–Lévy formula in (15.12), we get

$$
\begin{aligned}
J_i = & E\{V(\tilde{x}_i(0)) - V(\tilde{x}_i(t_p))\} + E\{\int_0^{t_p} [(\tilde{x}_i(t)^T Q_i \tilde{x}_i(t) \\
& + u_i^T(t) R_i u_i(t) - \rho_i u_{-i}^T(t) u_{-i}(t))dt + \left(\frac{\partial V(\tilde{x}_i(t))}{\partial \tilde{x}_i(t)}\right)^T f(\tilde{x}_i(t))dt \\
& + \left(\frac{\partial V(\tilde{x}_i(t))}{\partial \tilde{x}_i(t)}\right)^T g_i(\tilde{x}_i(t)) u_i(t)dt + \left(\frac{\partial V(\tilde{x}_i(t))}{\partial \tilde{x}_i(t)}\right)^T g_{-i}(\tilde{x}_i(t)) u_{-i}(t)dt \\
& + \frac{1}{2} l(\tilde{x}_i(t))^T \frac{\partial^2 V(\tilde{x}_i(t))}{\partial \tilde{x}_i^2(t)} l(\tilde{x}_i(t))dt + \left(\frac{\partial V(\tilde{x}_i(t))}{\partial \tilde{x}_i(t)}\right)^T l(\tilde{x}_i(t))dwt \\
& + [V(\tilde{x}_i(t) + n(\tilde{x}_i(t))) - V(\tilde{x}_i(t))]dp(t)]\},
\end{aligned} \tag{15.42}
$$

Completing the square for $u_i(t)$ and $u_{-i}(t)$, i.e.,

$$
\begin{aligned}
& u_i^T(t) R_i u_i(t) + \left(\frac{\partial V(\tilde{x}_i(t))}{\partial \tilde{x}_i(t)}\right)^T g_i(\tilde{x}_i(t)) u_i(t) \\
& = \left(u_i(t) + \frac{1}{2} R_i^{-1} g_i^T(\tilde{x}_i(t)) \left(\frac{\partial V(\tilde{x}_i(t))}{\partial \tilde{x}_i(t)}\right)\right)^T R_i \left(u_i(t) + \frac{1}{2} R_i^{-1} g_i^T(\tilde{x}_i(t)) \left(\frac{\partial V(\tilde{x}_i(t))}{\partial \tilde{x}_i(t)}\right)\right) \\
& \quad - \frac{1}{4} \left(\frac{\partial V(\tilde{x}_i(t))}{\partial \tilde{x}_i(t)}\right)^T g_i(\tilde{x}_i(t)) R_i^{-1} g_i^T(\tilde{x}_i(t)) \left(\frac{\partial V(\tilde{x}_i(t))}{\partial \tilde{x}_i(t)}\right),
\end{aligned} \tag{15.43}
$$

and

$$
\begin{aligned}
& -\rho_i u_{-i}^T(t) u_{-i}(t))dt + \left(\frac{\partial V(\tilde{x}_i(t))}{\partial \tilde{x}_i(t)}\right)^T g_{-i}(\tilde{x}_i(t)) u_{-i}(t) \\
& = -\rho_i \left(u_{-i}(t) + \frac{1}{2\rho_i} g_{-i}^T(\tilde{x}_i(t)) \left(\frac{\partial V(\tilde{x}_i(t))}{\partial \tilde{x}_i(t)}\right)\right)^T \left(u_{-i}(t) + \frac{1}{2\rho_i} g_{-i}^T(\tilde{x}_i(t)) \left(\frac{\partial V(\tilde{x}_i(t))}{\partial \tilde{x}_i(t)}\right)\right) \\
& \quad + \frac{1}{4\rho_i} \left(\frac{\partial V(\tilde{x}_i(t))}{\partial \tilde{x}_i(t)}\right)^T g_{-i}(\tilde{x}_i(t)) g_{-i}^T(\tilde{x}_i(t)) \left(\frac{\partial V(\tilde{x}_i(t))}{\partial \tilde{x}_i(t)}\right),
\end{aligned} \tag{15.42}
$$

By the fact that $Edw = 0$, $Edp = \lambda dt$ and substituting (15.43)–(15.44) into (15.42), we get

$$
\begin{aligned}
J_1^* &= \min_{u_i(t)} \max_{u_{-i}(t)} E\{V(\tilde{x}_i(0)) - V(\tilde{x}_i(t_p))\} + \min_{u_i(t)} \max_{u_{-i}(t)} E\{\int_0^{t_p} [(\tilde{x}_i(t)^T Q_i \tilde{x}_i(t) \\
&+ \left(\frac{\partial V(\tilde{x}_i(t))}{\partial \tilde{x}_i(t)}\right)^T f(\tilde{x}_i(t)) + \frac{1}{2} l(\tilde{x}_i(t))^T \frac{\partial^2 V(\tilde{x}_i(t))}{\partial \tilde{x}_i^2(t)} l(\tilde{x}_i(t)) \\
&+ \lambda[V(\tilde{x}_i(t) + n(\tilde{x}_i(t))) - V(\tilde{x}_i(t))]) \\
&+ \left(u_i(t) + \frac{1}{2} R_i^{-1} g_i^T(\tilde{x}_i(t)) \left(\frac{\partial V(\tilde{x}_i(t))}{\partial \tilde{x}_i(t)}\right)\right)^T R_i \left(u_i(t) + \frac{1}{2} R_i^{-1} g_i^T(\tilde{x}_i(t)) \left(\frac{\partial V(\tilde{x}_i(t))}{\partial \tilde{x}_i(t)}\right)\right) \\
&- \frac{1}{4} \left(\frac{\partial V(\tilde{x}_i(t))}{\partial \tilde{x}_i(t)}\right)^T g_i(\tilde{x}_i(t)) R_i^{-1} g_i^T(\tilde{x}_i(t)) \left(\frac{\partial V(\tilde{x}_i(t))}{\partial \tilde{x}_i(t)}\right) \\
&- \rho_i \left(u_{-i}(t) + \frac{1}{2\rho_i} g_{-i}^T(\tilde{x}_i(t)) \left(\frac{\partial V(\tilde{x}_i(t))}{\partial \tilde{x}_i(t)}\right)\right)^T \left(u_{-i}(t) + \frac{1}{2\rho_i} g_{-i}^T(\tilde{x}_i(t)) \left(\frac{\partial V(\tilde{x}_i(t))}{\partial \tilde{x}_i(t)}\right)\right) \\
&+ \frac{1}{4\rho_i} \left(\frac{\partial V(\tilde{x}_i(t))}{\partial \tilde{x}_i(t)}\right)^T g_{-i}(\tilde{x}_i(t)) g_{-i}^T(\tilde{x}_i(t)) \left(\frac{\partial V(\tilde{x}_i(t))}{\partial \tilde{x}_i(t)}\right)] dt\}.
\end{aligned} \tag{15.45}
$$

It is obvious that the optimal strategies $u_i^*(t)$ and $u_{-i}^*(t)$ should be defined as the following equations to minimize the function and maximize the function, respectively:

$$
\begin{aligned}
u_i^*(t) &= -\frac{1}{2} R_i^{-1} g_i^T(\tilde{x}_i(t)) \left(\frac{\partial V(\tilde{x}_i(t))}{\partial \tilde{x}_i(t)}\right), \\
u_{-i}^*(t) &= \frac{1}{2\rho_i} g_{-i}^T(\tilde{x}_i(t)) \left(\frac{\partial V(\tilde{x}_i(t))}{\partial \tilde{x}_i(t)}\right).
\end{aligned} \tag{15.46}
$$

substituting (15.46) into (15.45), we have

$$
\begin{aligned}
J_1^* &= \min_{u_i(t)} \max_{u_{-i}(t)} E\{V(\tilde{x}_i(0)) - V(\tilde{x}_i(t_p))\} + E\{\int_0^{t_p} [(\tilde{x}_i(t)^T Q_i \tilde{x}_i(t) \\
&+ \left(\frac{\partial V(\tilde{x}_i(t))}{\partial \tilde{x}_i(t)}\right)^T f(\tilde{x}_i(t)) + \frac{1}{2} l(\tilde{x}_i(t))^T \frac{\partial^2 V(\tilde{x}_i(t))}{\partial \tilde{x}_i^2(t)} l(\tilde{x}_i(t)) \\
&+ \lambda[V(\tilde{x}_i(t) + n(\tilde{x}_i(t))) - V(\tilde{x}_i(t))]) \\
&- \frac{1}{4} \left(\frac{\partial V(\tilde{x}_i(t))}{\partial \tilde{x}_i(t)}\right)^T g_i(\tilde{x}_i(t)) R_i^{-1} g_i^T(\tilde{x}_i(t)) \left(\frac{\partial V(\tilde{x}_i(t))}{\partial \tilde{x}_i(t)}\right) \\
&+ \frac{1}{4\rho_i} \left(\frac{\partial V(\tilde{x}_i(t))}{\partial \tilde{x}_i(t)}\right)^T g_{-i}(\tilde{x}_i(t)) g_{-i}^T(\tilde{x}_i(t)) \left(\frac{\partial V(\tilde{x}_i(t))}{\partial \tilde{x}_i(t)}\right)] dt\}.
\end{aligned} \tag{15.47}
$$

By the HJIIs in (15.22), we get

$$
J_1^* = \min_{u_i(t)} \max_{u_{-i}(t)} J_i \le \min_{u_i(t)} \max_{u_{-i}(t)} EV(\tilde{x}_i(0)) = EV(\tilde{x}_i(0)). \tag{15.48}
$$

By the inequalities in (15.19), we get (15.23).

Q.E.D.

15.8.2 Appendix B: Proof of Lemma 15.3

It must be proven by contradiction. According to the definition of Nash equilibrium solution of noncooperative H_∞ game problem, the inequalities in (15.10) must be satisfied. If any inequality in (15.10) is violated, for example:

$$(\rho_1^*, \rho_2^*, \ldots, \rho_i^*, \ldots, \rho_m^*) \geq (\rho_1^*, \rho_2^*, \ldots \rho_{i-1}^*, \rho_i, \rho_{i+1}^* \ldots, \rho_m^*), i = 1, 2, \ldots, m, \quad (15.49)$$

it will violate the Pareto domination of $(\rho_1^*, \ldots, \rho_i^*, \ldots, \rho_m^*)$ of the multiobjective optimization in (15.24). Therefore, when MOP in (15.24) is solved, the inequalities in (15.10) all hold and $(\rho_1^*, \rho_2^*, \ldots, \rho_i^*, \ldots, \rho_m^*)$ is the Nash equilibrium solution.

Q.E.D

15.8.3 Appendix C: Proof of Theorem 15.2

Based on Proposition 1, we select the Lyapunov function $V(\tilde{x}_i(t)) = \tilde{x}_i^T(t) P \tilde{x}_i(t)$. By the global linearization technique in (15.25)–(15.27), we get

$$\begin{aligned}
f(\tilde{x}_i(t)) &= \sum_{j=1}^{J} \alpha_j(\tilde{x}_i(t)) A_{ij} \tilde{x}_i(t), \\
g_i(\tilde{x}_i(t)) &= \sum_{j=1}^{J} \alpha_j(\tilde{x}_i(t)) B_{ij}, \\
g_{-i}(\tilde{x}_i(t)) &= \sum_{j=1}^{J} \alpha_j(\tilde{x}_i(t)) B_{-ij}, \\
l(\tilde{x}_i(t)) &= \sum_{j=1}^{J} \alpha_j(\tilde{x}_i(t)) L_{ij} \tilde{x}_i(t), \\
n(\tilde{x}_i(t)) &= \sum_{j=1}^{J} \alpha_j(\tilde{x}_i(t)) N_{ij} \tilde{x}_i(t).
\end{aligned} \quad (15.50)$$

Then,

$$\begin{aligned}
u_i^*(t) &= -\frac{1}{2} R_i^{-1} g_i^T(\tilde{x}_i(t)) \left(\frac{\partial V(\tilde{x}_i(t))}{\partial \tilde{x}_i(t)} \right), \\
&= \sum_{j=1}^{J} \alpha_j(\tilde{x}_i(t)) R_i^{-1} B_{ij}^T P \tilde{x}_i(t),
\end{aligned} \quad (15.51)$$

and

$$\begin{aligned}
u_{-i}^*(t) &= \frac{1}{2\rho_i} g_{-i}^T(\tilde{x}_i(t)) \left(\frac{\partial V(\tilde{x}_i(t))}{\partial \tilde{x}_i(t)} \right) \\
&= \frac{1}{\rho_i} \sum_{j=1}^{J} \alpha_j(\tilde{x}_i(t)) B_{-ij}^T P \tilde{x}_i(t),
\end{aligned} \quad (15.46)$$

which are the equations (15.29) and (15.30), respectively. The HJIIs in (15.22) have the following inequality after using Lemma 15.4:

$$\begin{aligned}
&\left(\frac{\partial V(\tilde{x}_i(t))}{\partial \tilde{x}_i(t)}\right)^T f(\tilde{x}_i(t)) + \tilde{x}_i(t)^T Q_i \tilde{x}_i(t) \\
&- \frac{1}{4}\left(\frac{\partial V(\tilde{x}_i(t))}{\partial \tilde{x}_i(t)}\right)^T g_i(\tilde{x}_i(t)) R_i^{-1} g_i^T(\tilde{x}_i(t)) \left(\frac{\partial V(\tilde{x}_i(t))}{\partial \tilde{x}_i(t)}\right) \\
&+ \frac{1}{4\rho_i}\left(\frac{\partial V(\tilde{x}_i(t))}{\partial \tilde{x}_i(t)}\right)^T g_{-i}(\tilde{x}_i(t)) g_{-i}^T(\tilde{x}_i(t)) \left(\frac{\partial V(\tilde{x}_i(t))}{\partial \tilde{x}_i(t)}\right) \\
&+ \frac{1}{2} l^T(\tilde{x}_i(t)) \frac{\partial^2 V(\tilde{x}_i(t))}{\partial \tilde{x}_i^2(t)} l(\tilde{x}_i(t)) + \lambda[V(\tilde{x}_i(t) + n(\tilde{x}_i(t))) - V(\tilde{x}_i(t))] \\
&\leq \sum_{j=1}^{J} \alpha_j(\tilde{x}_i) \tilde{x}_i^T(t)(Q_i + A_{ij}^T P + P A_{ij} - P B_{ij}^T R_i^{-1} B_{ij} P + \frac{1}{\rho_i} P B_{-ij} B_{-ij}^T P \\
&+ L_{ij}^T P L_{ij} + \lambda(N_{ij}^T P N_{ij} + N_{ij}^T P + P N_{ij})) \tilde{x}_i(t) \leq 0.
\end{aligned} \tag{15.53}$$

According to (15.53), once the Riccati-like inequalities in (15.31) hold, the HJIIs in (15.22) can also be satisfied. For the inequalities in (15.23), we have

$$0 \leq E\{\tilde{x}_i^T(0) P \tilde{x}_i(0)\} \leq \rho_i E\{\tilde{x}_i^T(0) P \tilde{x}_i(0)\}, \tag{15.54}$$

The sufficient and necessary conditions to satisfy (15.54) are the following LMIs:

$$0 < P \leq \rho_i I, \text{ for } i = 1, 2, \ldots, m. \tag{15.55}$$

which are the LMIs in (15.32).

Q.E.D.

Part V

Stochastic Game Strategies in Biological Systems

16

Robust Reference Therapeutic Control of Immune Systems under Environmental Disturbances via Dynamic H_∞ Game Strategy

16.1 Introduction

After a continuous intrusion of exogeneous pathogens, a dynamic response of the immune system includes innate immune system and adaptive immune system, which are induced by infectious microbes or environmental disturbances. The innate immune system could provide a tactical response, signaling the presence of "non-self" organisms and activating B cells to produce antibodies to bind to the intruders' epitopic sites. The antibodies could identify targets for scavenging cells that engulf and consume the microbes, reducing them to non-functioning units [348]. The antibodies can also stimulate the production of cytokines, complement factors, and acute-phase response proteins that either damage an intruder's plasma membrane directly or trigger the second phase of immune response. Obviously, the innate immune system could protect against many extracellular bacteria or free viruses found in blood plasma, lymph, tissue fluid, or interstitial space between cells, but it cannot clean out microbes that burrow into cells, such as viruses, intracellular bacteria, and protozoa [348–350].

On the other hand, activated by the innate immune response, the adaptive immune system could provide strategic response to invading microbe and yield protective cells. These protective cells could remember specific antigens and produce antibodies to counter the antigens, and seek for epitopes of antigens on the surfaces of infected cells. It is also found that adaptive immune mechanisms could depend on the actions of B- and T-lymphocytes that become dedicated to a single antibody type through clonal selection. Meanwhile, killer T-cells (or cytotoxic T-lymphocytes) bind to infected cells and kill them by initiating programmed cell death (apoptosis). Further, helper T-cells could also assist naive B-cells in maturing into plasma cells to produce the needed antibody type. Then, immune cells with narrowly focused memory are generated, ready to respond rapidly if invading microbes with the same antigen epitopes are encountered again. Finally, we also found some elements of the innate and adaptive immune systems are shared, and response mechanisms are coupled, even though distinctive modes of operation can be identified [348–350].

For clinical treatment of infectious diseases, the current therapeutic control methods focus on killing the invading microbes, neutralizing their response, and providing palliative or healing care to affected organs of the body. Few biological or chemical agents have adverse side effects; for example, an agent that kills a virus may also damage healthy "self" cells. A critical function of drug discovery and development is hence to identify new compounds that have the maximum intended efficacy with the minimal side effect on the general population. These therapeutic treatment examples include antibiotics as microbe killers; interferons as microbe neutralizers; interleukins, antigens from killed (i.e. non-toxic) pathogens, and pre-formed and monoclonal antibodies as immunity enhancers; and anti-inflammatory and anti-histamine compounds as palliative drugs [348].

Recently, there are many dynamic models of immune response to infection [351–354] with a special emphasis on the human-immunodeficiency virus [355–358]. Some papers have discussed immune defense models with moving target strategy [359]. Wiener [360] and Bellman [361] appreciated and anticipated the application of mathematical analysis to treatment in a broad sense, and Swan [362] surveys early optimal control applications to biomedical treatment problems. Notably, Kirschner et al. [363] offer an optimal control approach to HIV treatment, and intuitive therapeutic control approaches are presented [364–368].

The system dynamics of drug response (pharmacokinetics) have been modeled in several works [369,370] and control theory is applied to drug delivery in other studies [371–379]. Recently, Stengel et al. [380] presented a simple model for the dynamic response of the innate immune system to infection and therapeutic intervention, reviewed the prior methods and results of optimization, and introduced a significant extension to the H_2 quadratic optimal therapeutic control of enhancing immune response by solving a two-point boundary-value problem via an iterative method. Their results show not only the treatment progression from an initially life-threatening state to a controlled or cured condition but also the H_2 quadratic optimal control history of therapeutic agents that produce that condition. In their study, the therapeutic control method is extended by adding linear-optimal feedback control to the nominal optimal solution. However, the therapeutic treatment performance of H_2 quadratic optimal control for immune systems may be decayed by the continuous exogenous pathogen input, which is considered as an environmental disturbance of the immune system. Furthermore, some over-shoots may occur in the optimal control process and may lead to organ failure because the H_2 quadratic optimal control only minimizes a quadratic cost function that is only the integration of squares of states and allows the existence of overshoot [381]. A series of dynamic control optimization methods are therefore proposed to design the H_2 quadratic optimal therapeutic schedule for host defense, immune memory, and post-infection pathogen levels in mammals [382–385].

In this chapter, a robust model reference control of immune response is proposed for the therapeutic enhancement to achieve a desired immune response under uncertain exogenous pathogen input, environmental disturbances, and uncertain initial states. Because of the uncertainties of these factors in the therapeutic treatment mentioned above, in order to attenuate their detrimental effects, their worst-case effects should be considered in the reference therapeutic control procedure from the robust design perspective. The worst-case effect of all possible uncertain factors on the reference error to a desired immune response is minimized for the enhanced immune systems, i.e. the proposed robust model reference control is designed from the minimax H_∞ game perspective. This minimax H_∞ game problem could be transformed to an equivalent nonlinear dynamic game problem [280]. The exogenous pathogen input is considered as a player to maximize (worsen) the reference error, while the therapeutic control agent is considered as another player to minimize the reference matching error. Since the innate immune system is highly nonlinear, it is not easy to solve the robust model reference control problem by the nonlinear dynamic game method directly. Recently, fuzzy systems have been employed to efficiently approximate nonlinear dynamic systems to solve the nonlinear control problem [74,105,386,387]. A Takagi–Sugeno (T–S) fuzzy model is proposed to interpolate several local linearized immune systems at different operating points to approximate the innate immune system via smooth fuzzy membership functions. Then, with the help of fuzzy approximation method, a fuzzy dynamic H_∞ game scheme is developed so that the minimax reference therapeutic control of immune systems could be easily solved by the linear dynamic H_∞ game method, which can be subsequently solved by a constrained optimization scheme via the linear matrix inequality (LMI) technique [23] with the help of Robust Control Toolbox in Matlab. Because the fuzzy dynamic model can approximate any nonlinear dynamic system, the proposed model reference matching method via fuzzy H_∞ game theory can be applied to the robust therapeutic control design of any model of immune system that can be interpolated T–S fuzzy modeled. Finally, the computational simulation examples considering the side effect of agents are given to illustrate the design procedure and to confirm the efficiency and efficacy of the proposed minimax match control method for immune systems.

16.2 Nonlinear Dynamic Model of Immune Response to Infectious Disease

For the principal goals to study the general infection course of a disease and to clarify some observational results, a simple four-nonlinear, ordinary differential equation for the dynamic model of infectious disease is introduced as the following equations to describe rates of change of pathogen, immune cell, and antibody concentrations and of an indicator of organic health [351,380]. A more general nonlinear dynamic model will be given next:

$$\begin{aligned}
\dot{x}_1 &= (a_{11} - a_{12}x_3)x_1 + b_1u_1 + w_1, \\
\dot{x}_2 &= a_{21}(x_4)a_{22}x_1x_3 - a_{23}(x_2 - x_2^*) + b_2u_2 + w_2, \\
\dot{x}_3 &= a_{31}x_2 - (a_{32} + a_{33}x_1)x_3 + b_3u_3 + w_3, \\
\dot{x}_4 &= a_{41}x_1 - a_{42}x_4 + b_4u_4 + w_4, \\
a_{21}(x_4) &= \begin{cases} \cos(\pi x_4), & 0 \le x_4 \le 1/2 \\ 0, & 1/2 \le x_4, \end{cases}
\end{aligned} \tag{16.1}$$

where x_1 denotes the concentration of a pathogen that expresses a specific foreign antigen, x_2 denotes the concentration of immune cells that are specific to the foreign antigen, x_3 denotes the concentration of antibodies that bind to the foreign antigen, and x_4 denotes the characteristic of a damaged organ ($x_4 = 0$: healthy, $x_4 \ge 1$: dead). The combined therapeutic control agents and the exogenous inputs are described as follows: u_1 denotes the pathogen killer's agent, u_2 denotes the immune cell enhancer, u_3 denotes the antibody enhancer, u_4 denotes the organ healing factor (or health enhancer), and w_1 denotes the rate of continuing introduction of exogenous pathogens. $w_2 - w_4$ denote the environmental disturbances or unmodeled errors and residues. $a_{21}(x_4)$ is a nonlinear function that describes the mediation of immune cell generation by the damaged cell organ. And if there is no antigen, then the immune cell maintains the steady equilibrium value of x_2^*. The system parameters have been chosen to produce a system that recovers naturally from the pathogen infections (without treatment) as a function of initial conditions during a period of times. For the benchmark example in (16.1), both system parameters and time units are abstractions, as no specific disease is addressed. The state and control are always positive because concentrations cannot go below zero, and organ death is indicated when $x_4(t) \geq 1$. The structural relationship of system variables in (16.1) is illustrated in Fig. 16.1. Organ health mediates immune cell production, inferring a relationship between immune response and fitness of the individual. Antibodies bind to the attacking antigens, thereby killing pathogenic microbes directly, activating complement proteins, or triggering an attack by phagocytic cells, e.g. macrophages and neutrophils. Each element of the state is subject to independent therapeutic control, and new microbes may continue to enter the immune response system.

Several typical uncontrolled responses to increasing levels of initial pathogen concentration under sub-clinical, clinical, chronic, and lethal conditions are shown in Fig. 16.2 [380]. In general, the sub-clinical response would not require medical examination, while the clinical case warrants medical consultation but is self-healing without intervention. Pathogen concentration stabilizes at non-zero values in the chronic case, which is characterized by permanently degraded organ health, and it diverges in the lethal case and expires the organ. The "lethal" simulation of Fig. 16.2 is allowed to continue to past the point at which x_4 exceeds 1 [348]. Finally, a more general nonlinear disease dynamic model could be represented as

$$\dot{x}(t) = f(x(t)) + Bu(t) + Dw(t), \quad x(0) = x_0, \tag{16.2}$$

where $x(t) \in \Re^{n\times 1}$ is the state vector, $u(t) \in \Re^{m\times 1}$ is the control agent input, $w(t) \in \Re^{n\times 1}$ includes exogenous pathogens and environmental disturbances or uncertainty vector. $f(x(t))$ denotes all possible nonlinear interactions in the immune system.

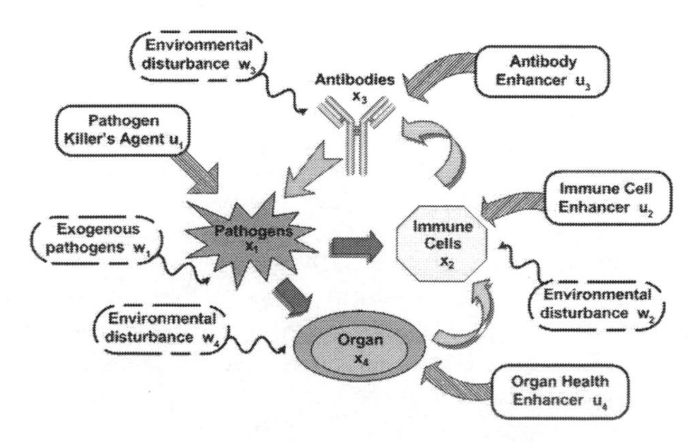

FIG 16.1
Innate and enhanced immune response to a pathogenic attack under exogenous pathogens and environmental disturbances.

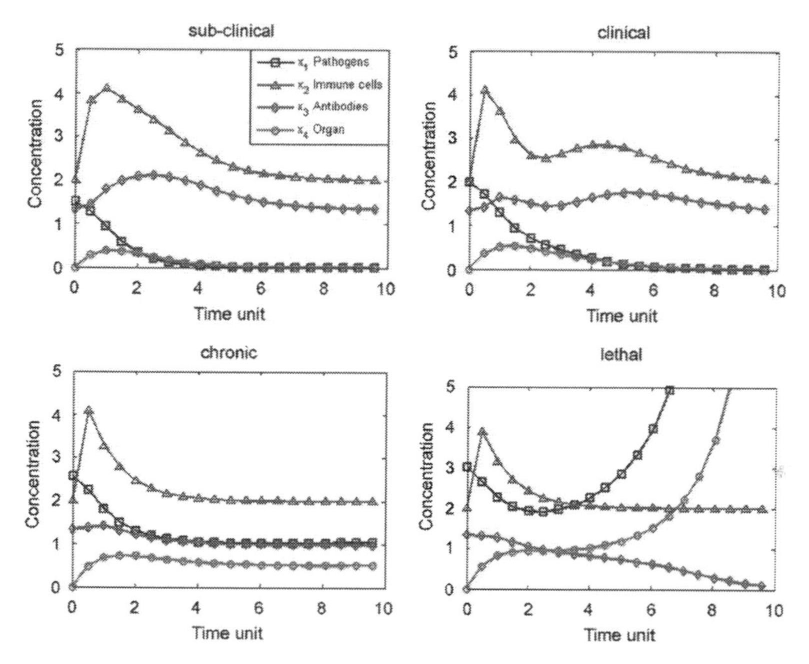

FIGURE16.2
Native immune responses to pathogens which are under sub-clinical, clinical, chronic, and lethal conditions [380].

16.3 Robust H_∞ Nonlinear Therapeutic Control of Immune Response by Nonlinear Dynamic H_∞ Game Strategy

The H_2 quadratic optimal control is to specify $u(t)$ such that the following cost function is minimized [380]:

$$J = \frac{1}{2}\left\{ x^T(t_f)Px(t_f) + \int_0^{t_f} [x^T(t)Qx(t) + u^T(t)Ru(t)]dt \right\}, \tag{16.3}$$

where P, Q, and R are weighting matrices to be specified by designer. Because the quadratic control is only to minimize J in (16.3), i.e. the integration of $x^T(t)Qx(t) + u^T(t)Ru(t)$ to be minimized, a control leading to large overshoot of $x(t)$ but with small integration of $x^T(t)Qx(t)$ may be specified in the H_2 quadratic optimal control design [381]. This therapeutic control will lead to organ failure because $x_4(t) \geq 1$. Furthermore, the cost function does not include exogenous pathogens and environmental disturbances $w(t)$, which may degrade the performance of the H_2 quadratic optimal control. Therefore, it is more appealing to prescribe a desired reference immune response of the disease dynamic in (16.2) beforehand. Next, we design therapeutic control agents $u(t)$ to optimally track the desired time response and at the same time the influence of exogenous pathogens and environmental disturbances $w(t)$ on the reference response tracking should be eliminated as much as possible.

Consider a reference model of immune system with a desired reference immune response prescribed as follows:

$$\dot{x}_r(t) = A_r x_r(t) + r(t), \tag{16.4}$$

where $x_r(t) \in \Re^{n\times 1}$ is the reference state vector, $A_r \in \Re^{n\times n}$ is a specific asymptotically stable matrix, and $r(t)$ is a desired reference signal. It is assumed that $x_r(t), \forall t>0$ represents a desired immune response for (16.2) to follow, i.e. the therapeutic control is to specify $u(t)$ such that the tracking error $\tilde{x}(t) = x(t) - x_r(t)$ must be as small as possible under the influence of uncertain exogenous pathogens and environmental disturbances $w(t)$. Since the exogenous pathogens and environmental disturbances $w(t)$ and the initial state $x(0)$ are uncertain and reference signal $r(t)$ could be arbitrarily assigned, the robust model matching control design should be specified so that the worst-case effect of three uncertainties $w(t)$, $x(0)$, and $r(t)$ on the tracking error could be minimized and set below a prescribed value ρ^2, i.e. both the minimax reference matching and H_∞ robustness against uncertainties $w(t)$, $x(0)$, and $r(t)$ should be achieved simultaneously from the minimax H_∞ game perspective [23,280]:

$$\rho_0^2 = \min_{u(t)} \max_{\substack{w(t),r(t)\\ \tilde{x}(0)}} \frac{\int_0^{t_f} (\tilde{x}^T(t)Q\tilde{x}(t) + u^T(t)Ru(t))dt}{\int_0^{t_f} (w^T(t)w(t) + r^T(t)r(t))dt + \tilde{x}^T(0)\tilde{x}(0)} \le \rho^2, \quad \forall \tilde{x}(0), \tag{16.5}$$

where the weighting matrices Q and R are assumed diagonal as follows:

$$Q = \begin{bmatrix} q_{11} & 0 & 0 & 0 \\ 0 & q_{22} & 0 & 0 \\ 0 & 0 & q_{33} & 0 \\ 0 & 0 & 0 & q_{44} \end{bmatrix}, R = \begin{bmatrix} r_{11} & 0 & 0 & 0 \\ 0 & r_{22} & 0 & 0 \\ 0 & 0 & r_{33} & 0 \\ 0 & 0 & 0 & r_{44} \end{bmatrix}$$

.

The diagonal element q_{ii} of Q denotes the punishment on the corresponding tracking error and the diagonal element r_{ii} of R denotes the relative therapeutic cost. Since the worst-case effect of $w(t)$, $r(t)$, and uncertain initial state $x(0)$ on tracking error $\tilde{x}(t)$ and control $u(t)$ is minimized from the energy point of view, the minimax problem of (16.5) is suitable for the minimax reference model matching problem under unknown initial $x(0)$, uncertain environmental disturbances $w(t)$, and changeable reference $r(t)$, which are always met in practical design cases. Because it is not easy to solve the minmax H_∞ dynamic game problem in (16.5) subject to (16.2) and (16.4) directly, we provide an upper bound ρ^2 of ρ_0^2 to solve the minimax H_∞ game problem indirectly from the suboptimal perspective.

Remark 16.1 Actually, the design idea is the same as the model adaptive control (MRAC) [388]. The desired immune reference response in (16.4) is the model reference in [388]. The difficulty of the H_∞ model reference control design of immune system is that all the immune systems are nonlinear and external disturbances are uncertain. Therefore, the minimax H_∞ game theory in (16.5) and fuzzy interpolation method are employed to simplify the design procedure of the nonlinear MRAC design problem of immune systems in the next approach.

We will first solve the above sub-minimax game problem and then decrease the upper bound ρ^2 as small as possible to get the real minimax H_∞ game problem. Since $w(t)$ and $r(t)$ in the denominator in (16.5) are independent of $u(t)$ and is not zero, (16.5) is equivalent to [23,280]:

$$\begin{aligned}&\min_{u(t)}\max_{w(t),r(t)}\int_0^{t_f}(\tilde{x}^T(t)Q\tilde{x}(t)+u^T(t)Ru(t)-\rho^2w^T(t)w(t)\\&-\rho^2r^T(t)r(t))dt\leq\rho^2\tilde{x}^T(0)\tilde{x}(0),\forall\tilde{x}(0).\end{aligned}\tag{16.6}$$

Let us denote

$$\begin{aligned}&\min_{u(t)}\max_{w(t),r(t)}J(u(t),w(t),r(t))\\&=\min_{u(t)}\max_{w(t),r(t)}\int_0^{t_f}(\tilde{x}^T(t)Q\tilde{x}(t)+u^T(t)Ru(t)-\rho^2w^T(t)w(t)-\rho^2r^T(t)r(t))dt.\end{aligned}$$

From the above analysis, the dynamic game problem in (16.5) or (16.6) is equivalent to finding the worst-case disturbance $w^*(t)$ and reference signal $r^*(t)$ which maximize $J(u(t),w(t),r(t))$ and then a minimax control $u^*(t)$ which minimizes $J(u(t),w^*(t),r^*(t))$ such that the minimax value $J(u^*(t),w^*(t),r^*(t))$ is less than $\rho^2\tilde{x}^T(0)\tilde{x}(0)$, i.e.

$$\begin{aligned}&J(u^*(t),w^*(t),r^*(t))\\&=\min_{u(t)}J(u(t),w^*(t),r^*(t))\\&=\min_{u(t)}\max_{w(t),r(t)}J(u(t),w(t),r(t))\leq\rho^2\tilde{x}^T(0)\tilde{x}(0),\quad\forall\tilde{x}(0).\end{aligned}\tag{16.7}$$

Hence, if there exist $u^*(t)$, $w^*(t)$, and $r^*(t)$ such that minimax H_2 matching game problem in (16.7) is solved, then they can satisfy the robust model matching performance in (16.5) as well. Therefore, the first step of robust reference matching therapeutic control design of therapeutic agents for immune systems is to solve the following minimax H_2 quadratic dynamic game problem:

$$\min_{u(t)}\max_{w(t),r(t)}J(u(t),w(t),r(t))\tag{16.8}$$

subject to the disease dynamic model in (16.2) and the desired reference model in (16.4). After that, the next step is to check whether the condition $J(u^*(t),w^*(t),r^*(t))\leq\rho^2\tilde{x}^T(0)\tilde{x}(0)$ is satisfied or not for any $\tilde{x}(0)$.

In general, it is not easy to solve the minimax H_2 reference matching game problem directly; it should be transformed to an equivalent minimax regulation problem. Let us denote

$$F(\bar{x}(t))=\begin{bmatrix}f(x(t))\\A_rx_r(t)\end{bmatrix},\bar{x}(t)=\begin{bmatrix}x(t)\\x_r(t)\end{bmatrix}\in\Re^{2n\times1},u(t)\in\Re^{m\times1}\text{ and }v(t)=\begin{bmatrix}w(t)\\r(t)\end{bmatrix}\in\Re^{2n\times1}.$$

Then we can rewrite the minimax H_2 reference matching game problem as

$$\min_{u(t)} \max_{v(t)} J(u(t), v(t))$$
$$= \min_{u(t)} \max_{v(t)} \int_0^{t_f} (\bar{x}^T(t)\mathbf{Q}\bar{x}(t) + u^T(t)Ru(t) - \rho^2 v^T(t)v(t))dt, \forall \tilde{x}(0) \tag{16.9}$$

subject to the following augmented system of (16.2) and (16.4):

$$\dot{\bar{x}}(t) = F(\bar{x}(t)) + \mathbf{B}u(t) + \mathbf{C}v(t), \tag{16.10}$$

where

$$\mathbf{Q} = \begin{bmatrix} Q & -Q \\ -Q & Q \end{bmatrix}, \mathbf{B} = \begin{bmatrix} B \\ 0 \end{bmatrix}, \mathbf{C} = \begin{bmatrix} D & 0 \\ 0 & I \end{bmatrix}, \mathbf{I} = \begin{bmatrix} I & -I \\ -I & I \end{bmatrix},$$

in which **I** denotes the 4×4 identity matrix. Then the minimax H_2 reference matching game problem in (16.8) is equivalent to the following minimax H_2 regulation game problem of the augmented system in (16.10):

$$\min_{u(t)} \max_{v(t)} J(u(t), v(t))$$
$$= \min_{u(t)} \max_{v(t)} \int_0^{t_f} (\bar{x}^T(t)\mathbf{Q}\bar{x}(t) + u^T(t)Ru(t) - \rho^2 v^T(t)v(t))dt \leq \rho^2 \tilde{x}^T(0)\mathbf{I}\tilde{x}(0), \forall \tilde{x}(0) \tag{16.11}$$

subject to (16.10).

Theorem 16.1

The minimax H_2 quadratic game problem for robust reference matching therapeutic control of immune response in (16.11) could be solved by the following minimax matching therapeutic control $u^*(t)$ and the worst-case disturbance $v^*(t)$:

$$u^*(t) = -\frac{1}{2}R^{-1}\mathbf{B}^T \frac{\partial V(\bar{x}(t))}{\partial \bar{x}(t)}, \tag{16.12}$$

$$v^*(t) = \frac{1}{2\rho^2}\mathbf{C}^T \frac{\partial V(\bar{x}(t))}{\partial \bar{x}(t)}, \tag{16.13}$$

where $V(\bar{x}(t)) > 0$ is the positive solution of the following Hamilton–Jacobi inequality (HJII):

$$\left(\frac{\partial V(\bar{x}(t))}{\partial \bar{x}(t)}\right)^T F(\bar{x}(t)) + \bar{x}^T(t)\mathbf{Q}\bar{x}(t) - \frac{1}{4}\left(\frac{\partial V(\bar{x}(t))}{\partial \bar{x}(t)}\right)^T \mathbf{B}R^{-1}\mathbf{B}^T \left(\frac{\partial V(\bar{x}(t))}{\partial \bar{x}(t)}\right)$$
$$+ \frac{1}{4\rho^2}\left(\frac{\partial V(\bar{x}(t))}{\partial \bar{x}(t)}\right)^T \mathbf{C}\mathbf{C}^T \left(\frac{\partial V(\bar{x}(t))}{\partial \bar{x}(t)}\right) < 0 \tag{16.14}$$

with

$$V(\bar{x}(0)) \leq \rho^2 \tilde{x}^T(0) \mathrm{I} \tilde{x}(0), \quad \forall \tilde{x}(0). \tag{16.15}$$

Proof: See Appendix A.

Since ρ^2 is the upper bound of ρ_0^2 for the minimax H_∞ game problem in (16.5), based on the analysis above, the minimax H_∞ reference matching control $u^*(t)$ and the worst-case disturbance $v^*(t)$ still need to minimize the upper bound ρ^2 as follows:

$$\begin{aligned} &\rho_0^2 = \min_{V(\bar{x}(t))>0} \rho^2 \\ &\text{subject to (16.4) and (16.5).} \end{aligned} \tag{16.16}$$

After solving a $V_0(\bar{x}(t))$ and ρ_0^2. from the constrained optimization in (16.16), we substitute this solution $V_0(\bar{x}(t))$ to obtain the minimax H_∞ matching control $u^*(t)$ in (16.12), i.e. $u^*(t) = -\frac{1}{2} R^{-1} \mathrm{B}^T \frac{\partial V_0(\bar{x}(t))}{\partial \bar{x}(t)}$.

16.4 Robust H_∞ Reference Matching Therapeutic Control of Innate Immune System via Fuzzy Dynamic H_∞ Game Method

Because it is very difficult to solve the nonlinear HJII in (16.14), no simple approach is available to solve the constrained optimization problem in (16.16) for robust H_∞ model matching control of innate immune system. Recently [4,74,105], the fuzzy T–S model has been widely applied to approximate the nonlinear system via interpolating several local linearized systems at different operating points so that the nonlinear dynamic H_∞ game problem could be transformed to a fuzzy dynamic H_∞ game problem. Using such approach, the HJII in (16.14) can be replaced by a set of LMIs. In this situation, the nonlinear dynamic game problem in (16.5) could be easily solved by fuzzy H_∞ dynamic game method for the design of robust model matching control for innate immune response systems.

Suppose the augmented system in (16.10) can be represented by the T–S fuzzy model [4]. The T–S fuzzy model is a piecewise interpolation of several local linearized models through membership functions. The fuzzy model is described by fuzzy "If–Then" rules and will be employed to deal with the nonlinear dynamic H_∞ game problem for robust reference model matching control to achieve a desired immune response under exogenous pathogens input and environmental disturbances. The ith rule of fuzzy model for nonlinear system in (16.10) is of the following form [74,105]:

- Rule i

If $x_1(t)$ is F_{i1} and $x_g(t)$ is F_{ig},
then,

$$\dot{\bar{x}}(t) = \mathrm{A}_i \bar{x}(t) + \mathrm{B}u(t) + Cv(t), \quad i = 1, 2, 3, \ldots, L, \tag{16.17}$$

in which

$$\mathrm{A}_i = \begin{bmatrix} A_i & 0 \\ 0 & A_r \end{bmatrix}, \mathrm{B} = \begin{bmatrix} B \\ 0 \end{bmatrix}, \mathrm{C} = \begin{bmatrix} D & 0 \\ 0 & I \end{bmatrix},$$

and F_{ij} is the fuzzy set; A_i, B and C are known constant matrices; L is the number of If–Then rules, g is the number of premise variables and $x_1(t)$, $x_2(t)$, …, $x_g(t)$ are the premise variables. The fuzzy system is inferred as follows [4,74,105]:

$$\begin{aligned} \dot{\bar{x}}(t) &= \frac{\sum_{i=1}^{L} \mu_i(x(t))[\mathrm{A}_i\bar{x}(t) + \mathrm{B}u(t) + \mathrm{C}v(t)]}{\sum_{i=1}^{L} \mu_i(x(t))} \\ &= \sum_{i=1}^{L} h_i(x(t))[\mathrm{A}_i\bar{x}(t) + \mathrm{B}u(t) + \mathrm{C}v(t)], \end{aligned} \tag{16.18}$$

where $\mu_i(x(t)) = \prod_{j=1}^{g} F_{ij}(x_j(t))$, $h_i(x(t)) = \mu_i(x(t))/\sum_{i=1}^{L} \mu_i(x(t))$, $x(t) = [x_1(t), x_2(t), \ldots, x_g(t)]$ and $F_{ij}(x_j(t))$ is the grade of membership of $x_j(t)$ in F_{ij}.

We assume

$$\mu_i(x(t)) \geq 0 \text{ and } \sum_{i=1}^{L} \mu_i(x(t)) > 0. \tag{16.19}$$

Therefore, we get

$$h_i(x(t)) \geq 0 \text{ and } \sum_{i=1}^{L} h_i(x(t)) = 1. \tag{16.20}$$

The T–S fuzzy model in (16.18) is to interpolate L local linear systems to approximate the nonlinear system in (16.10) via the fuzzy basis function $h_i(x(t))$. We specify the parameter A_i easily so that $\sum_{i=1}^{L} h_i(x(t))\mathrm{A}_i\bar{x}(t)$ in (16.18) can approximate $F(\bar{x}(t))$ in (16.10) by the fuzzy identification method [4].

After the nonlinear system in (16.10) is approximated as the T–S fuzzy system in (16.18), the nonlinear dynamic H_2 quadratic game problem in (16.10) and (16.11) is replaced by solving the fuzzy dynamic H_2 quadratic game problem in (16.18) and (16.11).

Theorem 16.2

The minimax control and the worst-case disturbance for the fuzzy dynamic H_2 quadratic game problem in (16.11) subject to (16.18) are solved respectively as follows:

$$u^*(t) = -R^{-1}\mathrm{B}^T P\bar{x}(t) \text{ and } v^*(t) = \frac{1}{\rho^2}\mathrm{C}^T P\bar{x}(t), \tag{16.21}$$

where P is the positive definite symmetric matrix solution of the following Riccati-like inequality:

$$\begin{aligned} &P\mathrm{A}_i + \mathrm{A}_i^T P + \mathrm{Q} - P^T\mathrm{B}R^{-1}\mathrm{B}^T P + \frac{1}{\rho^2}P^T\mathrm{C}\mathrm{C}^T P \leq 0, \\ &i = 1, \ldots, L, \;\; P \leq \rho^2 I. \end{aligned} \tag{16.22}$$

Proof: See Appendix B.

By fuzzy approximation, obviously, the HJII in (16.14) can be approximated by a set of algebraic inequalities in (16.22).

Since ρ^2 is the upper bound of ρ_0^2 for the minimax H_∞ game problem in (16.5), the minimax H_∞ game problem still needs to minimize ρ^2 as follows:

$$\rho_0^2 = \min_{P>0} \rho^2 \quad \text{subject to (16.22).} \tag{16.23}$$

In order to solve the above-constrained optimization in (16.23) by the conventional LMI method, we let $W = P^{-1} > 0$. Then, (16.22) can be equivalent to

$$A_i W + W A_i^T + WQW - BR^{-1}B^T + \frac{1}{\rho^2} CC^T \leq 0, \quad i = 1, \dots, L,$$

$$\text{or } A_i W + W A_i^T + W \begin{bmatrix} Q & -Q \\ -Q & Q \end{bmatrix} W - BR^{-1}B^T + \frac{1}{\rho^2} CC^T \leq 0, \; i = 1, \dots, L,$$

$$\text{or } A_i W + W A_i^T + W \begin{bmatrix} Q^{1/2} \\ -Q^{1/2} \end{bmatrix} I \begin{bmatrix} Q^{1/2} & -Q^{1/2} \end{bmatrix} W - BR^{-1}B^T + \frac{1}{\rho^2} CC^T \leq 0, \; i = 1, \dots, L, \tag{16.24}$$

By the Schur complements [23], the constrained optimization in (16.22) and (16.23) for minimax H_∞ game in (16.5) is equivalent to the following LMI-constrained optimization:

$$\rho_0^2 = \min_{W>0} \rho^2$$

subject to

$$\begin{bmatrix} A_i W + W A_i^T - BR^{-1}B^T + \frac{1}{\rho^2} CC^T & W \begin{bmatrix} Q^{1/2} \\ -Q^{1/2} \end{bmatrix} \\ \begin{bmatrix} Q^{1/2} & -Q^{1/2} \end{bmatrix} W & -I \end{bmatrix} \leq 0, \; i = 1, \dots, L. \tag{16.25}$$

$$\rho^2 W \geq I.$$

Remark 16.2

(1) The fuzzy basis function $h_i(x(t))$ in (16.18) and (16.20) can be replaced by other interpolation functions, for example, cubic spline functions.

(2) By the fuzzy approximation, the HJII in (16.14) of nonlinear dynamic H_2 quadratic game problem is replaced by a set of inequalities in (16.22), which can be easily solved by LMI-constrained optimization in (16.25).

(3) The constrained optimization to solve ρ_0 and $W_0 = P_0^{-1}$ in (16.24) and (16.25) can be easily solved by decreasing ρ^2 until there exists no $W > 0$ solution in (16.25). After solving W_0 and then $P_0 = W_0^{-1}$ from the constrained optimization problem in (16.24) and (16.25), the minimax H_∞ game control can be obtained from (16.21).

(4) The solution $W_0>0$ in LMI-constrained optimization (16.25) can be solved by Robust Control Toolbox in Matlab efficiently.

(5) If the conventional H_2 quadratic optimal control in (16.3) is considered [348], i.e. the effect of disturbance is not considered in the design procedure, the H_2 quadratic optimal tracking control problem is equivalent to letting $\rho^2 = \infty$ in (16.5) [381]. Then the H_2 optimal control design $u^*(t) = -R^{-1}BP\bar{x}(t)$ can be solved by a common positive definite symmetric matrix P from (16.22) with $\rho^2 = \infty$, i.e. solving a common positive definite symmetric matrix $P>0$ from the following constrained inequalities: $PA_i + A_i^T P + Q - P^T BR^{-1}B^T P \leq 0, i = 1, \ldots, L$ [23]. In order to solve the H_2 quadratic optimal tracking control by LMI technique, the H_2 quadratic optimal tracking control is equivalent to solving a common $W = P^{-1}$ from the following constrained inequalities:

$$A_i W + WA_i^T + W\begin{bmatrix} Q^{1/2} \\ -Q^{1/2} \end{bmatrix} I \begin{bmatrix} Q^{1/2} & -Q^{1/2} \end{bmatrix} W - BR^{-1}B^T \leq 0, \quad i = 1, \ldots, L \tag{16.26}$$

or equivalently,

$$\begin{bmatrix} A_i W + WA_i^T - BR^{-1}B^T & W\begin{bmatrix} Q^{1/2} \\ -Q^{1/2} \end{bmatrix} \\ \begin{bmatrix} Q^{1/2} & -Q^{1/2} \end{bmatrix} W & -I \end{bmatrix} \leq 0, \quad i = 1, \ldots, L, \tag{16.27}$$

which is equivalent to (16.25) with $\rho^2 = \infty$.

According to the above analysis, the robust H_∞ model matching control of innate immune system via fuzzy interpolation method is summarized as follows.

Design procedure of robust minimax H_∞ therapeutic control:

(1) Give a desired reference model in (16.4) of immune system.
(2) Select membership functions and construct fuzzy plant rules in (16.17).
(3) Give weighting matrices Q and R in (16.5).
(4) Solve the LMI-constrained optimization in (16.25) to obtain W_0 (thus $P_0 = W_0^{-1}$ can also be obtained) and ρ_0^2.
(5) Construct the minimax H_∞ game controller $u^*(t) = -R^{-1}BP_0\bar{x}(t)$ under the worst-case disturbance $v^*(t)$ in (16.21).

Remark 16.3 The software packages such as Robust Control Toolbox in Matlab can be employed to solve the LMI-constrained optimization problem in (16.25) easily.

16.5 Computational Examples

Example 16.1. We consider the innate immune system in (16.1) and in Fig. 16.1. The values of the parameters are in Table 16.1. The environmental disturbances $w_1 - w_4$ are unknown but bounded signals. Under infectious situation, the microbes infect the organ

TABLE 16.1

Model parameters of dynamic innate immune system [348,390].

Parameter	Value	Description
a_{11}	1	Pathogens reproduction rate coefficient
a_{12}	1	The suppression by pathogens coefficient
a_{22}	3	Immune reactivity coefficient
a_{23}	1	The mean immune cell production rate coefficient
x_2^*	2	The steady-state concentration of immune cells
a_{31}	1	Antibodies production rate coefficient
a_{32}	1.5	The antibody mortality coefficient
a_{33}	0.5	The rate at which antibodies suppress pathogens
a_{41}	0.5	The organ damage depends on the pathogen's damage possibilities coefficient
a_{42}	1	Organ recovery rate
b_1	−1	Pathogen killer's agent coefficient
b_2	1	Immune cell enhancer coefficient
b_3	1	Antibody enhancer coefficient
b_4	−1	Organ health enhancer coefficient

not only by an initial concentration at the beginning but also by the continuous pathogens input. For the convenience of computer simulation, suppose the continuous pathogens input to the immune system is viewed as an environmental disturbance w_1 shown in Fig. 16.3. For the convenience of simulation, $w_2 - w_4$ are assumed zero mean white noises with standard deviations all equal to 2. The stochastic noises of immune systems are mainly due to measurement errors, modeling errors, and process noises [389]. The dynamic model of innate immune system in (16.1) under exogenous pathogens input and environmental disturbances is controlled by a combined therapeutic control shown as follows [380]:

$$\begin{aligned}
\dot{x}_1 &= (1 - x_3)x_1 - u_1 + 3w_1, \\
\dot{x}_2 &= a_{21}(x_4)3x_1x_3 - (x_2 - 2) + u_2 + 3w_2, \\
\dot{x}_3 &= x_2 - (1.5 + x_1)x_3 + u_3 + 3w_3, \\
\dot{x}_4 &= 0.5x_1 - x_4 - u_4 + 3w_4 \\
a_{21}(x_4) &= \begin{cases} \cos(\pi x_4), & 0 \le x_4 \le 1/2, \\ 0, & 1/2 \le x_4, \end{cases}
\end{aligned} \tag{16.28}$$

with the set of the initial condition $x(0) = [x_1(0) \quad x_2(0) \quad x_3(0) \quad x_4(0)]^T = [3 \quad 3.1 \quad 1 \quad 0.98]^T$. In this example, therapeutic controls $u_1 - u_4$ are combined to enhance the immune system.

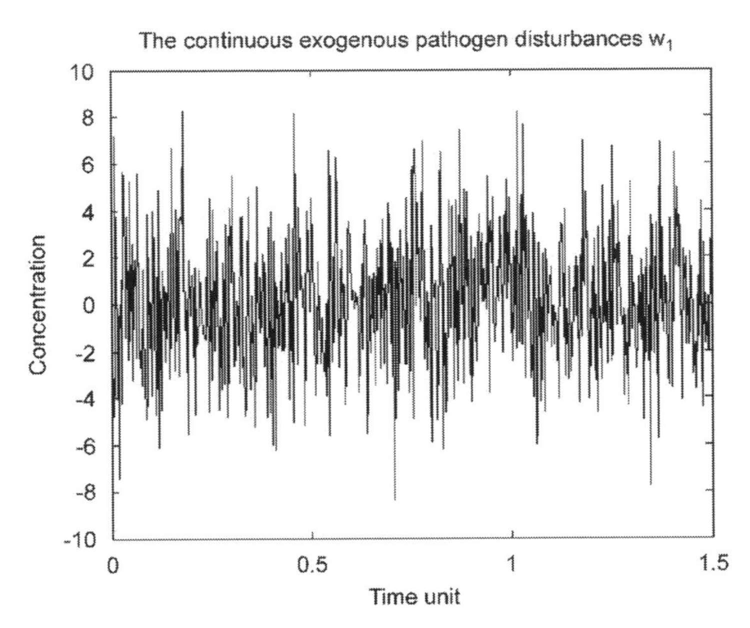

FIGURE 16.3
The continuous exogenous pathogens intrusion.

Our reference model design objective is that system matrix A_r and in (16.4) should be specified beforehand so that its transient responses and steady state of reference system for innate immune response system are desired. If the real parts of eigenvalues of A_r are more negative (i.e. more robust stable), the tracking system will be more robust to environmental disturbances. After some numerical simulations for clinical treatment, the desired reference signals are obtained by the following reference model (see Fig. 16.4):

$$\dot{x}_r(t) = \begin{bmatrix} -12 & 0 & 0 & 0 \\ 0 & -2.3 & 0 & 0 \\ 0 & 0 & -10 & 0 \\ 0 & 0 & 0 & -3 \end{bmatrix} \times x_r(t) + B_r \times u_{step}(t), \tag{16.29}$$

where $B_r = [0 \quad 4.6 \quad 13.3333 \quad 0]^T$ and $u_{step}(t)$ is the unit step function.

From the investigation of the uncontrolled innate immune response (lethal case) in Fig. 16.5, the pathogen concentration is increasing rapidly and causes organ failure. We try to administrate a treatment after a period of pathogens infection to enhance the immune system. The cutting line (black solid line) in Fig. 16.5 is a proper time to take drugs. Suppose the set of the initial condition of the desired reference model is about $x_r(0) = [2.9 \quad 3.2 \quad 1.1 \quad 0.9]^T$. The time response of the desired reference model in (16.29) is shown in Fig. 16.4.

To minimize the design effort and complexity for this nonlinear innate immune system in (16.28), we employ the T–S fuzzy model in (16.7) to construct fuzzy rules to approximate the nonlinear innate immune system with the innate immune system's state variables as premise variables in the following:

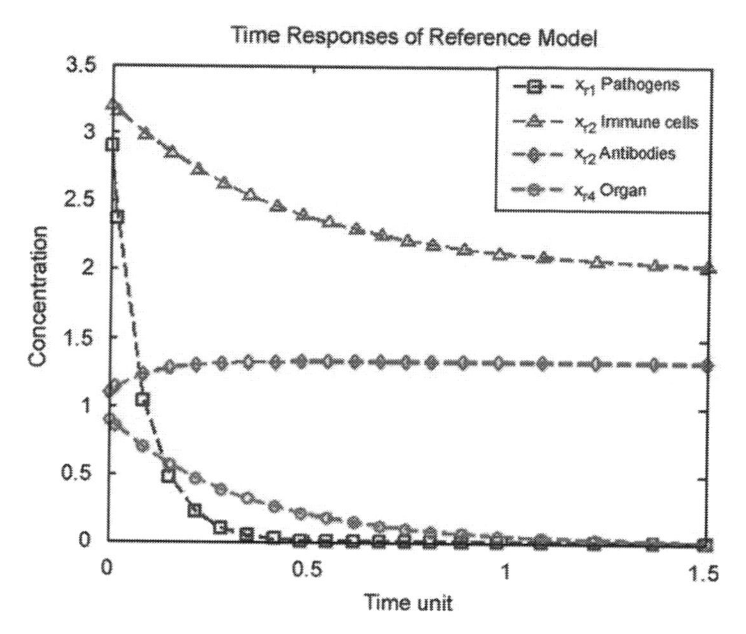

FIGURE 16.4
The desired reference model with four desired states in (16.29): pathogens (x_{r1}), immune cells (x_{r2}), antibodies (x_{r3}), and organ (x_{r4}). The initial conditions of the reference model are $x_r(0) = [2.9 \quad 3.2 \quad 1.1 \quad 0.9]^T$.

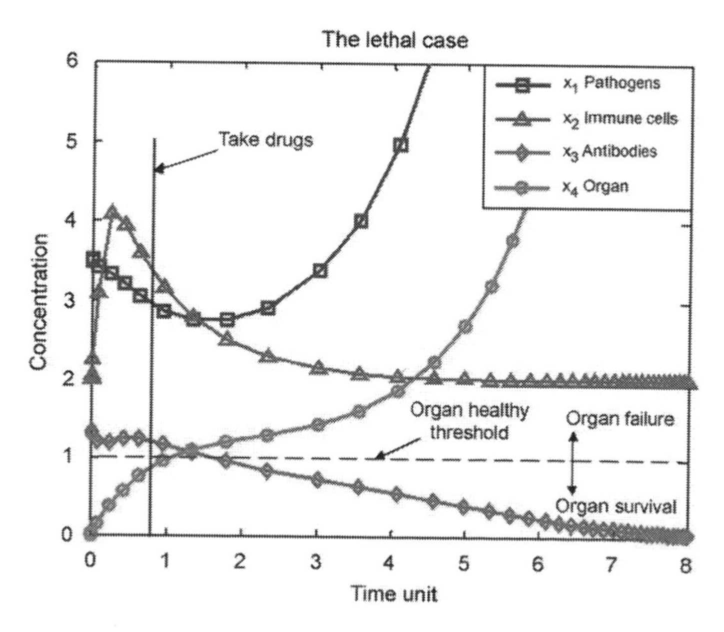

FIGURE 16.5
The uncontrolled immune responses (lethal case) in (16.28) are shown to increase the level of pathogen concentration at the beginning of the time period. In this case, we try to administrate a treatment after a short period of pathogens infection. The cutting line (solid line) is an optimal time point to give drugs. The organ will survive or fail based on the organ health threshold (horizontal dashed line) ($x_4 < 1$: survival; $x_4 \geq 1$: failure).

- *Rule i*

If $x_1(t)$ is F_{i1}, $x_2(t)$ is F_{i2}, $x_3(t)$ is F_{i3}, and $x_4(t)$ is F_{i4},
then

$$\dot{\bar{x}}(t) = \mathrm{A}_i\bar{x}(t) + \mathrm{B}u(t) + Cv(t), \quad i = 1, 2, 3, \ldots, L, \tag{16.30}$$

where $\bar{x} = [x_1 \ x_2 \ x_3 \ x_4 \ x_{r1} \ x_{r2} \ x_{r3} \ x_{r4}]^T$, $u = [u_1 u_2 u_3 u_4]^T$, $v = [w_1 w_2 w_3 w_4 r_1 r_2 r_3 r_4]^T$, the parameters $\mathrm{A}_i, \mathrm{B}, C$ are described in Appendix C and the number of the fuzzy rules is $L = 16$. To construct the fuzzy model in (16.30), we need to find the operating points of the innate immune response. Suppose the operating points for x_1 are at $\bar{x}_{11} = 0$ and $\bar{x}_{12} = 4$. Similarly, the operating points of x_2, x_3, and x_4 are at $\bar{x}_{21} = 0$, $\bar{x}_{22} = 10$, $\bar{x}_{31} = 0$, $\bar{x}_{32} = 5$, $\bar{x}_{41} = 0$, and , respectively. For the convenience of design, triangle-type membership functions are taken for Rule 1 through Rule 16. We create two triangle-type membership functions for each state at these operating points (see Fig. 16.6). In order to accomplish the robust minimax H_∞ game performance in (16.5), we should tune up a set of the weighting matrices Q and R of the cost function in (16.5) as follows:

$$Q = \begin{bmatrix} 1 & 0 & 0 & 0 \\ 0 & 1 & 0 & 0 \\ 0 & 0 & 1 & 0 \\ 0 & 0 & 0 & 1 \end{bmatrix}, R = \begin{bmatrix} 0.003 & 0 & 0 & 0 \\ 0 & 0.003 & 0 & 0 \\ 0 & 0 & 0.003 & 0 \\ 0 & 0 & 0 & 0.003 \end{bmatrix}.$$

After specifying the desired reference model, we need to solve the constrained optimization in (16.25) for the robust minimax H_∞ reference control in (16.21) by employing Matlab Robust Control Toolbox. Finally, we obtain a minimum attenuation level $\rho_0^2 = 0.98$ and a common positive definite symmetric matrix $P_0 = W_0^{-1}$ for (16.25) as follows:

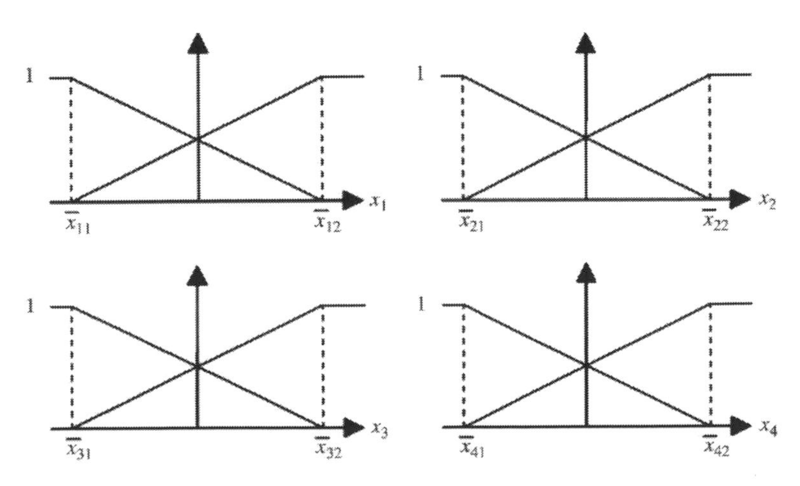

FIGURE 16.6
Membership functions for four states x_1, x_2, x_3, and x_4.

$$P_0 = \begin{bmatrix} 0.43313 & 0 & 0 & 0 & -0.43313 & 0 & 0 & 0 \\ 0 & 0.56172 & 0 & 0 & 0 & -0.56172 & 0 & 0 \\ 0 & 0 & 0.42678 & 0 & 0 & 0 & -0.42678 & 0 \\ 0 & 0 & 0 & 0.28482 & 0 & 0 & 0 & -0.28482 \\ -0.43313 & 0 & 0 & 0 & 0.50151 & 0 & 0 & 0 \\ 0 & -0.56172 & 0 & 0 & 0 & 0.62738 & 0 & 0 \\ 0 & 0 & -0.42678 & 0 & 0 & 0 & 0.49526 & 0 \\ 0 & 0 & 0 & -0.28482 & 0 & 0 & 0 & 0.34567 \end{bmatrix}.$$

Figures 16.7 and 16.8 present the simulation results for the robust minimax H_∞ model matching control. Figure 16.7 shows the responses of the controlled immune system by minimax H_∞ model matching control with the concentrations of the pathogens x_1, immune cells x_2, antibodies x_3, and organ index x_4 to track the desired reference states x_{r1}, x_{r2}, x_{r3} and x_{r4}, respectively. From the simulation results, the tracking performance of the robust minimax H_∞ reference model matching therapeutic control via T–S fuzzy interpolation is quite satisfactory. Figure 16.8 shows the four combined therapeutic control signals. Obviously, from Figs. 16.8 and 16.9, it is seen that the effect of stochastic external disturbances on the robust minimax H_∞ reference model tracking of immune system is attenuated significantly by the proposed robust minimax H_∞ game therapeutic control design.

Example 16.2. It is sometimes unavoidable for drugs to have adverse side effects. Limiting the impact of side effects should be considered in this minimax H_∞ reference therapeutic control design. Suppose the pathogen killer's agent u_1 (i.e. drugs or antibiotics) also degrades organ health. The modified model of (16.28) of the organ state x_4 becomes

$$\dot{x}_4 = 0.5x_1 - x_4 - u_4 + 0.2u_1 + 3w_4, \tag{16.31}$$

i.e. an increase of 0.2 side effect of the pathogen killer's agent u_1 on the organ state x_4. The detailed design of fuzzy approximation is similar to the previous procedure and the parameters are the same except B, i.e.

$$\mathrm{B} = \begin{bmatrix} -1 & 0 & 0 & 0 \\ 0 & 1 & 0 & 0 \\ 0 & 0 & 1 & 0 \\ 0.2 & 0 & 0 & -1 \\ 0 & 0 & 0 & 0 \\ 0 & 0 & 0 & 0 \\ 0 & 0 & 0 & 0 \\ 0 & 0 & 0 & 0 \end{bmatrix}.$$

In this design case, we choose

$$Q = \begin{bmatrix} 1 & 0 & 0 & 0 \\ 0 & 1 & 0 & 0 \\ 0 & 0 & 1 & 0 \\ 0 & 0 & 0 & 2 \end{bmatrix}, R = \begin{bmatrix} 0.002 & 0 & 0 & 0 \\ 0 & 0.002 & 0 & 0 \\ 0 & 0 & 0.002 & 0 \\ 0 & 0 & 0 & 0.002 \end{bmatrix}.$$

Since the organ health state suffers from the impact of the side effects, we should increase the weighting q_{44} to impose a penalty on the organ health tracking error. Furthermore, we relax the weighting on the control signals to improve the tracking efficiency. By the same design procedure, we obtain a minimum attenuation level $\rho_0^2 = 0.9$ and a common positive definite symmetric matrix P_0 as follows:

$$P_0 = \begin{bmatrix} 0.29932 & 0 & 0 & 0 & -0.29932 & 0 & 0 & 0 \\ 0 & 0.39859 & 0 & 0 & 0 & -0.39859 & 0 & 0 \\ 0 & 0 & 0.28408 & 0 & 0 & 0 & -0.28408 & 0 \\ 0 & 0 & 0 & 0.30502 & 0 & 0 & 0 & -0.30502 \\ -0.29932 & 0 & 0 & 0 & 0.35217 & 0 & 0 & 0 \\ 0 & -0.39859 & 0 & 0 & 0 & 0.43261 & 0 & 0 \\ 0 & 0 & -0.28408 & 0 & 0 & 0 & 0.33404 & 0 \\ 0 & 0 & 0 & -0.30502 & 0 & 0 & 0 & 0.33088 \end{bmatrix}.$$

The control design with considered drug's side effects is simulated in Figs. 16.9 and 16.10. Figure 16.9 shows the responses of the controlled immune system by the proposed minimax H_∞ reference model game control with undergoing the side effects of pathogen killer's

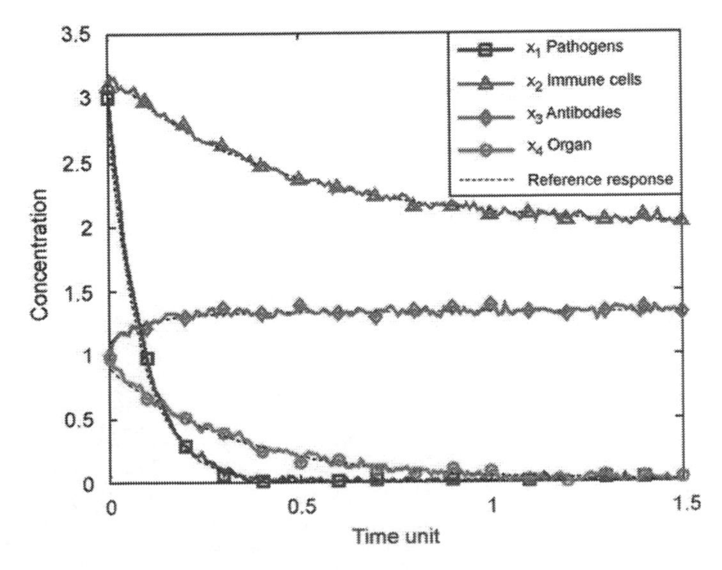

FIGURE 16.7
The tracking of innate immune system to the desired reference model by the robust minimax H_∞ game control under the continuous exogenous pathogens and environmental disturbances.

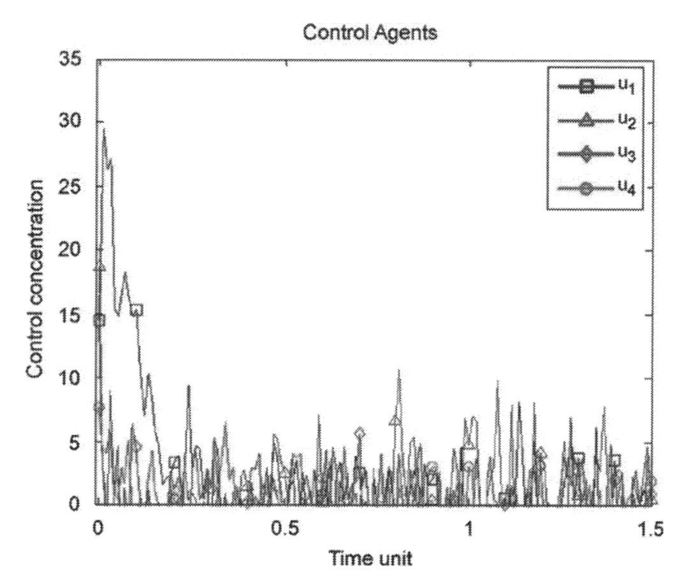

FIGURE 16.8
The minimax H_∞ game controls in the simulation example. The drug controls u_1 for pathogens, u_2 for immune cells, u_3 for antibodies, and u_4 for organ.

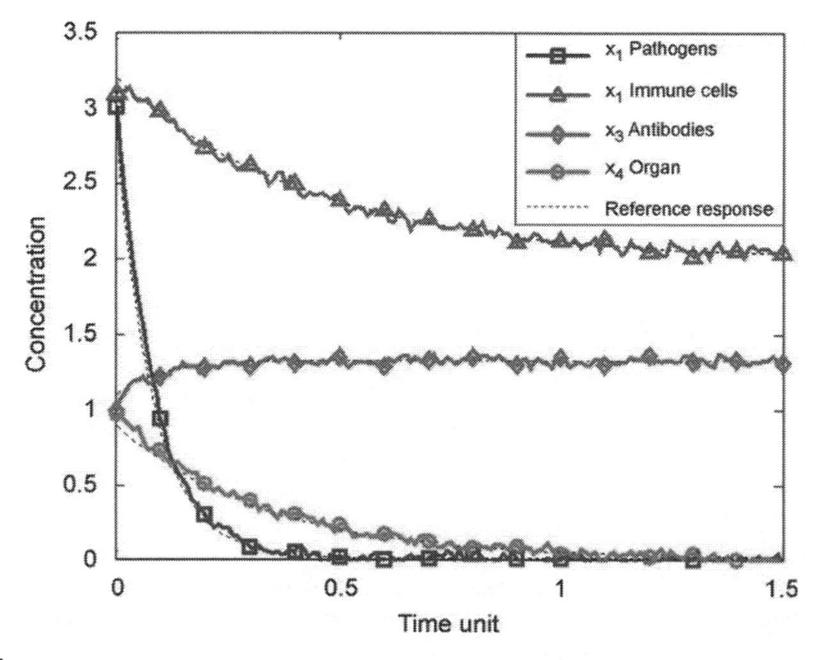

FIGURE 16.9
The tracking of innate immune system to the desired reference model by the robust minimax H_∞ game control under the continuous exogenous pathogens, environmental disturbances, and pathogen killer's agent side effects.

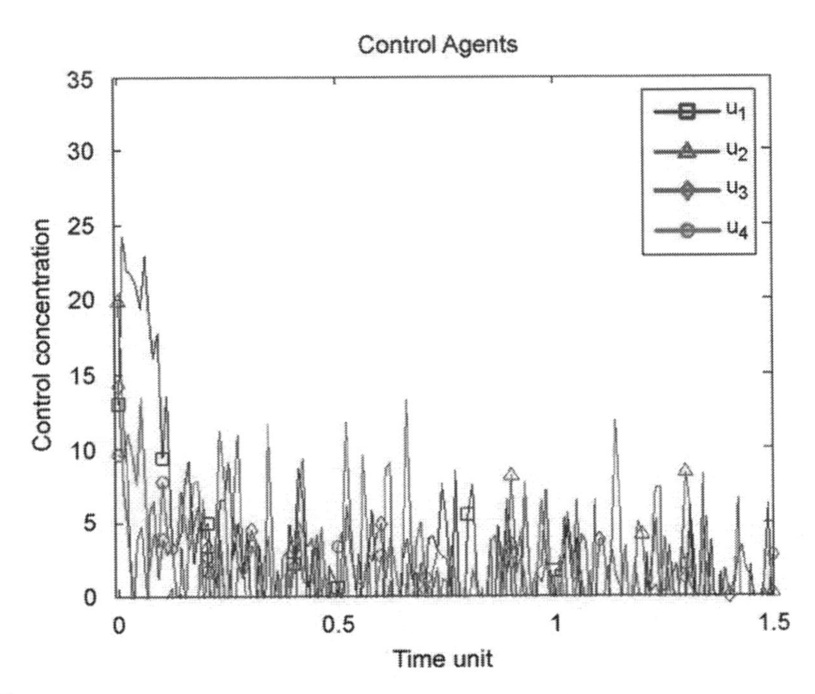

FIGURE 16.10
The minimax H_∞ game controls with considering the pathogen killer's agent side effects. The drug controls u_1 for pathogens, u_2 for immune cells, u_3 for antibodies, and u_4 for organs.

agent u_1. From the simulation result, the organ health state x_4 tracks the desired immune response leisurely at the beginning but it totally matches the desired response at the end. Figure 16.10 shows the four combined therapeutic control signals with the side effects of pathogen killer's agent u_1. At the beginning, the drug of organ health enhancer u_4 needs to increase to minimize the side effects from the pathogen killer's agent u_1.

16.6 Discussion

From the simulation results (Figs. 16.7 and 16.8), it is shown that the innate immune system under the continuous intrusion of exogenous pathogens and the corruption of environmental disturbances can be controlled by a robust model matching control design to achieve the desired time response. If we consider the conventional optimal H_2 quadratic control in (16.3), the effect of the environmental disturbances is not included in the cost function; the optimal H_2 quadratic tracking control problem is equivalent to letting $\rho^2 = \infty$ in (16.5) and (16.22) [23]. From the simulation results (Figs. 16.11 and 16.12), the four states of optimal H_2 quadratic tracking control of the immune system are overshooting and diverging without tracking the desired immune time response. Obviously, exogenous pathogens and the environmental disturbances have deteriorated the optimal H_2 quadratic tracking performance and therefore their effects should be considered in the robust minimax H_∞ game therapeutic control design procedure. In the situation, the proposed minimax H_∞ reference robust game control design is necessary to achieve a desired time response.

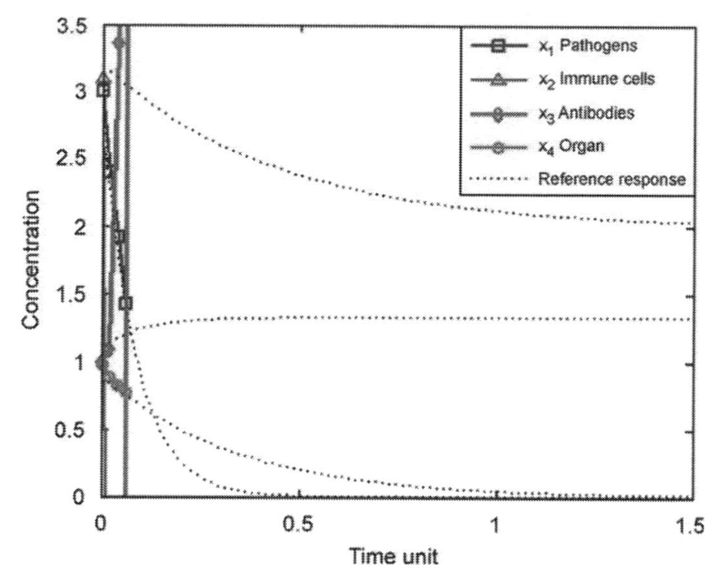

FIGURE 16.11
In the case of conventional optimal H_2 quadratic matching control (i.e. $\rho = \infty$ in (16.5)), since the effect of continuous exogenous pathogens intrusion and environmental disturbances is not considered in the design procedure, the states of innate immune system overshoot and diverge and cannot track the desired reference responses.

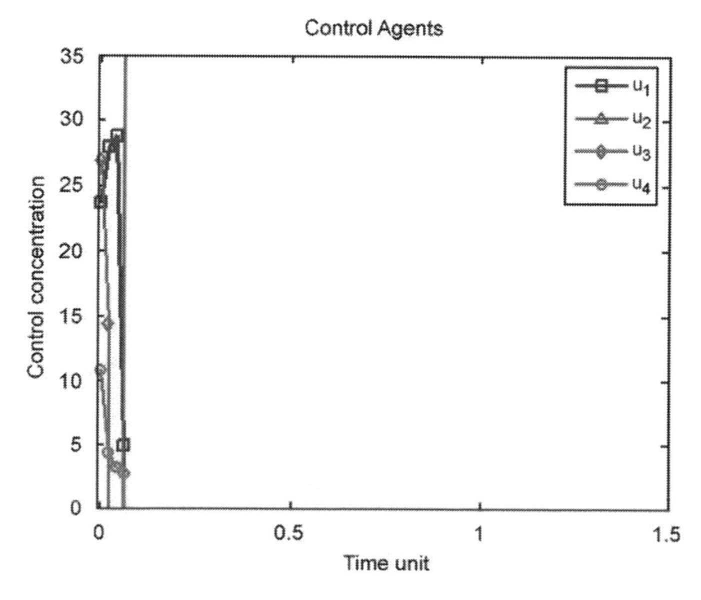

FIGURE 16.12
The controls of conventional optimal H_2 quadratic matching control (i.e. $\rho = \infty$ in (16.5)) without considering the effect of continuous exogenous pathogens intrusion and environmental disturbances in the design procedure; the drug controls u_1 for pathogens, u_2 for immune cells, u_3 for antibodies, and u_4 for organs that are all overshooting and divergent.

Pathogen killer's agent u_1 has toxic side effect that can damage the organ during the treatment. So, we should consider its detrimental impacts into the innate immune system in the model in (16.31). In the robust minimax H_∞ reference model game control design procedure, the element of weighting matrix q_{44} should be increased to impose a penalty on the reference tracking error of organ health state to eliminate the side effect of u_1 on x_4. From the simulation results (Figs. 16.9 and 16.10), the organ health state x_4 tracks the desired response slowly at the beginning but it almost matches the desired immune response at the end. It is also observed that the innate immune system can be controlled by a combined therapies design to achieve the desired time response under the continuous intrusion of exogenous pathogens, environmental disturbances, and side effects.

The combined therapies design is an important issue for all human diseases [391]. For a long period, the treatment of inflammatory skin diseases such as psoriasis, contact dermatitis, and atopic dermatitis has included agents that alleviate symptoms, but these agents have not been aimed at any specific molecular targets involved in the pathogenesis of the disease. Insights into this immune mechanism may facilitate the development of combination therapies that take advantage of the robust minimax H_∞ model reference game design, with the aim of achieving higher efficacy at a lower drug dosage and with a reduced probability of side effect on organ health. The proposed robust minimax H_∞ reference model game design has used four control variables, i.e. pathogen killer's agent u_1, immune cell enhancer u_2, antibody enhancer u_3, and health enhancer u_4, to achieve a minimax H_∞ matching game performance and to efficiently attenuate the effect of exogenous pathogens and environmental disturbances on the immune system.

In this chapter, the nonlinear dynamic model of innate immune response system is taken from the literature, which still needs to compare quantitatively with empirical evidence in practical application. For practical implementation, more accurate biodynamic models are required for therapeutic treatment application. However, model identification is not the topic of this chapter. Furthermore, we have made an assumption that the four states $(x_1 - x_4)$ of the concentrations or indices can be measured accurately by the medical equipment. With these detectable signals, we can solve these nonlinear dynamic minimax H_∞ game problems for robust reference tracking therapeutic control design of innate immune system to obtain the drug administration values in real time through medical instrument readout. If measurement is corrupted by noises in the measurement process, some filter designs should be employed to attenuate these noises to estimate the state variables for control in (16.21) [392]. Nevertheless, the implementation of filter will increase the complexity of the design problem [393]. Since the proposed robust minimax H_∞ reference model game control design can provide an efficient way to create a real-time therapeutic regime to protect suspected patients from the pathogens infection, in the future, we will focus on applications of robust minimax H_∞ reference game control design to therapy and drug design incorporating with nanotechnology and metabolic engineering scheme.

As a comparison, the similarity and difference between our minimax H_∞ game control method and the optimal control method [384] are given in the following. We all want to minimize the total cost of design, i.e. the weighted sum of the damage caused by pathogens and the cost paid by specific immune cells. On the other hand, the differences are given in the following: (i) A desired model reference is given to be optimally tracked for the enhancement of the immune system in our proposed minimax H_∞ robust matching therapeutic control method and Shudo and Iwasa have designed an optimal quadratic control to minimize the cost. (ii) The effect of external disturbances has not been considered in Shudo and Iwasa, but the worst-case effect of external disturbances has been considered and minimized in our design via the minimax H_∞ Nash game method.

(iii) Fuzzy interpolation technique is employed by our method so that LMIs technique is used to efficiently solve the nonlinear minimax H_∞ tracking optimization problem in our design procedure. However, Shudo and Iwasa have used a dynamic programming method to derive an optimal schedule to solve the nonlinear optimization problem of host defense, immune memory, and post-infection pathogen levels in mammals.

16.7 Conclusion

Stability robustness is a significant cellular property that allows the innate immune system to maintain its cellular function despite exogenous pathogens, environmental disturbances, and system uncertainties. Based on the nonlinear dynamic minimax H_∞ game theory, the robust H_∞ reference tracking therapeutic control is formulated as a minimax nonlinear dynamic H_∞ game problem for an innate immune system to achieve a desired time response prescribed prior under environmental disturbances, unknown initial conditions, and side effects. In general, the robust minimax nonlinear H_∞ reference model matching therapeutic control design for innate immune system needs to solve nonlinear HJI, which is generally difficult to solve for this H_∞ robust therapeutic control design. Based on the proposed fuzzy dynamic minimax H_∞ game scheme, the design of nonlinear dynamic robust H_∞ reference matching control problem for innate immune system is transformed to solve an equivalent constrained dynamic minimax H_2 quadratic game problem. Such transformation can then allow us an easier approach by solving an LMI-constrained optimization problem for robust dynamic minimax H_∞ reference game control design. With the help of the Robust Control Toolbox in Matlab instead of the HJI, we could solve these nonlinear dynamic H_∞ game problems for robust H_∞ reference matching control of innate immune system efficiently. From the in silico simulation examples, the proposed minimax H_∞ reference game therapeutic control of immune system could track the prescribed reference time response robustly, which may lead to potential application in therapeutic drug design for a desired immune response during an infection episode.

16.8 Appendix

16.8.1 Appendix A. Proof of Theorem 16.1

Let us denote a Lyapunov energy function $V(\bar{x}(t))>0$. Then, (16.9) is equivalent to the following minimax H_2 game problem:

$$\begin{aligned} &\min_{u(t)}\max_{v(t)} J \\ &= \min_{u(t)}\max_{v(t)}\{V(\bar{x}(0)) - V(\bar{x}(t_f)) + \int_0^{t_f} (\bar{x}^T(t)Q\bar{x}(t) + u^T(t)Ru(t) \\ &- \rho^2 v^T(t)v(t))dt + \frac{dV(\bar{x}(t))}{dt}]dt\}, \quad \forall\bar{x}(0). \end{aligned} \tag{16.A.1}$$

By the chain rule, we get

$$\begin{aligned}\frac{dV(\bar{x}(t))}{dt} &= \left(\frac{\partial V(\bar{x}(t))}{\partial \bar{x}(t)}\right)^T \frac{d\bar{x}(t)}{dt} \\ &= \left(\frac{\partial V(\bar{x}(t))}{\partial \bar{x}(t)}\right)^T F(\bar{x}(t)) + Bu(t) + Cv(t)).\end{aligned} \tag{16.A.2}$$

Substituting (16.A.2) into (16.A.1), we get

$$\begin{aligned}\min_{u(t)}\max_{v(t)} J &= \min_{u(t)}\max_{v(t)}\{V(\bar{x}(0)) - V(\bar{x}(t_f)) + \int_0^{t_f} (\bar{x}^T(t)Q\bar{x}(t) + u^T(t)Ru(t) \\ &- \rho^2 v^T(t)v(t))dt + \left(\frac{\partial V(\bar{x}(t))}{\partial \bar{x}(t)}\right)^T F(\bar{x}(t)) + \left(\frac{\partial V(\bar{x}(t))}{\partial \bar{x}(t)}\right)^T Bu(t) \\ &+ \left(\frac{\partial V(\bar{x}(t))}{\partial \bar{x}(t)}\right)^T Cv(t))]dt\}, \quad \forall \bar{x}(0).\end{aligned}$$

$$\begin{aligned}\min_{u(t)}\max_{v(t)} J &= \min_{u(t)}\max_{v(t)}\left\{V(\bar{x}(0)) - V(\bar{x}(t_f)) + \int_0^{t_f}\left[\left(\frac{\partial V(\bar{x}(t))}{\partial \bar{x}(t)}\right)^T F(\bar{x}(t))\right.\right. \\ &+ \bar{x}^T(t)Q\bar{x}(t) - \frac{1}{4}\left(\frac{\partial V(\bar{x}(t))}{\partial \bar{x}(t)}\right)^T BR^{-1}B^T\frac{\partial V(\bar{x}(t))}{\partial \bar{x}(t)} \\ &+ \frac{1}{4\rho^2}\left(\frac{\partial V(\bar{x}(t))}{\partial \bar{x}(t)}\right)^T CC^T\frac{\partial V(\bar{x}(t))}{\partial \bar{x}(t)} \\ &+ \left(u^T(t) + \frac{1}{2}R^{-1}B^T\frac{\partial V(\bar{x}(t))}{\partial \bar{x}(t)}\right)^T R\left(u^T(t) + \frac{1}{2}R^{-1}B^T\frac{\partial V(\bar{x}(t))}{\partial \bar{x}(t)}\right) \\ &\left.\left.- \left(\rho v(t) - \frac{1}{2\rho}C^T\frac{\partial V(\bar{x}(t))}{\partial \bar{x}(t)}\right)^T\left(\rho v(t) - \frac{1}{2\rho}C^T\frac{\partial V(\bar{x}(t))}{\partial \bar{x}(t)}\right)\right]dt\right\}, \quad \forall \bar{x}(0).\end{aligned}$$

Therefore, the minimax H_2 quadratic game solution is given as follows:

$$\begin{aligned}J(u^*(t), v^*(t)) &= \left\{V(\bar{x}(0)) - V(\bar{x}(t_f)) + \int_0^{t_f}\left[\left(\frac{\partial V(\bar{x}(t))}{\partial \bar{x}(t)}\right)^T F(\bar{x}(t))\right.\right. \\ &+ \bar{x}^T(t)Q\bar{x}(t) - \frac{1}{4}\left(\frac{\partial V(\bar{x}(t))}{\partial \bar{x}(t)}\right)^T BR^{-1}B^T\frac{\partial V(\bar{x}(t))}{\partial \bar{x}(t)} \\ &\left.\left.+ \frac{1}{4\rho^2}\left(\frac{\partial V(\bar{x}(t))}{\partial \bar{x}(t)}\right)^T CC^T\frac{\partial V(\bar{x}(t))}{\partial \bar{x}(t)}\right]dt\right\}, \quad \forall \bar{x}(0)\end{aligned}$$

with

$$u^*(t) = -\frac{1}{2}R^{-1}B^T\frac{\partial V(\bar{x}(t))}{\partial \bar{x}(t)}, v^*(t) = \frac{1}{2\rho^2}C^T\frac{\partial V(\bar{x}(t))}{\partial \bar{x}(t)}.$$

If (16.14) holds, then

$$J(u^*(t), v^*(t) \leq V(\bar{x}(0)) - V(\bar{x}(t_f)).$$

From the inequality in (16.11), this minimax H_2 quadratic game solution should be less than $\rho^2\bar{x}^T(0)\mathrm{I}\bar{x}(0)$, and then we get the inequality in (16.15):

$$\min_{u(t)}\max_{v(t)} J \leq V(\bar{x}(0)) - V(\bar{x}(t_f)) \leq V(\bar{x}(0)) \leq \rho^2\bar{x}^T(0)\mathrm{I}\bar{x}(0), \ \forall\bar{x}(0).$$

16.8.2 Appendix B. Proof of Theorem 16.2

Let us denote a Lyapunov energy function $V(\bar{x}(t)) = \bar{x}^T(t)P\bar{x}(t) > 0$. Then, (16.9) is equivalent to the following:

$$\begin{aligned}
\min_{u(t)}\max_{v(t)} J &= \min_{u(t)}\max_{v(t)}\Big\{\bar{x}^T(0)P\bar{x}(0) - \bar{x}^T(t_f)P\bar{x}(t_f) + \int_0^{t_f}(\bar{x}^T(t)Q\bar{x}(t) \\
&\quad + u^T(t)Ru(t) - \rho^2 v^T(t)v(t)) + \frac{dV(\bar{x}(t))}{dt})dt\Big\}, \forall\bar{x}(0) \\
&= \min_{u(t)}\max_{v(t)}\{\bar{x}^T(0)P\bar{x}(0) - \bar{x}^T(t_f)P\bar{x}(t_f) + \int_0^{t_f}(\bar{x}^T(t)Q\bar{x}(t) \\
&\quad + u^T(t)Ru(t) - \rho^2 v^T(t)v(t)) + 2\bar{x}^T(t)P\dot{\bar{x}}(t)dt)\}, \forall\bar{x}(0) \\
&= \min_{u(t)}\max_{v(t)}\{\bar{x}^T(0)P\bar{x}(0) - \bar{x}^T(t_f)P\bar{x}(t_f) + \int_0^{t_f}(\bar{x}^T(t)Q\bar{x}(t) \\
&\quad + u^T(t)Ru(t) - \rho^2 v^T(t)v(t))dt + 2\bar{x}^T(t)P\Big(\sum\nolimits_{i=1}^{L} h_i(x(t))\mathrm{A}_i\bar{x}(t)\Big) \\
&\quad + 2\bar{x}^T(t)P\mathrm{B}u(t) + 2\bar{x}^T(t)PCv(t)]dt\}, \forall\bar{x}(0) \\
&= \min_{u(t)}\max_{v(t)}\{\bar{x}^T(0)P\bar{x}(0) - \bar{x}^T(t_f)P\bar{x}(t_f) + \int_0^{t_f}(\bar{x}^T(t)Q\bar{x}(t) \\
&\quad + u^T(t)Ru(t) - \rho^2 v^T(t)v(t))dt + \sum\nolimits_{i=1}^{L} h_i(x(t))(2\bar{x}^T(t)P\mathrm{A}_i\bar{x}(t) \\
&\quad + 2\bar{x}^T(t)P\mathrm{B}u(t) + 2\bar{x}^T(t)PCv(t)]dt\}, \forall\bar{x}(0) \\
&= \min_{u(t)}\max_{v(t)}\{\bar{x}^T(0)P\bar{x}(0) - \bar{x}^T(t_f)P\bar{x}(t_f) + \int_0^{t_f}[(\bar{x}^T(t)Q\bar{x}(t) \\
&\quad + \sum\nolimits_{i=1}^{L} h_i(x(t))\bar{x}^T(t)[P\mathrm{A}_i + \mathrm{A}_i^T P - P^T\mathrm{B}R^{-1}\mathrm{B}^T P + \frac{1}{\rho^2}P^T CC^T P]\bar{x}(t) \\
&\quad + [u(t) + R^{-1}\mathrm{B}^T P\bar{x}(t)]^T R[u(t) + R^{-1}\mathrm{B}^T P\bar{x}(t)] \\
&\quad - \Big(\rho v(t) - \frac{1}{\rho}C^T P\bar{x}(t)\Big)^T\Big(\rho v(t) - \frac{1}{\rho}C^T P\bar{x}(t)\Big)\Big]dt\Big\}, \forall\bar{x}(0).
\end{aligned}$$

The minimax H_2 quadratic game solution is given as follows:

$$J(u^*(t),v^*(t)) = \{\bar{x}^T(0)P\bar{x}(0) - \bar{x}^T(t_f)P\bar{x}(t_f) + \int_0^{t_f} [(\bar{x}^T(t)Q\bar{x}(t) + \sum_{i=1}^{L} h_i(x(t))\bar{x}^T(t)[PA_i + A_i^TP - P^TBR^{-1}B^TP + \frac{1}{\rho^2}P^TCC^TP]\bar{x}(t)]dt\}, \forall\bar{x}(0)$$

with

$$u^*(t) = -R^{-1}B^TP\bar{x}(t), v^*(t) = \frac{1}{\rho^2}C^TP\bar{x}(t).$$

In order to simplify the above equation, suppose the inequality in (16.22) holds, then

$$\min_{u(t)}\max_{v(t)} J \leq \bar{x}(0)P\bar{x}(0), \forall\bar{x}(0).$$

From the inequality in (16.11), this minimax H_2 quadratic game should be less than $\rho^2\bar{x}^T(0)I\bar{x}(0)$, and then

$$\min_{u(t)}\max_{v(t)} J \leq \bar{x}(0)P\bar{x}(0) \leq \rho^2\bar{x}^T(0)I\bar{x}(0), \forall\bar{x}(0)$$
$$\text{i.e., } P \leq \rho^2 I.$$

Since we assume ρ^2 is the upper bound in (16.5), the minimax H_∞ game model reference control becomes how to design $u^*(t)$ in (16.21) by solving the constrained optimization problem in (16.22) and (16.23).

16.8.3 Appendix C. Parameters of the T–S fuzzy system

The nonlinear innate immune system in (16.28) could be approximated by a T–S fuzzy system. By the fuzzy modeling method [4], the matrices of the local linear system , the parameters B, and C are calculated as follows:

$$A_1 = \begin{bmatrix} \begin{matrix} 0.3462 & 0 & 0 & 0 \\ 3.3258 & -1.1155 & 0.3433 & 2.1987 \\ -0.2674 & 1.1816 & -1.7724 & 0.1108 \\ 0.6295 & 0 & 0 & -1.3994 \end{matrix} & 0 \\ 0 & \begin{matrix} -12 & 0 & 0 & 0 \\ 0 & -2.3 & 0 & 0 \\ 0 & 0 & -10 & 0 \\ 0 & 0 & 0 & -3 \end{matrix} \end{bmatrix},$$

$$A_2 = \begin{bmatrix} \begin{matrix} 0.348 & 0 & 0 & 0 \\ -2.8398 & -0.9076 & -1.4305 & -1.9826 \\ -0.235 & 1.1644 & -1.6635 & 0.1108 \\ 0.295 & -0.0902 & -0.1643 & -1.3994 \end{matrix} & 0 \\ 0 & \begin{matrix} -12 & 0 & 0 & 0 \\ 0 & -2.3 & 0 & 0 \\ 0 & 0 & -10 & 0 \\ 0 & 0 & 0 & -3 \end{matrix} \end{bmatrix},$$

$$A_3 = \begin{bmatrix} \begin{matrix} -2.3807 & 0 & 0 & 0 \\ 10.616 & -0.9107 & 0.3194 & -1.6078 \\ -2.4789 & 0.5007 & -1.7724 & -0.4134 \\ 0.4117 & 0 & 0 & -0.5537 \end{matrix} & 0 \\ 0 & \begin{matrix} -12 & 0 & 0 & 0 \\ 0 & -2.3 & 0 & 0 \\ 0 & 0 & -10 & 0 \\ 0 & 0 & 0 & -3 \end{matrix} \end{bmatrix},$$

$$A_4 = \begin{bmatrix} \begin{matrix} -2.2996 & 0 & 0 & 0 \\ -8.4642 & -0.9535 & -0.4587 & 1.1711 \\ -0.23919 & 0.4936 & -1.6635 & -0.4134 \\ 0.1957 & -0.0781 & -0.1643 & -0.5537 \end{matrix} & 0 \\ 0 & \begin{matrix} -12 & 0 & 0 & 0 \\ 0 & -2.3 & 0 & 0 \\ 0 & 0 & -10 & 0 \\ 0 & 0 & 0 & -3 \end{matrix} \end{bmatrix},$$

$$A_5 = \begin{bmatrix} \begin{matrix} 0.1987 & 0 & 0 & 0 \\ 0.6956 & -0.933 & 0.3733 & -3.5255 \\ 0.3935 & 1.1816 & -0.4106 & 0.1732 \\ 0.2311 & 0 & 0 & -0.3611 \end{matrix} & 0 \\ 0 & \begin{matrix} -12 & 0 & 0 & 0 \\ 0 & -2.3 & 0 & 0 \\ 0 & 0 & -10 & 0 \\ 0 & 0 & 0 & -3 \end{matrix} \end{bmatrix},$$

$$A_6 = \begin{bmatrix} \begin{matrix} 0.1981 & 0 & 0 & 0 \\ -2.9804 & -1.0591 & -2.0136 & 3.3337 \\ 0.3963 & 1.1644 & -0.3837 & 0.1732 \\ 0.1101 & -0.0902 & -0.0619 & -0.3611 \end{matrix} & 0 \\ 0 & \begin{matrix} -12 & 0 & 0 & 0 \\ 0 & -2.3 & 0 & 0 \\ 0 & 0 & -10 & 0 \\ 0 & 0 & 0 & -3 \end{matrix} \end{bmatrix},$$

$$A_7 = \begin{bmatrix} \begin{matrix} -1.0821 & 0 & 0 & 0 \\ 5.3534 & -0.7742 & -0.6384 & -0.7012 \\ -0.7249 & 0.5007 & -0.4106 & -0.0417 \\ 0.1682 & 0 & 0 & 0.1482 \end{matrix} & 0 \\ 0 & \begin{matrix} -12 & 0 & 0 & 0 \\ 0 & -2.3 & 0 & 0 \\ 0 & 0 & -10 & 0 \\ 0 & 0 & 0 & -3 \end{matrix} \end{bmatrix},$$

$$A_8 = \begin{bmatrix} \begin{matrix} -1.057 & 0 & 0 & 0 \\ -2.8391 & -1.532 & 0.7299 & 0.2814 \\ -0.7023 & 0.4936 & -0.3837 & -0.0417 \\ 0.0812 & -0.0781 & -0.0619 & -0.1482 \end{matrix} & 0 \\ 0 & \begin{matrix} -12 & 0 & 0 & 0 \\ 0 & -2.3 & 0 & 0 \\ 0 & 0 & -10 & 0 \\ 0 & 0 & 0 & -3 \end{matrix} \end{bmatrix},$$

$$A_9 = \begin{bmatrix} \begin{matrix} 0.3462 & 0.2243 & -2.2977 & 0.087 \\ 3.0103 & -0.0287 & 7.6732 & -2.1137 \\ -0.2674 & 0.9206 & -2.1832 & -0.0587 \\ 0.6295 & 0.1473 & 0.1675 & 0.0737 \end{matrix} & 0 \\ 0 & \begin{matrix} -12 & 0 & 0 & 0 \\ 0 & -2.3 & 0 & 0 \\ 0 & 0 & -10 & 0 \\ 0 & 0 & 0 & -3 \end{matrix} \end{bmatrix},$$

$$A_{10} = \begin{bmatrix} \begin{matrix} 0.348 & 0.2223 & -2.224 & 0.087 \\ -2.4956 & -1.8301 & -5.6794 & 0.7798 \\ -0.235 & 0.9101 & -2.1142 & -0.0587 \\ 0.295 & 0.0728 & 0.0811 & 0.0737 \end{matrix} & 0 \\ 0 & \begin{matrix} -12 & 0 & 0 & 0 \\ 0 & -2.3 & 0 & 0 \\ 0 & 0 & -10 & 0 \\ 0 & 0 & 0 & -3 \end{matrix} \end{bmatrix},$$

$$A_{11} = \begin{bmatrix} \begin{matrix} -2.3807 & -0.9739 & -2.2977 & -0.5749 \\ 12.767 & 2.4941 & 11.649 & -1.196 \\ -2.4789 & -0.2104 & -2.1832 & -0.598 \\ 0.4117 & 0.1305 & 0.1675 & 0.0489 \end{matrix} & 0 \\ 0 & \begin{matrix} -12 & 0 & 0 & 0 \\ 0 & -2.3 & 0 & 0 \\ 0 & 0 & -10 & 0 \\ 0 & 0 & 0 & -3 \end{matrix} \end{bmatrix},$$

$$A_{12} = \begin{bmatrix} \begin{matrix} -2.2996 & -0.9641 & -2.224 & -0.5749 \\ -11.901 & -3.8172 & -11.21 & -3.8954 \\ -2.3919 & -0.2087 & -2.1142 & -0.598 \\ 0.1957 & 0.0646 & 0.0811 & 0.0489 \end{matrix} & 0 \\ 0 & \begin{matrix} -12 & 0 & 0 & 0 \\ 0 & -2.3 & 0 & 0 \\ 0 & 0 & -10 & 0 \\ 0 & 0 & 0 & -3 \end{matrix} \end{bmatrix},$$

$$A_{13} = \begin{bmatrix} \begin{matrix} 0.1987 & 0.2243 & -1.0582 & 0.0495 \\ 0.5515 & -0.6453 & 3.8238 & 2.1633 \\ 0.3935 & 0.9206 & -0.7957 & 0.0991 \\ 0.2311 & 0.1473 & 0.0772 & 0.0275 \end{matrix} & 0 \\ 0 & \begin{matrix} -12 & 0 & 0 & 0 \\ 0 & -2.3 & 0 & 0 \\ 0 & 0 & -10 & 0 \\ 0 & 0 & 0 & -3 \end{matrix} \end{bmatrix},$$

$$A_{14} = \begin{bmatrix} \begin{matrix} 0.1981 & 0.2223 & -1.0349 & 0.0495 \\ 0.1057 & -1.2516 & -3.148 & -2.882 \\ 0.3963 & 0.9101 & -0.7759 & 0.0991 \\ 0.1101 & 0.0728 & 0.0377 & 0.0275 \end{matrix} & 0 \\ 0 & \begin{matrix} -12 & 0 & 0 & 0 \\ 0 & -2.3 & 0 & 0 \\ 0 & 0 & -10 & 0 \\ 0 & 0 & 0 & -3 \end{matrix} \end{bmatrix},$$

$$A_{15} = \begin{bmatrix} \begin{matrix} -1.0821 & -0.9739 & -1.0582 & -0.2642 \\ 3.6655 & 3.8523 & 4.0278 & 4.4254 \\ -0.7249 & -0.2104 & -0.7957 & -0.1756 \\ 0.1682 & 0.1305 & 0.0772 & 0.0203 \end{matrix} & 0 \\ 0 & \begin{matrix} -12 & 0 & 0 & 0 \\ 0 & -2.3 & 0 & 0 \\ 0 & 0 & -10 & 0 \\ 0 & 0 & 0 & -3 \end{matrix} \end{bmatrix},$$

$$A_{16} = \begin{bmatrix} \begin{matrix} -1.057 & -0.9641 & -1.0349 & -0.2642 \\ -6.4434 & -4.5651 & -4.4027 & -6.7461 \\ -0.7023 & -0.2087 & -0.7759 & -0.1756 \\ 0.0812 & 0.0646 & 0.0377 & 0.0203 \end{matrix} & 0 \\ 0 & \begin{matrix} -12 & 0 & 0 & 0 \\ 0 & -2.3 & 0 & 0 \\ 0 & 0 & -10 & 0 \\ 0 & 0 & 0 & -3 \end{matrix} \end{bmatrix},$$

$$B = \begin{bmatrix} -1 & 0 & 0 & 0 \\ 0 & 1 & 0 & 0 \\ 0 & 0 & 1 & 0 \\ 0 & 0 & 0 & -1 \\ 0 & 0 & 0 & 0 \\ 0 & 0 & 0 & 0 \\ 0 & 0 & 0 & 0 \\ 0 & 0 & 0 & 0 \end{bmatrix}, C = \begin{bmatrix} \begin{matrix} 3 & 0 & 0 & 0 \\ 0 & 3 & 0 & 0 \\ 0 & 0 & 3 & 0 \\ 0 & 0 & 0 & 3 \end{matrix} & 0 \\ 0 & \begin{matrix} 1 & 0 & 0 & 0 \\ 0 & 1 & 0 & 0 \\ 0 & 0 & 1 & 0 \\ 0 & 0 & 0 & 1 \end{matrix} \end{bmatrix}.$$

17

Robust Synthetic Genetic Circuit Design: Stochastic H_∞ Game Approach

17.1 Introduction

Development of synthetic gene circuit design in synthetic biology will revolutionize how we conceptualize and approach the engineering of biological systems. The vision and applications of this emerging field will influence many other scientific and engineering disciplines, as well as affect various aspects of daily life and society [394]. Synthetic biology builds living machines from the off-the-shelf chemical ingredients, utilizing many of the same strategies that electrical engineers employ to make computer chips [395]. The main goal of the nascent field of synthetic biology is to design and construct biological systems with the desired behavior [394–403]. By the development of powerful techniques for the automated synthesis of DNA molecules and their assembly into genes and microbial genomes, synthetic biology envisions the redesign of natural biological systems for greater efficiency as well as the construction of functional "genetic circuit" and metabolic pathways for practical purposes [394,401,404–409]. Synthetic biology is foreseen to have important applications in biotechnology and medicine in near future [394].

Though the engineering of genetic networks of inter-regulating genes, so-called synthetic gene networks, has demonstrated the feasibility of synthetic biology [406], the design of gene networks is still a difficult problem and most of the newly designed gene networks cannot work properly and duratively. These design failures in genetic circuits are mainly due to intrinsic perturbations such as gene expression noises, splicing, mutation, uncertain initial states, and disturbances such as changing extra-cellular environments and interactions with cellular context. Therefore, how to design a robust synthetic gene network, which could tolerate uncertain initial conditions and intrinsic perturbations, attenuate the effect of all disturbances, and function properly on the host cell, will be an important research topic for synthetic biology [394–403,410–416]. Previously, sensitivity analysis has been used for analysis of the dynamic properties of gene networks either in qualitative simulations of coarse-grained models or in extensive numerical simulations of nonlinear differential equation models or stochastic dynamic models [417,418]. For applications in synthetic biology, these approaches are not satisfying. The local sensitivity analysis can provide only a partial description of all possible behaviors of a nonlinear gene network to external disturbance. In particular, it cannot guarantee that a synthetic gene network behaves as expected for all uncertain initial conditions and disturbances. Moreover, obtaining all convergences of states and parameters by extensive numerical simulations quickly becomes computationally intractable when the size of the synthetic network grows [410].

Recently, an approach has been developed using semidefinite programming to partition the parameter spaces of polynomial differential equation models into the so-called feasible and infeasible regions [419]. Following that, a robustness analysis and tuning approach of synthetic gene networks was proposed to provide a means to assess the stability robustness of the expected behavior of a synthetic gene network in spite of parameter variations [410]. This approach has the capability to search for parameter sets for which a given robustness property is satisfied through a publicly available tool called RoVerGeNe. Several gene circuit design networks have been introduced to implement or delete some circuits from an existing gene network so as to modify its system structure for improving its robust stability or filtering ability [420–422]. However, robust synthetic gene network design is a different topic. It needs to design a complete man-made gene network to be inserted into a host cell. Therefore, the synthetic gene networks should be designed with enough stability robustness to tolerate uncertain initial conditions and intrinsic perturbations and to resist all possible external disturbances on the host cell so that they can function properly in a desired steady state. This is a so-called robust regulation design that can achieve a desired steady state of synthetic gene networks despite uncertain initial conditions intrinsic perturbations and disturbances on the host cell.

In this chapter, a robust regulation design of synthetic gene network is proposed based on stochastic H_∞ game strategy to achieve a desired steady state in spite of uncertain initial conditions, parameter variations, and disturbances on the host cell. Because most information of these uncertain factors on the host cell is unavailable, in order to attenuate their detrimental effects, their worst-case effect should be considered by the designer in the regulation design procedure from the stochastic H_∞ game perspective. The worst-case effect of all possible initial conditions and disturbances on the regulation error to a desired steady state is minimized for the robust synthetic gene networks, i.e. the proposed robust synthetic gene network is designed from the minimax H_∞ stochastic game perspective. The minimax H_∞ game design scheme is on the safe side being a simple robust synthetic gene network design method because we do not need the precise information of the initial conditions, parameter variations, and disturbances on the synthetic gene network in the host cell, which are not easy to measure in the design procedure. This minimax H_∞ regulation game design problem for robust synthetic gene networks could be transformed to an equivalent dynamic Nash quadratic game problem [9,280]. Dynamic Nash quadratic game methods have been widely applied to many fields of robust engineering design problems with external disturbances. Recently, the application of dynamic Nash quadratic game theory has been used for robust model matching control of immune systems under environmental disturbances [15]. A robust drug administration (control input) is designed by Nash quadratic game method to obtain a prescribed immune response under uncertain initial states and environmental disturbances. In this chapter, the stochastic H_∞ game theory will be used for robust synthetic gene network design so that the engineered gene network can work properly under uncertain initial conditions and environmental disturbances on the host cell. The uncertain initial states and disturbances are considered as a player doing his/her best to deteriorate the regulation performance from the stochastic H_∞ game point of view, while the system parameters to be designed are considered as another player optimizing the regulation performance under the worst-case deterioration of a former player. Since the synthetic gene networks are highly nonlinear, it is not easy to solve the robust synthetic gene network design problem directly by the nonlinear dynamic H_∞ game method. Recently, fuzzy systems have been employed to efficiently approximate nonlinear dynamic systems to solve the nonlinear control problem [4,74,105,386,387,423].

A Takagi–Sugeno (T–S) fuzzy model [4] is proposed to interpolate several linearized genetic networks at different operating points to approximate the nonlinear gene network via some smooth fuzzy membership functions. Then with the help of the fuzzy approximation method, a fuzzy dynamic H_∞ game scheme [9] is developed so that the minimax regulation design of robust synthetic gene networks could be easily solved by the techniques of the linear dynamic H_∞ game theory, which can be subsequently solved by a constrained optimization scheme via the linear matrix inequality (LMI) technique [23] that can be efficiently solved by the Robust Control Toolbox in Matlab [424]. Because the fuzzy model can approximate any nonlinear system, the proposed robust regulation design method developed from the fuzzy stochastic H_∞ game theory can be applied to the robust regulation design problem of any synthetic gene network that can be interpolated by a T–S fuzzy model. For comparison, the conventional stochastic optimal H_2 regulation design method without considering the effect of disturbances is also proposed for the synthetic gene network. Because the effect of disturbances is not attenuated efficiently, the optimal regulation design method of synthetic gene networks is much influenced by the disturbances on the host cell. Finally, an in silico example is given to illustrate the design procedure and to confirm the efficiency and efficacy of the proposed stochastic minimax H_∞ regulation game design method for robust synthetic gene networks.

17.2 Systems and Robust Synthetic Gene Circuit Design Methods

First, for the convenience of design problem description, a simple design example of a four-gene network in [410] is provided to give an overview of the design problem of robust synthetic gene networks. A more general design problem of robust synthetic gene networks will be given in the sequel. Let us consider a robust regulation design problem of a cascade loop of transcriptional inhibitions built in *Escherichia coli.* [425]. The synthetic gene network is represented in Fig. 17.1. It consists of four genes: *tetR, lacI, cI,* and *eyfp* that code, respectively, three repressor proteins, TetR, LacI, and CI, and the fluorescent protein EYFP (enhanced yellow fluorescent protein) [410]. aTc (anhydrotetracycline) is the input to the system. The fluorescence of the system, due to the protein EYFP, is the measured output. The protein CI inhibits gene *eyfp*. The protein TetR inhibits gene *lacI*. The protein LacI inhibits gene *cI*. The regulatory dynamic equations of the synthetic transcriptional gene cascade in Fig. 17.1 are given as follows [410]:

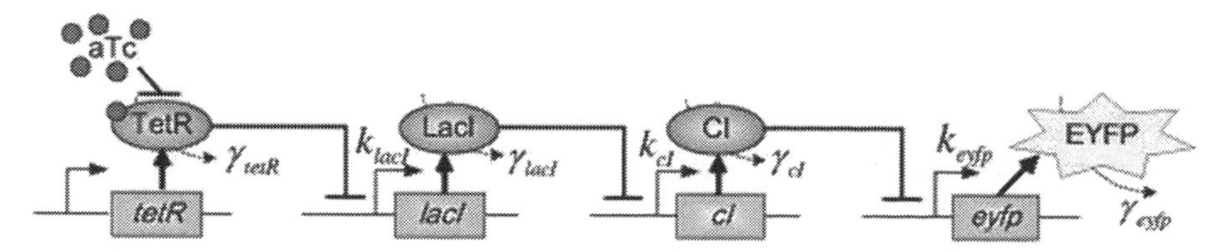

FIGURE 17.1
Synthetic transcription cascade loop in silico design example. aTc represses TetR, TetR represses lacI, LacI represses cI, and CI represses eyfp. aTc is the system input and the fluorescent protein EYFP is the output.

$$
\begin{aligned}
\dot{x}_{tetR} &= k_{tetR,0} - \gamma_{tetR} x_{tetR} + w_1 \\
\dot{x}_{lacI} &= k_{lacI,0} + k_{lacI}(r_{lacI}(x_{tetR}) + a_{lacI}(u_{aTc}) - r_{lacI}(x_{tetR}) a_{lacI}(u_{aTc})) - \gamma_{lacI} x_{lacI} + w_2 \\
\dot{x}_{cI} &= k_{cI,0} + k_{cI} r_{cI}(x_{lacI}) - \gamma_{cI} x_{cI} + w_3 \\
\dot{x}_{eyfp} &= k_{eyfp,0} + k_{eyfp} r_{eyfp}(x_{cI}) - \gamma_{eyfp} x_{eyfp} + w_4
\end{aligned}
\tag{17.1}
$$

with the uncertain initial conditions $x_{tetR}(0)$, $x_{lacI}(0)$, $x_{cI}(0)$, and $x_{eyfp}(0)$ in the host cell. $k_{tetR,0}$, $k_{lacI,0}$, $k_{cI,0}$, and $k_{eyfp,0}$ are basal production rates of the corresponding proteins, which are assumed to be given constants. k_{lacI}, k_{cI}, and k_{eyfp} are the production rate parameters, while γ_{tetR}, r_{lacI}, γ_{cI}, and γ_{eyfp} are decay rate parameters of the corresponding proteins. The regulatory functions r_{lacI}, γ_{cI}, and γ_{eyfp} are the Hill functions for repressors of a gene and a_{lacI} for an activator of a gene.

The Hill function can be derived from considering the equilibrium binding of the transcription factor to its site on the promoter region of a gene. For a repressor, Hill function is an S-shaped curve which can be described in the form $r(x) = \beta_r/(1 + x/K_r)^n)$. β_r is the maximal expression level of promoter. K_r is the repression coefficient. The Hill coefficient n governs the steepness of the input function. For an activator, Hill function can be described in the form $a(x) = \beta_a x^n/(K_a^n + x^n)$. β_a is the maximal expression level of promoter. K_a is the activation coefficient. n determines the steepness of the input function [397]. w_1, w_2, w_3, and w_4 are the disturbances of the synthetic gene network, which denote the total of environmental noises, modeling residuals, and intrinsic parameter fluctuations on four genes in the host cell. Therefore, w_i, $i = 1 - 4$ are assumed uncertain but bounded disturbances. The synthetic gene network design is to specify k_{lacI}, k_{cI}, k_{eyfp} and γ_{tetR}, r_{lacI}, γ_{cI}, γ_{eyfp} such that the system states x_{tetR}, x_{lacI}, x_{cI}, and x_{eyfp} can approach the desired states x_{d1}, x_{d2}, x_{d3}, and x_{d4}, respectively, in spite of uncertain initial conditions and disturbances on synthetic gene network.

If a synthetic gene network consists of n genes, then Equation (17.1) can be extended to the following n-gene network dynamics:

$$
\dot{x} = k_0 + f(x, k, \gamma) + g(u) + w, x(0) = x_0 \tag{17.2}
$$

where the state vector x denotes the concentrations of proteins in the synthetic gene network. k_0 denotes the vector of basal production rates of the corresponding proteins. $f(x, k, \gamma)$ denotes the regulation vector of synthetic gene network, which is the function of production rate parameters k and decay rate parameters γ to be designed. $g(u)$ denotes the input function to the synthetic gene network. w denotes the vector of stochastic disturbances on the host cell, whose statistics may be unavailable. The initial condition x_0 is assumed stochastic with unknown covariance. The robust synthetic gene network design is to select parameters k and γ from feasible ranges so that the state vector x can approach a desired state vector x_d in spite of uncertain initial condition $x(0)$ and disturbances w on the host cell, i.e. $x \to x_d$ at the steady state despite uncertain $x(0)$ and w. This is a robust regulation problem of synthetic gene networks, i.e. the state vector x of synthetic gene networks is robustly regulated to x_d in the host cell.

Let us denote the regulation error as

$$
\tilde{x} = x - x_d \tag{17.3}
$$

Then the regulation error dynamic system is given by

$$\dot{\tilde{x}} = f(\tilde{x} + x_d, k, \gamma) + v, \tilde{x}(0) = \tilde{x}_0 \tag{17.4}$$

where $v = k_0 + g(u) + w$ denotes the total uncertain disturbance in the regulation error system because these terms always fluctuate in the host cell and are not easily measured correctly. Because of the uncertainty of v and $\tilde{x}(0)$, the minimax regulation design method is an efficient but simple design scheme for robust synthetic gene network. The uncertainty of disturbance v and initial condition $\tilde{x}(0)$ in the following minimax stochastic H_∞ game design can be considered as a player maximizing their effects on the regulation error in the robust design problem of synthetic gene networks [9,280].

$$g_*^2 = \min_{\substack{k \in [k_1, k_2] \\ \gamma \in [\gamma_1, \gamma_2]}} \max_{\tilde{x}(0), v} \frac{E\left[\int_0^{t_f} \tilde{x}^T Q \tilde{x} dt\right]}{E\left[\int_0^{t_f} v^T v dt + \tilde{x}^T(0)\tilde{x}(0)\right]} \tag{17.5}$$

where Q is the weighting matrix. In general, Q is a diagonal weighting matrix with $Q = diag([q_{11}, q_{22}, \ldots, q_{nn}])$ to denote the punishment on regulation error. If only the last state x_n is required to be regulated to achieve the desired steady state x_{dn}, then we can let $q_{nn} = 1$ and $q_{11} = q_{22} = \ldots = q_{n-1n-1} = 0$. $[k_1, k_2]$ and $[\gamma_1, \gamma_2]$ denote the allowable ranges of production rate vector k and decay rate vector γ, respectively. The allowable ranges are determined by the engineering biotechnologies of synthetic biology. k and γ to be designed can be considered as another player minimizing the worst-case effect of $\tilde{x}(0)$ and v on the regulation error. If the disturbances v and initial condition $\tilde{x}(0)$ are deterministic, then the expectation operation $E[\,]$ in (17.5) could be neglected.

The physical meaning of stochastic minimax H_∞ game in (17.5) is that the worst-case effect of uncertain $\tilde{x}(0)$ and v on the regulation error $\tilde{x}$ must be minimized in safety from the mean energy perspective by k and γ, which are chosen from the allowable ranges. Therefore, for uncertain $\tilde{x}(0)$ and v, the robust synthetic gene network design is to solve the minimax H_∞ game problem in (17.5) subject to the regulation error dynamic system in (17.4). This is the so-called stochastic H_∞ game problem in the robust synthetic gene network design [280].

In general, it is not easy to solve the nonlinear stochastic minmax H_∞ game problem in (17.5) subject to (17.4) directly. It is always solved by a sub-minimax method. First, let the upper bound g^2 of g_*^2 in (17.5) [9,280]

$$g_*^2 = \min_{\substack{k \in [k_1, k_2] \\ \gamma \in [\gamma_1, \gamma_2]}} \max_{\tilde{x}(0), v} \frac{E\left[\int_0^{t_f} \tilde{x}^T Q \tilde{x} dt\right]}{E\left[\int_0^{t_f} v^T v dt + \tilde{x}^T(0)\tilde{x}(0)\right]} \le g^2 \tag{17.6}$$

We will first solve the sub-minimax problem in (17.6) and then decrease the upper bound g^2 as much as possible to approach its stochastic minimax H_∞ game solution. In general, the stochastic minimax H_∞ game problem in (17.6) is equivalent to the following minimax Nash quadratic game problem [9,280]:

$$\min_{\substack{k \in [k_1, k_2] \\ \gamma \in [\gamma_1, \gamma_2]}} \max_{v} E\left[\int_0^{t_f} (\tilde{x}^T Q \tilde{x} - g^2 v^T v) dt\right] \le g^2 E[\tilde{x}^T(0)\tilde{x}(0)], \ \forall \tilde{x}(0) \tag{17.7}$$

where g^2 is to be minimized because it is the upper bound in (17.6) and should be as small as possible to approach g_*^2 of the minimax H_∞ game solution. Let us denote the cost function as

$$J(k, r, v) = E\left[\int_0^{t_f} (\tilde{x}^T Q \tilde{x} - g^2 v^T v) dt\right] \tag{17.8}$$

17.3 Results of Minimax Stochastic H$_\infty$ Synthetic Gene Networks

17.3.1 Sub-Minimax H$_\infty$ Game Design for Robust Synthetic Gene Networks

From the above analysis, the stochastic game problem in (17.6) or (17.7) is equivalent to finding the worst-case disturbance v^* which maximizes $J(k, \gamma, v)$ and then the minimax k^* and γ^* which minimize $J(k, \gamma, v^*)$ such that the minimax Nash quadratic game value $J(k^*, \gamma^*, v^*)$ is less than $g^2 E[\tilde{x}^T(0)\tilde{x}(0)]$, i.e.

$$\begin{aligned} J(k^*, \gamma^*, v^*) &= \min_{\substack{k \in [k_1, k_2] \\ \gamma \in [\gamma_1, \gamma_2]}} J(k, r, v^*) \\ &= \min_{\substack{k \in [k_1, k_2] \\ \gamma \in [\gamma_1, \gamma_2]}} \max_{v} J(k, r, v) \leq g^2 E[\tilde{x}^T(0)\tilde{x}(0)], \ \forall \tilde{x}(0) \end{aligned} \tag{17.9}$$

Hence, if there exist k^*, γ^*, and v^* such that the minimax Nash quadratic game design problem in (17.9) is solved, then they can satisfy the sub-minimax H_∞ tracking performance of the robust synthetic gene network design in (17.6) as well. Therefore, the first step of robust synthetic gene network design is to solve the following dynamic Nash quadratic game problem:

$$\min_{\substack{k \in [k_1, k_2] \\ \gamma \in [\gamma_1, \gamma_2]}} \max_{v} J(k, r, v) \tag{17.10}$$

subject to the error dynamic equation in (17.4). Since $J(k^*, \gamma^*, v^*) \leq g^2 E[\tilde{x}^T(0)\tilde{x}(0)]$ according to (17.9) and g^2 is the upper bound of the stochastic minimax H_∞ game in (17.6), the sub-minimax has to make g^2 as small as possible, too.

From the above analysis, we obtain the following sub-minimax H_∞ game result for robust synthetic gene network design.

Proposition 17.1 The sub-minimax stochastic H_∞ game design of synthetic gene network is equivalent to solving the following constrained optimization for k^* and γ^*

$$g_*^2 = \min_{\substack{k \in [k_1, k_2] \\ \gamma \in [\gamma_1, \gamma_2]}} g^2 \tag{17.11}$$

subject to the following Hamilton–Jacobi–Issac inequality (HJII):

$$\left(\frac{\partial V(\tilde{x})}{\partial \tilde{x}}\right)^T f(\tilde{x}+x_d,k,\gamma)+\tilde{x}^T Q\tilde{x}+\frac{1}{4g^2}\left(\frac{\partial V(\tilde{x})}{\partial \tilde{x}}\right)^T\left(\frac{\partial V(\tilde{x})}{\partial \tilde{x}}\right)<0$$
$$E[V(\tilde{x}(0))]\leq g^2 E[\tilde{x}^T(0)\tilde{x}(0)] \tag{17.12}$$

with $V(\tilde{x})>0$, and the worst-case disturbance is given by

$$v^* = \frac{1}{2g^2}\frac{\partial V(\tilde{x})}{\partial \tilde{x}} \tag{17.13}$$

Proof: see Appendix A

Remark 17.1

(1) From (17.6), g^2 is the upper bound of g_*^2 stochastic minimax H_∞ game. In (17.11), we minimize the upper bound g^2 to achieve the sub-minimax solution for robust synthetic gene networks.

(2) The physical meaning of the constrained minimization in (17.11) and (17.12) is that we want to specify k^* and γ^* from the allowable parameter ranges such that the upper bound g^2 is as small as possible until no positive solution $V(\tilde{x})>0$ of HJII in (17.12) exists.

(3) At present, there exists no efficient analytic or numerical method to solve the HJII in (17.12) for nonlinear stochastic system control or filtering designs [275,276].

17.3.2 Minimax Stochastic H_∞ Game Strategy for Robust Synthetic Gene Networks via Fuzzy Interpolation Method

Because it is very difficult to solve the nonlinear HJII in (17.12), no simple approach is available for solving the constrained optimization problem in (17.11) for the stochastic minimax H_∞ game of the synthetic gene network design problem. Recently, the T–S fuzzy model has been widely employed [4,74,105,423] to approximate the nonlinear system via interpolating several linearized systems at different operating points so that the nonlinear minimax stochastic H_∞ game problem could be transformed to a fuzzy stochastic minimax H_∞ game problem [9]. By using such approach, the HJII in (17.12) can be replaced by a set of LMIs. In this situation, the nonlinear stochastic minimax H_∞ game problem in (17.10) could be easily solved by the fuzzy dynamic method for the robust design of sub-minimax stochastic H_∞ game design problem.

Suppose the nonlinear system in (17.4) could be approximated by a T–S fuzzy system [4]. The T–S fuzzy model is a piecewise interpolation of several linearized models through fuzzy membership functions. The fuzzy model is described by fuzzy if–then rules and will be employed to deal with the nonlinear stochastic minimax H_∞ game problem for robust synthetic gene network design under uncertain initial conditions and disturbances. The ith rule of fuzzy model for nonlinear systems in (17.4) is of the following form [4,105]:

Rule i:

If $\tilde{x}_1(t)$ is F_{i1} and $\tilde{x}_q(t)$ is F_{iq},

then

$$\dot{\tilde{x}} = \mathrm{A}_i(k,\gamma)\tilde{x}+v, i=1,2,\ldots,L \tag{17.14}$$

where F_{ij} is the fuzzy set. $\mathrm{A}_i(k,\gamma)$ is constant matrix with the elements of k and γ contained in its entries. q is the number of premise variables and $\tilde{x}_1,\ldots,\tilde{x}_q$ are the premise variables. The fuzzy system is inferred as follows [4,74,105,386,387]:

$$\dot{\tilde{x}}(t)=\frac{\sum_{i=1}^{L}\mu_i(\tilde{x}(t))[\mathrm{A}_i(k,\gamma)\tilde{x}(t)+v]}{\sum_{i=1}^{L}\mu_i(\tilde{x}(t))}=\sum_{i=1}^{L}h_i(\tilde{x}(t))[\mathrm{A}_i(k,\gamma)\tilde{x}(t)+v],\tilde{x}(0)=\tilde{x}_0 \quad (17.15)$$

where $\mu_i(\tilde{x}(t))=\prod_{j=1}^{q}F_{ij}(\tilde{x}_j(t)),h_i(\tilde{x}(t))=\mu_i(\tilde{x}(t))/\sum_{i=1}^{L}\mu_i(\tilde{x}(t))$ and $F_{ij}(\tilde{x}_j(t))$ are the grade of membership of $\tilde{x}_j(t)$ in F_{ij}.

We assume

$$\mu_i(\tilde{x}(t))0 \text{ and } \sum_{i=1}^{L}\mu_i(\tilde{x}(t))>0 \quad (17.16)$$

Therefore, we get the following fuzzy basis functions:

$$h_i(\tilde{x}(t))0 \text{ and } \sum_{i=1}^{L}h_i(\tilde{x}(t))=1 \quad (17.17)$$

The T–S fuzzy model in (17.15) is to interpolate L linear systems to approximate the nonlinear system in (17.4) via the fuzzy basis functions $h_i(\tilde{x}(t))$. We could specify system parameter $\mathrm{A}_i(k,\gamma)$ easily so that $\sum_{i=1}^{L}h_i(\tilde{x}(t))\mathrm{A}_i(k,\gamma)\tilde{x}$ can approximate $f(\tilde{x}+x_d,k,\gamma)$ in (17.4) by the fuzzy identification method [4].

After the nonlinear system in (17.4) is approximated by the T–S fuzzy system in (17.15), the nonlinear dynamic game problem in (17.10) is replaced by solving a dynamic game problem in (17.6) subject to the fuzzy system (17.15).

Proposition 17.2 The sub-minimax stochastic H_∞ game design of synthetic gene network is to solve k^* and γ^* by the following constraint optimization:

$$g_*^2=\min_{\substack{k\in[k_1,k_2]\\ \gamma\in[\gamma_1,\gamma_2]}} g^2 \quad (17.18)$$

subject to

$$\begin{aligned}&P\mathrm{A}_i(k,\gamma)+\mathrm{A}_i^T(k,\gamma)P+Q+\frac{1}{g^2}PP\le 0, i=1,2,\ldots,L\\ &P\le g^2I, P>0\end{aligned} \quad (17.19)$$

and the worst-case disturbance v^* is given by

$$v^*=\frac{1}{g^2}\sum_{i=1}^{L}h_i(\tilde{x}(t))P\tilde{x} \quad (17.20)$$

Proof: see Appendix B.

By the fuzzy approximation, the HJII in (17.12) can be approximated by a set of algebraic Riccati-like inequalities in (17.19). By Schur complement [23], the constrained optimization problem in (17.18) and (17.19) is equivalent to the following LMI-constrained optimization problem:

$$g_* = \min_{\substack{k \in [k_1, k_2] \\ \gamma \in [\gamma_1, \gamma_2]}} g^2 \tag{17.21}$$

subject to

$$\begin{aligned} &\begin{bmatrix} P\mathrm{A}_i(k,\gamma) + \mathrm{A}_i^T(k,\gamma)P + Q & P \\ P & -g^2 I \end{bmatrix} \leq 0, i = 1, 2, \ldots, L \\ &P \leq g^2 I, P > 0 \end{aligned} \tag{17.22}$$

Remark 17.2

(1) The fuzzy basis functions $h_i(\tilde{x})$ in (17.15) and (17.17) can be replaced by other interpolation functions, for example, cubic spline functions.

(2) By the fuzzy approximation, the HJII in (17.12) of nonlinear stochastic minimax H_∞ game problem can be solved by Robust Control Toolbox in Matlab efficiently [424]. The constrained optimization problem in (17.18) and (17.19) can be solved by decreasing g^2 until there is no positive definite solution $P>0$ in (17.22) with $k^* \in [k_1, k_2]$ and $\gamma^* \in [\gamma_1, \gamma_2]$.

(3) In the LMI-constrained optimization in (17.22) for the robust synthetic gene network design, we do not need the statistics of initial conditions and disturbances on the host cell, which are not easy to be measured. Therefore, the proposed stochastic minimax H_∞ game method is simple but robust for synthetic gene networks.

Remark 17.3 For comparison, the conventional optimal regulation design is also proposed for synthetic gene networks. If the effect of external disturbances and uncertain initial conditions on the regulation error is not considered as (17.5) in the design procedure, i.e. only the following stochastic H_2 optimal regulation design is considered:

$$\min_{\substack{k \in [k_1, k_2] \\ \gamma \in [\gamma_1, \gamma_2]}} E\left[\int_0^{t_f} \tilde{x}^T Q \tilde{x} dt\right] \tag{17.23}$$

subject to (17.4), then we obtain the following sub-optimal regulation design for synthetic gene networks.

Proposition 17.3 The stochastic sub-optimal H_2 synthetic gene network design in (17.23) is to solve the following constrained optimization:

$$\min_{\substack{k \in [k_1, k_2] \\ \gamma \in [\gamma_1, \gamma_2]}} E[V(\tilde{x}(0))] \tag{17.24}$$

subject to

$$V(\tilde{x})>0, \frac{\partial V(\tilde{x})}{\partial \tilde{x}} f(\tilde{x} + x_d, k, \gamma) + \tilde{x}^T Q \tilde{x} + \frac{1}{2}\left(\frac{\partial V(\tilde{x})}{\partial \tilde{x}}\right)^T \left(\frac{\partial V(\tilde{x})}{\partial \tilde{x}}\right) < 0 \tag{17.25}$$

Proof: see Appendix C.

Because it is not easy to solve the above HJI-constrained optimization for the stochastic sub-optimal H_2 regulation design in (17.24) and (17.25), the fuzzy approximation method is needed to simplify the design procedure. If the nonlinear error dynamic equation in (17.4) is represented by the fuzzy interpolation system in (17.15), then the stochastic optimal H_2 synthetic gene network design in (17.23) is equivalent to the following stochastic optimal H_2 regulation design problem:

$$\min_{\substack{k\in[k_1,k_2]\\ \gamma\in[\gamma_1,\gamma_2]}} E\left[\int_0^{t_f} \tilde{x}^T Q\tilde{x}dt\right] \quad \text{subject to } \dot{\tilde{x}} = \sum_{i=1}^{L} h_i(\tilde{x})A_i(k,\gamma)\tilde{x} + v \tag{17.26}$$

Proposition 17.4 The stochastic sub-optimal H_2 regulation design problem in (17.26) becomes how to solve the following LMIs-constrained optimization problem:

$$\min_{\substack{k\in[k_1,k_2]\\ \gamma\in[\gamma_1,\gamma_2]}} \operatorname{Tr}(PR_0) \tag{17.27}$$

subject to

$$P>0, \begin{bmatrix} A_i^T(k,\gamma)P + PA_i(k,\gamma) + Q & P \\ P & -2I \end{bmatrix} \leq 0, i = 1,2,\ldots,L \tag{17.28}$$

where R_0 denotes the covariance matrix $E[\tilde{x}(0)\tilde{x}^T(0)]$.

Proof: Similar to the proof of Proposition 17.2.

Since the effect of stochastic disturbances on $E[\tilde{x}(0)\tilde{x}^T(0)]$ is not considered as (17.5) in the above stochastic sub-optimal H_2 synthetic gene network design, the synthesized gene networks will be more sensitive to the external disturbances or other uncertain factors. They will be compared with the sub-minimax stochastic H_∞ game method for robust synthetic gene network in the simulation example.

Remark 17.4 Since the effect of the disturbance v on the regulation error has not been attenuated efficiently on the design procedure of the stochastic sub-optimal H_2 regulation in Propositions 17.3 and 17.4, the disturbance will have much effect on the sub-optimal H_2 regulation design of synthetic gene network. This property will be discussed and compared with the proposed stochastic minimax H_∞ game design for robust synthetic gene network in the design example in the following section.

Design procedure: According to the above analyses, a design procedure is developed for the proposed robust synthetic gene network via stochastic minimax H_∞ game method.

(1) Give feasible parameter ranges $[k_1, k_2]$ and $[\gamma_1, \gamma_2]$ for production rate parameters k and decay rate parameters γ, respectively, according to the biotechnology ability.

(2) Give the desired steady state x_d according to the design purpose and develop a regulation error dynamic (17.4) for a synthetic gene network.

(3) Construct a T–S fuzzy model in (17.15) to approximate the regulation error dynamic in (17.4).

(4) Solve the constrained optimization problem from the ranges $k \in [k_1, k_2]$ and $\gamma \in [\gamma_1, \gamma_2]$ in (17.21) and (17.22) for the robust synthetic gene network design k^* and γ^*, respectively according to the sub-minimax stochastic H_∞ game scheme or solve the constrained optimization problem in (17.27) and (17.28) for the sub-optimal regulation design.

17.3.3 Design Example In Silico for the Proposed Robust Design by Stochastic Minimax H_∞ Game Method

Consider the man-made synthetic gene network in the dynamic equation (17.1) [410]. The synthetic gene network is shown in Fig. 17.1, where $k_{tetR,0}$, $k_{lacI,0}$, $k_{cI,0}$, and$k_{eyfp,0}$ are basal production rates of the corresponding proteins, which are assumed to be 5000, 587, 210, and 3487, respectively [410,425]. k_{lacI}, k_{cI}, andk_{eyfp} are the production rate parameters while γ_{tetR}, r_{lacI}, γ_{cI}, and γ_{eyfp} are the decay rate parameters of the corresponding proteins in the host cell (i.e. *E. coli.*). In the robust synthetic gene network design via stochastic minimax H_∞ game, we should select the parameters k and γ from feasible ranges so that the state of synthetic gene network x_i could approach a desired steady state $x_{d,i}$ for some biotechnical purpose. r_{lacI}, γ_{cI}, andγ_{eyfp} are the decreasing Hill functions for regulations of repressors. a_{lacI} is an increasing function since aTc is an activator. The Hill function is an S-shaped curve [397]. u_{aTc} is the input to the synthetic gene network system. We assume anhydrotetracycline input concentration to be a constant value 10,000 (i.e. $u_{aTc} = 10000$). For the convenience of simulation, we assume that extrinsic disturbances $w_1 - w_4$ are $w_i = [500n_1\ 10000n_2 100n_3 100000n_4\ ]^T$, where n_i, $i = 1,2,3,4$ are independent Gaussian white noises with zero mean and unit variance.

From the stochastic minimax H_∞ game for the robust design of synthetic gene network, we give the feasible parameter ranges of production rate parameters k and decay rate parameters γ as follows [410]:

$$\begin{array}{cc} k_{lacI} \subset [70,7000] & \gamma_{tetR} \in [0.05,5] \\ k_{cI} \in [75,8000] & \gamma_{lacI} \in [0.01314,0.1517] \\ k_{eyfp} \in [30,30000] & \gamma_{cI} \in [0.7617,7.2815] \\ & \gamma_{eyfp} \in [0.007,0.067] \end{array} \tag{17.29}$$

Then, we give the desired steady states of the synthetic gene network as $x_{d,i} = [1000, 50000, 300, 500000]^T$, $i = tetR, lacI, cI, eyfp$. Then, the regulation error dynamic equation in (17.4) is developed for the synthetic gene network. Because it is very difficult to solve the nonlinear HJI in (17.12), no simple approach is available to solve the constrained optimization problem in (17.11) for robust parameters k_i^* and γ_i^*. We construct the T–S fuzzy model in (17.15) to approximate the regulation error dynamic in (17.4) with the regulation error dynamic system's state variables as the premise variables in the following:

Rule i:

If $\tilde{x}_1(t)$ is F_{i1} and $\tilde{x}_2(t)$ is F_{i2} and $\tilde{x}_3(t)$ is F_{i3} and $\tilde{x}_4(t)$ is F_{i4}, then

$$\dot{\tilde{x}} = \mathrm{A}_i(k,\gamma)\tilde{x} + v, i = 1,2,\ldots,L$$

where the parameters $\mathrm{A}_i(k,\gamma)$ and the number of fuzzy rules is $L = 16$. To construct the fuzzy model, we need to find the operating points of the regulation error dynamic system. The operating points for $\tilde{x}_1$ are chosen at $\tilde{x}_{11} = -40$ and $\tilde{x}_{12} = 4040$. Similarly, the operating points of $\tilde{x}_2, \tilde{x}_3, \tilde{x}_4$ are chosen at $\bar{x}_{21} = -38510, \bar{x}_{22} = 381$, $\bar{x}_{31} = -16.7, \bar{x}_{32} = 1686, \bar{x}_{41} = -441590$ and $\bar{x}_{42} = 4372$, respectively. For the convenience of design, triangle-type membership functions are taken for Rule 1 through Rule 16. We create two triangle-type membership functions for each state (Fig. 17.2).

In order to simplify the nonlinear stochastic minimax H_∞ game problem of the robust synthetic gene network, we just solve only the sub-minimax H_∞ game problem in (17.6) instead. With the help of fuzzy approximation method and LMI technique, we can easily

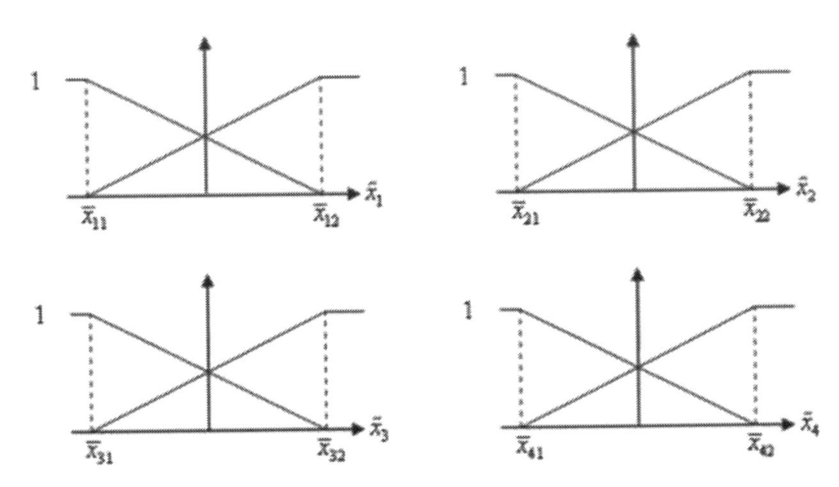

FIGURE 17.2
Membership functions for four states $\tilde{x}_1$, $\tilde{x}_2$, $\tilde{x}_3$, and $\tilde{x}_4$.

solve the constrained optimization problem in (17.21) and (17.22) instead of the nonlinear constrained optimization problem in (17.11) and (17.12) for the stochastic minimax H_∞ game design for the robust synthetic gene network. Finally, we obtain the upper bound of the stochastic minimax H_∞ game in (17.6) $g_*^2 = 0.847536$ and a common positive definite symmetric matrix P_* for (17.22) as follows:

$$P_* = \begin{bmatrix} 0.4582 & -0.0079 & 0.0143 & -0.00068 \\ -0.0079 & 0.07186 & -0.000557 & 0.00268 \\ 0.0143 & -0.000557 & 0.04847 & 0.000718 \\ -0.00068 & 0.00268 & 0.000718 & 0.0578 \end{bmatrix}$$

with the specified robust production rate parameters $k_{lacI}^* = 7000, k_{cI}^* = 4037.5$ and $k_{eyfp}^* = 30000$ and robust decay rate parameters $\gamma_{tetR}^* = 5, \gamma_{lacI}^* = 0.1517, \gamma_{cI}^* = 4.0216$ and $\gamma_{eyfp}^* = 0.067$ of the synthetic gene network. With these design parameters, the parameters A_i of fuzzy model are described in Appendix D.

Figure 17.3 presents the simulation result for robust synthetic gene networks by using Monte–Carlo method with 50 rounds and with the uncertain initial values. $x_1(0) - x_4(0)$ are assumed normal-distributed random numbers with means 5000, 8000, 2000, 10000 and SDs 500, 800, 200, 1000, respectively. As can be seen, the synthetic gene network has robust regulation ability to achieve the desired steady state (black dashed line) in spite of uncertain initial states and the disturbances on the host cell. Obviously, the robust synthetic gene network by the proposed sub-minimax H_∞ regulation game design method has robust stability to the uncertain initial conditions and enough filtering ability to attenuate the disturbances on the host cell and can approach the desired steady states.

For comparison, we solve the stochastic sub-optimal H_2 regulation design problem in (17.27) and (17.28) for the specified production rate parameters $k_{lacI}^* = 70, k_{cI}^* = 4037.5$, and $k_{eyfp}^* = 15015$ and decay rate parameters $\gamma_{tetR}^* = 2.525, \gamma_{lacI}^* = 0.1517, \gamma_{cI}^* = 7.2815$, and $\gamma_{eyfp}^* = 0.067$ of the synthetic gene network. The simulation result of conventional optimal regulation design is also shown in Fig. 17.4. As can be seen, the conventional stochastic optimal H_2 regulation design of the synthetic gene network is more sensitive

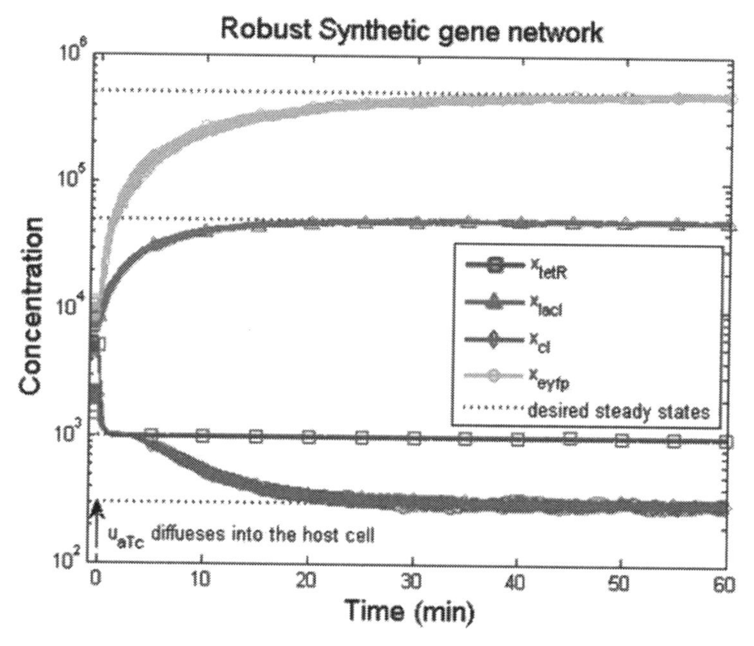

FIGURE 17.3
The robust synthetic gene network design by the stochastic minimax H_∞ game method with uncertain initial values and the desired steady states $x_d = [1000, 50000, 300, 500000]^T$. And with specified robust production rate parameters $k^*_{lacI} = 7000, k^*_{cI} = 4037.5$ and $k^*_{eyfp} = 30000$ while the specified robust decay rate parameters are $\gamma^*_{tetR} = 5, \gamma^*_{lacI} = 0.1517, \gamma^*_{cI} = 4.0216$ and $\gamma^*_{eyfp} = 0.067$ of the synthetic gene network. The Monte–Carlo simulation method is used with 50 rounds.

to the initial conditions and disturbances, and cannot achieve the desired steady state under the uncertain initial conditions and disturbances.

Remark 17.5 The experimental systems in the above example may not be fully observable. If we want to know whether all state variables can approach to the desired states x_d, several fluorescent proteins (red, green, and cyan color) should be necessary to observe their protein expressions of all state variables in the experimental design.

17.4 Discussion and Conclusion

Because the initial conditions and disturbances on the host cell are uncertain, to simplify the design problem and to be on the safe side, a robust synthetic biology design is formulated as a stochastic minimax H_∞ game problem in this chapter. The uncertain initial conditions and disturbances due to intrinsic and extrinsic molecular noises on the host cell are considered as a player maximizing the regulation error and the design parameters are considered as another player minimizing the regulation error. In order to avoid solving HJII in the stochastic minimax H_∞ game theory-based design problem, a T–S fuzzy interpolation method is introduced to simplify the design procedure of robust synthetic gene networks via only solving a set of LMIs, which can be efficiently solved by Robust Control Toolbox in Matlab.

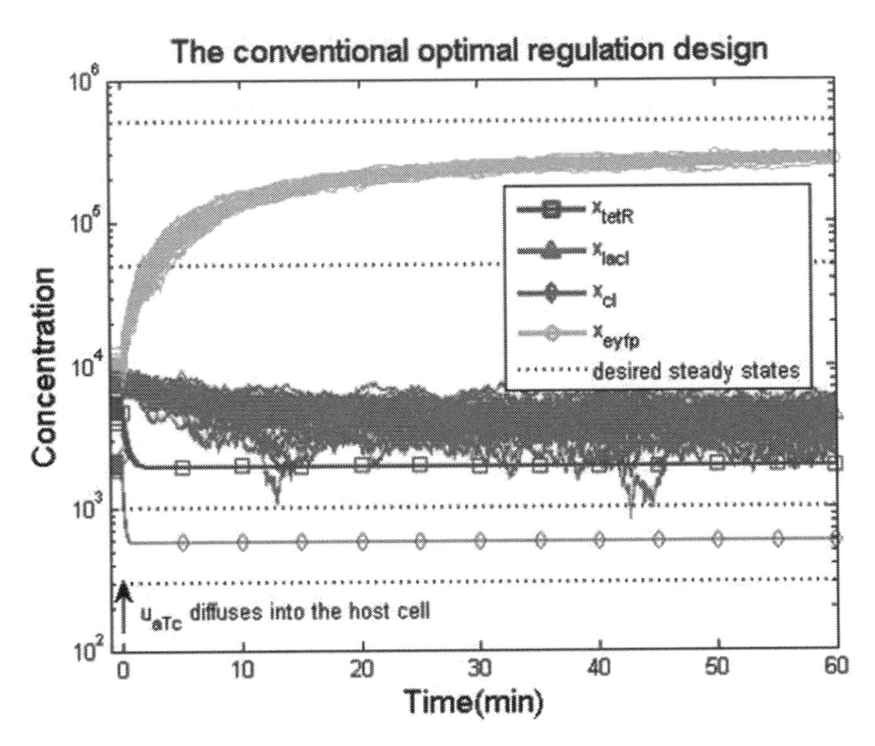

FIGURE 17.4
The conventional stochastic optimal H_2 regulation design with uncertain initial values and the desired steady states $x_d = [1000, 50000, 300, 500000]^T$. And with specified production rate parameters $k^*_{lacI} = 70, k^*_{cI} = 4037.5$ and $k^*_{eyfp} = 15015$, while the specified decay rate parameters are $\gamma^*_{tetR} = 2.525, \gamma^*_{lacI} = 0.1517, \gamma^*_{cI} = 7.2815$ and $\gamma^*_{eyfp} = 0.067$ of the synthetic gene network. It is seen that the conventional stochastic optimal H_2 regulation design of the synthetic gene network is sensitive to the initial conditions and disturbances and cannot achieve the desired steady states. The Monte–Carlo simulation method is used with 50 rounds.

In this study, we can select the weighting matrix $Q = diag([q_{11}, q_{22}, q_{33}, q_{44}])$ which denotes the punishment on the corresponding tracking error $\tilde{x}$. If we only need to achieve a desired steady state x_{d4} (EYFP), we just assign a value to the fourth diagonal element q_{44} of the weighting matrix Q and set $q_{11} = q_{22} = q_{33} = 0$. The states $x_1 - x_3$ will not approach to the given steady state $x_{d1} - x_{d3}$ because of no punishment. However, in this case, some infeasible steady states of x_1, x_2 and x_3 may be obtained, and even an optimal x_4 can be achieved. In this chapter, the desired steady states of x_1, x_2 and x_3 are given because we can avoid obtaining infeasible steady states in x_1, x_2 and x_3 when an optimal x_4 is achieved. Further, the undesired steady states of x_1, x_2 and x_3 may also have metabolic toxicity on host cell and should be avoided. Since the steady states of $x_1 - x_3$ are not that important, the desired steady states $x_{d1} - x_{d3}$ can be adjusted within feasible ranges, so that the desired steady state x_{d4} can still achieve some optimization as possible. This kind of design can avoid hampering the optimization of x_4 when x_1, x_2, and x_3 achieve some feasible steady states.

In our in silico design example, we can design the specified robust production rate parameters k^*_i and decay rate parameters γ^*_i within the feasible parameter ranges to achieve the desired steady states of the synthetic gene network. As for the biological implementation, we could refer to standard biological parts in biological device datasheets to construct the genetic circuits with the fine-tuned production rate parameters k_i^* and decay rate parameters γ_i^*. In this way, synthetic biologists can increase the efficiency of gene circuit design through registries of biological parts and standard datasheets, which are developed concerned with proper packing and characterizing of "modular" biological activities so that these biological parts or devices with some desired characteristics may be efficiently assembled into gene circuits [426].

Quantitative descriptions of devices in the form of standardized, comprehensive datasheets are widely used in many engineering disciplines. A datasheet is intended to allow an engineer to quickly determine whether the behavior of a device will meet the requirements of a system in which a device might be used [426]. Such a determination is based on a set of standard characteristics of device behavior, which are the product of engineering theory and experience. In the datasheets of engineering, the characteristics typically reported are common across a wide range of device types, such as sensors, logic elements, and actuators. Recently, biological datasheets have been set as standards for characterization, manufacture, and sharing of information about modular biological devices for a more efficient, predictable, and design-driven genetic engineering science [426,427]. Because datasheets of biological parts or devices are an embodiment of engineering standard for synthetic biology [426], a good device standard should define sufficient information about biological parts or devices to allow the design of gene circuit systems with the optimal parameters. Datasheets contain a formal set of input–output transfer functions, dynamic behaviors, compatibility, requirements, and other details about a particular part or device [426,427]. Since parameters k_i are combinations of transcription and translation, they could be measured from the input–output transfer functions and dynamic behaviors of biological parts or devices in biological device datasheets. From properly characterized input–output transfer functions and dynamic behaviors of parts or devices in biological device datasheets, an engineer can estimate the corresponding parameters of biological parts or devices. When the biological parts and devices in datasheets become more complete in future, we can rapidly select from a vast list the parts that will meet our design parameters k_i. Therefore, we can ensure that devices selected from datasheets can fit the optimal parameters and systems synthesized from them can satisfy the requirements of design specifications for robust synthetic gene networks.

In order to guarantee the biological feasibility of the calculated optimal parameters, the ranges $[k_1, k_2]$ and $[\gamma_1, \gamma_2]$ of parameters should be determined by the whole parameters of biological parts repositories (http://partsregistry.org/) so that the optimal parameters selected within these ranges to minimize g^2 in Equations (17.21) and (17.22) have biological meaning, or equivalently from the whole biological parts in biological device datasheets, we can find a set of biological parts whose parameters can minimize the g^2 in Equations (17.21) and (17.22) to achieve the robust optimal design of synthetic gene network from the stochastic minimax H_∞ game perspective.

In synthetic gene networks, there is much uncertainty about what affects the behavior of biological circuitry and systems. For example, devices will perturb the cellular functions and there are also likely to be parasitic and unpredictable interactions among components as well as with the host. Since k_i is a combination of promoter strength, ribosome binding site, and degradation of the transcript, there are some variations or uncertainties on the parameter value k_i. These variations or uncertainties of k_i can be transformed to an equivalent uncertain disturbance w_i in Equation (17.1) from the viewpoint of mathematic model. The proposed robust minimax synthetic biology design method can predict the most robust value of k_i from the perspective of stochastic minimax H_∞ game. In our robust design method, we do not need the statistics of these parameter uncertainties because the proposed synthetic genetic network not only can achieve the desired steady state but also can tolerate the worst-case effect due to these uncertain parameter variations and external noises on the host cell.

For comparison, a stochastic sub-optimal H_2 regulation design for synthetic gene network is also developed for synthetic gene network. Because the stochastic sub-optimal H_2 regulation design cannot efficiently attenuate the effect of uncertain initial conditions and disturbances on the regulation, it is not suitable for robust synthetic gene networks with uncertain initial conditions and disturbances on the host cell. As seen in the example in silico, the proposed robust synthetic gene network through the stochastic minimax H_∞ game strategy can function properly in spite of uncertain initial conditions and disturbances on the host cell. Design of more robust and complex genetic circuits is foreseen to have important applications in biotechnology, medicine, and biofuel production, and to revolutionize how we conceptualize and approach the engineering of biological systems [394]. Therefore, it has much potential for the robust synthetic gene network design in the near future.

17.5 Appendix

17.5.1 Appendix A. Proof of Proposition 17.1

Let us consider a Lyapunov energy function $V(\tilde{x})>0$, then the cost function in equation (17.8) is equivalent to

$$J(k,r,v)=E\left[V(\tilde{x}(0))-V(\tilde{x}(t_f))+\int_0^{t_f}(\tilde{x}^TQ\tilde{x}-g^2v^Tv+\frac{dV(\tilde{x})}{dt})dt\right] \tag{17.A1}$$

By the chain rule, we get

$$\frac{dV(\tilde{x}(t))}{dt}=\left(\frac{\partial V(\tilde{x}(t))}{\partial \tilde{x}(t)}\right)^T\cdot\frac{d\tilde{x}(t)}{dt}=\left(\frac{\partial V(\tilde{x}(t))}{\partial \tilde{x}(t)}\right)^T\cdot(f(\tilde{x}(t)+x_d(t),k,r)+v(t)) \tag{17.A2}$$

Substituting (17.A2) into (17.A1), we maximize $J(k,r,v)$ by the uncertain disturbance v

$$\begin{aligned}&\max_v J(k,r,v)\\&=\max_v E\left[V(\tilde{x}(0))-V(\tilde{x}(t_f))+\int_0^{t_f}(\tilde{x}^TQ\tilde{x}-g^2v^Tv+\left(\frac{\partial V(\tilde{x})}{\partial\tilde{x}}\right)^T(f(\tilde{x}+x_d,k,r)+\left(\frac{\partial V(\tilde{x})}{\partial\tilde{x}}\right)^T v)dt\right]\\&=\max_v E\left[V(\tilde{x}(0))-V(\tilde{x}(t_f))+\int_0^{t_f}(\tilde{x}^TQ\tilde{x}+\left(\frac{\partial V(\tilde{x})}{\partial\tilde{x}}\right)^T(f(\tilde{x}+x_d,k,r)\right.\end{aligned}$$

$$-\left(gv-\frac{1}{2g}\frac{\partial V(\tilde{x})}{\partial \tilde{x}}\right)^T\left(gv-\frac{1}{2g}\frac{\partial V(\tilde{x})}{\partial \tilde{x}}\right)+\frac{1}{4g^2}\left(\frac{\partial V(\tilde{x})}{\partial \tilde{x}}\right)^T\left(\frac{\partial V(\tilde{x})}{\partial \tilde{x}}\right)\Bigg)dt\Bigg]$$

$$=E\left[V(\tilde{x}(0))-V(\tilde{x}(t_f))+\int_0^{t_f}(\tilde{x}^TQ\tilde{x}+\left(\frac{\partial V(\tilde{x})}{\partial \tilde{x}}\right)^T(f(\tilde{x}+x_d,k,r)+\frac{1}{4g^2}\left(\frac{\partial V(\tilde{x})}{\partial \tilde{x}}\right)^T\left(\frac{\partial V(\tilde{x})}{\partial \tilde{x}}\right)\Bigg)dt\right] \tag{17.A3}$$

with the worst-case disturbance $v^*=\frac{1}{2g^2}\frac{\partial V(\tilde{x})}{\partial \tilde{x}}$.

By the inequality in (17.12), it is seen that $V(\tilde{x}(0))$ is the upper bound of (17.A3) i.e., the sub-minimax H_∞ game problem becomes how to solve the following constrained Nash quadratic game problem:

$$\min_{\substack{k\in[k_1,k_2]\\ \gamma\in[\gamma_1,\gamma_2]}}\max_{v} J(k,r,v)\le \min_{\substack{k\in[k_1,k_2]\\ \gamma\in[\gamma_1,\gamma_2]}} E[V(\tilde{x}(0))] \tag{17.A4}$$
$$\text{subject to (17.12) and } V(\tilde{x})>0.$$

By the fact in (17.9), $g^2E[\tilde{x}^T(0)\tilde{x}(0)]$ is the upper bound of $\min_{\substack{k\in[k_1,k_2]\\ \gamma\in[\gamma_1,\gamma_2]}}\max_{v} J(k,r,v)$.

Therefore, $E[V(\tilde{x}(0))]$ in (17.A4) should be bounded by $g^2E[\tilde{x}^T(0)\tilde{x}(0)]$, i.e. $E[V(\tilde{x}(0))]\le g^2E[\tilde{x}^T(0)\tilde{x}(0)]$. Therefore, the suboptimal solution is to minimize its upper bound. Hence, the sub-minimax problem in (17.A4) could be replaced by

$$\min_{\substack{k\in[k_1,k_2]\\ \gamma\in[\gamma_1,\gamma_2]}}\max_{v} J(k,r,v)\le \min_{\substack{k\in[k_1,k_2]\\ \gamma\in[\gamma_1,\gamma_2]}} E[V(\tilde{x}(0))]\le \min_{\substack{k\in[k_1,k_2]\\ \gamma\in[\gamma_1,\gamma_2]}} g^2[\tilde{x}^T(0)\tilde{x}(0)]=\min_{\substack{k\in[k_1,k_2]\\ \gamma\in[\gamma_1,\gamma_2]}} g^2\text{Tr}(R_0) \tag{17.A5}$$

where $\text{Tr}(R_0)$ denotes the trace of R_0 and R_0 denotes the covariance of the initial condition $\tilde{x}(0)$ i.e., $g^2E[\tilde{x}^T(0)\tilde{x}(0)]$, which is independent of the choice of k and γ. Therefore, the stochastic sub-minimax H_∞ game design problem is equivalent to solving the following constrained optimization:

$$\min_{\substack{k\in[k_1,k_2]\\ \gamma\in[\gamma_1,\gamma_2]}} g^2$$
$$\text{subject to (17.12) and } V(\tilde{x})>0.$$

17.5.2 Appendix B. Proof of Proposition 17.2

We replace error dynamic system in (17.4) by its fuzzy interpolation system in (17.15). Then, HJII in (17.12) can be represented by

$$\left(\frac{\partial V(\tilde{x})}{\partial \tilde{x}}\right)^T\left(\sum_{i=1}^{L} h_i(\tilde{x})A_i(k,\gamma)\tilde{x}\right)+\tilde{x}^TQ\tilde{x}+\frac{1}{4g^2}\left(\frac{\partial V(\tilde{x})}{\partial \tilde{x}}\right)^T\left(\frac{\partial V(\tilde{x})}{\partial \tilde{x}}\right)<0 \tag{17.B1}$$

Let us choose the Lyapunov function $V(\tilde{x})$ as $V(\tilde{x}) = \tilde{x}^T P \tilde{x}$ for some positive definite symmetric matrix P and substitute it into (17.B1). Then we get

$$\sum_{i=1}^{L} h_i(\tilde{x})\left\{\tilde{x}^T\left(PA_i(k,r) + A_i^T(k,r)P + Q + \frac{1}{g^2}PP\right)\tilde{x}\right\} \leq 0 \qquad (17.B2)$$

where the property in (17.17) is used.

It is seen that the inequalities in (17.19) imply (17.B2). Therefore, the stochastic sub-minimax H_∞ game design for the fuzzy equivalent system becomes how we solve the constrained optimization in (17.18) and (17.19). By substituting $V(\tilde{x}) = \tilde{x}^T P \tilde{x}$ into (17.13), we get the worst-case disturbances v^* in (17.20).

17.5.3 Appendix C. Proof of Proposition 17.3

Again, let us consider a Lyapunov energy function $V(\tilde{x})>0$, then Equation (17.23) is equivalent to

$$\begin{aligned}
&\min_{\substack{k\in[k_1,k_2]\\ \gamma\in[\gamma_1,\gamma_2]}} E[\int_0^{t_f} \tilde{x}^T Q \tilde{x} dt] \\
&= \min_{\substack{k\in[k_1,k_2]\\ \gamma\in[\gamma_1,\gamma_2]}} E\left[V(\tilde{x}(0)) - V(\tilde{x}(t_f)) + \int_0^{t_f} (\tilde{x}^T Q \tilde{x} + \left(\frac{dV(\tilde{x})}{dt}\right)dt\right] \\
&= \min_{\substack{k\in[k_1,k_2]\\ \gamma\in[\gamma_1,\gamma_2]}} E\left[V(\tilde{x}(0)) - V(\tilde{x}(t_f)) + \int_0^{t_f} \left(\tilde{x}^T Q \tilde{x} + \left(\frac{\partial V(\tilde{x})}{\partial \tilde{x}}\right)^T (f(\tilde{x}+x_d,k,r) + \left(\frac{\partial V(\tilde{x})}{\partial \tilde{x}}\right)^T v\right)dt\right]
\end{aligned}$$

By the fact that $2a^T b \leq a^T a + b^T b$ for any two-vectors a and b, we get

$$\begin{aligned}
\min_{\substack{k\in[k_1,k_2]\\ \gamma\in[\gamma_1,\gamma_2]}} E[\int_0^{t_f} \tilde{x}^T Q \tilde{x} dt] = &\min_{\substack{k\in[k_1,k_2]\\ \gamma\in[\gamma_1,\gamma_2]}} E[V(\tilde{x}(0)) - V(\tilde{x}(t_f)) + \int_0^{t_f} (\tilde{x}^T Q \tilde{x} \\
&+ \left(\frac{\partial V(\tilde{x})}{\partial \tilde{x}}\right)^T (f(\tilde{x}+x_d,k,r) + \frac{1}{2}\left(\frac{\partial V(\tilde{x})}{\partial \tilde{x}}\right)^T \left(\frac{\partial V(\tilde{x})}{\partial \tilde{x}}\right) + \frac{1}{2}v^T v)dt]
\end{aligned}$$

By the inequality in (17.25), we get the stochastic sub-optimal H_2 regulation problem as follows:

$$\min_{\substack{k\in[k_1,k_2]\\ \gamma\in[\gamma_1,\gamma_2]}} E[\int_0^{t_f} \tilde{x}^T Q \tilde{x} dt] \leq \min_{\substack{k\in[k_1,k_2]\\ \gamma\in[\gamma_1,\gamma_2]}} E[V(\tilde{x}(0)) + \int_0^{t_f} v^T v dt$$

Since disturbance v is independent of the choice of parameters k and γ, and only the choice of $V(\tilde{x})$ will influence the above minimization, the stochastic sub-optimal H_2 regulation design becomes how to solve the constrained optimization problem in (17.24) and (17.25).

17.5.4 Appendix C. Parameters of the T–S Fuzzy Model with the Specified Kinetic Parameters K^* and Decay Rates γ^*

$$A_1 = \begin{bmatrix} -1.6879 & -0.060601 & 0.11879 & -0.0092833 \\ 0.38914 & -0.093297 & 0.010249 & -0.0065119 \\ 0.10826 & -0.02841 & -1.4996 & -0.0060343 \\ 0.00097167 & -0.0025457 & 0.0053402 & -0.066832 \end{bmatrix}, A_2 = \begin{bmatrix} -3.5629 & -0.12704 & 0.25074 & -0.0092833 \\ 0.20138 & -0.193 & 0.021476 & -0.0065119 \\ 0.22906 & -0.11458 & -3.1644 & -0.0060447 \\ 0.0014069 & -0.00054073 & 0.0071284 & -0.066833 \end{bmatrix}$$

$$A_3 = \begin{bmatrix} -1.5351 & -0.060529 & 0.11879 & -0.0092832 \\ 0.40408 & -0.092851 & 0.010249 & -0.0065573 \\ 0.18285 & 0.0322 & -1.4996 & -0.0041303 \\ 0.0012516 & -0.0027325 & -0.0017594 & -0.066801 \end{bmatrix}, A_4 = \begin{bmatrix} -3.2403 & -0.12689 & 0.25074 & -0.0092832 \\ 0.23298 & -0.19466 & 0.021741 & -0.0065573 \\ -0.38598 & 0.039126 & -3.1671 & 0.0041304 \\ 0.0019632 & 0.00067731 & -0.00013562 & -0.066801 \end{bmatrix}$$

$$A_5 = \begin{bmatrix} -3.5287 & -0.060601 & 0.24784 & -0.0093278 \\ 0.19497 & -0.093286 & 0.017006 & 0.0019312 \\ 0.22212 & -0.080273 & -3.1614 & -0.0042428 \\ 0.001744 & -0.0025529 & 0.0072707 & -0.067233 \end{bmatrix}, A_6 = \begin{bmatrix} -7.4489 & -0.12704 & 0.52318 & -0.0092833 \\ -0.18351 & -0.1982 & 0.095548 & -0.0014778 \\ 0.21344 & -0.11298 & -7.2861 & -0.00040939 \\ -0.012439 & -0.0026832 & 0.025864 & -0.066952 \end{bmatrix}$$

$$A_7 = \begin{bmatrix} -3.2061 & -0.060529 & 0.24784 & -0.0093277 \\ 0.22649 & -0.092851 & 0.01727 & 0.0018016 \\ -0.38483 & -0.019544 & -3.1642 & 0.0068314 \\ 0.0023517 & -0.0027325 & 6.7334e-0.06 & -0.067149 \end{bmatrix}, A_8 = \begin{bmatrix} -6.768 & -0.12689 & 0.52318 & -0.0093277 \\ -0.14191 & -0.19465 & -0.023172 & -0.0018026 \\ -0.81178 & -0.012738 & -6.0679 & -0.0068211 \\ 0.0043172 & 0.00067013 & 0.040657 & -0.06715 \end{bmatrix}$$

$$A_9 = \begin{bmatrix} -1.6879 & 0.12793 & -0.25078 & 0.019598 \\ -0.727 & -0.07319 & -0.026022 & -0.003619 \\ 0.10826 & -0.031432 & -0.80806 & -0.005567 \\ 0.00097182 & -0.0027504 & 0.0047284 & -0.066801 \end{bmatrix}, A_{10} = \begin{bmatrix} -3.5629 & 0.26819 & -0.52934 & 0.019598 \\ -0.91465 & -0.15344 & -0.05495 & -0.003619 \\ 0.22793 & -0.094274 & -1.7058 & -0.005567 \\ 0.0013385 & 0.00063963 & 0.0057541 & -0.066801 \end{bmatrix}$$

$$A_{11} = \begin{bmatrix} -1.5351 & 0.12778 & -0.25078 & 0.019598 \\ -0.71206 & -0.073303 & -0.026022 & -0.0036189 \\ -0.18285 & -0.034225 & -0.80806 & 0.0041294 \\ 0.0012516 & -0.0026058 & -0.0023716 & -0.066797 \end{bmatrix}, A_{12} = \begin{bmatrix} -3.2403 & 0.26787 & -0.52934 & 0.019598 \\ -0.88316 & -0.15367 & -0.054951 & -0.0036189 \\ -0.38597 & 0.043382 & -1.7058 & 0.0041294 \\ 0.0019634 & 0.00094337 & -0.0013455 & -0.066797 \end{bmatrix}$$

$$A_{13} = \begin{bmatrix} -3.5287 & 0.12793 & -0.52322 & 0.019692 \\ -0.92106 & -0.07319 & -0.058507 & 0.0047537 \\ 0.22099 & -0.083177 & -1.7026 & 0.0029125 \\ 0.0016756 & -0.0027503 & 0.0059171 & -0.067149 \end{bmatrix}, A_{14} = \begin{bmatrix} -7.4489 & 0.26819 & -1.1045 & 0.019692 \\ -1.3492 & -0.15343 & -0.12363 & 0.0047547 \\ 0.72194 & -0.14614 & -3.593 & -0.0029229 \\ 0.018292 & 0.00063245 & 0.0083449 & -0.06715 \end{bmatrix}$$

$$A_{15} = \begin{bmatrix} -3.2061 & 0.12778 & -0.52322 & 0.019692 \\ -0.88965 & -0.073303 & -0.058507 & 0.0047076 \\ -0.38483 & -0.01752 & -1.7026 & 0.0073033 \\ 0.0023519 & -0.0026058 & -0.0011826 & -0.067117 \end{bmatrix}, A_{16} = \begin{bmatrix} -6.768 & 0.26787 & -1.1045 & 0.019692 \\ -1.2579 & -0.15367 & -0.12336 & 0.0047076 \\ -0.81291 & -0.0083629 & -3.5957 & 0.0073033 \\ 0.0042487 & 0.00094338 & 0.0010809 & -0.067117 \end{bmatrix}$$

18

On the Natural Selection Strategy in Stochastic H_∞ Evolutionary Game of a Population of Biological Networks

18.1 Introduction

In general, biological network evolution process is the physical, genetic, or behavioral changing process in populations of biological networks over time. Natural selection is an innovation process that causes biological networks to adapt to their environments. Network robustness is important in ensuring the systematic stability of phenotypic traits of biological networks that are constantly exposed to genetic variations and non-genetic environmental disturbances. A better understanding of network robustness is paramount for understanding biological network evolution. Network robustness can be discussed sensibly only if two cardinal questions have been resolved: what is the phenotypic trait of interest in network evolution, and what is the network perturbation of interest [428–431]? There are three principal kinds of network perturbations to which a biological network needs to be robust: stochastic intrinsic molecular noise, environmental change, and genetic variation [432–434]. Stochastic intrinsic molecular noise refers to the stochastic fluctuations that occur in any biological network, for example, in the concentration of a biological molecule. Environmental change is the variation in the external environment, for example, a change in temperature, salinity, or nutrient availability. Genetic variation is genetic change, either through de novo mutation or through recombination in the evolutionary process [428]. The systematic ability to buffer genetic variation is called genetic robustness, the systematic filtering ability to attenuate the effect of intrinsic molecular noise is called noise robustness, and the systematic ability to resist environmental change is called environmental robustness. The ability to maintain a desired phenotypic trait is called phenotypic robustness. In order to maintain the desired phenotypic trait, a biological network needs an evolutionary strategy with high network robustness to confer genetic robustness (buffering genetic variation), noise robustness (filtering stochastic intrinsic molecular noise), and environmental robustness to buffer environmental change in the evolutionary process.

Network evolvability is defined as the systematic ability to deviate from a phenotypic trait to adapt to an environmental change. The network evolvability of a biological network appears to be the reverse of network robustness. If phenotypic traits are robust against genetic variations and external disturbances, a network population may be expected to have difficult adaptation to an environmental change, as several studies have suggested [429,430,437]. However, other studies contend that robust networks are more adaptable [431–434]. Obviously, the relationship between network robustness and network evolvability is complex, because robust network populations harbor a large diversity of neutral genotypes that may be important in adaptation. At present, the study of the evolvability and robustness of biological networks is still in its infancy. In fact, network robustness allows changes in the structure and components of biological networks, due to genetic variations, while still allowing the desired phenotypic trait to be maintained. Network evolvability allows some environmental changes to influence a biological network, so that a new phenotypic trait of the biological network may evolve by natural selection, to adapt to the new environment [435–438].

In the evolutionary process, a population of evolving biological networks can be represented by a nonlinear stochastic system with heritable random genetic variations and non-heritable random environmental disturbances. A phenotypic trait of a biological network can be represented by an equilibrium point of the nonlinear stochastic dynamic system in the fitness landscape (see Fig. 18.1). There are many variations of phenotypes (population) around this equilibrium point, due to intrinsic random genetic variations and random environmental disturbances [435–438]. When a phenotypic trait is perturbed so much that it can transit from the domain of its equilibrium point into the domains of other equilibrium points, a new phenotypic trait is generated and presented. A heritable genetic variation that does not directly help a biological network to evolve a new phenotypic trait is called neutral genetic variations. In the evolutionary process, a robust biological network can harbor a large number of neutral genetic variations, before a phenotypic trait is pushed from one equilibrium point to another one. The capacity to harbor these neutral genetic variations is a measure of the genetic robustness of the biological network in evolution. If the amount of stochastic neutral genetic variations is larger than the genetic robustness of a biological network, so that a phenotypic trait can transit from one equilibrium point to another, a new phenotype is presented to start another period of network evolution around the new equilibrium point. Although it has been shown that many features of biological networks can arise by non-adaptive processes [439], this chapter will offer a different possible understanding of biological network adaptive processes by natural selection in evolution.

In this chapter, the network evolvability in response to stochastic environmental changes and genetic variations of a nonlinear stochastic network can be formulated for the first time as a stochastic minimax H_∞ game problem, because the environmental changes in the evolutionary process are unpredictable. In this situation, the minimization of the worst-case phenotypic deviation for all possible environmental changes is therefore considered as the phenotypic robustness strategy used by a biological network to resist the effect of environmental changes and to maintain its phenotype in the evolutionary process. Therefore, both the network evolvability and the phenotypic robustness of a nonlinear stochastic biological network in the evolutionary process of biological network can be discussed from the nonlinear stochastic game perspective. In other words, the stochastic biological network wants to minimize its network evolvability (or maximize its network robustness) while the stochastic environmental disturbances try to maximize the network evolvability (or minimize the network robustness). The interplay between genetic robustness, environmental robustness, and network robustness in the phenotypic robustness of an evolutionary biological network [440–442] can also be investigated from the point of view of a stochastic minimax H_∞ game.

In this chapter, using the stochastic minimax H_∞ evolutionary game strategy, we find that some genetic variations in the evolutionary process are selected to construct new negative feedback loops to efficiently improve network robustness, thereby maximizing

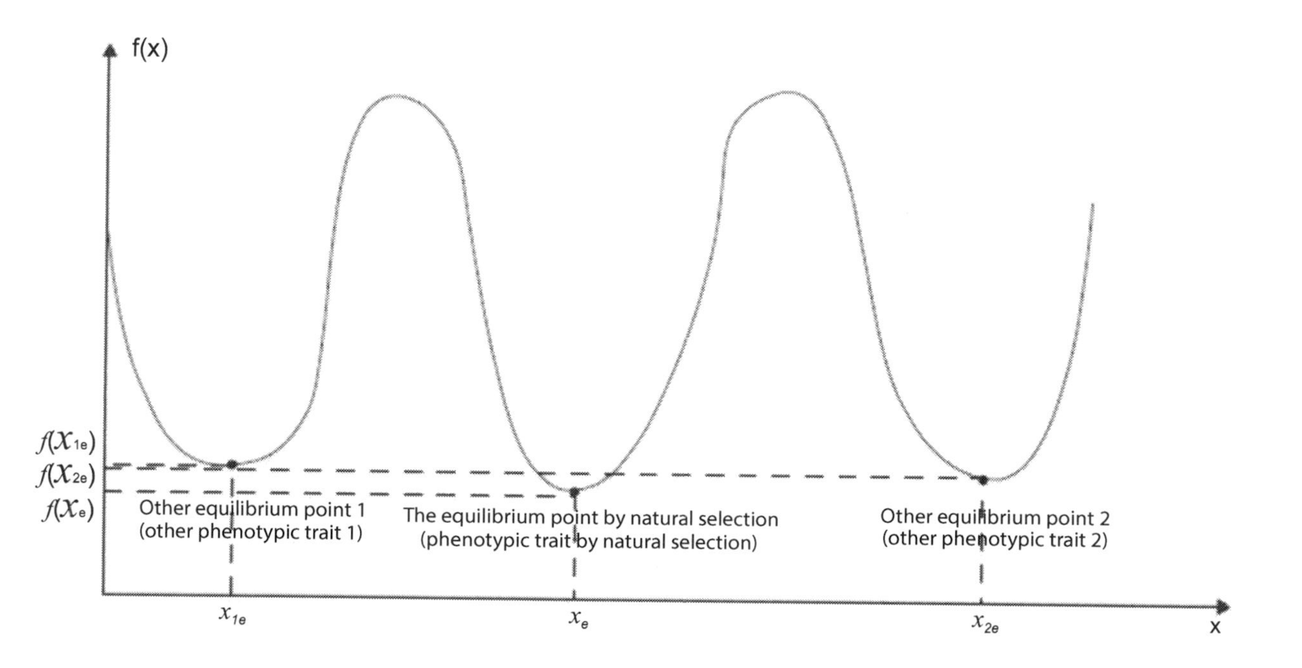

FIGURE 18.1
The equilibrium points of a nonlinear biological network in a phenotypic landscape. The nonlinear stochastic biological network has many local equilibrium points (phenotypic traits). The phenotypic landscape of three stable equilibrium points with a vertical scale illustrating the relative network robustness of the equilibrium points (phenotypic traits) of the nonlinear stochastic biological network in evolution, i.e., the equilibrium point at x_e with deeper and wider basin, is more robust. Here, x_e is the equilibrium point (phenotypic trait) of interest by natural selection.

the fitness (or minimizing the evolution level) of the biological network in evolution. We also find that with the stochastic minimax H_∞ evolutionary game strategy, a biological network has enough network robustness and that the stochastic evolutionary biological network becomes more robust near the equilibrium point. This is the so-called "phenotypic robustness criterion" of biological networks in the evolutionary process. However, when random neutral genetic variations harbored by network robustness are accumulated to the extent of violating the phenotypic robustness criterion, new material is provided for more evolutionary paths to other phenotypes of the gene network by random drift. In the face of a large environmental change, the phenotypic trait of the biological network cannot be maintained and a phenotype transition may eventually occur by a shift to another equilibrium point. In other words, while the stochastic minimax H_∞ evolutionary game strategy can improve the network robustness for the phenotype at the beginning, the accumulated neutral genetic variations harbored by the improved network robustness will finally violate the network phenotypic robustness criterion to break down the network phenotype and make a new phenotypic adaptation in network evolutionary process.

Based on nonlinear stochastic minimax H_∞ game theory, the study of network robustness and network evolvability needs to solve a very difficult Hamilton–Jacobi–Isaac inequality (HJII) [9]. At present, no good way exists to efficiently solve the HJII analytically or numerically. In this chapter, a global linearization method is employed to interpolate several local linear stochastic systems, at different operation points, to approximate the nonlinear stochastic biological network in evolution. In this situation, linear stochastic minimax H_∞ game techniques can be employed to solve the nonlinear stochastic minimax H_∞ evolutionary game problem of the biological network, in which the HJII is replaced by a set of linear matrix inequalities (LMIs). Based on the global linearization technique and linear system theory, the network evolvability and robustness are related to the locations of the eigenvalues of the local linearized stochastic evolutionary biological networks. If the eigenvalues are located in the far left-hand side of the s-complex domain, then the nonlinear stochastic biological network is more robust with the phenotypic trait (or the equilibrium point) and less adaptive to other equilibrium points (or other phenotypic traits) in the evolutionary process. On the other hand, if these eigenvalues of these local linearized biological networks are near the image axis of the s-complex domain, then they are more easily perturbed to the right-hand side of the s-complex domain. In this situation, the phenotypic trait of the nonlinear stochastic evolutionary biological network is less robustly stable but more adaptable to environmental change in the evolutionary process. In order to maintain the phenotype robustness of the evolutionary biological network, the evolutionary game strategy can select some genetic variations to generate adequate negative feedback loops, so that the eigenvalues of the local linearized biological network can be shifted to the far left-hand side of the s-complex domain. Obviously, the H_∞ evolutionary game strategy can efficiently enhance network robustness by developing adequate negative feedback loops to buffer neutral genetic variations and to resist environmental disturbance in the evolutionary process of biological networks.

In this chapter, we investigate quantitative measures of network evolvability and network robustness from a stochastic minimax H_∞ game theory point of view. The phenotype robustness criterion for a nonlinear stochastic evolutionary biological network can be measured by solving an evolutionary minimax H_∞ game strategy based on the LMIs of local linearized stochastic biological networks. That is, if the genetic robustness plus the noise robustness and environmental robustness is less than the network robustness in each local linearized stochastic biological network, then the phenotypic trait of the nonlinear stochastic biological network is robustly stable in the evolutionary process. If not, the biological network might evolve to another phenotypic trait with the release of accumulated cryptic neutral genetic variations and the aid of strong environmental changes. In this chapter, we first reveal the evolutionary strategy for adaptive fitness of a biological network in evolution from a stochastic minimax H_∞ game perspective, and then we investigate the tradeoffs between genetic robustness and environmental robustness and the antagonistic properties between network robustness and network evolvability from a stochastic system theory perspective. If there exist m competitive populations and k environmental changes in the network evolutionary process, then the proposed evolutionary minimax H_∞ game strategy can be easily modified to the n-tuple evolutionary minimax H_∞ game problem of biological network with m players (competitive populations) and k environmental dynamics. Finally, an in silico example is given to illustrate the stochastic minimax H_∞ evolutionary game of stochastic biological networks and to determine their network evolvability and fitness in evolution.

18.2 The Stochastic Minimax H_∞ Evolutionary Game of the Linear Biological Network under Natural Selection

For simplicity of analysis, we first consider a linear biological network as follows:

$$\frac{dx}{dt} = Ax(t), \tag{18.1}$$

where the state vector $x(t) = [x_1(t) \quad x_2(t) \quad \cdots \quad x_n(t)]^T$ denotes the concentrations of n genes in a genetic regulatory network, or protein concentrations of n proteins in a protein–protein interaction network, or the population densities of an ecological network with n different species, and $x_i(t)$ represents the number, density, or biomass of individuals of species i. Although we have assumed that there are n different genes, proteins, or species, these numbers need not be fixed. In fact, one feature of Darwinian dynamics is that the evolutionary process may determine the number n of genes, proteins, or species as a product of the evolutionary biological network. In the evolutionary biological network, the interaction matrix $A \in \mathbb{R}^{n\times n}$ among species is denoted as

$$A = \begin{bmatrix} a_{11} & \cdots & \cdots & a_{1n} \\ \vdots & a_{22} & & \vdots \\ \vdots & & \ddots & \vdots \\ a_{n1} & \cdots & \cdots & a_{nn} \end{bmatrix}, \tag{18.2}$$

where a_{ij} denotes the interaction from species j to species i, i.e., A denotes the dynamic interactions between species in a biological network from the present generation to the next generation.

Remark 18.1 (i) The linear biological network in (18.1) can be considered as a local linearized system of a nonlinear biological network $\frac{dx}{dt} = f(x(t))$ at an equilibrium point (phenotypic trait) x_e of interest by natural selection in Fig. 18.1, i.e., $A = \left.\frac{\partial f(x)}{\partial x}\right|_{x=x_e}$. Without loss of generality, the origin of the nonlinear biological network is always shifted to the equilibrium point (phenotypic trait) x_e, i.e., $x_e = 0$. This detail will be discussed later. (ii) The interaction matrix A between species may suffer from environmental disturbances due to climate and salinity change, and from random genetic variations due to heritable DNA mutation, genetic transition, and recombination in the evolutionary process.

In the evolutionary process, the biological network (18.1) suffers from extrinsic disturbances due to environmental changes as well as intrinsic phenotypic variations due to heritable random genetic variations. In order to maintain the favorable phenotypic trait at $x_e = 0$ by natural selection (i.e., the phenotypic trait robustness in evolution), the biological network needs to select some heritable genetic variations (the evolutionary strategy of the biological network) to improve network robustness so that intrinsic variations and environmental disturbances can be resisted. In this situation, the evolutionary biological network can be represented as follows:

$$\frac{dx(t)}{dt} = \left(A + \sum_{k=1}^{N(t)} A_k P(t - t_k)\right) x(t) + Bv(t), \tag{18.3}$$

where the weighted Poisson point (counting) process [443]

$$s(t) = \sum_{k=1}^{N(t)} A_k P(t - t_k) = \sum_{k=1}^{N(t)} A_k, t \geq 0, \tag{18.4}$$

denotes heritable network interaction variations due to some genetic variations occurring at $t_1, t_2, \ldots, t_p$; $N(t)$ denotes a Poisson point (counting) process of genetic variation to be selected by natural selection at t_k, where $k = 0, 1, \ldots, p$. It should be noted that N(t) is the same as p(t) in (2.1). Here we employ the notations of Poisson processing in [443] for more suitable representation of discrete genetic variation in the evolutionary process. Among these, $s_1(t) \triangleq \sum_{k=1}^{N_1(t)} A_k P(t - t_k)$ are selected by natural selection to change the interaction matrix A, in order to improve network robustness to resist an environmental disturbance $v(t)$ due to the surrounding change and interference from other competitive populations. In addition, $s_2(t) \triangleq \sum_{k=1}^{N_2(t)} A_k P(t - t_k)$ are due to neutral genetic variations, which have no effect on A to directly change a phenotypic trait, i.e., $s(t) = s_1(t) + s_2(t)$ in the evolutionary process (Fig. 18.2). The Poisson counting process $N(t)$ is denoted by

$$N(t) = \sum_{k=1}^{N(t)} P(t - t_k) = \sum_{k=1}^{N(t)} 1 \tag{18.5}$$

with mean λt and variance λt [443]. The matrix $A_k \in \mathbb{R}^{n \times n}$ denotes the random phenotypic variation of the biological network due to the heritable genetic point variation $P(t - t_k)$, occurring at $t = t_k$; $v(t) \in \mathbb{R}^m$ denotes the effect of environmental disturbances; and $B \in \mathbb{R}^n$ denotes the coupling matrix between the environmental disturbance and the biological network,

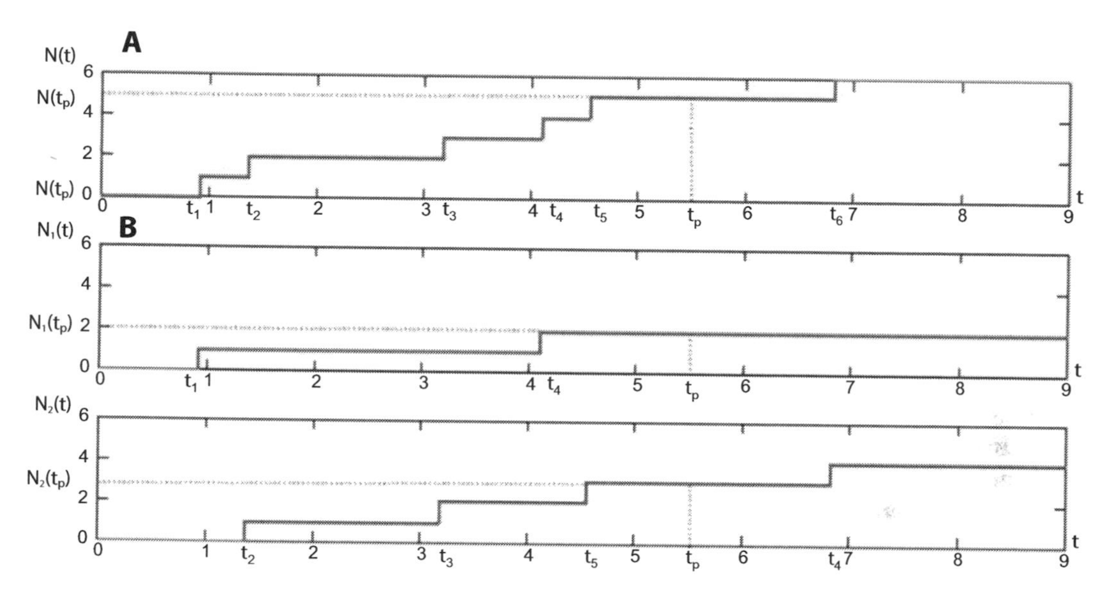

FIGURE 18.2
(A) The Poisson counting process of genetic variations occurring at t_k in the evolutionary process, where t_p denotes the present time. (B) The Poisson counting process $N(t)$ of genetic variations divided into two parts: (top) genetic variations $N_1(t)$ selected by natural selection and (bottom) neutral genetic variations $N_2(t)$ in the evolutionary process.

$$A_k = \begin{bmatrix} a_{k11} & \cdots & \cdots & a_{k1n} \\ \vdots & a_{22} & & \vdots \\ \vdots & & \ddots & \vdots \\ a_{kn1} & \cdots & \cdots & a_{knn} \end{bmatrix}, B = \begin{bmatrix} b_{11} & \cdots & \cdots & b_{1m} \\ \vdots & b_{22} & & \vdots \\ \vdots & & \ddots & \vdots \\ b_{m1} & \cdots & \cdots & b_{mn} \end{bmatrix}, v(t) = \begin{bmatrix} v_1(t) \\ v_2(t) \\ \vdots \\ v_m(t) \end{bmatrix}, \tag{18.6}$$

where $a_{kij} = 0$ if the kth genetic variation has no phenotypic effect on the interaction a_{ij} between species i and species j in the biological network; $b_{ij} = 0$ if the $v_j(t)$ has no effect on the species i.

Remark 18.2 (i) The weighted Poisson point process $s(t) \triangleq \sum_{k=1}^{N(t)} A_k P(t - t_k) = \sum_{k=1}^{N(t)} A_k$, $t \geq 0$, denotes the heritable network interaction variations accumulated by the biological network, some of which are to be selected by natural selection to resist environmental disturbances in evolution. Since the weighted Poisson point process $\sum_{k=1}^{N(t)} A_k P(t - t_k)$ is a random process due to random genetic variations, the stochastic evolutionary network in (18.3) can represent a population of biological networks over all possible Poisson random genetic variations in the evolutionary process. In other words, one possible Poisson genetic variation represents one possible biological network in the network population. (ii) The accumulated phenotypic variations $s(t)$ may lead to permanent changes in the phenotypic trait. (iii) $v(t)$ denotes external disturbances due to environmental changes or stresses, such as temperature or salinity or interferences from other competitive populations. Environmental disturbance can perturb $x(t)$ away from its phenotypic trait at the equilibrium point $x_e = 0$. Since the environmental disturbance $v(t)$ cannot affect the interaction matrix A of the biological network directly, its effect on the biological network is non-heritable. In order to resist the effect of the environmental disturbance, some phenotypic variations $s_1(t) = \sum_{k=1}^{N_1(t)} A_k P(t - t_k)$ are to be selected for the biological network to change the network interaction matrix A by natural selection, to improve the network robustness (i.e., the so-called evolutionary strategy of a biological network for phenotypic trait robustness in the evolutionary process). The remaining genetic variations $s_2(t) = \sum_{k=1}^{N_2(t)} A_k P(t - t_k)$, which have no direct phenotypic effect on the network interaction matrix A, are called neutral genetic variations, i.e., $s(t) = s_1(t) + s_2(t)$ and $N(t) = N_1(t) + N_2(t)$, where the Poisson processes $N_1(t)$ and $N_2(t)$ have means $\lambda_1 t$ and $\lambda_2 t$, respectively, and $\lambda = \lambda_1 + \lambda_2$ (see Fig. 18.2(b)).

Therefore, the stochastic gene network under environmental disturbances and natural selection in (18.3) can be represented as the following Poisson point process:

$$
\begin{aligned}
\frac{dx(t)}{dt} &= Ax(t) + \sum_{k=1}^{N(t)} A_k P(t - t_k)x(t) + Bv(t) \\
&= Ax(t) + s_1(t)x(t) + s_2(t)x(t) + Bv(t) \\
&= Ax(t) + u(t) + s_2(t)x(t) + Bv(t),
\end{aligned} \tag{18.7}
$$

where $u(t) = s_1(t)x(t) = \sum_{k=1}^{N_1(t)} A_k P(t - t_k)x(t)$ denotes the evolutionary strategy of the biological network to select adequate phenotypic variations by natural selection to modify the network interaction matrix A, thus improving network robustness to tolerate the neutral genetic variations and resist the environmental disturbance, maintaining the phenotypic trait in the evolutionary process. The main evolutionary problem of the stochastic biological network in (18.7) for natural selection is how to determine an evolutionary strategy $u(t) = s_1(t)x(t) = \sum_{k=1}^{N_1(t)} A_k P(t - t_k)x(t)$, i.e., how to select some adequate genetic variations to enhance system matrix A to improve the network robustness of biological network. This enables the biological network to effectively tolerate the accumulated neutral genetic variations $s_2(t) = \sum_{k=1}^{N_2(t)} A_k P(t - t_k)$ and efficiently resist the environmental disturbance $v(t)$ to achieve the phenotypic trait robustness of biological network in the evolutionary process. Since environmental disturbances are stochastic and unpredictable, the phenotypic robustness strategy $u(t)$ of the biological network to resist the environmental disturbances is based on the worst-case effect of all possible environmental disturbances. Let us denote the set of all possible bounded environmental disturbances as $L_2[0, t_p]$, i.e., $v(t) \in L_2[0, t_p]$ if $\int_0^{t_p} v(t)^T v(t)dt < \infty$, where t_p denotes the present time, i.e., $L_2[0, t_p]$ denotes the set of all possible bounded environmental disturbances to the present time. (Unbounded environmental disturbances are not considered here because they will lead to the extinction of all biological networks.)

Remark 18.3 (i) The stochastic biological system in (18.7) can be considered a Darwinian dynamic to describe a linear dynamic biological network underlying natural selection. Darwinian dynamics are biological systems of dynamical equations that satisfy Darwin's conditions of variability, heritability, and the strategy to survive [444]. (ii) If the stochastic biological network in (18.7) is robustly stable at $x_e = 0$ (if $x_e \neq 0$, then the origin of the linear dynamic network in (18.7) must be shifted to x_e for the simplicity of analysis, i.e., the phenotypic trait survives at $x_e = 0$), then the evolutionary strategy $u(t) = s_1(t)x(t)$ is a robust evolutionary strategy of biological network by natural selection. The Poisson genetic variations $\sum_{k=1}^{N_2(t)} A_k P(t - t_k)x(t)$ are not selected by natural selection, and are called neutral genetic variations without direct phenotypic effect on the system matrix A of the genetic network. These variations are further accumulated and harbored in the deeper basin around the phenotypic trait x_e of the robust biological network (see Fig. 18.1).

Let us denote the phenotypic deviation around x_e as $\tilde{x}(t) = x(t) - x_e$. Then, the evolution level of the stochastic biological network due to the effect of all possible bounded environmental disturbances in the evolutionary process is defined as

$$e = \max_{v(t)\in L_2[0,t_p]} \frac{E\int_0^{t_p}(\tilde{x}(t)^T Q\tilde{x}(t) + u(t)^T u(t))dt}{E\tilde{x}(0)^T\tilde{x}(0) + E\int_0^{t_p} v(t)^T v(t)dt}, \tag{18.8}$$

where $E(\cdot)$ denotes the expectation of $(\cdot)$, Q is a positive definite weighting matrix, and t_p denotes the present time. The biological meaning of (18.8) is that the worst-case (maximum) effect of all possible bounded but unpredictable environmental disturbances on both phenotypic deviation around x_e and the evolutionary strategy $u(t)$ is considered as the evolution level of the biological network on the basis of the average energy. Since the initial phenotypic deviation $\tilde{x}(0)$ also has an effect on $\tilde{x}(t)$ and $u(t)$, its effect on the evolution level is considered in the denominator of (18.8). If $x(0) = x_e$ or $\tilde{x}(0) = 0$, i.e., the initial phenotypic trait $x(0)$ is at the equilibrium point x_e, then the term $E\tilde{x}(0)^T\tilde{x}(0)$ in the denominator should be neglected.

The fitness (phenotypic robustness) of the biological network around the phenotypic trait at the equilibrium point x_e is inversely correlated to the evolution level of the biological network and is defined as

$$f = \frac{1}{e}. \tag{18.9}$$

A biological network with a low evolution level will have a high fitness to the phenotypic trait, and vice versa. Therefore, the fitness maximization of a biological network to the phenotypic trait x_e is equivalent to the minimization of the evolution level as follows:

$$\max f = \frac{1}{\min e}. \tag{18.10}$$

Combining (18.8) and (18.10), the evolutionary strategy $u(t)$ for the maximization of the fitness of the biological network to the phenotypic trait x_e is formulated for the first time as the following stochastic minimax H_∞ evolution (H_∞ evolutionary game) problem:

$$e_0 \triangleq \min_{u(t)} e = \min_{u(t)} \max_{v(t)\in L_2[0,t_p]} \frac{E\int_0^{t_p}(\tilde{x}(t)^T Q\tilde{x}(t) + u(t)^T u(t))dt}{E\tilde{x}(0)^T\tilde{x}(0) + E\int_0^{t_p} v(t)^T v(t)dt}. \tag{18.11}$$

The minimum evolution level e_0 in (18.11) is called the network evolvability of the biological network in the evolutionary process. The physical meaning of network evolvability is the minimum worst-case effect of environmental disturbances on the biological network with a minimax stochastic H_∞ evolutionary game strategy $u(t)$. The minimax stochastic H_∞ evolutionary game strategy is performed for the stochastic biological network by natural selection to resist intrinsic neutral genetic variations and unpredictable environmental disturbances from the point of view of the stochastic minimax H_∞ game in the evolutionary process. The minimax stochastic H_∞ evolution performance in (18.11) has been used for robust control design under intrinsic perturbations and external disturbances [51,275]. Therefore, the maximum phenotypic robustness (or fitness maximization) problem of the evolutionary biological network in (18.7) can be formulated as a stochastic H_∞ evolutionary game problem in (18.11), i.e., how to select $u(t) = \sum_{k=1}^{N_1(t)} A_k P(t - t_k)x(t) = s_1(t)x(t)$ to achieve the fitness maximization in (18.10) for the evolutionary biological network in (18.7) with the phenotype at x_e. The evolutionary strategy of a biological network to maintain its phenotypic trait against intrinsic genetic fluctuations and environmental disturbances is to develop a feasible evolutionary strategy $u(t) = s_1(t)x(t)$ to enhance the biological network with the maximum fitness at x_e, or equivalently, to select some genetic variations $s_1(t)$ to modify the interaction matrix A of the biological network to minimize the worst-case (maximum) effect of environmental disturbance $v(t)$ on both the variation of the phenotypic trait $\tilde{x}(t)$ and the H_∞ evolutionary game strategy $u(t)$ in (18.11) on the basis of mean energy.

Why is $u(t)$ also considered in the numerator of (18.11) or (18.8)? The reason is that the biological network aims to achieve the maximum fitness f_0 (or minimum evolution level e_0) with a minimum effort (or cost). This effort or cost can be described as a change $\sum_{k=1}^{N_1(t)} A_k P(t - t_k)$ on system matrix A of the biological network. There are two players, $u(t)$ and $v(t)$, in the evolutionary game problem in (18.11). In the minimax H_∞ evolutionary game problem in (18.11), the player $v(t)$ will try to maximize the effect of environmental disturbances on the deviation of the phenotype from its equilibrium point x_e, and the other player $u(t)$ will try to select some genetic variations to improve network robustness in order to maintain the phenotype around x_e by minimizing the worst-case phenotypic deviation due to all possible environmental disturbances. In general, it is very difficult to solve the stochastic H_∞ evolutionary game problem in (18.11) for the stochastic gene network in (18.7) directly. Therefore, the following indirect method is employed to solve the stochastic minimax H_∞ evolutionary game problem [9]:

$$e_0 = \min_{u(t)} \max_{v(t) \in L_2[0,t_p]} \frac{E\int_0^{t_p} (\tilde{x}(t)^T Q\tilde{x}(t) + u(t)^T u(t))dt}{E\tilde{x}(0)^T \tilde{x}(0) + E\int_0^{t_p} v(t)^T v(t)dt} \leq e, \tag{18.12}$$

where the positive value e denotes the upper bound of the evolvability e_0. The upper bound e of e_0 will be given beforehand and then decreased to as small a value as possible to approach e_0 and attain the real minimax H_∞ game problem, which is the so-called suboptimal approach to the stochastic evolutionary game problem in (18.11). The suboptimal stochastic H_∞ evolutionary game in (18.12) is equivalent to the following constrained stochastic Nash quadratic (H_2) game because $u(t)$ is independent on $v(t)$ [6,9,15]

$$\min_{u(t)} \max_{v(t) \in L_2[0,t_p]} E\int_0^{t_p} (\tilde{x}(t)^T Q\tilde{x}(t) + u(t)^T u(t) - ev(t)^T v(t))dt \leq eE\tilde{x}(0)^T \tilde{x}(0). \tag{18.13}$$

Let us denote

$$J \triangleq E\int_0^{t_p} (\tilde{x}(t)^T Q\tilde{x}(t) + u(t)^T u(t) - ev(t)^T v(t))dt. \tag{18.14}$$

We then need two steps to solve the constrained stochastic Nash H_2 (quadratic) game problem in (18.13). The first step is to solve the following stochastic Nash H_2 (quadratic) game problem:

$$J_0 \triangleq \min_{u(t)} \max_{v(t)\in L_2[0,t_p]} J = \min_{u(t)} \max_{v(t)} E\int_0^{t_p} (\tilde{x}(t)^T Q\tilde{x}(t) + u(t)^T u(t) - ev(t)^T v(t))dt, \tag{18.15}$$

and the second step is to solve the following constraint problem:

$$J_0 \leq E\tilde{x}(0)^T \tilde{x}(0). \tag{18.16}$$

By solving (18.15) and (18.16) for the constrained stochastic Nash H_2 game in (18.13), we get the following result.

Proposition 18.1 The stochastic evolutionary H_2 quadratic game problem of the evolutionary gene network in (18.13) for the sub-optimal H_∞ evolutionary game strategy can be solved by the following evolutionary strategy $u^*(t)$ and the worst-case environmental disturbance $v^*(t)$:

$$u^*(t) = s^*(t)x = \sum\nolimits_{k=1}^{N_1(t)} A_k P(t - t_k)x = -Px, v^*(t) = \frac{1}{e}B^T Px, \tag{18.17}$$

where the positive definite matrix P is the solution of the following Riccati-like inequality:

$$A^T P + PA - PP + \frac{1}{e}PBB^T P + Q + \lambda_2 \sum\nolimits_{k=1}^{N_{2p}} A_k^T PA_k \leq 0 \text{ and} \tag{18.18}$$

$$0 < P \leq eI \tag{18.19}$$

where $N_{2p} \triangleq N_2(t_p)$ represents the total number of neutral genetic variations to the present time t_p.

Proof. See Appendix A.

The term $\lambda_2 \sum_{k=1}^{N_{2p}} A_k^T PA_k$ is the effect of random genetic variations that are not selected by evolutionary strategy, i.e., the effect of neutral genetic mutations on the gene network in the evolutionary process. The solution of the evolutionary quadratic game in (18.17) is that based on the worst-case environmental disturbance $v^*(t)$ on the biological network (due to the uncertainty and unpredictability of environmental disturbances, the evolutionary strategy $u^*(t)$ needs to consider the effect of the worst-case disturbance $v^*(t)$ from the Nash quadratic game's perspective), thus the minimax quadratic game strategy is to select adequate genetic variations as new network interactions amounting to $s^*(t) = \sum_{k-1}^{N_1(t)} A_k P(t - t_k) = \sum_{k-1}^{N_1(t)} A_k = -P$ to enhance the system matrix A, which can minimize the evolution level or maximize the fitness of the biological network with the phenotype at x_e.

Since e is the upper bound of the evolvability e_0 in the stochastic minimax H_∞ evolutionary game in (18.12), the minimum value of e in (18.12) will approach the evolvability $e_0 = \frac{1}{f_0}$, where f_0 is the maximum fitness of the biological network by the minimax H_∞ evolutionary strategy. Therefore, we need to solve the following constrained optimization problem for e_0 or f_0 of the stochastic minimax H_∞ evolutionary game strategy in (18.12) in the stochastic biological network in evolution

$$\frac{1}{f_0} = e_0 \stackrel{\Delta}{=} \min_{P>0} e \quad \text{subject to (18.18) and (18.19).} \tag{18.20}$$

Remark 18.4. (i) The solution of e_0 in (18.20) can be solved by decreasing e until no positive solution $P>0$ exists in (18.18) and within $0<P\leq eI$ in (18.19). (ii) After solving e_0 and f_0 from (18.20), then the Riccati-like inequality in (18.18) becomes

$$A^TP + PA - PP + \frac{1}{e_0}PBB^TP + Q + \lambda_2 \sum_{k=1}^{N_{2p}} A_k^TPA_k \leq 0. \tag{18.21}$$

The above Riccati-like inequality can be considered to be the phenotypic trait robustness criterion for the stochastic biological network (18.7) under the minimax H_∞ evolutionary game strategy in the evolutionary process. For the nominal linear biological network in (18.1), if $A^TP + PA \leq 0$ for some $P>0$, then the biological network is stable based on Lyapunov stability, i.e., all the eigenvalues of A have to be in the left-hand side of the complex s-domain (see Fig. 18.3). Therefore, $-(A^TP + PA)$ is a measure of network robustness. The term $\lambda_2 \sum_{k=1}^{N_{2p}} A_k^TPA_k$ is the effect $\sum_{k=1}^{N_2(t)} A_kp(t-t_k)$ on the phenotypic trait robustness of the biological network. The term $\frac{1}{e_0}PBB^TP + Q$ is the effect of an environmental disturbance on the phenotypic trait robustness and a measure of the

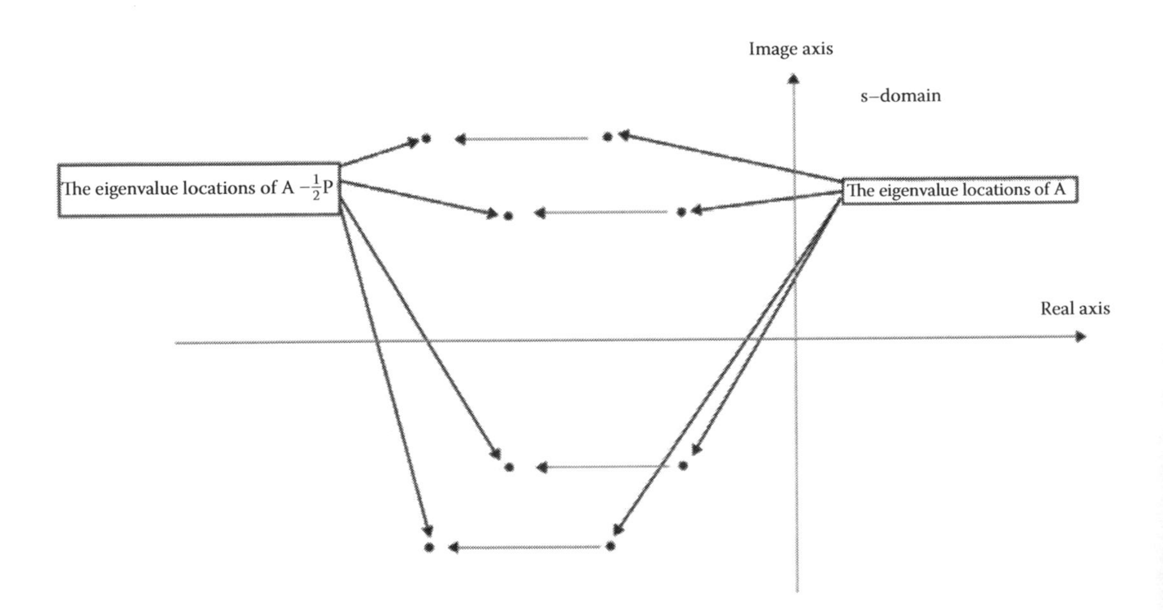

FIGURE 18.3
The eigenvalues of system matrix A in the evolution network are shifted further to the left-hand side of the s-domain by the minimax H_∞ evolutionary game strategy $u^*(t) = -Px$, i.e., the eigenvalues of $A - \frac{1}{2}P$ are farther left than those of A.

environmental robustness of the biological network. If the evolvability e_0 is small, then the environmental robustness of the biological network should be large enough to resist the phenotypic effect of environmental changes and vice versa. The term $-PP$ due to the evolutionary game strategy $u^*(t) = -Px$ (i.e., $s^*(t) = \sum_{k=1}^{N_1(t)} A_k P(t - t_k) = -P$) could improve the phenotypic trait robustness in (18.21) by enhancing network robustness through negative feedback loops $-Px$ as follows:

$$\left(A - \frac{1}{2}P\right)^T P + P\left(A - \frac{1}{2}P\right) + Q + \lambda_2 \sum_{k=1}^{N_{2p}} \left(A_k^T P + PA_k + A_k^T PA_k\right) + f_0 PBB^T P \leq 0. \tag{18.22}$$

The eigenvalues of $\left(A - \frac{1}{2}P\right)$ are farther to the left of the complex s-domain than those of A, such that the network robustness of the biological network is improved (see Fig. 18.3). Although the selected phenotypic variations up to the amount $-Px$ caused by the minimax H_∞ evolutionary strategy can improve the network robustness, the accumulated random neutral genetic variations $\sum_{k=1}^{N_2(t)} A_k P(t - t_k)$ harbored in the robust genetic network during the evolutionary process can also lead to a negative effect on the phenotypic trait robustness of the biological network (i.e., the term $\lambda_2 \sum_{k=1}^{N_{2p}} A_k^T PA_k$ will make the network robustness inequality in (18.22) more difficult to satisfy). In this situation, the phenotypic trait robustness criterion in (18.22) for the stochastic biological network in (18.3) under the minimax (H_∞) evolutionary game strategy in (11) can be reformulated as follows:

$$\underbrace{\lambda_2 \sum_{k=1}^{N_{2p}} \left(A_k^T P + PA_k + A_k^T PA_k\right)}_{\text{genetic robustness}} + \underbrace{Q + \frac{1}{e_0} PBB^T P_0}_{\text{environmental robustness}} \leq \underbrace{-\left[\left(A - \frac{1}{2}P\right)^T P + P\left(A - \frac{1}{2}P\right)\right]}_{\text{enhanced network robustness through the minimax } H_\infty \text{ evolutionary game strategy}}. \tag{18.23}$$

The physical meaning of (18.23) is that if the enhanced network robustness of the biological network through the stochastic minimax H_∞ evolutionary game strategy in (18.11) can confer both genetic robustness to tolerate random Poisson phenotypic variations due to the accumulated neutral genetic variations and environmental robustness to resist environmental disturbance, then the phenotypic trait of the biological network is maintained in the evolutionary process. Otherwise, the phenotypic trait may not be maintained under environmental disturbance and random genetic variations. In de Visser et al. [445], the evolutionary causes of genetic robustness were discussed using different evolutionary criteria. The correlation with environmental robustness is considered to be the most probable cause of genetic robustness in evolution. From (18.23), based on the congruence scenario, it can be seen that the genetic robustness of a biological network is a by-product of the environmental robustness of the biological network. This is because environmental changes are more frequent than genetic variations in evolution, i.e., when a biological network has developed network robustness in (18.23) to resist environmental disturbances, it would simultaneously provide genetic robustness to tolerate genetic variations. Therefore, the perspective of de Visser et al. in [445] could be verified by (18.23).

From the phenotypic trait robust criterion in (18.22), it is seen that if the system coupling matrix B is small, then the phenotypic trait is more robust. That is why the biological systems are always with membrane or semi-transparent membrane to isolate them from the environment and only with some channels or receptors to communicate with environment.

(iii) By the stochastic minimax (H_∞) evolutionary game strategy in (18.11), the total accumulated network interaction variations $\sum_{k=1}^{N_1(t)} A_k P(t-t_k) = \sum_{k=1}^{N_1(t)} A_k$ are chosen by natural selection to the amount of $-P$ to construct negative feedback loops $-Px(t)$ in (18.17). The stochastic evolutionary biological network is thus of the following form:

$$\frac{dx(t)}{dt} = (A-P)x(t) + \sum_{k=1}^{N_2(t)} A_k P(t-t_k)x(t) + Bv(t). \tag{18.24}$$

Since the eigenvalues of $A-P$ are in the farther left-hand side of the complex s-domain, the biological network has been made more robust by the minimax H_∞ evolutionary game strategy to resist all possible environmental disturbances in (18.24). (iv) An evolutionary biological network in (18.24) with greater network robustness in (18.23) will provide a greater buffer to harbor numerous neutral genetic variations simultaneously in the evolutionary process [428,429,431] (these neutral genetic variations $\sum_{k=1}^{N_2(t)} A_k P(t-t_k)$ have no obvious phenotypic effect on the system matrix A of the biological network). However, a large amount of neutral genetic variations may provide raw material for new evolutionary possibilities. Therefore, a robust biological network may finally lead to network evolution in the long run when the accumulated neutral genetic variations $\sum_{k=1}^{N_2(t)} A_k P(t-t_k)$ are large enough, i.e., the term $\lambda_2 \sum_{k=1}^{N_{2p}} A_k^T P A_k$ in (18.22) or (18.23) is large enough that the phenotypic trait robustness criterion is violated. (v) In order to provide a buffer against the accumulated neutral genetic variations and all possible environmental disturbances, a biological network needs to develop negative feedback loops to improve its network robustness. This is the minimax H_∞ evolutionary game strategy $u^*(t) = -Px$, providing negative feedback loops to enhance network robustness in (18.22) or (18.23). Furthermore, hubs and redundant mechanisms are frequently found in biological networks to provide efficient responses to external stimuli and to attenuate the effect of genetic variations, respectively. They are therefore also favored by natural selection to improve the phenotypic trait robustness of the biological network in the evolutionary process. The minimax H_∞ evolutionary game strategy through negative feedback interactions $u^*(t) = -Px$ in a genetic network might result in these hubs and redundant mechanisms, in order to improve phenotypic trait robustness in the evolutionary process.

18.3 The Stochastic Minimax H_∞ Evolutionary Game Strategy of Nonlinear Biological Network

In real biological networks, the system dynamics in (18.1) are always nonlinear and should be modified as follows:

$$\frac{dx(t)}{dt} = f(x), \tag{18.25}$$

where $f(x)$ denotes the nonlinear interactive vector among n different species in the biological network. For a nonlinear stochastic biological network with stochastic heritable random parameter fluctuations and non-heritable environmental disturbances in the evolutionary process, the stochastic biological network in (18.7) is modified as in the following Poisson point process:

$$\begin{aligned}\frac{dx(t)}{dt} &= f(x) + \sum_{k=1}^{N(t)} f_k(x)p(t - t_k) + Bv(t)\\ &= f(x) + \sum_{k=1}^{N_1(t)} f_k(x)p(t - t_k) + \sum_{k=1}^{N_2(t)} f_k(x)p(t - t_k) + Bv(t)\\ &= f(x) + u(t) + \sum_{k=1}^{N_2(t)} f_k(x)p(t - t_k) + Bv(t)\end{aligned} \tag{18.26}$$

with

$$\sum_{k=1}^{N(t)} f_k(x)p(t - t_k) = \sum_{k=1}^{N_1(t)} f_k(x)p(t - t_k) + \sum_{k=1}^{N_2(t)} f_k(x)p(t - t_k),$$

where $f_k(x)$, for $k = 1, 2, \ldots, N(t)$, denotes the nonlinear stochastic heritable phenotypic variations due to Poisson random genetic variations in the evolutionary process. Sources of these random genetic variations include DNA mutation, deletion, recombination, duplication, inversion, and translocation of chromosomes in the evolutionary process. The nonlinear stochastic system in (18.26) represents a nonlinear biological network under the evolutionary strategy $u(t) = \sum_{k=1}^{N_1(t)} f_k(x)p(t - t_k)$ by natural selection, the neutral genetic variations $\sum_{k=1}^{N_2(t)} f_k(x)p(t - t_k)$, and the environmental disturbances $v(t)$ in the evolutionary process.

For the nonlinear stochastic biological network in (18.26), there exist many equilibrium points (phenotypic traits) (see Fig. 18.1). Suppose that the phenotypic trait near a stable equilibrium point x_e is of interest in natural selection. For convenience of analysis, the origin of the nonlinear stochastic biological network in (18.26) is shifted to the equilibrium point (phenotypic trait) x_e. In this case, if the shifted nonlinear stochastic network is robustly stable at the origin, then the equilibrium point (phenotypic trait) of interest in natural selection is also robustly stable. Therefore, the analysis of the minimax H_∞ evolutionary strategy in (18.11) of the nonlinear stochastic biological network in evolution is simplified. Let us denote the phenotypic deviation around the phenotype at x_e as $\tilde{x}(t) = x(t) - x_e$, so then the following shifted nonlinear stochastic biological network is obtained:

$$\begin{aligned}\frac{d\tilde{x}(t)}{dt} &= f(\tilde{x}) + Bv(t) + \sum_{k=1}^{N(t)} f_k(\tilde{x})p(t - t_k)\\ &= f(\tilde{x}) + Bv(t) + u(t) + \sum_{k=1}^{N_2(t)} f_k(\tilde{x})p(t - t_k).\end{aligned} \tag{18.27}$$

For convenience of notation, we denote $f(\tilde{x}+x_e)$ and $f_k(\tilde{x}+x_e)$ as $f(\tilde{x})$ and $f_k(\tilde{x})$, respectively. In this situation, the origin $\tilde{x}(t)=0$ of the nonlinear stochastic biological network in (18.27) is at the equilibrium point x_e of the original nonlinear stochastic network in (18.26), and $u(t)=s_1(\tilde{x})=\sum_{k=1}^{N_1(t)} f_k(\tilde{x})p(t-t_k)$ denotes the minimax H_∞ evolutionary strategy of the biological network to select some phenotypic variations by natural selection to improve the network robustness, and thus tolerate the effect of neutral genetic variations $\sum_{k=1}^{N_2(t)} f_k(\tilde{x})p(t-t_k)$, as well as to resist the environmental disturbances $v(t)$. To maintain the robustness of the phenotypic trait at x_e, which is favored by natural selection, the robust evolution of the nonlinear stochastic biological network is to develop an evolutionary strategy $u(t)$ to select some genetic variations to solve the stochastic minimax H_∞ evolutionary game problem in (18.11) or (18.12). By considering the nonlinear stochastic H_∞ minimax evolution problem (18.11) of the nonlinear stochastic biological network in (18.27), we get the following result.

Proposition 18.2 For the nonlinear stochastic biological network in (18.27), the nonlinear stochastic Nash H_2 (quadratic) evolutionary game problem in (18.13) (the suboptimal minimax H_∞ evolutionary game problem in (8.12)) is solved by the following minimax H_∞ evolutionary game strategy $u^*(t)$ and the worst-case environmental disturbance $v^*(t)$

$$u^*(t)=s^*(\tilde{x})=\sum_{k=1}^{N_1(t)} f_k(\tilde{x})p(t-t_k)=-\frac{1}{2}\left(\frac{\partial V(\tilde{x})}{\partial \tilde{x}}\right), v^*(t)=\frac{1}{2e}B^T\left(\frac{\partial V(\tilde{x})}{\partial \tilde{x}}\right), \tag{18.28}$$

where the positive function $V(\tilde{x})>0$ is the positive solution of the following (HJII):

$$\begin{aligned}&\left(\frac{\partial V(\tilde{x})}{\partial \tilde{x}}\right)^T f(\tilde{x})+\tilde{x}^TQ\tilde{x}-\frac{1}{4}\left(\frac{\partial V(\tilde{x})}{\partial \tilde{x}}\right)^T\left(\frac{\partial V(\tilde{x})}{\partial \tilde{x}}\right)\\&+\frac{1}{4e}\left(\frac{\partial V(\tilde{x})}{\partial \tilde{x}}\right)^T BB^T\left(\frac{\partial V(\tilde{x})}{\partial \tilde{x}}\right)+\frac{1}{2}\lambda_2\sum_{k=1}^{N_{2p}}\left[V(\tilde{x}+f_k(\tilde{x}))-V(\tilde{x})\right]\le 0\end{aligned} \tag{18.29}$$

with

$$V(\tilde{x}(0))\le e_0\tilde{x}(0)^T\tilde{x}(0). \tag{18.30}$$

Proof: See Appendix B.

The last term in (18.29) is the effect of random neutral genetic variations, which are not selected by natural selection but are harbored by network robustness in the evolutionary process [437,438]. The biological meaning of the Nash H_2 (quadratic) evolutionary game strategy in (18.28) is that the worst-case environmental disturbance $v^*(t)$ represents the worst-case effect of all possible uncertain environmental disturbances on both the phenotypic deviation and the evolutionary strategy of the biological network in the stochastic game (18.12). In addition, the Nash H_2 (quadratic) evolutionary game strategy $u^*(t)$ in (18.28) denotes that, if the number of phenotypic variations selected by natural selection is equal to the number of negative feedback loops $-\frac{1}{2}\left(\frac{\partial V(\tilde{x})}{\partial \tilde{x}}\right)$, then the worst-case effect of all possible environmental disturbances on the phenotypic trait will be minimized; i.e., the phenotypic trait of the biological network will achieve the maximum fitness or the minimum evolution of x_e in (18.12) or (18.20). Since e is the upper bound of the evolvability e_0 of the minimax H_∞ stochastic evolutionary network in (18.12), i.e. $u^*(t)$ in (18.28) is only the sub-optimal solution of minimax H_∞ evolutionary game, e_0 could be approached by the minimax H_∞ game strategy in (18.12) for the nonlinear stochastic evolutionary network in (18.26) by minimizing its upper bound e, i.e. we could solve the minimax stochastic H_∞ evolutionary game strategy problem in (18.11) as follows:

$$\frac{1}{f_0} = e_0 = \min_{V(\tilde{x})>0} e \quad \text{subject to (18.29) and (18.30).} \tag{18.31}$$

Remark 18.5 (i) In general, the solution of the network evolvability e_0 in (18.31) could be obtained by decreasing e until the HJII in (18.29) has no positive solution $V(\tilde{x})>0$. (ii) After solving f_0 and e_0 for the minimax stochastic H_∞ evolutionary game strategy in (18.31), then the HJII in (18.29) can be modified as

$$\begin{aligned}&\left(\frac{\partial V(\tilde{x})}{\partial \tilde{x}}\right)^T f(\tilde{x}) + \tilde{x}^T Q\tilde{x} - \frac{1}{4}\left(\frac{\partial V(\tilde{x})}{\partial \tilde{x}}\right)^T\left(\frac{\partial V(\tilde{x})}{\partial \tilde{x}}\right)\\&+\frac{1}{4e_0}\left(\frac{\partial V(\tilde{x})}{\partial \tilde{x}}\right)^T BB^T\left(\frac{\partial V(\tilde{x})}{\partial \tilde{x}}\right) + \frac{1}{2}\lambda_2\sum_{k=1}^{N_{2p}}[V(\tilde{x}+f_k(\tilde{x})) - V(\tilde{x})] \le 0\end{aligned} \tag{18.32}$$

with $V(\tilde{x}(0)) \le e_0\tilde{x}(0)^T\tilde{x}(0)$. This HJII can be considered as the phenotypic trait robustness criterion of the nonlinear stochastic biological network in (18.27) with the minimax H_∞ evolutionary game strategy $u^*(t)$ in the evolutionary process. The term $\frac{1}{4e_0}\left(\frac{\partial V(\tilde{x})}{\partial \tilde{x}}\right)^T BB^T\left(\frac{\partial V(\tilde{x})}{\partial \tilde{x}}\right)$ in (18.32) is due to the effect of environmental disturbances on the biological network, and is considered as a measure of environmental robustness. The term $\frac{1}{2}\lambda_2\sum_{k=1}^{N_{2p}} f_k^T(\tilde{x})\frac{\partial^2 V(\tilde{x})}{\partial \tilde{x}^2}f_k(\tilde{x})$ in (18.32) is the random effect of the accumulated random neutral genetic variations on the biological network in the evolutionary process. The term $-\frac{1}{4}\left(\frac{\partial V(\tilde{x})}{\partial \tilde{x}}\right)^T\left(\frac{\partial V(\tilde{x})}{\partial \tilde{x}}\right)$ in (18.32) is due to the minimax stochastic H_∞ evolutionary game strategy $u^*(t)$, i.e., the robust effect of phenotypic variations $\sum_{k=1}^{N_1(t)} f_k(\tilde{x})p(t-t_k)$, which are selected by natural selection to construct the negative feedback loops $-\frac{1}{2}\left(\frac{\partial V(\tilde{x})}{\partial \tilde{x}}\right)$ to minimize the worst-case effect of all possible environmental disturbances. When the biological network is free of genetic fluctuations, environmental disturbances, and the absence of evolutionary game strategy as in (18.25), i.e., in the nominal nonlinear network case in (18.25), the Lyapunov stability criterion at the equilibrium x_e is $\left(\frac{\partial V(\tilde{x})}{\partial \tilde{x}}\right)^T f(\tilde{x}) \le 0$, and $\left(\frac{\partial V(\tilde{x})}{\partial \tilde{x}}\right)^T f(\tilde{x})$ can be considered as the measure of phenotypic stability of x_e for the nominal biological network in (18.25). With the minimax stochastic H_∞ evolutionary game strategy $u^*(t)$ to improve the network robust stability of the biological network in (18.26) and to maintain the phenotype at x_e in order to resist the effects of genetic variations and environmental disturbances, the phenotypic trait robustness criterion of the evolutionary biological network in (18.32) can be rewritten as follows:

$$\underbrace{\frac{1}{2}\lambda_2 \sum_{k=1}^{N_{2p}} [V(\tilde{x}+f_k(\tilde{x})) - V(\tilde{x})]}_{\text{genetic robustness}} + \underbrace{\tilde{x}^T Q \tilde{x} + \frac{1}{4e_0}\left(\frac{\partial V(\tilde{x})}{\partial \tilde{x}}\right)^T BB^T \left(\frac{\partial V(\tilde{x})}{\partial \tilde{x}}\right)}_{\text{environmental robustness}} \\ \leq \underbrace{-\left(\frac{\partial V(\tilde{x})}{\partial \tilde{x}}\right)^T \left(f(\tilde{x}) - \frac{1}{4}\left(\frac{\partial V(\tilde{x})}{\partial \tilde{x}}\right)\right)}_{\text{network robustness through the minimax H}_\infty\text{ evolutionary game strategy}} \tag{18.33}$$

From the right-hand side of (18.33), it is apparent that the minimax evolutionary H_∞ game strategy can improve the network robustness in the evolutionary process by making $\left(\frac{\partial V(\tilde{x})}{\partial \tilde{x}}\right)^T \left(f(\tilde{x}) - \frac{1}{4}\left(\frac{\partial V(\tilde{x})}{\partial \tilde{x}}\right)\right)$ more negative, letting the basin of the phenotypic trait at x_e become much deeper and wider in Fig. 18.1. The physical meaning of the phenotypic trait robustness criterion in (18.33) is that if the enhanced network robustness by the minimax evolutionary game strategy can confer enough genetic robustness to buffer random neutral genetic variations and enough environmental robustness to resist environmental disturbances, then the phenotypic trait of the nonlinear stochastic biological network in (18.27) is maintained in the evolutionary process. By substituting the minimax stochastic H_∞ evolutionary game strategy $u^*(t)$ into the nonlinear stochastic biological network in (18.27), we get

$$\frac{d\tilde{x}(t)}{dt} = \left(f(\tilde{x}) - \frac{1}{2}\frac{\partial V(\tilde{x})}{\partial \tilde{x}}\right) + Bv(t) + \sum_{k=1}^{N_2(t)} f_k(\tilde{x})p(t-t_k). \tag{18.34}$$

Since the term $f(\tilde{x}) - \frac{1}{2}\frac{\partial V(\tilde{x})}{\partial \tilde{x}}$ will make the nonlinear stochastic biological network more robustly stable than $f(\tilde{x})$ due to the negative feedback $-\frac{1}{2}\frac{\partial V(\tilde{x})}{\partial \tilde{x}}$, the biological network with the minimax stochastic H_∞ evolutionary strategy will be more robust to tolerate the random neutral genetic variations and to resist the environmental disturbances $v(t)$ in (18.34) in the evolutionary process. (iii) If the network robustness on the right-hand side of (18.33) is improved by the minimax stochastic H_∞ evolutionary game strategy so that the genetic robustness is increased, then the nonlinear biological network can harbor a large amount of neutral genetic variations. If the stochastic neutral genetic variations are accumulated in a sufficiently large amount in (18.27) or (18.34) in the evolutionary process, so that the phenotypic trait robustness criterion in (18.33) may finally be violated, then they may provide raw material for a new evolutionary possibility, causing a transition from one phenotypic trait to another through genetic drift [433,434]; for example, from the basin x_e to another basin x_{1e} in Fig. 18.1. At first sight, one might expect an evolutionary strategy to significantly improve network robustness to slow or even stop the evolution of a phenotypic trait in a biological network, but, because the more robust phenotypic trait can harbor a large number of neutral genetic variations, the robust phenotypic trait of the stochastic biological network might show increased rather than decreased evolutionary potential, in the long term of the biological network's evolution. The reasons why increasing network robustness may eventually lead to increasing network evolution include the accumulation of hidden neutral genetic variations that may be useful for later evolution. This is why network robustness is intrinsic to evolution and can eventually improve evolution [432,433]. This chapter extends these evolution results from the genetic level to the nonlinear stochastic gene network level, using the systematic methods in (18.32) and (18.33) from the point of the view of a stochastic minimax H_∞ evolutionary game. (iv) The tradeoff between network robustness and the evolvability of the nonlinear stochastic biological network in evolution can be seen from the phenotypic trait robustness criterion in (18.33); i.e., genetic robustness plus environmental robustness is less than or equal to network robustness. In this phenotypic trait robustness criterion, if the biological network has less evolvability (a small e_0 or a large fitness f_0) in evolution, then its environmental robustness becomes larger. In this situation, the network robustness has to be large enough to confer a large environmental robustness to resist environmental disturbances in order to maintain the phenotypic trait robustness, i.e., less evolvability implies a larger network robustness, and vice versa. (v) In general, the cost of the minimax stochastic H_∞ evolutionary strategy in selecting phenotypic variations as negative feedback loops $u^*(t) = -\frac{1}{2}\frac{\partial V(\tilde{x})}{\partial \tilde{x}}$ in order to improve network robustness and resist environmental disturbances is much higher [446]. In the case of the stress-avoidance strategy ($u(t) = 0$), the basin of the equilibrium point x_e in Fig. 18.1 becomes shallow, i.e., the term $\left(\frac{\partial V(\tilde{x})}{\partial \tilde{x}}\right)^T \left(f(\tilde{x}) - \frac{1}{4}\left(\frac{\partial V(\tilde{x})}{\partial \tilde{x}}\right)\right)$ on the right-hand side of (18.33) is reduced to $\left(\frac{\partial V(\tilde{x})}{\partial \tilde{x}}\right)^T f(\tilde{x})$ and is less negative. In this case, network robustness becomes smaller and environmental robustness must also be smaller in evolution, so that the evolvability e_0 will become larger and the biological network will be more adaptive in responding to environmental stresses and rapid environmental changes [446].

18.4 The Minimax Stochastic H_∞ Evolutionary Game Strategy of the Nonlinear Stochastic Biological Network Based on the Global Linearization Method

In general, it is still very difficult to solve the HJII-constrained optimization problem in (18.31) for the stochastic minimax H_∞ evolutionary strategy in a nonlinear stochastic biological network in evolution or the HJII problem in (18.33) for the phenotypic trait robustness criterion of a nonlinear stochastic biological network under natural selection in evolution. At present, there is no good method to solve the nonlinear partial differential HJII problem either analytically or numerically [32,275,276]. In this case, in order to simplify the stochastic evolutionary game problem, the global linearization technique [23,448] is employed to interpolate a nonlinear stochastic biological network with a set of local stochastic linearized biological networks.

Using the global linearization method, we suppose that all the global linearizations of the nonlinear stochastic biological network in (18.27) are bounded by a polytope, consisting of M vertices as [23,448]

$$\begin{bmatrix} \frac{\partial f(\tilde{x})}{\partial \tilde{x}} \\ \frac{\partial f_1(\tilde{x})}{\partial \tilde{x}} \\ \vdots \\ \frac{\partial f_{N(t)}(\tilde{x})}{\partial \tilde{x}} \end{bmatrix} \in C_0 \left[\begin{bmatrix} A_1 \\ A_{11} \\ \vdots \\ A_{N(t)1} \end{bmatrix} \cdots \begin{bmatrix} A_i \\ A_{1i} \\ \vdots \\ A_{N(t)i} \end{bmatrix} \cdots \begin{bmatrix} A_M \\ A_{1M} \\ \vdots \\ A_{N(t)M} \end{bmatrix} \right], \forall \tilde{x}(t), \tag{18.35}$$

where C_0 denotes the convex hull of the polytope with M vertices. That is, if the local linearized systems at all $\tilde{x}$ are inside the convex hull C_0, then the evolutionary trajectory $\tilde{x}$ of the nonlinear stochastic evolutionary network in (18.27) will belong to the convex combination of the stable trajectories of the following M local linearized networks, defined at M vertices of the polytope [23,448]

$$\frac{d\tilde{x}(t)}{dt} = A_i\tilde{x} + Bv(t) + u(t) + \sum_{k=1}^{N_2(t)} A_{ki}p(t-t_k)\tilde{x}, i = 1, \ldots, M. \tag{18.36}$$

Based on the global linearization theory [448], if (18.35) holds, then every evolutionary trajectory of the nonlinear stochastic biological network in (18.27) can be represented by a convex combination of the M local linearized biological networks in (18.36). This combination can be represented by

$$\frac{d\tilde{x}(t)}{dt} = \sum_{i=1}^{M} \alpha_i(\tilde{x})\left(A_i\tilde{x} + Bv(t) + u(t) + \sum_{k=1}^{N_2(t)} A_{ki}p(t-t_k)\tilde{x}\right), \tag{18.37}$$

where the interpolation function $\alpha_i(\tilde{x})$ satisfies $0 \leq \alpha_i(\tilde{x}) \leq 1$ and $\sum_{i=1}^{M} \alpha_i(\tilde{x}) = 1$, i.e., the evolutionary trajectory of the nonlinear stochastic biological network in (18.27) can be represented by the evolutionary trajectory of the interpolated biological network in (18.37). In (18.37), the evolutionary strategy $u(t) = \sum_{k=1}^{N_1(t)} f_k(\tilde{x})p(t-t_k)$ can be constructed as $\sum_{i=1}^{M} \sum_{k=1}^{N_1(t)} \alpha_i(\tilde{x})A_{ki}p(t-t_k)\tilde{x}$ by the global linearization technique. For the minimax stochastic H_∞ evolutionary strategy problem (18.11) of the interpolated evolutionary biological network in (18.37), we get the following result.

Proposition 18.3 For the interpolated evolutionary biological network in (18.37), the suboptimal solution of stochastic minimax H_∞ evolutionary game problem in (18.13) is solved by the following minimax evolutionary strategy $u^*(t)$ and the worst-case environmental disturbance $v^*(t)$

$$u^*(t) = \sum_{i=1}^{M} \sum_{k=1}^{N_1(t)} \alpha_i(\tilde{x})A_{ki}p(t-t_k)\tilde{x} = -P\tilde{x}, v^*(t) = \frac{1}{e}B^TP\tilde{x}, \tag{18.38}$$

where the positive definite matrix P is the solution of the following Riccati-like inequalities:

$$A_i^TP + PA_i + Q - PP + \frac{1}{e}PBB^TP + \lambda_2\sum_{k=1}^{N_{2p}} (A_{ki}^TP + PA_{ki} + A_{ki}^TPA_{ki}) \leq 0, \tag{18.39}$$

$$0 < P \leq eI \tag{18.40}$$

Proof: See Appendix C.

Remark 18.6 (i) In comparison with Proposition 18.1 of the stochastic linear biological network, the Riccati-like inequality in (18.18) is based only on the local linearized network at $\tilde{x}(t) = 0$, while the Riccati-like inequalities in (18.39) are based on M local linearized networks in (18.36) at the M vertices of the polytope in (18.35). Therefore, the result of the minimax evolutionary game strategy in Proposition 18.3 is more suitable for nonlinear stochastic biological networks in evolution. (ii) The Riccati-like inequalities in (18.39) can be considered as the local linearized HJII in (18.29) at M vertices of the polytope in (18.35). In general, it is very difficult to solve the HJII. However, it is very easy to solve the Riccati-like inequalities in (18.39) by transforming (18.39) into equivalent LMIs

$$\begin{bmatrix} X(A_i^T + \sum_{k=1}^{N_{2p}} \lambda_2 A_{ki}^T) + (A_i + \sum_{k=1}^{N_{2p}} \lambda_2 A_{ki})X - I + \frac{1}{e}BB^T & X & XA_{1i}^T & \cdots & XA_{N_{2p}i}^T \\ X & -Q^{-1} & 0 & \cdots & 0 \\ A_{1i}X & 0 & -\frac{1}{\lambda_2}X & \ddots & \vdots \\ \vdots & \vdots & \ddots & \ddots & 0 \\ A_{N_{2p}i}X & 0 & \cdots & 0 & -\frac{1}{\lambda_2}X \end{bmatrix} \leq 0 \tag{18.41}$$

and $X \triangleq P^{-1} > 0$, for $i = 1, \ldots, M$, using the Schur complement transformation method [448]. These can be easily solved by the LMI toolbox in Matlab.

Since e in (18.39) is the upper bound of the minimax stochastic H_∞ evolvability of the evolutionary game in (18.12), the minimum value of e will approach the evolvability $e_0 = \frac{1}{f_0}$. Therefore, we need to solve the following constrained optimization problem for e_0 or f_0 for the minimax H_∞ evolutionary game strategy in the nonlinear stochastic biological network in evolution

$$\frac{1}{f_0} = e_0 = \min_{P>0} e \tag{18.42}$$

subject to (18.40) and (18.41)

Remark 18.7 (i) The e_0 can be solved by decreasing e until (18.41) has no positive solution P within $0 < P \leq eI$ with the help of the LMI toolbox in Matlab. (ii) After solving e_0 and f_0 from (18.42), the Riccati-like inequalities in (18.39) and (18.40) then need to be modified as

$$A_i^T P + PA_i + Q - PP + \frac{1}{e_0} PBB^T P + \lambda_2 \sum_{k=1}^{N_{2p}} (A_{ki}^T P + PA_{ki} + A_{ki}^T PA_{ki}) \leq 0, \quad i = 1, \ldots, M \tag{18.43}$$

$$0 < P \leq e_0 I. \tag{18.44}$$

After solving P in (18.43), then $u^*(t) = -P\tilde{x}$ and $v^*(t) = \frac{1}{e_0} B^T P\tilde{x}$ in (18.38) form the minimax stochastic H_∞ evolutionary game strategy in (18.11) for the nonlinear stochastic biological network in evolution. (iii) With the minimax H_∞ evolutionary game strategy, the Riccati-like inequalities in (18.43) and (18.44) can be considered as the phenotypic trait robustness criteria of the biological network in evolution. If (18.43) and (18.44) are satisfied, then the phenotypic trait of the nonlinear stochastic biological network in (18.27) or (18.37) is maintained in the basin around the equilibrium point x_e under environmental disturbances and neutral genetic variations in the evolutionary process. If (18.43) and (18.44) have no positive solution P with $0 < P \leq e_0 I.$, then the phenotypic trait at x_e might not exist again or might transit to another phenotypic trait at another equilibrium point under environmental disturbances and genetic variations in evolution. (iii) The minimax stochastic H_∞ evolutionary game strategy $u^*(t)$ in (18.38) can enhance network robustness by shifting the eigenvalues of the local linear networks farther to the far left-hand side of the s-complex domain as follows:

$$(A_i-\frac{1}{2}P)^TP+P(A_i-\frac{1}{2}P)+Q+\frac{1}{e_0}PBB^TP+\lambda_2\sum_{k=1}^{N_{2p}}(A_{ki}^TP+PA_{ki}+A_{ki}^TPA_{ki})\leq 0, \tag{18.45}$$

$i=1,\ldots,M$

That is, every interaction matrix A_i of the local linearized biological networks is shifted by $-\frac{1}{2}P$ so that eigenvalues of $(A_i-\frac{1}{2}P)$ are on the farther left-hand side of the s-complex domain, and thus the evolutionary biological network is more robust to simultaneously tolerate local genetic variations and resist environmental disturbances. In general, the minimax stochastic H_∞ evolutionary game strategy is to select the total network variations $\sum_{i=1}^{M}\sum_{k=1}^{N_1(t)}\alpha_i(\tilde{x})A_{ki}p(t-t_k)=-P$, which consist of negative feedback loops for some positive matrix P. This is in order to select adequate genetic variations to form negative feedback circuits in the biological network, so as to improve network robustness to resist environmental disturbances and attenuate the effects of neutral genetic variation in the evolutionary process. (iv) The phenotypic trait robustness criterion in (18.45) can be reformulated as follows:

$$\underbrace{\lambda_2\sum_{k=1}^{N_{2p}}(A_{ki}^TP+PA_{ki}+A_{ki}^TPA_{ki})}_{\text{local genetic robustness}}+\underbrace{Q+\frac{1}{e_0}PBB^TP}_{\text{local environmental robustness}}\leq\underbrace{-[(A_i-\frac{1}{2}P)^TP+P(A_i-\frac{1}{2}P)]}_{\text{local network robustness through the minimax } H_\infty \text{ evolutionary game strategy}},i=1,\ldots,M \tag{18.46}$$

The biological meaning in (18.46) is that if the enhanced local network robustness of each local linearized biological network by the minimax H_∞ evolutionary game strategy can confer both local neutral genetic robustness to buffer local genetic variations and local environmental robustness to resist the local environmental disturbance in evolution, then the phenotypic trait of the nonlinear stochastic biological network with the minimax stochastic H_∞ evolutionary game strategy can be maintained in the basin of the equilibrium point at x_e in the evolutionary process. (v) Since the minimax stochastic H_∞ evolutionary game strategy can improve the local network robustness of local linearized networks on the right-hand side of (18.46), the nonlinear stochastic biological network will harbor more neutral genetic variations and will have greater environmental robustness to decrease the evolvability e_0 (or increase the fitness f_0) to attenuate the effect of environmental stimulus. However, when the harbored neutral genetic variations are accumulated to the extent that the first term on the left-hand side of (18.46) becomes very large and the phenotypic robustness criterion in (18.46) cannot be guaranteed in the evolutionary process, the phenotype of the biological network might be shifted, with the help of an environmental disturbance, from one equilibrium point to another equilibrium point in Fig. 18.1

and another period of network evolution will start at the other equilibrium point (phenotype) favored by natural selection.

Remark 18.8 If we consider that network evolution can only be plausibly studied at a population level, the evolutionary game should naturally invoke an n-tuple game with m-players under k environmental dynamics. In this n-tuple evolutionary game problem, the linear stochastic biological network in (18.7) should be modified as

$$\frac{dx(t)}{dt} = Ax(t) + u(t) + s_2(t)x(t) + \sum_{i=1}^{m} B_i u_i(t) + \sum_{j=1}^{k} D_j v_j(t), \tag{18.47}$$

where $u_i(t)$ denotes the ith decision strategy from other players (i.e., other competitive populations) and $v_j(t)$ denotes the jth environmental dynamics. Further, the evolutionary game problem in (8.11) should be modified as the following n-tuple minimax stochastic H_∞ evolutionary game

$$\min_{u(t)} \max_{\substack{u_i(t),i=1,\ldots,m \\ v_j(t),j=1,\ldots,k}} \frac{E\int_0^{t_p} (\tilde{x}(t)^T Q\tilde{x}(t) + u(t)^T u(t))dt}{E\tilde{x}(0)^T\tilde{x}(0) + E\sum_{i=1}^{m}\int_0^{t_p} u_i(t)^T u_i(t)dt + E\sum_{j=1}^{k}\int_0^{t_p} v_j(t)^T v_j(t)dt}, \tag{18.48}$$

The decision strategies $u_i(t)$ from other players and $v_j(t)$ from environmental dynamics are decided to maximize their effect on the phenotypic deviation $\tilde{x}(t)$ and evolutionary strategy $u(t)$ of the evolutionary biological network while the evolutionary strategy $u(t)$ is decided by the evolutionary network to minimize the worst-case phenotypic deviation and the effort (or cost) of evolutionary strategy. For simplicity of solving this n-tuple minimax stochastic H_∞ evolutionary game problem in the evolutionary biological network, the stochastic network dynamics in (18.47) can be represented by

$$\frac{dx(t)}{dt} = Ax(t) + u(t) + s_2(t)x(t) + BV(t), \tag{18.49}$$

where

$$B \triangleq [B_1\ B_2\ \cdots\ B_m\ D_1\ D_2\ \cdots\ D_k],\ V(t) \triangleq [u_1(t)\ u_2(t)\ \cdots\ u_m(t)\ v_1(t)\ v_2(t)\ \cdots\ v_k(t)].$$

In this situation, the n-tuple game in (18.48) is reduced to the 2-tuple game in (18.11), i.e., the decision variable $v(t)$ of the 2-tuple game in (18.11) can represent $V(t)$ of m-players and k environmental dynamics in (18.47) and the n-tuple H_∞ evolutionary game problem in (18.48) can be replaced by the 2-tuple evolutionary game in (18.11). Similarly, the result of the nonlinear stochastic minimax H_∞ evolutionary game problems in (18.26) and (18.11) can be easily extended to an n-tuple game problem of nonlinear stochastic network with m-players (populations) and k environmental dynamics. Therefore, the proposed theoretical approach to evolutionary game strategy naturally invokes an n-tuple game with m competitive populations and k environmental disturbances if B and $V(t)$ in (18.7) can be replaced by $B \triangleq [B_1\ B_2\ \cdots\ B_m\ D_1\ D_2\ \cdots\ D_k]$ and $V(t) \triangleq [u_1(t)\ u_2(t)\ \cdots\ u_m(t)\ v_1(t)\ \ v_2(t)\ \cdots\ v_k(t)]$ of the evolutionary network in (18.49), respectively. More details will be discussed in the following chapter.

18.5 Simulation Example

Consider the generic inhibition and activation biological model in evolution in Fig. 18.4, which is a known metabolic pathway and has been widely used for imitating systematic characteristics of a real metabolic pathway [16,449]. In this metabolic pathway, the metabolite X_3 converted from the metabolite X_2 inhibits an early step in its own production pathway, which is the synthesis of X_1. The metabolite X_2 is converted from the metabolite X_1, which is a divergence branching point. The degradation processes of X_1 into X_2 or X_4 are independent of each other. Then, the metabolite X_4 modulates downstream to activate the transformation of X_3. Let's denote $x_1(t)$, $x_2(t)$, $x_3(t)$, and $x_4(t)$ as the concentrations of metabolites X_1, X_2, X_3, and X_4, respectively. Based on the metabolic pathway [16,449] in evolution in Fig. 18.4, we have the following metabolic reaction network.

$$\begin{bmatrix} \frac{dx_1(t)}{dt} \\ \frac{dx_2(t)}{dt} \\ \frac{dx_3(t)}{dt} \\ \frac{dx_4(t)}{dt} \end{bmatrix} = \begin{bmatrix} f_1(x) \\ f_2(x) \\ f_3(x) \\ f_4(x) \end{bmatrix} + \sum_{k=1}^{N_1(t)} \begin{bmatrix} \lambda_{11k} & \lambda_{12k} & \cdots & \lambda_{14k} \\ \lambda_{21k} & \lambda_{22k} & \cdots & \lambda_{24k} \\ \vdots & \vdots & \ddots & \vdots \\ \lambda_{41k} & \lambda_{42k} & \cdots & \lambda_{44k} \end{bmatrix} \begin{bmatrix} x_1(t) \\ x_2(t) \\ x_3(t) \\ x_4(t) \end{bmatrix} p(t-t_k)$$

$$+ \sum_{k=1}^{N_2(t)} \begin{bmatrix} \lambda_{1k} \times \alpha_1 x_3^{g_{13}}(t) \\ \lambda_{2k} \times \alpha_2 x_3^{g_{21}}(t) \\ \lambda_{3k} \times \alpha_3 x_3^{g_{32}}(t) \\ \lambda_{4k} \times \alpha_4 x_1^{g_{41}}(t) \end{bmatrix} p(t-t_k) + \begin{bmatrix} v_1(t) \\ v_2(t) \\ v_3(t) \\ v_4(t) \end{bmatrix} \tag{18.50}$$

with

$$f(x) \triangleq \begin{bmatrix} f_1(x) \\ f_2(x) \\ f_3(x) \\ f_4(x) \end{bmatrix} \triangleq \begin{bmatrix} \alpha_1 x_3^{g_{13}}(t) - \beta_1 x_1^{h_{11}}(t) \\ \alpha_2 x_3^{g_{21}}(t) - \beta_2 x_2^{h_{22}}(t) \\ \alpha_3 x_3^{g_{32}}(t) - \beta_3 x_3^{h_{33}}(t) x_4^{h_{34}}(t) \\ \alpha_4 x_1^{g_{41}}(t) - \beta_4 x_4^{h_{44}}(t) \end{bmatrix},$$

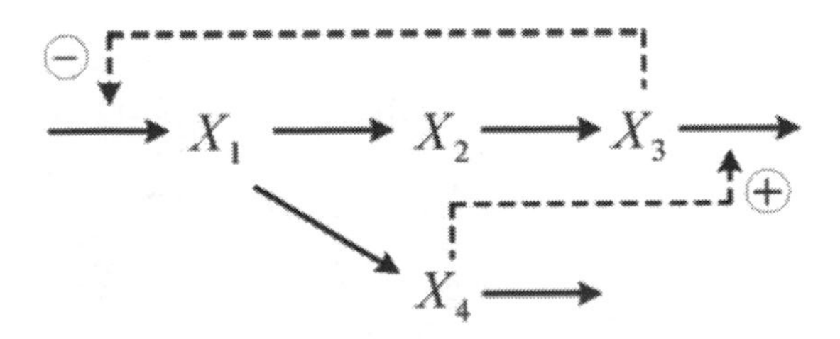

FIGURE 18.4
A real branched metabolic pathway of the generic inhibition and activation model where X_1, ..., X_4 are metabolites.

where $\alpha_1, \ldots, \alpha_4$ and $\beta_1, \ldots, \beta_4$ are the rate constants of net influxes and effluxes with kinetic orders g_{ij} and h_{ij}, $i, j \in S$ and $S = \{1, 2, 3, 4\}$. The parameters $\lambda_{1k}, \ldots, \lambda_{4k}$ are represented as weights of the random phenotypic variations of the biological network, which have no direct phenotypic effect on the network. The parameters of this metabolic reaction network are chosen as follows:

$$\alpha_1 = 12; \alpha_2 = 8; \alpha_3 = 3; \alpha_4 = 2; \beta_1 = 10; \beta_2 = 3; \beta_3 = 5; \beta_4 = 6; g_{13} = -0.8;$$
$$g_{21} = 0.5; g_{32} = 0.75; g_{41} = 0.5; h_{11} = 0.5; h_{22} = 0.75; h_{34} = 0.2; h_{44} = 0.8;$$
$$EV_{ik} = V = 0.1; EV_{ik}^2 = 0.04; E\lambda_{ik} = 0.1; E\lambda_{ik}^2 = 0.04.$$

The mean of the Poisson process $p(t)$ is imposed on $\lambda = 0.01$. The equilibrium point x_e of the nominal metabolic reaction network is $[x_{e1}, x_{e2}, x_{e3}, x_{e4}]^{eT} \simeq [0.400, 2.006, 2.228, 0.143]^T$ simulated from the initial condition $x(0) \triangleq [1.4, 2.25, 1.25, 0.45]^T$. For simplicity of analysis, the state of the network is shifted via the form $\tilde{x}(t) = x(t) - x_e$. Thus, the new equilibrium point of the phenotype concerned is at $\tilde{x}_e = 0_{4\times1}$ throughout this simulation example.

Based on the above evolutionary game strategy, the following phenotypic variations in the network in (18.50)

$$\sum_{k=1}^{N_1(t)} \begin{bmatrix} \lambda_{11k} & \lambda_{12k} & \cdots & \lambda_{14k} \\ \lambda_{21k} & \lambda_{22k} & \cdots & \lambda_{24k} \\ \vdots & \vdots & \ddots & \vdots \\ \lambda_{41k} & \lambda_{42k} & \cdots & \lambda_{44k} \end{bmatrix} \begin{bmatrix} \tilde{x}_1(t) \\ \tilde{x}_2(t) \\ \tilde{x}_3(t) \\ \tilde{x}_4(t) \end{bmatrix} p(t - t_k) \tag{18.51}$$

are selected by natural selection to construct negative feedback circuits to modify the metabolic reaction network, in order to resist environmental disturbance and tolerate the following effect due to neutral genetic variations

$$\sum_{k=1}^{N_2(t)} \begin{bmatrix} \lambda_{1k} \times \alpha_1 x_3^{g_{13}}(t) \\ \lambda_{2k} \times \alpha_2 x_3^{g_{21}}(t) \\ \lambda_{3k} \times \alpha_3 x_3^{g_{32}}(t) \\ \lambda_{4k} \times \alpha_4 x_1^{g_{41}}(t) \end{bmatrix} p(t - t_k), \tag{18.52}$$

where the Poisson counting process $N(t) = N_1(t) + N_2(t)$ for genetic variations in the evolutionary process is described in Fig. 18.6. The minimax H_∞ evolutionary game strategy based on Proposition 18.3 and the constrained optimization (18.42) is to select genetic variations to construct the following negative feedback loops to improve the network robustness:

$$u^*(t) = \sum_{k=1}^{N_1(t)} \begin{bmatrix} \lambda_{11k} & \lambda_{12k} & \cdots & \lambda_{14k} \\ \lambda_{21k} & \lambda_{22k} & \cdots & \lambda_{24k} \\ \vdots & \vdots & \ddots & \vdots \\ \lambda_{41k} & \lambda_{42k} & \cdots & \lambda_{44k} \end{bmatrix} \begin{bmatrix} \tilde{x}_1(t) \\ \tilde{x}_2(t) \\ \tilde{x}_3(t) \\ \tilde{x}_4(t) \end{bmatrix} p(t - t_k) = -P\tilde{x}(t), \tag{18.53}$$

where P is solved from the constrained optimization problem in (18.42). Based on the minimax stochastic H_∞ evolutionary game strategy, the global linearization technique and the LMI toolbox in Matlab, the following result is obtained from the constrained optimization problem (18.42):

$$P \simeq 10^{-4} \times \begin{bmatrix} 1.352 & 0.039 & -0.290 & 0.116 \\ * & 0.003 & -0.003 & -0.008 \\ * & * & 0.074 & -0.050 \\ * & * & * & 0.064 \end{bmatrix}. \tag{18.54}$$

We also find the evolvability $e_0 \simeq 0.7843$ and the maximum fitness $f_0 \simeq 1.2750$. The simulation result of the metabolic reaction network with the minimax stochastic H_∞ evolutionary game strategy is shown in Fig. 18.5. It can be seen that the phenotypic trait of the metabolic reaction network is maintained at the equilibrium point $\tilde{x}_e = 0_{4\times 1}$ or $x_e \simeq [0.400, 2.006, 2.228, 0.143]^T$ for a period of time from $t = 0$ to $t = 383$ in Fig. 18.5, and the evolvability of the metabolic reaction network by the minimax H_∞ evolutionary game strategy in this period is estimated with $v_i(t)$, $i = 1, \ldots, 4$, standard white noises, by a Monte–Carlo simulation with 10,000 runs, as follows:

$$\frac{E\int_0^{383} (\tilde{x}(t)^T Q \tilde{x}(t) + u(t)^T u(t))dt}{E\tilde{x}(0)^T \tilde{x}(0) + E\int_0^{383} v(t)^T v(t)dt} \simeq 0.7066 \leq e_0 = 0.7843. \tag{18.48}$$

From the calculation of the evolvability of the metabolic reaction network in (18.48), it can be seen that the estimated evolvability e_0 by the proposed method is conservative.

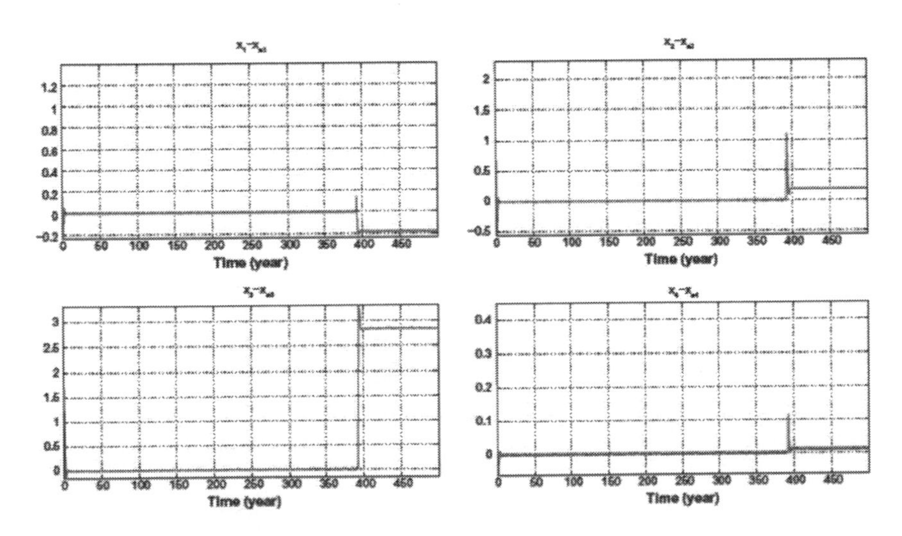

FIGURE 18.5
The trajectories of the metabolic reaction network with states $\tilde{x}_1, \ldots, \tilde{x}_4$ where $\tilde{x}_i(t) = x_i(t) - x_{ei}$, for $i = 1, \ldots, 4$, suffering from environmental disturbances $v_1(t), \ldots, v_4(t)$ and the Poisson genetic variations with the Poisson counting processes $N(t) = N_1(t) + N_2(t)$ are shown in Fig. 18.6. It can be seen that the minimax H_∞ evolutionary game strategy can maintain phenotype stability around x_e for a period of time from $t = 0$ to $t = 383$. However, when neutral genetic variations are continually accumulated by the robust network, it is seen that the phenotype robustness criterion in (18.46) is violated, and the phenotype of the metabolic reaction network becomes perturbed by neutral genetic variations after $t = 383$, finally shifting to another equilibrium point to start another evolutionary period.

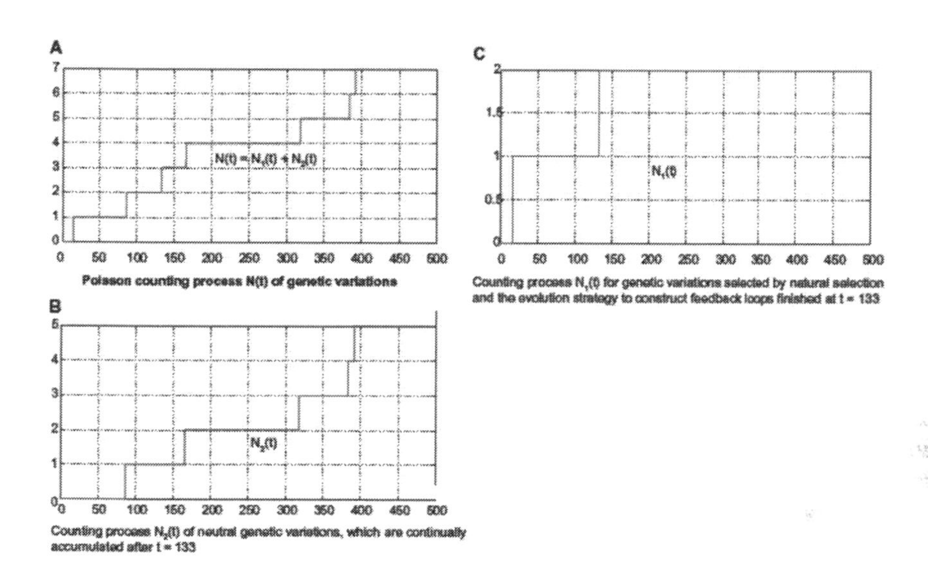

FIGURE 18.6
Poisson counting processes of genetic variations in the evolutionary process, for (A) $N(t) = N_1(t) + N_2(t)$. (B) $N_1(t)$, denoting the counting process of genetic variations selected for constructing negative feedback loops $u^* = -P\tilde{x}$ by natural selection. (C) $N_2(t)$, denoting the neutral genetic variations harbored by the robust network, without direct effect on the network phenotype.

The main reasons are the conservativeness of the global linearization and the LMIs used to approximate the nonlinear stochastic reaction network and the HJIIs, respectively. Furthermore, the convex optimization method employed by the LMI toolbox in Matlab to solve the LMI constraints for evolvability in (18.42) can also lead to conservative results.

From the simulation results in Fig. 18.5, it can be seen that by the evolutionary game strategy, as the neutral genetic variations are continually accumulated, after the negative feedback loops in (18.53) have been constructed by the evolutionary game strategy (see Fig. 18.6), the environmental disturbance can be buffered and neutral genetic variations can be tolerated by the metabolic reaction network only for a certain period of evolution time (from $t = 0$ to $t = 383$). However, when the neutral genetic mutations accumulate to the extent that the phenotypic trait robustness criterion of the evolutionary biological network in (18.33) or (18.46) is violated at approximately $t = 383$, the network stability will be destroyed and the phenotype transits to another equilibrium to begin another period of network evolution.

18.6 Discussion

In this chapter, a population of evolutionary biological networks with genetic variations selected by natural selection to resist intrinsic genetic variations and buffer environmental disturbances, maintaining its phenotypic trait in the evolutionary process, is modeled as a robust nonlinear stochastic stabilization system. Since environmental disturbances are uncertain and unpredictable, in order to maintain the robust stability of the phenotypic trait of interest by natural selection, the minimax stochastic H_∞ evolutionary game strategy for phenotypic trait robustness of stochastic biological network in (18.11) is developed to select adequate phenotypic variations by natural selection for improving network robustness. This minimax stochastic H_∞ evolutionary game strategy minimizes the worst-case effect of uncertain environmental disturbances on the phenotypic deviation from its phenotypic trait in the evolutionary process. The minimax stochastic H_∞ game design scheme has been widely used for robust control in robot systems [6,9], robust synthetic biology design [15,415], and robust reference control of economic systems [9] when some system parameters are perturbative and external disturbances are uncertain and unpredictable, so that their worst-case effect on system performance has to be as small a value as possible [6,9,15,415,420,422,444,445,450]. With the minimax stochastic H_∞ evolutionary game strategy, the phenotypic trait robustness criterion of the biological network in evolution is found to be as follows: genetic robustness plus environmental robustness is less than or equal to enhanced network robustness. That is, if the enhanced network robustness by the minimax stochastic H_∞ evolutionary game strategy can confer both genetic robustness to buffer random neutral genetic variations and environmental robustness to resist environmental disturbances in the evolutionary process, then the network phenotypic trait is robustly stable in the basin around the equilibrium point at x_e of interest by natural selection. The minimax stochastic H_∞ evolutionary game strategy can improve network robustness to enhance phenotypic trait robustness by modifying the network interaction matrix A by $\left(A - \frac{1}{2}P\right)$ in (18.23) (or $f(\tilde{x})$ by $f(\tilde{x}) - \frac{1}{4}\left(\frac{\partial V(\tilde{x})}{\partial x}\right)$ in (33)) through negative feedback loops, which could decrease the evolvability e_0 of the biological network in the evolutionary process. This result extends the idea of Waddington [451] from the gene level to the biological network level. Based on the phenotypic trait robustness criterion in (18.23), (18.33), or (18.46), the correlation between genetic robustness, environmental robustness, and network robustness by recent genomic experiments in yeast [452] (genes with similar robustness to genetic, environmental, and phenotypic perturbations in evolution) can be rationally explained from the network robustness perspective. That is, if the network robustness of a biological network is enhanced by the minimax stochastic H_∞ evolutionary game strategy (i.e., the basin of x_e in Fig. 18.1 becomes larger and deeper), then more neutral genetic variations and environmental disturbances can be overcome singly or simultaneously, and the phenotypic trait can be maintained by the minimax stochastic H_∞ evolutionary game strategy through natural selection during the evolutionary process. The fitness f_0 of the biological network, which is inversely proportional to the evolvability e_0, is increased by the minimax stochastic H_∞ evolutionary game strategy via natural selection in Proposition 18.2, so that the phenotypic trait is more robust in (18.33) due to the larger and deeper basin at the equilibrium point x_e caused by the adaptive negative feedback loops $u^*(t) = \sum_{k=1}^{N_1(t)} f_k(\tilde{x})p(t-t_k) =-\frac{1}{2}\left(\frac{\partial V(\tilde{x})}{\partial \tilde{x}}\right)$ in the evolutionary process. However, when large amounts of neutral genetic variations are accumulated continuously to enhance network robustness through the minimax stochastic H_∞ evolutionary game strategy, they may provide more raw material for new evolutionary possibilities through random genetic drift [451]. Therefore, the minimax stochastic H_∞ evolutionary game strategy may improve the network robustness to maintain the phenotypic trait in a period of evolutionary process, and may eventually improve the evolution of the biological network in the long term. This phenomenon has been illustrated by the in silico example of the evolutionary genetic regulatory network in Fig. 18.5.

From the phenotypic trait robustness criteria in (18.23) and (18.33), it is observed that there are two ways to improve phenotypic trait robustness in evolution. One way is to improve the network robustness (i.e., to make the right-hand side of (18.23) or (18.33) as large a value as possible so that the phenotypic trait robustness criterion always holds), and the other way is to reduce the effect of neutral genetic variations and environmental disturbances on the biological network (i.e., to make the left-hand side of (18.23) or (18.33) as small a value as possible). The minimax H_∞ evolutionary game strategy is to select heritable phenotypic variations such that $u^*(t) = -Px$ in (18.17) for linear biological networks or $u^*(t) = -\frac{1}{2}\left(\frac{\partial V(\tilde{x})}{\partial \tilde{x}}\right)$ in (18.28) for nonlinear biological networks, in order to develop negative feedback loops as the first way to improve the network robustness on the right-hand side of (18.23) or (18.33) to resist environmental disturbances. This is why there are many negative feedback loops in biological networks – to improve the network robustness. Furthermore, there are numerous redundant, modular, and scale-free structures in biological networks, which are helpful in attenuating the effect of intrinsic neutral genetic variations and environmental disturbances on the left-hand side of (18.23) or (18.33), which is the second way to maintain the phenotypic trait in the evolutionary process. Therefore, based on the phenotypic trait robustness criterion of the evolutionary biological network, we can gain a deeper evolutionary insight into the development of negative feedback, redundant, modular, and scale-free structures for biological networks by adaptive or non-adaptive processes [439] in evolution. In different levels of biological networks, two favored strategies can improve phenotypic trait robustness in evolution. One is to improve the network robustness to tolerate more intrinsic neutral genetic variations and to resist environmental disturbances. Negative feedback is the systematic mechanism to improve network robustness, and it is favored by natural selection at different levels of the biological network in evolution, which matches the minimax stochastic H_∞ evolutionary game strategy in this chapter. Another strategy is to reduce the effect of intrinsic neutral genetic variations and environmental disturbances on different biological networks. Redundancies and repairs are the mechanisms of this strategy, and are also favored by natural selection. This is why there are many different redundancies from duplicated genes in gene regulatory networks, redundant pathways in biological chemical networks, and species redundancies in ecological networks [445].

Recently, synthetic biology has focused on the rational construction of biological systems based on engineering principles to fulfill a particular purpose and to allow a better understanding of biological design, using synthetic circuits [400,453–460]. High throughput techniques are necessary to determine what is evolved and how engineered genetic networks in vivo mimic the evolutionary processes of genetic networks. An approach involving network evolution has been used to rationally design a synthetic network with a desired behavior in the host cell [458]. Unfortunately, synthetic biologists are faced with the uncertainties of modeling variation, evolutionary genetic variation, and environmental disturbances in vivo (i.e., the phenotypic trait robustness problem of the synthetic biological network) [455–457]. Given this situation, a robust synthetic network design is currently an important topic in the field of synthetic biology [400,420,458–460]. As the intrinsic genetic perturbations and environmental disturbances are unpredictable for synthetic biologists, the proposed minimax stochastic H_∞ evolutionary game strategy should be an efficient robust design method for a synthetic gene network with evolutionary genetic variations, intrinsic uncertainties, and environmental disturbances [400,420,458–460].

The contents of this chapter address the following: (i) The Poisson point process in (18.7) is used to model genetic variations in stochastic gene network; (ii) the evolution level (18.8) is defined for the stochastic biological network in the evolutionary process, i.e., the worst-case effect of all possible environmental disturbances on the phenotypic deviation around x_e and the effect (cost) of evolutionary strategy $u(t)$. Then, the fitness of the biological network around the phenotypic trait is inversely correlated to the evolution level of the biological network in (18.9). Therefore, the problem of evolvability or fitness can be considered as the minimax stochastic H_∞ evolutionary game problem in (18.11). Based on the minimax stochastic H_∞ evolutionary game strategy, we also find some accumulated genetic variations can provide negative feedback via natural selection to maximize the robust stability of networks; (iii) then the global linearization method is proposed to solve the nonlinear stochastic game problem for the minimax stochastic H_∞ evolutionary game strategy of nonlinear stochastic biological network in the evolutionary process; (iv) based on the minimax stochastic H_∞ evolutionary game strategy in (18.17) or (18.28), we could obtain the phenotypic trait robustness criterion in (18.23) or (18.33) for stochastic biological networks in evolution. Finally, the phenotypic trait robustness criterion is further analyzed for the evolutionary network to get insight into the minimax stochastic H_∞ evolutionary game strategy from the systematic point of view.

In the conventional evolution network [444], the evolvability and fitness of biological networks are not well defined. Therefore, the evolutionary game problem is not easily formulated for evolutionary biological networks. Actually, they only solve some evolutionary optimization problems without considering random genetic variations and environmental disturbances. Further, these methods could not provide an efficient method to solve Hamilton–Jacobi equations (HJE) for evolutionary game problems. Hence, it is not easy to get insight into the evolutionary mechanisms of nonlinear biological networks from the systematic point of view.

18.7 Conclusion

In this chapter, biological networks with random genetic variations and uncertain environmental disturbances in the evolutionary process are described as nonlinear stochastic systems. In order to maintain the phenotypic trait favored by natural selection, a biological network needs to select some heritable phenotypic variations to improve network robustness, so as to tolerate neutral genetic variations and to buffer environmental disturbances in the evolutionary process. Since the environmental disturbances are uncertain and unpredictable, the evolutionary strategy of a biological network needs to minimize the worst-case effect of all possible bounded environmental disturbances (i.e., to perform the minimax stochastic H_∞ evolutionary game strategy) in the evolutionary process. We have found that the minimax stochastic H_∞ evolutionary game strategy of a biological network is equivalent to accumulating some phenotypic variations due to random genetic variations to construct robust negative feedback loops, in order to efficiently improve the network robustness, resist the uncertain environmental disturbances, and tolerate neutral genetic variations. However, those harbored random neutral genetic variations also have an effect on network robustness when they accumulate to a large enough amount, serving to provide raw material for new adaptation and evolutionary innovation. If network robustness can be improved by the minimax stochastic H_∞ evolutionary game strategy, conferring environmental robustness against the environmental disturbance as well as genetic robustness to tolerate the neutral genetic variations, then the phenotypic trait of the biological network can be maintained in the evolutionary process. Otherwise, the phenotypic trait at an equilibrium point might be destroyed and shifted to another phenotypic trait at a neighboring equilibrium point to commence another period of network evolution.

In general, to gain insight into the minimax stochastic H_∞ evolutionary game strategy of a nonlinear evolutionary biological network, it is necessary to solve a difficult optimization problem constrained by a HJII. A global linearization technique is employed to simplify the minimax H_∞ evolutionary game problem, i.e., the HJII-constrained optimization problem for the minimax stochastic H_∞ evolutionary game strategy can be simplified to an equivalent optimization problem constrained by LMIs. Based on the global linearization technique, the tradeoff between evolvability, genetic robustness, and network robustness based on the minimax stochastic H_∞ game strategy in evolution can be easily discussed from the local stability robustness and local filtering ability of a set of local evolutionary biological networks on the basis of linear system theory. Finally, the proposed minimax stochastic H_∞ evolutionary game strategy is also extended to an n-tuple noncooperative stochastic H_∞ evolutionary game problem of a population of biological networks with m players (competitive populations) and k environmental dynamics in the evolutionary process. The detailed discussion of n-player noncooperative and cooperative game strategies in biological networks is given in the next chapter.

18.8 Appendix

18.8.1 Appendix A. Proof of Proposition 18.1

From

$$J \triangleq E\int_0^{t_p} (\tilde{x}(t)^T Q\tilde{x}(t) + u(t)^T u(t) - ev(t)^T v(t))dt.$$

and by the fact $\tilde{x}(t) = x(t)$ i.e. $x_e = 0$ in the linear biological case, we get

$$\begin{aligned} J &= E\int_0^{t_p} (\tilde{x}(t)^T Q\tilde{x}(t) + u(t)^T u(t) - ev(t)^T v(t))dt \\ &= Ex(0)^T Px(0) - Ex(t_p)^T Px(t_p) + E\int_0^{t_p} (\tilde{x}(t)^T Q\tilde{x}(t) + u(t)^T u(t) \\ &- ev(t)^T v(t) + \frac{d(x(t)Px(t))}{dt})dt \end{aligned} \tag{18.A1}$$

where $V(x(t),t) \triangleq x^T(t)Px(t), P = P^T \geq 0$. By using the Itô formula [16,32,275,276, 420,422,458,461], we can get

$$\begin{aligned}
J &= Ex(0)^T Px(0) - Ex(t_p)^T Px(t_p) + E\int_0^{t_p} (x^T Qx + u^T u - ev^T v + \left(\frac{\partial V}{\partial x}\right)^T \\
&\times (Ax + Bv + u) + \lambda_2 \left[\sum_{k=1}^{N_{2p}} x^T A_k^T\right] P \left[\sum_{l=1}^{N_{2p}} A_l x\right] \}dt \\
&= Ex(0)^T Px(0) - Ex(t_p)^T Px(t_p) + E\int_0^{t_p} (x^T[Q + A^T P + PA]x + u^T Px + x^T Pu \\
&+ u^T u - ev^T v + x^T PBv + v^T B^T Px + \lambda_2 \sum_{k=1}^{N_{2p}} x^T(A_k^T P + PA_k + A_k^T PA_k)x\}dt \\
&= Ex(0)^T Px(0) - Ex(t_p)^T Px(t_p) + E\int_0^{t_p} (x^T[Q + A^T P + PA - PP + \frac{1}{e}PBB^T P \\
&+ \sum_{k=1}^{N_{2p}} \lambda_2 A_k^T PA_k]x + (Px + u)^T(Px + u) - (\frac{1}{\sqrt{e}}B^T Px - \sqrt{e}v)^T(\frac{1}{\sqrt{e}}B^T Px - \sqrt{e}v)\}dt.
\end{aligned} \tag{18.A2}$$

$$\begin{aligned}
\min_{u(t)} \max_{v(t)} J &= Ex(0)^T Px(0) - Ex(t_p)^T Px(t_p) + \min_{u(t)} \max_{v(t)} E\int_0^{t_p} \{x^T[Q + A^T P \\
&+ PA - PP + \frac{1}{e}PBB^T P + \sum_{k=1}^{N_{2p}} \lambda_2(A_k^T P + PA_k + A_k^T PA_k)]x + (Px + u)^T(Px + u) \\
&- (\frac{1}{\sqrt{e}}B^T Px - \sqrt{e}v)^T(\frac{1}{\sqrt{e}}B^T Px - \sqrt{e}v)\}dt.
\end{aligned} \tag{18.A3}$$

Then, we get the solution of minimax Nash stochastic H_2 (quadratic) game as

$$u^*(t) = -Px \tag{18.A4}$$

$$v^*(t) = \frac{1}{e}B^T Px \tag{18.A5}$$

$$\begin{aligned}
\min_{u(t)} \max_{v(t)} J &= Ex(0)^T Px(0) - Ex(t_p)^T Px(t_p) + E\int_0^{t_p} \{x^T[Q + A^T P \\
&+ PA - PP + \frac{1}{e}PBB^T P + \sum_{k=1}^{N_{2p}} \lambda_2(A_k^T P + PA_k + A_k^T PA_k)]x\}dt.
\end{aligned} \tag{18.A6}$$

By the inequality

$$A^T P + PA - PP + \frac{1}{e}PBB^T P + \sum_{k=1}^{N_{2p}} A_k^T PA_k \leq 0,$$

we get

$$J_o = \min_{u(t)} \max_{v(t)} J \leq Ex(0)^T Px(0) - Ex(t_p)^T Px(t_p) \leq Ex(0)^T Px(0).$$

By the inequality $J_o \leq eE\tilde{x}(0)^T \tilde{x}(0)$, we get

$$P \leq eI \tag{18.A7}$$

which is the inequality

$$0 < P \leq eI.$$

Q.E.D.

18.8.2 Appendix B. Proof of Proposition 18.2

From

$$J \triangleq E \int_0^{t_p} (\tilde{x}(t)^T Q\tilde{x}(t) + u(t)^T u(t) - ev(t)^T v(t))dt.$$

for some positive function $V(\tilde{x})$

$$J = EV(\tilde{x}(0)) - EV(\tilde{x}(t_p)) + E\int_0^{t_p} (\tilde{x}^T Q\tilde{x} + u^T u - ev^T v)dt + EdV(\tilde{x}). \tag{18.A8}$$

By the Itô formula [16,32,275,276,420,422,458,461] for nonlinear stochastic network

$$\begin{aligned}\frac{d\tilde{x}(t)}{dt} &= f(\tilde{x}) + Bv(t) + \sum_{k=1}^{N(t)} f_k(\tilde{x})p(t - t_k) \\ &= f(\tilde{x}) + Bv(t) + u(t) + \sum_{k=1}^{N_2(t)} f_k(\tilde{x})p(t - t_k),\end{aligned}$$

and by the fact that $u(t) = \sum_{k=1}^{N_1(t)} f_k(\tilde{x})p(t - t_k)$, we can get

$$EdV(\tilde{x}) = \left[\left(\frac{\partial V(\tilde{x})}{\partial \tilde{x}}\right)^T (f(\tilde{x}) + u + Bv) + \frac{1}{2}\lambda_2 \sum_{k=1}^{N_{2p}} [V(\tilde{x} + f(\tilde{x})) - V(\tilde{x})\right] dt \tag{18.A9}$$

Substituting (18.A9) into (18.A8), we get

$$\begin{aligned}J = EV(\tilde{x}(0)) - EV(\tilde{x}(t_p)) + E\int_0^{t_p} \{\tilde{x}^T Q\tilde{x} + u^T u - ev^T v + \left(\frac{\partial V(\tilde{x})}{\partial \tilde{x}}\right)^T f(\tilde{x}) \\ + \left(\frac{\partial V(\tilde{x})}{\partial \tilde{x}}\right)^T u + \left(\frac{\partial V(\tilde{x})}{\partial \tilde{x}}\right)^T Bv + \frac{1}{2}\lambda_2 \sum_{k=1}^{N_{2p}} V(\tilde{x} + f(\tilde{x})) - V(\tilde{x})\}dt.\end{aligned} \tag{18.A10}$$

By the fact that

$$u^T u + \left(\frac{\partial V(\tilde{x})}{\partial \tilde{x}}\right)^T u = \left(u + \frac{1}{2}\frac{\partial V(\tilde{x})}{\partial \tilde{x}}\right)^T \left(u + \frac{1}{2}\frac{\partial V(\tilde{x})}{\partial \tilde{x}}\right) - \frac{1}{4}\left(\frac{\partial V(\tilde{x})}{\partial \tilde{x}}\right)^T \frac{\partial V(\tilde{x})}{\partial \tilde{x}} \tag{18.A11}$$

$$\begin{aligned} ev^T v + \left(\frac{\partial V(\tilde{x})}{\partial \tilde{x}}\right)^T Bv = &- \left(\sqrt{e}v - \frac{1}{2\sqrt{e}}B^T\left(\frac{\partial V(\tilde{x})}{\partial \tilde{x}}\right)\right)^T \left(\sqrt{e}v - \frac{1}{2\sqrt{e}}B^T\left(\frac{\partial V(\tilde{x})}{\partial \tilde{x}}\right)\right) \\ &+ \frac{1}{4e}\left(\frac{\partial V(\tilde{x})}{\partial \tilde{x}}\right)^T BB^T\left(\frac{\partial V(\tilde{x})}{\partial \tilde{x}}\right). \end{aligned} \tag{18.A12}$$

Substituting (18.A11) and (18.A12) into (18.A10), we get

$$\begin{aligned} J = &EV(\tilde{x}(0)) - EV(\tilde{x}(t_p)) + E\int_0^{t_p} \{\left(\frac{\partial V(\tilde{x})}{\partial \tilde{x}}\right)^T f(\tilde{x}) + \tilde{x}^T Q\tilde{x} \\ &+ \frac{1}{2}\lambda_2 \sum_{k=1}^{N_{2p}} [V(\tilde{x} + f(\tilde{x})) - V(\tilde{x})] + \frac{1}{4e}\left(\frac{\partial V(\tilde{x})}{\partial \tilde{x}}\right)^T BB^T\left(\frac{\partial V(\tilde{x})}{\partial \tilde{x}}\right) \\ &- \left(\sqrt{e}v - \frac{1}{2\sqrt{e}}B^T\left(\frac{\partial V(\tilde{x})}{\partial \tilde{x}}\right)\right)^T \left(\sqrt{e}v - \frac{1}{2\sqrt{e}}B^T\left(\frac{\partial V(\tilde{x})}{\partial \tilde{x}}\right)\right) \\ &- \frac{1}{4}\left(\frac{\partial V(\tilde{x})}{\partial \tilde{x}}\right)^T \frac{\partial V(\tilde{x})}{\partial \tilde{x}} \left(u + \frac{1}{2}\frac{\partial V(\tilde{x})}{\partial \tilde{x}}\right)^T \left(u + \frac{1}{2}\frac{\partial V(\tilde{x})}{\partial \tilde{x}}\right)\}dt. \end{aligned} \tag{18.A13}$$

Then,

$$\begin{aligned} &\min_{u(t)}\max_{v(t)} J = EV(\tilde{x}(0)) - EV(\tilde{x}(t_p)) + \min_{u(t)}\max_{v(t)} E\int_0^{t_p} \{\left(\frac{\partial V(\tilde{x})}{\partial \tilde{x}}\right)^T f(\tilde{x}) + \tilde{x}^T Q\tilde{x} \\ &+ \frac{1}{2}\lambda_2 \sum_{k=1}^{N_{2p}} [V(\tilde{x} + f(\tilde{x})) - V(\tilde{x})] + \frac{1}{4e}\left(\frac{\partial V(\tilde{x})}{\partial \tilde{x}}\right)^T BB^T\left(\frac{\partial V(\tilde{x})}{\partial \tilde{x}}\right) \\ &- \left(\sqrt{e}v - \frac{1}{2\sqrt{e}}B^T\left(\frac{\partial V(\tilde{x})}{\partial \tilde{x}}\right)\right)^T \left(\sqrt{e}v - \frac{1}{2\sqrt{e}}B^T\left(\frac{\partial V(\tilde{x})}{\partial \tilde{x}}\right)\right) \\ &- \frac{1}{4}\left(\frac{\partial V(\tilde{x})}{\partial \tilde{x}}\right)^T \frac{\partial V(\tilde{x})}{\partial \tilde{x}} \left(u + \frac{1}{2}\frac{\partial V(\tilde{x})}{\partial \tilde{x}}\right)^T \left(u + \frac{1}{2}\frac{\partial V(\tilde{x})}{\partial \tilde{x}}\right)\}dt. \end{aligned} \tag{18.A14}$$

Then, we get the minimax stochastic H_2 (quadratic) game solution

$$u^*(t) = s^*(\tilde{x}) = \sum_{k=1}^{N_1(t)} f_k(\tilde{x})p(t - t_k) = -\frac{1}{2}\left(\frac{\partial V(\tilde{x})}{\partial \tilde{x}}\right), v^*(t) = \frac{1}{2e}B^T\left(\frac{\partial V(\tilde{x})}{\partial \tilde{x}}\right).$$

Further, by the HJII

$$\begin{aligned} &\left(\frac{\partial V(\tilde{x})}{\partial \tilde{x}}\right)^T f(\tilde{x}) + \tilde{x}^T Q\tilde{x} - \frac{1}{4}\left(\frac{\partial V(\tilde{x})}{\partial \tilde{x}}\right)^T \frac{\partial V(\tilde{x})}{\partial \tilde{x}} \\ &+ \frac{1}{4e}\left(\frac{\partial V(\tilde{x})}{\partial \tilde{x}}\right)^T BB^T\left(\frac{\partial V(\tilde{x})}{\partial \tilde{x}}\right) + \frac{1}{2}\lambda_2 \sum_{k=1}^{N_{2p}} [V(\tilde{x} + f(\tilde{x})) - V(\tilde{x})] \le 0, \end{aligned}$$

we get

$$J_o = \min_{u(t)} \max_{v(t)} J \leq EV(\tilde{x}(0)) - EV(\tilde{x}(t_p)) \leq EV(\tilde{x}(0)). \tag{18.A15}$$

By the inequality constraint

$$J_o \leq eE\tilde{x}(0)^T \tilde{x}(0),$$

we get

$$EV(\tilde{x}(0)) \leq eE\tilde{x}(0)^T \tilde{x}(0) \tag{18.A16}$$

which in the inequality

$$V(\tilde{x}(0)) \leq e\tilde{x}(0)^T \tilde{x}(0).$$

Q.E.D.

18.8.3 Appendix C. Proof of Proposition 18.3

Since we replace the nonlinear stochastic biological network

$$\begin{aligned}\frac{d\tilde{x}(t)}{dt} &= f(\tilde{x}) + Bv(t) + \sum\nolimits_{k=1}^{N(t)} f_k(\tilde{x})p(t - t_k) \\ &= f(\tilde{x}) + Bv(t) + u(t) + \sum\nolimits_{k=1}^{N_2(t)} f_k(\tilde{x})p(t - t_k)\end{aligned}$$

by the interpolated biological network

$$\frac{d\tilde{x}(t)}{dt} = \sum\nolimits_{i=1}^{M} \alpha_i(\tilde{x})\Big(A_i\tilde{x} + Bv(t) + u(t) + \sum\nolimits_{k=1}^{N_2(t)} A_{ki}(\tilde{x})p(t - t_k)\tilde{x}\Big).$$

can be replaced by

$$\begin{aligned}&\left(\frac{\partial V(\tilde{x})}{\partial \tilde{x}}\right)^T \left(\sum\nolimits_{i=1}^{M} A_i\tilde{x}\right) + \tilde{x}^T Q\tilde{x} - \frac{1}{4}\left(\frac{\partial V(\tilde{x})}{\partial \tilde{x}}\right)^T \frac{\partial V(\tilde{x})}{\partial \tilde{x}} + \frac{1}{4e}\left(\frac{\partial V(\tilde{x})}{\partial \tilde{x}}\right)^T BB^T \left(\frac{\partial V(\tilde{x})}{\partial \tilde{x}}\right) \\ &+ \lambda_2 \sum_{i=1}^{M}\sum_{i=1}^{M}\sum_{k=1}^{N_{2p}} \alpha_i(\tilde{x})\alpha_j(\tilde{x})\tilde{x}^T (A_{ki}^T P + PA_{ki} + A_{ki}^T PA_{ki})\tilde{x} \\ &\leq \left(\frac{\partial V(\tilde{x})}{\partial \tilde{x}}\right)^T \left(\sum\nolimits_{i=1}^{M} A_i\tilde{x}\right) + \tilde{x}^T Q\tilde{x} - \frac{1}{4}\left(\frac{\partial V(\tilde{x})}{\partial \tilde{x}}\right)^T \frac{\partial V(\tilde{x})}{\partial \tilde{x}} + \frac{1}{4e}\left(\frac{\partial V(\tilde{x})}{\partial \tilde{x}}\right)^T BB^T \left(\frac{\partial V(\tilde{x})}{\partial \tilde{x}}\right) \\ &+ \lambda_2 \sum_{i=1}^{M}\sum_{k=1}^{N_{2p}} \alpha_i(\tilde{x})\alpha_j(\tilde{x})\tilde{x}^T (A_{ki}^T P + PA_{ki} + A_{ki}^T PA_{ki})\tilde{x} \leq 0\end{aligned} \tag{18.A17}$$

with $V(\tilde{x}(0)) \leq e\tilde{x}(0)^T\tilde{x}(0)$. If we choose $V(\tilde{x}) \leq \tilde{x}^T P\tilde{x}$, by the fact $\sum_{i=1}^{M} \alpha_i \tilde{x} = 1$ and $\frac{\partial V(\tilde{x})}{\partial \tilde{x}} = 2P\tilde{x}$ in (18.A17), then we get

$$\sum_{i=1}^{M} \alpha_i \tilde{x}^T [A_i^T P + PA_i + Q - PP + \frac{1}{e} PBB^T P + \lambda_2 \sum_{k=1}^{N_{2p}} (A_{ki}^T P + PA_{ki} + A_{ki}^T PA_{ki})]\tilde{x} \leq 0. \quad (18.A18)$$

The above inequality holds if the Riccati-like inequalities

$$A_i^T P + PA_i + Q - PP + \frac{1}{e} PBB^T P + \lambda_2 \sum_{k=1}^{N_{2p}} (A_{ki}^T P + PA_{ki} + A_{ki}^T PA_{ki}) \leq 0, i = 1, \ldots, M$$

hold. Further,

$$u^*(t) = -\frac{1}{2}\left(\frac{\partial V(\tilde{x})}{\partial \tilde{x}}\right) = -P\tilde{x}, v^*(t) = \frac{1}{e} B^T p\tilde{x} \quad (18.A19)$$

which are the solution of minimax stochastic H_2 (quadratic) evolutionary game problem

$$u^*(t) = \sum_{i=1}^{M} \sum_{k=1}^{N_1(t)} \alpha_i(\tilde{x}) A_{ki} p(t - t_k)\tilde{x} = -P\tilde{x}, v^*(t) = \frac{1}{e} B^T P\tilde{x}.$$

Q.E.D.

19

On the Nonlinear Stochastic H_∞ Game Strategy of Evolutionary Biological Network in Carcinogenic Process

19.1 Introduction

In the real life, biological systems are always subject to intrinsic random genetic mutations, epigenetic alterations, and noises arising from internal and external stochastic fluctuations [454,456]. However, it is very important that the most potential perturbative effects of stochastic noise [462], genetic mutations, and epigenetic alterations are buffered at the systemic phenotypic level. Indeed, biological systems are expected to display phenotypes that are robust to these genetic and epigenetic variations and noise [428,430–432,463]. However, phenotypic changes under the evolutionary fitness pressure require that biological systems are able to adapt by producing heritable phenotypic variants. This ability has been termed as evolvability. Recently, a paradoxical question about the intuitively expected trade-off between robustness and evolvability in the biological systems has attracted much attention [415,440–442]: how can biological systems generate phenotypes that are resistant to genetic mutations, epigenetic alterations, and noise, yet also retain their ability to evolve as a result of these effects? One answer is that the ability of a phenotype to resist mutations might enable genetic diversity, which could then be translated to phenotypic diversity as a result of subsequent genetic and epigenetic alterations, or environmental changes under selection pressure [464] through what has been termed the neutral network [428,435,437,465].

The onset of cancer is characterized by an accumulation of genetic mutations and epigenetic alterations that are caused by different environmental stresses, including tobacco, chemical agents, radiation, and viruses. These mutations and alterations could typically modify the structures of DNA and chromatin, and consequently alter the gene products or the regulation of gene expression [466]. Molecular biologists have long recognized carcinogenesis as a somatic evolutionary process that involves genetic mutations, epigenetic alterations and natural selection. Indeed, cancer is driven by the somatic evolution of cell lineages [438,467–479]. In this chapter, the somatic evolution of cell lineages in an organ was modeled as a nonlinear stochastic biological network to study natural selection in the evolutionary carcinogenesis. The somatic evolutionary biological network of an organ is driven by intrinsic random fluctuations because of genetic and epigenetic variations and external disturbances attributed to carcinogens and other stressors in the ambient microenvironment. Therefore, we modeled the intrinsic random fluctuations because of genetic and epigenetic variations as a Poisson counting process, and the external disturbances were described by an uncertain signal. Therefore, an organ with different cell species undergoing carcinogenesis was modeled by a nonlinear stochastic system with an intrinsic Poisson counting process and external random disturbances during somatic evolutionary process [17,443].

A natural selection scheme develops a robust strategy to efficiently attenuate intrinsic fluctuations and withstand external disturbances at the beginning of carcinogenesis. This robust strategy maintains the normal function (i.e., phenotype) of an evolutionary biological network under intrinsic random fluctuations because of genetic and epigenetic variations and external disturbances attributed to environmental stresses. A robust feedback scheme [440–442] may exist in natural selection to maintain the healthy phenotype of a biological network at the desired equilibrium point x_e (attractor; Fig. 19.1). Because the environmental disturbances and the intrinsic random Poisson somatic genetic and epigenetic fluctuations are unpredictable, the minimax H_∞ evolutionary game strategy was employed to attenuate the effect of environmental disturbance and intrinsic fluctuation. This strategy utilized a robust optimal feedback scheme that minimized the worst-case effects of external disturbances and intrinsic fluctuations on the robust stability of the phenotype at an equilibrium point favored by natural selection. Therefore, the strategy of natural selection on the evolutionary biological network during carcinogenesis was formulated as a nonlinear stochastic H_∞ evolutionary game problem. Based on the stochastic H_∞ game strategy, effective feedback loops could be developed by natural selection for a cancer-associated biological network in the somatic evolutionary process. This evolutionary strategy can significantly improve the network's robustness to buffer genetic and epigenetic variations and to attenuate extrinsic disturbances, as well as maintain the organ's normal phenotype during the somatic evolutionary process. However, neutral genetic mutations and epigenetic alterations could be gradually accumulated, and eventually overrode the network's robustness developed through natural selection during the somatic evolutionary process. The robust stability of the biological network is thereby deteriorated, which leads to a shift from the healthy phenotypic state at the equilibrium point x_e to the cancerous phenotypic state at another equilibrium point x_{1e}. Hence, cancer could initiate another new somatic evolutionary process (Fig. 19.1). As time increased, the above mentioned shift of phenotypic state always occurred when the accumulated genetic mutations and epigenetic alterations were large enough. Therefore, cancer can be considered an aging disease.

In this chapter, based on the stochastic H_∞ game strategy for the robust stabilization of evolutionary nonlinear biological networks in carcinogenesis [6,9,46,444,445,457], both natural selection and external disturbances affected network evolution. Natural selection minimized network evolution (or maximized network fitness), while external disturbances, including genetic mutations and epigenetic alterations, maximized network evolution (or minimized network fitness). We solved a Hamilton–Jacobi–Issac inequality (HJII)-constrained optimization problem using the stochastic H_∞ game strategy to obtain a robust evolutionary feedback scheme for natural selection during carcinogenesis. Furthermore, the phenotype robustness criterion was also derived from the stochastic H_∞ game strategy to obtain insights into the evolvability of the cancer biological network during somatic evolutionary process. In order to simplify the solving procedure of stochastic H_∞ game strategy, the global linearization method was employed to interpolate several local linear stochastic systems and to approximate the nonlinear stochastic system of an organ undergoing carcinogenesis. Therefore, the HJII-constrained optimization problem could be replaced by a linear matrix inequality

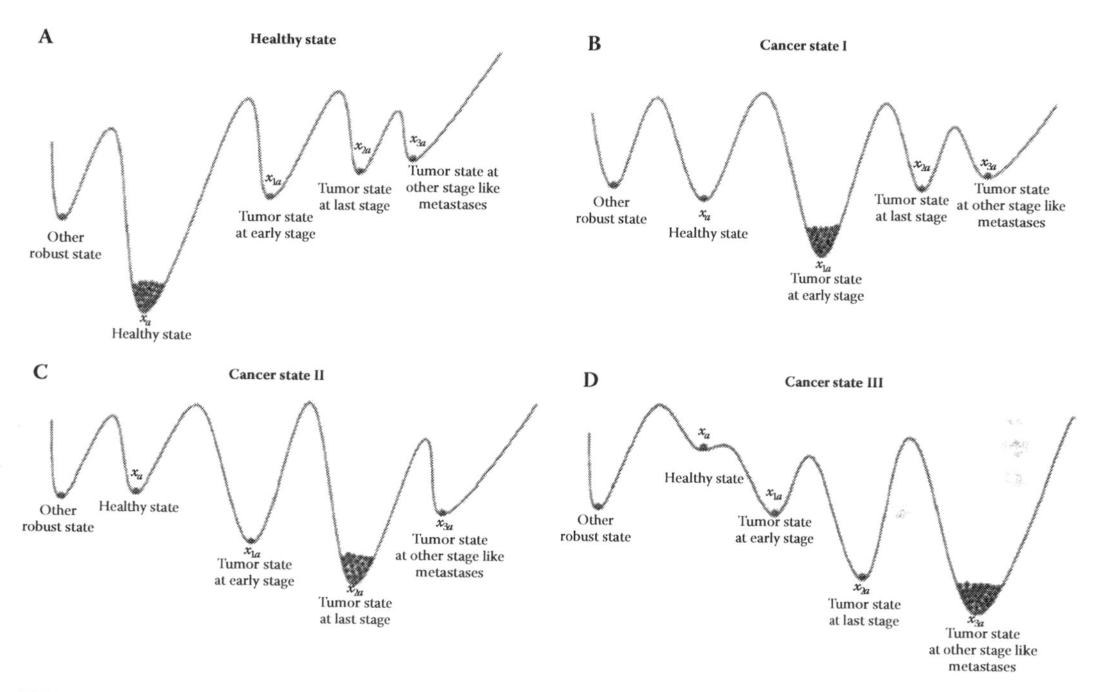

FIGURE 19.1
Four equilibrium points in the phenotypic landscape of the nonlinear stochastic biological network of an organ in (19.10) or (19.11) in carcinogenesis. The dots in the basin of the equilibrium points denote the accumulation of Poisson phenotypic fluctuations (population) of biological networks because of genetic and epigenetic changes in carcinogenesis. The vertical states illustrate the relative stability of robust states, healthy state, and tumor state of different stages at different equilibrium points, in the multiple dimensional state space. (A) At the beginning, the phenotypic trait of biological network is near the equilibrium point x_e at healthy state. (B) By suffering the accumulation of neutral phenotypic variations and the influence of environmental stresses in carcinogenesis, the robust stability of healthy state cannot be maintained through the evolutionary strategy $u(t)$ with the effect of natural selection, and then, the phenotypic trait may be shifted from x_e to x_{1e} of tumor state at the early stage to start another somatic evolutionary process (carcinogenesis). (C) Similarly, if the accumulated neutral phenotypic variations are large enough and microenvironmental disturbances $v(t)$ are strong enough, the evolutionary strategy $u(t)$ cannot confer enough robust network stability to maintain the network local stability at x_{1e}. Then, the phenotypic trait of biological network is shifted from cancer stage i at the equilibrium point x_{1e} to cancer stage ii at the equilibrium point x_{2e} to begin another stage of evolutionary process (carcinogenesis). (D) Finally, if the accumulation of neutral phenotypic variations is large enough to violate the phenotypic robustness criterion at x_{2e}, cancer is transited from cancer stage ii to cancer stage iii. Since the equilibrium points x_e in (B) and (C) are still locally stable, they are curable (treatable), i.e., the curable stages in (B) and (C) may be reversible to (A) by adequate surgical or medical treatment (one kind of enhanced negative feedback). If the healthy state x_e is incurable (saddle point) as shown in (D), then it is impossible that this somatic evolutionary process is revertible to (A).

(LMI)-constrained optimization problem, which was solved easily using the LMI Control Toolbox in MATLAB [8,23,275].

Finally, an in silico example of evolutionary oncogenesis in lung cancer has been provided to illustrate natural selection based on the stochastic H_∞ game strategy. The nonlinear stochastic biological network of lung cancer-related tissue contains parenchymal cells, stromal cells, immature vascular endothelial cells (VECs), and VECs [472]. The lung cancer-related network was constructed using Poisson genetic and epigenetic variations and environmental disturbances during evolutionary oncogenesis. By using a robust feedback scheme with the natural selection property based on the proposed nonlinear stochastic H_∞ game strategy, we found that the tissue network maintained a normal phenotype for ~54.5 years in the aging process under genetic and epigenetic variations and environmental disturbances. When the network robustness of the tissue network was violated by the accumulated neutral genetic and epigenetic variations and environmental disturbances during somatic evolutionary (aging) process, the normal phenotype of the tissue shifted from its original equilibrium point to another equilibrium point with a new cancerous phenotype. This new phenotype [472,480] took about 1.5 years to progress from stage I to stage II cancer, and about 2.5 years to progress from stage II to stage III cancer.

19.2 Stochastic Biological Network of Organ in the Carcinogenic Process

First, we considered a simplified analysis by using the linear biological network of a tissue or organ in carcinogenesis as the following random dynamic process:

$$\frac{dx(t)}{dt} = \left(A + \sum_{k=1}^{N(t)} A_k p(t - t_k)\right) x(t) + Bv(t) \tag{19.1}$$

where $x(t) = [x_1(t)\ x_2(t)\ x_3(t), \ldots, x_n(t)]^T$ denotes the masses at time t of n cell species in the biological network of a tissue or organ (in the liquid tumor case, i.e., lymphomas and leukemias, the masses of cell species are changed to the densities of cancer cells in blood or lymph), $v(t)$ denotes the environmental disturbances (signals or mutagens) because of different kinds of damages or carcinogens. $p(t)$ is a Poisson point process and A_k denotes the effects of random genetic mutations and epigenetic alterations at $t - t_k$ on the phenotype of the tissue. $N(t)$ denotes a Poisson counting process of phenotypic variations in the tissue level because of random cellular genetic and epigenetic variations [466] in carcinogenesis as follows:

$$N(t) = \sum\nolimits_{k=1}^{N(t)} p(t - t_k) = \sum\nolimits_{k=1}^{N(t)} 1 \tag{19.2}$$

with mean λt and variance λt [443], and the weighting Poisson counting process

$$\sum\nolimits_{k=1}^{N(t)} A_k p(t - t_k) = \sum\nolimits_{k=1}^{N(t)} A_k, \text{ for } t \geq 0 \tag{19.3}$$

denotes the total phenotype effect of random genetic and epigenetic variations on the system matrix A. Poisson point processes are useful for modeling a wide range of event-driven phenomena involving the occurrence of events at random points in time, such as emission of subatomic particles from radioactive material, the detection of photons by the receiver of a weak optical signal, genetic mutations, neural discharges, lightning discharges, seismic events, and the arrival of customers always used to model a dynamic system exposed to sudden, infrequent, highly localized changes [443,481]. Therefore, it is suitable to model discontinuous genetic and epigenetic variations of biological network in the carcinogenesis process. The system matrix A denotes the interactions among these n cell species in the tissue or organ, and B denotes the coupling matrix between environmental disturbances (signals) and cell species in the organ, i.e., it denotes the effect of environmental signals on the biological network of the organ, where

$$A = \begin{bmatrix} a_{11} & a_{12} & \cdots & a_{1n} \\ a_{21} & a_{22} & \cdots & a_{2n} \\ \vdots & \vdots & \ddots & \vdots \\ a_{n1} & a_{n2} & \cdots & a_{nn} \end{bmatrix}; A_k = \begin{bmatrix} a_{k11} & a_{k12} & \cdots & a_{k1n} \\ a_{k21} & a_{k22} & \cdots & a_{k2n} \\ \vdots & \vdots & \ddots & \vdots \\ a_{kn1} & a_{kn2} & \cdots & a_{knn} \end{bmatrix};$$
$$B = \begin{bmatrix} b_{11} & b_{12} & \cdots & b_{1n} \\ b_{21} & b_{22} & \cdots & b_{2n} \\ \vdots & \vdots & \ddots & \vdots \\ b_{n1} & b_{n2} & \cdots & b_{nn} \end{bmatrix}; v(t) = \begin{bmatrix} v_1(t) \\ v_2(t) \\ \vdots \\ v_m(t) \end{bmatrix}; \tag{19.4}$$

$a_{ij} = 0$ if there exists no interaction between cell species i and cell species j; $a_{kij} = 0$ if the kth genetic or epigenetic variation has no phenotype effect on a_{ij}; and $b_{ij} = 0$ if the disturbance $v_j(t)$ has no effect on the cell species i.

In general, the biological processes in the cancer are time-varying, i.e., $A(t)$ and $B(t)$ in (19.1). In this chapter, because of discontinuous time-varying parameter variations due to genetic and epigenetic variations, $A(t)$ is replaced by $A + \sum_{k=1}^{N(t)} A_k p(t - t_k)$ in (19.1). $B(t)$ is replaced by time-invariant B but with varying part merged to uncertain disturbance $v(t)$. In this chapter, $v(t)$ is any bounded environmental stress. Therefore, in this chapter, we are with time-invariant system models A and B, but with random process to represent the time-varying process across time.

In order to operate successfully in the organ within the context of the organism, all the cells in the organ must be integrated into an architecture as system matrix A in the biological network of the organ in (19.1).

From (19.1), we get a linear stochastic evolutionary biological network in carcinogenesis as

$$\frac{dx(t)}{dt} = Ax(t) + \sum_{k=1}^{N(t)} A_k x(t) p(t - t_k) + Bv(t) \tag{19.5}$$

Since the weighted Poisson counting process $\sum_{k=1}^{N(t)} A_k p(t - t_k)$ in (19.1) or (19.5) is an accumulating process of phenotypic changes of biological network in the tissue level because of random genetic and epigenetic variations in the cellular level, the stochastic biological network in (19.5) can represent a population of biological networks over all possible Poisson genetic and epigenetic variations in the somatic evolutionary process, i.e., one possible Poisson genetic or epigenetic variation represents one possible biological network in a population of the organs in carcinogenesis. The accumulated phenotypic variations $\sum_{k=1}^{N(t)} A_k x(t) p(t - t_k)$ in (19.5) may lead to permanent changes in the phenotypic trait of the organ. $v(t)$ denotes the environmental stresses because of environmental changes or any agent capable of causing cancer. Therefore, the phenotypic trait $x(t)$ of biological network will be perturbed away from the equilibrium point $x_e = 0$. Recently [482], the lifetime risk of cancers of many different types is shown to be strongly correlated (0.81) with the total number of divisions of the normal self-renewing cells maintaining that tissue's homeostasis, i.e., to be strongly related to the accumulated Poisson phenotypic changes $\sum_{k=1}^{N(t)} A_k x(t) p(t - t_k)$ in the aging process.

Under the phenotypic variations and environmental disturbances in (19.5), in order to maintain its normal function and homeostasis, i.e., to operate successfully around the equilibrium point x_e, the natural selection strategy of biological network in (19.5) needs to select some adequate phenotypic variations to improve the biological network robustness to resist the effects of genetic and epigenetic variations and environmental changes in carcinogenesis.

Suppose that the natural selection strategy of evolutionary biological network in the organ is to select the following phenotypic variations to improve the network robustness in carcinogenesis

$$u(t) = \sum_{k=1}^{N_1(t)} A_k x(t) p(t - t_k^{'}) \tag{19.6}$$

and the remaining (or neutral) phenotype variations are

$$\begin{aligned} R(t) &= \sum_{k=1}^{N_2(t)} A_k x(t) p(t - t_k^*) \\ &= \sum_{k=1}^{N(t)} A_k x(t) p(t - t_k) - \sum_{k=1}^{N_1(t)} A_k x(t) p(t - t_k^{'}) \end{aligned} \tag{19.7}$$

which have not been selected by natural selection and are accumulated in the biological network. The temporal parameters $t_k^{'}$ and t_k^* were defined as $t_{N_1(t)}^{'} \geq, \ldots, \geq t_1^{'}$ and $t_{N_2(t)}^* \geq, \ldots, \geq t_1^*$, respectively. It is worth noting that the $t_k^{'}$ is not equal to t_k^*. From (19.5)–(19.7), it is seen that

$$N(t) = N_1(t) + N_2(t), \sum_{k=1}^{N(t)} A_k x(t) p(t - t_k) = u(t) + R(t) \tag{19.8}$$

where the Poisson counting processes $1 \sim N_1(t)$ and $1 \sim N_2(t)$ have means $\lambda_1 t$ and $\lambda_2 t$, respectively, and $\lambda = \lambda_1 + \lambda_2$.

Therefore, the stochastic evolutionary biological network of an organ under somatic genetic and epigenetic variations, environmental disturbances, and natural selection in (19.5) can be represented as follows:

$$\begin{aligned} \frac{dx(t)}{dt} &= Ax(t) + \sum_{k=1}^{N_1(t)} A_k x(t) p(t - t_k^{'}) + \sum_{k=1}^{N_2(t)} A_k x(t) p(t - t_k^*) + Bv(t) \\ &= Ax(t) + u(t) + \sum_{k=1}^{N_2(t)} A_k x(t) p(t - t_k^*) + Bv(t) \end{aligned} \tag{19.9}$$

where $u(t) = \sum_{k=1}^{N_1(t)} A_k x(t) p(t - t_k^{'})$ denotes the evolutionary strategy of biological network by natural selection, i.e., how to select adequate phenotypic variations to improve the network robustness to phenotypic variations and environmental disturbance $v(t)$ and to finally maintain the normal function of biological network (or organ) in the network evolutionary process (carcinogenesis) was considered.

In the real organ, the stochastic evolutionary biological networks for n cell species, $x_1, \ldots, x_n$, in carcinogenesis are always nonlinear, and the linear stochastic evolution network in (19.5) is thus modified as

$$\frac{dx(t)}{dt} = f(x) + \sum_{k=1}^{N(t)} f_k(x) p(t - t_k) + Bv(t) \tag{19.10}$$

where $f(x)$ denotes the nonlinear interactive vector among n different cell species in the biological network, and $f_k(x)$, for $k = 1, \ldots, N(t)$, denote the nonlinear phenotypic variations because of Poisson somatic genetic and epigenetic variations in carcinogenesis.

Suppose that the phenotypic variations in (19.10) could be separated into the part of evolutionary strategy $u(t) = \sum_{k=1}^{N_1(t)} f_k(x) p(t - t_k^{'})$ by natural selection and the remaining neutral phenotypic variations $\sum_{k=1}^{N_2(t)} f_k(x) p(t - t_k^{*})$, which are accumulated as random dots in the basin of healthy state x_e in Fig. 19.1A. The above nonlinear stochastic evolutionary biological network in carcinogenesis was then modified as

$$\frac{dx(t)}{dt} = f(x) + u(t) + \sum_{k=1}^{N_2(t)} f_k(x) p(t - t_k^{*}) + Bv(t) \tag{19.11}$$

where

$$\begin{aligned} \sum_{k=1}^{N(t)} f_k(x) p(t - t_k) &= \sum_{k=1}^{N_1(t)} f_k(x) p(t - t_k^{'}) + \sum_{k=1}^{N_2(t)} f_k(x) p(t - t_k^{*}) \\ &= u(t) + \sum_{k=1}^{N_2(t)} f_k(x) p(t - t_k^{*}) \end{aligned} \tag{19.12}$$

The nonlinear stochastic biological system in (19.11) represents an evolutionary organ under evolutionary strategy $u(t) = \sum_{k=1}^{N_1(t)} f_k(x) p(t - t_k^{'})$ by natural selection, and the neutral phenotypic variations $\sum_{k=1}^{N_2(t)} f_k(x) p(t - t_k^{*})$ and environmental disturbance $Bv(t)$ are accumulated in the basin of the healthy state at the equilibrium point x_e in Fig. 19.1A during the carcinogenesis.

The nonlinear stochastic biological network in (19.10) and (19.11) has many equilibrium points (phenotypic traits) (Fig. 19.1). Suppose that the phenotypic trait of the biological network near the stable equilibrium point x_e is the healthy state, which is the phenotypic trait with normal function of interest by natural selection. For the convenience of analysis, the origin of nonlinear stochastic biological network in (19.11) is shifted to the equilibrium point (phenotypic trait) x_e. In this case, if the shifted nonlinear stochastic network is robustly stable at the origin, the equilibrium point (phenotype trait) x_e of interest is also robustly stable. Let us denote the phenotypic deviation around the phenotype at x_e as $\tilde{x}(t) = x(t) - x_e$ so that the following shifted nonlinear stochastic biological network of an organ is obtained as

$$\begin{aligned}\frac{d\tilde{x}(t)}{dt} &= f(\tilde{x}) + \sum_{k=1}^{N(t)} f_k(x)p(t-t_k) + Bv(t) \\ &= f(\tilde{x}) + u(t) + \sum_{k=1}^{N_2(t)} f_k(x)p(t-t_k^*) + Bv(t)\end{aligned} \quad (19.13)$$

where the notations $f(\tilde{x}+x_e)$ and $f_k(\tilde{x}+x_e)$ are, respectively, simplified by $f(\tilde{x})$ and $f_k(\tilde{x})$. $u(t) = \sum_{k=1}^{N_1(t)} f_k(x)p(t-t_k^{'})$ denotes the evolutionary strategy of nonlinear stochastic biological network by natural selection to maintain the normal function of evolutionary biological network.

In this situation, the origin $\tilde{x}(t) = 0$ of the shifted nonlinear stochastic biological network in (19.13) is the equilibrium point x_e of the original biological network in (19.11).

Remark 19.1

i. The linear stochastic biological network of an organ in (19.9) was considered as the linearized system of (19.13) at the equilibrium point $\tilde{x} = 0$ [or at the equilibrium point x_e in the nonlinear stochastic biological network in (19.11)] with $A = \frac{\partial f(\tilde{x})}{\partial \tilde{x}}|_{\tilde{x}=0}$ and $A_k = \frac{\partial f_k(\tilde{x})}{\partial \tilde{x}}|_{\tilde{x}=0}$ since $x(t) = x_e$ [i.e., $\tilde{x}(t) = x(t) - x_e$] is the only equilibrium point in the linear biological system in (19.5) or (19.9).

ii. The nonlinear stochastic biological system in (19.13) was considered as a Darwinian dynamic equation that satisfies Darwin's conditions of variability, heritability, and the strategy to survive [444]. However, the heritability in this chapter implies the heritability of somatic cell evolution in carcinogenesis.

19.3 Natural Selection Scheme in Carcinogenic Process

In carcinogenesis, the evolutionary biological network of an organ was modeled as the nonlinear stochastic system in (19.13). Under the effects of random somatic genetic and epigenetic variations and environmental disturbance $v(t)$, in order to maintain the normal function of evolutionary biological network to be in the healthy state at the equilibrium point x_e in Fig. 19.1, an evolutionary strategy $u(t) = \sum_{k=1}^{N_1(t)} f_k(x)p(t-t_k^{'})$ must be selected by natural selection to improve the robust stability of the healthy state or the equilibrium point at $x(t) = x_e$ (or $\tilde{x}(t) = 0$) of somatic evolutionary biological network. According to the robust stochastic stabilization strategy, the evolutionary strategy needs to make the system state (phenotype) of evolutionary biological network deviate from the healthy state as small as possible with a parsimonious effort of evolutionary strategy $u(t)$ selected by natural selection in response to environmental disturbances, somatic genetic and epigenetic variations, and initial state. Because the environmental disturbance $v(t)$ and neutral somatic genetic and epigenetic variations are unpredictable, the evolution level e of the biological network, because of the worst-case of all possible bounded disturbances, neutral somatic genetic and epigenetic variations, and initial conditions, was defined as [435,465]

$$e = \max_{\substack{v(t)\in L_2[0,t_p] \\ R(t)=\sum_{k=1}^{N_2(t)} f_k(\tilde{x})p(t-t_k^*)}} \frac{E\int_0^{t_p}\{\tilde{x}^T(t)Q\tilde{x}(t)+u^T(t)u(t)\}dt}{E\{\tilde{x}^T(0)\tilde{x}(0)\}+E\int_0^{t_p}\{\sum_{k=1}^{N_2(t)} f_k^T(\tilde{x})f_k(\tilde{x})p(t-t_k^*)+v^T(t)v(t)\}dt} \tag{19.14}$$

where $L_2[0,t_p]$ denotes the set of all possible bounded environmental disturbances, i.e., $v(t)\in L_2[0,t_p]$, if $\int_0^{t_p} v^T(t)v(t)dt<\infty$. $E\{*\}$ denotes the expectation of *. The integration $E\int_0^{t_p}\{\tilde{x}^T(t)Q\tilde{x}(t)+u^T(t)u(t)\}dt$ in the numerator denotes the performance index of natural selection strategy of biological network in carcinogenesis. The positive symmetric weighting matrix Q is a trade-off between the state deviation $\tilde{x}(t)$ and the effort $u(t)$, i.e., a large Q will lead to a small state deviation in $\tilde{x}(t)$ but with large effort $u(t)$, and vice versa. If Q is selected as the diagonal matrix, $Q=diag[q_1,\ldots,q_i,\ldots,q_n]$, in the situation, the selection of a larger q_i will make the corresponding ith state $x_i(t)$ be not easier to deviate from the equilibrium point, i.e., a less $\tilde{x}(t)$ in carcinogenesis, and vice versa.

The physical meaning of stochastic H_∞ evolution performance in (19.14) is that the worst-case effect of the initial condition, all possible environmental disturbance $v(t)$, neutral somatic genetic and epigenetic variations on the robust stability performance of $\tilde{x}(t)$, and control strategy effort of $u(t)$ were considered as the somatic evolution level of the biological network because the environmental disturbances and neutral genetic and epigenetic variations are unpredictable in carcinogenesis. The deviation $\tilde{x}(t)$ and evolutionary strategy $u(t)$ are simultaneously considered in the evolutionary performance in (19.14) because natural selection employs a parsimonious strategy to achieve the robust stability with a less effort.

The phenotypic fitness (or phenotypic robustness) of the biologic network around the healthy phenotypic trait at the equilibrium point x_e is inversed to the evolution level of biological network, [457] i.e.,

$$f\alpha\frac{1}{e} \tag{19.15}$$

In other words, a biological network with a low evolution level e in carcinogenesis will lead to a high fitness. Therefore, the maximization of the biological network fitness to the phenotypic trait at equilibrium point x_e under the effect of somatic genetic and epigenetic variations and environmental disturbances is equivalent to the minimization of the evolution level as

$$\max f \sim \min e \tag{19.16}$$

Therefore, by combining (19.14) and (19.16), the evolutionary strategy $u(t)$ for the biological network suffers from the effect of natural selection to tolerate neutral phenotypic variations, to resist environmental disturbances, and to finally maintain the normal phenotype of biological network in carcinogenesis. It was to maximize the fitness of the biological network to the healthy state x_e as the following minimax stochastic H_∞ evolutionary game problem:

$$e_0 = \min_{u(t)\in\sum_{k=1}^{N_1(t)} f_k(\tilde{x})p(t-t_k')} e$$
$$= \min_{u(t)\in\sum_{k=1}^{N_1(t)} f_k(\tilde{x})p(t-t_k')} \left[\max_{v(t)\in L_2[0,t_p]} \left(\frac{E\int_0^{t_p}\{\tilde{x}^T(t)Q\tilde{x}(t)+u^T(t)u(t)\}dt}{E\{\tilde{x}^T(0)\tilde{x}(0)\}+E\int_0^{t_p}\{\sum_{k=1}^{N_2(t)} f_k^T(\tilde{x})f_k(\tilde{x})p(t-t_k^*)\}dt+E\int_0^{t_p}\{v^T(t)v(t)\}dt} \right) \right] \tag{19.17}$$

where $u(t)$ was selected as $u(t)=\sum_{k=1}^{N_1(t)} f_k(\tilde{x})p(t-t_k')$ from the total phenotypic variations $\sum_{k=1}^{N_1(t)} f_k(\tilde{x})p(t-t_k')$ to maximize the fitness (or minimize the evolution level) of biological network in carcinogenesis. Since the initial phenotypic deviation $\tilde{x}(0)$ also affects $\tilde{x}(t)$ and $u(t)$, its effect on the evolution level was also considered in the denominator of (19.17). e_0 in (19.17) was called the network evolvability of biological network with the minimax(game) strategy $u(t)$ by considering natural selection, i.e., the H_∞ stochastic game is performed for biological network of the organ by natural selection to resist the unpredictable environmental disturbances and tolerate the neutral phenotypic variations because of somatic genetic and epigenetic variations in carcinogenesis. Therefore, the maximum phenotypic robustness (fitness maximization) strategy of biological network by natural selection in carcinogenesis was formulated as a stochastic H_∞ game problem in (19.17) for nonlinear stochastic biological network of (19.13), i.e., how to select $u(t)=\sum_{k=1}^{N_1(t)} f_k(\tilde{x})p(t-t_k')$ from the total phenotypic variations $\sum_{k=1}^{N(t)} f_k(\tilde{x})p(t-t_k)$ because of the somatic genetic and epigenetic variations in carcinogenesis to solve the minimum evolution (or maximum fitness) problem of the biological network at the healthy state x_e was considered.

The stochastic H_∞ evolutionary game in (19.17) has two players $u(t)$ and $v(t)$. The environmental disturbance $v(t)$ and neutral phenotype variations $\sum_{k=1}^{N_2(t)} f_k(\tilde{x})p(t-t_k^*)$ will maximize their effect on the deviation of phenotype $\tilde{x}(t)$ from its equilibrium point x_e (i.e., the worst-case effect of all possible disturbances $v(t)\in L_2[0,t_p]$ and neutral phenotypic variations $\sum_{k=1}^{N_2(t)} f_k(\tilde{x})p(t-t_k^*)$ were considered in the evolution game) and the other player $u(t)$ selected phenotypic variations to significantly improve the network robustness (fitness) of the phenotype x_e by minimizing the worst-effect of all possible neutral phenotypic variations, environmental disturbances, and initial conditions on the phenotype deviation and evolution strategy effort in carcinogenesis.

In general, it is very difficult to solve the stochastic H_∞ game problem in (19.17) directly for the evolutionary strategy of nonlinear stochastic biological network of organs in carcinogenesis. In this chapter, a suboptimal method was introduced to solve the stochastic H_∞ evolutionary game problem by minimizing the upper bound e of e_0 indirectly as follows:

$$\min_{u(t)\in\sum_{k=1}^{N_1(t)} f_k(\tilde{x})p(t-t_k')} \left[\max_{v(t)\in L_2[0,t_p]} \left(\frac{E\int_0^{t_p}\{\tilde{x}^T(t)Q\tilde{x}(t)+u^T(t)u(t)\}dt}{E\{\tilde{x}^T(0)\tilde{x}(0)\}+E\int_0^{t_p}\{\sum_{k=1}^{N_2(t)} f_k^T(\tilde{x})f_k(\tilde{x})p(t-t_k^*)\}dt+E\int_0^{t_p}\{v^T(t)v(t)\}dt} \right) \right] \le e \tag{19.18}$$

The upper bound e will be given beforehand to solve the suboptimal H_∞ evolution game problem in (19.18) at first and e will be decreased as small as possible to approach to the solution e_0 of the stochastic H_∞ evolutionary game problem in (19.17).

The suboptimal stochastic H_∞ evolutionary game problem in (19.18) is equivalent to the following constrained stochastic Nash quadratic (H_2) game problem [6,9,16]:

$$\begin{aligned}&\min_{u(t)\in\sum_{k=1}^{N_1(t)} f_k(\tilde{x})p(t-t_k')}\ \max_{v(t)\in L_2[0,t_p]} E\int_0^{t_p}\{\tilde{x}^T(t)Q\tilde{x}(t)+u^T(t)u(t)-ev^T(t)v(t)\\&-e\sum_{k=1}^{N_2(t)} f_k^T(\tilde{x})f_k(\tilde{x})p(t-t_k^*)\}dt\le eE\{\tilde{x}^T(0)\tilde{x}(0)\}\end{aligned} \tag{19.19}$$

Let us denote

$$J=E\int_0^{t_p}\{\tilde{x}^T(t)Q\tilde{x}(t)+u^T(t)u(t)-ev^T(t)v(t)-e\sum_{k=1}^{N_2(t)} f_k^T(\tilde{x})f_k(\tilde{x})p(t-t_k^*)\}dt \tag{19.20}$$

Therefore, we developed two steps to solve the constrained Nash stochastic quadratic (H_2) game in (19.18) or (19.19). The first step was to solve the following minimax Nash quadratic (H_2) game problem

$$\begin{aligned}J_0&=\min_{u(t)\in\sum_{k=1}^{N_1(t)} f_k(\tilde{x})p(t-t_k')}\ \max_{v(t)\in L_2[0,t_p]} J\\&=\min_{u(t)\in\sum_{k=1}^{N_1(t)} f_k(\tilde{x})p(t-t_k')}\ \max_{v(t)\in L_2[0,t_p]} E\int_0^{t_p}\{\tilde{x}^T(t)Q\tilde{x}(t)+u^T(t)u(t)-ev^T(t)v(t)\\&\quad-e\sum_{k=1}^{N_2(t)} f_k^T(\tilde{x})f_k(\tilde{x})p(t-t_k^*)\}dt\end{aligned} \tag{19.21}$$

The second step is to solve the following constraint problem:

$$J_0\le eE\{\tilde{x}^T(0)\tilde{x}(0)\} \tag{19.22}$$

19.4 Stochastic H_∞ Game Strategy of Linear Evolutionary Biological Network in Carcinogenesis

For the convenience of illustration, we first solve the stochastic H_∞ evolutionary game strategy of linear biological network in carcinogenesis. By solving the constrained stochastic Nash quadratic (H_2) game problem in (19.21) and (19.22) [or (19.19)] for the natural selection strategy of the linear evolutionary stochastic biological network in (19.9) with the effect of the natural selection in carcinogenesis, we get the following result.

19.4.1 Proposition 19.1

For the linear stochastic evolutionary biological network (19.9) in somatic evolutionary process, the suboptimal stochastic H_∞ evolutionary game problem in (19.18) or (19.19) can be solved by the following somatic evolutionary game strategy $u^*(t)$ and worst-case environmental disturbance $v^*(t)$:

$$u^*(t) = \sum_{k=1}^{N_1(t)} A_k x(t) p(t - t_k') = -Px(t) \tag{19.23}$$

$$v^*(t) = \frac{1}{e} B^T P x(t) \tag{19.24}$$

where the positive-definite matrix $P>0$ is the solution of the following Riccati-like inequalities:

$$\begin{aligned} &PA + A^T P - PP + Q \\ &+ \sum_{k=1}^{N_{2p}} \lambda_2 (PA_k + A_k^T P + A_k^T P A_k - e A_k^T A_k) + \frac{1}{e} PBB^T P \leq 0 \end{aligned} \tag{19.25}$$

and

$$P \leq eI \tag{19.26}$$

where $N_{2p} \triangleq N_2(t_p)$, and N_{2p} represents the total number of selected phenotypic variations and neutral phenotypic variations to the present time t_p, respectively.

Proof: see Appendix A.

The Proposition 19.1 is the solution of the suboptimal H_∞ stochastic evolutionary game in (19.18) or (19.19). We need to minimize the upper bound e to approach the network evolvability e_0 in (19.17) of the linear biological network (19.9) in the somatic evolutionary process. The stochastic H_∞ game in (19.17) for network evolvability e_0 of the biological network in (19.9) needs to solve the following constrained optimization problem:

$$\begin{aligned} &e_0 = \min_{P>0} e \\ &\text{subject to (19.25) and (19.26)} \end{aligned} \tag{19.27}$$

Remark 19.2

i. The solution of the above network evolvability e_0 of linear biological network (19.9) in carcinogenesis can be obtained by decreasing e until no positive solution P in (19.25) within $0<P \leq eI$ in (19.27).

ii. After solving e_0 from (19.27), the Riccati-like inequality in (19.25) becomes

$$\begin{aligned} &PA + A^T P - PP + Q \\ &+ \sum_{k=1}^{N_{2p}} \lambda_2 (PA_k + A_k^T P + A_k^T P A_k - e_0 A_k^T A_k) + \frac{1}{e_0} PBB^T P \leq 0 \end{aligned} \tag{19.28}$$

which can be considered as the phenotypic robustness criterion of stochastic evolutionary biological network in (19.9) because the inequality in (19.28) is required for the robust stability of the evolutionary biological network under genetic and epigenetic variations and environmental disturbances in the somatic evolutionary process.

iii. The term $\sum_{k=1}^{N_{2p}} \lambda_2(PA_k + A_k^T P + A_k^T PA_k - e_0 A_k^T A_k)$ in the phenotypic robustness criterion (19.28) is the effect of neutral phenotypic variations that are not selected by natural selection but are accumulated randomly at the biological network in carcinogenesis.

iv. After solving e_0 and $P>0$ from the constrained optimization problem in (19.27), the solution of stochastic H_∞ evolutionary game strategy in (19.17) for biological network of an organ by natural selection in the somatic evolutionary process is obtained by modifying (19.23) and (19.24) as

$$u^*(t) = \sum_{k=1}^{N_1(t)} A_k x(t) p(t - t_k^{'}) = -Px(t) \tag{19.29}$$

$$v^*(t) = \frac{1}{e_0} B^T Px(t) \tag{19.30}$$

where $P>0$ is the solution of (19.28). The somatic H_∞ evolutionary game strategy $u^*(t)$ by natural selection in (19.29) is to select an adequate amount of phenotypic variations $\sum_{k-1}^{N_1(t)} A_k p(t - t_k^{'}) = \sum_{k=1}^{N_1(t)} A_k$ to approach $-P$ to enhance the robust stability of the biological system and minimize the evolution level e (or maximize the phenotypic fitness of the biological network of an organ via $f_0 = \frac{1}{e_0}$). With the negative feedback $-Px$ because of stochastic H_∞ evolutionary game evolution strategy by natural selection, the phenotypic robustness criterion in (19.28) was rearranged as

$$\begin{aligned} &P(A - \frac{1}{2}P) + (A - \frac{1}{2}P)^T P + Q \\ &+ \sum_{k=1}^{N_{2p}} \lambda_2(PA_k + A_k^T P + A_k^T PA_k - eA_k^T A_k) + \frac{1}{e_0} PBB^T P \le 0 \end{aligned} \tag{19.31}$$

The eigenvalues of $(A - \frac{1}{2}P)$ are farther to the left-hand side of the complex s-domain than those of A, such that the network robustness of biological network is improved by negative feedback loops of the stochastic H_∞ evolutionary game strategy $u^*(t) = \sum_{k=1}^{N_1(t)} A_k x(t) p(t - t_k^{'}) = -Px(t)$ in the somatic evolutionary process.

Further, the phenotypic robustness criterion in (19.31) of evolutionary biological network in carcinogenesis was reformulated as

$$\begin{aligned} &\underbrace{Q + \frac{1}{e_0} PBB^T P}_{\text{environmental robustness}} \quad \underbrace{+ \sum_{k=1}^{N_{2p}} \lambda_2(PA_k + A_k^T P + A_k^T PA_k - eA_k^T A_k)}_{\text{phenotypic robustness}} \\ &\underbrace{\le -[P(A - \frac{1}{2}P) + (A - \frac{1}{2}P)^T P]}_{\text{enhanced network robustness}} \end{aligned} \tag{19.32}$$

The physical meaning of network phenotypic robustness criterion in (19.32) is that if the enhanced network robustness of the biological network through negative feedback loops of the stochastic H_∞ evolutionary strategy in (19.17) can confer the environmental robustness to resist the environmental disturbances and phenotypic robustness and tolerate the neutral Poisson phenotypic variations, the phenotypic trait of biological network of an organ is maintained in carcinogenesis. Otherwise, the phenotypic trait of biological network of an organ may not be maintained under environmental disturbances and random Poisson genetic and epigenetic variations in carcinogenesis [445]. According to the statistical results [482], the lifetime risk of many different cancers is strongly correlated (0.81) with the total number of divisions of the normal self-renewing cells maintaining that tissue's homeostasis. Also, the violation of phenotypic robustness criterion in (19.32) can be explained by the number of cell divisions, and a third of cancer risk among tissues is attributable to environmental factors. Namely, the cause of the violation of phenotypic robustness criterion in (19.32) during carcinogenesis is attributed 1/3 risk (i.e., $Q+\frac{1}{e_0}PBB^TP$) to environmental factor and 2/3 risk to intrinsic phenotypic factor (i.e., $\sum_{k=1}^{N_{2p}}\lambda_2(PA_k+A_k^TP+A_k^TPA_k-e_0A_k^TA_k)$) in (19.32).

v. Through the stochastic H_∞ evolutionary game strategy of biological network of an organ in carcinogenesis, the accumulated phenotypic variations $\sum_{k=1}^{N_1(t)} A_kp(t-t_k')=\sum_{k=1}^{N_1(t_p)} A_kp(t-t_k')=\sum_{k=1}^{N_{1p}} A_k$, where $N_{1p}\triangleq N_1(t_p)$ are chosen by natural selection to the amount of $-P$ to construct negative feedback loops $-Px(t)$ in (19.29). This is why many feedback loops, and the epigenetic methylations and miRNA regulations in caner are always found to be with negative feedback (inhibition). The stochastic biological network of an organ in (19.9) was thus of the following form:

$$\frac{dx(t)}{dt}=(A-P)x(t)+\sum_{k=1}^{N_2(t)}A_kx(t)p(t-t_k^*)+Bv(t) \tag{19.33}$$

Because the eigenvalues of $(A-P)$ of biological network are in the far left-hand side of the complex s-domain, the biological network of an organ is more robust by the stochastic H_∞ game strategy of natural selection to resist environmental disturbances and to tolerate neutral phenotypic variations.

vi. The stochastic H_∞ evolutionary game strategy by natural selection can provide negative feedback loop $-Px(t)$ for a greater network stability robustness improvement to buffer more neutral phenotypic variations $\sum_{k=1}^{N_2(t)} A_kx(t)p(t-t_k^*)$ in carcinogenesis. However, when the neutral phenotypic variations because of Poisson genetic and epigenetic variations are accumulated in carcinogenesis, i.e., the term $\sum_{k=1}^{N_{2p}}\lambda_2(PA_k+A_k^TP+A_k^TPA_k-eA_k^TA_k)$ is large enough, and the network phenotypic robustness criterion of biological network in (19.32) is violated eventually, and the biological network will become unstable at the healthy state, $x(t)=x_e$. Finally, the organ will become cancer in the somatic evolutionary process.

19.5 Stochastic H_∞ Evolutionary Game Strategy of Nonlinear Biological Network in Carcinogenesis

After solving the stochastic H_∞ evolutionary game strategy of linear biological network of an organ by natural selection in carcinogenesis, we will solve the stochastic H_∞ game problem in (19.17) for the evolutionary strategy of nonlinear biological network in (19.13) by natural selection in carcinogenesis. First, we solved the suboptimal stochastic H_∞ evolutionary game problem in (19.18), or equivalently in (19.21) and (19.22), for nonlinear stochastic biological network in (19.17) operating near the healthy state x_e in Fig. 19.1A.

Position 19.2 For the nonlinear stochastic biological network (19.13) in somatic evolutionary process, the suboptimal stochastic Nash game problem in (19.18) or (19.21) and (19.22) can be solved by the following evolutionary game strategy $u^*(t)$ and worst-case environmental disturbance $v^*(t)$:

$$u^*(t) = \sum_{k=1}^{N_1(t)} f_k(\tilde{x}(t))p(t-t_k) = -\frac{1}{2}\frac{\partial V(\tilde{x}(t))}{\partial \tilde{x}} \tag{19.34}$$

$$v^*(t) = \frac{1}{2e}B^T\frac{\partial V(\tilde{x}(t))}{\partial \tilde{x}} \tag{19.35}$$

where the Lyapunov function $V(\tilde{x}(t)) > 0$ is the solution of the following constrained HJII:

$$\begin{aligned}
&\left(\frac{\partial V(\tilde{x}(t))}{\partial \tilde{x}}\right)^T f(\tilde{x}) + \sum_{k=1}^{N_2(t)} (V(\tilde{x} + f_k(\tilde{x})p(t-t_k^*) - V(\tilde{x})) \\
&-\frac{1}{4}\left(\frac{\partial V(\tilde{x}(t))}{\partial \tilde{x}}\right)^T\left(\frac{\partial V(\tilde{x}(t))}{\partial \tilde{x}}\right) + \frac{1}{4}\left(\frac{\partial V(\tilde{x}(t))}{\partial \tilde{x}}\right)^T B^TB\left(\frac{\partial V(\tilde{x}(t))}{\partial \tilde{x}}\right) \\
&+\tilde{x}^T(t)Q\tilde{x}(t) - e\sum_{k=1}^{N_2(t)} f_k^T(\tilde{x})f_k(\tilde{x})p(t-t_k^*) \leq 0
\end{aligned} \tag{19.36}$$

with

$$E\{V(\tilde{x}(t))\} \leq eE\{\tilde{x}^T(0)\tilde{x}(0)\} \tag{19.37}$$

Proof: see Appendix B.

After solving the above suboptimal H_∞ evolutionary game problem in (19.18) or (19.21) and (19.22), for the stochastic biological network in (19.13), the stochastic H_∞ evolutionary strategy by natural selection in (19.17) was solved by minimizing the upper bound e as follows:

$$\begin{aligned}
e_0 = &\min_{V(\tilde{x}(t))>0} e \\
&\text{subject to (19.36) and (19.37)}
\end{aligned} \tag{19.38}$$

If $V(\tilde{x})$ and e_0 are obtained by solving the HJII-constrained optimization in (19.38), the network robust stability criterion in (19.36) and (19.37) in the somatic evolutionary process was modified as

$$\begin{aligned}&\left(\frac{\partial V(\tilde{x}(t))}{\partial \tilde{x}}\right)^T f(\tilde{x}(t)) + \sum_{k=1}^{N_2(t)} (V(\tilde{x}(t) + f_k(\tilde{x}(t))p(t-t_k^*) - V(\tilde{x}(t)))\\&-\frac{1}{4}\left(\frac{\partial V(\tilde{x}(t))}{\partial \tilde{x}}\right)^T\left(\frac{\partial V(\tilde{x}(t))}{\partial \tilde{x}}\right) + \frac{1}{4e_0}\left(\frac{\partial V(\tilde{x}(t))}{\partial \tilde{x}}\right)^T B^T B\left(\frac{\partial V(\tilde{x}(t))}{\partial \tilde{x}}\right)\\&+\tilde{x}^T(t)Q\tilde{x}(t) - e_0\sum_{k=1}^{N_2(t)} f_k^T(\tilde{x})f_k(\tilde{x})p(t-t_k^*) \leq 0\end{aligned} \tag{19.39}$$

with $V(\tilde{x}(t)) \leq e_0\tilde{x}^T(0)\tilde{x}(0)$. Since the HJI in (19.39) is necessary for the phenotype at $\tilde{x}(t) = 0$ or $x(t) = x_e$ to be robust, it can be considered as the phenotypic trait robustness criterion of nonlinear stochastic network in (19.13) at the phenotypic state x_e in the somatic evolutionary process.

In this situation, the evolutional game strategy $u^*(t)$ in (19.34) and the worst-case environmental disturbance $v^*(t)$ in (19.35) were modified, respectively, as follows:

$$u^*(t) = \sum_{k=1}^{N_1(t)} f_k(\tilde{x}(t))p(t-t_k') = -\frac{1}{2}\frac{\partial V(\tilde{x}(t))}{\partial \tilde{x}} \tag{19.40}$$

$$v^*(t) = \frac{1}{2e_0}B^T\frac{\partial V(\tilde{x}(t))}{\partial \tilde{x}} \tag{19.41}$$

where the Lyapunov function $V(\tilde{x}(t))$ is the solution of the network robust stability criterion in (19.39). The stochastic H_∞ evolutionary game strategy of $u^*(t)$ is to select $\sum_{k=1}^{N_1(t)} f_k(\tilde{x}(t))p(t-t_k')$ to approach the nonlinear negative feedback $-\frac{\partial V(\tilde{x}(t))}{\partial \tilde{x}}$ to minimize the evolution level e, and v^* denotes the worst-case disturbance to maximize the evolution level e.

The robust stability criterion in (19.39) was rewritten as the following phenotypic robustness criterion in carcinogenesis:

$$\begin{aligned}&\underbrace{\sum_{k=1}^{N_2(t)} (V(\tilde{x}(t) + f_k(\tilde{x}(t))p(t-t_k^*) - V(\tilde{x}(t))) - e_0\sum_{k=1}^{N_2(t)} f_k^T(\tilde{x})f_k(\tilde{x})p(t-t_k^*)}_{\text{phenotypic robustness}}\\&\underbrace{+\tilde{x}^T(t)Q\tilde{x}(t) + \frac{1}{4e_0}\left(\frac{\partial V(\tilde{x}(t))}{\partial \tilde{x}}\right)^T B^T B\left(\frac{\partial V(\tilde{x}(t))}{\partial \tilde{x}}\right)}_{\text{environmental robustness}}\\&\leq \underbrace{-\frac{1}{2}\left(\frac{\partial V(\tilde{x}(t))}{\partial \tilde{x}}\right)^T\left(f(\tilde{x}(t)) - \frac{1}{4}\left(\frac{\partial V(\tilde{x}(t))}{\partial \tilde{x}}\right)\right)}_{\text{enhanced network robustness}}\\&\underbrace{-\frac{1}{2}\left(f(\tilde{x}(t)) - \frac{1}{4}\left(\frac{\partial V(\tilde{x}(t))}{\partial \tilde{x}}\right)\right)^T\left(\frac{\partial V(\tilde{x}(t))}{\partial \tilde{x}}\right)}_{\text{enhanced network robustness}}\end{aligned} \tag{19.42}$$

The first two terms are because of neutral phenotypic variations from the somatic genetic and epigenetic variations in carcinogenesis, the last two terms on the left-hand side of (19.42) are by the worst-case environmental disturbance, and the terms on the right-hand side of (19.42) are because of the negative feedback through evolutionary stochastic H_∞ evolutionary game strategy $u^*(t) = -\frac{1}{2}\frac{\partial V(\tilde{x}(t))}{\partial \tilde{x}}$. The biological meaning of phenotypic robustness criterion in (19.42) is that if the enhanced network robustness by stochastic H_∞ evolutionary strategy $u^*(t)$ can confer phenotypic robustness to tolerate the neutral phenotypic variations $\sum_{k=1}^{N_2(t)} f_k(\tilde{x}(t))p(t-t_k^*)$ and environmental disturbances, the healthy state (or phenotypic trait) x_e was maintained in the somatic evolutionary process. By substituting the stochastic H_∞ evolutionary game strategy $u^*(t)$ in (19.40) into the nonlinear stochastic biological network in (19.13), we get

$$\frac{d\tilde{x}(t)}{dt} = \left(f(\tilde{x}(t)) - \frac{1}{2}\left(\frac{\partial V(\tilde{x}(t))}{\partial \tilde{x}}\right)\right) + \sum_{k=1}^{N_2(t)} f_k(\tilde{x}(t))p(t-t_k^*) + Bv(t) \tag{19.43}$$

Form (19.43), it is seen that $\left(f(\tilde{x}(t)) - \frac{1}{2}\left(\frac{\partial V(\tilde{x}(t))}{\partial \tilde{x}}\right)\right)$ can make the nonlinear stochastic biological network of an organ more robustly stable than $f(\tilde{x}(t))$ because of the negative feedback loops of stochastic H_∞ game evolutionary strategy by considering natural selection in carcinogenesis.

If the network robustness on the right-hand side of (19.42) is improved by the stochastic H_∞ game evolutionary strategy, from the landscape of phenotype in Fig. 19.1A, the healthy phenotypic trait is with deeper basin and steep cliff at the equilibrium point x_e, and the nonlinear biological network can harbor much more neutral phenotypic variations $\sum_{k=1}^{N_2(t)} f_k(\tilde{x}(t))p(t-t_k^*)$, i.e., much more dots in the basin of x_e. However, as time increases, the neutral phenotypic variations are accumulated to a sufficiently large amount in carcinogenesis that the enhanced network robustness on the right-hand side of (19.42) cannot provide enough phenotypic robustness to tolerate these accumulated neutral phenotypic variations and resist environmental disturbances, and eventually the phenotypic robustness criterion in (19.42) is violated. This is why cancer is an aging-associated disease and the accumulating process of network phenotypic variations can be considered as an aging process of the tissue. In this situation, these accumulated neutral phenotypic variations may provide raw materials for biological network evolution to cause a possible phenotypic transition to the early stage of tumor state at the equilibrium point x_{e1}, i.e., a phenotypic transition may occur from the basin x_e to another basin x_{e1} in Fig. 19.1B to start another cancer phenotypic evolutionary process at the equilibrium point x_{e1} in carcinogenesis.

Remark 19.3 For the early stage of tumor at the equilibrium point x_{e1} of stage I in Fig. 19.1B, by a similar analysis as above, we could shift the origin of nonlinear stochastic biological network of an organ to x_{e1} as follows:

$$\frac{d\tilde{x}'(t)}{dt} = f(\tilde{x}'(t) + u(t) + \sum_{k=1}^{N_2'(t)} f_k(\tilde{x}'(t))p(t-t_k^*) + Bv'(t) \tag{19.44}$$

where $\tilde{x}'(t) = x(t) - x_{e1}$.

As the stochastic H_∞ evolutionary game strategy in (19.40) is chosen for nonlinear stochastic network in (19.43) by natural selection in the healthy stage of carcinogenesis, the biological network of an organ with early cancer stage is of the following form through the stochastic H_∞ evolutionary game strategy at x_{e1}:

$$\frac{d\tilde{x}'(t)}{dt} = \left(f(\tilde{x}'(t) - \frac{1}{2}\left(\frac{\partial V(\tilde{x}'(t))}{\partial \tilde{x}'}\right)\right) + \sum_{k=1}^{N_2'(t)} f_k(\tilde{x}'(t))p(t - t_k^*) + Bv'(t) \tag{19.45}$$

As the neutral phenotypic variations $\sum_{k=1}^{N_2'(t)} f_k(\tilde{x}'(t))p(t - t_k^*)$ are accumulated to a sufficiently large amount in the early stage of carcinogenesis so that the basin of x_{e1} could not tolerate these neutral phenotypic variations and has not enough ability to resist environmental disturbances simultaneously, the biological network transits from the equilibrium point x_{e1} to the equilibrium point x_{e2} of another phenotype of cancer to begin the second stage of carcinogenesis in Fig. 19.1C. Finally, the phenotype of cancer stage II will be transited to the phenotype of cancer stage III in Fig. 19.1D when the phenotypic robustness criterion at x_{2e} is violated by Poisson neutral genetic and epigenetic variations in carcinogenesis. In the late stage of carcinogenesis, the network evolutionary strategy that maintains network robustness can be hijacked by cancer biological network to maintain dysfunction at the other equilibrium point in Fig. 19.1, as that occurs in cancer drug resistance [473,474].

Remark 19.4

i. The feedback loop $-\frac{1}{2}\left(\frac{\partial V(\tilde{x}(t))}{\partial \tilde{x}}\right) = \sum_{k=1}^{N_1(t)} f_k(\tilde{x}(t))p(t - t_k')$ provides the orderly tissue structure of organ by considering natural selection. However, the accumulated neutral phenotypic variations $\sum_{k=1}^{N_2(t)} f_k(\tilde{x}(t))p(t - t_k^*)$ are the source of random and complex tissue structures of cancer and provide raw materials to increase the heterogeneity of tumor cells for a new evolutionary possibility to cause a transition from one cancer state to another cancer state in carcinogenesis.

ii. In Fig. 19.1B and C, the cancer states at x_{1e} and x_{2e} are still curable, i.e., the carcinogenesis is still reversible to Fig. 19.1A by an adequate surgical or medical treatment (one kind of enhanced negative feedback). If the healthy state is incurable (saddle point) as shown in Fig. 19.1D, then it is impossible that this process is reversible to the healthy state in Fig. 19.1A.

19.6 The Stochastic H_∞ Evolutionary Game of Nonlinear Cancer-Associated Network in Carcinogenesis

In general, it is very difficult to solve the HJII-constrained optimization problem in (19.38) for the stochastic H_∞ evolutionary game strategy $u^*(t)$ of the somatic evolutionary biological network in (19.11) in carcinogenesis. At present, there is no efficient method to solve the nonlinear partial differential HJII in (19.39) either analytically or numerically. Therefore, in order to simplify the stochastic H_∞ evolutionary game strategy, the global linearization technique [23] is employed to interpolate the nonlinear stochastic biological network in (19.13) with a set of local stochastic linearized biological networks like (19.9) at different operation points. Using the global linearization technique [23], we suppose that all the global linearization of the nonlinear stochastic biological network in (19.13) are bounded by the following L vertices as [23]

$$\begin{bmatrix} \frac{\partial f(\tilde{x})}{\partial \tilde{x}} \\ \frac{\partial f_1(\tilde{x})}{\partial \tilde{x}} \\ \vdots \\ \frac{\partial f_{N(t)}(\tilde{x})}{\partial \tilde{x}} \end{bmatrix} = C_0 \left[\begin{bmatrix} A_1 \\ A_{11} \\ \vdots \\ A_{N(t)1} \end{bmatrix} \cdots \begin{bmatrix} A_i \\ A_{1i} \\ \vdots \\ A_{N(t)i} \end{bmatrix} \cdots \begin{bmatrix} A_L \\ A_{1L} \\ \vdots \\ A_{N(t)L} \end{bmatrix} \right], \forall \tilde{x}(t) \tag{19.46}$$

where C_0 denotes the convex hull of the polytope with L vertices. That is, if the local linearized biological systems at all $\tilde{x}(t)$ are inside the convex hull C_0, the trajectory $\tilde{x}(t)$ of the nonlinear evolutionary biological network in (19.13) can be represented by the convex combination of the stable trajectories of the following L local linearized biological networks at L vertices of the polytope in (19.46):

$$\frac{d\tilde{x}(t)}{dt} = A_i\tilde{x}(t) + Bv(t) + u(t) + \sum_{k=1}^{N_2(t)} A_{ki}p(t - t_k^*)\tilde{x}(t), \text{for} i = 1, \ldots, L \tag{19.47}$$

Based on the global linearization theory [23], if (19.46) holds, every trajectory of nonlinear stochastic evolutionary biological network in (19.13) can be represented by a convex combination of L local linear biological networks in (19.47), i.e.,

$$\frac{d\tilde{x}(t)}{dt} = \sum_{i=1}^{L} \alpha_i(\tilde{x})(A_i\tilde{x}(t) + Bv(t) + u(t) + \sum_{k=1}^{N_2(t)} A_{ki}p(t - t_k^*)\tilde{x}(t)) \tag{19.48}$$

where the interpolation functions $\alpha_i(\tilde{x})$ satisfy the following constraints: $0 \leq \alpha_i(\tilde{x}) \leq 1$ and $\sum_{i=1}^{L} \alpha_i(\tilde{x}) = 1$. The physical meaning of (19.48) is that the evolutionary trajectory of the nonlinear stochastic biological network in (19.13) can be represented by the evolutionary trajectory of the interpolated biological network in (19.48). In (19.48), the evolutionary strategy $u(t) = \sum_{k=1}^{N_1(t)} f_k(\tilde{x})p(t - t_k')$ in (19.13) can be represented as $u(t) = \sum_{i=1}^{L} \sum_{k=1}^{N_1(t)} \alpha_i(\tilde{x})A_{ki}p(t - t_k')\tilde{x}(t)$ by the global linearization technique. For the stochastic H_∞ evolutionary game strategy problem in (19.17) in the carcinogenesis of the nonlinear stochastic network in (19.48), we get the following results for Nash quadratic (H_2) game problem in (19.19), or (19.21), and (19.22).

Proposition 19.3 For the interpolated evolutionary biological network (19.48) in carcinogenesis, the stochastic Nash quadratic (H_2) evolutionary game problem in (19.19) is solved for the suboptimal H_∞ evolutionary game in (19.18) by the following evolutionary strategy $u^*(t)$ and the worst-case environmental disturbance $v^*(t)$:

$$u^*(t) = \sum_{i=1}^{L} \sum_{k=1}^{N_1(t)} \alpha_i(\tilde{x})A_{ki}p(t - t_k')\tilde{x}(t) = -P\tilde{x} \tag{19.49}$$

$$v^*(t) = \frac{1}{e_0} B^T P \tilde{x}(t) \tag{19.50}$$

where the positive-definite matrix P is the solution of the following Riccati-like inequalities:

$$\begin{aligned} &PA_i + A_i^T P - PP + Q \\ &+ \sum_{k=1}^{N_{2p}} \lambda_2 (PA_{ki} + A_{ki}^T P + A_{ki}^T P A_{ki} - e A_{ki}^T A_{ki}) + \frac{1}{e} PBB^T P \leq 0, i = 1, \ldots, L \end{aligned} \tag{19.51}$$

$$0 < P < eI \tag{19.52}$$

Proof: see Appendix C.

Remark 19.5

i. In comparison with Proposition 19.1 of a stochastic linear biological network in carcinogenesis in (19.9), the Riccati-like inequality in (19.25) is based on the local linearized network at $\tilde{x}(t) = 0$, while the L Riccati-like inequalities in (19.51) are based on L local linearized biological networks in (19.48). Hence, the result of the Nash quadratic (H_2) evolutionary game strategy in Proposition 19.3 is more suitable for the suboptimal H_∞ evolutionary game of nonlinear stochastic biological system in carcinogenesis.

ii. The Riccati-like inequalities in (19.51) can be considered as the local linearization of HJI in (19.36) at the L vertices of polytope in (19.46).

iii. For solving P in (19.51), the L Riccati-like inequalities in (19.52) can be transformed to the following equivalent LMIs by the Schur complement transformation method [23]:

$$\begin{bmatrix} PA_i + A_i^T P + Q + \sum_{k=1}^{N_{2p}} \lambda_2 (PA_{ki} + A_{ki}^T P + A_{ki}^T P A_{ki} - e A_{ki}^T A_{ki}) & P \\ P & -(\frac{1}{e} BB^T - I) \end{bmatrix} \leq 0 \tag{19.53}$$

for$i = 1, 2, \ldots, L$.

These LMIs in (19.53) can be easily solved by using the LMI toolbox in MATLAB. However, e in (19.51) and (19.52) is only the upper bound of the network evolvability e_0. The network evolvability e_0 of the nonlinear biological network under evolutionary game strategy can be obtained by solving the following LMI-constrained optimization problem:

$$\begin{aligned} e_0 &= \min_{P>0} e \\ &\text{subject to (19.52) and (19.53)} \end{aligned} \tag{19.54}$$

Remark 19.6

i. The e_0 in (19.54) could be solved by decreasing e until (19.53) has no positive solution for $0 < P \leq eI$ with the help of LMI toolbox in MATLAB.

ii. After solving e_0 in (19.54), the Riccati-like inequalities for the network robustness criteria of local linearized biological networks in (19.51) become

$$\begin{aligned}&PA_i + A_i^T P - PP + Q + \sum\nolimits_{k=1}^{N_{2p}} \lambda_2 (PA_{ki} + A_{ki}^T P \\ &+A_{ki}^T PA_{ki} - e_0 A_{ki}^T A_{ki}) + \frac{1}{e_0} PBB^T P \leq 0, i = 1, \ldots, L\end{aligned} \tag{19.55}$$

iii. After solving P and e_0 from (19.54), the evolutionary stochastic H_∞ game strategy $u^*(t) = -P\tilde{x}(t)$ and the worst-case disturbance $v^*(t) = \frac{1}{e_0} B^T P\tilde{x}(t)$ are solved for the nonlinear stochastic network in (19.13).

iv. The Riccati-like inequalities for local network robustness criteria in (19.51) can be rearranged as

$$\begin{aligned}&P(A_i - \frac{1}{2}P) + (A_i - \frac{1}{2}P)^T P + Q \\ &+ \sum\nolimits_{k=1}^{N_{2p}} \lambda_2 (PA_{ki} + A_{ki}^T P + A_{ki}^T PA_{ki} - eA_{ki}^T A_{ki}) + \frac{1}{e_0} PBB^T P \leq 0, i = 1, \ldots, L\end{aligned} \tag{19.56}$$

Obviously, every interaction matrix A_i of local linearized biological networks is shifted by $-\frac{1}{2}P$ so that the eigenvalues of $(A_i - \frac{1}{2}P)$ are on the farther left-hand side of the s-complex domain, and thus the evolutionary biological network is with more network robustness to tolerate local genetic and epigenetic variations and resist environmental disturbances in carcinogenesis.

v. The stochastic H_∞ evolutionary game strategy $u^*(t)$ is to select the total amount of phenotypic variations $\sum_{i=1}^{L} \sum_{k=1}^{N_1(t)} \alpha_i(\tilde{x}) A_{ki} p(t - t_k') = -P$. This is so as to select adequate genetic and epigenetic variations to form negative feedback loops $-P\tilde{x}(t)$ in the nonlinear biological network, which can resist environmental disturbances and buffer neutral genetic and epigenetic variations in carcinogenesis. Therefore, the increase of inhibitive epigenetic methylation and miRNA regulation can be observed [466] in carcinogenesis [483].

vi. The phenotypic robustness criterion in (19.56) can be rearranged as

$$\begin{aligned}&\underbrace{Q + \frac{1}{e_0} PBB^T P}_{\text{localenvironmental robustness}} \quad + \underbrace{\sum\nolimits_{k=1}^{N_{2p}} \lambda_2 (PA_{ki} + A_{ki}^T P + A_{ki}^T PA_{ki} - eA_{ki}^T A_{ki})}_{\text{local phenotypic robustness}} \\ &\leq \underbrace{-[P(A_i - \frac{1}{2}P) + (A_i - \frac{1}{2}P)^T P]}_{\text{enhanced local network robustness}}, \text{for} i = 1, \ldots, L\end{aligned} \tag{19.57}$$

The biological meaning in (19.57) is that if the enhanced local network robustness of each local linearized biological network by using stochastic H_∞ evolutionary game strategy $u^*(t)$ under natural selection can confer both local phenotypic robustness to buffer neutral local somatic genetic and epigenetic variations and local environmental robustness and to resist the local environmental disturbances in carcinogenesis, the phenotype of nonlinear stochastic network with stochastic H_∞ game evolutionary strategy can be maintained in the basin of the equilibrium point in carcinogenesis.

vii. From the phenotype robustness criterion in (19.57) in the cancer evolutionary process, it is more appealing to make the left-hand side of (19.57) as small as possible and to make the right-hand side of (19.57) as large as possible. In this situation, the coupling matrix B between biological network and environmental disturbances, as well as phenotypic effects A_{ki} (because of genetic mutations and epigenetic alterations), should be as small as possible. In order to maintain network robustness, there always exists some membrane or transparent membrane to make B smaller to isolate the biological network from environmental disturbance, and some redundant and modular structures make the phenotypic effect A_{ki} of somatic genetic mutations and epigenetic alterations as small as possible. Further, the negative feedback loop $-P\tilde{x}(t)$ mediated by the stochastic H_∞ evolutionary game strategy $u^*(t)$ makes the right-hand side of (19.57) as large as possible. These networks are also hijacked by cancer to maintain dysfunction in the late stage of carcinogenesis. This always occurs in cancer drug resistance [473,474].

viii. Because the stochasticH_∞ evolutionary game strategy can improve significantly the local network robustness of each local linearized biological network, the nonlinear stochastic biological network will harbor more neutral genetic and epigenetic variations. However, as the harbored neutral genetic and epigenetic variations are accumulated to the extent that the third term in (19.57) becomes very large and the phenotypic robustness criterion of biological network in (19.57) cannot be guaranteed in carcinogenesis, the phenotype of biological network might be shifted to cancer state at another equilibrium point in Fig. 19.1 with the help of the environmental disturbances and start another period of cancer network evolution at the other equilibrium point favored by natural selection.

19.7 Simulation Result

Considering the stochastic H_∞ evolutionary game strategy in the oncogenesis of lung cancer, some cell lineages related to lung cancer are described as follows. The tissue contains parenchymal cells and stromal cells. The stroma typically contains VECs, pericytes, and smooth muscle cells in its blood vessels [475,476,478,479]. Along with fibroblasts and other cell types, parenchymal cells tend to exhibit considerable phenotypic variations because of genetic and epigenetic variations during oncogenesis [472].

Let $x_1(t)$ and $x_2(t)$ be the mass of parenchymal cells with phenotypes 1 and 2, respectively. Further, let $y(t)$ be the mass of immature VECs within the tumor and $z(t)$ be the mass of VECs, which can be represented by the length of existing microvessel in 1 g of undiseased tissue (i.e., one unit is equal to microvessel length in 1 g of undiseased tissue). The dynamic system of cells related to lung cancer is thus described as follows [472]:

$$\begin{aligned}
\frac{dx_1(t)}{dt} &= \Phi_1(v)x_1(t) - 0.01x_1(t) \\
\frac{dx_2(t)}{dt} &= \Phi_2(v)x_2(t) - 0.01x_2(t) \\
\frac{dy(t)}{dt} &= (0.06H(x_1, x_2, z) - 0.04)y(t) \\
\frac{dz(t)}{dt} &= 3y(t) - 0.004m(t)z(t)
\end{aligned} \tag{19.58}$$

where
$b_1(m) = 0.28c(m)e^{-0.06c(m)}, b_2(m) = 0.4c(m)e^{-0.06c(m)}, m(t) = z(t)/[x_1(t) + x_2(t)],\ c(m) = 95m(t)/[1.375 + m(t)]$, and $H(x_1, x_2, z) = [x_1(t)b_1(m) + x_2(t)b_2(m)]/[x_1(t) + x_2(t)]$.

The variable $m(t)$ represents tumor vascularization (perfusion) in microvessel units per gram of parenchyma [$m(t)$, to a physiologist, is proportional to tumor microvessel length density]. The functions $\Phi_i(m)$ express per capita growth rate of cell type i as a function of blood supply and are represented by the following equations, respectively [472,479]:

$$\Phi_1(m) = \frac{0.05c^2(m)}{0.76^2 + c^2(m)} - 0.05(1 - \frac{c^2(m)}{0.4^2 + c^2(m)}) \tag{19.59}$$

and

$$\Phi_2(m) = \frac{0.05c^2(m)}{0.53^2 + c^2(m)} - 0.03(1 - \frac{c^2(m)}{0.1^2 + c^2(m)}) \tag{19.60}$$

In carcinogenesis, the tissue dynamic system in (19.58) suffers from phenotypic variations because of somatic genetic and epigenetic variations and environmental disturbances as the following nonlinear stochastic Poisson system:

$$\begin{bmatrix} \frac{dx_1(t)}{dt} \\ \frac{dx_2(t)}{dt} \\ \frac{dy(t)}{dt} \\ \frac{dz(t)}{dt} \end{bmatrix} = f(x) + \sum_{k=1}^{N_1(t)} f_k(\tilde{x}(t))p(t - t_k^{'}) + \sum_{k=1}^{N_2(t)} f_k(\tilde{x}(t))p(t - t_k^{*}) + v(t) \tag{19.61}$$

where $x(t) = \begin{bmatrix} x_1(t) \\ x_2(t) \\ y(t) \\ z(t) \end{bmatrix}, f(x) = \begin{bmatrix} \Phi_1(m)x_1(t) - 0.01x_1(t) \\ \Phi_2(m)x_2(t) - 0.01x_2(t) \\ (0.06H(x_1, x_2, z) - 0.04)y(t) \\ 3y(t) - 0.004m(t)z(t) \end{bmatrix}, f_k(x) = 0.5f(x)$ and the mean of Poisson process is imposed as $\lambda_2 = 0.05$.

The equilibrium point x_e of the nominal lung tissue system is at $x_e = [x_{e1}\ x_{e2}\ x_{e3}\ x_{e4}]^T = [6.19 \times 10^{-11} 2.20 \times 10^{-3}\ 0\ 9.88 \times 10^{-6}]^T$ of the nominal system in (19.58) simulated from the initial condition $x(0) = [10^{-2}\ \ -1.20 \times 10^{-2}\ 0\ 1.16 \times 10^{-7}]^T$. For the simplicity, the state of the lung tissue in (19.61) is shifted to the form $\tilde{x}(t) = x(t) - x_e$. Thus, the new equilibrium point of the phenotype concerned is at $\tilde{x}_e = O_{4\times1}$, through the following in silico example.

Based on the above evolutionary game strategy, the phenotypic variations $\sum_{k=1}^{N_1(t)} f_k(\tilde{x})p(t - t_k^{'})$ are selected by considering natural selection to construct negative feedback loops to modify the lung cancer-associated cell system in (19.58), in order to resist environmental disturbances and to tolerate the effect of the neutral phenotypic variations $\sum_{k=1}^{N_2(t)} f_k(\tilde{x})p(t - t_k^{*})$. According to the global linearization scheme in (19.46)–(19.48), the nonlinear stochastic Poisson system in (19.61) can be approximated by interpolating the following l local linear stochastic Poisson systems:

$$\begin{bmatrix} \frac{d\tilde{x}_1(t)}{dt} \\ \frac{d\tilde{x}_2(t)}{dt} \\ \frac{d\tilde{x}_3(t)}{dt} \\ \frac{d\tilde{x}_4(t)}{dt} \end{bmatrix} = \sum_{i=1}^{l} \phi_i(\tilde{x}) \left(A_i\tilde{x}(t) + \sum_{k=1}^{N_1(t)} A_{ki}(\tilde{x}(t))p(t - t_k^{'}) + \sum_{k=1}^{N_2(t)} A_{ki}(\tilde{x}(t))p(t - t_k^{*}) \right) + v(t) \tag{19.62}$$

where A_i are given in Appendix D, $A_{ki} = 0.5A_i$ and $l = 5$.

Based on the stochastic H_∞ evolutionary game strategy in Proposition 9.3, natural selection is to select the phenotypic variations to construct the following negative feedback loops to improve the system robustness:

$$u^*(t) = \sum_{i=1}^{l} \sum_{k=1}^{N_1(t)} \phi_i(\tilde{x}(t))A_{ki}\tilde{x}(t)p(t - t_k^{'}) = -P\tilde{x}(t)$$

where P is the solution of the constrained optimization problem in (19.54) as follows:

$$P = 10^{-4} \times \begin{bmatrix} 1.59 & -0.07 & -0.69 & -0.86 \\ * & 23.34 & 7.37 & -4.55 \\ * & * & 191.87 & -41.90 \\ * & * & * & 11.97 \end{bmatrix}$$

We also find the evolvability $e_0 = 0.4688$ and fitness $f_0 = 2.1331$ for the lung tissue system with the phenotype at the equilibrium point $x_e = [6.19 \times 10^{-11} 2.20 \times 10^{-3}\ 0\ 9.88 \times 10^{-6}]^T$ in (19.61). From the computer simulation in Fig. 19.2, it is seen that the phenotype of the lung tissue is maintained at equilibrium point x_e under the Poisson somatic genetic and epigenetic variations and environmental disturbances in the period of time $t = 0 \sim 19,906$(days) ≈ 54.5(years).

The neutral phenotypic variations $\sum_{k=1}^{N_2(t)} f_k(\tilde{x})p(t - t_k^*)$ are accumulated large enough as shown in Fig. 19.3 so that the phenotypic robustness criterion in (19.42) or (19.57) of nonlinear stochastic biological system in (19.61) or (19.62) is eventually violated in carcinogenesis. The network robust stability is thus broken down and the phenotypic trait of the lung tissue is shifted to another equilibrium point $x_{2e} = \tilde{x}_{2e} + x_e = [7.09 \times 10^{-4} 1.60 \times 10^{-3}\ 1.30 \times 10^{-4}\ 3.60 \times 10^{-4}]^T$, which can be considered as the phenotype at the first stage of lung cancer, to start another evolutionary process in carcinogenesis. The simulation result in Fig. 19.2 indicates that the phenotypic shift of the lung cancer-associated cell network takes 54.5 years from the normal state to stage I cancer, 1.5 years from stage I to stage II cancer, and 2.5 years from stage II to stage III cancer. For clarity, the evolutionary simulation in the phase plane is given in Fig. 19.4 to illustrate the transition of lung cancer from the healthy state at $x_e = [6.19 \times 10^{-11} 2.20 \times 10^{-3}\ 0\ 9.88 \times 10^{-6}]^T$ to stage I cancer at $x_{1e} = [1.46 \times 10^{-4} 1.80 \times 10^{-3}\ 1.02 \times 10^{-4}\ 2.75 \times 10^{-4}]^T$, then to stage II cancer at $x_{2e} = [7.09 \times 10^{-4} 1.60 \times 10^{-3} 1.30 \times 10^{-4}\ 3.60 \times 10^{-4}]^T$, and finally to stage III cancer at $x_{3e} = [1.01 \times 10^{-3} 1.35 \times 10^{-3}\ 1.80 \times 10^{-4}\ 5.12 \times 10^{-4}]^T$. Based on the results of simulation, we found that the average age of lung cancer is more than 50 years, the

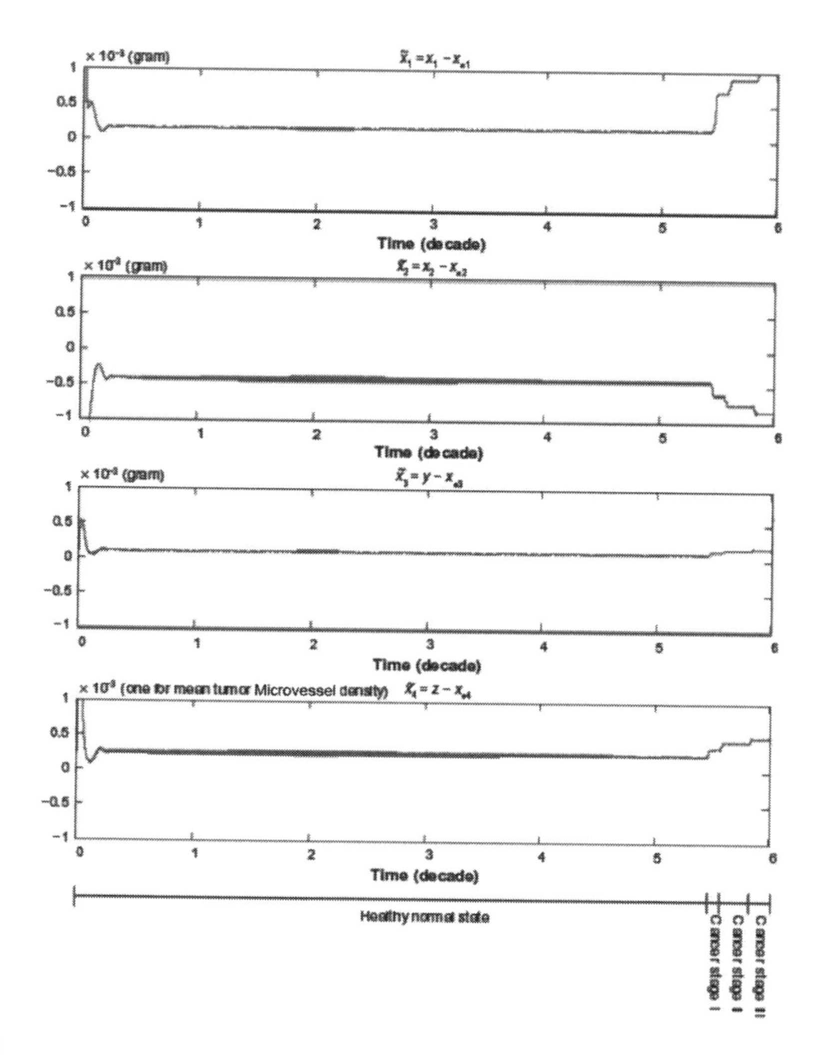

FIGURE 19.2
The trajectories of the nonlinear stochastic biological network of lung cancer cells in carcinogenesis with states $\tilde{x} = [x_1 - x_{e1}, x_2 - x_{e2}, y - x_{e3}, z - x_{e4}]^T$, suffering from Poisson genetic and epigenetic variations with Poisson counting process as shown in Fig. 19.3 and environmental disturbance $v(t)$. Even the network robustness is enhanced by the stochastic H_∞ evolutionary game strategy; however, it is seen that the phenotype will shift to another phenotype when the neutral phenotype variations are accumulated to a sufficiently large amount such that the phenotypic robustness criterion in (19.42) or (19.57) is violated (i.e., the normal phenotype of the lung cancer-associated network is shifted from the equilibrium point at $x_e = [6.19 \times 10^{-11} 2.20 \times 10^{-3}\ 0\ 9.88 \times 10^{-6}]^T$ to the phenotype of first-stage lung cancer state at the equilibrium point $x_{1e} = \tilde{x}_{1e} + x_e = [1.46 \times 10^{-4} 1.80 \times 10^{-3}\ 1.02 \times 10^{-4}\ 2.75 \times 10^{-4}]^T$ at 54.5 years of age, and then the second-stage lung cancer state at the equilibrium point $x_{2e} = \tilde{x}_{2e} + x_e = [7.09 \times 10^{-4} 1.60 \times 10^{-3}\ 1.30 \times 10^{-4}\ 3.60 \times 10^{-4}]^T$ at 58.5 years of age and so on). Obviously, the cancer network evolution seems to increase its step in the late stages of carcinogenesis. From the literature in [484], lung cancer occurs mostly in men over 50 years, often at age of 60–75 years. Moreover, the units of state x_1, x_2, and y are grams, and the unit of state z represents the length of existing microvessel in 1 g of undiseased tissue (i.e., one unit is equal to microvessel length in 1 g of undiseased tissue).

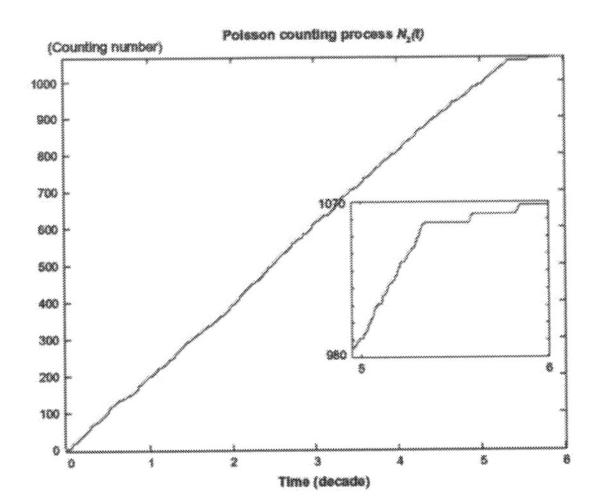

FIGURE 19.3
The Poisson counting process of the neutral genetic and epigenetic variations $N_2(t)$ that are accumulated in carcinogenesis of lung cancer cell lineages in the example. The vertical axis denotes the counting number of phenotypic changes of Poisson counting process $N_2(t)$ in carcinogenesis.

transition from stage I cancer to stage II cancer is about 1.5 years, and the period from stage II to stage III lung cancer is about 2.5 years. The literature [484] seems to give a statistical support to these results.

19.8 Discussion

For ecologists and evolutionary biologists, natural selection and evolution are usually viewed as the domains of peppered moths and finches, driven to adapt by predators and competition. A long time ago, few evolutionary biologists could conceive that their field of biology would have a pivotal role in the understanding and combat of complex diseases such as cancer [485]. Recently, molecular biologists have acknowledged carcinogenesis as an evolutionary process involving natural selection that buffers neutral phenotypic variations and withstands environmental disturbances in the aging process [486]. In order to maintain the phenotypic stability of interest by natural selection, a stochastic H_∞ evolutionary game strategy was developed using natural selection to improve the stability of a cancer-associated biological network. This stochastic H_∞ evolutionary strategy minimized the worst-case effects of uncertain neutral genetic and epigenetic variations and environmental disturbances on the variations from its normal carcinogenic phenotype. With the stochastic H_∞ evolutionary game strategy, the robust phenotypic stability criterion for evolutionary biological networks in carcinogenesis is found in (19.42) or (19.57). This criterion states that the phenotypic robustness plus the environmental robustness should be less than or equal to the enhanced network robustness. Essentially, if the enhanced network robustness by the stochastic H_∞evolutionary game strategy can confer both phenotypic robustness to buffer neutral Poisson epigenetic and genetic variations, and environmental

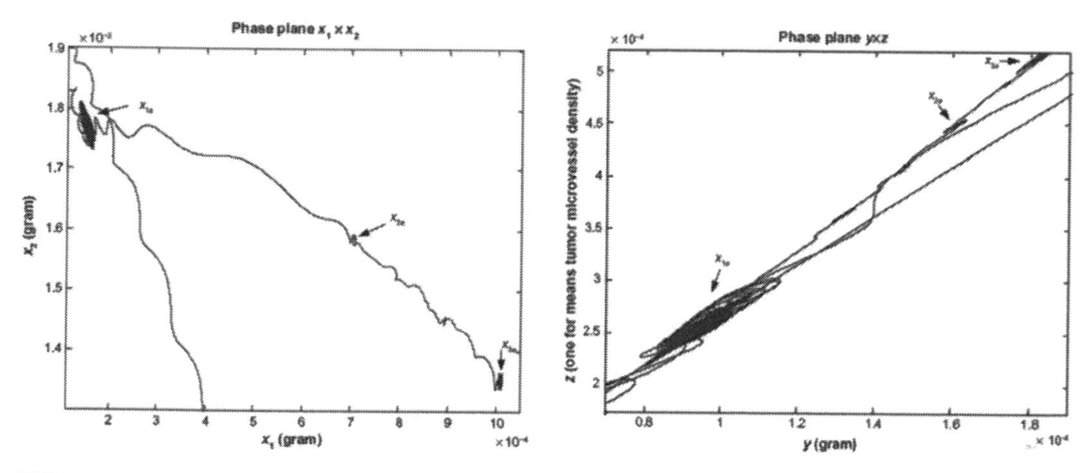

FIGURE 19.4
The evolutionary simulation in phase plane for the biological network of in silico example in carcinogenesis. This evolutionary simulation in phase plane could support the phenotypic transition of lung cancer from the healthy state at the equilibrium point $x_e = [6.19 \times 10^{-11} 2.20 \times 10^{-3}\ 0\ 9.88 \times 10^{-6}]^T$, to the cancer stage I state at the equilibrium point $x_{1e} = [1.46 \times 10^{-4} 1.80 \times 10^{-3}\ 1.02 \times 10^{-4}\ 2.75 \times 10^{-4}]^T$, then to the cancer stage II state at the equilibrium point $x_{2e} = [7.09 \times 10^{-4} 1.60 \times 10^{-3}\ 1.30 \times 10^{-4}\ 3.60 \times 10^{-4}]^T$, and finally to the cancer stage III state at the equilibrium point $x_{3e} = [1.01 \times 10^{-3} 1.35 \times 10^{-3}\ 1.80 \times 10^{-4}\ 5.12 \times 10^{-4}]^T$. Moreover, the units of state x_1, x_2, and y are grams, and the unit of state z represents the length of existing microvessel in 1 g of undiseased tissue (i.e., one unit is equal to microvessel length in 1 g of undiseased tissue).

robustness to withstand environmental disturbances in carcinogenesis, by considering natural selection, the network phenotype of the cancer-associated biological network can be robustly stable in the basin around the equilibrium point x_e. In the evolutionary period, the linear network interaction matrix A is modified by $(A - \frac{1}{2}P)$ in (19.31) [or $f(\tilde{x}(t))$ is modified by $f(\tilde{x}(t)) - \frac{1}{4}\frac{\partial V(\tilde{x}(t))}{\partial \tilde{x}}$ in (19.42) for a nonlinear network case] through negative feedback loops of the stochastic H_∞ evolutionary game strategy. This could then decrease the evolvability e_0 of the biological network in carcinogenesis or increase the fitness f_0 of the biological network in (19.15). In this situation, the phenotype is more robust because of the larger and deeper basin in the phenotypic landscape, as presented in Fig. 19.1. However, when large amounts of neutral genetic and epigenetic variations are continuously accumulated in a larger and deep basin in a period of carcinogenesis, they may provide more raw material for new evolutionary heterogenic possibilities through random genetic drift [450,472,473]. Therefore, the stochastic H_∞ evolutionary game strategy may improve the network robustness in order to maintain the phenotypic trait in a somatic evolutionary period and may eventually improve the evolution of the cancer biological network, with a phenotypic shift from the normal phenotype at the equilibrium point x_e to the cancer phenotype at the equilibrium point x_{1e}. This evolutionary phenomenon has been illustrated by the in silico example of the evolutionary network of lung cancer cells in Fig. 19.2. From the phenotypic robustness criterion in (19.42) or (19.57), the results reveal two ways to improve the phenotypic robustness of the evolutionary biological network in carcinogenesis. One way is to improve the stability and robustness of the network by making the right-hand side of (19.42) or (19.57) as large a value as possible so that the phenotypic robustness criterion always holds. The other way to improve the stability and robustness of the network is to reduce the effect of neutral genetic variations A_{ki} and the coupling B of environmental disturbances $v(t)$ on the biological network. This has the effect of making the left-hand side of (19.42) or (19.57) as small as possible. The stochastic H_∞ evolutionary game strategy involves selecting certain somatic phenotypic variations to develop negative feedback loops $-\frac{1}{2}\frac{\partial V(\tilde{x}(t))}{\partial \tilde{x}}$ in (19.40) as the first way to improve the robustness of nonlinear biological networks and thus maintain their normal phenotypic function. This is why many negative feedback loops and inhibitive epigenetic methylations and miRNA regulations in biological networks have been found to improve the robustness of the network and prevent a phenotypic shift to the cancerous state. Furthermore, there are numerous redundant, membranous, modular, and complex structures in cancer biological networks, which are helpful in attenuating the effect of intrinsic neutral genetic and epigenetic variations and environmental disturbances on the left-hand side of (19.42) or (19.57). This is the second way to maintain the phenotype of the biological network in carcinogenesis. After enduring several evolutionary stages, the biological network phenotype is shifted to late-stage cancer at certain equilibrium points. These robust H_∞ network game strategies are always hijacked by cancer in order to resist anticancer drugs. This is why many anticancer treatments have limited success and cures remain elusive [473,474]. Therefore, based on the phenotype robustness criterion of the evolutionary network in (19.42) or (19.57), we could gain deep insight into the development of complex negative feedback, and redundant, modular, membranous, and scale-free structures of cancer tissue by natural selection during carcinogenesis.

Recently, it has been reported that the phenotype of a synthetic gene network always evolves because of genetic mutations from one generation to another generation, thus making it difficult to properly design a robust synthetic gene network [449,450,453,455,480]. Therefore, the robust phenotypic criterion in (19.42) and (19.57), based on the proposed stochastic H_∞ evolutionary strategy, may be a suitable remedy for the limitations of conventional synthetic design methods.

19.9 Conclusion

In this chapter, evolutionary biological networks with unpredictable somatic genetic variations and uncertain environmental disturbances in carcinogenesis were modeled as nonlinear Poisson dynamic systems. An evolutionary H_∞ game strategy was developed to select certain phenotypic somatic variations to construct negative feedback loops. This stochastic H_∞ evolutionary game strategy buffered neutral genetic and epigenetic variations and unpredictable environmental disturbances to efficiently improve network robustness and to avoid shifting to the cancerous state, thus maintaining the phenotype favored by natural selection during carcinogenesis. The phenotypic robustness criterion for each cancer-related state at an equilibrium point of the evolutionary biological network during carcinogenesis was also derived from this chapter. We found that the evolutionary H_∞ game strategy was able to select certain phenotypic variations because of somatic Poisson genetic and epigenetic variations. Negative feedback loops were constructed to make the basin of the phenotype equilibrium point significantly deeper and wider in the phenotypic landscape (Fig. 19.1), so that the biological network could buffer more neutral genetic and epigenetic variations and withstand additional environmental disturbances during carcinogenesis. However, as enough random neutral genetic and epigenetic variations accumulated over a long period of time, they provided the raw heterogenic material for new adaptation and evolutionary innovation. This enabled the shift to a new phenotype at a neighboring equilibrium point to commence another period of network evolution during carcinogenesis. Taken together, our results demonstrate why cancer is a disease associated with aging.

The global linearization technique was also employed to simplify the solution procedure for the nonlinear biological network in carcinogenesis. In this manner, a difficult HJII-constrained optimization problem for the stochastic H_∞ evolutionary game strategy could be simplified to an equivalent LMIs-constrained optimization problem. Therefore, the tradeoff between evolvability, genetic and epigenetic robustness, and network phenotypic robustness can be easily discussed in terms of the local stability robustness and local environmental disturbance-filtering ability of a set of local linear evolutionary biological networks. In the example of a dynamic system of cells related to lung cancer, the results of the simulation support the proposed stochastic H_∞ evolutionary game strategy of an evolutionary biological network during carcinogenesis.

19.10 Appendix

19.10.1 Appendix A. Proof of Proposition 19.1

For the linear stochastic biological network in (19.9), since $x_e = 0$, $\tilde{x}(t) = x(t) - x_e = x(t)$.

$$
\begin{aligned}
J &= E\int_0^{t_p} \{x^T(t)Qx(t) + u^T(t)u(t) - ev^T(t)v(t) - e\sum_{k=1}^{N_2(t)} x^T(t)A_k^T A_k x(t)p(t-t_k^*)\}dt \\
&= E\{x^T(0)Px(0)\} - E\{x^T(t_p)Px(t_p)\} + E\int_0^{t_p} \{x^T(t)Qx(t) + u^T(t)u(t) - ev^T(t)v(t) \\
&\quad - e\sum_{k=1}^{N_2(t)} x^T(t)A_k^T A_k x(t)p(t-t_k^*) + \frac{d}{dt}x(t)^T Px(t)\}dt \\
&= E\{x^T(0)Px(0)\} - E\{x^T(t_p)Px(t_p)\} + E\int_0^{t_p} \{x^T(t)Qx(t) + u^T(t)u(t) - ev^T(t)v(t) \\
&\quad + (x^T(t)A^T + u^T(t) + v^T(t)B^T)P(x(t) + x^T(t)P(Ax(t) + u(t) + Bv(t) \\
&\quad + \left(\sum_{k=1}^{N_2(t)} (x^T(t) + x^T(t)A_k^T p(t-t_k^*))P(x(t) + A_k x(t)p(t-t_k^*)) - x^T(t)Px(t)\right) \\
&\quad - e\sum_{k=1}^{N_2(t)} x^T(t)A_k^T A_k x(t)p(t-t_k^*)\}dt \\
&= E\{x^T(0)Px(0)\} - E\{x^T(t_p)Px(t_p)\} + E\int_0^{t_p} \{x^T(t)Qx(t) + u^T(t)u(t) - ev^T(t)v(t)
\end{aligned}
$$

$$+x^T(t)A^TPx(t)+u^T(t)Px(t)+v^T(t)B^TPx(t)+x^T(t)PAx(t)+x^T(t)Pu(t)$$
$$+x^T(t)PBv(t)+\sum_{k=1}^{N_2(t)}x^T(t)Px(t)\sum_{k=1}^{N_2(t)}x^T(t)A_k^TPA_kx(t)p(t-t_k^*)-\sum_{k=1}^{N_2(t)}x^T(t)Px(t)$$
$$-e\sum_{k=1}^{N_2(t)}x^T(t)A_k^TA_kx(t)p(t-t_k^*)\}dt$$
$$=E\{x^T(0)Px(0)\}-E\{x^T(t_p)Px(t_p)\}+E\int_0^{t_p}\{x^T(t)Qx(t)+u^T(t)u(t)-ev^T(t)v(t)$$
$$+x^T(t)A^TPx(t)+u^T(t)Px(t)+v^T(t)B^TPx(t)+x^T(t)PAx(t)+x^T(t)Pu(t)$$
$$+x^T(t)PBv(t)+\sum_{k=1}^{N_2(t)}x^T(t)A_k^TPx(t)p(t-t_k^*)+\sum_{k=1}^{N_2(t)}x^T(t)PA_kx(t)p(t-t_k^*)$$
$$+\sum_{k=1}^{N_2(t)}x^T(t)A_k^TPA_kx(t)p(t-t_k^*)-e\sum_{k=1}^{N_2(t)}x^T(t)A_k^TA_kx(t)p(t-t_k^*)\}dt$$
$$=E\{x^T(0)Px(0)\}-E\{x^T(t_p)Px(t_p)\}+E\int_0^{t_p}\{x^T(t)Qx(t)+x^T(t)A^TPx(t)$$
$$+x^T(t)PAx(t)+\sum_{k=1}^{N_2(t)}x^T(t)A_k^TPx(t)p(t-t_k^*)+\sum_{k=1}^{N_2(t)}x^T(t)PA_kx(t)p(t-t_k^*)$$
$$+\sum_{k=1}^{N_2(t)}x^T(t)A_k^TPA_kx(t)p(t-t_k^*)-e\sum_{k=1}^{N_2(t)}x^T(t)A_k^TA_kx(t)p(t-t_k^*)\}dt$$
$$+(u(t)+Px(t))^T(u(t)+Px(t))-x^T(t)PPx(t)$$
$$-\left(\sqrt{e}v(t)-\frac{1}{\sqrt{e}}B^TPx(t)\right)^T\left(\sqrt{e}v(t)-\frac{1}{\sqrt{e}}B^TPx(t)\right)+\frac{1}{e}x^T(t)PBB^TPx(t)\}dt$$

Then, the stochastic Nash quadratic (H_2) game in (19.21) becomes

$$J_0=\min_{u\in\sum_{k=1}^{N_1(t)}A_kx(t)p(t-t_k')}\max_{v(t)}J$$
$$=E\{x^T(0)Px(0)\}-E\{x^T(t_p)Px(t_p)\}+E\int_0^{t_r}\{x(t)^T(PA+A^TP-PP+Q$$
$$+\sum\nolimits_{k=1}^{N_2(t)}\lambda_2(PA_k+A_k^TP+A_k^TPA_k-eA_k^TA_k)+\frac{1}{e}PBB^TP)x(t)\}dt$$

By the Riccati-like inequality in (19.25), we get J_0 which has the upper bound as follows:

$$J_0\le E\{x^T(0)Px(0)\}$$

By the second step of stochastic Nash game in (19.22), we get

$$P\le eI$$

which is the inequality in (19.26). Q.E.D.

19.10.2 Appendix B. Proof of Proposition 19.2

For the nonlinear stochastic biological network in (19.13), we get

$$J = E\int_0^{t_p} \{x^T(t)Qx(t) + u^T(t)u(t) - ev^T(t)v(t) - e\sum_{k=1}^{N_2(t)} f_k^T(\tilde{x})f_k(\tilde{x})p(t-t_k^*)\}dt \tag{19.B1}$$

For some Lyapunov positive $V(\tilde{x}(t))$, get the following equivalent equation:

$$\begin{aligned} J &= E\{V(\tilde{x}(0))\} - E\{V(\tilde{x}(t_p))\} \\ &+ E\int_0^{t_p} \{x^T(t)Qx(t) + u^T(t)u(t) - ev^T(t)v(t) \\ &- e\sum_{k=1}^{N_2(t)} f_k^T(\tilde{x})f_k(\tilde{x})p(t-t_k^*)\}dt + E\int_0^{t_p} \{dV(\tilde{x}(t))\} \end{aligned} \tag{19.B2}$$

By the Itô's formula of Poisson process in (19.13), we get

$$\begin{aligned} E\int_0^{t_p} \{dV(\tilde{x}(t))\} = E\int_0^{t_p} \{&\left(\frac{\partial V(\tilde{x}(t))}{\partial \tilde{x}}\right)^T (f(\tilde{x}) + u(t) + Bv(t) \\ &+ \sum_{k=1}^{N_2(t)} (V(\tilde{x} + f_k(\tilde{x})p(t-t_k^*)) - V(\tilde{x})))\}dt \end{aligned} \tag{19.B3}$$

Substituting (19.B3) into (19.B2), we get

$$\begin{aligned} J &= E\{V(\tilde{x}(0))\} - E\{V(\tilde{x}(t_p))\} \\ &+ E\int_0^{t_p} \{x^T(t)Qx(t) + u^T(t)u(t) - ev^T(t)v(t) \\ &- e\sum_{k=1}^{N_2(t)} f_k^T(\tilde{x})f_k(\tilde{x})p(t-t_k^*)) + \left(\frac{\partial V(\tilde{x}(t))}{\partial \tilde{x}}\right)^T (f(\tilde{x}) + u(t) + Bv(t) \\ &+ \sum_{k=1}^{N_2(t)} (V(\tilde{x} + f_k(\tilde{x})p(t-t_k^*)) - V(\tilde{x})))\}dt \end{aligned} \tag{19.B4}$$

By the fact that

$$\begin{aligned} &u^T(t)u(t) + \left(\frac{\partial V(\tilde{x}(t))}{\partial \tilde{x}}\right)^T u(t) \\ &= \left(u(t) + \frac{1}{2}\left(\frac{\partial V(\tilde{x}(t))}{\partial \tilde{x}}\right)\right)^T \left(u(t) + \frac{1}{2}\left(\frac{\partial V(\tilde{x}(t))}{\partial \tilde{x}}\right)\right) - \frac{1}{4}\left(\frac{\partial V(\tilde{x}(t))}{\partial \tilde{x}}\right)^T \left(\frac{\partial V(\tilde{x}(t))}{\partial \tilde{x}}\right) \\ &- ev(t)v(t) + \left(\frac{\partial V(\tilde{x}(t))}{\partial \tilde{x}}\right)^T Bv(t) \end{aligned} \tag{19.B5}$$

$$= -\left(\sqrt{e}v(t) - \frac{1}{2\sqrt{e}}B^T\left(\frac{\partial V(\tilde{x}(t))}{\partial \tilde{x}}\right)\right)^T\left(\sqrt{e}v(t) - \frac{1}{2\sqrt{e}}B^T\left(\frac{\partial V(\tilde{x}(t))}{\partial \tilde{x}}\right)\right)$$
$$+\frac{1}{4e}\left(\frac{\partial V(\tilde{x}(t))}{\partial \tilde{x}}\right)^T BB^T\left(\frac{\partial V(\tilde{x}(t))}{\partial \tilde{x}}\right)$$

Substituting (19.B.5) into (19.B.4), we get

$$\begin{aligned}
J = {} & E\{V(\tilde{x}(0))\} - E\{V(\tilde{x}(t_p))\} + E\int_0^{t_p}\{x^T(t)Qx(t) + \left(\frac{\partial V(\tilde{x}(t))}{\partial \tilde{x}}\right)^T f(\tilde{x}) \\
& + \sum_{k=1}^{N_2(t)}(V(\tilde{x}+f_k(\tilde{x})p(t-t_k^*)) - V(\tilde{x})) - \frac{1}{4}\left(\frac{\partial V(\tilde{x}(t))}{\partial \tilde{x}}\right)^T\left(\frac{\partial V(\tilde{x}(t))}{\partial \tilde{x}}\right) \\
& + \left(u(t) + \frac{1}{2}\left(\frac{\partial V(\tilde{x}(t))}{\partial \tilde{x}}\right)\right)^T\left(u(t) + \frac{1}{2}\left(\frac{\partial V(\tilde{x}(t))}{\partial \tilde{x}}\right)\right) \\
& - \left(\sqrt{e}v(t) - \frac{1}{2\sqrt{e}}B^T\left(\frac{\partial V(\tilde{x}(t))}{\partial \tilde{x}}\right)\right)^T\left(\sqrt{e}v(t) - \frac{1}{2\sqrt{e}}B^T\left(\frac{\partial V(\tilde{x}(t))}{\partial \tilde{x}}\right)\right) \\
& + \frac{1}{4e}\left(\frac{\partial V(\tilde{x}(t))}{\partial \tilde{x}}\right)^T BB^T\left(\frac{\partial V(\tilde{x}(t))}{\partial \tilde{x}}\right) - e\sum_{k=1}^{N_2(t)} f_k^T(\tilde{x})f_k(\tilde{x})p(t-t_k^*))\}dt
\end{aligned} \tag{19.B6}$$

Then,

$$\begin{aligned}
\min_{u(t)}\max_{v(t)} J = {} & E\{V(\tilde{x}(0))\} - E\{V(\tilde{x}(t_p))\} + \min_{u(t)}\max_{v(t)} E\int_0^{t_p}\{x^T(t)Qx(t) \\
& + \left(\frac{\partial V(\tilde{x}(t))}{\partial \tilde{x}}\right)^T f(\tilde{x}) + \sum_{k=1}^{N_2(t)}(V(\tilde{x}+f_k(\tilde{x})p(t-t_k^*)) - V(\tilde{x})) \\
& - \frac{1}{4}\left(\frac{\partial V(\tilde{x}(t))}{\partial \tilde{x}}\right)^T\left(\frac{\partial V(\tilde{x}(t))}{\partial \tilde{x}}\right) \\
& + \left(u(t) + \frac{1}{2}\left(\frac{\partial V(\tilde{x}(t))}{\partial \tilde{x}}\right)\right)^T\left(u(t) + \frac{1}{2}\left(\frac{\partial V(\tilde{x}(t))}{\partial \tilde{x}}\right)\right) \\
& - \left(\sqrt{e}v(t) - \frac{1}{2\sqrt{e}}B^T\left(\frac{\partial V(\tilde{x}(t))}{\partial \tilde{x}}\right)\right)^T\left(\sqrt{e}v(t) - \frac{1}{2\sqrt{e}}B^T\left(\frac{\partial V(\tilde{x}(t))}{\partial \tilde{x}}\right)\right) \\
& + \frac{1}{4e}\left(\frac{\partial V(\tilde{x}(t))}{\partial \tilde{x}}\right)^T BB^T\left(\frac{\partial V(\tilde{x}(t))}{\partial \tilde{x}}\right) - e\sum_{k=1}^{N_2(t)} f_k^T(\tilde{x})f_k(\tilde{x})p(t-t_k^*))\}dt \\
= {} & E\{V(\tilde{x}(0))\} - E\{V(\tilde{x}(t_p))\} + E\int_0^{t_p}\{x^T(t)Qx(t) + \left(\frac{\partial V(\tilde{x}(t))}{\partial \tilde{x}}\right)^T f(\tilde{x})
\end{aligned} \tag{19.B7}$$

$$+\sum_{k=1}^{N_2(t)}(V(\tilde{x}+f_k(\tilde{x})p(t-t_k^*))-V(\tilde{x}))-\frac{1}{4}\left(\frac{\partial V(\tilde{x}(t))}{\partial \tilde{x}}\right)^T\left(\frac{\partial V(\tilde{x}(t))}{\partial \tilde{x}}\right)$$
$$+\frac{1}{4e}\left(\frac{\partial V(\tilde{x}(t))}{\partial \tilde{x}}\right)^T BB^T\left(\frac{\partial V(\tilde{x}(t))}{\partial \tilde{x}}\right)-e\sum_{k=1}^{N_2(t)}f_k^T(\tilde{x})f_k(\tilde{x})p(t-t_k^*)\}dt$$

By the HJI in (19.36) and the fact $E\{V(\tilde{x}(t_p))\}\geq 0$, we get

$$\min_{u(t)}\max_{v(t)} J\leq E\{V(\tilde{x}(0))\}\leq eE\{x^T(0)x(0)\} \tag{19.B8}$$

which is the inequality (19.37). Q.E.D.

19.10.3 Appendix C. Proof of Proposition 19.3

Since we replace the nonlinear stochastic network in (19.13) by the interpolated local linear biological network in (19.48), the HJI in Proposition 19.2 can be replaced by

$$\left(\frac{\partial V(\tilde{x}(t))}{\partial \tilde{x}}\right)\left(\sum_{i=1}^{L}\alpha_i(\tilde{x})A_i\tilde{x}(t)\right)+\tilde{x}^T(t)Q\tilde{x}(t)-\frac{1}{4}\left(\frac{\partial V(\tilde{x}(t))}{\partial \tilde{x}}\right)^T\left(\frac{\partial V(\tilde{x}(t))}{\partial \tilde{x}}\right)$$
$$+\frac{1}{4e}\left(\frac{\partial V(\tilde{x}(t))}{\partial \tilde{x}}\right)^T BB^T\left(\frac{\partial V(\tilde{x}(t))}{\partial \tilde{x}}\right)+\sum_{k=1}^{N_2(t)}\left(V\left(\sum_{i=1}^{L}\alpha_i(\tilde{x})A_{ki}\tilde{x}(t)p(t-t_k^*)\right)-V(\tilde{x})\right) \tag{19.C1}$$
$$-e\sum_{k=1}^{N_2(t)}\sum_{i=1}^{L}\sum_{j=1}^{L}\alpha_i(\tilde{x})\alpha_j(\tilde{x})A_{ki}A_{kj}p(t-t_k^*)\leq 0$$

with $V(\tilde{x}(0))\leq e\tilde{x}^T(0)\tilde{x}(0)\}$. If we choose $V(\tilde{x}(t))\leq \tilde{x}^T(t)P\tilde{x}(t)\}$, by the fact $\sum_{i=1}^{L}\alpha_i(\tilde{x})=1$ and $\left(\frac{\partial V(\tilde{x}(t))}{\partial \tilde{x}}\right)=2P\tilde{x}(t)$, we get

$$\begin{aligned}&\sum_{i=1}^{L}\alpha_i(\tilde{x})\tilde{x}^T(t)(PA_i+A_i^TP-PP+Q+\frac{1}{e}PBB^TP\\&+\sum\nolimits_{k=1}^{N_{2p}}\lambda_2(PA_{ki}+A_{ki}^TP+A_{ki}^TPA_{ki}-eA_{ki}^TA_{ki}))\tilde{x}(t)\leq 0\end{aligned} \tag{19.C2}$$

The above inequality holds if the Riccati-like inequalities in (19.51) hold. Further,

$$u^*(t)=-\frac{1}{2}\frac{\partial V(\tilde{x}(t))}{\partial \tilde{x}}=-P\tilde{x}(t), v^*(t)=\frac{1}{e}B^TPx(t) \tag{19.C3}$$

which are the solutions of Nash quadratic (H_2) game evolutionary strategy

$$u^*(t)=\sum_{i=1}^{L}\sum_{k=1}^{N_1(t)}\alpha_i(\tilde{x})A_ip(t-t_k^{'})=-P\tilde{x}(t), v^*(t)=\frac{1}{e}B^TP\tilde{x}(t) \tag{19.C4}$$

Q.E.D.

19.10.4 Appendix D. Parameters A_i of Global Linearization Scheme In Silico Example.

$$A_1 = \begin{bmatrix} -0.0534782 & -0.0254467 & 0.0520144 & 0.0003126 \\ 0.0005168 & -0.0038916 & -0.0135070 & -0.0001828 \\ 0.0000006 & 0.0000004 & -0.0237363 & -4.9803948 \times 10^{-9} \\ -0.0000367 & -0.0000331 & -0.0000629 & -0.0237372 \end{bmatrix}$$

$$A_2 = \begin{bmatrix} -0.0596919 & 0.0036681 & 0.0000219 & 0.5821825 \\ -0.0001722 & -0.0417919 & -0.0003613 & -0.2061930 \\ -0.0003118 & -0.0006342 & -0.0395600 & 0.6118400 \\ 1.2427067 \times 10^{-9} & -2.6932285 \times 10^{-9} & 3.0000000 & -0.0000029 \end{bmatrix}$$

$$A_3 = \begin{bmatrix} -0.0599859 & 0.0000641 & 0.0001830 & 0.3217975 \\ -0.0000170 & -0.0395879 & -0.0000082 & -0.0324759 \\ -0.0000328 & 0.0001248 & -0.0398643 & 0.1280925 \\ 2.0485833 \times 10^{-10} & -2.7951458 \times 10^{-10} & 3.0000000 & -0.0000008 \end{bmatrix}$$

$$A_4 = \begin{bmatrix} -0.0599820 & 0.0004016 & -0.0000326 & 0.2173435 \\ -0.0000171 & -0.0400054 & 0.0000678 & -0.0178187 \\ -0.0000197 & 0.0000895 & -0.0398498 & 0.0719197 \\ 1.1501642 \times 10^{-10} & -1.9750672 \times 10^{-9} & 3.0000000 & -0.0000005 \end{bmatrix}$$

$$A_5 = \begin{bmatrix} -0.0600004 & 0.0005292 & 0.0002048 & 0.1781146 \\ -0.0000175 & -0.0401022 & 0.0000980 & -0.0078165 \\ 0.00000255 & 0.0000218 & -0.0397513 & 0.0722756 \\ 1.7267656 \times 10^{-10} & -1.3279752 \times 10^{-9} & 3.0000000 & -0.0000005 \end{bmatrix}$$

20

Noncooperative and Cooperative Stochastic H_∞ Evolutionary Game Strategies of a Population of Evolutionary Biological Networks

20.1 Introduction

In the last decades, evolutionary game theory is well established in both biology and economics. In the biological version of von Neumann's game theory, the concept of fitness is treated as an equivalent of "payoff". Evolutionary game theory has demonstrated that the basic ideas of game theory can be applied even to situations in which no individual is overtly reasoning or making explicit decisions [27, 28, 487–489]. In these cases, game-theoretic analysis can rather be applied to settings in which individuals can exhibit different forms of behavior which may include some that are not the result of conscious choices. This also allows exploration of which forms of behavior have the ability to persist in a population and which have a tendency to be driven out by others. In the evolution of biological systems, the key insight of evolutionary game theory is that many behaviors involve the interaction of multiple organisms in a population, and the success of any one of these organisms depends on how its behavior interacts with that of others. In this situation, the fitness of an individual organism cannot be measured in isolation; rather, it has to be evaluated in the context of the full population. In other words, an organism's genetically determined characteristics and behaviors are like its strategy in a game, its fitness is like its payoff, and this payoff depends on the strategies (characteristics) of the organisms with which it interacts [487, 490, 491].

In the game theory, the concept of the Nash equilibrium is central in analyzing the outcome of a game [28]. A Nash equilibrium for a two-player game is a choice of strategies that tends to persist once the players are using it because neither player has an incentive to deviate from their strategy [28, 492]. In an evolutionary setting, this corresponds to genetically determined evolutionary game strategies that tend to persist once prevalent in a population. A given evolutionary strategy can therefore be called evolutionarily stable when the entire population is using it, and any small group of invaders using a different evolutionary strategy will eventually die off over subsequent generations [490, 493, 494]. The evolutionarily stable strategies of the Prisoner's Dilemma game and Hawk–Dove game have been widely discussed in the context of the evolution of populations [495].

By comparing the evolutionarily stable strategy with the Nash equilibrium, we can see that a strict Nash equilibrium is an evolutionarily stable strategy and that an evolutionarily stable strategy is a Nash equilibrium [490, 496]. Nash developed decision rules to determine the game strategies that rational players should choose in a game, but how do individuals involved in actual game-theoretical settings find a Nash equilibrium without a knowledge of game theory [487, 495, 496]? Further, if there is more than one Nash equilibrium, which one is the true solution of the game? Answering these questions also requires knowledge of the dynamics of evolutionarily stable strategies; if a small proportion of a stable population mutates, will it eventually evolve back to the original state, or will populations, starting from some random initial condition, evolve to an evolutionarily stable strategy (i.e., a Nash equilibrium) [490, 494]? How populations learn to achieve a Nash equilibrium and to select among different Nash equilibria are thus important issues in the evolution of populations and their strategies.

By the fact that evolutionary game theory provides a readily applicable mathematical framework for understanding the selective pressures, which could affect the evolution of the game strategies of agents engaged in interactions with potential conflicts [497], in simple settings, the costs and benefits of decisions can be used to predict the optimal strategy. More realistic settings such as finite populations, non-vanishing mutation rates, stochastic decision-making, communication between agents, and spatial interactions require agent-based methods where each agent is modeled as an individual making decisions and where the evolutionary outcome can only be ascertained by the evolution of agents. Recently, evolutionary game theory using agent-based methods was covered in a recent review [498]. Another study introduced an analytical framework for polymorphic evolutionary games suitable for explicitly modeling evolutionary processes in diploid populations with sexual reproduction [499]. The principal aspect of the proposed approach to an evolutionary game is to add diploid genetics with sexual recombination to a traditional evolutionary game, and to switch strategies from acting on phenotypes to acting on haplotypes. For the network-based evolutionary games, players reside in the vertices of a graph and games are played between neighboring vertices. As the complexity of the system model increases with the number of vertices in the graph, a "lumping" approach was introduced to reduce the graph and simplify analysis [500]. Evolutionary game theory examines frequency-dependent strategy interactions and thus usually omits the life history and demographic structure of the population. However, a recent study introduced the modeling of evolutionary games in populations with a demographic structure [501]. Therefore, individuals have two basic strategic behaviors, interacting in pairwise games. This allows the diversity of life-stage structures and life-stage-dependent strategies to promote each other. Using the relevance of evolutionary game theory to situations where choices are affected by the choices of others, whether actors are rational or not, another study applied the concept to strategic innovation to highlight the rising importance of bandwagon pressure to developing the generation and evolution of strategic innovation [502].

At the beginning, for the convenience of reviewing evolutionary game strategies, two important dynamic models are introduced in some mathematical detail. The following replicator equation was the first and most successful example of an evolutionary game dynamics [503, 504]

$$\dot{x}_i = x_i[(Wx)_i - x^T Wx] \tag{20.1}$$

where $x = [x_1 \cdots x_i \cdots x_n]^T$ is the state of the population, individuals (players) x_i, $i = 1, \ldots, n$ meet randomly, engaging in a game with payoff matrix W, $(Wx)_i$ is the expected payoff for an individual (players) using strategy S_i, and $x^T Wx$ is the average payoff of the population state x.

The replicator equation (20.1) could translate these mathematical terms into the elementary principle of natural selection: strategies, or individuals using a given strategy, that reproduce with more efficiently, spread, and displace those with lesser fitness. Based on the dynamic equation in (20.1), the tools of dynamical system theory can be used to derive the fundamental characteristics of an evolutionary game. Since every rest point of the replicator dynamics is a Nash equilibrium, (20.1) can provide an evolutionary mechanism for achieving an equilibrium. Note that the replicator equation can be transformed by appropriate changes in variables into the Lotka–Volterra equation [503, 505]. From (20.1), it is noted also that (i) the replicator equation contains no mutations – strategies increase or decrease only due to reproduction; (ii) variation in population size has a linear relation to payoff difference; and (iii) the influence of environmental changes is not considered in the evolutionary mechanism [506]. Since random genetic variations and environmental disturbances play an important role in real biological evolution, the omission of these factors constitutes a deficiency of the replicator equation.

In the evolutionary process, the emergence of cooperation is one of the most important events to which evolutionary game theory is being applied. At the basis of many biological phenomena from mutualism to the appearance of multicellular organisms is one individual helping another to achieve greater fitness, implying more reproductive success [507, 508]. When we consider the cooperation problem in the framework of the dynamic replicator equation in (20.1), cooperation is simply not possible. Because defection is the only Nash equilibrium of the Prisoner's Dilemma, for any initial condition with a positive fraction of defectors, replicator dynamics in (20.1) will inevitably take the population to a final state only consisting of defectors [503, 505]. To remedy the above two deficiencies of replicator dynamics (lack of intrinsic genetic mutations and environmental changes, and cooperation), the following general stochastic dynamic model is proposed, incorporating an intrinsic Poisson process to model genetic variations and external environmental disturbances [443, 504, 506].

$$\dot{x}(t) = f(x) + u(t) + \sum_{k=1}^{N(t)} f_k(x)p(t - t_k) + Bv(t) \tag{20.2}$$

where $x(t) = [x_1(t) \cdots x_i(t) \cdots x_n(t)]^T$ denotes the state variables of n species in the biological network, $f(x)$ denotes nonlinear interaction vector among n different species in the biological network, the Poisson counting process $N(t)$ denotes the sequence of random event times $\{t_k\}$ of genetic variations with mean λt and variance λt in which λ is the mean of Poisson process in unit time, and $p(\cdot)$ denotes the unit step function [443]

$$p(t) = \begin{cases} 1, & t \geq 0, \\ 0, & t<0. \end{cases}, \; p(t - t_k) = \begin{cases} 1, & k \leq N(t), \\ 0, & k>N(t), \end{cases}$$

since $t_k \leq t$ for $k \leq N(t)$, and $t_k > t$ for $k > N(t)$, and t_k is the random time of occurrence of the kth genetic variation. And $f_k(x)$ denotes the nonlinear stochastic heritable phenotypic variation due to Poisson random genetic variations occurring at $t = t_k$ randomly in the evolutionary process [443]. Sources of these random genetic variations include DNA mutation, deletion, recombination, duplication, inversion, and translocation of chromosomes in the evolution process. The stochastic term $\sum_{k=1}^{N(t)} f_k(x)p(t - t_k) = \sum_{k=1}^{N(t)} f_k(x)$

denotes the random phenotypic effect of genetic variations on the biological network during evolution. The stochastic dynamics in (20.2) could represent a population of stochastic evolutionary biological networks. $v(t)$ denotes the random environmental disturbances (stimuli) and B denotes the coupling matrix from environmental disturbances to the biological network. $u(t) = [u_1(t) \cdots u_i(t) \cdots u_m(t)]^T$ denotes the cooperative strategy of m individuals in the biological network, i.e., $u_1(t) \cdots u_m(t)$ are considered as a common strategy $u(t)$ to optimize a common goal.

Darwin's theory of evolution was based on the idea of natural selection through genetic variations resulting in the adaptation of individuals and species to their environment. Clearly, the nonlinear stochastic model in (20.2) is more suitable for modeling a stochastic evolutionary game within a population of biological networks because it incorporates individuals' strategies to resist the effect of random genetic variations and environmental disturbances. At present, there exists no efficient method to solve the stochastic cooperative evolutionary game problem of the nonlinear stochastic system in (20.2) [444, 509]. Network evolvability is defined as the ability to deviate from a phenotypic trait to adapt to environmental changes. In [506], the evolvability of the nonlinear stochastic network in (20.2) is discussed from the vantage of the worst-case effect of an environmental disturbance on phenotypic change at an equilibrium point of interest. The network evolvability of a biological network appears to be inverse to its network robustness. A phenotypic trait of a biological network can be represented by an equilibrium point of the nonlinear stochastic system in the fitness landscape (see Fig. 20.1). In [506] and [503], the cooperative evolution game of the stochastic biological network in (20.2) was played by cooperative species at an equilibrium point of interest through natural selection. The interplay between the genetic robustness, environmental robustness, and network robustness of an evolutionary stochastic biological network was also investigated from the point of view of a stochastic Nash game [428, 429, 431, 432, 445]. In the cooperative evolution game strategy, we find that some genetic variations are selected to construct new negative feedback loops to efficiently improve network robustness, thereby maximizing the fitness of the network [1, 9, 415]. Using this

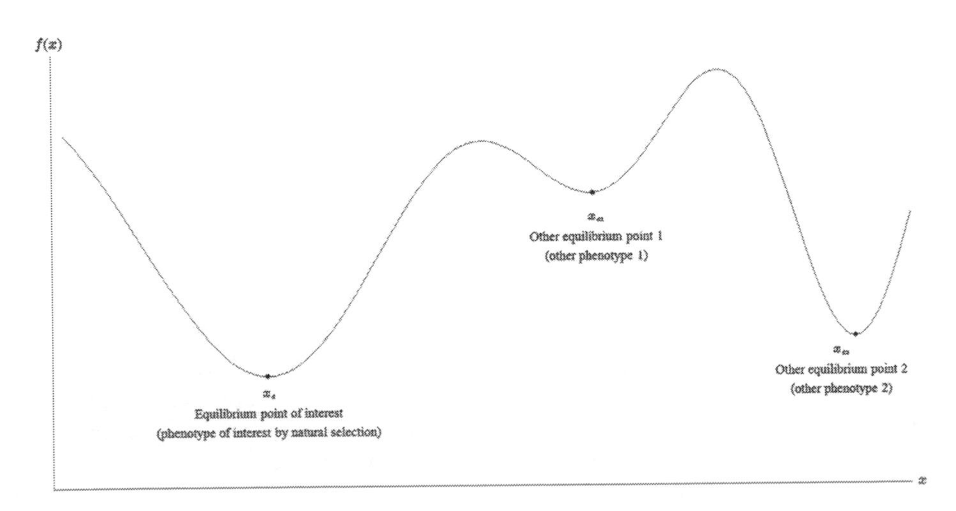

FIGURE 20.1
The landscape of three stable equilibrium points (phenotypes) of nonlinear stochastic biological network in (20.2).

strategy, the stochastic biological network has enough network robustness near the equilibrium point x_e to efficiently buffer neutral genetic variations and resist environmental disturbance. This is the so-called "phenotypic robustness criterion" of biological networks in the evolutionary process [506]. However, when the remaining (neutral) genetic variations accumulate to the extent of violating the phenotypic robustness criterion, new materials for further evolutionary paths to other phenotypes are provided by random drift; i.e., the phenotypic trait of x_e cannot be maintained, and a phenotype transition may eventually occur by a shift to another equilibrium point (x_{e1} or x_{e2} in Fig. 20.1) [444, 509]. While the cooperative evolutionary game strategy can thus improve network robustness for the phenotype at the beginning, accumulating neutral genetic variations will finally break down the network phenotype and force new phenotypic adaptations [467, 438, 464, 510–513].

In the literature, molecular biologists have long recognized that carcinogenesis and aging constitute an evolutionary process of somatic cells that involves natural selection. In [18], the evolution of somatic cancer cell lineages during carcinogenesis was modeled as a shifting process of the equilibrium point (i.e., the phenotype of the attractor) of a nonlinear stochastic biological network. This process is subject to intrinsic random fluctuations due to somatic genetic and epigenetic variations, as well as environmental disturbances due to carcinogens and stressors (i.e., changes to the cancer-microenvironment). Aging is caused by the antagonistically pleiotropic effects of cellular responses that protect the organism from cancer. A cooperative evolutionary game strategy was proposed in [18] and [504] to select certain genetic and epigenetic variations to modify the network structure with the aim of improving network robustness to attenuate intrinsic fluctuation and to resist extrinsic disturbance in order to maintain the phenotype of the evolutionary biological network at an equilibrium point (attractor). During the carcinogenic process, however, the remaining (or neutral) genetic and epigenetic variations accumulate too strongly to maintain the normal phenotype at the desired equilibrium point of the evolutionary cellular network [469, 504]. The network is thus shifted to a cancer phenotype at a new equilibrium point, beginning a new evolutionary process at the later stages of carcinogenesis.

In the real biology, each individual in a biological network might pursue its own interests, which are partly in conflict with those of others. Assuming that network evolution can only be plausibly studied in a population of realistic biological networks, the following noncooperative evolutionary game of m individuals (players) is posited [1, 9, 31]:

$$\dot{x}(t) = f(x) + \sum_{i=1}^{m} u_i(t) + \sum_{k=1}^{N(t)} f_k(x)p(t-t_k) + Bv(t) \tag{21.3}$$

where $u_i(k)$ denotes the strategy of the ith individual or player $x_i(t)$ in the biological network.

In the noncooperative H_∞ evolutionary game with m players and m strategies $u_i(t)$, $i = 1, \ldots, m$, each player pursues an individual goal that may partly conflict with those of others. At present, there still exists no method to efficiently solve the noncooperative H_∞ evolutionary game problem in (20.3) for the nonlinear stochastic biological network with Poisson intrinsic genetic variations and environmental disturbances in the evolution process [1, 9]. In this chapter, we will discuss the solution of the noncooperative stochastic H_∞ evolution game problem of the nonlinear stochastic biological network in (20.3). For the m-player stochastic evolution game, an m-tuple evolutionary game strategy provides a feedback Nash equilibrium solution for the stochastic noncooperative H_∞ evolutionary game. In fact, this stochastic H_∞ noncooperative evolutionary game is equivalent to a multi-objective optimization problem of the network in (20.3). This requires solving m Hamilton–Jacobi–Isaac inequalities (HJIIs)-constrained multi-objective optimization problems. We found that this not only allows all players in the biological network to simultaneously and optimally achieve their goals, but also improves network robustness to tolerate more genetic variations and buffer more strongly against environmental disturbances. Further, we will show that the noncooperative stochastic H_∞ evolutionary game strategy based on multi-objective optimization is a Nash equilibrium solution. However, compared with the stochastic cooperative H_∞ evolutionary game strategy, the stochastic noncooperative H_∞ evolutionary strategy required more effort and engendered greater difficulties in striking compromises between players.

In order to overcome the difficulty to solve the cooperative evolutionary game of nonlinear stochastic biological network, the global linearization technique is employed to interpolate several local linear biological networks to approximate a nonlinear evolutionary biological network. This results in a simplification by replacing the HJIIs-constrained optimization problem to solve nonlinear stochastic noncooperative H_∞ evolutionary game with a linear matrix inequalities (LMIs)-constrained problem. Finally, a simple evolutionary metabolic pathway is simulated to illustrate the solution procedure of noncooperative and cooperative stochastic H_∞ evolutionary strategies and to confirm and compare the results of the two evolutionary game strategies in the evolutionary process.

20.2 Noncooperative Stochastic H$_\infty$ Evolutionary Game for Stochastic Biological Networks

a. Noncooperative H$_\infty$ evolutionary game in linear stochastic biological network

For simplicity of analysis and discussion, we first consider the following linear evolutionary biological network implementing the noncooperative evolutionary game [1, 443]:

$$\dot{\tilde{x}}(t) = A\tilde{x}(t) + \sum_{k=1}^{N(t)} A_k p(t - t_k)\tilde{x}(t) + \sum_{i=1}^{m} u_i(t) + Bv(t) \tag{20.4}$$

where the dynamic interaction matrix $A \in \mathbb{R}^{n \times n}$ among species, whose element a_{ij} denotes the interaction between species j and species i. The weighted Poisson counting process $s(t) = \sum_{k=1}^{N(t)} A_k p(t - t_k) = \sum_{k=1}^{N(t)} A_k$ denotes heritable network interaction variations due to some genetic variations occurring randomly at $t_1, \ldots, t_p$. $\tilde{x}(t) = x(t) - x_e(t)$ denotes the phenotypic deviation around the phenotypic trait at the equilibrium point x_e. $u_i(t)$ denotes the game strategy of player i.

Remark 20.1 The linear evolutionary biological network in (20.4) could be considered as the linearization of the nonlinear evolutionary biological network in (20.3) at an equilibrium point of interest, for example at x_e [514], i.e.,

$$A = \frac{\partial f(x)}{\partial x}\bigg|_{x=x_e}. = \frac{\partial f(\tilde{x})}{\partial \tilde{x}}\bigg|_{\tilde{x}=0}., A_k = \frac{\partial f_k(x)}{\partial x}\bigg|_{x=x_e}. = \frac{\partial f_k(\tilde{x})}{\partial \tilde{x}}\bigg|_{\tilde{x}=0}..$$

In the noncooperative evolutionary game of the biological network in (20.4), every evolutionary game strategy $u_i(t)$ selects some beneficial genetic variations [506]

$$u_i(t) = \sum_{k=1}^{N(t)} A_k p(t - t_k)\tilde{x}(t), \text{for } i = 1, 2, \ldots, m \tag{20.5}$$

from genetic variations $\sum_{k=1}^{N(t)} A_k p(t - t_k)$ as the noncooperative evolutionary strategy. The total selections of m noncooperative evolutionary strategies are given by

$$u(t) = \sum_{i=1}^{m} u_i(t) = \sum_{i=1}^{m} \sum_{k=1}^{N(t)} A_k p(t - t_k)\tilde{x}(t) \tag{20.6}$$

and the remaining neutral genetic variations $\sum_{k=1}^{N_r(t)} A_k p(t - t_k)$ after the beneficial selection of noncooperative evolutionary strategies in (20.5) are given by

$$\sum_{i=1}^{N_r(t)} A_k p(t - t_k) = \sum_{k=1}^{N(t)} A_k p(t - t_k) - \sum_{i=1}^{m} \sum_{k=1}^{N_i(t)} A_k p(t - t_k) \tag{20.7}$$

For each evolutionary game strategy $u_i(t)$, all other game strategies $u_1(t) \cdots u_{i-1}(t)$ $u_{i+1}(t) \cdots u_m(t)$ and environmental disturbances $v(t)$ are unavailable for the ith player and for simplicity of analysis can be bundled together as a competitive game strategy to $u_i(t)$. Since environmental disturbance plays an important role in the biological evolutionary process and its information is unavailable to each player, its effect on the biological system should be considered from the worst-case perspective and each player tries to minimize the worst-case effect. Therefore, environment should be considered as a player in the biological evolutionary game. Therefore, after the selections in (20.6), the evolutionary biological system in (20.4) can be represented by

$$\dot{\tilde{x}}(t) = A\tilde{x}(t) + \sum_{k=1}^{N_r(t)} A_k p(t - t_k)\tilde{x}(t) + u_i(t) + \bar{B}_i \bar{v}_i(t) \tag{20.8}$$

where $\bar{B}_i = [B \quad I \quad \cdots \quad I], \bar{v}_i(t) = \begin{bmatrix} v(t) \\ u_1(t) \\ \vdots \\ u_{i-1}(t) \\ u_{i+1}(t) \\ \vdots \\ u_m(t) \end{bmatrix}$

Since all other strategies and environmental disturbances are unavailable in the noncooperative system, each evolutionary strategy $u_i(t)$ attempts to simultaneously minimize the worst-case effects of the remaining neutral genetic variation, the combined competitive strategy, and environmental disturbances $\bar{v}_i(t)$ on individual goals (i.e., minimize phenotypic deviation $\tilde{x}(t)$ with minimal effort $u_i(t)$) as the following stochastic n-person noncooperative stochastic H_∞ evolutionary game [30]:

$$e_i^* = \min_{u_i(t)} \max_{\bar{v}_i(t)} \frac{E\{\int_0^{t_p} (\tilde{x}^T(t)Q_i\tilde{x}(t) + u_i^T(t)R_iu_i(t))dt\}}{E\{(\tilde{x}^T(0)\tilde{x}(0) + \int_0^{t_p} \bar{v}_i^T(t)\bar{v}_i(t))dt\}}, \text{for } i = 1, 2, \ldots, m \quad (20.9)$$

where $Q_i$andR_i denote the corresponding weighting matrices for each player to tradeoff between state deviation $\tilde{x}(t)$ from x_e and the strategy effort of $u_i(t)$ towards the goal of player i; t_p denotes the present time; the evolvability e_i^* of the ith strategy $u_i(t)$ is inversely proportional to the network fitness f_i^*, i.e., $e_i^* \propto \frac{1}{f_i^*}$ or the ith strategy of noncooperative stochastic H_∞ evolutionary game is equivalent to maximize network fitness f_i^* from the perspective of the ith player. $(e_1^* \cdots e_i^* \cdots e_n^*)$ denotes the solution vector of the m-person noncooperative game. Selection of different weighting matrices $Q_i$andR_i in (20.9) could lead to different game strategies with different levels of cooperative behavior within these players. However, there is still a certain level of cooperation present in n-person noncooperative stochastic H_∞ game strategies, because the individual goals of these strategies are not completely in conflict with one another in (20.9). Each player needs to specify two weighting matrices Q_i and R_i so that game strategies have their desired levels of cooperative behavior to achieve their fitness within a given game.

The physical meaning of the noncooperative stochastic H_∞ evolutionary game strategy in (20.9) is twofold: (i) the primary goal of each player is to minimize state deviation $\tilde{x}(t)$ from the equilibrium point due to genetic variations (i.e., to maximize their fitness) by parsimonious effort $u_i(t)$; (ii) since random environmental disturbance acts as a player in the evolutionary process, the worst-case effects of both random environmental disturbances and the strategies of other players on the goal of each player are to be simultaneously minimized. The noncooperative H_∞ evolutionary game strategy $u_i(t)$ of the ith player in (20.9) is to select a smaller amount of beneficial genetic variations in Eq. (21.5) to modify systematic cooperative behavior of the biological network in (20.8) according to the noncooperative stochastic H_∞ evolutionary game in (20.9) with two specified weighting matrices $Q_i$andR_i; the aim is to minimize both the effect of the remaining neutral genetic variation and the worst-case effects of the unpredictable combined competitive strategies and environmental disturbances $\bar{v}_i(t)$ on phenotypic deviation $\tilde{x}(t)$, to achieving their fitness using parsimonious effort $u_i(t)$. $E\{(\tilde{x}^T(0)\tilde{x}(0)\}$ in (20.9) denotes the effect of initial condition $\tilde{x}(0)$ on the game, while e* denotes the solution of the stochastic noncooperative H_∞ evolutionary game and can be considered the evolvability of the biological network under evolutionary strategy $u_i(t)$.

In general, it is still very difficult to solve the m-person noncooperative stochastic H_∞ evolutionary game problem in (20.9) directly. Therefore, the following indirect suboptimal method is employed which simultaneously minimizes the corresponding upper bounds:

$$e_i^* = \min_{u_i(t)} \max_{\bar{v}_i(t)} \frac{E\{\int_0^{t_p} (\tilde{x}^T(t)Q_i\tilde{x}(t) + u_i^T(t)R_iu_i(t))dt\}}{E\{(\tilde{x}^T(0)\tilde{x}(0) + \int_0^{t_p} \bar{v}_i^T(t)\bar{v}_i(t))dt\}} \leq e_i, \text{for } i = 1, 2, \ldots, m \quad (20.10)$$

where positive value e_i denotes the upper bound of evolvability e_i^* by the ith strategy. An initial upper bound e_i evolutionary of e_i^* is provided and then minimized as far as possible to simultaneously approach e_i^* for all i. This is done to attain the real m minimax stochastic H_∞ evolutionary game problems of the m-person noncooperative suboptimal evolutionary game in (20.10) and constitutes the so-called suboptimal approach to the m-person noncooperative stochastic H_∞ evolutionary game problem in (20.9). In other words, we need to solve the following multi-objective optimization problems to simultaneously minimize their corresponding upper bounds in (20.10) [1, 9, 515]

$$(e_1^* \cdots e_i^* \cdots e_m^*) = \min(e_1 \cdots e_i \cdots e_m) \tag{20.11}$$

$$\text{subject to} \min_{u_i(t)} \max_{\bar{v}_i(t)} \frac{E\{\int_0^{t_p} (\tilde{x}^T(t)Q_i\tilde{x}(t) + u_i^T(t)R_iu_i(t))dt\}}{E\{(\tilde{x}^T(0)\tilde{x}(0) + \int_0^{t_p} \bar{v}_i^T(t)\bar{v}_i(t))dt\}} \leq e_i, \text{for } i = 1, 2, \ldots, m. \tag{20.12}$$

Lemma 20.1 *The indirect multi-objective problem in (20.11) and (20.12) for the suboptimal noncooperative game strategy is equivalent to the noncooperative stochastic H_∞ evolutionary game problem in Eq. (20.9) or Eq. (20.10) when the optimization solution in Eq. (20.11) is achieved.*

Proof: The proof of Lemma 20.1 is straightforward. It is only required to prove that the inequalities in (20.10) or (20.12) become equalities for Pareto optimal solutions. This can be shown through proof by contradiction. Given Pareto optimal solution $(e_1^* \cdots e_i^* \cdots e_m^*)$ of (20.11), we assume that any one of the inequalities in (20.10) or (20.12) remains a strict inequality at the optimal solution. Without loss of generality, suppose $\min_{u_i(t)} \max_{\bar{v}_i(t)} \frac{E\{\int_0^{t_p} (\tilde{x}^T(t)Q_i\tilde{x}(t)+u_i^T(t)R_iu_i(t))dt\}}{E\{(\tilde{x}^T(0)\tilde{x}(0)+\int_0^{t_p} \bar{v}_i^T(t)\bar{v}_i(t))dt\}} = e_i^{'}$ such that $e_i^{'} < e_i^*$. In this situation, $(e_1^* \cdots e_i^{'} \cdots e_m^*)$ dominates Pareto optimal solution $(e_1^* \cdots e_i^* \cdots e_m^*)$, leading to a contradiction. This implies that m inequalities in (20.12) indeed become equalities when the Pareto optimal solution is achieved. Therefore, the optimization problem in (20.11) is equivalent to the multi-objective problem in (20.9). Q.E.D.

The m constrained suboptimal stochastic noncooperative H_∞ evolutionary games in (20.12) are equivalent to the following m-constrained stochastic Nash quadratic (H_2) evolution games [1, 9]

$$\min_{u_i(t)} \max_{\bar{v}_i(t)} E\{\int_0^{t_p} (\tilde{x}^T(t)Q_i\tilde{x}(t) + u_i^T(t)R_iu_i(t) - e_i\bar{v}_i^T(t)\bar{v}_i(t))dt\} \leq e_iE\{(\tilde{x}^T(0)\tilde{x}(0)\}, \quad \text{for } i = 1, 2, \ldots, m. \tag{20.13}$$

Let us denote

$$J_i = E\{\int_0^{t_p} (\tilde{x}^T(t)Q_i\tilde{x}(t) + u_i^T(t)R_iu_i(t) - e_i\bar{v}_i^T(t)\bar{v}_i(t))dt\}, \text{ for } i = 1, 2, \ldots, m. \tag{20.14}$$

Two steps are required to solve the constrained stochastic noncooperative Nash quadratic evolutionary game problem in (20.13). The first step is to solve the following problem:

$$\begin{aligned} J_i^* &= \min_{u_i(t)} \max_{\bar{v}_i(t)} J_i \\ &= \min_{u_i(t)} \max_{\bar{v}_i(t)} E\{\int_0^{t_p} (\tilde{x}^T(t)Q_i\tilde{x}(t) + u_i^T(t)R_iu_i(t) - e_i\bar{v}_i^T(t)\bar{v}_i(t))dt\}, i = 1, 2, \ldots, m. \end{aligned} \tag{20.15}$$

and the second step is to solve the following inequality constraint problem:

$$J_i^* \leq e_i E\{(\tilde{x}^T(0)\tilde{x}(0)\}, \text{for } i = 1, 2, \ldots, m. \tag{20.16}$$

Solving (20.15) and (20.16) for the constrained stochastic Nash evolution game in (20.13) yields the following result in preparation for the multi-objective optimization problem in (20.11) and (20.12).

Proposition 20.1 *The constrained suboptimal stochastic noncooperative H_∞ evolutionary games of the evolutionary biological network in (20.13) can be solved by the following n-person noncooperative evolutionary game strategies $u_i^*(t)$, $i = 1, 2, \ldots, m$ and the worst-case combined competitive strategies and environmental disturbances $\bar{v}_i^*(t)$, $i = 1, 2, \ldots, m$*

$$u_i^*(t) = \sum_{k=1}^{N_i(t)} A_k p(t - t_k)\tilde{x}(t) = -R_i^{-1}P\tilde{x}(t), \bar{v}_i^*(t) = \frac{1}{e_i}\bar{B}_i^T P\tilde{x}(t) \tag{20.17}$$

where the positive definite matrix P is the solutions of the following Riccati-like inequalities:

$$A^T P + PA - PR_i^{-1}P + \frac{1}{e_i}P\bar{B}_i\bar{B}_i^T P + Q_i + \sum_{k=1}^{N_{rp}} (A_k^T P + PA_k + A_k^T PA_k) \leq 0, \quad \text{for } i = 1, 2, \ldots, m \tag{20.18}$$

$$\text{with } 0 < P \leq e_i I, \text{ for } i = 1, 2, \ldots, m \tag{20.19}$$

where $N_{rp} = N_r(t_p)$ represents the total number of remaining neutral genetic variations to the present time t_p after selection by m players.

Proof: See Appendix A.

The physical meaning of the suboptimal noncooperative H_∞ evolutionary game strategy in (20.17) is that each strategy $u_i^*(t)$ selecting genetic variations $\sum_{k=1}^{N_i(t)} A_k p(t - t_k)$ that are equal to $-R_i^{-1}P$ in amount could minimize the worst-case effects of the combined competitive strategies and environmental disturbances $\bar{v}_i^*(t)$ on phenotypic deviation with less effort to achieve each player's goal. The final term $\sum_{k=1}^{N_{rp}} (A_k^T P + PA_k + A_k^T PA_k)$ in (20.18) is the effect of random genetic variations that are not selected by the m evolutionary game strategies of m players, i.e., of neutral genetic variations. Equations (20.17)–(20.19) implement the fact that due to the unavailability of other strategies and the unpredictability of environmental disturbances, strategies $u_i^*(t)$, $i = 1, 2, \ldots, m$ need to consider the worst-case effects of other competitive strategies and environmental disturbances $\bar{v}_i^*(t)$ on their goal, based on the stochastic Nash quadratic game perspective in (20.12) or (20.13). Through the selection of adequate tradeoff weighting matrices $Q_i$andR_i in (20.9), players can obtain adequate network robustness through adequate strategy effort in the noncooperative game and pass on their improved strategies in the population.

After solving the constrained suboptimal evolutionary games in (20.12) or (20.13) by Proposition 20.1, the multi-objective optimization problem for the m-person noncooperative evolutionary game in (20.11) and (20.12) can be transformed into the following multi-objective optimization problem [30, 515]:

$$(e_1^* \cdots e_i^* \cdots e_m^*) = \min_P (e_1 \cdots e_i \cdots e_m) \tag{20.20}$$

$$\text{subject to } A^TP + PA - PR_i^{-1}P + \frac{1}{e_i}P\bar{B}_i\bar{B}_i^TP + Q_i + \sum_{k=1}^{N_{rp}}(A_k^TP + PA_k + A_k^TPA_k) \le 0, \tag{20.21}$$

for $i = 1, 2, \ldots, m$

$$0 < P \le e_iI, \text{ for } i = 1, 2, \ldots, m. \tag{20.22}$$

Some further discussions about the multi-objective optimization problem in (20.20) are given as follows:

Remark 20.2

(i) The multi-objective optimal solution in (20.20)–(20.22) for m-person noncooperative H_∞ evolution game dominates any other solution, i.e., $(e_1^* \cdots e_i^* \cdots e_m^*) \le (e_1^* \cdots e_i \cdots e_m^*)$ for any i. According to Theorem 2.3, it is a Nash equilibrium solution, i.e., no player could benefit by changing his/her strategy.

(ii) Since Riccati-like inequalities in (20.21) are bilinear inequalities, they cannot easily be solved directly. Performing $W = P^{-1} > 0$ on both sides of the inequalities, the multi-objective optimization problem in (20.20)–(20.22) is equivalent to the following LMIs-constrained multi-objective problem through the Schur complement method [23]

$$(e_1^* \cdots e_i^* \cdots e_m^*) = \min_{W>0}(e_1 \cdots e_i \cdots e_m) \tag{20.23}$$

subject to

$$\begin{bmatrix} WA^T + AW - R_i^{-1} + \frac{1}{e_i}\bar{B}_i\bar{B}_i^T + \sum_{k=1}^{N_{rp}}(WA_k^T + A_kW) & W & (A_1W)^T & \cdots & (A_{N_{rp}}W)^T \\ W & -Q_i^{-1} & 0 & \cdots & 0 \\ A_1W & 0 & -W & & \vdots \\ \vdots & \vdots & & \ddots & 0 \\ A_{N_{rp}}W & 0 & \cdots & 0 & -W \end{bmatrix} \le 0, \tag{20.24}$$

for $i = 1, 2, \ldots, m$,

$$\begin{bmatrix} W & I \\ I & e_iI \end{bmatrix} \ge 0, \text{ for } i = 1, 2, \ldots, m. \tag{20.25}$$

(iii) In contrast to the single-objective optimization problem, a global solution that is optimal for all conflicting objectives in the multi-objective problem in (20.20) or (20.23) may not exist [30]; it is thus desirable to seek a set of good compromise solutions that are no worse than other solutions. This set is called the Pareto optimal solutions for the noncooperative stochastic H_∞ evolutionary game from the multi-objective perspective [30]. Recently, a multi-objective evolutionary algorithm (MOEA)-based searching algorithm has been proposed to efficiently compute Pareto optimal solutions using the LMI toolbox in MATLAB [40]. In the proposed MOEA-based searching algorithm, we begin by choosing an initial parent population $(e_1 \cdots e_i \cdots e_m)$ in the feasible set (i.e., it should satisfy the LMIs in (20.24) and (20.25)). Pareto optimal solutions are then found via an MOEA-based non-dominant searching algorithm, which searches for the non-dominant level of simultaneous minimization in the initial population. Next, a child population is generated by crossover and mutation, and the search for the non-dominant level in the feasible set is carried out to find a compromise among m noncooperative players. This procedure is conducted iteratively to find the final population that approaches the Pareto front to simultaneously achieve the optimal individual targets [40].

(iv) MOEA is a stochastic searching method that simulates natural selection. It is particularly suited for solving the multi-objective optimization problem in the noncooperative H_∞ evolutionary game of nonlinear stochastic biological systems due to its population-based nature, permitting a set of Pareto optimal solutions to be obtained in a single run. In the multi-objective problem in (20.23)–(20.25), we employ the evolution algorithm (EA) to search for multi-objective solutions $(e_1^* \cdots e_i^* \cdots e_m^*)$ and their corresponding $W = P^{-1} > 0$ (Pareto optimal solutions). However, in the searching process, each candidate solution should satisfy the LMI constraints in (20.24) and (20.25), and no candidate solution should be dominated by other solutions to guarantee simultaneous optimization; that is, candidates dominated by other candidates should be deleted in the non-dominant search process because they are not solutions of the multi-objective optimization problem. The remaining non-dominant candidates are selected for crossover for the new population in the next generation of EA.

(v) Similar to Theorem 2.3, the solution of MOP in (20.20)-(20.22) could be proven the Nash equilibrium solution of the noncooperative evolution game in (20.9)

The MOEA algorithm for the multi-objective optimization problem in (20.23)–(20.25) is given as follows:

Step 1: Initialization

Set the searching region $[\underline{e}_1 \cdots \underline{e}_i \cdots \underline{e}_m] \times [\bar{e}_1 \cdots \bar{e}_i \cdots \bar{e}_m]$, where $\underline{e}_i$ and $\bar{e}_i$ denote the lower and upper bounds of e_i, respectively, the maximum number of individuals N_g, the iteration number N, the crossover rate M_c, and the mutation rate M_r.

Step 2: Updating

Perform EA (including crossover and the mutation operator) and examine N-feasible processes to produce $2N_g$-feasible individuals for the multi-objective optimization problem in (20.23).

Select N_g elitist individuals from $2N_g$ feasible individuals via the non-dominated sorting scheme.

Step 3: Stop criterion

If iteration index $t > N$, then stop, obtain the Pareto front as the Pareto optimal solutions, and solve the corresponding $W^* = P^{*^{-1}}$.

Step 4: Obtain the stochastic H_∞ evolutionary game strategies $u_i^*(t) = -R_i^{-1}P^*\tilde{x}(t) = -R_i^{-1}W^{*^{-1}}\tilde{x}(t)$ for each player.

(v) Once the set of Pareto optimal solutions in the Pareto front $(e_1^* \cdots e_i^* \cdots e_m^*)$ is obtained, we can solve LMIs in (20.24) and (20.25) to obtain the optimal W^* in Eq. (20.23) and then the optimal $P^* = W^{*^{-1}}$ in (20.20). Finally, the optimal evolutionary game strategies $u_i^*(t) = -R_i^{-1}P^*\tilde{x}(t)$ of the noncooperative stochastic H_∞ evolutionary game in (20.17) are obtained; that is, each player i selects their game strategy $\sum_{k=1}^{N_i(t)} A_k p(t-t_k)\tilde{x}(t) = -R_i^{-1}P^*\tilde{x}(t)$, and the total strategies of m players are $\sum_{i=1}^{m} u_i^*(t) = \sum_{i=1}^{m}\sum_{k=1}^{N_i(t)} A_k p(t-t_k)\tilde{x}(t) = -\sum_{i=1}^{m} R_i^{-1}P^*\tilde{x}(t)$.

Then, the remaining neutral genetic variations are $\sum_{i=1}^{N_r(t)} A_k p(t-t_k) = \sum_{k=1}^{N(t)} A_k p(t-t_k) - \sum_{i=1}^{m} R_i^{-1} P^*$.

(vi) After solving the m-person noncooperative stochastic H_∞ evolutionary game through the indirect multi-objective optimization problem in (20.23)–(20.25), we substitute e_i^*, $i = 1, \ldots, m$ into the Riccati-like inequalities in (20.18), which are necessary for the phenotypic trait to be robustly maintained at equilibrium point x_e, to obtain the phenotypic criterion under the m-person noncooperative H_∞ evolutionary game

$$(A - \frac{1}{2} R_i^{-1} P^*)^T P^* + P^* (A - \frac{1}{2} R_i^{-1} P^*) \frac{1}{e_i} P^* \bar{B}_i \bar{B}_i^T P^* + Q_i$$
$$+ \sum_{k=1}^{N_{rp}} (A_k^T P^* + P^* A_k + A_k^T P^* A_k) \le 0, \text{ for } i = 1, 2, \ldots, m$$

Under the noncooperative stochastic H_∞ evolutionary game strategy in (20.9), the phenotypic robustness criterion for each game strategy in (20.21) or the above inequalities could be reformulated as

$$\underbrace{\sum_{k=1}^{N_{rp}} (A_k^T P^* + P^* A_k + A_k^T P^* A_k)}_{\text{robustness to tolerate the effect of neutral genetic variations}} + \underbrace{Q_i + \frac{1}{e_i} P^* \bar{B}_i \bar{B}_i^T P^*}_{\text{robustness to buffer the effect of other strategies and environmental disturbance}}$$
$$\le \underbrace{-\left[(A - \frac{1}{2} R_i^{-1} P^*)^T P^* + P^* (A - \frac{1}{2} R_i^{-1} P^*) \right]}_{\text{network robustness enhanced by noncooperative evolutionary game strategies}}, \quad \text{for } i = 1, 2, \ldots, m \tag{20.26}$$

It is apparent that if the network robustness enhanced by the noncooperative H_∞ evolutionary game can confer enough stability robustness to both tolerate the effect of neutral genetic variations and to buffer the effect of other conflicting strategies and environmental disturbances, then the phenotypic trait will be maintained. Otherwise, the phenotypic robustness criterion of the biological system in (20.26) is violated and the trait may not be maintained.

Substituting noncooperative H_∞ evolutionary game strategies $u_i(t)$, $i = 1, 2, \ldots, m$ in (20.17) into the linear biological system in (20.4) yields

$$\dot{\tilde{x}}(t) = (A - \sum_{i=1}^{m} R_i^{-1} P^*) + \sum_{k=1}^{N_{rp}} A_k p(t-t_k) \tilde{x}(t) + Bv(t) \tag{20.27}$$

where $\sum_{k=1}^{N_r(t)} A_k p(t-t_k) =$ denotes the neutral (remaining) genetic variations after being selected by m players, that is,

$$\sum_{i=1}^{N_r(t)} A_k p(t-t_k) = \sum_{k=1}^{N(t)} A_k p(t-t_k) - \sum_{i=1}^{m} \sum_{k=1}^{N_i(t)} A_k p(t-t_k) \tilde{x}(t).$$

Clearly, the eigenvalues of the linear biological network are shifted to the left complex domain (i.e., shifted left by $-\sum_{i=1}^{m} R_i^{-1} P^*$ in the s-domain) to improve stability robustness by greater control effort $\sum_{i=1}^{m} u_i^*(t) = -\sum_{i=1}^{m} R_i^{-1} P^* \tilde{x}(t)$.

The phenotypic robustness criterion in Eq. (21.26) is applicable to many feedback loops in real biological networks, as the negative state feedback from noncooperative H_∞ evolutionary game strategies can enhance network robustness. Many modular and parallel structures as well as repair mechanisms to efficiently attenuate A_k may work (by minimizing the impact of genetic variations in the first term of (20.26)) to guarantee the phenotypic robustness criterion in (20.26). This would buffer the effect of genetic variations on the biological system to improve phenotypic robustness.

20.2.1 Noncooperative Stochastic H∞ Evolutionary Game in Nonlinear Stochastic Biological Network

In real biological networks, the stochastic network dynamics are always nonlinear as in (21.3). For this type of network, there always exist many equilibrium points (possible phenotypic traits) (see Fig. 20.1). Suppose that the phenotypic trait near stable equilibrium point x_e is of evolutionary consequence. For simplicity of analysis, let us denote the phenotypic deviation around the phenotype at x_e as $\tilde{x}(t) = x(t) - x_e$. In this situation, (20.3) becomes

$$\dot{\tilde{x}}(t) = f(\tilde{x}) + \sum_{i=1}^{m} u_i(t) + \sum_{k=1}^{N(t)} f_k(\tilde{x})p(t - t_k)\tilde{x}(t) + Bv(t) \tag{20.28}$$

In the noncooperative H_∞ evolutionary game of the above biological network, every strategy selects some beneficial genetic variations

$$u_i(t) = \sum_{k=1}^{N_i(t)} f_k(\tilde{x})p(t - t_k) = \sum_{k=1}^{N_i(t)} f_k(\tilde{x}(t_k)), \text{ for } i = 1, 2, \ldots, m \tag{20.29}$$

to simultaneously solve the stochastic m-person noncooperative stochastic H_∞ evolutionary game in (20.9). After beneficial genetic variations are selected by these noncooperative stochastic H_∞ evolutionary game strategies, the linear evolutionary biological system in (20.8) is modified for the nonlinear stochastic case in (20.28) as follows:

$$\dot{\tilde{x}}(t) = f(\tilde{x}) + u_i(t) + \sum_{k=1}^{N_r(t)} f_k(\tilde{x})p(t - t_k) + \bar{B}_i\bar{v}_i(t) \tag{20.30}$$

where $\bar{v}_i(t)$ denotes the combined competitive strategies and environmental disturbances as shown in (20.8). Similarly, the suboptimal approach to the problem in (20.11) and (20.12) of the network in (20.28) or (20.30) is equivalent to solving the following constrained multi-objective optimization problem:

$$(e_1^* \cdots e_i^* \cdots e_m^*) = \min(e_1 \cdots e_i \cdots e_m) \tag{20.31}$$

$$\text{subject to} \min_{u_i(t)} \max_{\bar{v}_i(t)} \frac{E\{\int_0^{t_p} (\tilde{x}^T(t)Q_i\tilde{x}(t) + u_i^T(t)R_iu_i(t))dt\}}{E\{(\tilde{x}^T(0)\tilde{x}(0) + \int_0^{t_p} \bar{v}_i^T(t)\bar{v}_i(t))dt\}} \le e_i, \text{ for } i = 1, 2, \ldots, m.$$

The m-constrained suboptimal H_∞ noncooperative evolutionary games in the multi-objective optimization problem in (20.31) are equivalent to the following n-constrained stochastic Nash evolution game problems [1, 9]:

$$e_i = \min_{u_i(t)} \max_{\bar{v}_i(t)} E\{\int_0^{t_p} (\tilde{x}^T(t)Q_i\tilde{x}(t) + u_i^T(t)R_iu_i(t) - e_i\bar{v}_i^T(t)\bar{v}_i(t))dt\} \leq e_iE\{(\tilde{x}^T(0)\tilde{x}(0)\},$$
$$\text{for } i = 1,2,\ldots,m \tag{20.32}$$

Let us denote

$$J_i = E\{\int_0^{t_p} (\tilde{x}^T(t)Q_i\tilde{x}(t) + u_i^T(t)R_iu_i(t) - e_i\bar{v}_i^T(t)\bar{v}_i(t))dt\}, \text{ for } i = 1,2,\ldots,m.$$

Two steps are required to solve the constrained suboptimal stochastic H_∞ evolutionary game problems in (20.32). The first step is to solve the following stochastic Nash evolution game problems:

$$J_i^* = \min_{u_i(t)} \max_{\bar{v}_i(t)} E\{\int_0^{t_p} (\tilde{x}^T(t)Q_i\tilde{x}(t) + u_i^T(t)R_iu_i(t) - e_i\bar{v}_i^T(t)\bar{v}_i(t))dt\}, \ i = 1,2,\ldots,m. \tag{20.33}$$

and the second step is to solve the following constraint problem:

$$J_i^* \leq e_iE\{(\tilde{x}^T(0)\tilde{x}(0)\}, \text{ for } i = 1,2,\ldots,m. \tag{20.34}$$

Solving the suboptimal H_∞ evolutionary game in (20.32) by the two-step method in (20.33) and (20.34) yields the following result for the suboptimal noncooperative H_∞ evolution game problem in (20.32) or (20.31).

Proposition 20.2 *The suboptimal H_∞ evolutionary games in (20.33) and (20.34) or the suboptimal H_∞ evolutionary games in (20.32) can be solved by the following m-person noncooperative suboptimal H_∞ evolutionary strategies $u_i^*(t)$ and the worst-case combined competitive strategies and environmental disturbances $\bar{v}_i^*(t)$*

$$u_i^*(t) = \sum_{k=1}^{N_i(t)} f_k(\tilde{x})p(t - t_k) = -\frac{1}{2}R_i^{-1}\left(\frac{\partial V(\tilde{x})}{\partial \tilde{x}}\right), \text{ for } i = 1,2,\ldots,m, \tag{20.35}$$

$$\bar{v}_i^*(t) = \frac{1}{2e_i}\bar{B}_i^T\left(\frac{\partial V(\tilde{x})}{\partial \tilde{x}}\right), \text{ for } i = 1,2,\ldots,m \tag{20.36}$$

where the positive Lyapunov function $V(\tilde{x})>0$ is the positive solution of the following Hamilton–Jacobi–Isaac inequalities (HJIIs)

$$\left(\frac{\partial V(\tilde{x})}{\partial \tilde{x}}\right)^T f(\tilde{x}) + \tilde{x}^TQ_i\tilde{x} - \frac{1}{4}\left(\frac{\partial V(\tilde{x})}{\partial \tilde{x}}\right)^T R_i^{-1}\left(\frac{\partial V(\tilde{x})}{\partial \tilde{x}}\right) + \frac{1}{4e_i}\left(\frac{\partial V(\tilde{x})}{\partial \tilde{x}}\right)^T \bar{B}_i\bar{B}_i^T\left(\frac{\partial V(\tilde{x})}{\partial \tilde{x}}\right)$$
$$+ \sum_{k=1}^{N_{rp}} (V(\tilde{x} + f_k(\tilde{x})) - V(\tilde{x})) \leq 0, \text{ for } i = 1,2,\ldots,m \tag{20.37}$$

$$V(\tilde{x}(0)) \leq e_i \tilde{x}^T(0)\tilde{x}(0), \text{ for } i = 1, 2, \ldots, m. \tag{20.38}$$

Proof: See Appendix B.

The last term $\sum_{k=1}^{N_{rp}} (V(\tilde{x} + f_k(\tilde{x})) - V(\tilde{x}))$ in (20.37) is the effect of the neutral Poisson genetic variations that are not selected by the noncooperative H_∞ evolutionary strategies of m players [31]. Solutions $u_i^*(t)$ in (20.35) and $\bar{v}_i^*(t)$ in (20.36) are the solutions of the constrained Nash game problem in (20.32). Therefore, the multi-objective optimization problem in (20.31) for the noncooperative stochastic H_∞ evolutionary game in (20.9) of the nonlinear stochastic biological system in (20.30) can be transformed into the following multi-objective optimization problem:

$$(e_1^* \cdots e_i^* \cdots e_m^*) = \min_{V(\tilde{x})>0} (e_1 \cdots e_i \cdots e_m) \tag{20.39}$$

subject to Eq. (20.37) and Eq. (20.38).

Similar to Theorem 3.4, the solution of MOP in (20.39) could be shown to be the Nash equilibrium solution of noncooperative evolutionary game strategy in (20.30) and (20.31) of nonlinear biological system.

The multi-objective optimization solution in (20.39) for noncooperative evolutionary game strategy (20.9) of nonlinear stochastic biological system dominates any other solution, i.e. $(e_1^* \cdots e_i^* \cdots e_m^*) \leq (e_1^* \cdots e_i \cdots e_m^*)$. According to Theorem 2.2, it is a Nash equilibrium solution, i.e. no player could benefit by changing his/her strategy.

After solving $e_1^*, \cdots, e_m^*$, and $V^*(\tilde{x})$ for the multi-objective optimization problem in (20.39), the minimax strategies in (20.9) for system in (20.30) or (20.28) are $u_i^*(t) = \sum_{k=1}^{N_i(t)} f_k(\tilde{x})p(t - t_k) = -\frac{1}{2}R_i^{-1}\left(\frac{\partial V^*(\tilde{x})}{\partial \tilde{x}}\right)$, $i = 1, 2, \ldots, m$. Substituting e_1^*, $i = 1, 2, \ldots, m$ into (20.37) yields the following network phenotypic robustness criterion:

$$\underbrace{\sum_{k=1}^{N_{rp}} (V^*(\tilde{x} + f_k(\tilde{x})) - V^*(\tilde{x}))}_{\substack{\text{genetic robustness to tolerate the} \\ \text{effect of neutral genetic variations}}} + \underbrace{\tilde{x}^T Q_i \tilde{x} + \frac{1}{4e_i^*}\left(\frac{\partial V^*(\tilde{x})}{\partial \tilde{x}}\right)^T \bar{B}_i \bar{B}_i^T \left(\frac{\partial V^*(\tilde{x})}{\partial \tilde{x}}\right)}_{\substack{\text{environmental robustness to buffer the effect} \\ \text{of other strategies and environmental disturbance}}}$$
$$\underbrace{\leq \left(\frac{\partial V^*(\tilde{x})}{\partial \tilde{x}}\right)^T \left(f(\tilde{x}) - \frac{1}{4}R_i^{-1}\left(\frac{\partial V^*(\tilde{x})}{\partial \tilde{x}}\right)\right)}_{\text{network robustness enhanced by noncooperative evolutionary game}}, \quad \text{for } i = 1, 2, \ldots, m, \tag{21.40}$$

Based on this criterion, the stochastic noncooperative H_∞ game strategies provide each player with enhanced network robustness by negative feedback $-\frac{1}{4}R_i^{-1}\left(\frac{\partial V^*(\tilde{x})}{\partial \tilde{x}}\right)$; i.e., the enhanced term $\left(\frac{\partial V^*(\tilde{x})}{\partial \tilde{x}}\right)^T \times \left(f(\tilde{x}) - \frac{1}{4}R_i^{-1}\left(\frac{\partial V^*(\tilde{x})}{\partial \tilde{x}}\right)\right)$ is lower than $\left(\frac{\partial V^*(\tilde{x})}{\partial \tilde{x}}\right)^T f(\tilde{x})$ without the noncooperative stochastic H_∞ evolutionary game strategy, which is equivalent to causing the basin of the phenotypic trait at x_e to become much deeper and wider in Fig. 20.1, conferring additional network robustness. This provides greater tolerance for neutral genetic variations and for buffering the effect of competitive strategies and environmental disturbances. If the criterion holds for all noncooperative stochastic H_∞ evolutionary game strategies, then the network phenotypic trait at x_e will be maintained in the evolutionary process. If stochastic neutral genetic variations are accumulated to a sufficiently large amount in the first term of (20.40) that the criterion is eventually violated, they may provide the raw material for further evolutionary processes to cause a transition from one phenotypic trait to another through genetic drift; for example, from basin x_e to another basin x_{e1} in Fig. 20.1.

Substituting the noncooperative stochastic H_∞ evolutionary game strategies $u_i^*(t)$ in (20.35) into (20.28), we get

$$\dot{\tilde{x}}(t) = f(\tilde{x}) + \sum_{i=1}^{m} \frac{1}{2} R_i^{-1} \left(\frac{\partial V^*(\tilde{x})}{\partial \tilde{x}} \right) + \sum_{k=1}^{N_r(t)} f_k(\tilde{x}) p(t - t_k) + Bv(t) \tag{20.41}$$

where neutral genetic variations $\sum_{k=1}^{N_r(t)} f_k(\tilde{x}) p(t - t_k)$ are defined as the remaining genetic variations $\sum_{k=1}^{N_r(t)} f_k(\tilde{x}) p(t - t_k) = \sum_{k=1}^{N(t)} f_k(\tilde{x}) p(t - t_k) - \sum_{i=1}^{m} \sum_{k=1}^{N_i(t)} f_k(\tilde{x}) p(t - t_k)$. The negative feedback terms $\sum_{i=1}^{m} \frac{1}{2} R_i^{-1} \left(\frac{\partial V^*(\tilde{x})}{\partial \tilde{x}} \right)$ will deepen and widen the basin of equilibrium point x_e of $f(\tilde{x})$, enhancing the phenotypic stability robustness at x_e.

Compared to the cooperative H_∞ evolutionary game strategy, the noncooperative stochastic H_∞ evolutionary game strategy is more able to improve the network robustness of the nonlinear biological network through the selection of more beneficial genetic variations, creating additional negative feedback loops. This will be discussed in detail in the sequel.

20.2.2 Global Linearization Approach to Nonlinear Stochastic Noncooperative H_∞ Evolutionary Game

In general, it is very difficult to solve the HJII-constrained multi-objective optimization problem in (20.39) for the noncooperative H_∞ evolutionary game of the nonlinear stochastic biological network in (20.28). At present, there is no good method for solving it either analytically or numerically. In this chapter, we used the global linearization technique [23] to interpolate a nonlinear stochastic biological network with a set of local linearized networks. This results in bounding the nonlinear stochastic biological network in (20.28) by a polytope, consisting of M vertices as [23, 275]

$$\begin{bmatrix} \frac{\partial f(\tilde{x})}{\partial \tilde{x}} \\ \frac{\partial f_1(\tilde{x})}{\partial \tilde{x}} \\ \vdots \\ \frac{\partial f_{N(t)}(\tilde{x})}{\partial \tilde{x}} \end{bmatrix} \in C_0 \left[\begin{bmatrix} A_1 \\ A_{11} \\ \vdots \\ A_{N(t)1} \end{bmatrix} \cdots \begin{bmatrix} A_1 \\ A_{1j} \\ \vdots \\ A_{N(t)j} \end{bmatrix} \cdots \begin{bmatrix} A_M \\ A_{1M} \\ \vdots \\ A_{N(t)M} \end{bmatrix} \right], \forall \tilde{x}(t) \tag{20.42}$$

where C_0 denotes the convex hull of the polytope with M vertices. In other words, if the local linearized systems at all $\tilde{x}(t)$ are inside convex hull C_0, then evolutionary trajectory $\tilde{x}(t)$ of the nonlinear stochastic network in (20.30) will belong to the convex combination of the stable trajectories of the following M local linearized biological networks, at M vertices of the polytope [23]

$$\dot{\tilde{x}}(t) = A_j \tilde{x} + u_i(t) + \sum_{k=1}^{N_r(t)} A_{kj} p(t - t_k) \tilde{x}(t) + \bar{B}_i \bar{v}_i(t), \text{ for } j = 1, 2, \ldots, M \tag{20.43}$$

where $\bar{B}_i$ and $\bar{v}_i(t)$ are in (20.8).

Based on global linearization theory [23], every evolutionary trajectory of the network in (20.30) can be represented by a convex combination of the M local linearized networks in (20.43). This combination is represented by [31] and [18]

$$\dot{\tilde{x}}(t) = \sum_{j=1}^{M} \alpha_j(\tilde{x})\left(A_j\tilde{x}(t) + u_i(t) + \sum_{k=1}^{N_r(t)} A_{kj}p(t-t_k)\tilde{x}(t) + \bar{B}_i\bar{v}_i(t)\right) \tag{20.44}$$

where $\alpha_j(\tilde{x})$ are interpolation functions with $0 \le \alpha_j(\tilde{x}) \le 1$ and $\sum_{j=1}^{M}\alpha_j(\tilde{x}) = 1$; that is, the evolutionary trajectory of the network in (20.30) can be represented by the trajectory of the interpolated network in (20.44).

In (20.44), the noncooperative stochastic H_∞ evolutionary game strategy $u_i(t)$ is to select the following genetic variations based on the global linearization technique:

$$u_i(t) = \sum_{j=1}^{M} \alpha_j(\tilde{x}) \sum_{k=1}^{N_i(t)} A_{kj}p(t-t_k)\tilde{x}(t), \text{ for } i = 1,2,\ldots,m. \tag{20.45}$$

This yields the following result:

Proposition 20.3 *For the interpolated evolutionary biological network in (20.44), the suboptimal noncooperative H_∞ evolutionary game problem in (20.32)–(20.34) is solved by the following:*

$$u_i^*(t) = \sum_{j=1}^{M}\sum_{k=1}^{N_i(t)} \alpha_j(\tilde{x})A_{kj}p(t-t_k)\tilde{x}(t) = -R_i^{-1}P\tilde{x}(t), \text{ for } i = 1,2,\ldots,m, \tag{20.46}$$

$$\bar{v}_i^*(t) = \tfrac{1}{e_i}\bar{B}_i^T P\tilde{x}(t), \text{ for } i = 1,2,\ldots,m$$

where the positive matrix P is the solution of the following Riccati-like inequalities:

$$\begin{aligned} &A_j^TP + PA_j + Q_i - PR_i^{-1}P + \frac{1}{e_i}P\bar{B}_i\bar{B}_i^TP + \sum_{k=1}^{N_{rp}}(A_{kj}^TP + PA_{kj} + A_{kj}^TPA_{kj}) \le 0,\\ &\text{for } i = 1,2,\ldots,m,\\ &\text{for j} = 1,2,\ldots,M\\ &\text{with } 0<P \le e_iI, \text{ for } i = 1,2,\ldots,m. \end{aligned} \tag{20.47}$$

Proof: See Appendix C.

Remark 20.3

(i) By comparing Proposition 20.3 with Proposition 20.1, the Riccati-like inequality in (20.18) is based on the localized biological network in (20.4) at $\tilde{x}(t) = 0$ when the inequalities in (20.47) are based on the M local linearized networks in (20.43) at the M vertices of the polytope in (20.42). Therefore, the noncooperative strategy based on the global linearization theory in Proposition 20.3 is more suitable for nonlinear biological networks in the evolutionary process.

(ii) The Riccati-like inequalities in (20.47) can be considered as the local linearized HJIIs in (20.37) at M vertices of the polytope in (20.42). While it is very difficult to solve the HJIIs in (20.37), the Riccati-like inequalities in (20.47) are readily solved by transforming them into the following equivalent LMIs using the Schur complement transformation method [23]

$$\begin{bmatrix} WA_j^T+A_jW-R_i^{-1}+\frac{1}{e_i}\bar{B}_i\bar{B}_i^T+\sum_{k=1}^{N_{rp}}(WA_{kj}^T+A_{kj}W) & W & (A_{1j}W)^T & \cdots & (A_{N_{rpj}}W)^T \\ W & -Q_i^{-1} & 0 & \cdots & 0 \\ A_{1j}W & 0 & -W & & \vdots \\ \vdots & \vdots & & \ddots & 0 \\ A_{N_{rpj}}W & 0 & \cdots & 0 & -W \end{bmatrix} \leq 0, \tag{20.48}$$

for $i=1,2,\ldots,m$,
for $j=1,2,\ldots,M$.

which can be easily solved by the LMI toolbox in MATLAB.

Interpolating the nonlinear network in (20.30) with the M local linear networks in (20.44) transforms the HJIIs-constrained multi-objective optimization problem in (20.39) for the noncooperative stochastic H_∞ evolutionary game of the global linearized network in (20.44) into the following LMIs-constrained multi-objective problem

$$(e_1^* \cdots e_i^* \cdots e_m^*) = \min_{W>0}(e_1 \cdots e_i \cdots e_m) \tag{20.49}$$

subject to LMIs in Eq. (20.48) and LMIs in Eq. (20.25).

Remark 20.4

(i) The proposed LMIs-constrained MOEA searching algorithm could be employed to efficiently solve the Pareto optimal solution $(e_1^* \cdots e_i^* \cdots e_m^*)$ and W^* of the multi-objective problem in (20.49) (Chen and Ho, 2016).

(ii) The noncooperative H_∞ evolutionary strategies $u_i^*(t)$ select genetic variations $\sum_{j=1}^{M}\sum_{k=1}^{N_i(t)} \alpha_j(\tilde{x})A_{kj}p(t-t_k)\tilde{x}(t)$ to produce negative feedback loops $-R_i^{-1}P^*\tilde{x}(t) = -R_i^{-1}W^{*^{-1}}\tilde{x}(t)$ with the aim of improving network robustness. This allows resisting the effect of other strategies and environmental disturbances, and attenuating the effects of neutral genetic variations to maintain the local network phenotypes as follows:

$$\begin{aligned} &\underbrace{\sum_{k=1}^{N_{rp}}(A_{kj}^TP^* + P^*A_{kj} + A_{kj}^TP^*A_{kj})}_{\text{genetic robustness to tolerate the local genetic variations}} + \underbrace{Q_i + \frac{1}{e_i^*}P^*\bar{B}_i\bar{B}_i^TP^*}_{\text{environmental robustness to buffer the effect of other strategies and environmental disturbance}} \\ &\leq \underbrace{-\left[\left(A_j - \frac{1}{2}R_i\right)^T P^* + P^*\left(A_j - \frac{1}{2}R_i\right)\right]}_{\text{local network robustness enhanced by noncooperative evolutionary game}}, \text{ for } i=1,2,\ldots,m, \text{ for } j=1,2,\ldots,M. \end{aligned} \tag{20.50}$$

In other words, if the enhanced local network robustness of each local linearized biological network in (20.43) by each noncooperative H_∞ evolutionary game strategy can confer both local neutral genetic robustness and local environmental robustness as shown in (20.50), then the phenotype trait can be maintained in the basin of the equilibrium point at x_e.

\(iii) Since the noncooperative stochastic H_∞ evolutionary strategy could improve the local network robustness of the local linearized networks on the right-hand side of (20.50), the nonlinear network will contain more neutral genetic variations and offer more robustness to eliminate the effect of competitive strategies and environmental disturbances, decreasing evolvability e_i^*. However, when the neutral genetic variations are accumulated to the extent that the first term in (20.50) becomes very large and the local phenotypic robustness criteria in (20.50) are violated, the phenotypic trait of the biological network may shift from one equilibrium point to another (Fig. 20.1) and start another period of network evolution at another equilibrium point (phenotypic trait) favored by natural selection.

(iv) If we substitute the noncooperative stochastic H_∞ evolutionary strategies $u_i^*(t)$, $i = 1, 2, \ldots, m$, in Proposition 20.3 into the biological system in (20.44), we get

$$\dot{x}(t) = \sum_{j=1}^{M} \alpha_j(\tilde{x}) \left[\left(A_j - \sum_{i=1}^{m} R_i^{-1} P^* \right) \tilde{x}(t) + \sum_{k=1}^{N_r(t)} A_{kj} p(t - t_k) \tilde{x}(t) + Bv(t) \right] \quad (20.51)$$

where the remaining neutral genetic variation is defined as $\sum_{k=1}^{N_r(t)} A_{kj} p(t - t_k) = \sum_{k=1}^{N(t)} A_{kj} \times p(t - t_k) - \sum_{i=1}^{m} sum_{k=1}^{N_i(t)} A_{kj} p(t - t_k)$.

Evidently, the noncooperative H_∞ evolutionary game strategy is able to significantly improve network robustness with more strategical use of effort and more compromises made between players to achieve individual optimal goals.

20.3 Cooperative Stochastic H_∞ Evolutionary Game Strategy for Stochastic Biological Networks

In the noncooperative H_∞ evolutionary game, each player pursues an individual goal which may partly conflict with others, therefore much strategy efforts $\sum_{i=1}^{m} u_i(t)$ must be paid to maximize the network fitness (or minimize the evolvability) of every player by the biological networks to maintain their robust stability of the phenotypic trait at an equilibrium point x_e in the evolutionary process. In the emergence of the cooperative H_∞ evolutionary strategy, an individual helps others achieve more fitness, implying more reproductive success and less effort, which are at the roots of many biological phenomena from mutualism to the appearance of multicellular organisms [27, 444]. In the following we discuss the cooperative H_∞ evolutionary game of the linear stochastic biological network at first.

20.3.1 Cooperative Stochastic H_∞ Evolutionary Game Strategy in Linear Stochastic Biological Network

Consider the following linear evolutionary biological network with the cooperative stochastic H_∞ evolutionary game strategy:

$$\dot{\tilde{x}}(t) = A\tilde{x}(t) + u(t) + \sum_{k=1}^{N(t)} A_k p(t - t_k)\tilde{x}(t) + Bv(t). \tag{20.52}$$

The cooperative stochastic H_∞ evolutionary strategy $u(t)$ of the linear stochastic biological network in (20.52) is how to select beneficial genetic variations

$$u(t) = \sum_{k=1}^{N_1(t)} A_k p(t - t_k)\tilde{x}(t) = \sum_{k=1}^{N_1(t)} A_k \tilde{x}(t_k)$$

from genetic variations $\sum_{k=1}^{N_1(t)} A_k p(t - t_k)\tilde{x}(t)$ to achieve the following cooperative (common) H_∞ evolutionary game strategy:

$$e_0 = \min_{u(t)} \max_{v(t)} \frac{E\{\int_0^{t_p} (\tilde{x}^T(t)Q\tilde{x}(t) + u^T(t)Ru(t))dt\}}{E\{(\tilde{x}^T(0)\tilde{x}(0) + \int_0^{t_p} v^T(t)v(t))dt\}} \leq e \tag{20.53}$$

where e_0 denotes the evolvability of the cooperative evolutionary network and e denotes the upper bound of e_0, i.e., all n species in $\tilde{x}(t) = [\tilde{x}_1 \cdots \tilde{x}_i \cdots \tilde{x}_n]^T$ stochastic biological network in (20.52) are cooperative with the same strategy $u(t)$ for minimizing both the effect of genetic variations and the worst-case effect of unpredictable environmental disturbance on the phenotypic deviation around x_e with a least strategy effort $u(t)$, i.e. to maximize the fitness of phenotypic trait of the biological network at x_e.

The remaining (neutral) genetic variations after the selection in (20.52) are denoted as

$$\sum_{k=1}^{N_2(t)} A_k p(t - t_k) = \sum_{k=1}^{N(t)} A_k p(t - t_k) - \sum_{k=1}^{N_1(t)} A_k \tilde{x}(t_k). \tag{20.54}$$

In general, it is still difficult to select some beneficial genetic variations in (20.52) to solve the cooperative stochastic H_∞ evolutionary game in (20.53) for e_0 directly. The following indirect suboptimal method is employed by minimizing its upper bound:

$$\begin{aligned} &e_0 = \min_{u(t)} e \\ &\text{subject to} \min_{u(t)} \max_{v(t)} \frac{E\{\int_0^{t_p} (\tilde{x}^T(t)Q\tilde{x}(t) + u^T(t)Ru(t))dt\}}{E\{(\tilde{x}^T(0)\tilde{x}(0) + \int_0^{t_p} v^T(t)v(t))dt\}} \leq e \end{aligned} \tag{20.55}$$

The minimax constraint on the suboptimal cooperative stochastic H_∞ evolutionary game problem in (20.55) is equivalent to the following constrained stochastic quadratic (H_2) game [1, 9]:

$$\min_{u(t)} \max_{v(t)} E\{\int_0^{t_p} (\tilde{x}^T(t)Q\tilde{x}(t) + u^T(t)Ru(t) - ev^T(t)v(t))dt\} \leq eE\{\tilde{x}^T(0)\tilde{x}(0)\}. \tag{20.56}$$

Let us denote

$$J = E\{\int_0^{t_p} (\tilde{x}^T(t)Q\tilde{x}(t) + u^T(t)Ru(t) - ev^T(t)v(t))dt\}.$$

Two steps are needed to solve the constrained quadratic stochastic game problem in (20.56). The first step is to solve the following stochastic Nash game:

$$\begin{aligned} J^0 &= \min_{u(t)} \max_{v(t)} J \\ &= \min_{u(t)} \max_{v(t)} E\{\int_0^{t_p} (\tilde{x}^T(t)Q\tilde{x}(t) + u^T(t)Ru(t) - ev^T(t)v(t))dt\} \end{aligned} \quad (20.57)$$

and the second step is to solve the following constraint problem:

$$J^0 \leq eE\{\tilde{x}^T(0)\tilde{x}(0)\}. \quad (20.58)$$

By solving (20.57) and (20.58), for the constrained Nash game problem in (20.56), we get the following result:

Proposition 20.4 *The constrained stochastic quadratic game problem for the suboptimal H_∞ evolutionary game of the cooperative evolutionary biological network in (20.56) could be solved by the following evolutionary strategy $u^0(t)$ and the worst-case environmental disturbance $v^0(t)$*

$$u^0(t) = \sum\nolimits_{k=1}^{N_1(t)} A_k p(t - t_k)\tilde{x}(t) = -R^{-1}P\tilde{x}(t),\ v^0(t) = \frac{1}{e}B^T P\tilde{x}(t) \quad (20.59)$$

where P is the solution of the following Riccati-like inequality

$$A^T P + PA - PR^{-1}P + \frac{1}{e}PBB^T P + Q + \sum_{k=1}^{N_{2p}} (A_k^T P + PA_k + A_k^T PA_k) \leq 0 \quad (20.60)$$

with

$$0 < P \leq eI \quad (20.61)$$

where $N_{2p} = N_2(t_p)$ represents the number of neutral genetic variations to the present time t_p.

Proof: The proof follows a similar procedure to that in Appendix A.

The last term $\sum_{k=1}^{N_{2p}} (A_k^T P + PA_k + A_k^T PA_k)$ in (20.60) is the effect of neutral genetic variations on the evolutionary network under the constrained stochastic quadratic H_∞ evolutionary game strategy (i.e. the suboptimal solution of the cooperative stochastic H_∞ evolutionary game problem) in (20.56). After solving the suboptimal solution of the cooperative stochastic H_∞ evolutionary game problem, the n-person cooperative stochastic H_∞ evolutionary game problem in (20.55) can be transformed to the following constrained optimization problem:

$$\begin{aligned} e_0 &= \min_P e \\ &\text{subject to (20.60) and (20.61).} \end{aligned} \quad (20.62)$$

By performing $W = P^{-1}$ to both sides of (20.60) and by the Schur complement method [23], the constrained optimization problem for the linear cooperative evolution game in (20.62) is equivalent to the following LMIs-constrained optimization problem [23]:

$$e_0 = \min_{W>0} e \quad (20.63)$$

subject to

$$\begin{bmatrix} WA^T + AW - R^{-1} + \frac{1}{e}BB^T + \sum_{k=1}^{N_{2p}}(WA_k^T + A_kW) & W & (A_1W)^T & \cdots & (A_{N_{2p}}W)^T \\ W & -Q^{-1} & 0 & \cdots & 0 \\ A_1W & 0 & -W & & \vdots \\ \vdots & \vdots & & \ddots & 0 \\ A_{N_{2p}}W & 0 & \cdots & 0 & -W \end{bmatrix} \leq 0, \tag{20.64}$$

$$\begin{bmatrix} W & I \\ I & eI \end{bmatrix} \geq 0. \tag{20.65}$$

Remark 20.5

(i) The solution of e_0 and $W_0 = P_0^{-1}$ in (20.63) can be found by decreasing e until no positive solution $W > 0$ exists in (20.64) and (20.65).

(ii) After solving e_0 and P_0 from (20.63)–(20.65), the phenotypic trait robustness criterion in (20.60) for linear stochastic biological network in (20.52) under the cooperative stochastic H_∞ evolutionary game can be reformulated as follows:

$$\underbrace{\sum_{k=1}^{N_{2p}}(A_k^TP_0 + P_0A_k + A_k^TP_0A_k) +}_{\text{robustness to tolerate neutral genetic variations}} \underbrace{Q + \frac{1}{e_0}P_0BB^TP_0}_{\text{robustness to buffer environmental disturbance}}$$
$$\leq \underbrace{-\left[(A - \frac{1}{2}R^{-1}P_0)^TP_0 + P_0(A - \frac{1}{2}R^{-1}P_0)\right]}_{\text{network robustness enhanced by cooperative evolutionary game}}. \tag{20.66}$$

The cooperative stochastic H_∞ evolutionary game strategy $u^0(t)$ selects some beneficial genetic variations $\sum_{k=1}^{N_1(t)} A_k \times p(t - t_k) = -R^{-1}P_0$ as negative feedback loops to improve the network robustness in the right-hand side of (20.66). If the improved network robustness could confer enough network robustness in the right-hand side of (20.66) to tolerate neutral genetic variations and to buffer environmental disturbance in the left-hand side of (20.66) simultaneously, then the phenotypic trait of the linear biological network could be maintained in the evolutionary process. Otherwise, the phenotypic trait may not be maintained under environmental disturbance and genetic variations in evolution. This is why so many negative feedback loops and hubs are found in real biological networks. If the network robustness in the right-hand side of (20.66) is fixed, the small coupling matrix B and small impact A_k due to the kth genetic variation will make network evolvability e_0 small (or make network fitness $f_0 = \frac{1}{e_0}$ large) according to the phenotypic trait robustness criterion in (20.66). These are why we find so many membranes, transparent membranes, and compartments to separate biological systems from their environment, only with some receptors and channels to communicate with their environment to make coupling matrix B small, which could attenuate environmental disturbance to cause network evolvability e_0 small. In this situation, small evolvability e_0 could not violate the phenotypic trait robustness criterion in (20.66) to maintain a phenotypic trait in evolution because of small B. We have found so many modules and redundant mechanisms in cell to buffer genetic variations to attenuate their effect on phenotypic variations A_k, i.e. modules and redundant mechanisms could attenuate the phenotypic variation A_k due to the impact of genetic variation A_k occurring at $p(t-t_k)$. Therefore, we could predict the resulting structure of a biological network in the evolutionary process from the phenotypic trait robustness criterion under the cooperative stochastic H_∞ evolutionary game strategy in (20.66).

(iii) Substituting the cooperative strategy $u^0(t)$ of (20.59) into the linear biological network in (20.52), we get

$$\dot{\tilde{x}}(t) = (A - R^{-1}P_0)\tilde{x}(t) + \sum_{k=1}^{N_2(t)} A_k p(t-t_k)\tilde{x}(t) + Bv(t). \tag{20.67}$$

The cooperative stochastic H_∞ evolutionary game strategy can shift eigenvalues of the linear biological network to the deep left-hand s domain by $-R^{-1}P_0$ to improve the network stability robustness when the noncooperative stochastic H_∞ evolutionary game could shift eigenvalues more deeply by $-\sum_{i=1}^{m} R^{-1}P^*$. However, when compared with the noncooperative evolutionary strategy in (20.23) with more effort $\sum_{i=1}^{m} u_i(t) = \sum_{i=1}^{m}\sum_{k=1}^{N_i(t)} A_k p(t-t_k)\tilde{x}(t)$, the cooperative stochastic H_∞ evolutionary game is with a less strategy effort $u(t) = \sum_{k=1}^{N_1(t)} A_k p(t-t_k)\tilde{x}(t)$ to maintain network stability robustness.

(iv) In the noncooperative stochastic H_∞ evolutionary strategy, there are $2m$ LMI constraints on solving the m-tuple multi-objective optimization problem in (20.23)–(20.25). However, there are only 2 LMI constraints on solving the single-objective optimization problem for the cooperative stochastic H_∞ evolution game in (20.63)–(20.65). Therefore, it is more difficult and exhaustive to solve P^* for the noncooperative H_∞ evolutionary game strategy of a biological network than P_0 for the cooperative game strategy, especially the strategy effort of how to employ non-dominant MOEA searching algorithm to achieve simultaneous optimization in solving the multi-objective optimization problem in (20.23)–(20.25), which reflects the difficult compromise process among partly conflicting players of the noncooperative stochastic H_∞ evolutionary game with an infinite number of Pareto optimal solutions in the noncooperative evolutionary process.

20.3.2 Cooperative Stochastic H_∞ Evolutionary Game in Nonlinear Stochastic Biological Network

After discussing the cooperative H_∞ evolutionary game of a linear biological network, the cooperative stochastic H_∞ evolutionary game of nonlinear stochastic biological systems in (20.2) with phenotypic trait around the equilibrium point x_e will be discussed in the following.

The dynamic equation of phenotypic deviation $\tilde{x}(t) = x(t) - x_e$ is described in the following with a cooperative stochastic H_∞ evolutionary game strategy:

$$\dot{\tilde{x}}(t) = f(\tilde{x}) + u(t) + \sum_{k=1}^{N(t)} f_k(\tilde{x})p(t-t_k)\tilde{x}(t) + Bv(t). \tag{20.68}$$

The cooperative stochastic H_∞ evolutionary game strategy $u(t)$ is to select some beneficial genetic variations

$$u(t) = \sum_{k=1}^{N_1(t)} f_k(\tilde{x})p(t - t_k)\tilde{x}(t) \tag{20.69}$$

from genetic variations $\sum_{k=1}^{N(t)} f_k(\tilde{x})p(t - t_k)$ to achieve the following stochastic cooperative H_∞ evolution game:

$$e_0 = \min_{u(t)} \max_{v(t)} \frac{E\{\int_0^{t_p} (\tilde{x}^T(t)Q\tilde{x}(t) + u^T(t)Ru(t))dt\}}{E\{(\tilde{x}^T(0)\tilde{x}(0) + \int_0^{t_p} v^T(t)v(t))dt\}} \le e \tag{20.70}$$

where e_0 denotes the evolvability of a cooperative H_∞ evolutionary network with cooperative evolutionary game strategy and e denotes the upper bound of e_0.

Following the procedures of Propositions 20.4 and 20.2, we get the following suboptimal result for the cooperative stochastic H_∞ evolutionary game strategy in Eq. (20.70) for the nonlinear stochastic biological system in Eq. (20.68).

Proposition 20.5 *The constrained cooperative stochastic H_∞ evolutionary game in (20.70) could be solved for the suboptimal cooperative H_∞ evolutionary game by*

$$u^0(t) = \sum_{k=1}^{N_1(t)} f_k(\tilde{x})p(t - t_k)\tilde{x}(t) = -\frac{1}{2}R^{-1}\left(\frac{\partial V(\tilde{x})}{\partial \tilde{x}}\right), \tag{20.71}$$

$$v^0(t) = \frac{1}{2e}B^T\left(\frac{\partial V(\tilde{x})}{\partial \tilde{x}}\right) \tag{20.72}$$

where the positive Lyapunov function $V(\tilde{x}) > 0$ *is the positive solution of the following HJII:*

$$\begin{aligned}&\left(\frac{\partial V(\tilde{x})}{\partial \tilde{x}}\right)^T f(\tilde{x}) + \tilde{x}^T Q\tilde{x} - \frac{1}{4}\left(\frac{\partial V(\tilde{x})}{\partial \tilde{x}}\right)^T R^{-1}\left(\frac{\partial V(\tilde{x})}{\partial \tilde{x}}\right) + \frac{1}{4e}\left(\frac{\partial V(\tilde{x})}{\partial \tilde{x}}\right)^T \bar{B}\bar{B}^T\left(\frac{\partial V(\tilde{x})}{\partial \tilde{x}}\right)\\ &+ \sum_{k=1}^{N_{2p}} (V(\tilde{x} + f_k(\tilde{x})) - V(\tilde{x})) \le 0,\end{aligned} \tag{20.73}$$

with

$$V(\tilde{x}(0)) \le e_i\tilde{x}^T(0)\tilde{x}(0). \tag{20.74}$$

Proof: Similar to the procedure of Propositions 20.4 and 20.2.

The last term $\sum_{k=1}^{N_{2p}} (V(\tilde{x} + f_k(\tilde{x})) - V(\tilde{x}))$ in (20.73) is the effect of the remaining neutral Poisson genetic variations after selection by the cooperative stochastic H_∞ evolutionary game strategy. Therefore, based on Proposition 20.5, the suboptimal cooperative stochastic H_∞ evolution game problem in (20.70) is transformed to the following HJII-constrained optimization problem:

$$\begin{aligned}&e_0 = \min_{V(\tilde{x})>0} e\\ &\text{subject to Eqs(20.73) and (20.74).}\end{aligned} \tag{20.75}$$

After solving e_0 and $V^0(\tilde{x})$ from (20.75), the cooperative stochastic H_∞ evolutionary game strategy $u^0(t)$ is to select some beneficial genetic variations such that $u^0(t) = \sum_{k=1}^{N_1(t)} f_k(\tilde{x})p(t-t_k)\tilde{x}(t) = -\frac{1}{2}R^{-1}\left(\frac{\partial V^0(\tilde{x})}{\partial \tilde{x}}\right)$. Substituting e_0 into (20.73), we get the following network phenotypic robustness criterion under the cooperative stochastic H_∞ evolutionary game strategy:

$$\underbrace{\sum_{k=1}^{N_{2p}}(V^0(\tilde{x}+f_k(\tilde{x}))-V^0(\tilde{x}))}_{\text{robustness to tolerate neutral genetic variations}}+\underbrace{\tilde{x}^TQ\tilde{x}+\frac{1}{4e_0}\left(\frac{\partial V^0(\tilde{x})}{\partial \tilde{x}}\right)^T BB^T\left(\frac{\partial V^0(\tilde{x})}{\partial \tilde{x}}\right)}_{\text{robustness to buffer environmental disturbance}}$$
$$\leq \underbrace{-\left(\frac{\partial V^0(\tilde{x})}{\partial \tilde{x}}\right)^T\left(f(\tilde{x})-\frac{1}{4}R^{-1}\left(\frac{\partial V^0(\tilde{x})}{\partial \tilde{x}}\right)\right)}_{\text{network robustness enhanced by cooperative evolutionary game strategy}}. \tag{20.76}$$

From the network phenotypic trait robustness criterion of the cooperative stochastic H_∞ game strategy in (20.76), if the enhanced network robustness by negative feedback $-\frac{1}{4}R^{-1}\left(\frac{\partial V^0(\tilde{x})}{\partial \tilde{x}}\right)$ could confer enough network stability robustness to tolerate neutral genetic variations and to buffer environmental disturbances, then the phenotypic trait at the equilibrium point x_e will be maintained. However, if the stochastic neutral genetic variations are accumulated to a sufficiently large amount in the evolutionary process so that the phenotypic trait robustness criterion in (20.76) may be finally violated, then the phenotypic trait may cause a transition from an equilibrium point to another through genetic drift, i.e. from x_e to x_{e1} or x_{e2} in Fig. 20.1.

At first sight, one might expect a cooperative stochastic H_∞ evolutionary game strategy to significantly improve network robustness to slow or even stop the evolution of a phenotypic trait at an equilibrium point in a biological network, but, because the more robust phenotypic trait can harbor a larger amount of neutral genetic variations, the robust phenotypic trait of a stochastic biological network might show increased rather than decreased evolutionary potential, in the long term of biological network evolution. The reasons why increasing network robustness may eventually lead to increasing network evolution include the accumulation of hidden neutral genetic variations that may be useful for later evolution. This matches with the finding that network robustness is intrinsic to evolution and eventually improve network evolution [437, 464, 511].

Remark 20.6

(i) Comparing the HJII-constraints of the cooperative stochastic H_∞ evolutionary game in (20.73) and (20.74) with the HJII-constraints of the noncooperative stochastic H_∞ evolutionary game in (20.37) and (20.38), it is seen that more HJII constraints on solving the noncooperative H_∞ evolutionary game strategy than that on solving the cooperative stochastic H_∞ evolutionary game strategy make the noncooperative stochastic H_∞ evolutionary game problem in (20.39) more difficult to solve than the cooperative stochastic H_∞ evolutionary game in (20.75).

(ii) The solution of the cooperative stochastic H_∞ evolutionary game strategy in (20.75) is unique. However, there exist many Pareto optimal solutions to the multi-objective optimization problem for the noncooperative stochastic H_∞ evolutionary game strategy in (20.39). Obviously, there should exist some compromises among these partly conflicting evolutionary strategies for solution.

20.3.3 Global Linearization Approach to Cooperative Stochastic H_∞ Evolutionary Game in Nonlinear Stochastic Network

In general, it is still very difficult to solve the HJII-constrained optimization problem in (20.75) for the cooperative stochastic H_∞ game strategy of the nonlinear stochastic biological network in (20.68). For simplicity of solution, the global linearization technique in (20.42)–(20.44) is employed to represent the nonlinear stochastic biological network in (20.68) by the following interpolated biological network [23]:

$$\dot{x}(t) = \sum_{i=1}^{M} \alpha_i(\tilde{x})\left(A_i\tilde{x}(t) + u(t) + \sum_{k=1}^{N(t)} A_{ki}p(t-t_k)\tilde{x}(t) + Bv(t)\right) \tag{20.77}$$

and the cooperative H_∞ evolutionary strategy $u(t)$ is to select some beneficial genetic variations as follows:

$$u(t) = \sum_{i=1}^{M} \sum_{k=1}^{N_1(t)} \alpha_i(\tilde{x})A_{ki}p(t-t_k)\tilde{x}(t). \tag{20.78}$$

Then, we get the following result.

Proposition 20.6 *Based on the interpolated biological network in (20.77) to replace nonlinear biological network in (20.68) by the global linearization technique, the constrained cooperative stochastic H_∞ evolution game problem in Proposition 20.5 for the nonlinear stochastic biological network in (20.68) or (20.77) is solved by*

$$u^0(t) = \sum_{i=1}^{M} \sum_{k=1}^{N_1(t)} \alpha_i(\tilde{x})A_{ki}p(t-t_k)\tilde{x}(t) = -R^{-1}P\tilde{x}(t), \tag{20.79}$$

$$v^0(t) = \frac{1}{e}B^T P\tilde{x}(t) \tag{20.80}$$

where $P > 0$ is the solution of the following Riccati-like inequalities

$$A_i^T P + PA_i - PR^{-1}P + \frac{1}{e}PBB^T P + \sum_{k=1}^{N_{2p}} (A_{ki}^T P + PA_{ki} + A_{ki}^T PA_i) \le 0, i = 1, 2, \ldots, M \tag{20.81}$$

$$\text{with } 0 < P \le eI \tag{20.82}$$

Proof: The proof follows a similar procedure to that of Propositions 20.4 and 20.5.

Based on Proposition 20.6, the cooperative stochastic H_∞ evolution problem in (20.75) can be transformed to the following constrained optimization problem:

$$e_0 = \min_{P>0} e \quad \text{subject to (20.81) and (20.82)} \tag{20.83}$$

By a similar method to transforming the constrained optimization problem in (20.62) to the constrained optimization problem in (20.63)–(20.65), the constrained optimization problem in (20.83) for the cooperative stochastic H_∞ evolutionary game strategy of the nonlinear stochastic network could be transformed to the following LMIs-constrained optimization problem:

$$e_0 = \min_{W>0} e \tag{20.84}$$

subject to

$$\begin{bmatrix} WA_i^T + A_iW - R^{-1} + \frac{1}{e}BB^T + \sum_{k=1}^{N_{2p}}(WA_{ki}^T + A_{ki}W) & W & (A_{1i}W)^T & \cdots & (A_{N_{2p}i}W)^T \\ W & -Q^{-1} & 0 & \cdots & 0 \\ A_{1i}W & 0 & -W & & \vdots \\ \vdots & \vdots & & \ddots & 0 \\ A_{N_{2p}i}W & 0 & \cdots & 0 & -W \end{bmatrix} \leq 0, \tag{20.85}$$

$$i = 1, 2, \ldots, M, \tag{20.86}$$

$$\begin{bmatrix} W & I \\ I & eI \end{bmatrix} \geq 0. \tag{20.87}$$

where $W = P^{-1} > 0$.

Remark 20.7

(i) e_0 in (21.84) could be solved by decreasing e until (21.85) and (21.86) have no positive solution $W > 0$ with the help of the LMI toolbox in MATLAB [23].

(ii) After solving W_0 and $P_0 = W_0^{-1}$ from the LMIs-constrained optimization problem in (20.84)–(20.86), then $u^0(t) = -R^{-1}P_0\tilde{x}(t)$ and $v^0(t) = \frac{1}{e_0}B^TP_0\tilde{x}(t)$ form the suboptimal cooperative stochastic H_∞ evolutionary game solution for the nonlinear stochastic biological network in (20.68) or (20.77). With the cooperative H_∞ evolutionary game strategy, the total genetic variations $\sum_{i=1}^{M}\sum_{k=1}^{N_1(t)} \alpha_i(\tilde{x})A_{ki}p(t-t_k) = -R^{-1}P_0$ are selected to construct negative feedback circuits $-R^{-1}P_0\tilde{x}(t)$ in the biological network to improve local network robustness to resist environmental disturbances and to attenuate the effects of local neutral genetic variations in the evolutionary process.

(iii) The phenotypic trait robustness criterion of the cooperative stochastic H_∞ evolutionary game in the nonlinear stochastic biological network in (20.68) or (20.77) is given by

$$\begin{aligned} &\underbrace{\sum_{k=1}^{N_{2p}}(A_{ki}^TP_0 + P_0A_{ki} + A_{ki}^TP_0A_{ki})}_{\substack{\text{robustness to tolerate} \\ \text{local genetic variations}}} + \underbrace{Q + \frac{1}{e_0}P_0BB^TP_0}_{\substack{\text{robustness to buffer} \\ \text{environmental disturbance}}} \\ &\leq \underbrace{-\left[(A_i - \frac{1}{2}R^{-1}P_0)^T P_0 + P_0(A_i - \frac{1}{2}R^{-1}P_0)\right]}_{\substack{\text{local network robustness enhanced by} \\ \text{cooperative evolutionary game}}}, \text{ for } i = 1, 2, \ldots, M. \end{aligned} \tag{20.88}$$

If the enhanced local network robustness of each local linearized biological network by the cooperative stochastic H_∞ evolutionary game strategy can confer local network stability robustness to tolerate local genetic variations and attenuate environmental disturbance in evolution, then the phenotypic trait of the biological network with the cooperative evolutionary game strategy could be maintained in the basin of the equilibrium point at x_e in the evolutionary process.

(iv) Since the cooperative stochastic H_∞ evolutionary game strategy can improve the local network robustness of the local linearized networks on the right-hand side of Eq. (20.87), the nonlinear stochastic biological network will harbor more neutral genetic variations, and will have more environmental robustness to decrease the evolvability e_0 (or increase network fitness $f_0 = \frac{1}{e_0}$) to attenuate the effect of environmental stimuli. However, when the harbored neutral genetic variations are accumulated by the robust biological network to the extent that the first term in (20.87) becomes very large and the phenotypic criterion in (20.87) cannot be guaranteed in the evolutionary process, with the help of an environmental disturbance, the phenotype of the biological network can cause a transition from one equilibrium point to other equilibrium point to start another network evolution at the other equilibrium point (phenotype) favored by natural selection.

(v) Comparing Proposition 20.3 with Proposition 20.6, the number of constraints of Riccati-like inequalities of the noncooperative stochastic H_∞ evolutionary game problem in (20.47) is m-times more than the number of constraints of the cooperative H_∞ evolutionary game problem in (20.81) and (20.82), so it is more difficult and exhaustive to solve the noncooperative stochastic H_∞ evolutionary game problem in Eq. (20.49) than to solve the cooperative stochastic H_∞ evolutionary game problem in (20.83) through the global linearization method. Further, a set of Pareto solutions exist for the noncooperative stochastic H_∞ evolutionary game problem in (20.49) but only one solution exists for the cooperative H_∞ evolutionary game problem in (20.83). Obviously, there should be some compromise among individual goals which may be partly in conflict with others in the noncooperative stochastic H_∞ evolutionary game process.

20.4 Computer Simulation Example

Consider the metabolic pathway in Fig. 20.2. This is a genetic inhibition and activation model in evolution and has been widely used for imitating characteristics of a real metabolic pathway. In this branched metabolic pathway, the metabolite x_2 is converted from the metabolite x_1 and then finally inhibits an early step in its own production pathway. The degradation process of metabolite x_1 is branched into the metabolite x_2 and the metabolite x_3, i.e., the metabolite x_3 is also converted from metabolite x_1 which is a divergence branching point. Then, the metabolite x_3 modulates downstream for the transformation of x_2 in an activation pathway.

Let us denote $x_1(t)$, $x_2(t)$, and $x_3(t)$ as the concentrations of metabolites x_1, x_2, and x_3, respectively. Based on the evolutionary metabolic pathway in Fig. 20.2, we have the following stochastic metabolic reaction network based on the S-system model [449]:

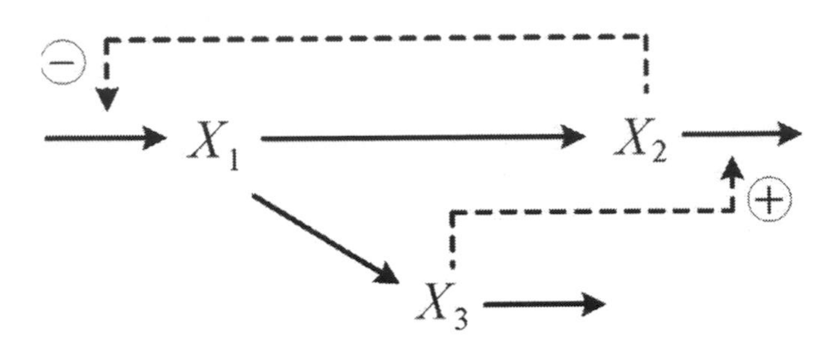

FIGURE 20.2
A branched metabolic pathway of genetic inhibition and activation model where x_1, x_2, and x_3 are metabolites.

$$\begin{bmatrix} \dot{x}_1(t) \\ \dot{x}_2(t) \\ \dot{x}_3(t) \end{bmatrix} = \begin{bmatrix} f_1(x) \\ f_2(x) \\ f_3(x) \end{bmatrix} + \sum_{k=1}^{N(t)} f_k(x)p(t-t_k) + \begin{bmatrix} v_1(t) \\ v_2(t) \\ v_3(t) \end{bmatrix} \tag{20.89}$$

where

$$f(x) = \begin{bmatrix} f_1(x) \\ f_2(x) \\ f_3(x) \end{bmatrix} = \begin{bmatrix} 12x_2^{-0.8}(t) - 10x_1^{0.5}(t) \\ 3x_1^{0.75}(t) - 5x_2^{0.5}(t)x_3^{0.2}(t) \\ 2x_1^{0.5}(t) - 6x_3^{0.8}(t) \end{bmatrix}, f_k(x) = \begin{bmatrix} 10x_1^{0.5}(t) \\ 5x_2^{0.5}(t)x_3^{0.2}(t) \\ 6x_3^{0.8}(t) \end{bmatrix}.$$

The Poisson counting process $N(t)$ with $\lambda = 0.01$ to indicate the occurring of genetic variations is shown in Fig. 21.3. The equilibrium point $x_e = \begin{bmatrix} x_{e1} & x_{e2} & x_{e3} \end{bmatrix}^T = \begin{bmatrix} 1.452 & 0.995 & 0.320 \end{bmatrix}^T$. For simplicity of analysis, the state of the metabolic pathway is shifted in the form $\tilde{x}(t) = x(t) - x_e$. Thus, the new equilibrium point of the phenotype concerned is at $\tilde{x}_e = 0_{3\times1}$ throughout this simulation example.

20.4.1 Noncooperative Stochastic H$_\infty$ Evolutionary Game Case

Suppose the 3-person noncooperative evolutionary game in (20.9) is performed by three individuals with $Q_i = I$, and $R_1 = I$, $R_2 = 3I$, $R_3 = 5I$, i.e. with different considerations of the cost or the effort for each individual. Based on the multi-objective optimization in (20.39) for the noncooperative stochastic H_∞ evolutionary game strategy, the following phenotypic variations are selected by natural selection:

$$u_i^*(t) = \sum_{k=1}^{N_i(t)} f_k(\tilde{x}(t))p(t-t_k)\tilde{x}(t), \text{ for } i = 1,2,3$$

to construct negative feedback loops to modify the metabolic pathway, in order to resist environmental disturbance and to tolerate the effect of neutral genetic variations.

According to the global linearization in (20.42) with M=3 and Proposition 20.3, we need to solve the LMIs-constrained optimization problem in (20.49) to get the noncooperative stochastic H_∞ evolutionary strategies for each player as follows:

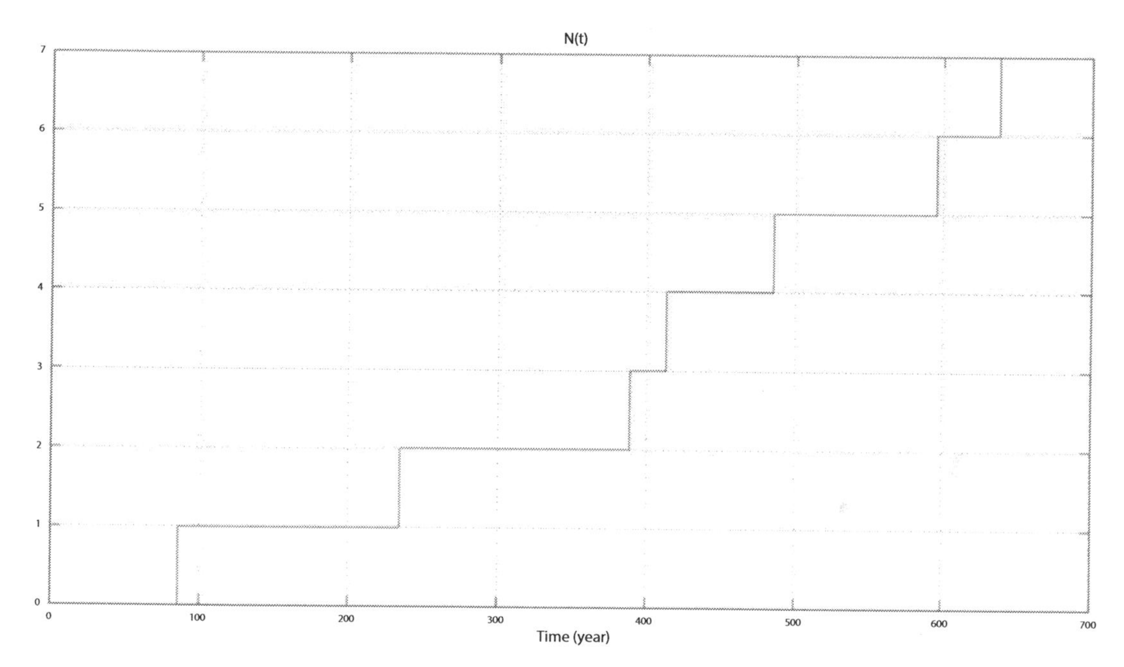

FIGURE 20.3
The Poisson counting process $N(t)$ of genetic variations with mean $\lambda = 0.01$.

$$u_i^*(t) = \sum_{j=1}^{M} \sum_{k=1}^{N_i(t)} \alpha_j(\tilde{x}) A_{kj} p(t - t_k)\tilde{x}(t) = -R_i^{-1} P^* \tilde{x}(t), \text{ for } i = 1, 2, 3.$$

We solve the LMIs-constrained multi-objective problem in Eq. (20.49) by the LMIs-constrained MOEA algorithm to obtain $W^* > 0$ and $(e_1^* \quad e_2^* \quad e_3^*)$ with computational time 349.88 s.

Since the noncooperative stochastic H_∞ evolutionary game solution $(e_1^* \quad e_2^* \quad e_3^*)$ is not unique, there exists a set of Pareto optimal solutions in the Pareto front as shown in Fig. 20.4, which are the result of compromise among 3 individual goals partly in conflict with each other. Suppose one Pareto optimal solution with a cross in Fig. 20.4 is selected for the following simulation, i.e. $(e_1^* \quad e_2^* \quad e_3^*) = (0.917 \quad 0.997 \quad 1.042)$ and the corresponding P^* and W^* are solved from (20.49) as

$$P^* = W^{*-1} = \begin{bmatrix} [c]ccc0.1359 & 0.0057 & -0.0544 \\ 0.0057 & 0.7565 & -0.3144 \\ -0.0544 & -0.3144 & 0.2933 \end{bmatrix}.$$

The trajectories of the noncooperative stochastic H_∞ evolutionary metabolic network with the above Pareto optimal solution are shown in Fig. 20.5. From the simulation result in Fig. 20.5, it is seen that by the noncooperative H_∞ evolutionary strategy, the environmental disturbance can be buffered and neutral genetic variations can be tolerated by the evolutionary metabolic pathway from $t = 0$ to $t = 638$ years. When the neutral genetic mutations accumulate to the extent that the phenotypic trait robustness criterion of the evolutionary metabolic pathway in (20.40) or (20.50) is violated at the approximate time $t = 638$ years,

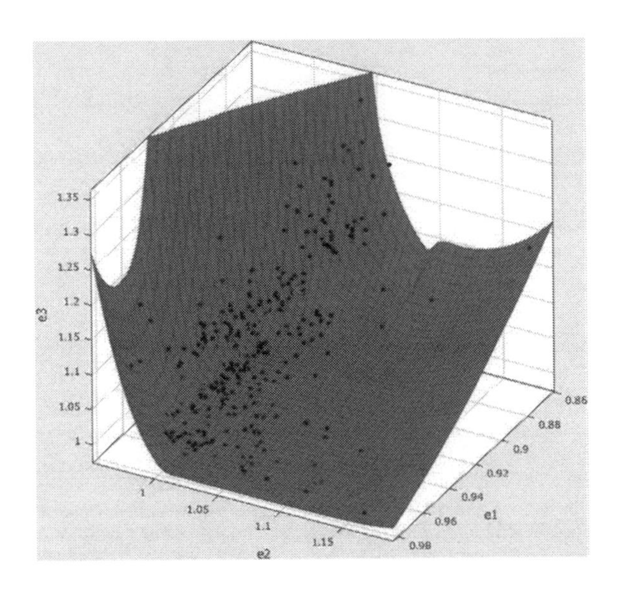

FIGURE 20.4
The Pareto front as the Pareto optimal solutions of the multi-objective optimization problem for the noncooperative stochastic H_∞ evolutionary game strategy of the stochastic metabolic pathway in Fig. 20.2. The Pareto optimal solution $(e_1^* \quad e_2^* \quad e_3^*) = (0.917 \quad 0.997 \quad 1.042)$ marked with cross X is used for simulations in Figs. 21.5–21.8.

the network stability at $x_e = [1.452 \quad 0.995 \quad 0.320]^T$ will be destroyed and the phenotypic trait will transit to another equilibrium point $x_{e1} = [1.595 \quad 1.101 \quad 0.484]^T$ to begin another period of network evolution. The noncooperative stochastic H_∞ evolutionary strategies $u_1^*(t), u_2^*(t)$ and $u_3^*(t)$ are shown in Figs. 20.6–20.8, respectively.

20.4.2 Cooperative Stochastic H$_\infty$ Evolutionary Game Case

Suppose the 3-person cooperative stochastic H_∞ evolutionary game in (20.70) is performed by the branched metabolic pathway with $Q = I$ and $R = 3I$ (the average of R_i, i = 1, 2, 3 in the noncooperative stochastic H_∞ evolutionary strategy). Based on the cooperative H_∞ evolutionary game strategy in Proposition 20.5, the following phenotypic variations are selected by natural selection

$$u^0(t) = \sum\nolimits_{k=1}^{N_1(t)} f_k(x)p(t-t_k)\tilde{x}(t)$$

to construct negative feedback circuits to modify the metabolic pathway to resist environmental disturbance and to tolerate the neutral genetic variations.

According to the global linearization in (20.42) and Proposition 20.6, we get

$$u^0(t) = \sum\nolimits_{j=1}^{3} \sum\nolimits_{k=1}^{N_i(t)} \alpha_i(\tilde{x})A_{ki}p(t-t_k)\tilde{x}(t) = -R^{-1}P_0\tilde{x}(t).$$

We need to solve the LMIs-constrained optimization problem in (20.84)–(20.86) to obtain W_0 and e_0 with computational time 4.33 s. Then, we get $e_0 = 0.993$ and

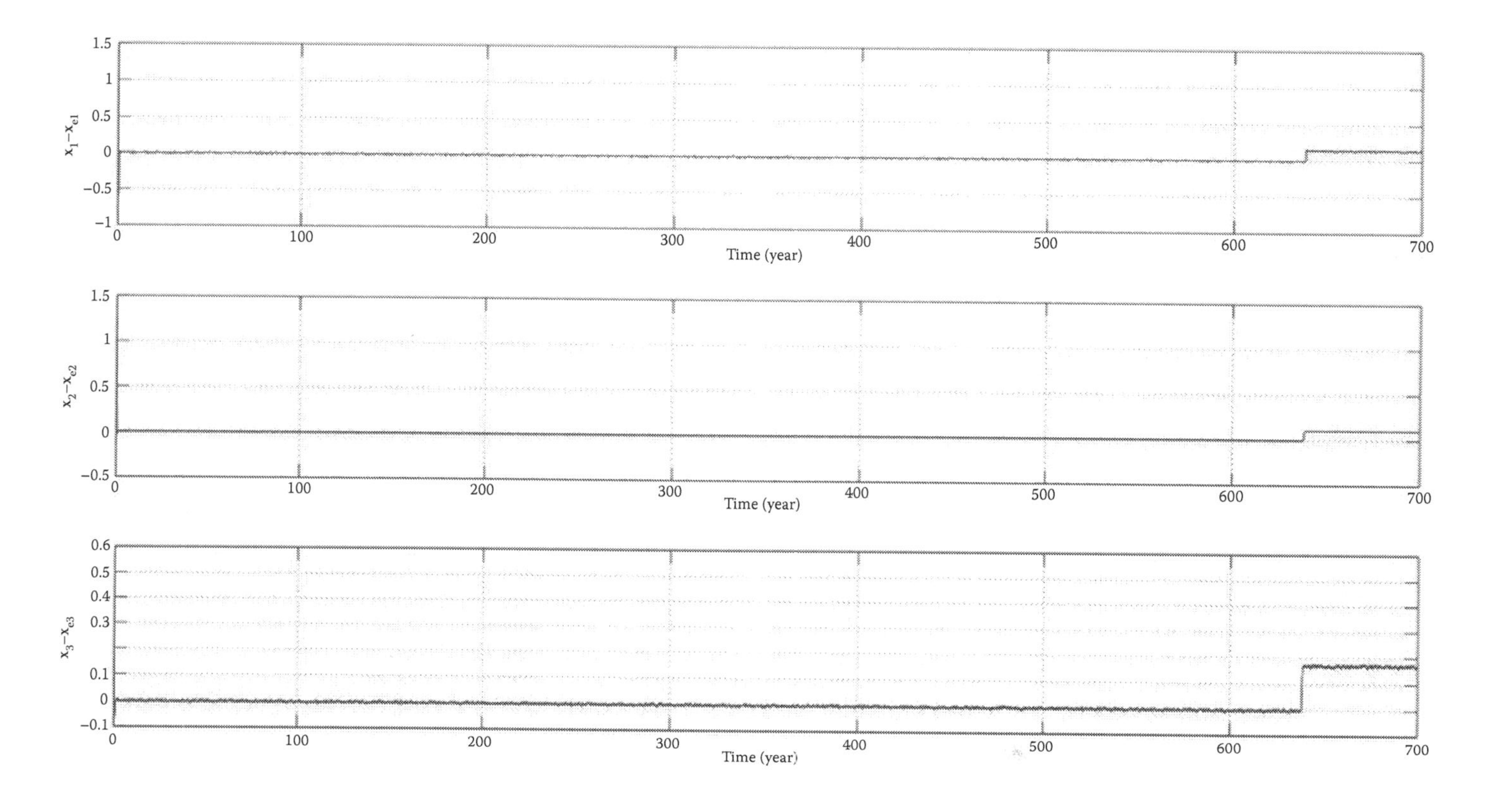

FIGURE 20.5
The trajectories of the noncooperative stochastic H_∞ evolutionary metabolic pathway in Fig. 20.2. The phenotypic trait at the equilibrium point $x_e = [1.452 \quad 0.995 \quad 0.320]^T$ is maintained by the noncooperative stochastic H_∞ evolutionary game strategies $u_1^*(t), u_2^*(t)$and$u_3^*(t)$ in Figs. 21.6–21.8, respectively, for 638 years, at which time the phenotypic robustness criterion in Eq.(21.40) or Eq. (21.50) is violated by the accumulation of neutral genetic variations and the influence of environmental disturbance so that the phenotypic trait is shifted to another equilibrium point $x_{e1} = [1.595 \quad 1.101 \quad 0.484]^T$ to begin another period of network evolution.

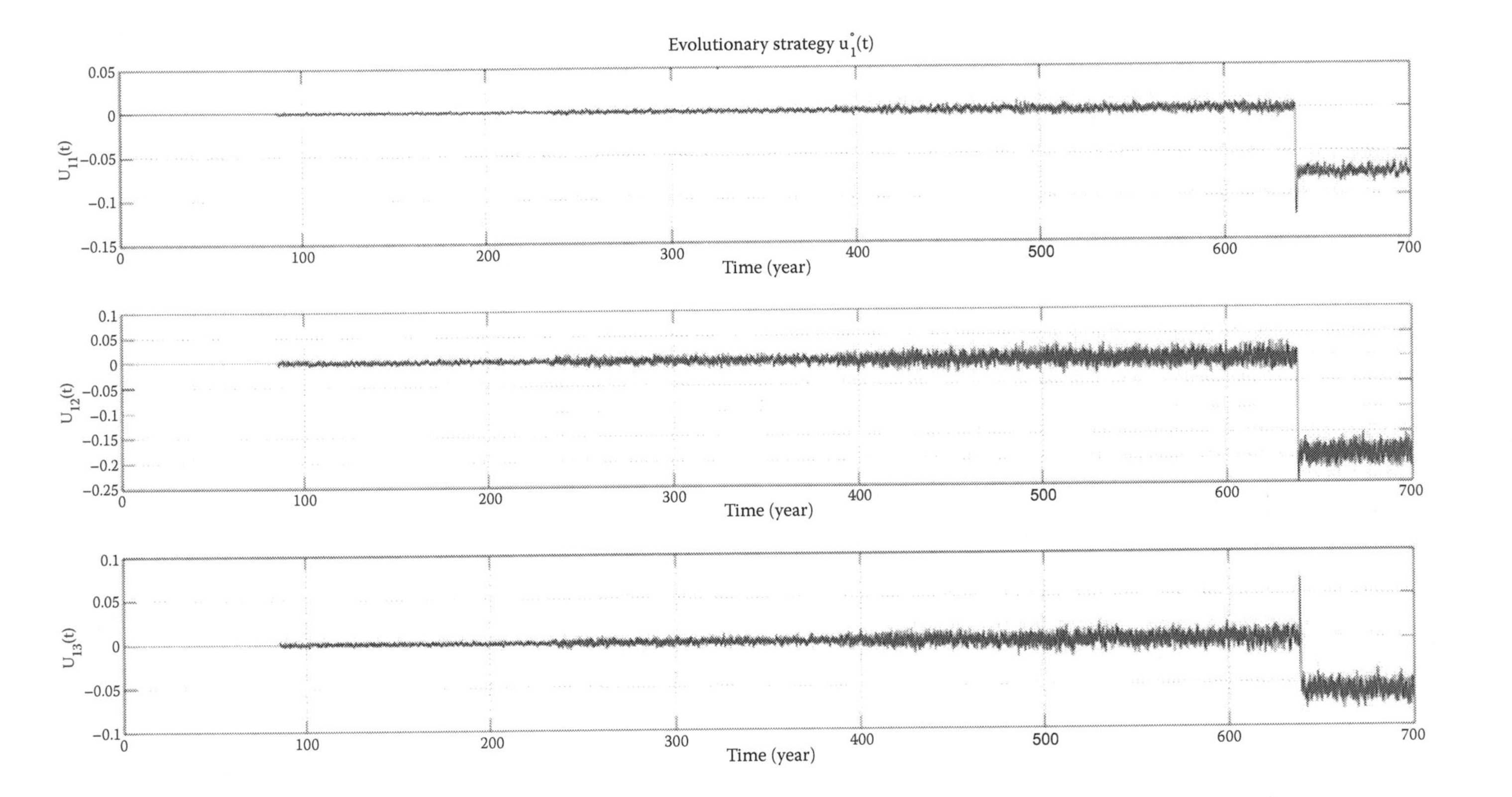

FIGURE 20.6
The noncooperative stochastic H_∞ evolutionary strategy $u_1^*(t)$ of the stochastic metabolic pathway in Fig. 20.2. In this figure, $u_1^*(t) = [\, u_{11}(t) \quad u_{12}(t) \quad u_{13}(t)\,]^T$. When compared with the cooperative stochastic H_∞ evolutionary strategy $u^0(t)$ in Fig. 20.10, it is seen that more effort and time are needed to maintain the phenotypic trait at x_e for the noncooperative stochastic H_∞ evolutionary strategies than the cooperative stochastic H_∞ evolutionary strategy $u^0(t)$ in Fig. 20.10.

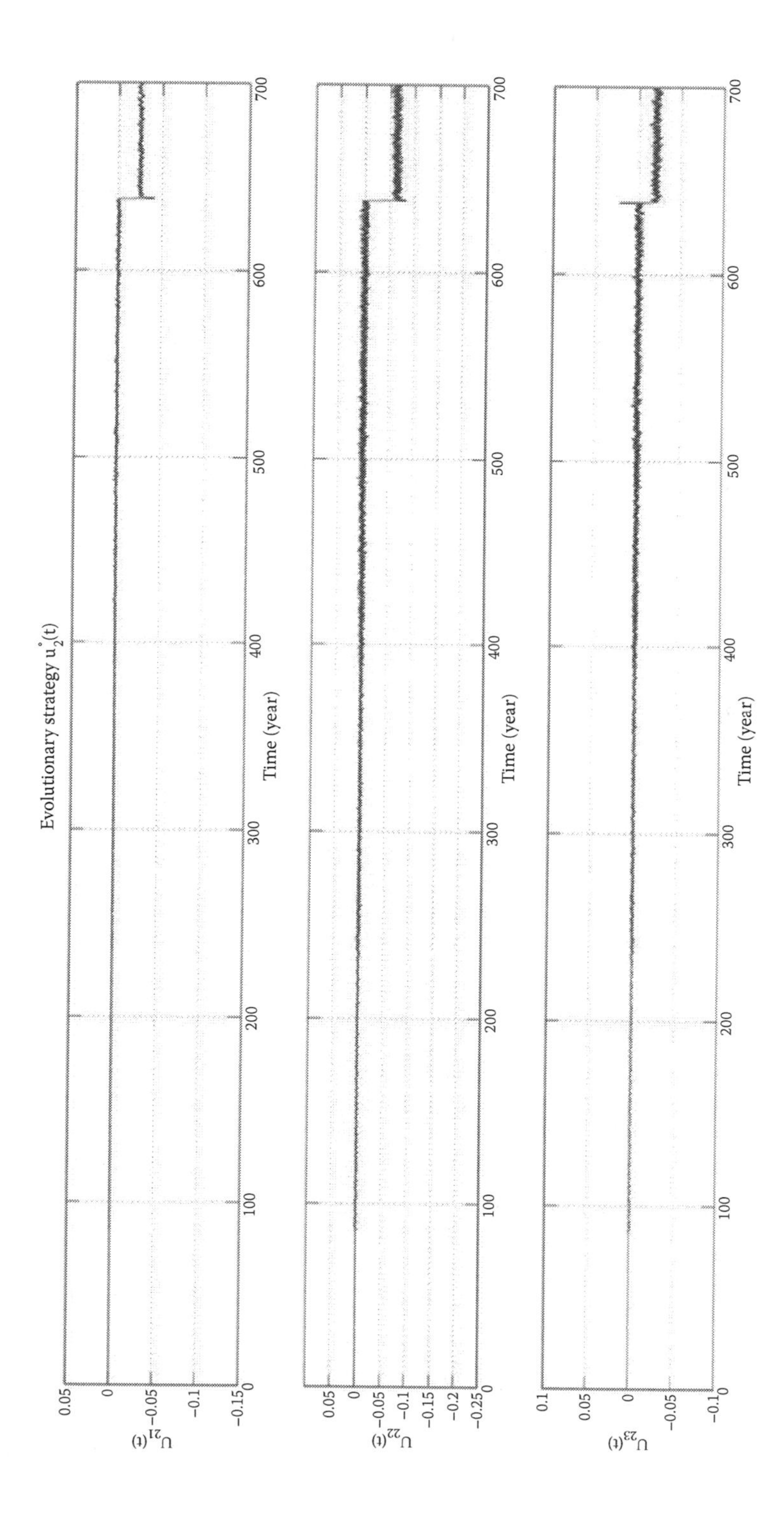

FIGURE 20.7
The noncooperative stochastic H_∞ evolutionary strategy $u_2^*(t)$ of the stochastic metabolic pathway in Fig. 20.2. In this figure, $u_2^*(t) = [\,u_{21}(t) \quad u_{22}(t) \quad u_{23}(t)\,]^T$. When compared with the cooperative stochastic H_∞ evolutionary strategy $u^0(t)$ in Fig. 20.10, it is seen that more effort and time are needed to maintain the phenotypic trait at x_e for the noncooperative H_∞ evolutionary strategies than the cooperative stochastic H_∞ evolutionary strategy $u^0(t)$ in Fig. 20.10.

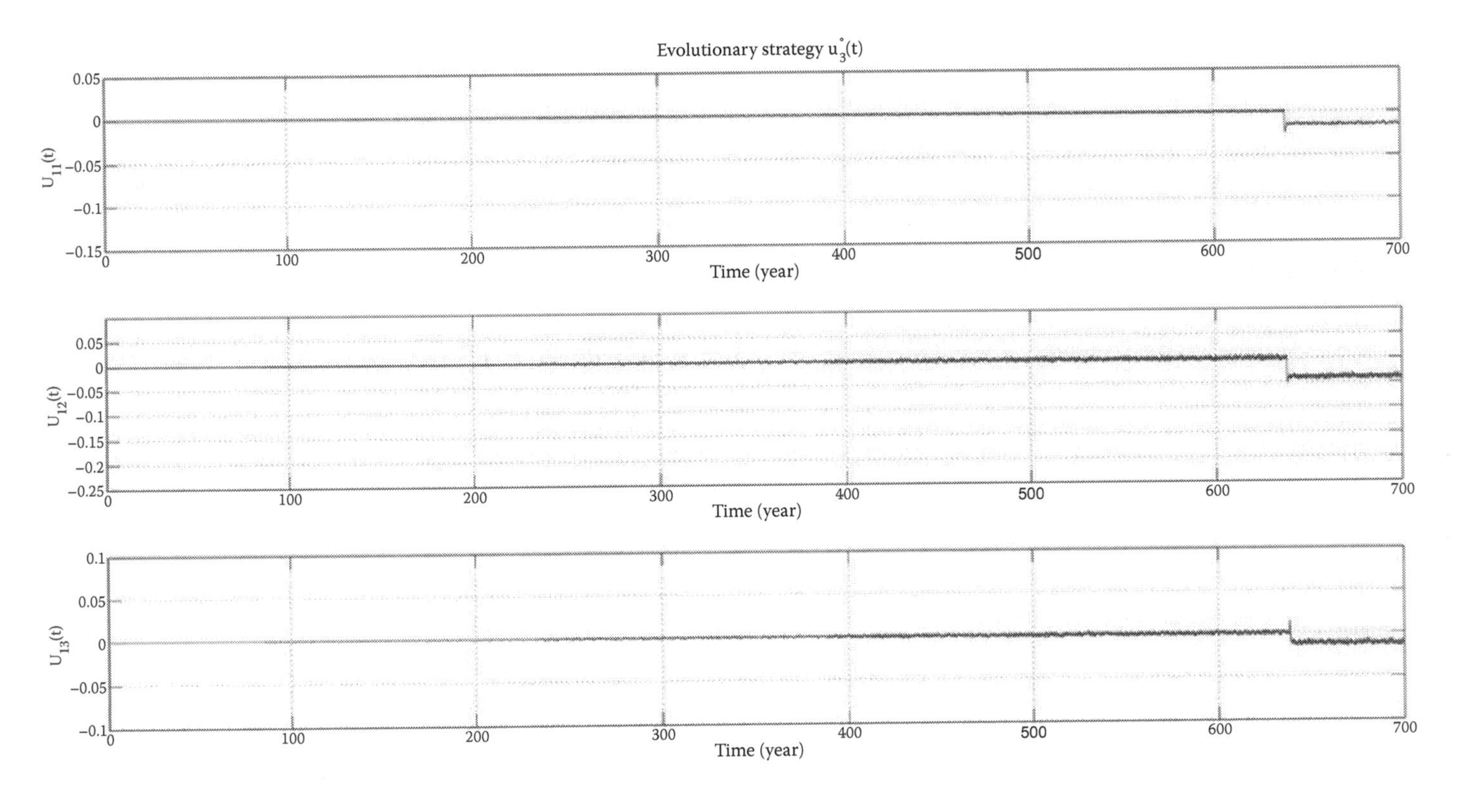

FIGURE 20.8
The noncooperative stochastic H_∞ evolutionary strategy $u_3^*(t)$ of the stochastic metabolic pathway in Fig. 20.2. In this figure, $u_3^*(t) = [u_{31}(t) \quad u_{32}(t) \quad u_{33}(t)]^T$. When compared with the cooperative stochastic H_∞ evolutionary strategy $u^0(t)$ in Fig. 20.10, it is seen that more effort and time are needed to maintain the phenotypic trait at x_e for the noncooperative stochastic H_∞ evolutionary strategies than the cooperative stochastic H_∞ evolutionary strategy $u^0(t)$ in Fig. 20.10.

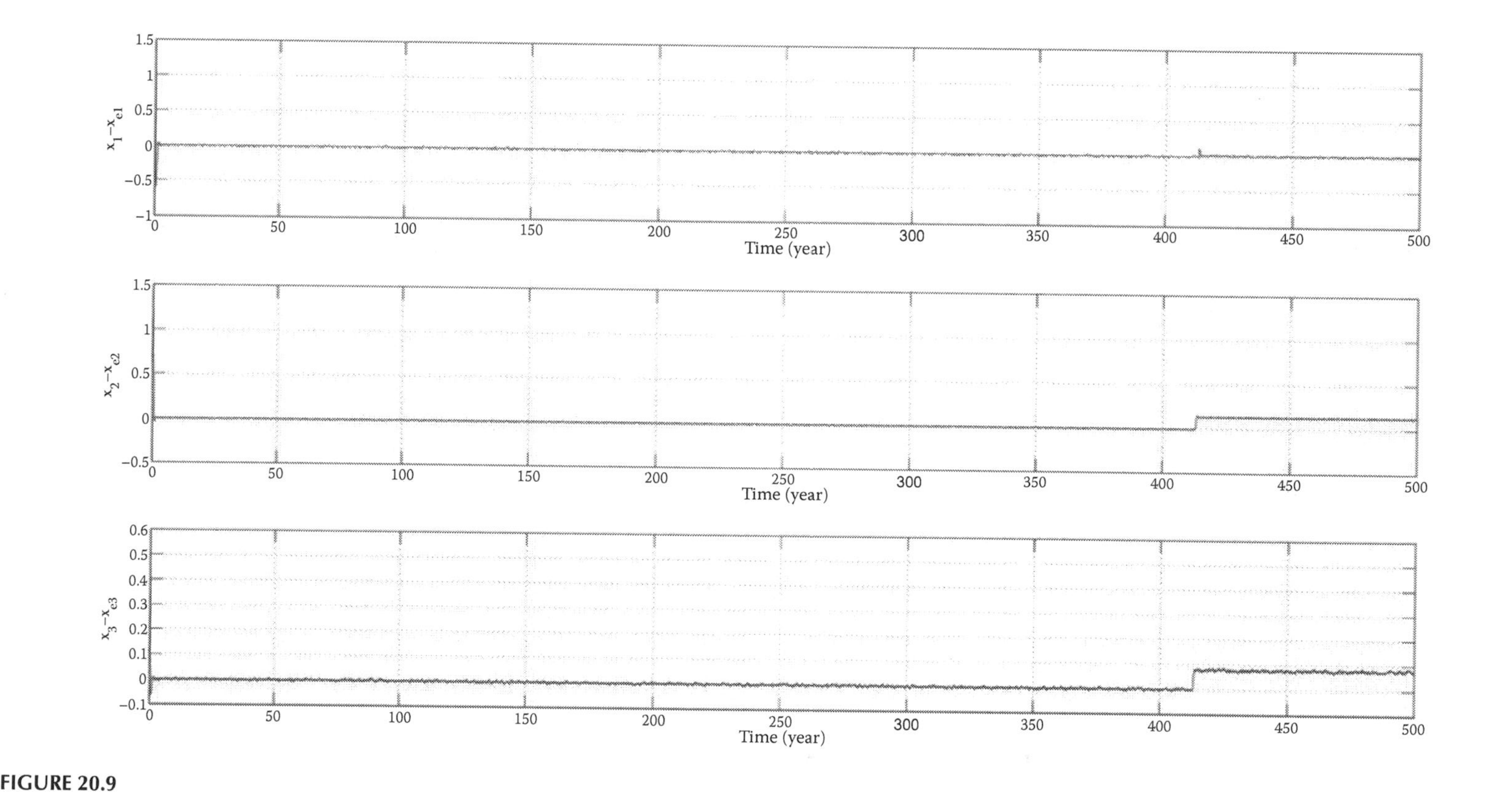

FIGURE 20.9
The trajectories of the cooperative stochastic H_∞ evolutionary metabolic pathway in Fig. 20.2. The phenotypic trait at the equilibrium point $x_e = [1.452 \quad 0.995 \quad 0.320]^T$ is maintained by the stochastic H_∞ cooperative game strategy $u^0(t)$ in Fig. 20.10 for 413 years, at which time the phenotypic robustness criterion in (20.76) or (20.87) is violated by the accumulation of neutral genetic variations and the influence of environmental disturbance so that the phenotypic trait is shifted to another equilibrium point $x_{e2} = [1.468 \quad 1.124 \quad 0.395]^T$ to begin another period of network evolution. Obviously, the cooperative stochastic H_∞ evolutionary game strategy is with less effort to maintain the phenotypic trait within a shorter period of time and maybe more adaptive and easier to survive under environmental changes.

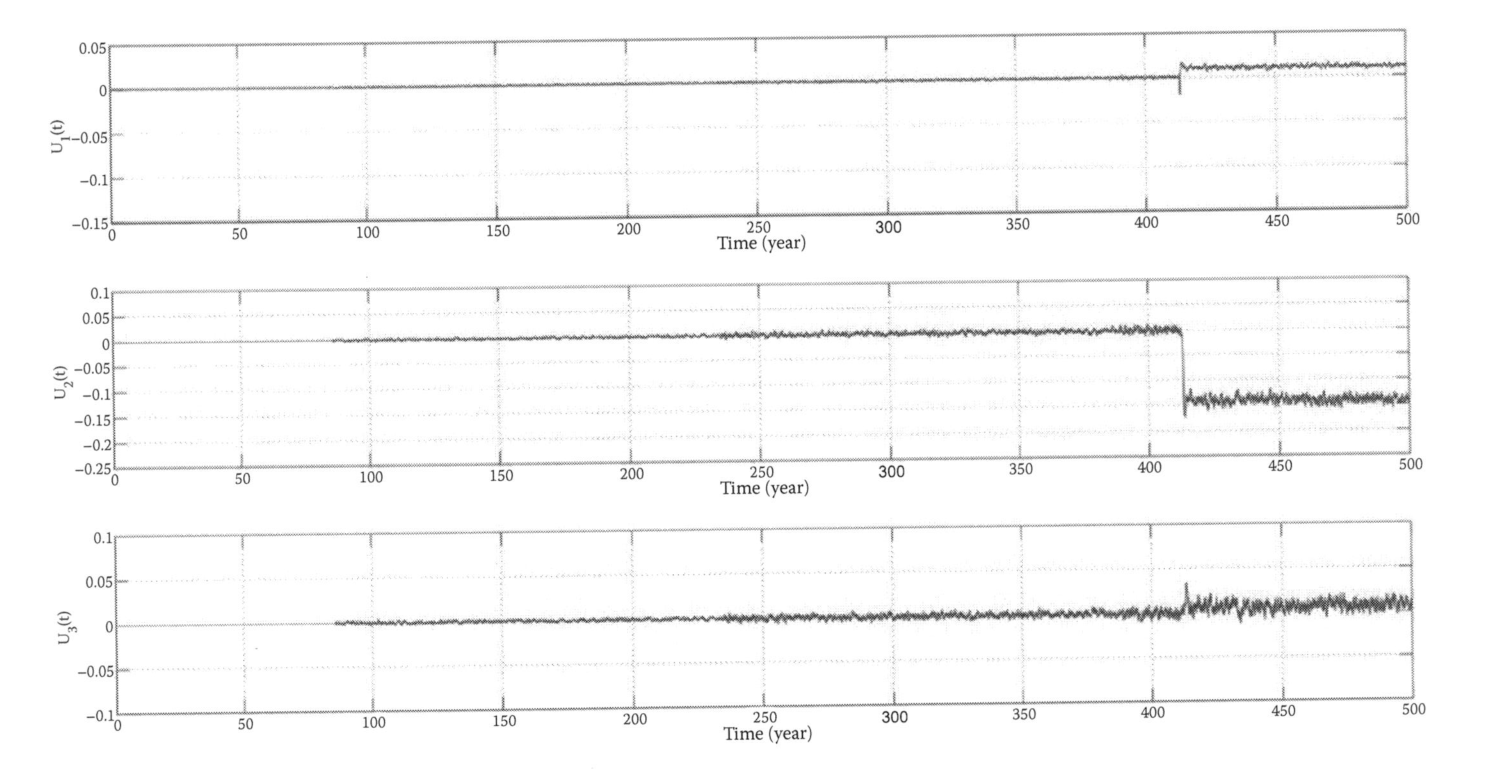

FIGURE 20.10
The strategy effort of the cooperative stochastic H_∞ evolutionary strategy $u^0(t)$ of stochastic metabolic pathway in Fig. 20.2. When compared with the noncooperative stochastic H_∞ evolutionary strategies in Figs. 21.6–21.8, it is seen that the phenotypic trait at x_e is maintained with less strategy effort and time by the cooperative stochastic H_∞ evolutionary strategy than the noncooperative stochastic H_∞ evolutionary strategies in Figs. 20.6–20.8.

$$P_0 = W_0^{-1} = \begin{bmatrix} 0.1946 & -0.0096 & -0.1258 \\ -0.0096 & 1.098 & -0.5634 \\ -0.1258 & -0.5634 & 0.9485 \end{bmatrix}.$$

From the simulation result as shown in Fig. 20.9, the cooperative H_∞ evolutionary game strategy could also effectively attenuate the environmental disturbance and tolerate the neutral genetic variations from $t = 0$ to $t = 413$ years with an effort $u^0(t)$ as shown in Fig. 20.10, which is less than that of the noncooperative H_∞ evolutionary game in Figs. 20.6–20.8. When the neutral genetic variations accumulate to the extent that the phenotypic trait robustness criterion in (20.76) or (20.87) is violated, the metabolic pathway stability at $x_e = [1.452 \quad 0.995 \quad 0.320]^T$ will be destroyed and the phenotypic trait will transit to another equilibrium point $x_{e2} = [1.468 \quad 1.124 \quad 0.395]^T$ to begin another period of network evolution. The cooperative evolutionary strategy $u^0(t)$ is shown in Fig. 20.10.

By comparing the cooperative stochastic H_∞ evolutionary strategy with the noncooperative stochastic H_∞ evolutionary strategy, obviously, the noncooperative stochastic H_∞ game strategy needs about 80 searching times of MOEA algorithm and more efforts to compromise with these conflicting objectives to solve the multi-objective problem in (20.49) for the noncooperative H_∞ evolutionary strategy than that of solving the single objective problem in (20.84) for the cooperative stochastic H_∞ evolutionary strategy. Hence, the noncooperative stochastic H_∞ evolutionary game strategy could perform with more effort to maintain the phenotypic trait longer at the equilibrium x_e. The cooperative stochastic H_∞ evolutionary strategy is easier to solve and with less effort to maintain the phenotypic trait within a shorter period of time at x_e so that it may be more adaptive and easier for biological systems to survive under environmental changes.

20.5 Discussions and Conclusions

In this chapter, we first reviewed existing evolutionary game strategies, especially the replicator equation, a pillar of modern evolutionary game theory that has produced many interesting and fruitful insights into different fields [488, 494, 507]. The evolutionary games based on the replicator equation, similar to those based on mean-field theories, have been very important in physics, but they cannot (nor are they intended to) describe all possible phenomena [504]. In discussing stochastic H_∞ evolutionary game strategies, we have mainly focused on network evolvability (or robustness) under the influence of random genetic variations and in response to the fluctuations of environmental disturbances in the course of evolution by natural selection. Evolutionary biological systems were modeled using a nonlinear stochastic dynamic equation incorporating a Poisson counting process to describe the effects of both genetic variations and random environmental disturbances. Both noncooperative and cooperative stochastic H_∞ evolutionary game strategies with, respectively, partly conflicting goals and a common goal were employed. These stochastic H_∞ evolutionary game strategies differ from conventional dynamic noncooperative and cooperative game strategies in that we considered both the natural selection of beneficial genetic variations and the worst-case effect of environmental disturbances. In addition, the Poisson point process is more suitable for modeling the random point process of genetic mutations than the more usually employed continuous Wiener process [510, 516].

At present, there exists no good way to solve the noncooperative stochastic H_∞ evolutionary game problem [1, 9]. Several iterative schemes based on an updating eigenvalue problem (EVP) algorithm have been developed to search Nash equilibrium point for noncooperative and cooperative game problems. However, these iterative schemes cannot guarantee to achieve Nash game solution. In this chapter, we proposed an indirect suboptimal method to transform the noncooperative stochastic H_∞ evolutionary game problem of linear stochastic biological networks into an equivalent multi-objective optimization problem with Nash equilibrium solution in (20.23)–(20.25), which can be efficiently solved by the proposed MOEA-based searching method. Using the global linearization technique in (20.42), the noncooperative stochastic H_∞ evolutionary game strategy problem of nonlinear stochastic biological networks could also be transformed into an equivalent multi-objective optimization problem with Nash equilibrium solution in (20.49), which can be solved by MOEA similarly. Since there are more m-time constraints on the multi-objective optimization problem of the noncooperative than for the cooperative stochastic H_∞ evolutionary game, and since the solutions to the problem are not unique (i.e., there are many Pareto solutions), it is more difficult to solve the multi-objective optimization problem of the noncooperative stochastic H_∞ evolutionary game in (20.23) than it is to solve the single-objective optimization problem of the cooperative stochastic H_∞ evolutionary game in (20.84). This is because of the need for more compromises among individuals (players) and correspondingly greater effort and time to solve the noncooperative stochastic H_∞ evolutionary game problem than is required for the cooperative stochastic H_∞ evolutionary game problem.

Since the cooperative stochastic H_∞ evolutionary game strategy includes a common goal, it can be transformed into an equivalent LMIs-constrained optimization problem in (20.63)–(20.65). Using global linearization, the same approach can be applied to the cooperative stochastic H_∞ evolutionary problem in the nonlinear network ((20.84)–(20.86)). Both transformed versions can be solved more easily and with less effort than the noncooperative stochastic H_∞ evolutionary game problems in (20.23)–(20.25) and (20.49) for linear and nonlinear networks, respectively. This is why species playing a cooperative stochastic H_∞ evolutionary strategy are more efficient than those playing a noncooperative stochastic H_∞ evolutionary game strategy when it comes to adapting to environmental changes, and have greater evolutionary stability. These results were confirmed by simulation.

The phenotypic trait robustness criterion of the cooperative stochastic H_∞ evolutionary game strategy in (20.66), (20.76), or (20.87) dictates that if network robustness as enhanced by the negative feedback loops of the cooperative strategy is able to confer sufficient robustness to tolerate neutral genetic variations and to buffer environmental disturbances, then the phenotypic trait can be maintained in the evolutionary process. Since a small coupling matrix B and small genetically driven phenotypic variations A_k help satisfy the criterion, they are favored by the cooperative stochastic H_∞ evolutionary strategy. This may explain why there exist so many membranes, permeable divisions, and compartments that separate biological systems from their environment: these allow only some receptors and channels to communicate with the environment, in effect decreasing the size of coupling matrix B. Similarly, it may explain that we found so many modules and redundant and repair mechanisms to maintain the phenotypic robustness in biological systems: these buffer genetic variation, attenuating its effect on phenotypic variation A_k. These phenotypic trait criteria as derived from cooperative stochastic H_∞ evolutionary game strategy may allow prediction of a variety of structures of evolutionary biological systems.

A comparison of the multi-objective optimization problems in (20.23), (20.39), and (20.49) of the noncooperative stochastic H_∞ evolutionary game strategy with the single-objective optimization problems of the cooperative stochastic H_∞ evolutionary game strategy in (20.62), (20.75), (20.84) showed that the former are more difficult to solve because of the m-time inequality constraints on the noncooperation evolutionary game problem. Further, there are many solutions for the multi-objective optimization problem. As more time, effort, and compromises are required to solve these noncooperative stochastic H_∞ evolutionary game strategies with partly conflicting goals, they may be less flexible in dealing with environmental changes than the cooperative stochastic H_∞ evolutionary game strategies. This may be a fundamental driver of many biological phenomena from mutualism to the appearance of multicellular organisms.

In both noncooperative and cooperative stochastic H_∞ evolutionary game strategies, negative feedback loops can be constructed that have the effect of widening and deepening the basin of the phenotypic equilibrium point (see Fig. 20.1), increasing robustness to buffer more neutral genetic variations and additional environmental disturbances. The accumulation of random neutral genetic variations over a long period will, however, eventually drive a shift to a new phenotype at a neighboring equilibrium point and provide raw heterogenetic materials to commence another period of network evolution.

The main aims of this chapter are described as follows:

(i) Poisson point process is modeled in a stochastic dynamic biological system to mimic genetic variations of biological network in the evolutionary process, and then the game strategies of how to select some beneficial genetic variations for noncooperative and cooperative stochastic H_∞ evolutionary game are also investigated for the first time to get an insight into the natural selection in a population of biological networks in the evolutionary process.

(ii) The noncooperative stochastic H_∞ evolutionary game problem of a biological system was transformed to an equivalent LMI-constrained MO problem with Nash equilibrium solution and an LMI-constrained MOEA algorithm was also proposed to efficiently solve the Pareto optimal solutions for the noncooperative stochastic H_∞ evolutionary game problem.

(iii) The global linearization technique was employed to simplify the nonlinear stochastic H_∞ evolutionary game problem by the interpolation of several local biological systems to approximate a nonlinear biological system so that a difficult HJI-constrained optimization problem of nonlinear stochastic H_∞ evolutionary game strategy problem could be replaced by a simple LMIs-constrained optimization problem.

20.5.1 Appendix A. Proof of Proposition 20.1

From Eq. (20.14), we get

$$\begin{aligned} J_i = & E\tilde{x}^T(0)P\tilde{x}(0) - E\tilde{x}^T(t_p)P\tilde{x}(t_p) \\ & + E\int_0^{t_p} (\tilde{x}^T(t)Q_i\tilde{x}(t) + u_i^T(t)R_iu_i(t) - e_i\bar{v}_i^T(t)\bar{v}_i(t) + \frac{d}{dt}(\tilde{x}^T(t)P\tilde{x}(t)))dt \end{aligned} \tag{20.A1}$$

where $V(\tilde{x}^T(t)) = \tilde{x}^T(t)P\tilde{x}(t)$ and $P = P^T > 0$.

By Levy formula [31, 443], we get

$$dV(\tilde{x}^T(t)) = \begin{bmatrix} \tilde{x}^T P(A\tilde{x}(t) + u_i(t) + \bar{B}_i v_i(t)) + (A\tilde{x}(t) + u_i(t) + \bar{B}_i v_i(t))^T P\tilde{x}(t) \\ + \sum_{k=1}^{N_r(t)} [\tilde{x}^T(t) + \tilde{x}^T(t) A_k^T p(t-t_k) P(\tilde{x}(t) + A_k \tilde{x}(t) p(t-t_k)) - \tilde{x}^T(t) P\tilde{x}(t) \end{bmatrix} dt. \tag{20.A2}$$

Substituting Eq. (20.A2) into Eq. (20.A1), we get

$$\begin{aligned} J_i &= E\tilde{x}^T(0)P\tilde{x}(0) - E\tilde{x}^T(t_p)P\tilde{x}(t_p) \\ &+ E\int_0^{t_p} \left\{ \begin{array}{l} \tilde{x}^T(t)Q_i\tilde{x}(t) + u_i^T(t)R_i u_i(t) - e_i \bar{v}_i^T(t)\bar{v}_i(t) \\ +\tilde{x}^T PA\tilde{x}(t) + \tilde{x}^T P u_i(t) + \tilde{x}^T P\bar{B}_i v_i(t) + \tilde{x}^T A^T P\tilde{x}(t) \\ +u_i^T(t)P\tilde{x}(t) + \bar{v}_i^T \bar{B}_i^T(t) P\tilde{x}(t) + \sum_{k=1}^{N_r(t)} \tilde{x}^T(t) A_k^T P\tilde{x}(t) p(t-t_k) \\ \sum_{k=1}^{N_r(t)} \tilde{x}^T(t) PA_k\tilde{x}(t)p(t-t_k) + \sum_{k=1}^{N_r(t)} \tilde{x}^T(t) A_k^T PA_k \tilde{x}(t) p(t-t_k) \end{array} \right\} dt \\ &= E\tilde{x}^T(0)P\tilde{x}(0) - E\tilde{x}^T(t_p)P\tilde{x}(t_p) \\ &+ E\int_0^{t_p} \left\{ \begin{array}{c} \tilde{x}^T(t) \begin{bmatrix} A^T P + PA - PR_i^{-1}P + \frac{1}{e_i} P\bar{B}_i\bar{B}_i^T P + Q_i \\ + \sum_{k=1}^{N_r(t)} (A_k^T P + PA_k + A_k^T PA_k) \end{bmatrix} \tilde{x}(t) \\ +(P\tilde{x}(t) + R_i u_i(t))^T R_i^{-1}(P\tilde{x}(t) + R_i u_i(t)) \\ -\left(\frac{1}{\sqrt{e_i}} \bar{B}_i^T P\tilde{x}(t) - \sqrt{e_i}\bar{v}_i(t)\right)^T \left(\frac{1}{\sqrt{e_i}} \bar{B}_i^T P\tilde{x}(t) - \sqrt{e_i}\bar{v}_i(t)\right) \end{array} \right\} dt. \end{aligned}$$

$$\begin{aligned} J_i^* &= \min_{u_i(t)} \max_{\bar{v}(t)} J_i \\ &= E\tilde{x}^T(0)P\tilde{x}(0) - E\tilde{x}^T(t_p)P\tilde{x}(t_p) \\ &+ \min_{u_i(t)} \max_{\bar{v}(t)} \int_0^{t_p} \left\{ \begin{array}{c} \tilde{x}^T(t) \begin{bmatrix} A^T P + PA - PR_i^{-1}P + \frac{1}{e_i} P\bar{B}_i\bar{B}_i^T P + Q_i \\ + \sum_{k=1}^{N_r(t)} (A_k^T P + PA_k + A_k^T PA_k) \end{bmatrix} \tilde{x}(t) \\ +(P\tilde{x}(t) + R_i u_i(t))^T R_i^{-1}(P\tilde{x}(t) + R_i u_i(t)) \\ -\left(\frac{1}{\sqrt{e_i}} \bar{B}_i^T P\tilde{x}(t) - \sqrt{e_i}\bar{v}_i(t)\right)^T \left(\frac{1}{\sqrt{e_i}} \bar{B}_i^T P\tilde{x}(t) - \sqrt{e_i}\bar{v}_i(t)\right) \end{array} \right\} dt. \end{aligned}$$

Therefore, the solution of the noncooperative evolutionary Nash quadratic evolutionary game in (20.15) is given by

$$u_i^*(t) = -R_i^{-1}P\tilde{x}(t), \text{ and } \bar{v}_i^*(t) = \frac{1}{e_i}\bar{B}_i^T P\tilde{x}(t)$$

and

$$J_i^* \le E\tilde{x}^T(0)P\tilde{x}(0) - E\tilde{x}^T(t_p)P\tilde{x}(t_p) \le E\tilde{x}^T(0)P\tilde{x}(0).$$

If the m Riccati-like inequalities in (20.18) hold, then we get

$$\begin{aligned}
J_i^* &= \min_{u_i(t)} \max_{\bar{v}(t)} J_i \\
&= E\tilde{x}^T(0)P\tilde{x}(0) - E\tilde{x}^T(t_p)P\tilde{x}(t_p) \\
&+ \min_{u_i(t)} \max_{\bar{v}(t)} \int_0^{t_p} \left\{ \tilde{x}^T(t) \begin{bmatrix} A^T P + PA - PR_i^{-1}P + \frac{1}{e_i} P\bar{B}_i\bar{B}_i^T P + Q_i \\ + \sum_{k=1}^{N_r(t)} (A_k^T P + PA_k + A_k^T PA_k) \end{bmatrix} \tilde{x}(t) \right\} dt.
\end{aligned}$$

From the constraint inequality in (20.16), we get the inequality constraint in (20.19).

Q.E.D.

20.5.2 Appendix B. Proof of Proposition 20.2

From

$$J_i = E\int_0^{t_p} (\tilde{x}^T(t)Q_i\tilde{x}(t) + u_i^T(t)R_iu_i(t) - e_i\bar{v}_i^T(t)\bar{v}_i(t))dt \tag{20.B1}$$

for some positive Lyapunov function $V(\tilde{x}(t))$

$$\begin{aligned}
J_i =& EV(\tilde{x}(0)) - EV(\tilde{x}(t_p)) \\
&+ E\int_0^{t_p} (\tilde{x}^T(t)Q_i\tilde{x}(t) + u_i^T(t)R_iu_i(t) - e_i\bar{v}_i^T(t)\bar{v}_i(t) + \frac{d}{dt}V(\tilde{x}(t)))dt.
\end{aligned} \tag{20.B2}$$

By Levy formula [31, 443], we get

$$EdV(\tilde{x}(t)) = E\begin{bmatrix} \left(\frac{\partial V(\tilde{x})}{\partial \tilde{x}}\right)^T (f(\tilde{x}) + u_i(t) + \bar{B}_i\bar{v}_i(t)) \\ + \sum_{k=1}^{N_r(t)} (V(\tilde{x}(t) + f_k(\tilde{x})) - V(\tilde{x}(t))) \end{bmatrix} dt. \tag{20.B3}$$

Substituting (20.B3) into (20.B2), we get

$$\begin{aligned}
J_i =& EV(\tilde{x}(0)) - EV(\tilde{x}(t_p)) \\
&+ E\int_0^{t_p} \begin{bmatrix} \tilde{x}^T(t)Q_i\tilde{x}(t) + u_i^T(t)R_iu_i(t) - e_i\bar{v}_i^T(t)\bar{v}_i(t) \\ + \left(\frac{\partial V(\tilde{x})}{\partial \tilde{x}}\right)^T f(\tilde{x}) + \left(\frac{\partial V(\tilde{x})}{\partial \tilde{x}}\right)^T u_i(t) + \left(\frac{\partial V(\tilde{x})}{\partial \tilde{x}}\right)^T \bar{B}_i\bar{v}_i(t) \\ + \sum_{k=1}^{N_r(t)} (V(\tilde{x}(t) + f_k(\tilde{x})) - V(\tilde{x}(t))) \end{bmatrix} dt.
\end{aligned} \tag{20.B4}$$

By the fact that

$$u_i^T(t)R_iu_i(t) + \left(\frac{\partial V(\tilde{x})}{\partial \tilde{x}}\right)^T u_i(t) = \left(R_iu_i(t) + \frac{1}{2}\frac{\partial V(\tilde{x})}{\partial \tilde{x}}\right)^T R_i^{-1}\left(R_iu_i(t) + \frac{1}{2}\frac{\partial V(\tilde{x})}{\partial \tilde{x}}\right) - \frac{1}{4}\left(\frac{\partial V(\tilde{x})}{\partial \tilde{x}}\right)^T R_i^{-1}\left(\frac{\partial V(\tilde{x})}{\partial \tilde{x}}\right) \tag{20.B5}$$

$$-e_i\bar{v}_i^T(t)\bar{v}_i(t) + \left(\frac{\partial V(\tilde{x})}{\partial \tilde{x}}\right)^T \bar{B}_i\bar{v}_i(t) = -\left[\sqrt{e_i}\bar{v}_i(t) - \frac{1}{2\sqrt{e_i}}\bar{B}_i^T\left(\frac{\partial V(\tilde{x})}{\partial \tilde{x}}\right)\right]^T \times \left[\sqrt{e_i}\bar{v}_i(t) - \frac{1}{2\sqrt{e_i}}\bar{B}_i^T\left(\frac{\partial V(\tilde{x})}{\partial \tilde{x}}\right)\right] + \frac{1}{4e_i}\left(\frac{\partial V(\tilde{x})}{\partial \tilde{x}}\right)^T \bar{B}_i\bar{B}_i^T\left(\frac{\partial V(\tilde{x})}{\partial \tilde{x}}\right). \tag{20.B6}$$

Substituting (20.B5) and (20.B6) into (20.B4), we get

$$J_i = EV(\tilde{x}(0)) - EV(\tilde{x}(t_p)) + E\int_0^{t_p}\left[\begin{array}{c} \left(\frac{\partial V(\tilde{x})}{\partial \tilde{x}}\right)^T f(\tilde{x}) + \tilde{x}^T(t)Q_i\tilde{x}(t) + \sum_{k=1}^{N_r(t)}\left(V(\tilde{x}(t) + f_k(\tilde{x})) - V(\tilde{x}(t))\right) \\ -\frac{1}{4}\left(\frac{\partial V(\tilde{x})}{\partial \tilde{x}}\right)^T R_i^{-1}\left(\frac{\partial V(\tilde{x})}{\partial \tilde{x}}\right) + \frac{1}{4e_i}\left(\frac{\partial V(\tilde{x})}{\partial \tilde{x}}\right)^T \bar{B}_i\bar{B}_i^T\left(\frac{\partial V(\tilde{x})}{\partial \tilde{x}}\right) \\ +\left(R_iu_i(t) + \frac{1}{2}\frac{\partial V(\tilde{x})}{\partial \tilde{x}}\right)^T R_i^{-1}\left(R_iu_i(t) + \frac{1}{2}\frac{\partial V(\tilde{x})}{\partial \tilde{x}}\right) \\ -\left[\sqrt{e_i}\bar{v}_i(t) - \frac{1}{2\sqrt{e_i}}\bar{B}_i^T\left(\frac{\partial V(\tilde{x})}{\partial \tilde{x}}\right)\right]^T\left[\sqrt{e_i}\bar{v}_i(t) - \frac{1}{2\sqrt{e_i}}\bar{B}_i^T\left(\frac{\partial V(\tilde{x})}{\partial \tilde{x}}\right)\right] \end{array}\right]dt. \tag{20.B7}$$

Then,

$$\begin{aligned} &\min_{u_i(t)}\max_{\bar{v}(t)} J_i \\ &= EV(\tilde{x}(0)) - EV(\tilde{x}(t_p)) \\ &\quad + E\int_0^{t_p}\left[\begin{array}{c} \left(\frac{\partial V(\tilde{x})}{\partial \tilde{x}}\right)^T f(\tilde{x}) + \tilde{x}^T(t)Q_i\tilde{x}(t) + \sum_{k=1}^{N_r(t)}\left(V(\tilde{x}(t) + f_k(\tilde{x})) - V(\tilde{x}(t))\right) \\ -\frac{1}{4}\left(\frac{\partial V(\tilde{x})}{\partial \tilde{x}}\right)^T R_i^{-1}\left(\frac{\partial V(\tilde{x})}{\partial \tilde{x}}\right) + \frac{1}{4e_i}\left(\frac{\partial V(\tilde{x})}{\partial \tilde{x}}\right)^T \bar{B}_i\bar{B}_i^T\left(\frac{\partial V(\tilde{x})}{\partial \tilde{x}}\right) \end{array}\right]dt \end{aligned} \tag{20.B8}$$

with the minimax solution

$$u_i^*(t) = \sum_{k=1}^{N_r(t)} f_k(\tilde{x})p(t - t_k) - \frac{1}{2}R_i^{-1}\left(\frac{\partial V(\tilde{x})}{\partial \tilde{x}}\right),$$

$$\bar{v}_i^*(t) = \frac{1}{2e_i}\bar{B}_i^T\left(\frac{\partial V(\tilde{x})}{\partial \tilde{x}}\right).$$

Further, if the HJIIs in (20.37) hold, then the inequality constraints in (20.38) also hold by the constraints in (20.34). Q.E.D.

20.5.3 Appendix C. Proof of Proposition 20.3

Based on global linearization method in (20.42)–(20.44), we use $\sum_{j=1}^{M}\alpha_j(\tilde{x})A_j\tilde{x}(t)$ to approximate $f(\tilde{x})$ and $\sum_{j=1}^{M}\alpha_j(\tilde{x})\sum_{k=1}^{N_r(t)}A_{kj}\tilde{x}(t)p(t-t_k)$ to approximate $\sum_{k=1}^{N_r(t)}f_k(\tilde{x})p(t-t_k)$. In Proposition 20.3, we replace $f(\tilde{x})$ by $\sum_{j=1}^{M}\alpha_j(\tilde{x})A_j\tilde{x}(t)$ and $\sum_{k=1}^{N_r(t)}f_k(\tilde{x})p(t-t_k)$ by $\sum_{j=1}^{M}\alpha_j(\tilde{x})\sum_{k=1}^{N_r(t)}A_{kj}\tilde{x}(t)p(t-t_k)$. Further, we choose Lyapunov function $V(\tilde{x})=\tilde{x}^T(t)P\tilde{x}(t)$. Then, based on Proposition 20.2, we get the noncooperative H_∞ evolutionary game strategy

$$u_i^*(t)=\sum_{j=1}^{M}\sum_{k=1}^{N_r(t)}\alpha_j(\tilde{x})A_{kj}\tilde{x}p(t-t_k)=-R_i^{-1}P\tilde{x}(t),\mathrm{i}=1,2,\ldots,\mathrm{m}$$
$$\bar{v}_i^*(t)=\frac{1}{e_i}\bar{B}_i^TP\tilde{x}(t).$$

And HJIIs in (20.37) become

$$(P\tilde{x}(t))^T\left(\sum_{j=1}^{M}\alpha_j(\tilde{x})A_j\tilde{x}(t)\right)+\left(\sum_{j=1}^{M}\alpha_j(\tilde{x})A_j\tilde{x}(t)\right)^TP\tilde{x}(t)$$
$$+\tilde{x}^T(t)Q_i\tilde{x}(t)-(P\tilde{x}(t))^TR_i(P\tilde{x}(t))+\frac{1}{e_i}(P\tilde{x}(t))^T\bar{B}_i\bar{B}_i^T(P\tilde{x}(t))$$
$$\leq 0,\ \text{for } i=1,2,\ldots,m$$

with $\tilde{x}^T(0)P\tilde{x}(0)\leq e_i\tilde{x}^T(0)\tilde{x}(0)$

which become

$$\sum_{j=1}^{M}\alpha_j(\tilde{x})\tilde{x}^T(t)\left(A_j^TP+PA_j+Q_i+PR_i^{-1}P+\frac{1}{e_i}P\bar{B}_i\bar{B}_i^TP+\sum_{k=1}^{N_{rp}}(A_{kj}^TP+PA_{kj}+A_{kj}^TPA_{kj})\right)\tilde{x}(t)$$
$$\leq 0,\ \text{for } i=1,2,\ldots,m,\ \text{for } j=1,2,\ldots,M$$

with $\tilde{x}^T(0)(P-e_iI)\tilde{x}(0)\leq 0$

which imply the inequalities in (20.47). Q.E.D.

References

1. Basar, T., Olsder, G. J., 1982. Dynamic Noncooperative Game Theory. Academic Press, New York.
2. Zhang, W., Xie, L., Chen, B. S., 2017. Stochastic H_2/H_∞ Control: A Nash Game Approach. CRC Press, Taylor & Francis Group, New York.
3. Liu, S. J., Krstic, M., 2011. Stochastic Nash Equilibrium Seeking for Game with General Nonlinear Payoffs. SIAM J. Cont. Optim. 49(4), 1659–1679.
4. Takagi, T., Sugeno, M., 1985. Fuzzy Identification of Systems and Its Applications to Modeling and Control. IEEE Trans. Syst. Man. Cyber. SMC-15(1), 116–132.
5. Tanaka, K., Wang, H. O., 2001. Fuzzy Control System Design and Analysis: Linear Matrix Inequality Approach. John Wiley & Sons. Inc.
6. Chen, B. S., Uang, H. J., Tseng, C. S., 1998. Robust Tracking Enhancement of Robot Systems Including Motor Dynamics: A Fuzzy-Based Dynamic Game Approach. IEEE Trans. Fuzzy Syst. 6, 538–552.
7. Chen, B. S., Wu, C. S., Uang, H. J., 2000. A Minimax Tracking Design for Wheeled Vehicles with Trailer Based on Adaptive Elimination Scheme. IEEE Trans. Control Syst. Technol. 8(2), 418–434.
8. Chen, B. S., Zhang, W., 2004. Stochastic H_2/H_∞ Control with State Dependent Noise. IEEE Trans. Automat. Control 49(1), 45–57.
9. Chen, B. S., Tseng, C. S., Uang, H. J., 2002. Fuzzy Differential Games for Nonlinear Stochastic Systems: Suboptimal Approach. IEEE Trans. Fuzzy Syst. 10(2), 222–233.
10. Chen, Y. L., Chen, B. S., 1994. Minimax Robust Deconvolution Filter under Stochastic Parametric and Noise Uncertainties. IEEE Trans. Signal Process. 42(1), 32–45.
11. Ho, J. H., Chen, B. S., 2010. Robust Minimax MSE Equalizer Design for MIMO Wireless Communications with Time-Varying Channel Uncertainties. IEEE Trans. Signal Process. 58 (11), 5835–5844.
12. Chiu, W. Y., Chen, B. S., 2012. Robust Relative Location Estimation in Wireless Sensor Networks with Inexact Position Problem. IEEE Trans. Mobile Comput. 11(6), 935–945.
13. Chen, B. S., Lin, Y. P., 2011. Robust Resource Management Control for CO_2 Emission and Reduction of Greenhouse Effect: Stochastic Game Approach. J. Environ. Protect. 2, 1172–1191.
14. Chen, B. S., Chen, W. Y., Young, C. T., Yan, Z., 2018. Noncooperative Game Strategy in Cyber-Financial Systems with Wiener and Poisson Random Fluctuation: LMIs-constrained MOEA Approach. IEEE Trans. Cybernetics. 48(12), 3323–3326.
15. Chen, B. S., Chang, C. H., Chung, Y. J., 2008. Robust Model Matching Control of Immune Systems under Environmental Disturbances: Dynamic Game Approach. J. Theoret. Biol. 253, 824–837.
16. Chen, B. S., Chang, C. H., Lee, H. C., 2009. Robust Synthetic Biology Design: Stochastic Game Theory Approach. Bioinformatics. 25(14), 1822–1830.
17. Chen, B. S., Ho, S. J., 2014. The Stochastic Evolutionary Game for a Population of Biological Networks under Natural Selection. Evolutionary Bioinform. 10(17), 17–38.
18. Chen, B. S., Tsai, K. W., Li, C. W., 2015. Using Nonlinear Stochastic Evolutionary Game Strategy to Model an Evolutionary Biological Network of Organ Carcinogenesis under Natural Selection Scheme. Evolutionary Bioinform. 11, 155–178.
19. Chen, B. S., Yeh, C. H., 2017. Stochastic Noncooperative and Cooperative Evolutionary Game Strategies of a Population of Biological Networks under Natural Selection. BioSystems. 162, 90–118.
20. Docknev, E., Jørgensen, S., 1984. Cooperative and Non-Cooperative Differential Game Solutions to an Investment and Pricing Problem. J. Oper. Res. Soc. 35(8), 731–739.
21. Engwerda, J., 2010. Necessary and Sufficient Condition for Pareto Optimal Solutions of Cooperative Differential Games. SIAM J. Control Optimiz. 48(6), 3859–3881.

22. Wu, C. F., Chen, B. S., Zhang, W., 2017. Multiobjective Investment Policy for a Nonlinear Stochastic Financial System: A Fuzzy Approach. IEEE Trans. Fuzzy Syst. 25(2), 460–474.
23. Boyd, S., etal., 1994. Linear Matrix Inequalities in System and Control Theory. SIAM, Philadelphia, PA.
24. Liu, G. P., Yang, J. B., Whidborne, J. F., 2002. Multiobjective Optimisation and Control. DSP Ltd, Hertfordshire, UK.
25. Abraham, A., Jain, L. C., Goldberg, R., 2005. Evolutionary Multiobjective Optimization: Theoretical Advances and Applications. Springer, New York.
26. Hardanyi, J. C., Selten, R., 1998. A General Theory of Equilibrium Selection in Games. The MIT Press, Cambridge, MA.
27. Hofbauer, J., Sigmund, K., 2003. Evolutionary Game Dynamics. Bull. Am. Math. Soc. 40, 479–519.
28. Vega-Redondo, F., 2003. Economics and Theory of Games. Cambridge University Press, Cambridge.
29. Chen, B. S., 2018. Systems Evolutionary Biology: Biological Network Evolution Theory, Stochastic Evolutionary Game Strategies, and Application to Systems Synthetic Biology. Academic Press, Elsevier, London.
30. Chen, B. S., Ho, S. J., 2016. Multiobjective Tracking Control Design of T-S Fuzzy System: Fuzzy Pareto Optimal Approach. Fuzzy Set. Syst. 290, 39–55.
31. Chen, B. S., Wu, C. F., 2015. Robust Scheduling Filter Design for a Class of Nonlinear Stochastic Poisson Signal System. IEEE Trans. Signal Process. 63, 6245–6257.
32. Chen, B. S., Chen, W. H., Wu, H. L., 2009. Robust H_2/H_∞ Global Linearization Filter Design for Nonlinear Stochastic Systems. IEEE Trans. Circuit Syst. I. 56(7), 1441–1454.
33. Chen, B. S., Chen, W. Y., Young, C. T., Yan, Z., 2018. Noncooperative Game Strategy in Cyber-Financial Systems with Wiener and Poisson Random Fluctuation: LMIs-constrained MOEA Approach. IEEE Trans. Cybernetics. 48,(12), 3323–3336.
34. Khasminskii, P., Milstein, G. N., 2011. Stochastic Stability of Differential Equations, 2nd ed. Springer, Berlin, Germany.
35. Del, K., 2001. Multi-Objective Optimization Using Evolutionary Algorithms. Wiley, Chichester, UK.
36. Del, K., Pratap, A., Agarwal, S., Meyarivan, T., A Fast and Elitist Multiobjective Genetic Algorithm: NSGA-II. IEEE Trans. Evol. Comput.
37. Spong, M. W., 1989. Robot Dynamics and Control. Wiley, New York.
38. Slotine, J. J. E., Li, W., 1991. Applied Nonlinear Control. Prentice-Hall, Englewood Cliffs, NJ.
39. Tarn, T. J., Bejcay, A. K., Yun, X., Li, Z., 1991. Effect of Motor Dynamics on Nonlinear Feedback Robot Arm Control. IEEE Trans. Robot. Automat. 7, 114–122.
40. Koivo, A. J., 1989. Fundamentals for Control of Robotic Manipulators. Wiley, New York.
41. Mahmoud, M. S., 1993. Robust Control of Robot Arms Including Motor Dynamic. Int. J. Control 58(4), 853–873.
42. Beekman, R., Lee, K., Jun. 1988. Nonlinear Robotic Control Including Driving Motor Interactions, In Proc. Amer. Control Conf., Atlanta, GA, pp. 1333–1338.
43. Pandian, S. R., Hanmandlu, M., 1993. Adaptive Generalized Model-Based Control of Robot Manipulators. Int. J. Control 58(4), 835–852.
44. Craig, J. J., 1988. Adaptive Control of Mechanical Manipulators. Addison-Wesley, New York.
45. Karakasoglu, A., Sudharsanan, S. I., Nov 1993. Identification and Decentralized Adaptive Control Using Dynamical Neural Network with Application to Robotic Manipulators. IEEE Trans. Neural Net. 4, 919–930.
46. Narendra, K. S., Parthusarathy, K., June 1990. Identification and Control of Dynamical Systems Using Neural Networks. IEEE Trans. Neural Net. 1, 4–27.
47. Kosko, B., 1992. Neural Network and Fuzzy Systems. Prentice-Hall, Englewood Cliffs, NJ.
48. Basar, T., 1991. H_∞-Optimal Control and Related Minimax Design Problems: A Dynamic Game Approach. Birkha¨user, Boston, MA.
49. Stoorvogel, A., 1992. The H_∞ Control Problem: A State Space Approach. Prentice-Hall, New York.

50. Doyle, J., Glover, K., Khargonekar, P. P., Francis, B. A., Aug. 1989. State Space Solution to Standard H_2 and H_∞ Control Problems. IEEE Trans. Automat. Control 34, 831–847.
51. Chen, B. S., Lee, T. S., Feng, J. H., 1994. A Nonlinear H_∞ Control Design in Robotic Systems under Parameter Perturbation and External Disturbance. Int. J. Control 59(2), 439–461.
52. Barmish, B. R., Corless, M., Leitmann, G., 1983. A New Class of Stabilizing Controllers for Uncertain Dynamical Systems. SIAM J. Control Optimiz. 21, 246–255.
53. Corless, M., Leitmann, G., 1981. Continuous State Feedback Guaranteeing Uniform Ultimate Boundedness of Uncertain Dynamic Systems. IEEE Trans. Automat. Control 26, 1139–1144.
54. Garofalo, F., Leitmann, G., 1989. Guaranteeing Ultimate Boundedness and Exponential Rate of Convergence for a Class of Nominally Linear Uncertain Systems. ASME J. Dynamic Systems, Measurements, Control 111, 584–588.
55. Driankov, D., Hellendoorn, H., Reinfrank, M., 1993. An Introduction to Fuzzy Control. Springer-Verlag, Berlin, Germany.
56. Jamshidi, M., Vadiee, N., Ress, T. J., 1993. Fuzzy Logic and Control. Prentice-Hall, Englewood Cliffs, NJ.
57. Narendra, K. S., Annaswamy, A. M., 1989. Stable Adaptive Systems. Prentice-Hall.
58. Sanner, R. M., Slotine, J. J. E., June 1992. Gaussian Networks for Direct Adaptive Control. IEEE Trans. Neural Net. 3, 837–864.
59. Ortega, R., Spong, M. W., 1989. Adaptive Motion Control of Rigid Robotics: A Tutorial. Automatica. 25, 877–888.
60. Wang, L. X., 1994. Adaptive Fuzzy Systems and Control: Design and Stability Analysis. Prentice-Hall, Englewood Cliffs, NJ.
61. Chen, B. S., Lee, C. H., Chang, Y. C., Feb. 1996. H_∞ Tracking Design of Uncertain Nonlinear SISO Systems: Adaptive Fuzzy Approach. IEEE Trans. Fuzzy Syst. 4, 32–43.
62. Isaacs, R., 1975. Differential Games. Kruger, Huntington, NY.
63. Chang, Y. C., Chen, B. S., Jan. 1997. A Nonlinear Adaptive H_∞ Tracking Control Design in Robotic Systems via Neural Networks. IEEE Trans. Control Syst. Technol. 5(1), 13–29.
64. Pappas, G. J., Kyriakopoulos, K. J., 1995. Stabilization of Nonholonomic Vehicles under Kinematic Constraints. Int. J. Control, 4, 933–947.
65. Pappas, G. J., Kyriakopoulos, K. J., 1992. Modeling and Feedback Control of Nonholonomic Mobile Vehicles, in Proc. 31st Conf. Decision Control, Tucson, AZ, pp. 2680–2685.
66. Pappas, G. J., Kyriakopoulos, K. J., 1993. Dynamic Modeling and Tracking Control Nonholonomic Wheeled Vehicles, in IFAC 12TH Triennial World Congr., Sydney, Australia, pp. 61–64.
67. Laumond, J.-P., 1993. Controllability of a Multibody Mobile Robot. IEEE Trans. Robot. Automat. 9, 755–763.
68. Kolmanovsky, I., Mcclamroch, N. H., 1995. Developments in Nonholonomic Control Problems. IEEE Trans. Control Syst. Technol. 5, 20–36.
69. Watanable, K., Tang, J., Nakamura, M., Koga, S., Fukuda, T., 1996. A Fuzzy- Gaussian Neural Network and Its Application to Mobile Robot Control. IEEE Trans. Control Syst. Technol. 4, 193–199.
70. Zeng, X. J., Singh, M. G., 1994. Approximation Theory of Fuzzy systems—SISO Case. IEEE Trans. Fuzzy Syst. 2, 162–176.
71. Limebeer, D. J. N., Anderson, B. D. O., Hendel, B., Jan. 1994. A Nash Game Approach to Mixed H_2/H_∞ Control. IEEE Trans. Automat. Control 39, 69–82.
72. Wu, C. S., Chen, B. S., 1999. Unified Design for H_2, H_∞ and Mixed Control of Spacecraft. J. Guid. Control Dyna. 22, 884–896.
73. Wu, C. S., Chen, B. S., 2001. Adaptive Attitude Control of Spacecraft: Mixed H_2/H_∞ Approach. J. Guidance, Control, Dyna. 24, 755–766.
74. ptimal Control and Related Minimax DesigChen, B. S., Tseng, C. S., Uang, H. J., 2002. Mixed H_2/H_∞ Fuzzy Output Feedback Control Design for Nonlinear Dynamic Systems: An LMI Approach. IEEE Trans. Fuzzy Syst. 8(3), 249–265.
75. Bernstein, D. S., Haddad, W. M., Feb. 1989. LQG Control with an H_2/H_∞ Performance Bound: A Riccati Equation Approach. IEEE Trans. Automat. Control 34, 293–305.

76. Hinrichsen, D., Pritchard, A. J., 1998. Stochastic H_∞. SIAM J. Control Optim. 36, 1504–1538.
77. Gershon, E., Limebeer, D. J. N., Shaked, U., Yaesh, I., Oct. 2001. Robust H_∞ Filtering of Stationary Continuous-Time Linear Systems with Stochastic Uncertainties. IEEE Trans. Automat. Control 46, 1788–1793.
78. Ugrinovskii, V. A., 1998. Robust H_∞ Control in the Presence of Stochastic Uncertainty. Int J. Cont. 71, 219–237.
79. de Farias, D. P., Geromel, J. C., Do Val, J. B. R., Costa, O. L. V., Aug. 2000. Output Feedback Control of Markov Jump Linear Systems in Continuous-Time. IEEE Trans. Automat. Control 45, 944–949.
80. Gershon, E., Shaked, U., Yaesh, I., 2001. H_∞ Control and Filtering of Discrete-Time Stochastic Systems with Multiplicative Noise. Automatica. 37, 409–417.
81. van der Schaft, A. J., June 1992. L2-Gain Analysis of Nonlinear Systems and Nonlinear State Feedback H_∞ Control. IEEE Trans. Automat. Control 37, 770–784.
82. Limebeer, D. J. N., Anderson, B. D. O., Khargonekar, P. P., Green, M., 1992. A Game Theoretic Approach to H_∞ Control for Time- Varying Systems. SIAM J. Control Optim. 30, 262–283.
83. Sweriduk, G. D., Calise, A. J., 1997. Differential Game Approach to the Mixed H_2-H_∞ Problem. J. Guid. Control Dyna. 20, 1229–1234.
84. Lin, W., 1995. Mixed H_2/H_∞ Control of Nonlinear Systems, in Proc. 34th Conf. Decision Control, New Orleans, LA, pp. 333–338.
85. Chen, X., Zhou, K., 2001. Multiobjective H_2/H_∞ Control Design. SIAM J. Control Optim. 40, 628–660.
86. Bernstein, D. S., Aug. 1987. Robust Static and Dynamic Output-Feedback Stabilization: Deterministic and Stochastic Perspectives. IEEE Trans. Automat. Control AC-32, 1076–1084.
87. Bernstein, D. S., Hyland, D. C., 1988. Optimal Projection Equations for Reduced-Order Modeling, Estimation and Control of Linear Systems with Multiplicative White Noise. J. Optim. Theory Appl. 58, 387–409.
88. Zhang, W., Chen, B. S., January 2004. On Stabilizability and Exact Observability of Stochastic Systems with Their Applications. Automatica. 40(1), 87–94.
89. Has'minskii, R. Z., 1980. Stochastic Stability of Differential Equations. Sijtjoff and Noordhoff, Alphen, The Netherlands.
90. Liu, Y., 1999. Backward Stochastic Differential Equation and Stochastic Control System, Ph. D. dissertation, Shandong University, China.
91. Fragoso, M. D., Costa, O. L. V., de Souza, C. E., 1998. A New Approach to Linearly Perturbed Riccati Equations Arising in Stochastic Control. Appl. Math. Optim. 37, 99–126.
92. Rami, M. A., Zhou, X. Y., Sept. 2000. Linear Matrix Inequalities, Riccati Equations, and Indefinite Stochastic Linear Quadratic Control. IEEE Trans. Automat. Control 45, 1131–1142.
93. Krylov, N. V., 1995. Introduction to the Theory of Diffusion Processes. In: Translations of Mathematical Monographs 142. Providence, RI, AMS.
94. Bensoussan, A., 1983. Lecture on Stochastic Control—Part I. Ser. Lecture Notes in Mathematics: Springer-Verlag. 972, 1–39.
95. Leibfritz, F., 2001. An LMI-based Algorithm for Designing Suboptimal Static Output Feedback Controllers. SIAM J. Control Optim. 39, 1711–1735.
96. Yaesh, I., Shaked, U., June 1997. Minimum Entropy Static Output Feedback Control with an Norm Performance Bound. IEEE Trans. Automat. Control 42, 853–858.
97. Bernstein, D. S., Haddad, W. M., 1990. Robust Stability and Performance Analysis for State Space Systems via Quadratic Lyapunov Bounds. SIAM J. Matrix Anal. Applicat. 11, 239–271.
98. Osburn, S. L., Bernstein, D. S., 1997. A Review and Comparison of Guaranteed Cost Bounds for Robust Stability and Performance, in Proc. 36th Conf. Decision Control, San Diego, CA, pp. 787–788.
99. Khargonekar, P. P., Petersen, I. R., Zhou, K., Mar. 1990. Robust Stabilization of Uncertain Linear Systems: Quadratic Stabilizability and H_∞ Control Theory. IEEE Trans. Automat. Control 35, 356–361.

100. Gahinet, P., Nemirovski, A., Laub, A. J., Chilali, M., 1995. LMI Control Toolbox. MathWorks, Natick, MA.
101. Jamshidi, M., 1982. Large-Scale Systems-Modeling and Control. North-Holland, Amsterdam, The Netherlands.
102. Altman, E., Basar, T., July 1998. Multiuser Rate-Based Flow Control. IEEE Trans. Commun. 46.
103. Anderson, B. D. O., Moore, J. B., 1990. Optimal Control: Linear Quadratic Methods. Prentice-Hall, Upper Saddle River, NJ.
104. Tanaka, K., Ikeda, T., Wang, H. O., Apr. 1998. Fuzzy Regulators and Fuzzy Observers: Relaxed Stability Conditions and LMI-based Designs. IEEE Trans. Fuzzy Syst. 6.
105. Chen, B. S., Tseng, C. S., Uang, H. J., Oct. 1999. Robustness Design of Nonlinear Dynamic Systems via Fuzzy Linear Control. IEEE Trans. Fuzzy Syst. 7, 571–585.
106. Sage, A. P., Melsa, J. L., 1971. Estimation Theory with Application to Communication and Control. McGraw-Hill, New York.
107. Scherer, C., Gahinet, P., July 1997. Multiobjective Output-Feedback Control via LMI Optimization. IEEE Trans. Automat. Control 42, 896–911.
108. Hwang, G. C., Lin, S. C., 1992. A Stability Approach to Fuzzy Control Design for Nonlinear Systems. Fuzzy Sets Syst. 48, 279–287.
109. Buckley, J. J., 1992. Theory of Fuzzy Controller: An Introduction. Fuzzy Sets Syst. 51, 249–258.
110. Tse, E., Dec. 1971. On the Optimal Control of Stochastic Linear Systems. IEEE Trans. Automat. Control AC-16, 776–784.
111. Sage, A. P., White, III, C. C., 1977. Optimal Systems Control, 2nd ed. Prentice-Hall, Upper Saddle River, NJ.
112. Siljak, D. D., 1978. Large-Scale Dynamic Systems-Stability and Structure. North-Holland, Amsterdam, The Netherlands.
113. Ho, Y. C., Bryson, Jr., A. E., Baron, S., 1965. Differential Games and Optimal Pursuit-Evasion Strategies. IEEE Trans. Autom. Cont. 10, 385–389.
114. Tolwinski, B., Haurie, A., Leitmann, G., Oct. 1986. Cooperative Equilibria in Differential Games. J Math Anal Appl. 119(Issues 1–2), 182–202.
115. Zhang, J., Gou, Q. L., Liang, L., Huang, Z. M., Apr. 2013. Supply Chain Coordination Through Cooperative Advertising With Reference Price Effect. Omega. 41(Issue 2), 345–353.
116. Gromova, E., Petrosian, O., Control of Information Horizon for Cooper-Ative Differential Game of Pollution Control, Stability and Oscillations of Nonlinear Control Systems (Pyatnitskiy's Conference), 2016 International Conference.
117. Lewis, F. L., Zhang, H. W., Hengster-Movric, K., Das, A., 2014. Cooperative Control of Multi-Agent Systems: Optimal and Adaptive Design Approaches, Springer Publishing Company, Incorporated.
118. Chen, H., Ye, R., Wang, X., Lu, R., 2015. Cooperative Control of Power System Load and Frequency by Using Differential Games. IEEE Trans. Cont. Syst. Technol. 23(3), 882–897.
119. Bruyne, G. D., Jul. 1979. Pareto Optimality of Non-Cooperative Equilibrium N a Time-Dependent Multi-Period Game. Eur Econ Rev. 12(3), 243–260.
120. Wu, H. Y., Feb. 2013. A Non-Cooperative Pareto-Efficient Solution to A One-Shot Prisoner's Dilemma, Munich Personal RePEc Archive (MPRA), University Library of Munich, Germany.
121. Maskery, M., et al, 2009. Decentralized Dynamic Spectrum Access for Cognitive Radios: Cooperative Design of a Non-Cooperative Game. IEEE Trans. Commun. 57(2), 459–469.
122. Han, Z., Ji, Z., Liu, K. J. R., Aug. 2007. Non-Cooperative Resource Competition Game by Virtual Referee in Multi-Cell OFDMA Networks. IEEE J. Sel. Areas Commun. 25(6), 1079–1090.
123. He, Z., Xu, H., Wang, J., Non-Cooperative Differential Game Based Load Balancing Algorithm in Radio-Over-Fibre System, Cyberspace Technology (CCT 2014), International Conference on. Nov. 2014.
124. Scutari, G., Palomar, D. P., Barbarossa, S., March 2008. Optimal Linear Precoding Strategies for Wideband Non-Cooperative Systems Based on Game Theory-Part I: Nash Equilibria. IEEE Trans. Signal Process. 56(3).

125. Osendal, B. K., Sulem, A., 2007. Applied Stochastic Control of Jump Diffusions, 2nd. ed. Springer, Berlin, Germany.
126. Hanson, F. B., 2007. Applied Stochastic Processes and Control for Jump Diffusions: Modeling, Analysis, and Computation. SIAM, Philadelphia, PA.
127. Jacobson, D., 1973. Optimal Stochastic Linear Systems with Exponential Performance Criteria and Their Relation to Deterministic Differential Games. IEEE Trans Automat Contr. 18(2), 124–131.
128. Cont, R., Tankov, P., 2004. Financial Modeling with Jump Processes. CRC Press, Boca Raton, FL.
129. Kamalapurkar, R., Klotz, J., Dixon, W., July 2014. Concurrent Learning-Based Online Approximate Feedback Nash Equilibrium Solution of N -Player Nonzero-Sum Differential Games. IEEE/CAA J. Autom. Sin. 1(3), 239–247.
130. Suijs, J., Borm, P., May 1999. Stochastic Cooperative Games: Superadditivity, Convexity, and Certainty Equivalents. Games Econ. Behav. 27(2), 331–345.
131. Chengguo, E., Gao, Z. F., Mao, A., 2007. Repeated N-Person Stochastic Cooperative Games: Superadditivity, Convexity, Control Conference, CCC.
132. Lin, Y. N., Zhang, W. H., 2017. Pareto Optimality in Finite Horizon LQ Stochastic Differential Games, in Control And Decision Conference (CCDC), 29th Chinese.
133. Isidori, A., Astolfi, A., 1992. Disturbance Attenuation and H_∞ Control via Measurement Feedback in Nonlinear Svstems. IEEE Trans Automat Contr. 37(9), 1283–1293.
134. Feng, C. S., Chen, B. S., Uang, H. J., Jan 2001. Fuzzy Tracking Control Design for Nonlinear Dynamic Systems via T-S Fuzzy Model. IEEE. Trans. Fuzzy Syst. 9(3), 381–392.
135. Zhang, H., Zhang, J., Yang, G.-H., Luo, Y., Feb. 2015. Leader-Based Optimal Coordination Control for the Consensus Problem of Multiagent Differential Games via Fuzzy Adaptive Dynamic Programming. IEEE Trans. Fuzzy Syst. 23(1), 152–163.
136. Del, K., Pratap, A., Agarwal, S., Meyarivan, T., Apr. 2002. A Fast and Elitist Multiobjective Genetic Algorithm: NSGA-II. IEEE Trans. Evol. Comput. 6(2), 182–197.
137. Wang, L. X., 1997. A Course in Fuzzy Systems and Control 1st Edition. Prentice-Hall, Upper Saddle River, NJ.
138. Liu, G. P., Yang, J. B., Whidborne, J. F., 2002. Multiobjective Optimizational Control. RSP Ltd, Hertfordshire, UK.
139. Rachmawati, L., Srinivasan, D., Aug. 2009. Multiobjective Evolutionary Algorithm with Controllable Focus on the Knees of the Pareto Front. IEEE Trans. Evol. Comput. 13(4), 810–824.
140. Haupt, R. L., Haupt, S. E., 2004. Practical Genetic Algorithms. 2nd Ed. Wiley, New York.
141. Chiu, W. Y., Nov. 2014. Multiobjective Controller Design by Solving a Multiobjective Matrix Inequality Problem. IET Cont. Theory Appl. 8(16), 1656–1665.
142. Ma, J. H., Chen, Y. S., 2001. Study for the Bifurcation Topological Structure and the Global Complicated Character of a Kind on Nonlinear Finance System. App. Math. Mech. 22(11), 1240–1251.
143. Chi, C. Y., Mendel, J. M., 1984. Performance of Minimum Variance Deconvolution Filter. IEEE Trans. Acoust. Speech Signal Process. ASSP-23(6), 1145–1153.
144. Lawrence, R. E., Kaufman, H., 1971. The Kalman Filter for Equalization of a Digital Communication Channel. IEEE Trans. Commun. COM-19, 1137–1141.
145. Miyoshi, M., Kaneda, Y., Feb. 1988. Inverse Filtering of Room Acoustics. IEEE Trans. Acoust. Speech Signal Process. ASSP-36, 145–152.
146. Ahlen, A., Stemad, M., Feb. 1989. Optimal Deconvolution Based on Polynomial Methods. IEEE Trans. Acoust. Speech Signal Process. 37, 217–226.
147. Chen, B. S., Peng, S. C., 1991. Optimal Deconvolution Filter Design Based on Orthogonal Principle. Signal Processing. 25, 361–372.
148. Chi, C. Y., June 1987. A Further Analysis for the Minimum-Variance Deconvolution Filter Performance. IEEE Trans. Acoust. Speech Signal Process. ASSP-35, 888–889.
149. Mahalanabis, A. K., Prasad, S., Mohandas, K. P., June 1983. Deconvolution of Nonstationary Seismic Data Using Adaptive Lattice Filters. IEEE Trans Acoust. Speech Signal Process. ASSP-31, 591–598.

150. Chiang, H.-H., Nikias, C. L., Jan. 1990. Adaptive Deconvolution and Identification of Non-minimum Phase FIR Systems Based on Cumulants. IEEE Trans. Automat. Control 35, 36–47.
151. Kassam, S. A., 1980. Two Dimensional Filters for Signal Processing under Modeling Uncertainties. IEEE Trans. Geosci. Electron. GE-18, 331–336.
152. Poor, H. V., Sep 1983. Robust Matched Filters. IEEE Trans. Inform. Theor. IT-29, 677–687.
153. Verdu, S., Poor, H. V., Feb. 1983. Minimax Robust Discrete-Time Matched Filters. IEEE Trans. Commun. COM-31, 208–215.
154. Poor, H. V., June 1980. On Robust Wiener Filtering. IEEE Trans. Automat. Control AC-25, 531–536.
155. Vastola, K. S., Poor, H. V., Mar 1984. Robust Wiener-Kolmogorov Theory. IEEE Trans. Inform. Theor. IT-30, 316–327.
156. D' Appolito, J. A., Hutchinson, C. E., 1972. A Minimax Approach to the Design of Low Sensitivity State Estimators. Automatica. 8, 599–608.
157. Morris, J. M., 1976. The Kalman Filter: A Robust Estimator for Some Classes of Linear Quadratic Problem. IEEE Trans. Inform. Theory IT-22, 526–534.
158. Poor, H. V., Looze, D. P., Aug 1981. Minimax State Estimation for Linear Stochastic Systems with Noise Uncertainty. IEEE Trans. Automat. Control AC-26, 902–906.
159. Verdu, S., Poor, H. V., June 1984. Minimax Linear Observers and Regulators for Stochastic System with Uncertain Second-Order Statistics. IEEE Trans Automat. Control AC-29, 499–510.
160. Moustakides, G. V., Kassam, S. A., Aug 1985. Minimax Equalization for Random Signals. IEEE Trans. Commun. COM-33, 820–825.
161. Desoer, C. A., Vidyasagar, M., 1975. Feedback Systems: Input-Output Properties. Academic Press, New York.
162. Bellman, R. E., 1970. Introduction to Matrix Analysis. McGrawHill, New York.
163. Roberts, A. W., Varberg, D. E., 1973. Convex Functions. Academic Press, New York.
164. Goodwin, G. C., Sin, K. S., 1984. Adaptive Filtering. Prediction and Control. Prentice-Hall, Englewood Cliffs, NJ.
165. Owen, G., 1968. Game Theory. Saunders, Philadelphia, PA.
166. Kalman, R. E., 1960. A New Approach to Linear Filtering and Prediction Problem. Trans. ASME. Series D. J. Basic Eng. 82, 35.
167. Koning, W. L. D., 1984. Optimal Estimation of Linear Discrete-Time Systems with Stochastic Parameters. Automatica. 20(1), 113–115.
168. Blackwell, D. H., Girshick, M. A., 1954. Theory of Game and Statistical Decisions. Wiley, New York.
169. Paulraj, A. J., et al., 2004. An Overview of MIMO Communications-A Key to Gigabit Wireless. Proc. IEEE. 92, 198–218.
170. Foschini, G. J., Gans, M. J., 1998. On Limits of Wireless Communications in a Fading Environment Using Multiple Antennas. Wireless Personal Commun. 6(3), 311–355.
171. Gesbert, D., et al., 2003. From Theory to Practice: An Overview of MIMO Space-Time Coded Wireless Systems. IEEE J. Sel. Areas Commun. 21(3), 281–302.
172. Yang, J., Roy, S., 1994. On Joint Transmitter and Receiver Optimization or Multiple-Input Multiple-Output (MIMO) Transmission Systems. IEEE Trans. Commun. 42(12), 3221–3231.
173. Li, Y., Liu, K. J., Nov. 1998. Adaptive Blind Source Separation and Equalization for Multiple-Input/Multiple-Output Systems. IEEE Trans. Inf. Theory. 44(7), 2864–2876.
174. Sampath, H., Stoica, P., Paulraj, A., 2001. Generalized Linear Precoder and Decoder Design for MIMO Channels Using the Weighted MMSE Criterion. IEEE Trans. Commun. 49, 2198–2206.
175. Scaglione, A., et al., May 2002. Optimal Designs for Space-Time Linear Precoders and Decoders. IEEE Trans. Signal Process. 50(5), 1051–1064.
176. Vikalo, H., Hassibi, B., 2002. Maximum-Likelihood Sequence Detection of Multiple Antenna Systems over Dispersive Channels via Sphere Decoding. EURASIP J. Appl. Signal Process. 525–531.
177. Komninakis, C., et al., May 2002. Multi-Input Multi-Output Fading Channel Tracking and Equalization Using Kalman Estimation. IEEE Trans. Signal Process. 50(5), 1065–1076.

178. Guo, Y. F., Levy, B., May 2006. Robust MSE Equalizer Design for MIMO Communication Systems in the Presence of Model Uncertainties. IEEE Trans. Signal Process. 54(5), 1840–1852.
179. Erdogan, A. T., Hassibi, B., Kailath, T., Mar. 2004. MIMO Decision Feedback Equalization from an H∞ Perspective. IEEE Trans. Signal Process. 52(3), 734–745.
180. Al-Dhahir, N., Sayed, A. H., Oct. 2000. The Finite-Length Multi-Input Multi-Output MMSE-DFE. IEEE Trans. Signal Process. 48(10), 2921–2936.
181. Lim, T. J., et al., Dec. 1998. An Asynchronous Multiuser CDMA Detector Base on the Kalman Filter. IEEE J. Sel. Areas Commun. 16, 1711–1722.
182. Nagpal, K. N., Khargonekar, P. P., 1991. Filtering and Smoothing in an Setting. IEEE Trans. Autom. Cont. 36(2), 152–166.
183. Verdu, S., Poor, H. V., 1984. On Minimax Robustness: A General Approach and Applications. IEEE Trans. Inf. Theory. 30(2), 328–340.
184. Kassam, S. A., Poor, H. V., Mar. 1985. Robust Techniques for Signal Processing: A Survey. Proc. IEEE. 73(3), 433–481.
185. Eldar, Y. C., Ben-Tal, A., Nemirovski, A., Jan. 2005. Robust Mean-Squared Error Estimation in the Presence of Model Uncertainties. IEEE Trans. Signal Process. 53(1), 168–181.
186. Ho, T.-J., Chen, B. S., Jan. 2007. Tracking of Dispersive DS-CDMA Channels: An AR-embedded Modified Interacting Multiple Model Approach. IEEE Trans. Wireless Commun. 6(1), 166–174.
187. Sadeghi, P., et al., Sep. 2008. Finite-State Markov Modeling of Fading Channels. IEEE Signal Process. Mag. 25(5), 57–80.
188. Ho, T.-J., Chen, B.-S., 2009. Robust Minimax Equalizer Design for Wire- Less MIMO Communications under Time-Varying Channel Uncertainties, in Proc. IEEE PIMRC, Tokyo, Japan, pp. 2666–2670.
189. Horn, R. A., Johnson, C. R., 1985. Matrix Analysis. Cambridge University Press, Cambridge, UK
190. Chan, S., et al., 2004. Asympotically Minimum BER Linear Block Precoders for MMSE Equalization. Proc. Inst. Electr. Eng. Commun. 151, 297–304.
191. Xu, F., et al., Mar. 2006. Design of Block Transceivers with Decision Feedback De- Tection. IEEE Trans. Signal Process. 54(3), 965–978.
192. Khaled, N. et al., 2004. A Robust Joint Linear Precoder and Decoder MMSE Design for Slowly Time-Varying MIMO Channels, presented at the IEEE Int. Conf. Acoustics, Speech and Signal Processing (ICASSP), Montreal, QC, Canada.
193. Guo, Y. F., Levy, B. C., Aug. 2005. Worst-Case MSE Precoder Design for Imper- Fectly Known MIMO Communications Channels. IEEE Trans. Signal Process. 53(8), 2918–2930.
194. Pascual-Iserte, A., et al., Jan. 2006. A Robust Maximin Approach for MIMO Communications with Imperfect Channel State Information Based on Convex Optimization. IEEE Trans. Signal Process 54(1), 346–362.
195. Jakes, W. C., Ed. 1994. Microwave Mobile Communications. IEEE Press, Piscataway, NJ.
196. Molisch, A., 2006. Wireless Communications. Wiley, Hoboken, NJ.
197. Ebner, A., et al., 2003. Performance of UTRA TDD Ad Hoc and IEEE 802.11b in Vehicular Environments, in Proc. IEEE Vehicular Technology Conf. (VTC-Spring), Jeju, Korea, pp. 960–964.
198. Akyildiz, I. F., Su, W., Sankarasubramaniam, Y., Cayirci, E., 2002. Wireless Sensor Networks: A Survey. Comp. Networks. 38(4), 393–422.
199. Moses, R. L., Krishnamurthy, D., Patterson, R., May 2002. An Auto- Calibration Method for Unattended Ground Sensors. Proc. IEEE Int. Conf. Acoustics, Speech, Signal Process. 3, 2941–2944.
200. Fleming, R., Kushner, C., July 1995. Low-Power, Miniature, Distributed Position Location and Communication Devices Using Ultra- Wideband, Nonsinusoidal Communication Technology, semi- annual technical report, Aetherwire, Inc., ARPA Contract J-FBI-94-058.
201. Patwari, N., Hero, III, A. O., Perkins, M., Correal, N. S., O'Dea, R. J., Aug. 2003. Relative Location Estimation in Wireless Sensor Networks. IEEE Trans. Signal Process. 51(8), 2137–2148.
202. Patwari, N., O'Dea, R. J., Wang, Y., May 2001. Relative Location in Wireless Networks. Proc. IEEE 53rd Vehicular Technology Conf. (VTC). 2, 1149–1153.

203. Biswas, P., Liang, T.-C., Toh, K.-C., Ye, Y., Wang, T.-C., Oct. 2006. Semidefinite Programming Approaches for Sensor Network Localization with Noisy Distance Measurements. IEEE Trans. Autom. Sci. Eng. 3(4), 360–371.
204. Zhang, L., Cheng, Q., Wang, Y., Zeadally, S., Feb. 2008. A Novel Distributed Sensor Positioning System Using the Dual of Target Tracking. IEEE Trans. Comput. 57(2), 246–260.
205. Tseng, P., 2007. Second-Order Cone Programming Relaxation of Sensor Network Localization. SIAM J. Optimiz. 18, 156–185.
206. Ji, X., Zha, H., Mar. 2004. Sensor Positioning in Wireless Ad-Hoc Sensor Networks Using Multidimensional Scaling. Proc. IEEE INFO- COM. 4, 2652–2661.
207. Niculescu, D., Nath, B., Nov. 2001. Ad-Hoc Positioning System (APS). Proc. IEEE GlobeCom. 5, 2926–2931.
208. Patwari, N., Hero, III, A. O., Sept. 2003. Using Proximity and Quantized RSS for Sensor Localization in Wireless Networks, in Proc. Second ACM Int'l Workshop Wireless Sensor Networks and Applications (WSNA' 03), pp. 20–29.
209. Priyantha, N. B., Chakraborty, A., Balakrishnan, H., Aug. 2000. The Cricket Location-Support System, in Proc. ACM MobiCom, pp. 32–43.
210. Savarese, C., Rabaey, J. M., Beutel, J., May 2001. Location in Distributed Ad-Hoc Wireless Sensor Networks, in Proc. IEEE Int'l Conf. Acoustics, Speech, and Signal Processing (ICASSP), pp. 2037–2040.
211. Albowicz, J., Chen, A., Zhang, L., Nov. 2001. Recursive Position Estimation in Sensor Networks, in Proc. IEEE Int'l Conf. Network Protocols, pp. 35–41.
212. Shah, R., Wolisz, A., Rabaey, J., May 2005. On the Performance of Geographical Routing in the Presence of Localization Errors, in Proc. IEEE Int'l Conf. Comm. (ICC'05), pp. 2979–2985.
213. Witt, M., Turau, V., 2006. The Impact of Location Errors on Geographic Routing in Sensor Networks, in Proc. Int'l Conf. Wireless and Mobile Comm. (ICWMC' 06), p. 76.
214. Kim, Y., Lee, J.-J., Helmy, A., Jan. 2004. Modeling and Analyzing the Impact of Location Inconsistencies on Geographic Routing in Wireless Networks. ACM SIGMOBILE Mobile Computing and Comm. Rev. 8(1), 48–60.
215. Manolakis, D., Cox, M., Jan. 1998. Effect in Range Difference Position Estimation Due to Stations' Position Errors. IEEE Trans. Aerosp. Electron. Syst. 34(1), 329–334.
216. Ho, K. C., Lu, X., Kovavisaruch, L., Feb. 2007. Source Localization Using TDOA and FDOA Measurements in the Presence of Receiver Location Errors: Analysis and Solution. IEEE Trans. Signal Process. 55(2), 684–696.
217. Yang, K., Wang, G., Luo, Z.-Q., July 2009. Efficient Convex Relaxation Methods for Robust Target Localization by a Sensor Network Using Time Differences of Arrivals. IEEE Trans. Signal Process. 57(7), 2775–2784.
218. Doherty, L., Pister, K. S. J., Ghaoui, L. E., Apr. 2001. Convex Position Estimation in Wireless Sensor Networks. Proc. IEEE INFOCOM. 3, 1655–1663.
219. Chang, C.-H., Liao, W., Jun. 2009. Revisiting Relative Location Estimation in Wireless Sensor Networks, in Proc. IEEE Int'l Conf. Comm. (ICC), pp. 1–5.
220. Sinha, K., Chowdhury, A. D., 2007. A Beacon Selection Algorithm for Bounded Error Location Estimation in Ad Hoc Networks, in Proc. Int'l Conf. Computing: Theory and Applications (ICCTA' 07), pp. 87–93.
221. Boyd, S., Vandenberghe, L., 2004. Convex Optimization. Cambridge University Press.
222. Rockah, Y., Schultheiss, P., Mar. 1987. Array Shape Calibration Using Sources in Unknown Locations-Part I: Far-Field Sources. IEEE Trans. Acoust. Speech Signal Process. 35(3), 286–299.
223. Rockah, Y., Schultheiss, P., June 1987. Array Shape Calibration Using Sources in Unknown Locations-Part II: Near-Field Sources and Estimator Implementation. IEEE Trans. Acoust. Speech Signal Process. 35(6), 724–735.
224. Noam, Y., Messer, H., June 2009. Notes on the Tightness of the Hybrid Cramer-Rao Lower Bound. IEEE Trans. Signal Process. 57(6), 2074–2084.
225. Kay, S. M., 1993. Fundamentals of Statistical Signal Processing. Prentice-Hall.

226. Hu, L., Evans, D., Sept. 2004. Localization for Mobile Sensor Networks, in Proc. ACM MobiCom, pp. 45–57.
227. Pathirana, P., Savkin, A., Jha, S., Bulusu, N., May/June 2005. Node Localization Using Mobile Robots in Delay-Tolerant Sensor Networks. IEEE Trans. Mobile Comput. 4(3), 285–296.
228. Chiu, W.-Y., Chen, B.-S., Jan. 2009. Mobile Location Estimation in Urban Areas Using Mixed Manhattan/Euclidean Norm and Convex Optimization. IEEE Trans. Wireless Comm. 8(1), 414–423.
229. Chiu, W.-Y., Chen, B.-S., Dec. 2009. Mobile Positioning Problem in Manhattan-Like Urban Areas: Uniqueness of Solution, Optimal Deployment of BSs, and Fuzzy Implementation. IEEE Trans. Signal Process. 57(12), 4918–4929.
230. Chiu, W.-Y., Chen, B.-S., May 2010. A Mixed-Norm Approach Using Simulated Annealing with Changeable Neighborhood for Mobile Location Estimation. IEEE Trans. Mobile Comput. 9(5), 633–642.
231. Chiu, W.-Y., Chen, B.-S., June 2009. Locating Mobiles in General Urban Areas Using Combined Convex Optimization and Weight-Product Tracking Method, in Proc. Int'l Conf. Wireless Comm. and Mobile Computing: Connecting the World Wirelessly (IWCMC), pp. 1085–1090.
232. Denis, B., Daniele, N., 2004. NLOS Ranging Error Mitigation in a Distributed Positioning Algorithm for Indoor UWB Ad-Hoc Networks, in Proc. Int'l Workshop Wireless Ad-Hoc Networks, pp. 356–360.
233. Venkatesh, S., Buehrer, R. M., Apr. 2006. A Linear Programming Approach to NLOS Error Mitigation in Sensor Networks, in Proc. Conf. Information Processing in Sensor Networks, pp. 301–308.
234. Shi, Q., Kyperountas, S., Correal, N. S., Niu, F., Apr. 2005. Performance Analysis of Relative Location Estimation for Multihop Wireless Sensor Networks. IEEE J. Sel. Areas Comm. 23(4), 830–838.
235. Savvides, A., Park, H., Srivastava, M. B., Sept. 2002. The Bits and Flops of the N-Hop Multilateration Primitive for Node Localization Problems, in Proc. Int'l Workshop Sensor Networks and Applications, pp. 112–121.
236. Capkun, S., Hamdi, M., Hubaux, J.-P., Jan. 2001. GPS-Free Positioning in Mobile Ad-Hoc Networks, in Proc. 34th IEEE Hawaii Int'l Conf. System Sciences.
237. Rahman, M. S., Park, Y., Kim, K.-D., Feb. 2009. Relative Location Estimation of Vehicles in Parking Management System. Proc. Int. Conf. Adv. Comm. Technol. 1, 729–732.
238. Bai, Y., Huang, Q., Chen, L., Sept. 2007. Segmentation-Aided and Density-Aware Hop-Count (SDH) Localization in Ad Hoc Networks, Proc. IEEE Int'l Symp. Personal, Indoor and Mobile Radio Comm. (PIMRC), pp. 1–5.
239. Montesano, L., Gaspar, J., Santos-Victor, J., Montano, L., Aug. 2005. Cooperative Localization by Fusing Vision-Based Bearing Mea- Surements and Motion, in Proc. IEEE/RSJ Int'l Conf. Intelligent Robots and Systems, pp. 2333–2338.
240. Savvides, A., Han, C., Srivastava, M., July 2001. Dynamic Fine-Grained Localization in Ad-Hoc Networks of Sensors, in Proc. ACM MobiCom, pp. 166–179.
241. Tian, H., Huang, C., Blum, B., Stankovik, J., Abdelzaher, T., Sept. 2003. Range-Free Localization Schemes for Large Scale Sensor Networks, in Proc. ACM MobiCom, pp. 81–95.
242. Yousef, N., Sayed, A., Oct. 1999. A New Adaptive Estimation Algorithm for Wireless Location Finding Systems, in Proc. IEEE Third Asilomar Conf. Signals, Systems, Computers, pp. 491–495.
243. Bahl, P., Padmanabhan, V., Apr. 2000. RADAR: An In-Building RF- Based User Location and Tracking System. Proc. IEEE INFOCOM. 2, 775–784.
244. Bechtler, T., Yenigun, H., May 2003. 2D Localization and Identification Based on SAW ID Tags at 2.5 GHz. IEEE Trans. Microw. Theory Tech. 51(5), 1584–1589.
245. Ma, C., Klukas, R., Lachapelle, G., May 2003. An Enhanced Two-Step Least Squared Approach for TDOA/AOA Wireless Location Communication, in Proc. IEEE Int'l Conf. Comm. (ICC), pp. 987–991.
246. Hashemi, H., July 1993. The Indoor Radio Propagation Channel. Proc. IEEE. 81(7), 943–968.
247. Chong, E. K. P., Zak, S. H., 2008. An Introduction to Optimization. Wiley- Interscience.

248. Ueda, K., Yamashita, N., 2010. On a Global Complexity Bound of the Levenberg-Marquardt Method. J. Optimiz. Theory Ap. 147(3), 443–453.
249. Monteiro, R. D., Tsuchiya, T., 2000. Polynomial Convergence of Primal-Dual Algorithms for the Second-Order Cone Program Based on the MZ-Family of Directions. Math. Programm. 88(1), 61–83.
250. Monteiro, R. D., Tsuchiya, T., 1997. Primal-Dual Path-Following Algorithms for Semidefinite Programming. SIAM J. Optimiz. 7(3), 663–678.
251. SeDuMi, 2011. http://sedumi.ie.lehigh.edu.
252. McGill, R., Tukey, J. W., Larsen, W. A., 1978. Variations of Boxplots. Am. Stat. 32(1), 12–16.
253. Kumar, S., Lai, T. H., Balogh, J., Sept. 2004. On k-Coverage in a Mostly Sleeping Sensor Network, in Proc. ACM MobiCom.
254. Zhou, Z., Das, S., Gupta, H., 2004. Connected K-Coverage Problem in Sensor Networks, in Proc. Int'l Conf. Computer Comm. and Networks, pp. 373–378.
255. Bejerano, Y., 2008, Simple and Efficient K-Coverage Verification without Location Information, in Proc. IEEE INFOCOM, pp. 291–295.
256. Yang, G., Qiao, D., Sept. 2008. Critical Conditions for Connected-K- Coverage in Sensor Networks. IEEE Comm. Lett. 12(9), 651–653.
257. Ammari, H. M., Giudici, J., 2009. On the Connected K-Coverage Problem in Heterogeneous Sensor Nets: The Curse of Randomness and Heterogeneity, in Proc. IEEE Int'l Conf. Distributed Computing Systems (ICDCS), pp. 265–272.
258. Wigley, T. M. L., 2005. The Climate Change Commitment. Science. 307(5716), 1766–1769. doi:10.1126/science.1103934
259. Friedlingstein, P., Solomon, S., 2005. Contributions of past and Present Human Generations to Committed Warming Caused by Carbon Dioxide. Proc. Natl. Acad. Sci. U.S.A. 102(31), 10832–10836. doi:10.1073/pnas.0504755102
260. Caetano, M. A. L., Gherardi, D. F. M., Yoneyama, T., 2008. Optimal Resource Management Control for CO2 Emission and Reduction of the Greenhouse Effect. Ecol Modell. 213(1), 119–126. doi:10.1016/j.ecolmodel.2007.11.014
261. Davis, S. J., Caldeira, K., Matthews, H. D., 2010. Future CO2 Emissions and Climate Change from Existing Energy Infrastructure. Science. 329(5997), 1330–1333. doi:10.1126/science.1188566
262. Salehi-Khojin, A., et al., 2010. Nonthermal Current-Stimulated Desorption of Gases from Car- Bon Nanotubes. Science. 329(5997), 1327–1330. doi:10.1126/science.1194210
263. Scheraga, J. D., Leary, N. A., 1992. Improving the Efficiency of Policies to Reduce CO2 Emissions. Energy Pol. 20(5), 394–404. doi:10.1016/0301-4215(92)90061-6
264. UNEP, United Nations Environment Program, www.unep.org/
265. Marland, G., Boden, T. A., Andres, R. J., et al., 2006. Global, Regional, and National Fossil Fuel CO_2 Emissions, Carbon Dioxide Information Analysis Center, Environ- mental Sciences Division. Oak Ridge National Laboratory, Oak Ridge, TN.
266. WWF, World Wildlife Fund, www.wwf.org.uk/researcher/issues/footprint/index.asp.
267. UNFCCC, United Nations Framework Convention on Climate Change, Status of Ratification of the Kyoto Protocol, http://unfccc.int/kyoto_protocol/status_of_ratification/items/2613.php.
268. Nordhaus, W. D., 1991. To Slow or Not to Slow: The Economics of the Greenhouse Effect. Econ. J. 101(407), 920–937. doi:10.2307/2233864
269. Nordhaus, W. D., 1991. A Sketch of the Economics of the Greenhouse Effect. Am Econ Rev. 81(2), 146–150.
270. Nordhaus, W. D., 1993. Reflections on the Economics of Climate Change. J. Econ. Perspect. 7(4), 11–25.
271. Poterba, J. M., 1993. Global Warming Policy: A Public Finance Perspective. J. Econ. Perspect. 7(4), 47–63.
272. Stollery, K. R., 1998. Constant Utility Paths and Irreversible Global Warming. Canadian J. Econ. 31(3), 730–742. doi:10.2307/136210
273. Caetano, M. A. L., et al., 2009. Reduction of CO2 Emission by Optimally Tracking a Pre-Defined Target. Ecol Modell. 220(19), 2536–2542. doi:10.1016/j.ecolmodel.2009.06.003

274. Oksendal, B. K., 2003. Stochastic Differential Equations: An Introduction with Applications. Springer Verlag, New York.
275. Zhang, W. H., Chen, B. S., 2006. State Feedback H_∞ Control for a Class of Nonlinear Stochastic Systems. SIAM. J. Control Optim. 44, 1973–1991. doi:10.1137/S0363012903423727
276. Zhang, W. H., Chen, B. S., Tseng, C. S., 2005. Robust H_∞ Filtering for Nonlinear Stochastic Systems. IEEE Trans. Signal Process. 53(2), 589–598. doi:10.1109/TSP.2004.840724
277. Agostini, P., Botteon, M., Carraro, C., 1992. A Carbon Tax to Reduce CO2 Emissions in Europe. Energy Econ. 14(4), 279–290. doi:10.1016/0140-9883(92)90034-B
278. Boyd, R., Krutilla, K., Viscusi, W. K., 1995. Energy Taxation as A Policy Instrument to Reduce CO_2 Emissions: A Net Benefit Analysis. J. Environ. Econ. Manage. 29(1), 1–24. doi:10.1006/jeem.1995.1028
279. McKitrick, R., 2001. What's Wrong with Regulating Carbon Dioxide Emissions, A Briefing at the United States Congress.
280. Basar, T., Olsder, G. J., 1999. Dynamic Noncooperative Game Theory, 2nd Edition ed. Society for Industrial and Applied Mathematics, Philadelphia, PA.
281. WRI, World Resources Institute, Earth Trends Environmental Information, www.earthtrends.wri.org.
282. FAOSTAT, Fao Corporate Document Repository, European Forest Sector Outlook Study 1960-2000-2020— Main Report, www.fao.org/docrep/008/ae428e/ae428e00.htm
283. Maddison, A., The World Economy, Available at: www.theworldeconomy.org/publications/worldeconomy/statistics.htm
284. USDA, United States Department of Agriculture, Economic Research Service, www.ers.usda.gov/Data/macroeconomics/Data/HistoricalGDPDeflatorValues.xls
285. HYDE, History Database of the Global Environment, Netherlands Environmental Assessment Agency, www.rivm.nl/hyde/bdf/gdp.
286. Grossman, G. M., 1995. Pollution and Growth: What Do We Know? In: Goldin, I., Winters, L. A. (eds.), The Economics of Sustainable Development. Cambridge University Press, Cambridge, pp. 19–50.
287. Jotzo, F., Pezzey, J. C. V., 2007. Optimal Intensity Targets for Greenhouse Gas Emissions Trading under Uncertainty. Environ. Resour. Econ. 38(2), 259–284. doi:10.1007/s10640-006-9078-z
288. Farhangi, H., Feb. 2010. The Path of the Smart Grid. IEEE Power Energy Mag. 8(1), 18–28.
289. Ipakchi, A., Albuyeh, F., 2009. Grid of the Future. IEEE Power Energy Mag. 7(2), 52–62.
290. Vojdani, A., 2008. Smart Integration. IEEE Power Energy Mag. 6(6), 71–79.
291. Tsoukalas, L. H., Gao, R., 2008. From Smart Grid to an Energy Internet: Assumptions, Architectures, and Requirements, Proc. 2008 3rd IEEE International Conference on Electric Utility Deregulation, pp. 94–98.
292. Zheng, J., Gao, D. W., Lin, L., Smart Meters in Smart Grid: An Overview, in 2013 IEEE Green Technologies Conference, 2013.
293. Lasseter, R. H., Paigi, P., Microgrid: A Conceptual Solution, in Proceedings of the 2004 Power Electronics Specialists Conference (PESC'04), 2004.
294. Lasseter, R. H., et al., Oct. 2003. White Paper on Integration of Distributed Energy Resources—The CERTS MicroGrid Concept, in Consort. Electr. Reliab. Technol. Solut., pp. 1–27.
295. Lopes, J. A. P., Moreira, C. L., Madureira, A. G., 2005. Defining Control Strategies for Analysing Microgrids Islanded Operation, in Proc. IEEE St. Petersburg PowerTech.
296. Kanellos, F. D., Tsouchnikas, A. I., Hatziargyriou, N. D., Member, S., 2005. Micro-Grid Simulation during Grid-Connected and Islanded Modes of Operation, in Presented at the International Conference on Power Systems Transients (IPST'05).
297. Hooshmand, A., Malki, H. A., Mohammadpour, J., Sep. 2012. Power Flow Management of Microgrid Networks Using Model Predictive Control. Comput. Math. Appl. 64(5), 869–876.
298. Nguyen, P., Kling, W., Georgiadis, G., Papatriantafilou, M., Tuan, L. A., Bertling, L., 2010. Distributed Routing Algorithms to Manage Power Flow in Agent-Based Active Distribution Network, in Proc. IEEE PES Conference on Innovative Smart Grid Technologies Europe.

299. Lu, M.-S., Chang, C.-L., Lee, W.-J., Wang, L., 2008. Combining the Wind Power Generation System with Energy Storage Equipments, in Proc. IEEE Industry Applications Society Annual Meeting.
300. Li, P., Dec. 2008. Energy Storage Is the Core of Renewable Technologies. IEEE Nano Tech. Mag. 2(4), 13–18.
301. Minciardi, R., Sacile, R., Mar. 2012. Optimal Control in a Cooperative Network of Smart Power Grids. IEEE Syst. J. 6(1), 126–133.
302. Chiu, W.-Y., Sun, H., Poor, H. V., 2012, Robust Power Flow Control in Smart Grids with Fluctuating Effects, in Proc. IEEE Int. Conference on Computer Communications Workshops.
303. Chiu, W.-Y., Sun, H., Poor, H. V., 2012. Demand-Side Energy Storage System Management Smartgrid, in Proc. 2012 IEEE 3rd Int. Conf. Smart Grid Commun.
304. Bruno, S., Lamonaca, S., Rotondo, G., Stecchi, U., La Scala, M., Oct. 2011. Unbalanced Three-Phase Optimal Power Flow for Smart Grids. IEEE Trans. Ind. Electron. 58(10), 4504–4513.
305. Xu, Y., Zhang, W., Hug, G., Kar, S., Li, Z., Jan. 2015. Cooperative Control of Distributed Energy Storage Systems in a Microgrid. IEEE Trans. Smart Grid. 6(1), 238–248.
306. Tanaka, K., Uchida, K., Ogimi, K., Goya, T., Yona, A., Senjy, T., Funabashi, T., Kim, C.-H., Sep. 2011. Optimal Operation by Controllable Loads Based on Smart Grid Topology considering Insolation Forecasted Error. IEEE Trans. Smart Grid. 2(3), 438–444.
307. Paudyal, S., Canizares, C. A., Bhattacharya, K., Oct. 2011. Optimal Operation of Distribution Feeders in Smart Grids. IEEE Trans. Ind. Electron. 58(10), 4495–4503.
308. Mohsenian-Rad, A., Wong, V., Jatskevich, J., Schober, R., LeonGarcia, A., Dec. 2010. Autonomous Demand-Side Management Based on Game Theoretic Energy Consumption Scheduling for the Future Smart Grid. IEEE Trans. Smart Grid. 1(3), 320–331.
309. Chiu, W.-Y., Sun, H., Poor, H. V., Sep. 2015. A Multiobjective Approach to Multimicrogrid System Design. IEEE Trans. Smart Grid. 6(5), 2263–2272.
310. Wang, Y., Saad, W., Han, Z., Poor, H. V., Basar, T., May 2014. A Game-Theoretic Approach to Energy Trading in the Smart Grid. IEEE Trans. Smart Grid. 5(3), 1439–1450.
311. Deb, K., Agrawal, S., Pratab, A., Meyarivan, T., Apr. 2002. A Fast and Elitist Multiobjective Genetic Algorithm: NSGA-II. IEEE Trans Evol Comput. 6(2), 681–695.
312. Chen, B. S., Lee, H. C., Wu, C. F., Apr. 2014. Pareto Optimal Filter Design for Nonlinear Stochastic Fuzzy Systems via Multiobjective Optimization. IEEE Trans. Fuzzy Syst. 23(2), 387–399.
313. Cohen, A., Shaked, U., 1998. Robust Discrete-Time H_∞-Optimal Tracking with Preview. Int. J. Robust Nonlin. Cont. 8(1), 29–37.
314. Joseph, A., Shahidehpour, M., 2006. Battery Storage Systems in Electric Power Systems, in IEEE Power Engineering Society General Meeting.
315. Sunnyfounder, Taipei, Taiwan, 2015. [Online]. Available at: www.sunnyfounder.com.
316. Bates, D. S., 1996. Jumps and Stochastic Volatility: Exchange Rate Processes Implicit in Deutschemark Options. Rev. Finance Stud. 9(1), 69–107.
317. Cai, N., Kou, S., 2011. Option Priceing under a Mixed-Exponential Jump Diffusion Model. Manage. Sci. 57, 2067–2081.
318. Leitner, Y., Dec. 2005. Financial Networks: Contagion, Commitment, and Private Sector Bailouts. J. Finance. 60(6), 2925–2953.
319. Castellacci, G., Choi, Y., 2014. Financial Instability Contagion: A Dynamic Systems Approach. Quant. Finance. 14(7), 1243–1255.
320. Melvin, M., Melvin, B. P., 2003. The Global Transmission of Volatility in the Foreign Exchange Market, Rev. Econ. Statist. 85(3), 670–679.
321. Griffith-Jones, S., 1999. Stabilising Capital Flows to Developing Countries. IDS Bull. 30(1), 33–49.
322. Choi, Y., Douady, R., 2012. Financial Crisis Dynamics: Attempt to Define a Market Instability Indicator. Quant. Finance. 12,(9), 1351–1365.
323. Castellacci, G., Choi, Y., 2015. Modeling Contagion in the Eurozone Crisis via Dynamical Systems. J. Bank. Financ. 50, 400–410.
324. Hu, Z. H., Chen, W., 2013. Modeling of Macroeconomics by a Novel Discrete Nonlinear Fractional Dynamical System. Discrete Dyn. Nat. Soc. 2013, 9, Art. no. 275134.

325. Xu, X., Zeng, Z., Xu, J., Zhang, M., 2017. Fuzzy Dynamical System Scenario Simulation-Based Cross-Border Contagion Analysis: A Perspective from International Capital Flows. IEEE Trans. Fuzzy Syst. 25(2), 439–459.
326. Cont, R., Tankov, P., 2004. Financial Modelling with Jump Processes. CRC Press, Boca Raton, FL.
327. Chen, B. S., Lee, T. S., Chang, W. S., 1996. A Robust H_∞ Model Reference Tracking Design for Nonholonomic Mechanical Control Systems. Int. J. Control. 63(2), 283–306.
328. Gulisashvili, A., 2012. Analytically Tractable Stochastic Stock Price Models. Springer, New York.
329. Stein, E. M., Stein, J. C., 1991. Stock Price Distributions with Stochastic Volatility: An Analytic Approach. Rev. Financial Stud. 4(4), 727–752.
330. Stoyanov, J., 2011. Stochastic Financial Models. J. Roy. Statist. Soc.: Ser. A (Statist. Soc.). 174,(2), 510–511.
331. Chen, W. C., 2008. Nonlinear Dynamics and Chaos in a Fractional-Order Financial System. Chaos, Solitons Fractals. 36(5), 1305–1314.
332. Schweitzer, F., Fagiolo, G., Sornette, D., Vega Redondo, F., Vespignani, A., White, D. R., July 2009. Economic Networks: The New Challenge. Science. 325(5939), 422–425.
333. Elloitt, M., Golul, B., Jackson, M. O., Oct. 2004. Financial Networks and Contagion. Amer. Econ. Rev. 104(10), 3115–3153.
334. Wang, L. X., Aug. 2015. Dynamic Model of Stock Prices Based on Technical Trading Rules Part I: The Models. IEEE Trans. Fuzzy Syst. 23(4), 787–801.
335. Wang, L. X., Aug. 2015. Dynamic Model of Stock Prices Based on Technical Trading Rules Part II: Analysis of the Model. IEEE Trans. Fuzzy Syst. 23(4), 1127–1141.
336. Wang, L. X., Oct. 2015. Dynamic Model of Stock Prices Based on Technical Trading Rules Part IIII: Application to Hong Kong Stocks. IEEE Trans. Fuzzy Syst. 23(5), 1680–1697.
337. Fratzscher, M., 2012. Capital Flows, Push versus Full Factors and the Global Financial Crsis. J Int Econ. 88(2), 341–356.
338. Gulisashuili, A., 2012. Analytically Tractable Stochastic Stock Price Models. Springer, New York.
339. Stein, E. M., Stein, J. C., 1991. Stock Price Distributions with Stochastic Volatillity: An Analytic Approach. Rev. Financial Stud. 4(4), 727–752.
340. Tan, X., Cruz, J. B., Sept. 2010. Adaptive Noncooperative N-Person Games With Unknown General Quadratic Objectives. IEEE Trans. Control Syst. Tech. 18(5), 1033–1043.
341. Ahmed, M., Mukaidani, H., Shima, T., Sept. 2017. H_∞-Constrained Incentive Stackelberg Games for Discrete-Time Stochastic Systems with Multiple Followers. IET Control Theory Appl. 11(15), 2475–2485.
342. Nourian, M., Caines, P. E., Malhame, R. P., Huang, M., Mar. 2013. Nash, Social and Centralized Solutions to Consensus Problems via Mean Field Control Theory. IEEE Trans. Auto. Control 58 (3), 639–653.
343. Liang, H., Zhou, Y., Ma, H., Zhou, Q., Jan. 2018. Adaptive Distributed Observer Approach for Cooperative Containment Control of Nonidentical Networks, in IEEE Trans. Syst. Man. Cyber, Syst. doi:10.1109/TSMC.2018.2791513
344. Liang, H., Li, H., Yu, Z., Li, P., Wang, W., July 2017. Cooperative Robust Containment Control for General Discrete-Time Multi-Agent Systems with External Disturbance. IET Control Appl. 11(12), 1928–1937.
345. Zhang, H., Feng, T., Liang, H., Luo, Y., Mar. 2017. LQR-Based Optimal Distributed Cooperative Design for Linear Discrete-Time Multi-Agent Systems. IEEE Trans. Neural Net. Learn. Syst. 28(3), 599–611.
346. Feng, T., Zhang, H., Luo, Y., Liang, H., Aug. 2016. Globally Optimal Distributed Cooperative Control for General Linear Multi-Agent Systems. Neurocomputing. 203, 12–21.
347. Zhang, J., Li, S., April 2017. Quanto European Option Pricing with Ambiguous Return Rate and Volatility. IEEE Trans. Fuzzy Syst. 25(2), 417–424.

348. Stengel, R. F., Ghigliazza, R. M., Kulkarni, N. V., 2002b. Optimal Enhancement of Immune Response. Bioinformatics. 18, 1227–1235.
349. Lydyard, P. M., Whelan, A., Fanger, M. W., 2000. Instant Notes in Immunology. Springer, New York.
350. Janeway, C., 2005. Immunobiology: The Immune System in Health and Disease. Garland, New York.
351. Asachenkov, A. L., 1994. Disease Dynamics. Birkha¨ user, Boston, MA.
352. Rundell, A., HogenEsch, H., DeCarlo, R., 1995. Enhanced Modeling of the Immune System to Incorporate Natural Killer Cells and Memory, in Proceedings of American Control Conference, vol. 1, Seattle, pp. 255–259.
353. Perelson, A. S., Weisbuch, G., 1997. Immunology for Physicists. Rev. Modern Phys. 69, 1219–1267.
354. Nowak, M. A., May, R. M., 2000. Virus Dynamics: Mathematical Principles of Immunology and Virology. Oxford University Press, Oxford.
355. Perelson, A. S., Kirschner, D. E., De Boer, R., 1993. Dynamics of HIV Infection of CD4+ T Cells. Math. Biosci. 114, 81–125.
356. Perelson, A. S., Neumann, A. U., Markowitz, M., Leonard, J. M., Ho, D. D., 1996. HIV-1 Dynamics in Vivo: Virion Clearance Rate, Infected Cell Life-Span, and Viral Generation Time. Science. 271, 1582–1586.
357. Nowak, M. A., May, R. M., Phillips, R. E., Rowland-Jones, S., Lalloo, D. G., McAdam, S., Klenerman, P., Koppe, B., Sigmund, K., Bangham, C. R., et al., 1995. Antigenic Oscillations and Shifting Immunodominance in HIV-1 Infections. Nature.375, 606–611.
358. Stafford, M. A., Corey, L., Cao, Y., Daar, E. S., Ho, D. D., Perelson, A. S., 2000. Modeling Plasma Virus Concentration during Primary HIV Infection. J. Theor. Biol. 203, 285–301.
359. Adler, F. R., Karban, R., 1994. Defended Fortresses or Moving Targets? Another Model of Inducible Defenses Inspired by Military Metaphors. Am. Nat. 144, 813–832.
360. Wiener, N., 1948. Cybernetics; Or, Control and Communication in the Animal and the Machine. Technology Press, Cambridge.
361. Bellman, R., 1983. Mathematical Methods in Medicine. World Scientific, Singapore.
362. Swan, G. W., 1981. Optimal-Control Applications in Biomedical-Engineering- a Survey. Optim. Control Appl. Meth. 2, 311–334.
363. Kirschner, D., Lenhart, S., Serbin, S., 1997. Optimal Control of the Chemotherapy of HIV. J. Math. Biol. 35, 775–792.
364. De Boer, R. J., Boucher, C. A., 1996. Anti-CD4 Therapy for AIDS Suggested by Mathematical Models. Proc. Biol. Sci. 263, 899–905.
365. Bonhoeffer, S., May, R. M., Shaw, G. M., Nowak, M. A., 1997. Virus Dynamics and Drug Therapy. Proc. Natl. Acad. Sci. USA. 94, 6971–6976.
366. Wein, L. M., D'Amato, R. M., Perelson, A. S., 1998. Mathematical Analysis of Antiretroviral Therapy Aimed at HIV-1 Eradication or Maintenance of Low Viral Loads. J. Theor. Biol. 192, 81–98.
367. Wodarz, D., Nowak, M. A., 1999. Specific Therapy Regimes Could Lead to Long-Term Immunological Control of HIV. Proc. Natl. Acad. Sci. USA. 96, 14464–14469.
368. Wodarz, D., Nowak, M. A., 2000. CD8 Memory, Immunodominance, and Antigenic Escape. Eur. J. Immunol. 30, 2704–2712.
369. Robinson, D. C., 1986. Topics in Clinical Pharmacology and Therapeutics, In: Maronde, R. F. (ed.), Principles of Pharmacokinetics. Springer, New York, pp. 1–12.
370. Van Rossum, J. M., Steyger, O., van Uem, T., Binkhorst, G. J., Maes, R. A. A., 1986. Pharmacokinetics by Using Mathematical Systems Dynamics, In: Eisenfeld, J., Witten, M. (eds.), Modelling of Biomedical Systems. Elsevier, Amsterdam, The Netherlands, pp. 121–126.
371. Bell, D. J., Katusiime, F., 1980. A Time-Optimal Drug Displacement Problem. Optim. Control Appl. Meth. 1, 217–225.
372. Carson, E. R., Cramp, D. G., Finkelstein, F., Ingram, D., 1985. Control System Concepts and Approaches in Clinical Medicine, In: Carson, E. R., Cramp, D. G. (eds.), Computers and Control in Clinical Medicine. Plenum Press, New York, pp. 1–26.

373. Chizeck, H. J., Katona, P. G., 1985. Closed-Loop Control, In: Carson, E. R., Cramp, D. G. (eds.), Computers and Control in Clinical Medicine. Plenum Press, New York, pp. 95–151.
374. Jelliffe, R. W., 1986. Clinical Applications of Pharmacokinetics and Control Theory: Planning, Monitoring, and Adjusting Regimens of Aminoglycosides, Lidocaine, Digitoxin, and Digoxin, In: Maronde, R. F. (ed.), Topics in Clinical Pharmacology and Therapeutics. Springer, New York, pp. 26–82.
375. Schumitzky, A., 1986. Stochastic Control of Pharmacokinetic Systems, In: Maronde, R. F. (ed.), Topics in Clinical Pharmacology and Therapeutics. Springer, New York, pp. 13–25.
376. Kwong, G. K., Kwok, K. E., Finegan, B. A., Shah, S. L., 1995. Clinical Evaluation of Long Range Adaptive Control for Mean Arterial Blood Pressure Regulation, in Proceedings of American Control Conference, vol. 1, Seattle, pp. 786–790.
377. Polycarpou, M. M., Conway, J. Y., 1995. Modeling and Control of Drug Delivery Systems Using Adaptive Neuralcontrol Methods, in Proceedings of American Control Conference, vol. 1, Seattle, pp. 781–785.
378. Parker, R. S., Doyle, III, J. F., Harting, J. E., Peppas, N. A., 1996. Model Predictive Control for Infusion Pump Insulin Delivery, in Proceedings of the 18th Annual International Conference of the IEEE Engineering in Medicine and Biology Society, vol. 5, Amsterdam, pp. 1822–1823.
379. Gentilini, A., Morari, M., Bieniok, C., Wymann, R., Schnider, T. W., 2001. Closed-Loop Control of Analgesia in Humans, in Proceedings of IEEE Conference on Decision and Control, vol. 1, Orlando, pp. 861–866.
380. Stengel, R. F., Ghigliazza, R., Kulkarni, N., Laplace, O., 2002a. Optimal Control of Innate Immune Response. Optim. Control Appl. Meth. 23, 91–104.
381. Zhou, K., Doyle, J. C., Glover, K., 1996. Robust and Optimal Control. Prentice-Hall, Upper Saddle River, NJ.
382. Shudo, E., Iwasa, Y., 2001. Inducible Defense against Pathogens and Parasites: Optimal Choice among Multiple Options. J. Theor. Biol. 209, 233–247.
383. Shudo, E., Iwasa, Y., 2002. Optimal Defense Strategy: Storage Vs. New Production. J. Theor. Biol. 219, 309–323.
384. Shudo, E., Iwasa, Y., 2004. Dynamic Optimization of Host Defense, Immune Memory, and Post-Infection Pathogen Levels in Mammals. J. Theor. Biol. 228, 17–29.
385. Shudo, E., Haccou, P., Iwasa, Y., 2003. Optimal Choice between Feedforward and Feedback Control in Gene Expression to Cope with Unpredictable Danger. J. Theor. Biol. 223, 149–160.
386. Lian, K. Y., Chiu, C. S., Chiang, T. S., Liu, P., 2001. LMI-based Fuzzy Chaotic Synchronization and Communications. IEEE Trans. Fuzzy Syst. 9, 539–553.
387. Li, T. H. S., Chang, S. J., Tong, W., 2004. Fuzzy Target Tracking Control of Autonomous Mobile Robots by Using Infrared Sensors. IEEE Trans. Fuzzy Syst. 12, 491–501.
388. Åström, K. J., Wittenmark, B., 1995. Adaptive Control. Addison-Wesley, Reading, MA.
389. Milutinovic, D., De Boer, R. J., 2007. Process Noise: An Explanation for the fluctuations in the Immune Response during Acute Viral Infection. Biophys. J. 92, 3358–3367.
390. Marchuk, G. I., 1983. Mathematical Models in Immunology. Optimization Software, Inc. Worldwide distribution rights by Springer, New York.
391. Villadsen, L. S., Skov, L., Baadsgaard, O., 2003. Biological Response Modifiers and Their Potential Use in the Treatment of Inflammatory Skin Diseases. Exp. Dermatol. 12, 1–10.
392. Stengel, R. F., Ghigliazza, R., 2004. Stochastic Optimal Therapy for Enhanced Immune Response. Math. Biosci. 191, 123–142.
393. Althaus, C. L., Ganusov, V. V., De Boer, R. J., 2007. Dynamics of CD8+T Cell Responses during Acute and Chronic Lymphocytic Choriomeningitis Virus Infection. J. Immunol. 179, 2944–2951.
394. Andrianantoandro, E., et al., 2006. Synthetic Biology: New Engineering Rules for an Emerging Discipline. Mol. Syst. Biol. 2, 1–14.
395. Tucker, J. B., Zilinskas, R. A., 2006. The Promise and Perils of Synthetic Biology. New Atlantis. 12, 25–45.
396. Alon, U., 2003. Biological Networks: The Tinkerer as an Engineer. Science. 301, 1866–1867.

397. Alon, U., 2007. An Introduction to Systems Biology: Design Principles of Biological Circuits. Chapman & Hall/CRC.
398. Church, G. M., 2005. From Systems Biology to Synthetic Biology. Mol. Syst. Biol. 1, 2005.0032.
399. Endy, D., 2005. Foundations for Engineering Biology. Nature. 438, 449–453.
400. Hasty, J., et al., 2002. Engineered Gene Circuits. Nature.420, 224–230.
401. Heinemann, M., Panke, S., 2006. Synthetic Biology - Putting Engineering into Biology. Bioinformatics. 22, 2790–2799.
402. Kobayashi, H., et al., 2004. Programmable Cells: Interfacing Natural and Engineered Gene Networks. Proc. Natl. Acad. Sci. 101, 8414–8419.
403. Pleiss, J., 2006. The Promise of Synthetic Biology. Appl. Microbiol. Biotechnol. 73, 735–739.
404. Ferber, D., 2004. Synthetic Biology: Microbes Made to Order. Science. 303, 158–161.
405. Forster, A. C., Church, G. M., 2007. Synthetic Biology Projects in Vitro. Genome Res. 17, 1–6.
406. Gardner, T. S., et al., 2000. Construction of a Genetic Toggle Switch in Escherichia Coli. Nature.403, 339–342.
407. Isaacs, F. J., et al., 2006. RNA Synthetic Biology. Nat. Biotechnol.24, 545–554.
408. Maeda, Y. T., Sano, M., 2006. Regulatory Dynamics of Synthetic Gene Networks with Positive Feedback. J. Mol. Biol. 359, 1107–1124.
409. Tucker, M., Parker, R., 2000. Mechanisms and Control of mRNA Decapping in Saccharomyces Cerevisiae. Ann. Rev. Biochem. 69, 571–595.
410. Batt, G., et al., 2007. Robustness Analysis and Tuning of Synthetic Gene Networks. Bioinformatics.23, 2415.
411. Goulian, M., 2004. Robust Control in Bacterial Regulatory Circuits. Curr. Opin. Microbiol. 7, 198–202.
412. Kaznessis, Y. N., 2006. Multi-Scale Models for Gene Network Engineering. Chem. Eng. Sci. 61, 940–953.
413. Kaznessis, Y. N., 2007. Models for Synthetic Biology. BMC Syst. Biol. 1, 47.
414. Kitano, H., 2002. Systems Biology: A Brief Overview. Science. 295, 1662–1664.
415. Kitano, H., 2004. Biological Robustness. Nat. Rev. Genet. 5, 826–837.
416. Salis, H., Kaznessis, Y. N., 2006. Computer-Aided Design of Modular Protein Devices: Boolean AND Gene Activation. Phys. Biol. 3, 295–310.
417. de Jong, H., 2002. Modeling and Simulation of Genetic Regulatory Systems: A Literature Review. J. Comput. Biol. 9, 67–103.
418. Szallasi, Z., et al., 2006. System Modeling in Cellular Biology: From Concepts to Nuts and Bolts. The MIT Press, Cambridge, MA.
419. Kuepfer, L., et al., 2007. Efficient Classification of Complete Parameter Regions Based on Semidefinite Programming. BMC Bioinformatics.8, 12.
420. Chen, B. S., Chen, P. W., 2008. Robust Engineered Circuit Design Principles for Stochastic Biochemical Networks with Parameter Uncertainties and Disturbances. IEEE Trans. Biomed. Circuits Syst. 2, 114–132.
421. Chen, B. S., Wu, W. S., 2008. Robust Filtering Circuit Design for Stochastic Gene Networks under Intrinsic and Extrinsic Molecular Noises. Math. Biosci. 211, 342–355.
422. Chen, B. S., et al., 2008b. Robust H_∞ Stabilization Design in Gene Networks under Stochastic Molecular Noises: Fuzzy-Interpolation Approach. IEEE Trans. Syst. Man Cybern. B.38, 25–42.
423. Hwang, C. L., 2004. A Novel Takagi-Sugeno-based Robust Adaptive Fuzzy Sliding-Mode Controller. IEEE Trans. Fuzzy Syst. 12, 676–687.
424. Balas, G., et al., 2008. Robust Control Toolbox User's Guide. The Math Works, Inc., Natick, MA.
425. Hooshangi, S., et al., 2005. Ultrasensitivity and Noise Propagation in a Synthetic Transcriptional Cascade. Proc. Natl Acad. Sci. USA.102, 3581–3586.
426. Canton, B., et al., 2008. Refinement and Standardization of Synthetic Biological Parts and Devices. Nat. Biotechnol. 26, 787–793.
427. Arkin, A., 2008. Setting the Standard in Synthetic Biology. Nat. Biotechnol. 26, 771–774.
428. Felix, M. A., Wagner, A., 2008. Robustness and Evolution: Concepts, Insights and Challenges from a Developmental Model System. Heredity. 100, 132–140.

429. Ancel, L. W., Fontana, W., 2000. Plasticity, Evolvability, and Modularity in RNA. J. Exp. Zool. 288, 242–283.
430. Carter, A. J. R., Hermisson, J., Hansen, T. F., 2005. The Role of Epistatic Gene Interactions in the Response to Selection and the Evolution of Evolvability. Theor Popul Biol. 68, 179–196.
431. Bloom, J. D., Labthavikul, S. T., Otey, C. R., Arnold, F. H., 2006. Protein Stability Promotes Evolvability. Proc. Natl. Acad. Sci. U.S.A. 103, 5869–5874.
432. Aldana, M., Balleza, E., Kauffman, S., Resendiz, O., 2007. Robustness and Evolvability in Genetic Regulatory Networks. J. Theor. Biol. 245, 433–448.
433. Elena, S. F., Sanjuan, R., 2008. The Effect of Genetic Robustness on Evolvability in Digital Organisms. BMC Evol. Biol. 8.
434. McBride, R. C., Ogbunugafor, C. B., Turner, P. E., 2008. Robustness Promotes Evolvability of Thermotolerance in an RNA Virus. BMC Evol. Biol. 8.
435. Chen, B. S., Lin, Y. P., 2011. On the Interplay between the Evolvability and Network Robustness in an Evolutionary Biological Network: A Systems Biology Approach. Evolutionary Bioinform. 7, 201–233.
436. Chen, B. S., Wu, W. S., Wu, W. S., Li, W. H., 2007. On the Adaptive Design Rules of Bio-Chemical Networks in Evolution. Evolutionary Bioinform. 3, 27–39.
437. Draghi, J. A., Parsons, T. L., Wagner, G. P., Plotkin, J. B., 2010. Mutational Robustness Can Facilitate Adaptation. Nature. 463, 353–355.
438. Tawfik, D. S., 2010. Messy Biology and the Origins of Evolutionary Innovations. Nat. Chem. Biol. 6, 692–696.
439. Lyuch, M., 2007. The Evolution of Genetic Networks by Non-Adaptive Processes. Nat. Rev. Genet. 8, 803–813.
440. Chen, B. S., Lin, Y. P., 2013. A Unifying Mathematical Framework for Genetic Robustness, Environmental Robustness, Network Robustness and Their Trade-Off on Phenotype Robustness in Biological Networks Part I: Gene Regulatory Networks in Systems and Evolutionary Biology. Evolutionary Bioinform. 9, 43–68.
441. Chen, B. S., Lin, Y. P., 2013. A Unifying Mathematical Framework for Genetic Robustness, Environmental Robustness, Network Robustness and Their Tradeoff on Phenotype Robustness in Biological Networks Part II: Ecological Networks. Evolutionary Bioinform. 9, 69–85.
442. Chen, B. S., Lin, Y. P., 2013. A Unifying Mathematical Framework for Genetic Robustness, Environmental Robustness, Network Robustness and Their Trade-Offs on Phenotype Robustness in Biological Networks. Part III: Synthetic Gene Networks in Synthetic Biology. Evolutionary Bioinform. 9, 87–109.
443. Gardner, W. A., 1990. Introduction to Random Processes, with Applications to Signals & Systems 2nd Edition. McGraw-Hill, New York.
444. Vincent, T. L., Brown, J. S., 2005. Evolutionary Game Theory, Natural Selection, and Darwinian Dynamics. Cambridge University Press, Cambridge.
445. de Visser, J. A. G. M., et al., 2003. Perspective: Evolution and Detection of Genetic Robustness. Evolution. 57, 1959–1972.
446. Badyaev, A. V., 2005. Stress-Induced Variation in Evolution: From Behavioural Plasticity to Genetic Assimilation. Proc. Royal Soc. B-Biol. Sci. 272, 877–886.
447. Chen, B. S., Wang, Y. C., 2006. On the Attenuation and Amplification of Molecular Noise in Genetic Regulatory Networks. BMC Bioinformatics. 7.
448. Sriyudthsak, K., Shiraishi, F., Hirai, M. Y., 2013. Identification of a Metabolic Reaction Network from Time-Series Data of Metabolite Concentrations. PLoS ONE. 8, 51212.
449. Voit, E. O., 2000. Computational Analysis of Biochemical Systems: A Practical Guide for Biochemists and Molecular Biologists. Cambridge University Press, UK, p. 531.
450. Waddington, C. H., 1942. Analization of Development and the Inheritance of Acquired Characters. Nature. 150, 563–565.
451. Lehner, B., 2010. Genes Confer Similar Robustness to Environmental, Stochastic, and Genetic Perturbations in Yeast. PLoS ONE. 5.

452. Basu, S., Gerchman, Y., Collins, C. H., Arnold, F. H., Weiss, R., 2005. A Synthetic Multicellular System for Programmed Pattern Formation. Nature. 434, 1130–1134.
453. Hasty, J., Dolnik, M., Rottschafer, V., Collins, J. J., 2002. Synthetic Gene Network for Entraining and Amplifying Cellular Oscillations. Phys. Rev. Lett. 88.
454. Kaern, M., Elston, T. C., Blake, W. J., Collins, J. J., 2005. Stochasticity in Gene Expression: From Theories to Phenotypes. Nat. Rev. Genet. 6, 451–464.
455. Isaacs, F. J., Hasty, J., Cantor, C. R., Collins, J. J., 2003. Prediction and Measurement of an Autoregulatory Genetic Module. Proc. Natl. Acad. Sci. U.S.A. 100, 7714–7719.
456. Kaern, M., Blake, W. J., Collins, J. J., 2003. The Engineering of Gene Regulatory Networks. Annu. Rev. Biomed. Eng. 5, 179–206.
457. Chen, B. S., Hsu, C. Y., Liou, J. J., 2011. Robust Design of Biological Circuits: Evolutionary Systems Biology Approach. J. Biomed. Biotechnol.
458. Chen, B. S., Wu, C. H., 2010. Robust Optimal Reference-Tracking Design Method for Stochastic Synthetic Biology Systems: T-S Fuzzy Approach. IEEE Trans. Fuzzy Syst. 18, 1144–1159.
459. Wu, C. H., Lee, H. C., Chen, B. S., 2011. Robust Synthetic Gene Network Design via Library-Based Search Method. Bioinformatics. 27, 2700–2706.
460. Chen, B. S., Chen, P. W., 2010. GA-based Design Algorithms for the Robust Synthetic Genetic Oscillators with Prescribed Amplitude, Period and Phase. Gene Regul Syst Bio. 4, 35–52.
461. Chen, B. S., Tseng, C. S., Uang, H. J., 2002. Fuzzy diCerential Games for Nonlinear Stochastic Systems: Suboptimal Approach. IEEE Trans. Fuzzy Syst. 10.
462. Chu, L. H., Chen, B. S., 2008. Construction of a Cancer-Perturbed Protein-Protein Interaction Network for Discovery of Apoptosis Drug Targets. BMC Syst. Biol. 2, 56.
463. Ao, P., 2005. Laws in Darwinian Evolutionary Theory. Phys Life Rev. 2(2), 117–156.
464. Wagner, A., 2008. Robustness and Evolvability: A Paradox Resolved. Proc. Biol. Sci. 275(1630), 91–100.
465. Chen, B. S., Wu, W. S., Wu, W. S., Li, W. H., 2007. On the Adaptive Design Rules of Biochemical Networks in Evolution. Evol. Bioinform. Online. 3, 27–39.
466. Barillot, E., 2013. Computational Systems Biology of Cancer. CRC Press, Boca Raton, FL.
467. Wong, Y. H., Chen, R. H., Chen, B. S., 2014. Core and Specific Network Markers of Carcinogenesis from Multiple Cancer Samples. J. Theor. Biol. 362, 17–34.
468. Ao, P., Galas, D., Hood, L., Zhu, X., 2008. Cancer as Robust Intrinsic State of Endogenous Molecular-Cellular Network Shaped by Evolution. Med. Hypotheses. 70(3), 678–684.
469. Weinberg, R. A., 2006. The Biology of Cancer. Taylor & Francis Group, New York.
470. Murgia, C., Pritchard, J. K., Kim, S. Y., Fassati, A., Weiss, R. A., 2006. Clonal Origin and Evolu-Tion of a Transmissible Cancer. Cell. 126(3), 477–487.
471. Vogelstein, B., Kinzler, K. W., 2004. Cancer Genes and the Pathways They Control. Nat. Med. 10(8), 789–799.
472. Nagy, J. D., 2004. Competition and Natural Selection in a Mathematical Model of Cancer. Bull Math Biol. 66(4), 663–687.
473. Kitano, H., 2004. Cancer as a Robust System: Implications for Anticancer Therapy. Nat. Rev. Cancer. 4(3), 227–235.
474. Kitano, H., 2003. Cancer Robustness: Tumour Tactics. Nature. 426(6963), 125.
475. Lobov, I. B., Brooks, P. C., Lang, R. A., 2002. Angiopoietin-2 Displays VEGF-dependent Modulation of Capillary Structure and Endothelial Cell Survival in Vivo. Proc. Natl. Acad. Sci. USA. 99(17), 11205–11210.
476. Yancopoulos, G. D., Davis, S., Gale, N. W., Rudge, J. S., Wiegand, S. J., Holash, J., 2000. Vascular- Specific Growth Factors and Blood Vessel Formation. Nature. 407(6801), 242–248.
477. Greaves, M. F., 2000. Cancer: The Evolutionary Legacy. Oxford University Press, Oxford.
478. Carmeliet, P., Jain, R. K., 2000. Angiogenesis in Cancer and Other Diseases. Nature. 407(6801), 249–257.
479. Weibel, E. R., 1984. The Pathway for Oxygen: Structure and Function in the Mammalian Respiratory System. Harvard University Press, Cambridge, MA.

480. Wang, Y. C., Chen, B. S., 2011. A Network-Based Biomarker Approach for Molecular Investigation and Diagnosis of Lung Cancer. BMC Med Genomics. 4, 2.
481. Chen, W.-H., Chen, B.-S., 2013. Robust Stabilization Design for Nonlinear Stochastic Sys- Tem with Poisson Noise via Fuzzy Interpolation Method. Fuzzy Set. Syst. 217(0), 41–61.
482. Tomasetti, C., Vogelstein, B., 2015. Variation in Cancer Risk among Tissues Can Be Explained by the Number of Stem Cell Divisions. Science. 347(6217), 78–81.
483. Shukla, S., Meeran, S. M., 2014. Epigenetics of Cancer Stem Cells: Pathways and Therapeu-Tics. Biochim. Biophys. Acta. 1840(12), 3494–3502.
484. Kozielski, J., Kaczmarczyk, G., Porębska, I., Szmygin-Milanowska, K., Gołecki, M., 2012. Lung Cancer in Patients under the Age of 40 Years. Contemp. Oncol (Pozn). 16, 413–415.
485. Crespi, B., Summers, K., 2005. Evolutionary Biology of Cancer. Trends Ecol. Evol. 20(10), 545–552.
486. Campisi, J., 2005. Aging, Tumor Suppression and Cancer: High Wire-Act!. Mech. Ageing Dev. 126(1), 51–58.
487. Kandori, M., Mailath, G. J., Rob, R., 1993. Learning, Mutation and Long-Run Equilibria in Games. Econometrica. 61(1), 29–56.
488. Maynard Smith, J., 1982. Evolution and the Theory of Games. Cambridge University Press, Cambridge.
489. Nowak, M. A., 2006. Five Rules for Evolution of Cooperation. Science. 314(5805), 1560–1563.
490. Harsanyi, J. C., Selten, R., 1988. A General Theory of Equilibrium Selection in Games. The MIT Press, Cambridge, MA.
491. Maynard Smith, J., Szathmáry, E., 1995. The Major Transition in Evolution. Oxford University Press, Oxford.
492. Page, K. M., Nowak, M. A., 2002. Unifying Evolutionary Dynamics. J. Theor. Biol. 219(1), 93–98.
493. Fogel, D. B., Fogel, G. B., Andrews, P. C., 1997. On the Instability of Evolutionary Stable Strategies. Biosystems. 44(2), 135–152.
494. Hofbauer, J., Sigmund, K., 1998. Evolutionary Games and Population Dynamics. Cambridge University Press, Cambridge.
495. Licht, A. N., 1999. Game Commissions Play: 2 × 2 Games of International Securities Regulation. Yale J. Int. Law. 24, 61–125.
496. Taylor, P., Jonker, L., 1978. Evolutionarily Stable Strategies and Game Dynamics. Math. Biosci. 40, 145–156.
497. Maynard Smith, J., Price, G. R., 1973. The Logic of Animal Conflict. Nature. 246, 15–18.
498. Adami, C., Schossau, J., Hintze, A., 2016. Evolutionary Game Theory Using Agent-Based Methods. Phys. Life Rev. 19, 1–26.
499. Li, X. Y., Giaimo, S., Baudisch, A., Traulsen, A., 2015. Modeling Evolutionary Games in Populations with Demographic Structure. J. Theor. Biol. 380, 506–515.
500. Fishman, M. A., 2016. Polymorphic Evolutionary Games. J. Theor. Biol. 398, 130–135.
501. Iacobelli, G., Madeo, D., Mocenni, C., 2016. Lumping Evolutionary Game Dynamics on Networks. J. Theor. Biol. 407, 328–338.
502. Ozkan-Canbolat, E., Beraha, A., Bas, A., 2016. Application of Evolutionary Game Theory to Strategic Innovation. Procedia Soc. Behav. Sci. 235, 685–693.
503. Luthi, L., Tomassini, M., Pestelacci, E., 2009. Evolutionary Game on Networks and Payoff Invariance under Replicator Dynamics. Biosystems. 96, 213–222.
504. Roca, C. P., Cuesta, J. A., Sánchez, A., 2009. Evolutionary Game Theory: Temporal and Spatial Effects beyond Replicator Dynamics. Phys. Life Rev. 6, 208–249.
505. Traulsen, A., Nowak, M. A., Pacheco, J. M., 2006. Stochastic Dynamic of Invasion and fixation. Phys. Rev. E. 74, 011909.
506. Chen, B. S., Ho, S. J., 2014. The Stochastic Evolutionary Game on Robust Biological Network by Natural Selection. Evol. Bioinform. 10, 17–38.
507. Axelrod, R., 1984. The Evolution of Cooperation. Basic Books, New York.
508. Fehr, E., Fischbacher, U., 2003. The Natural of Human Altruism. Nature. 425, 785–791.
509. Freeman, S., Herron, J. C., Payton, M., 2004. Evolutionary Analysis, 3rd ed ed. Prentice Hall, Upper Saddle River, NJ.

510. Chen, B. S., et al., 2005. A New Measure of the Robustness of Biochemical Networks. Bioinformatics.21, 2698–2705.
511. Gibson, G., Wagner, G., 2000. Canalization in Evolutionary Genetics: A Stabilizing Theory? BioEssays. 22(4), 372–380.
512. Lenski, R. E., Barrick, J. E., Ofria, C., 2006. Balancing Robustness and Evolvability. PLoS Biol. 4 (12), 2190–2192.
513. Veloz, T., Razeto-Barry, P., Dittrich, P., Fajardo, A., 2014. Reaction Networks and Evolutionary Game Theory. J. Math. Biol. 68, 181–206.
514. Chen, B. S., Cheng, Y. M., Lee, C. H., 1995. A Genetic Approach to Mixed H_2/H_∞ Optimal PID Control. IEEE Control Syst. Mag. 15(5), 51–60.
515. Chiu, W. Y., Chen, B. S., Poor, H. V., 2013. A Multiobjective Approach for Source Estimation in Fuzzy Networked Systems. IEEE Trans. Circuits Syst. I: Reg. Pap. 60(7), 1890–1900.
516. Chen, B. S., Wu, W. S., Li, W. H., 2007. On the Adaptive Rules of Biochemical Network in Evolution. Evol. Bioinform. 3, 27–39.

Index

F

G

H

I

K

L

T

U

V

W